AF679763

Epigenetics

Current Research and Emerging Trends

Edited by

Brian P. Chadwick

Department of Biological Science
Florida State University
Tallahassee, FL
USA

Caister Academic Press

Copyright © 2015

Caister Academic Press
Norfolk, UK

www.caister.com

British Library Cataloguing-in-Publication Data
A catalogue record for this book is available from the British Library

ISBN: 978-1-910190-07-4 (hardback)
ISBN: 978-1-910190-08-1 (ebook)

Description or mention of instrumentation, software, or other products in this book does not imply endorsement by the author or publisher. The author and publisher do not assume responsibility for the validity of any products or procedures mentioned or described in this book or for the consequences of their use.

All rights reserved. No part of this publication may be reproduced, stored in a retrieval system, or transmitted, in any form or by any means, electronic, mechanical, photocopying, recording or otherwise, without the prior permission of the publisher. No claim to original U.S. Government works.

Cover design adapted from images provided by Brian P. Chadwick and Zhuo Sun. The images show the distribution of the SMCHD1 protein (green) in female human nuclei (blue) by indirect immunofluorescence. The large staining foci in each nucleus corresponds to the territory of the inactive X chromosome.

Contents

Contributors

Aradhita Baral
Proteomics and Structural Biology Unit
CSIR-Institute of Genomics and Integrative Biology
Delhi
India

arybaral@gmail.com

Scott J. Bultman
Department of Genetics and Lineberger Comprehensive Cancer Center
University of North Carolina at Chapel Hill
Chapel Hill, NC
USA

Scott_Bultman@med.unc.edu

Brian P. Chadwick
Department of Biological Science
Florida State University
Tallahassee, FL
USA

chadwick@bio.fsu.edu

Shantanu Chowdhury
Proteomics and Structural Biology Unit; and
G. N. Ramachandran Knowledge Centre for Genome Informatics
CSIR-Institute of Genomics and Integrative Biology
Delhi
India

shantanuc@igib.in

Lauren A. Cole
Department of Biological Science
Florida State University
Tallahassee, FL
USA

lc13g@my.fsu.edu

Emily M. Darrow
Department of Biological Science
Florida State University
Tallahassee, FL
USA

edarrow@bio.fsu.edu

Sunny Das
Department of Biological Science
Florida State University
Tallahassee, FL
USA

sdas@bio.fsu.edu

Jonathan H. Dennis
Department of Biological Science
Florida State University
Tallahassee, FL
USA

dennis@bio.fsu.edu

Dana C. Dolinoy
Department of Environmental Health Sciences
University of Michigan School of Public Health
Ann Arbor, MI
USA

ddolinoy@umich.edu

Kenneth C. Ehrlich
Center for Bioinformatics and Genomics
Tulane Health Sciences Center
New Orleans, LA
USA

kehrlich@tulane.edu

Melanie Ehrlich
Program in Human Genetics; and
Tulane Cancer Center; and
Center for Bioinformatics and Genomics
Tulane Health Sciences Center
New Orleans, LA
USA

ehrlich@tulane.edu

Agustín F. Fernández
Cancer Epigenetics Laboratory
Instituto Universitario de Oncología del Principado de Asturias (IUOPA)
HUCA
Universidad de Oviedo
Oviedo
Spain

affernandez@hca.es

Mario F. Fraga
Cancer Epigenetics Laboratory
Instituto Universitario de Oncología del Principado de Asturias (IUOPA)
HUCA
Universidad de Oviedo
Oviedo
Spain; and
Department of Immunology and Oncology
Centro Nacional de Biotecnología (CNB-CSIC)
Madrid
Spain

mffraga@cnb.csic.es

María G. García
Cancer Epigenetics Laboratory
Instituto Universitario de Oncología del Principado de Asturias (IUOPA)
HUCA
Universidad de Oviedo
Oviedo
Spain

garciagarmaria@uniovi.es

Jaclyn M. Goodrich
Department of Environmental Health Sciences
University of Michigan School of Public Health
Ann Arbor, MI
USA

gaydojac@umich.edu

Ashleigh S. Hanner
Department of Biological Sciences
University at Buffalo
The State University of New York
Buffalo, NY
USA

ashanner@buffalo.edu

Justin M.H. Heltzel
Department of Biological Sciences
University at Buffalo
The State University of New York
Buffalo, NY
USA

jheltzel@buffalo.edu

Benjamin Hoffman
Department of Biological Science
Florida State University
Tallahassee, FL
USA

bhoffman@bio.fsu.edu

Kristen M. Humphrey
Department of Cancer Pathology and Prevention
Roswell Park Cancer Institute
Buffalo, NY
USA

kristen.humphrey@roswellpark.org

Luke Isbel
La Trobe Institute of Molecular Sciences
La Trobe University
Melbourne
Australia

l.isbel@latrobe.edu.au

Shivali Kapoor
Department of Biological Sciences
University at Buffalo
The State University of New York
Buffalo, NY
USA

shivalihm@gmail.com

Michelle Lacey
Tulane Cancer Center
Tulane Health Sciences Center
New Orleans, LA
USA; and
Department of Mathematics
Tulane University
New Orleans, LA
USA

mlacey@math.tulane.edu

Janine M. LaSalle
Medical Microbiology and Immunology; and
MIND Institute; and
Genome Center
University of California
Davis, CA
USA

jmlasalle@ucdavis.edu

Shannon M. McNulty
Department of Molecular Genetics and Microbiology
Division of Human Genetics
Duke University Medical Center
Durham, NC
USA

shannon.mcnulty@duke.edu

John H. Malone
Institute of Systems Genomics, Molecular and Cell Biology
University of Connecticut
Storrs, CT
USA

john.malone@uconn.edu

Harry Oey
La Trobe Institute of Molecular Sciences
La Trobe University
Melbourne
Australia

h.oey@latrobe.edu.au

Sriharsa Pradhan
New England Biolabs
Ipswich, MA
USA

pradhan@neb.com

Theodore P. Rasmussen
Department of Pharmaceutical Sciences; and
Department of Molecular and Cell Biology; and
University of Connecticut Stem Cell Institute
University of Connecticut
School of Pharmacy
Storrs, CT
USA

theodore.rasmussen@uconn.edu

Raed Rizkallah
Department of Biomedical Sciences
College of Medicine
Florida State University
Tallahassee, FL
USA

raed.rizkallah@med.fsu.edu

Justyne E. Ross
Department of Molecular Genetics and Microbiology
Division of Human Genetics
Duke University Medical Center
Durham, NC
USA

ross.justyne@gmail.com

Christopher B. Rupert
Department of Biological Sciences
University at Buffalo
The State University of New York
Buffalo, NY
USA

cbrupert@buffalo.edu

Laura N. Rusche
Department of Biological Sciences
University at Buffalo
The State University of New York
Buffalo, NY
USA

lrusche@buffalo.edu

Ingrid Sadler-Riggleman
Center for Reproductive Biology
School of Biological Sciences
Washington State University
Pullman, WA
USA

sadler@wsu.edu

Dhurjhoti Saha
Proteomics and Structural Biology Unit
CSIR-Institute of Genomics and Integrative Biology
Delhi
India

dhurjhotisaha@igib.in

Michael K. Skinner
Center for Reproductive Biology
School of Biological Sciences
Washington State University
Pullman, WA
USA

skinner@wsu.edu

Beth A. Sullivan
Department of Molecular Genetics and Microbiology
Division of Human Genetics
Duke University Medical Center
Durham, NC
USA

beth.sullivan@duke.edu

Zhuo Sun
Department of Biological Science
Florida State University
Tallahassee, FL
USA

zs13c@my.fsu.edu

Estela G. Toraño
Cancer Epigenetics Laboratory
Instituto Universitario de Oncología del Principado de Asturias (IUOPA)
HUCA
Universidad de Oviedo
Oviedo
Spain

UO179651@uniovi.es

Daniel L. Vera
Department of Biological Science
Florida State University
Tallahassee, FL
USA

dvera@bio.fsu.edu

Emma Whitelaw
La Trobe Institute of Molecular Sciences
La Trobe University
Melbourne
Australia

e.whitelaw@latrobe.edu.au

Guoqiang Zhang
New England Biolabs
Ipswich, MA
USA

zhangg@neb.com

Preface

What an exciting time in biology we find ourselves in. Many complex genomes are deciphered (Lander *et al.*, 2001; Mouse Genome Sequencing Consortium *et al.*, 2002), leaps and bounds are being made in stem cell biology (Takahashi *et al.*, 2007; Takahashi and Yamanaka, 2006; Thomson *et al.*, 1998), and molecular tools are being refined for precise editing and engineering of DNA (Carroll, 2011; Joung and Sander, 2013; Mali *et al.*, 2013). On top of all this, our understanding and appreciation for epigenetics has made phenomenal advances since the term 'epigenetics' was first put forward by Waddington in the 1940s to unite the fields of genetics and developmental biology (Waddington, 1942). Initially relating to developmental programming, today the designation of epigenetics is more focused and can be defined as the study of the mitotic and/or meiotic heritability of gene expression triggered through a transient initiating event that occurs in the absence of change to the DNA sequence (Berger *et al.*, 2009; Felsenfeld, 2014; Russo *et al.*, 1996).

Despite the many advances, there is so much more yet to be learned about this complex, yet fascinating topic. For example, as illustrated in Fig. 1, who directs and initiates whom? DNA methylation or histone modifications? What about the role of RNA, or DNA structural forms? In my opinion, we are but at the base of the mountain and have an exciting and long adventure of discovery as we clamber ahead.

This book begins with several chapters that focus on epigenetic processes. We start with a discussion of the multifunctional zinc finger protein YY1 (Chapter 1) that performs numerous central roles in epigenetic phenomena.

Figure 1 What is the epigenetic hierarchy? Which comes first, DNA methylation or histone modifications? Or is it RNA or the act of transcription, and what role does the DNA have in setting up the epigenome?

Next we transition to two chapters that relate to histone-modifying enzymes. The first focuses on the versatile role of the histone methyltransferase protein SETDB1 in establishing and maintaining gene silencing (Chapter 2), whereas the second discusses the multifaceted role of sirtuins in fungal biology (Chapter 3). The next chapter discusses the distribution and detection of the DNA modification, 5-hydroxymethylcytosine in the genome, and how this relates to development (Chapter 4). After this chapter, various approaches to identify proteins involved in epigenetic processes are discussed, including the remarkable power of an *N*-ethyl-*N*-nitrosourea mutagenesis screen to generate mouse lines with mutations in epigenetic factors that enhance or suppress variegated transgene expression (Chapter 5). This is followed by a chapter that focuses on the global response of chromatin in response to stimuli (Chapter 6), before a discussion of the complex nature of chromatin and epigenetics in defining the centromere of eukaryotic chromosomes (Chapter 7). The next chapter is a broad discussion of dosage compensation with a particular focus on what is known in frogs and toads and why these animals make ideal models to further investigate this process (Chapter 8). The first part of the book is rounded off with a discussion of long non-coding RNAs in epigenetic processes (Chapter 9).

The next section of the book highlights two human genetic disorders that are directly impacted by epigenetics. The first provides an in depth and current review of the autism spectrum disorder Rett syndrome, which is caused by mutations in the methyl-DNA binding protein MECP2, whose gene is located on the X-chromosome and the disease is therefore also impacted by the mammalian dosage compensation pathway, X-chromosome inactivation (Chapter 10). The second disease focused chapter provides and in depth look at the progressive muscle degenerative disorder facioscapulohumeral muscular dystrophy, a complex disease that is impacted by many epigenetic influences (Chapter 11).

Finally, the last section of the book focuses on relatively new aspects of epigenetics. We start with a discussion of challenges and approaches to reprogramming the epigenome (Chapter 12), followed by a discussion of the potential for G4 quadruplex structures as a means for epigenetic inheritance (Chapter 13). The book is then rounded off with four chapters that address current topics in epigenetics. The first discusses the application of epigenetics in cancer diagnosis, prognosis and therapy (Chapter 14). The second (Chapter 15) focuses on transgenerational inheritance: inheritance of phenotype in the absence of exposure (Fig. 2). The third discusses the influence of metabolites on the epigenome (Chapter 16), and the final chapter (Chapter 17) discusses the impact of the environment on our epigenome, an area of growing concern (Fig. 3).

I would like to thank all the contributors for their time and effort in making this book happen, and hope that you, the readers, will enjoy and be as enlightened as I was in its preparation.

Dr Brian P. Chadwick

Figure 2 Can we now blame (at least in part) our great grandmother or our grandfather for epimutations that impact us? What are we doing, unbeknownst to us, that will impact our lineage several generations removed?

Figure 3 We now know how some environmental toxicants impact our epigenome and increase susceptibility to disease. However, what about those natural or man-made agents that are not currently on our radar?

References

Berger, S.L., Kouzarides, T., Shiekhattar, R., and Shilatifard, A. (2009). An operational definition of epigenetics. Genes Dev. *23*, 781–783.

Carroll, D. (2011). Genome engineering with zinc-finger nucleases. Genetics *188*, 773–782.

Felsenfeld, G. (2014). A brief history of epigenetics. Cold Spring Harb. Perspect. Biol. *6*, 1–10.

Joung, J.K., and Sander, J.D. (2013). TALENs: a widely applicable technology for targeted genome editing. Nat. Rev. Mol. Cell. Biol. *14*, 49–55.

Lander, E.S., Linton, L.M., Birren, B., Nusbaum, C., Zody, M.C., Baldwin, J., Devon, K., Dewar, K., Doyle, M., FitzHugh, W., et al. (2001). Initial sequencing and analysis of the human genome. Nature *409*, 860–921.

Mali, P., Esvelt, K.M., and Church, G.M. (2013). Cas9 as a versatile tool for engineering biology. Nat. Methods *10*, 957–963.

Mouse Genome Sequencing Consortium, Waterston, R.H., Lindblad-Toh, K., Birney, E., Rogers, J., Abril, J.F., Agarwal, P., Agarwala, R., Ainscough, R., Alexandersson, M., et al. (2002). Initial sequencing and comparative analysis of the mouse genome. Nature *420*, 520–562.

Russo, V.E.A., Martienssen, R.A., and Riggs, A.D. (1996). Epigenetic mechanisms of gene regulation (Cold Spring Harbor Laboratory Press, Plainview, NY. USA).

Takahashi, K., Tanabe, K., Ohnuki, M., Narita, M., Ichisaka, T., Tomoda, K., and Yamanaka, S. (2007). Induction of pluripotent stem cells from adult human fibroblasts by defined factors. Cell *131*, 861–872.

Takahashi, K., and Yamanaka, S. (2006). Induction of pluripotent stem cells from mouse embryonic and adult fibroblast cultures by defined factors. Cell *126*, 663–676.

Thomson, J.A., Itskovitz-Eldor, J., Shapiro, S.S., Waknitz, M.A., Swiergiel, J.J., Marshall, V.S., and Jones, J.M. (1998). Embryonic stem cell lines derived from human blastocysts. Science *282*, 1145–1147.

Waddington, C.H. (1942). The epigenotype. Endeavour *1*, 18–20.

The Multifaceted Roles of YY1 in the Establishment of the Cellular Epigenetic Landscape

1

Raed Rizkallah

Abstract

Yin Yang 1 (YY1) is an essential multifunctional protein needed for the regulation of a very wide spectrum of biological processes. Since its initial identification as a transcription factor that can repress, activate, or initiate transcription, YY1 has been shown to regulate hundreds of mammalian genes. Our knowledge of the mechanisms governing epigenetic regulation has grown exponentially over the past two decades. Mounting evidence in the scientific literature clearly indicates that YY1 is a central player in the establishment of the cellular epigenetic landscape. This chapter will focus on the multilayered roles of YY1 in epigenetic regulation with emphasis on the polycomb pathway, genomic imprinting, and X-chromosome inactivation. In addition, the known mechanisms of YY1 regulation will be discussed, along with possible implications on human health and disease through epigenetic disturbance.

Introduction

In 1991–1992, several research groups reported the isolation of a new transcription factor acting in different gene contexts. This factor was named NF-E1 (for its interaction with the Igκ3′ enhancer and the IgH μE1 site) (Park and Atchison, 1991), UCRBP (for its binding to the upstream control region of retroviral LTR) (Flanagan *et al.*, 1992), delta/δ (for its binding to the delta motif in the promoters of ribosomal protein genes) (Hariharan *et al.*, 1991), and as Yin Yang 1/YY1 (for its dual activation and repression of the P5 promoter of adeno-associated virus) (Shi *et al.*, 1991). Because the 'YY1' name is most descriptive of this new protein's functional behaviour, it became adopted by the scientific community.

YY1 is a ubiquitously expressed multifunctional protein with multifaceted roles in cellular and developmental biology (for general reviews see Gordon *et al.*, 2006; Shi *et al.*, 1997). The essential role of YY1 is underscored by the consequences of its deletion. *Yy1*$^{-/-}$ knockout mice die quickly after peri-implanation, whereas *yy1*$^{-/+}$ mice survive until birth with a plethora of severe developmental defects (Donohoe *et al.*, 1999). Examination of conditional knockout mice, at different levels of the YY1 protein, revealed dosage-dependent roles of YY1 in the various biological pathways (Affar el *et al.*, 2006). Normal biological processes regulated by YY1 include embryogenesis and development, cell differentiation, cell growth and proliferation, DNA repair, and programmed cell death (apoptosis). Furthermore, the deregulation of YY1 has been associated with many human diseases, most notably cancer initiation and progression (Castellano *et al.*, 2009; Gordon *et al.*, 2006; Nicholson *et al.*, 2011).

YY1 is traditionally regarded as a transcription factor with a global role in gene regulation. It is estimated that approximately 10% of human promoters could be regulated by YY1 (Schug *et al.*, 2005). YY1's DNA consensus binding sites are overrepresented in core promoters regions, and most of these sites overlap translation or transcription start sites (Xi *et al.*, 2007). Dozens of genes are now demonstrated as bona fide physiological targets of YY1, and the number still mounts.

By directly binding DNA, YY1 can activate, repress, or initiate transcription (Thomas and Seto, 1999). Moreover, without directly binding DNA,

YY1 can function as a co-factor for other transcription factors, like the androgen receptor that regulates the expression of the prostate-specific antigen (PSA) (Deng *et al.*, 2009). It can also bind RNA and play a role in RNA metabolism (Belak *et al.*, 2008; Belak and Ovsenek, 2007). YY1 has also been shown to regulate other proteins by protein–protein interactions. For example, YY1 forms a ternary complex with Mdm2 and p53 leading to the down-regulation of p53 proteins levels (Gronroos *et al.*, 2004; Sui *et al.*, 2004). This means that the biological roles of YY1 can be exerted through many different mechanisms, most likely by subdividing into smaller populations of differentially modified YY1 pools. This behaviour of YY1 has made it particularly challenging to understand its specific roles and for teasing apart its different functions.

The past two decades have witnessed a revolution in our understanding of the molecular pathways underlying biological phenotypes. The sequencing of the human genome was a major landmark. However, it became apparent that the DNA sequence can only be as informative as it is expressed or not. And hence we entered the era of epigenetics and functional genomics. The knowledge of the various epigenetic mechanisms leading to gene activation or silencing has grown exponentially in the past 10 years. The transcription factor YY1 is now acknowledged as a key player in the establishment of the epigenetic landscape. This chapter will specifically focus on the accumulating evidence in the scientific literature connecting YY1 with epigenetic regulation.

YY1

The *yy1* gene

The human *yy1* gene is located on chromosome 14 band q32. A major gene transcription initiation site was mapped to 478 bp upstream of the ATG translation start site. The human *yy1* proximal promoter contains high G+C content, multiple Sp1 sites, but no TATA or CCAAT boxes, a typical feature of many housekeeping genes. The mouse *yy1* gene consists of five exons, with the first exon encoding 55% of the protein and the four zinc fingers are encoded by three different exons (Safrany and Perry, 1993; Yao *et al.*, 1998). Interestingly, the first intron of the *yy1* gene contains several YY1 binding sites, indicating the possibility of autoregulation (Kim *et al.*, 2009b). Recently, it was shown that the YY1 promoter and 5′-UTR regions contain G-quadruplex structures of the promoter are regulated by G4 resolvase 1 (Huang *et al.*, 2012).

The YY1 protein

The YY1 protein is highly conserved through evolution, with close homologues existing in most vertebrate species, and homologues and orthologues existing in some invertebrates (J.D. Kim *et al.*, 2007a). The highest conservation of YY1 at the amino acid level is in its DNA binding domain, which also mediates most of its protein–protein interactions.

The human *yy1* gene codes for a 414 amino acid protein belonging to the *GLI*–Krüppel family (Shi *et al.*, 1991). The main structural features of YY1 include four C2H2 zinc finger motifs at the C-terminal side and highly acidic regions at the N-terminal side. In between, there are several distinguishing features including a glycine/alanine-rich region (amino acids 154–201), proline/glutamine-rich region (amino acids 80–100), and a stretch of 11 histidine residues (amino acids 70–80).

Several biochemical studies have helped elucidate the primary functions of the YY1 structural domains (Austen *et al.*, 1997; Bushmeyer *et al.*, 1995; Bushmeyer and Atchison, 1998; Galvin and Shi, 1997). The DNA binding activity of YY1 has been shown to be mediated through the four zinc finger motifs. The N-terminal acidic region has been shown to act as a bipartite transcriptional activation domain, whereas the central (amino acids 170–200) and C-terminal regions have been shown to harbour two repression domains. The functional relevance of the histidine stretch is not yet understood. A considerable segment of YY1 is termed the ‘spacer’ region because of lack of clear functionality in the initial studies. However, this region has been shown to have some effects on YY1’s transcriptional activity, probably indirectly, by affecting the overall folding of the protein. This region also overlaps with the sites of direct interactions between YY1 and several other

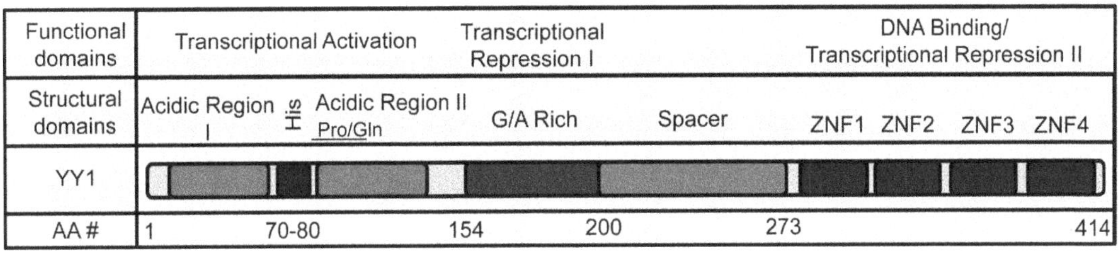

Figure 1.1 Illustration of the structural and functional domains of YY1.

proteins. Amino acids 331–399 have been shown to be critical for YY1's nuclear localization, but the exact localization signal within this region is not yet determined. The structural and functional domains of YY1 are portrayed in Fig. 1.1 (Austen *et al.*, 1997; Bushmeyer *et al.*, 1995).

YY1's DNA consensus binding site

There are thousands of YY1 potential binding sites throughout the human genome. Hundreds of them have been confirmed as real binding sites *in vivo* by ChIP studies and have been associated with specific functions.

The variable composition of the binding site can have significant bearing on the context-specific functionality of YY1. Analysis of YY1 binding sites from a panel of known activated or repressed genes in 1994 led to the extrapolation of the central consensus sequence CCATNTT (Shrivastava and Calame, 1994). The authors note that the YY1's conserved binding sequence in activated genes is slightly longer than the one in repressed genes. Also, the activating sequence is composed of higher G/C content in the flanking regions. From this observation, the authors hint on the possibility that stronger YY1 binding motifs correlate with activation. This fits well with a model where YY1 exerts repression by competing with an activator to occupy the binding site. Moderate or suboptimal binding of YY1 may be useful in this context. When the activator concentration increases to a certain threshold, it would displace YY1 and activate transcription (Shi *et al.*, 1997). The displacement model of YY1 activity been has been observed on the α-actin promoter by the serum response factor (SRF) (Lee *et al.*, 1992).

Another analysis of the YY1 consensus site was performed using bacterially expressed YY1 and the PCR-based binding site selection method. The derived consensus sequence from this analysis was 5′-(C/g/a)(G/t)(C/t/a)CATN(T/a)(T/g/c)-3′, with the upper case letters representing preferred bases (Hyde-DeRuyscher *et al.*, 1995). The authors note that the most important element is the core 5′-CAT–3′ tri-base sequence. These results were mostly confirmed by another independent analysis (Yant *et al.*, 1995). The work by Yant *et al.* (1995) revealed two core sequences (CCAT and ACAT) and observed that the CCAT core is associated with higher variability in the extended motif.

A recurrent element in the YY1 binding sequence is the CG motif 5′ to the CAT core. This element holds an important significance as it can be modified by methylation. It has been shown by some research groups that the CpG methylation can inhibit the YY1 DNA binding activity (Kim *et al.*, 2003). However, the methylation effect is still controversial because other groups did not observe this inhibition (Gaston and Fried, 1995a,b; Moseley *et al.*, 2012). It could also be a context-specific regulatory mechanism.

The base composition of a DNA sequence could affects its flexibility (Ortiz and de Pablo, 2011). There is evidence that YY1 might act, in some contexts, through bending the DNA strand (Natesan and Gilman, 1993). In this respect, the base composition of the binding site could modulate YY1's ability to bend DNA and significantly affect its function on that specific chromosomal region.

In 1998, Eliassen *et al.* (1998) showed that YY1 binds to an activating sequence (named the *alpha* element) within the coding region of replication-dependent histone genes (Bowman and Hurt, 1995; Bowman *et al.*, 1996; Eliassen *et al.*, 1998).

The composition of this element resembles the common YY1 consensus sequence; however, it greatly diverges at the 3′ end of the CAT core. In the more frequently occurring YY1 binding motif, the sequence 3′ to the core sequence is commonly composed of thymine bases. In the *alpha* element, it is largely occupied with guanine bases (5′-CGT**CAT**GGCG-3′). This enrichment could significantly increase the strength of binding. Importantly, this element presents a binding site for YY1 both on the plus and minus strands. This feature is not present in the more prevalent consensus site, where YY1 can either bind on minus or plus strands, thus determining its direction and possibly its function (Natesan and Gilman, 1993). The *alpha*-type motif was also identified in the macrosatellite DXZ4 on the inactive X-chromosome, and YY1 was shown to bind it both *in vitro* and *in vivo* (Moseley *et al.*, 2012). The biological implication of this variation in the binding motif is not yet clear and needs further investigation.

YY1's epigenetic partners

Most of YY1's functions have been shown to be mediated, and/or modulated, by its direct interactions with other proteins. As a transcription factor, YY1 has been shown to directly interact with the general factors of the transcription machinery, such as TBP, TFIIB, and $TAF_{II}55$ (Austen *et al.*, 1997; Usheva and Shenk, 1994). Moreover, YY1 directly interacts with numerous transcription factors including key tumour suppressors and oncoproteins, such as p53 (Gronroos *et al.*, 2004; Sui *et al.*, 2004), the retinoblastoma protein Rb (Delehouzee *et al.*, 2005), c-Myc (Shrivastava *et al.*, 1993; Shrivastava *et al.*, 1996). YY1's association with the transcription factor CTCF has particular significance to its role epigenetic regulation. CTCF is a zinc finger protein that has been described as the 'Master Weaver of the Genome' because of its numerous roles in regulating the epigenome (Phillips and Corces, 2009). Direct physical interaction between YY1 and CTCF has been documented. Moreover, clustering of YY1 and CTCF binding sites in close proximity and at high density has been observed in critical epigenetic regulatory domains. Several studies indicate a partnership of YY1 and CTCF in epigenetic regulation and chromatin shaping and organization (Donohoe *et al.*, 2007; Kim, 2008). The YY1-CTCF combinatory role will be discussed in more details later in this chapter.

Epigenetic regulation is primarily an alteration in the compaction state of the chromatin. YY1 has been shown to directly interact with multiple types of chromatin modifiers (Table 1.1). This interaction and recruitment has been mostly investigated in the context of transcriptional

Table 1.1 YY1 interacting chromatin modifiers

Interacting protein	Protein function	Reference
HDACs	Deacetylase	Coull *et al.* (2000), Liu *et al.* (2007), Luke *et al.* (2006), Ren *et al.* (2009), Sucharov *et al.* (2006), Yang *et al.* (1996), Yao *et al.* (2001)
P300/CBP	Acetyltransferase	Austen *et al.* (1997), Lee *et al.* (1995)
PRMT1	Methyltransferase (histone H4R3)	Rezai-Zadeh *et al.* (2003)
Ezh	Methyltransferase (PcG complex)	Caretti *et al.* (2004), Wang *et al.* (2004), Wilkinson *et al.* (2006)
PIASy	Sumoylation	Deng *et al.* (2007)
Ubc9	Sumoylation	Deng *et al.* (2007)
CtBP1	Corepressor/chromatin remodelling	Atchison *et al.* (2003), Srinivasan and Atchison (2004)
INO80	Chromatin remodelling	Cai *et al.* (2007), Wu *et al.* (2007)
RYBP	Component of the PcG repressor complex	Garcia *et al.* (1999)
CTCF	Transcription factor/insulator/chromatin remodelling.	Donohoe *et al.* (2007)

regulation of specific genes (Thomas and Seto, 1999). YY1 interacts with several histone-modifying enzymes, including histone acetyl transferases (HAT/CBP/p300) (Austen *et al.*, 1997; Lee *et al.*, 1995), deacetylases (HDACs) (Coull *et al.*, 2000; Liu *et al.*, 2007; Luke *et al.*, 2006; Ren *et al.*, 2009; Yang *et al.*, 1996), and methyltransferases (PRMT and Ezh1/2) (Baumeister *et al.*, 2005; Caretti *et al.*, 2004; Rezai-Zadeh *et al.*, 2003; Wilkinson *et al.*, 2006). Moreover, YY1 has been identified as a component of the INO80 chromatin remodelling complex and contributing to its functions, both in transcriptional regulation (Cai *et al.*, 2007) and homologous recombination-based repair (Wu *et al.*, 2007).

YY1 and the polycomb pathway

Polycomb group (PcG) was initially identified in *Drosophila* as a group of genes required for the maintenance, but not the initiation, of the repressive state of developmentally important homeotic genes (Kennison and Tamkun, 1988; Lewis, 1978; Paro, 1990; Ringrose and Paro, 2007). PcG proteins were shown to be necessary for proper anterior–posterior development (Simon *et al.*, 1992), and *PcG* mutant flies displayed severe homeotic transformations (Duncan, 1982). Although notable differences do exist, the general PcG pathway is functionally conserved throughout evolution and PcG components have been identified in vertebrates (Gould, 1997; Ringrose and Paro, 2004; Satijn and Otte, 1999; Schwartz and Pirrotta, 2013). In mammals, PcG proteins primarily regulate the expression of the developmental *hox* genes. The mammalian PcG pathway is important for skeletal-muscular development and haematopoiesis, but could also be involved in a plethora of recently identified pathways (Gould, 1997; Jacobs and van Lohuizen, 2002; Pirrotta, 1998; Schwartz and Pirrotta, 2013).

The silencing effects of PcG proteins are primarily mediated by repressive chromatin modifications like histone H3 deacetylation and H3K9 and H3K27 methylation (Cao *et al.*, 2002; Czermin *et al.*, 2002; Kuzmichev *et al.*, 2002; Muller *et al.*, 2002; Sewalt *et al.*, 1999; Tie *et al.*, 2003). PcG proteins function as high molecular weight complexes organized into two main complexes (PRC1 and PRC2) with differential roles within the PcG pathway [for review of canonical composition and newly discovered components see (Schwartz and Pirrotta, 2013)]. In *Drosophila* PcG protein complexes associate with target genes through sequences known as PRE (PcG response elements) (Brock and van Lohuizen, 2001; Pirrotta, 1997a,b, 1999; Ringrose and Paro, 2007; Satijn and Otte, 1999). In mammals, identification of PRE-like sequences is still an active field of investigation. Most PcG proteins lack direct sequence-specific DNA activity and recruitment of PcG complexes to their target genes remained enigmatic for a long time, even to date much of it is still not fully understood.

In 1998, Brown *et al.* (1998) identified PHO as the first *Drosophila* PcG protein that has sequence-specific DNA binding activity based on its interaction with the PRE of the *Drosophila engrailed* gene. Previous work had shown that *Drosophila pho* mutants display aberrant repression of homeotic genes, and that further depletion of the maternally deposited PHO results in defective phenotypes in cuticle formation and body segmentation (Breen and Duncan, 1986; Girton and Jeon, 1994). Analysis of the amino acid sequence of PHO showed that it is a homologue of the mammalian YY1 protein. PHO and YY1 share about 95% homology in the DNA-binding domain (100% homology in zinc fingers 2 and 3). Importantly, all the YY1 amino acids that have been shown by X-ray crystallography to make direct contact with DNA (Houbaviy *et al.*, 1996) are conserved in PHO. Consistent with this homology, PHO was able to efficiently bind the YY1-specific DNA consensus binding site in an *in vitro* binding assay. The remainder of the PHO amino acid sequence diverges significantly from that of YY1, with the peculiar exception of a small motif of about 26 amino acids (201–226) in what is termed the spacer region of YY1. This motif displays 82% homology between the two proteins. The significance of this motif will be discussed below.

Evidence for functional overlap between YY1 and PHO came from experiments demonstrating that YY1 can partially correct the phenotypic defects resulting from PHO loss in *pho* fly mutants. In addition, YY1 was able to repress a PcG responsive promoter in a PcG dependent manner, suggesting that YY1 is a vertebrate PcG protein (Atchison *et al.*, 2003). Moreover, exogenously expressed GAL-YY1 bound PREs *in vivo* and this binding was accompanied with PcG recruitment, deacetylation of histone H3K9 and H3K14, and methylation of histone H3K9 and H3K27 (Srinivasan and Atchison, 2004).

On the other hand, several points of functional divergence distinguish YY1 from PHO. YY1 represses transcription very early in fly embryogenesis, whereas PHO is known to act only at later stages of development (Atchison *et al.*, 2003). Furthermore, repressive activity of YY1 was shown to be dependent on CtBP, which physically interacts with YY1 (Atchison *et al.*, 2003; Srinivasan and Atchison, 2004). As a homodimer, CtBP could bind both YY1 and PcG proteins and bridge their interaction. YY1 contains the CtBP canonical binding motif, but PHO does not. Moreover, YY1 was able to repress a previously activated gene, a function that is not attributed to PcG. These results indicate that YY1 can act in multiple repressive pathways, in addition to the PcG pathway (Atchison *et al.*, 2003).

In vertebrates, insights into the specific roles of YY1 within the PcG pathway came from studies by Satijn *et al.* (2001). In their report, the authors show that YY1 can directly interact with the human EED protein (of the EED–EZH2 PcG complex/PRC1). However, YY1 did not interact with the HPC–HPH PcG/PRC2 protein complex, which contains HPC2 (human Polycomb 2), HPH (human Polyhomeotic), BMI1, and RING1. Specifically, EED was shown to interact with the C-terminal zinc finger domain of YY1 (amino acids 250–414). Functional analyses were performed in the *Xenopus* system by overexpressing XYY1 and XEED in embryos. The results of these experiments showed that EED and YY1 work synergistically and that the EED–YY1 axis is involved in neural induction and formation of the ectopic neural axis, but not in the induction of mesodermal tissues. Interestingly, the EED–EZH2, but not the HPC–HPH, complex is associated with deacetylase (HDAC) activity (Strouboulis *et al.*, 1999; van der Vlag and Otte, 1999). YY1 is known to interact with HDAC enzymes (Yao *et al.*, 2001). This suggests that YY1 could recruit deacetylase activity and contribute to the repressive histone code on a specific set of PcG target genes. It is worth mentioning that the *Drosophila* homologue of EED (esc – extra sex combs) has been shown to primarily function early in embryogenesis to establish stable and heritable repression (Gutjahr *et al.*, 1995; Struhl and Brower, 1982), and this is consistent with early YY1 effects. These findings help build a model of YY1's functional specificity within the PcG pathway, in addition to its sequence-specific targeting activity (Satijn *et al.*, 2001).

Knockout of YY1 in mice results in peri-implantation lethality, whereas reduced YY1 levels caused by heterozygote *yy1*$^{+/-}$ genotype result in pleiotropic effects (Donohoe *et al.*, 1999). These include severe developmental abnormalities that resemble the phenotypic abnormalities resulting from PcG mutations. These observations suggest a significant overlap between YY1 and PcG functions during mammalian development. Garcia *et al.* (1999) identified a novel murine protein, RYBP (Ring-1 and YY1 binding protein), as a new polycomb protein based on interaction with PcG components. Interestingly, the authors show that RYBP directly interacts with the first one and half zinc fingers (amino acids 274–333) of YY1. However, the combinatorial YY1–RYBP interaction with E2F2 and 3 on the CdC6 promoter was later shown to incur activation at the G1/S transcription of the cells cycle (Schlisio *et al.*, 2002). Thus, the role of RYBP in association with YY1 might be context specific and the activation/repression balance of this complex could depend on other interacting proteins. Further investigation is needed to tease apart these roles. A direct role for YY1 in mammalian PcG pathway was supported by the identification of YY1's interaction with Ezh2 and HDAC on genomic regions of silent muscle-specific genes of undifferentiated murine myoblasts (Caretti *et al.*, 2004). This association was correlated with H3K9 methylation. Moreover, YY1 knockdown abrogated the Ezh2 recruitment to DNA and resulted in reduction of H3K9 methylation. Furthermore, the YY1–Ezh2–HDAC complex was shown to dissociate from DNA upon activation the muscle-specific genes and progression towards differentiation (Carretti 2004). The role of YY1 as a PcG protein in mammals has been further reinforced with the recent identification of mammalian PRE-like sequences which contain the YY1 consensus binding site (Cuddapah *et al.*, 2012; Sing *et al.*, 2009; Woo *et al.*, 2010).

The YY1 protein contains multiple functional domains (Fig. 1.1). The DNA binding domain of YY1 serves an obvious essential function for the recruitment of PcG to specific DNA sequences. To understand what other YY1 domains are required for the PcG function, Wilkinson *et al.* (2006) analysed the ability of YY1 deletion mutants to recruit PcG complexes. Deletion of amino acids 201–226 abolished YY1's ability to repress a PcG-responsive reporter, but it did not affect the ability of YY1 to incur transient transcription repression on a non-PcG dependent reporter. Interestingly, this segment of YY1, fused to a heterologous DNA binding domain, was able to repress a PcG-dependent promoter, recruit PcG, and induce methylation of H3K27. This indicates that amino acids 201–226 of YY1 are sufficient and necessary for the recruitment of PcG complexes. However, histone deacetylation was not observed probably because another YY1 domain (170–200) is needed for HDAC recruitment. What is even more interesting is that amino acids 201–226 of YY1 represent the only motif that is conserved with PHO (apart from the DNA binding domain). The authors termed this motif the REPO domain (for REcruitment of POlycomb), and showed that it is conserved to a large extent in YY1 orthologues in 13 species. This is indicative of the highly conserved role of YY1 in PcG recruitment (Wilkinson *et al.*, 2006).

In summary, the PcG pathway is a critical regulatory pathway the significance of which in mammalian and human development is only beginning to surface (Schwartz and Pirrotta, 2013). Importantly, recent evidence indicates that mutations in PcG components are associated with human pathologies, like cancer (Radulovic *et al.*, 2013; Simon and Lange, 2008; Sparmann and van Lohuizen, 2006). Because of the complexity of the PcG pathway, particularly in mammals, we are only at the beginning of dissecting its roles and regulation. We have presented here solid evidence from the scientific literature which demonstrates that YY1 is a fundamental PcG component. In which specific PcG complex, on which specific gene sets, and at which stage of development and functional context is YY1 needed in the PcG pathway are open-ended questions that will hopefully be addressed in the coming years.

YY1 and genomic imprinting

Genomic imprinting is an epigenetic mechanism which regulates a specific subset of mammalian genes in a parent-of-origin manner (for reviews see Bartolomei, 2009; Fedoriw *et al.*, 2012). In this context, only one of the two alleles of an imprinted gene is expressed while the other allele is epigenetically silenced. The mode of expression or repression of each allele depends on the parental origin of its chromosomal locus. The imprints are established during gametogenesis in the parents, transmitted to the zygote, and later maintained in the somatic tissues of the adult offspring. On the other hand, these imprints are erased in germ cells to allow for new imprints to be established during gametogenesis, depending on the sex of the offspring.

A *cis*-acting regulatory element called the imprinted control region (ICR) is present in almost all imprinted genes. ICRs are essential for the establishment and maintenance of the epigenetic patterns associated with imprinting. These regions themselves are often the subject of multiple layers of epigenetic modifications in an allele specific manner. ICR often overlap with differentially methylated regions (DMRs) that play important roles in defining the parental origin.

Approximately 100 imprinted genes have been identified in the human and mouse genomes, but an additional few hundred genes are predicted to be controlled by imprinting mechanisms (Luedi *et al.*, 2005, 2007; Morison *et al.*, 2005). Imprinted genes are usually organized into chromosomal clusters that could span several hundreds of kilobases.

Because of its fixed establish–maintain–erase cycle through the generations, genomic imprinting has served as a robust experimental model to understand the mechanisms of epigenetic regulation. Moreover, abnormal establishment or maintenance of this mechanism could result in 'loss-of-imprinting' (LOI) on some of the imprinted genes. The significance of normal genomic imprinting to human health is underscored by the fact that LOI has been associated with several human diseases including Beckwith–Wiedemann syndrome, Silver–Russell syndrome, Prader–Willi syndrome, Angelman syndrome (Horsthemke and Wagstaff, 2008; Ideraabdullah *et al.*, 2008), and even cancer (Jelinic and Shaw, 2007; Lim and Maher, 2010b).

Initial indication of a role for YY1 in genomic imprinting came from the identification of multiple YY1 binding sites in the first intron of the imprinted *Peg3* gene (paternally expressed gene 3) (Kim *et al.*, 2003). The direct binding of YY1 to these sites was demonstrated in *in vitro* binding assays and *in vivo* by ChIP studies using human cells and mouse brain tissue (Kim *et al.*, 2003).

The YY1 sites in the *peg3* ICR are organized into tandem repeats. To identify similar YY1-regulated regions, Kim *et al.* (2006) searched the genome specifically for arrays of clustered repeats of YY1 binding sites. Thirty-five human and 21 mouse genomic regions were found containing tandem arrays of YY1 binding sites. Four of the 21 identified regions in the mouse genome turned out to be ICRs, which is a significant enrichment considering the low percentage of imprinted loci compared to the size of the genome (Kim *et al.*, 2006). Binding of YY1 to these imprinted sites *in vivo* was confirmed by ChIP studies. Furthermore, an independent unbiased investigation of the conserved elements in imprinted genes also showed enrichment of YY1 sites (Hutter *et al.*, 2010). The YY1 binding sites within these regions have been well maintained throughout evolution, further reinforcing their fundamental regulatory role (Hutter *et al.*, 2010; Kim *et al.*, 2003, 2006).

Interestingly, the majority of the identified YY1 binding sites are organized in one orientation relative to the transcriptional direction. Moreover, most are located in the first intron, exon, or promoter regions of individual genes (Kim *et al.*, 2006); indicating that the function of the YY1 sites is the transcriptional regulation of imprinted genes. In reporter assays, the region containing the YY1 binding sites in the *Peg3* gene acted as an insulator (Kim *et al.*, 2003). Therefore, the YY1 sites in *Peg3* could play a similar role to that of the CTCF binding sites in the H19/Igf2 imprinted locus (Bell *et al.*, 1999; Engel *et al.*, 2006; Hark *et al.*, 2000; Kurukuti *et al.*, 2006; Zhang *et al.*, 2011a). However, the insulator activity of the clustered YY1 sites observed in *peg3* was not observed in the promoters of the other imprinted loci that were tested (Kim *et al.*, 2006), which suggests locus-specific roles of the YY1 binding sites.

Evidence for the functional involvement of YY1 in the regulation of imprinted genes was demonstrated by RNAi-based knockdown of YY1 in cultured mouse cells (J.D. Kim *et al.*, 2007b). Reduction of YY1 protein levels caused alterations in the expression of the imprinted genes of the *Peg3* and *Gnas* domains. Upon YY1 knockdown, there was an increase in the expression of the *Peg3*, *Usp29*, *Zim1* and *Zim2* genes, within the *Peg3* domain. Interestingly, antagonistic effects were observed for the genes in the *Gnas* domain. Whereas the *Nespas* gene was up-regulated, the *Nesp* gene was down-regulated. These two genes are known to have sense/antisense regulation. The antagonistic effects of YY1 knockdown on the *Nespas/Nesp* genes are therefore in conformity with their physiological regulation patterns. This is supportive of a specific role for YY1 in this mechanism. Similar effects on the expression of these imprinted genes were observed in YY1 knockdown transgenic mice (Kim and Kim, 2008).

DNA methylation is a major regulatory mechanism in the imprinting process. Differentially methylated regions (DMRs) are a hallmark of ICRs of imprinted genes. The identified YY1 binding sites are located within the DMRs of the respective imprinted genes (Kim, 2008; Kim *et al.*, 2003, 2006); suggesting the involvement of YY1 in the regulation of DNA methylation. Indeed, reduced YY1 levels in cultured mice cells (J.D. Kim *et al.*, 2007b) and in transgenic mice (Kim and Kim, 2008) was correlated with disruption of the normal DNA methylation pattern in the ICRs of multiple imprinted genes. This was particularly evident for the ICRs of the genes that contain clustered YY1 binding sites.

As mentioned earlier, DNA methylation of DMR regions within imprinted genes is established during gametogenesis. Conditional knockdown of YY1 in mice greatly affected the *de novo* DNA methylation of *Peg3* and *Xist*, but not *Nespas* and *H19*, during oogenesis (Kim *et al.*, 2009a). The mechanism through which YY1 affects the *de novo* methylation during oogenesis is not yet known. However, it is hypothesized that the high numbers of YY1 binding sites in an intriguingly unidirectional pattern, would result in a high YY1 density at these loci. This might trigger a certain signal to methylate these regions (Kim, 2008; Kim and Kim, 2008). This hypothesis waits further challenging experimentation for proper validation.

Interestingly, YY1 knockdown had no significant effect on the establishment of DNA methylation during spermatogenesis (Kim and Kim, 2008). One possible explanation of this observation is that YY1's role in establishing *de novo* methylation on imprinted genes is specifically inhibited during spermatogenesis. In support of this view, the Rex1 (reduced expression gene 1) protein was found associated with YY1 consensus motifs in a manner that would competitively inhibit YY1 binding (Kim *et al.*, 2011). Rex1/Zfp42 is a protein closely related to YY1, with high sequence homology in the DNA binding domain. It is thought to have duplicated from YY1 via retrotransposition in placental mammals (J.D. Kim *et al.*, 2007a). In contrast to the ubiquitous expression of YY1, Rex1 is expressed very early in embryonic development. Rex1 levels are highest in pre-implantation embryos, with modest expression persisting later in the placenta and during spermatogenesis (Rogers *et al.*, 1991). In fact, Rex1 is a well-known marker of self-renewing stem cells, and its expression drops dramatically upon the induction of differentiation (Hosler *et al.*, 1989; Mongan *et al.*, 2006). Therefore, it seems that the emergence of Rex1 during mammalian evolution has established a context-specific regulatory mechanism of YY1's imprinting functions. This mechanism would inhibit YY1 from establishing *de novo* methylation specifically during paternal spermatogenesis but not maternal oogenesis. In line with this inhibitory effect, knockout of Rex1 resulted in hypermethylation of the regions containing YY1 binding sites in the *Peg3* and *Gnas* loci (Kim *et al.*, 2011). YY1 is known to be active during spermatogenesis, in the presence of Rex1, in establishing H3K27methylation in the pericentromeric chromosomal regions (Wu *et al.*, 2009). Therefore, the Rex1 competitive inhibition of YY1 appears to be specific to its role in imprinting.

Another regulatory mechanism of YY1's function in imprinting could lie in the methylation process itself. YY1 binding sites in the imprinted domains have a high conservation of a CpG element, the site for DNA methylation. Kim *et al.* (2003) showed that methylation of the CpG element within the YY1 consensus site can prevent the direct binding of YY1, when tested by *in vitro* binding assays. *In vivo*, ChIP studies, using human K562 cells and mouse brain tissue, showed that YY1 only binds the paternal allele of the *Peg3*. Interestingly, the YY1 binding sites in the *Peg3* gene are methylated only on the maternal chromosome.

Taken together these observations suggest a model in which YY1 plays a role in establishing the *de novo* methylation of the maternal, but not the paternal, *Peg3* gene. Later in development, this methylation itself inhibits YY1 binding to the maternal gene, and therefore YY1 is found only on the paternal gene. At this stage, the window of de novo methylation is closed and YY1 cannot induce DNA methylation on the paternal locus. In fact, it is very likely that YY1 might play a transcriptional activation role on the expressed paternal *Peg3* gene at this later stage.

The only caveat in this model is that the direct effects of CpG methylation on YY1 is controversial, because other reports in the literature demonstrate that DNA methylation does not inhibit the binding of YY1 to its target site (Gaston and Fried, 1995a,b; Moseley *et al.*, 2012). These studies could have used different assay conditions and therefore did not detect the subtle effects of methylation binding that was observed with Kim *et al.* (2003) Subtle effects on the binding affinity *in vitro* could have significant effects on the physiological binding *in vivo*. Alternatively, the chromatin associated with methylated DNA *in vivo* is probably modified by multiple mechanisms, and thus does not allow access to YY1 binding.

Genomic imprinting involves multiple layers of epigenetic regulation including several histone modifications. As discussed earlier, YY1 is now known to be an important part of the Polycomb pathway, particularly in the sequence-specific recruitment of the PcG complex. The role of the PcG pathway in genomic imprinting was first discovered by Mager *et al.* (2003). These pioneering experiments showed that *eed*$^{-/-}$ mice displayed abnormal imprinting in a specific subset of imprinted genes (Mager *et al.*, 2003). Notably, EED is the same component of the PRC2 that was shown to directly interact with YY1. Within this complex, YY1 is the component responsible for targeting EED (and the PRC2) to certain specific DNA sequences. Another part of the PRC2 complex is the Ezh2 histone methyltransferase and the YY1-interacting histone deacetylase HDAC. It is not currently known if YY1's polycomb recruiting activity plays a role in the establishment or maintenance of genomic imprinting. Moreover, YY1 could possibly play a PcG-independent role in establishing imprinting histone marks.

Finally, YY1 has been shown to be involved in the regulation of the imprinted *SNRPN* locus. The disruption for the normal paternal expression of this imprinted locus can lead to the Prader–Willi syndrome, a neurogenetic disorder. Abnormal regulation of *SNRPN* has been shown to be caused by deletion of control regions encompassing the YY1 binding sites (Rodriguez-Jato *et al.*, 2005).

In conclusion, a role for YY1 in regulating imprinted genes is well supported, but not yet fully understood. It is very important to note that the reports cited in this section analysed only a fraction of the imprinted genes. With YY1, there is always a large margin of context–specific variations and further research in the future is needed to help elucidate the multifaceted roles of YY1 in genomic imprinting.

YY1 and the inactivation of the X-chromosome

Mammalian dosage compensation, which involves the inactivation of one of the two X-chromosomes in the female, is probably the best-characterized experimental model for studying epigenetic regulation (for reviews see Deng *et al.*, 2014; Lucchesi *et al.*, 2005). Most of the currently known epigenetic regulatory mechanisms have been shown to be involved in the precise sequential initiation and then maintenance of X-chromosome inactivation (XCI). Non-coding RNAs, DNA methylation, multiple histone post-translational modifications, histone variant-specific substitutions, and large-scale heterochromatic changes are all needed for the XCI. Many of the implicated epigenetic regulators are now known but it is believed that many other factors are not yet discovered.

The main control programme of the inactivation process is found in one locus called the X-chromosome inactivation centre (Xic) (Clerc and Avner, 2003). The primary function of the Xic depends on a binary switch between two sense–antisense non-coding X-chromosome loci known as the *Xist* and *Tsix*. Expression of the *Xist* gene produces non-coding RNA that coats the future inactive chromosome in *cis* and is the central element in the XCI. However, when *Tsix* is expressed, it suppresses the *Xist* expression and the X-chromosome remains active. The choice is straightforward in males with only one X-chromosome that has to remain active. On the other hand, which of the two X-chromosomes is inactivated and which remains active appears to be due to a random process, and the regulation of this decision is not yet fully understood (for review see Navarro and Avner, 2010).

Over the past two decades, accumulating evidence showed a clear implication of YY1 in the regulation of the X-chromosome inactivation at multiple capacities. An initial clue came from the identification of a YY1 DNA consensus binding site in the human minimal *Xist* promoter (Hendrich *et al.*, 1997). The YY1 consensus site was shown to be conserved in the murine, lupine and equine *Xist* promoters. Hendrich *et al.* (1997) showed that YY1 can directly bind to this sequence in *in vitro* binding assays. In another report, clusters of YY1 binding sites were also identified in the first exon of *Xist* and second intron of *Tsix* (Kim, 2008). The tandem array organization of these clusters appears to be conserved throughout mammalian evolution. When tested in reporter assays, the YY1 binding sites increased the transcriptional strength of the promoters, suggesting that YY1 can play a transcriptional activation role on both *Xist* and *Tsix*.

A more functionally relevant role of YY1 was shown by Donohoe *et al.* (2007). In this study, the authors aimed at identifying transcription factors that could act as partners for CTCF in the XCI. The role of CTCF in XCI has long been established [reviewed in Ohlsson *et al.*, 2010; Phillips and Corces, 2009). Interestingly, YY1 binding sites were highly enriched near the CTCF sites. This was particularly true at in the 4-kb region at the 5′ end of *Tsix* and in the *DXPas34* repeat element (which works as an enhancer of *Tsix* expression) 1.5 kb downstream of the *Tsix* promoter. The pairing of CTCF-YY1 sites is conserved between mouse and human *Xic* (Donohoe *et al.*, 2007). Experimentally, the authors showed that YY1 can bind these sequences in *in vitro* binding assays and by ChIP studies in mouse ES cell. Further *in vivo* insight was provided by investigating the XCI process in mice embryos carrying homozygous ($yy1^{-/-}$) or heterozygous ($yy1^{+/-}$) deletions of YY1. The $yy1^{-/-}$ embryos can be monitored only during the pre-implantation period, because this mutation is lethal after peri-implanation. The heterozygous ($yy1^{+/-}$) embryos were observed through the peri-implantation blastocyst period at the ICM (Inner Cell Mass). In both cases, there was a clear aberration of the *Xist* expression in correlation with reduced YY1 levels. Furthermore, $yy1^{+/-}$ mouse embryonic stem cells (ES) – pre and post differentiation – had severely diminished *Tsix* levels and an increase in *Xist* levels. These effects were exacerbated after additional knockdown of YY1 levels with specific shRNA. The knockdown of CTCF showed similar changes of *Tsix* and *Xist* expression levels. Because of this and the pairing of YY1 binding sites to those of CTCF, in addition to the physical interactions between the two proteins, the authors proposed a YY1-CTCF binary switch to control XCI. In this switch, YY1 serves as a cofactor for CTCF.

The combined data of these reports indicate that YY1 is involved in the transcriptional regulation of both *Tsix* and *Xist*, and that proper balance of the *Tsix/Xist* circuit is disrupted in the absence of YY1. However, much remains to be learned. YY1 binds to both alleles in undifferentiated and differentiated mouse ES cells, pre and post XCI (Donohoe *et al.*, 2007; Navarro and Avner, 2010). Does YY1 have a yin–yang function of a repressor switching to an activator of *Xist* during XCI and how is this switch regulated?

Chromatin modifications along the *Xist* gene are known to be important for the regulation of *Xist* expression, and for the balance between the *Tsix/Xist* switch. In undifferentiated ES cells, the silenced *Xist* promoter is characterized by high levels of repressive marks (H3K9 and CpG methylation) and low levels of permissive marks (H3K4 methylation and H3K9 acetylation); whereas the *Tsix* promoter displays the opposite euchromatic features (Navarro *et al.*, 2006). During differentiation, the *Xist* promoter on the future inactive X assumes an active euchromatic pattern with the recruitment of activating marks like H3K4 methylation (Marks *et al.*, 2009). This is also accompanied by silencing of *Tsix* and the switching of its promoter to a repressive mode with enrichment of H3K27 methylation (Lee *et al.*, 1999; Marks *et al.*, 2009). YY1's ability to recruit chromatin modifiers is well documented, as detailed earlier in this chapter. Thus, it is very likely that YY1 plays a role in shaping the *Tsix/Xist* chromatic landscape. Surprisingly, there is currently no evidence supporting (or refuting) this hypothesis.

Intriguingly, in the Donohoe *et al.* (2007) report, even with the increase of *Xist* expression in $yy1^{+/-}$ ES cells, the authors did not observe coating of the X-chromosome with Xist RNA. In an interesting and clever set of experiments, Jeon and Lee demonstrated that YY1 binds the *Xist* gene locus at three sites that are needed to the proper tethering of the Xist RNA to the X-chromosome (Jeon and Lee, 2011). Moreover, the authors showed that YY1 can directly and specifically bind to the Xist RNA. This led to the proposal that YY1 can tether the Xist RNA to the future inactive X-chromosome and serve as a docking receptor protein for Xist particles. This finding addresses a long standing question of the mechanism of RNA tethering to the inactive X. However, although dual binding properties of YY1 to both DNA and RNA were demonstrated, it is not fully understood if YY1 can simultaneously bind both molecules. Although this is one possibility, another plausible hypothesis could be envisaged based on recent findings that YY1 molecules can oligomerize and form a ring structure (Lopez-Perrote *et al.*, 2014). This newly discovered feature of YY1 could be important for its DNA–RNA bridging role. Jeon and Lee show that the binding of YY1 to the *Xist* locus is the primary nucleation site. Recently, YY1 has been shown to associate with other loci on the inactive X. Moseley *et al.* (2012) provided evidence that YY1 binds the macrosatellite DXZ4, specifically in the inactive X-chromosome. This additional association is probably not needed for the nucleation process; however, it could serve to spread the coating along the inactive X.

During development, X-chromosome inactivation occurs independently at two different stages. An early imprinted X inactivation takes place at the two- to four-cell pre-implantation stage by specifically inactivating the paternal X (Okamoto *et al.*, 2004). This inactivation persists until the blastocyst stage, when the X is reactivated in the inner cell mass (ICM), but remains inactive in the extraembryonic tissue (Takagi and Sasaki, 1975). X-chromosome inactivation is re-established during the differentiation of the ICM, but this time it appears to be a random selection. From all the accumulating data, YY1 binds X-chromosome sites at all these stages, pre and post differentiation of ES cells. It is currently not clear whether YY1's roles in genomic imprinting and as a PcG protein could lead to differential YY1 roles in the imprinted versus the random X inactivation.

Careful experimentation and innovative approaches will have to devised in future studies to address the complicated and multifaceted roles of YY1 in this fundamental biological process.

Finally, how does the research in mice translate to human biology? The X-chromosome inactivation mechanism does have a considerable homology between the two species, but also significant differences. The discovered YY1 binding sites and their roles seem to be conserved between mice and humans and it is very likely that most of YY1's functions, discovered in mice, occur similarly in humans.

YY1 and heterochromatin establishment during spermatogenesis

Meiotic recombination is a critical step of normal mammalian spermatogenesis. This process is heavily dependent on the proper execution of induced, or programmed, DNA double strand breaks and the subsequent homologous repair. This process by itself is greatly affected by epigenetic modifications and the status of chromatin compaction. Current evidence reveals that the heterochromatin is 'refractory' to gamma H2AX modification and is mostly excluded from meiotic recombination events (J.A. Kim *et al.*, 2007).

Because of previous finding of YY1 involvement of DNA repair, Wu *et al.* (2009) investigated if YY1 is needed for meiotic recombination during spermatogenesis. Immunocytostaining of mouse spermatocytes revealed an enrichment of YY1 localization in the heterochromatin fraction, co-localizing with HP1-gamma and histone H3K9me3. Moreover, conditional deletion of YY1 resulted in a defective heterochromatic status, with reduced HP1-gamma and H3K9me3 staining. No changes were detected in the H3K9me2 modification, which is associated with repressive states of the euchromatin (Peters *et al.*, 2003). This phenotype was accompanied by an increase in DNA double strand breaks and abnormal progression through meiosis.

However, it was not clear whether the perturbation of meiosis resulting from YY1 deletion is mainly due to its role in heterochromatin regulation or its direct involvement in DNA repair. A plausible scenario is that YY1 is needed for both layers, and possibly others yet to be discovered.

Regulation of YY1

Compared with what is known about YY1's downstream functions, target genes, and pathways, very little is known about the upstream regulation of YY1.

Regulation of YY1 protein levels

YY1 is ubiquitously expressed in most tissues. Fluctuations in the YY1 mRNA levels have not been reported as a major regulatory mechanism for YY1 during development. The *yy1* promoter contains multiple Sp1 sites, but no TATA or CCAAT boxes, reminiscent of constitutively expressed housekeeping genes. However, the expression of the *yy1* gene has been shown to be modulated by several factors. It has been proposed that YY1 can regulate its own expression (Flanagan, 1995). This was later supported by the detection of multiple YY1 binding sites in the first intron of the *yy1* gene (Kim *et al.*, 2009b). In this model, at low levels, YY1 can act as an activator.

At higher levels and higher occupancy of the YY1 sites, YY1 turns into a repressor of its own expression. Another interesting element in the *yy1* gene is the presence of DNA G-quadruplex structures. The resolving of these structures by G4 resolvase 1 can increase expression (Huang *et al.*, 2012). In many cases, the increase in G4 resolvase 1 has been associated with tumorigenesis, which correlates well with the increase of YY1 levels in cancer cells (Huang *et al.*, 2012).

Specific cleavage of the YY1 protein has been associated with certain biological contexts. It has been shown that YY1 is specifically cleaved by caspases during apoptosis (Krippner-Heidenreich *et al.*, 2005), by calpain II during muscle differentiation (Walowitz *et al.*, 1998), and by a cathepsin B-like protease in NT2 (human teratocarcinoma) cells as they were differentiated with retinoic acid treatment (Pizzorno, 2001). Interestingly, in all these mechanisms, the N-terminal activation domain is cleaved and the C-terminal DNA binding/repression domain remains intact. In some of these examples, YY1 is eventually completely degraded. However, this transient cleavage mechanism converting YY1 to only a repressor protein could have significant implications. Further research is needed to elucidate this mechanism.

Regulation of YY1 by post-translational modifications

The YY1 protein has been reported to be the subject of many post-translational modifications. An intricate interplay of dynamic acetylation/deacetylation has been demonstrated on YY1 (Yao *et al.*, 2001). The association between YY1 and the p300 acetyltransferase lead to YY1's acetylation in the region comprising residues 170–200. This region represents one of the two repression domains of YY1. PCAF (p300-CBP associated factor) was also shown to have the ability to acetylate that region of YY1. Moreover, PCAF can also acetylate the DNA binding domain region; which contains the second repression domain. Acetylation of YY1 enhances its interaction with HDAC, which then deacetylates the 170–200 region, but not the DNA-binding domain. The authors also show that acetylation of YY1 in the 170–200 region increase its repressive activity whereas acetylation of the DNA-binding domain decrease its DNA binding activity *in vitro* (Yao *et al.*, 2001). The physiological relevance of this interplay is not fully understood. Another group showed that acetylation of YY1 is actually required for its DNA-binding ability to the promoter region of the mouse homeobox gene Otx2 *in vivo*, whereas it has no effect on binding of YY1 to an enhancer region upstream of the same gene (Takasaki *et al.*, 2007).

YY1 has been proposed to be a phosphoprotein by multiple research groups and that phosphorylation can affect its DNA binding activity (Becker *et al.*, 1994; Kaludov *et al.*, 1996; Patten *et al.*, 2000). The first direct evidence for YY1 phosphorylation was demonstrated by Rizkallah and Hurt in 2009. YY1 was shown to be phosphorylated in its DNA binding domain during cell division (mitosis), which results in the inactivation of its DNA binding activity (Rizkallah and Hurt, 2009). Interestingly, this phosphorylation occurs in the linker sequences between the zinc finger domains, and was later shown to be a common mechanism to inactivate the DNA binding activity of all C2H2 zinc finger proteins during mitosis (Rizkallah *et al.*, 2011a). Recently, the oncogenic kinase TOPK/PBK was identified as the master regulator that phosphorylates YY1 and other C2H2 zinc finger proteins, leading to their dissociation from the mitotic chromosomes (Rizkallah *et al.*, 2014). Furthermore, Plk1 (Polo-like kinase 1) and AurA (Aurora Kinase B) have been shown to phosphorylate YY1 at the entry into mitosis. Plk1 and AurB phosphorylate YY1 in the activation and repression domains, respectively (Kassardjian *et al.*, 2012; Rizkallah *et al.*, 2011b). The functional implications of these phosphorylation events need further investigation. On the other hand, phosphorylation of YY1 by CKII (Casein kinase 2) in the caspase cleavage site was shown to protect YY1 from cleavage during early stages of apoptosis (Riman *et al.*, 2012). Currently, there are multiple known phosphorylation sites on YY1, mapped in large-scale proteomics analyses (Beausoleil *et al.*, 2004; Molina *et al.*, 2007; Nousiainen *et al.*, 2006). However, many of these phosphorylation events are not yet correlated with specific biological contexts or functional effects.

Furthermore, several other types of post-translational modifications have been shown on

YY1, such as O-linked-glycosylation (Hiromura *et al.*, 2003), sumoylation (Deng *et al.*, 2007), ubiquitination (Jeong *et al.*, 2014), poly(ADP-ribosyl) ation (Oei and Shi, 2001), and S-nitrosation (Hongo *et al.*, 2005). Interestingly, many of these modifications occur on residues that are in close proximity within YY1. A possible interplay between the various modifications on these sites has not yet been properly investigated.

YY1 and cancer

Cancer has long been considered primarily a genetic disease. With the recent advancements in deciphering the epigenome, it became clear that epigenetic regulation plays a major role in the initiation, development, and progression of cancer. Cancer cells display both localized and global alterations in their chromatin configuration and nuclear architecture. Today, the epigenetics of cancer cells is under wide exploration at the level of basic understanding of cancer aetiology, but also at the level of diagnostics and therapeutics (Easwaran *et al.*, 2014; Rodriguez-Paredes and Esteller, 2011).

The involvement of YY1 in cancer biology was initially proposed because of its role in cell proliferation. In addition, an array of oncoproteins and tumour suppressor proteins have been shown to be regulated by YY1, such as p53 (Gronroos *et al.*, 2004; Sui *et al.*, 2004), c-Myc (Riggs *et al.*, 1993; Shrivastava *et al.*, 1993, 1996), and Rb (Delehouzee *et al.*, 2005; Petkova *et al.*, 2001). Furthermore, YY1 has been shown to confer resistance to tumour cells at multiple levels. For example, YY1's repression of death receptors, such as Fas and DR5, revealed roles in resistance to external apoptotic stimuli (Garban and Bonavida, 2001; Huerta-Yepez *et al.*, 2009). Also, YY1's role in regulating p53 and its function in DNA repair with the INO80 complex is needed for genomic integrity and resistance to genotoxic stress (Gronroos *et al.*, 2004; Sui *et al.*, 2004; Wu *et al.*, 2007). YY1 can exert tumorigenic effect at multiple stages of cancer development. de Nigris *et al.* (2008) showed that depletion of YY1 can significantly decrease bone cancer invasion and metastasis. The various proposed roles of YY1 in tumour development and resistance to therapy, especially in different cancer types, are numerous (for comprehensive reviews analysing the roles of YY1 in cancer see Atchison *et al.*, 2011; Gordon *et al.*, 2006; Nicholson *et al.*, 2011; Zaravinos and Spandidos, 2009, 2010; Zhang *et al.*, 2011b).

Both up- and down-regulation of YY1 have been observed in different types of tumours. Overall, the mechanisms of YY1 deregulation are not fully understood. However, some tissue-specific mechanisms have been documented. For example, the 3′-untranslated region (3′-UTR) of YY1 mRNA was shown to be the target of microRNA-34 (Wang *et al.*, 2014). Down-regulation of this microRNA in gastric cancer tissues was correlated with up-regulation of YY1 and contribution to gastric carcinogenesis. In addition, the promoter of YY1 was found to be a frequent target for the insertion of the Graffi murine leukaemia virus, an event known to induce myeloid leukaemia. This insertion results in an increase in YY1 expression levels (Erkeland *et al.*, 2003). However, with the high diversity of mechanisms involved in tumorigenesis of different types of cancer, it is likely that YY1 expression is altered by many different pathways in a tissue-specific manner.

With the exception of one point mutation in the DNA-binding domain of YY1 in sporadic insulinomas (Cao *et al.*, 2013), no other mutations have been detected on YY1 in cancer cells. However, a summation of YY1 studies clearly indicates that tight regulation of YY1 levels is critical for normal cellular functions and for development. YY1 cellular concentration could affect the functionality of YY1, and could even switch its yin yang activation/repression effects (Bushmeyer *et al.*, 1995). Fluctuations in YY1 protein levels could also impact the normal imprinting of its target genes discussed earlier (Thiaville and Kim, 2011). Slight changes in the expression of imprinted genes could have dire consequences on human health (Hirasawa and Feil, 2010). Loss of imprinting (LOI) on several of YY1's target genes has also been clearly correlated with tumorigenesis (Jelinic and Shaw, 2007; Lim and Maher, 2010a).

Importantly, the normal function of the polycomb complexes depends on the finely tuned stoichiometry of its components. Therefore, alteration of YY1 levels in cancer cells can have significant effects on the polycomb pathway. As discussed earlier, YY1 recruits the Ezh2 methyltransferase as part of the PRC2 polycomb complex (Caretti *et al.*, 2004; Satijn *et al.*, 2001). Disruption of normal Ezh2 function through differential expression and/or mutation has been shown to have broad implications in tumorigenesis. Moreover, Ezh2 inhibitors are showing great promise as novel therapeutic strategies (Lund *et al.*, 2014; McCabe and Creasy, 2014; Simon and Lange, 2008). A correlation of high expression of YY1 and Ezh2 has been documented in brain tumours (Crea *et al.*, 2010). Moreover, YY1 recruitment of Ezh2 to the promoter of the tumour suppressor APC (Adenomatous Polyposis Coli) causes an increase in trimethylation of histone H3K37 and repression of APC expression. YY1-mediated repression of APC is a critical factor in the progression of human endometrioid endometrial adenocarcinoma (Yang *et al.*, 2013). Moreover, p53 has been shown to suppress the expression of Ezh2 (Tang *et al.*, 2004). YY1 overexpression in tumour cells could indirectly contribute to the overexpression of Ezh2 by down-regulation of p53 protein levels.

On the other hand, down-regulation of YY1 was observed in the invasive human breast cancer cell line HS578T, compared with the non-invasive MCF7 cells. YY1 was shown to positively regulate the expression of heterochromatin protein 1 (HP1), and that down-regulation of YY1 in these cells was correlated with down-regulation of HP1. Reduced levels of HP1 could lead to impaired heterochromatic organization and contribute to the invasiveness of the tumour (Lieberthal *et al.*, 2009).

In summary, deregulation of YY1 could have significant impact on the epigenetic landscape and aggressiveness of cancer cells. Although YY1 has been proposed as a prognostic marker and a promising target in some cancers, much more research is required to better understand its diverse tumorigenic effects.

Conclusions

Today, there is strong and ample evidence in the scientific literature to support the conclusion that the YY1 protein is a central player in epigenetic regulation. YY1 can function directly and indirectly at the DNA, RNA, and protein levels to modulate the epigenome. It is specifically this multifunctionality that makes the study of YY1 particularly challenging. Currently, it is believed that cellular YY1 is divided into numerous subpopulations that are differentially modified and interact with various combinations of other proteins. A comprehensive approach to fractionate these populations would provide most valuable insights into the biochemistry of YY1 and its effects on human biology.

The largest gap in our knowledge of YY1, at the moment, is in its regulation. Because in most biological contexts YY1 levels do not fluctuate significantly, it is believed that post-translational modifications play a major role in this regulation. Although many modifications have been reported to occur on YY1, their functional effects and physiological relevance remain obscure. Importantly, the interplay of all these context-specific modifications and effects on the interactions between YY1 and other proteins has not been properly investigated. For a long period of time, we lacked the adequate tools to study these modifications. The recent development of YY1 modification-specific antibodies is a step in the right direction.

The current knowledge of the significance of YY1 in human biology and disease dictates a better understanding of its roles and regulation. It is believed that we have only seen the tip of the YY1 iceberg; much further research is needed to dissect the 'YY1 code'.

References

Affar el, B., Gay, F., Shi, Y., Liu, H., Huarte, M., Wu, S., Collins, T., and Li, E. (2006). Essential dosage-dependent functions of the transcription factor yin yang 1 in late embryonic development and cell cycle progression. Mol. Cell. Biol. *26*, 3565–3581.

Atchison, L., Ghias, A., Wilkinson, F., Bonini, N., and Atchison, M.L. (2003). Transcription factor YY1 functions as a PcG protein *in vivo*. EMBO J. *22*, 1347–1358.

Atchison, M., Basu, A., Zaprazna, K., and Papasani, M. (2011). Mechanisms of Yin Yang 1 in oncogenesis: the importance of indirect effects. Crit. Rev. Oncog. *16*, 143–161.

Austen, M., Luscher, B., and Luscher-Firzlaff, J.M. (1997). Characterization of the transcriptional regulator YY1. The bipartite transactivation domain is independent of interaction with the TATA box-binding protein, transcription factor IIB, TAFII55, or cAMP-responsive element-binding protein (CPB)-binding protein. J. Biol. Chem. *272*, 1709–1717.

Bartolomei, M.S. (2009). Genomic imprinting: employing and avoiding epigenetic processes. Genes Dev. *23*, 2124–2133.

Baumeister, P., Luo, S., Skarnes, W.C., Sui, G., Seto, E., Shi, Y., and Lee, A.S. (2005). Endoplasmic reticulum stress induction of the Grp78/BiP promoter: activating mechanisms mediated by YY1 and its interactive chromatin modifiers. Mol. Cell. Biol. *25*, 4529–4540.

Beausoleil, S.A., Jedrychowski, M., Schwartz, D., Elias, J.E., Villen, J., Li, J., Cohn, M.A., Cantley, L.C., and Gygi, S.P. (2004). Large-scale characterization of HeLa cell nuclear phosphoproteins. Proc. Natl. Acad. Sci. U.S.A. *101*, 12130–12135.

Becker, K.G., Jedlicka, P., Templeton, N.S., Liotta, L., and Ozato, K. (1994). Characterization of hUCRBP (YY1, NF-E1, delta): a transcription factor that binds the regulatory regions of many viral and cellular genes. Gene *150*, 259–266.

Belak, Z.R., and Ovsenek, N. (2007). Assembly of the Yin Yang 1 transcription factor into messenger ribonucleoprotein particles requires direct RNA binding activity. J. Biol. Chem. *282*, 37913–37920.

Belak, Z.R., Ficzycz, A., and Ovsenek, N. (2008). Biochemical characterization of Yin Yang 1-RNA complexes. Biochem. Cell Biol. *86*, 31–36.

Bell, A.C., West, A.G., and Felsenfeld, G. (1999). The protein CTCF is required for the enhancer blocking activity of vertebrate insulators. Cell *98*, 387–396.

Bowman, T.L., and Hurt, M.M. (1995). The coding sequences of mouse H2A and H3 histone genes contains a conserved seven nucleotide element that interacts with nuclear factors and is necessary for normal expression. Nucleic Acids Res. *23*, 3083–3092.

Bowman, T.L., Kaludov, N.K., Klein, M., and Hurt, M.M. (1996). An H3 coding region regulatory element is common to all four nucleosomal classes of mouse histone-encoding genes. Gene *176*, 1–8.

Breen, T.R., and Duncan, I.M. (1986). Maternal expression of genes that regulate the bithorax complex of *Drosophila melanogaster*. Dev. Biol. *118*, 442–456.

Brock, H.W., and van Lohuizen, M. (2001). The Polycomb group – no longer an exclusive club? Curr. Opin. Genet. Dev. *11*, 175–181.

Brown, J.L., Mucci, D., Whiteley, M., Dirksen, M.L., and Kassis, J.A. (1998). The Drosophila Polycomb group gene pleiohomeotic encodes a DNA binding protein with homology to the transcription factor YY1. Mol. Cell *1*, 1057–1064.

Bushmeyer, S.M., and Atchison, M.L. (1998). Identification of YY1 sequences necessary for association with the nuclear matrix and for transcriptional repression functions. J. Cell. Biochem. *68*, 484–499.

Bushmeyer, S., Park, K., and Atchison, M.L. (1995). Characterization of functional domains within the multifunctional transcription factor, YY1. J. Biol. Chem. *270*, 30213–30220.

Cai, Y., Jin, J., Yao, T., Gottschalk, A.J., Swanson, S.K., Wu, S., Shi, Y., Washburn, M.P., Florens, L., Conaway, R.C., *et al.* (2007). YY1 functions with INO80 to activate transcription. Nat. Struct. Mol. Biol. *14*, 872–874.

Cao, R., Wang, L., Wang, H., Xia, L., Erdjument-Bromage, H., Tempst, P., Jones, R.S., and Zhang, Y. (2002). Role of histone H3 lysine 27 methylation in Polycomb-group silencing. Science *298*, 1039–1043.

Cao, Y., Gao, Z., Li, L., Jiang, X., Shan, A., Cai, J., Peng, Y., Li, Y., Huang, X., Wang, J., *et al.* (2013). Whole exome sequencing of insulinoma reveals recurrent T372R mutations in YY1. Nat. Commun. *4*, 2810.

Caretti, G., Di Padova, M., Micales, B., Lyons, G.E., and Sartorelli, V. (2004). The Polycomb Ezh2 methyltransferase regulates muscle gene expression and skeletal muscle differentiation. Genes Dev. *18*, 2627–2638.

Castellano, G., Torrisi, E., Ligresti, G., Malaponte, G., Militello, L., Russo, A.E., McCubrey, J.A., Canevari, S., and Libra, M. (2009). The involvement of the transcription factor Yin Yang 1 in cancer development and progression. Cell Cycle *8*, 1367–1372.

Clerc, P., and Avner, P. (2003). Multiple elements within the Xic regulate random X inactivation in mice. Semin. Cell Dev. Biol. *14*, 85–92.

Coull, J.J., Romerio, F., Sun, J.M., Volker, J.L., Galvin, K.M., Davie, J.R., Shi, Y., Hansen, U., and Margolis, D.M. (2000). The human factors YY1 and LSF repress the human immunodeficiency virus type 1 long terminal repeat via recruitment of histone deacetylase 1. J. Virol. *74*, 6790–6799.

Crea, F., Hurt, E.M., and Farrar, W.L. (2010). Clinical significance of Polycomb gene expression in brain tumors. Mol. Cancer *9*, 265.

Cuddapah, S., Roh, T.Y., Cui, K., Jose, C.C., Fuller, M.T., Zhao, K., and Chen, X. (2012). A novel human polycomb binding site acts as a functional polycomb response element in Drosophila. PLoS ONE *7*, e36365.

Czermin, B., Melfi, R., McCabe, D., Seitz, V., Imhof, A., and Pirrotta, V. (2002). Drosophila enhancer of Zeste/ESC complexes have a histone H3 methyltransferase activity that marks chromosomal Polycomb sites. Cell *111*, 185–196.

Delehouzee, S., Yoshikawa, T., Sawa, C., Sawada, J., Ito, T., Omori, M., Wada, T., Yamaguchi, Y., Kabe, Y., and Handa, H. (2005). GABP, HCF-1 and YY1 are involved in Rb gene expression during myogenesis. Genes Cells *10*, 717–731.

Deng, X., Berletch, J.B., Nguyen, D.K., and Disteche, C.M. (2014). X-chromosome regulation: diverse patterns in development, tissues and disease. Nat. Rev. Genet. *15*, 367–378.

Deng, Z., Wan, M., and Sui, G. (2007). PIASy-mediated sumoylation of Yin Yang 1 depends on their interaction but not the RING finger. Mol. Cell. Biol. *27*, 3780–3792.

Deng, Z., Wan, M., Cao, P., Rao, A., Cramer, S.D., and Sui, G. (2009). Yin Yang 1 regulates the transcriptional activity of androgen receptor. Oncogene *28*, 3746–3757.

Donohoe, M.E., Zhang, X., McGinnis, L., Biggers, J., Li, E., and Shi, Y. (1999). Targeted disruption of mouse Yin

Yang 1 transcription factor results in peri-implantation lethality. Mol. Cell. Biol. *19*, 7237–7244.

Donohoe, M.E., Zhang, L.F., Xu, N., Shi, Y., and Lee, J.T. (2007). Identification of a Ctcf cofactor, Yy1, for the X-chromosome binary switch. Mol. Cell *25*, 43–56.

Duncan, I.M. (1982). Polycomblike: a gene that appears to be required for the normal expression of the bithorax and antennapedia gene complexes of Drosophila melanogaster. Genetics *102*, 49–70.

Easwaran, H., Tsai, H.C., and Baylin, S.B. (2014). Cancer epigenetics: tumor heterogeneity, plasticity of stem-like states, and drug resistance. Mol. Cell *54*, 716–727.

Eliassen, K.A., Baldwin, A., Sikorski, E.M., and Hurt, M.M. (1998). Role for a YY1-binding element in replication-dependent mouse histone gene expression. Mol. Cell. Biol. *18*, 7106–7118.

Engel, N., Thorvaldsen, J.L., and Bartolomei, M.S. (2006). CTCF binding sites promote transcription initiation and prevent DNA methylation on the maternal allele at the imprinted H19/Igf2 locus. Hum. Mol. Genet. *15*, 2945–2954.

Erkeland, S.J., Valkhof, M., Heijmans-Antonissen, C., Delwel, R., Valk, P.J., Hermans, M.H., and Touw, I.P. (2003). The gene encoding the transcriptional regulator Yin Yang 1 (YY1) is a myeloid transforming gene interfering with neutrophilic differentiation. Blood *101*, 1111–1117.

Fedoriw, A., Mugford, J., and Magnuson, T. (2012). Genomic imprinting and epigenetic control of development. Cold Spring Harb. Perspect. Biol. *4*, a008136.

Flanagan, J.R. (1995). Autologous stimulation of YY1 transcription factor expression: role of an insulin-like growth factor. Cell Growth Differ. *6*, 185–190.

Flanagan, J.R., Becker, K.G., Ennist, D.L., Gleason, S.L., Driggers, P.H., Levi, B.Z., Appella, E., and Ozato, K. (1992). Cloning of a negative transcription factor that binds to the upstream conserved region of Moloney murine leukemia virus. Mol. Cell. Biol. *12*, 38–44.

Galvin, K.M., and Shi, Y. (1997). Multiple mechanisms of transcriptional repression by YY1. Mol. Cell. Biol. *17*, 3723–3732.

Garban, H.J., and Bonavida, B. (2001). Nitric oxide inhibits the transcription repressor Yin–Yang 1 binding activity at the silencer region of the Fas promoter: a pivotal role for nitric oxide in the up-regulation of Fas gene expression in human tumor cells. J. Immunol. *167*, 75–81.

Gaston, K., and Fried, M. (1995a). CpG methylation and the binding of YY1 and ETS proteins to the Surf-1/Surf-2 bidirectional promoter. Gene *157*, 257–259.

Gaston, K., and Fried, M. (1995b). CpG methylation has differential effects on the binding of YY1 and ETS proteins to the bi-directional promoter of the Surf-1 and Surf-2 genes. Nucleic Acids Res. *23*, 901–909.

Girton, J.R., and Jeon, S.H. (1994). Novel embryonic and adult homeotic phenotypes are produced by pleiohomeotic mutations in Drosophila. Dev. Biol. *161*, 393–407.

Gordon, S., Akopyan, G., Garban, H., and Bonavida, B. (2006). Transcription factor YY1: structure, function, and therapeutic implications in cancer biology. Oncogene *25*, 1125–1142.

Gould, A. (1997). Functions of mammalian Polycomb group and trithorax group related genes. Curr. Opin. Genet. Dev. *7*, 488–494.

Gronroos, E., Terentiev, A.A., Punga, T., and Ericsson, J. (2004). YY1 inhibits the activation of the p53 tumor suppressor in response to genotoxic stress. Proc. Natl. Acad. Sci. U.S.A. *101*, 12165–12170.

Gutjahr, T., Frei, E., Spicer, C., Baumgartner, S., White, R.A., and Noll, M. (1995). The Polycomb-group gene, extra sex combs, encodes a nuclear member of the WD-40 repeat family. EMBO J. *14*, 4296–4306.

Hariharan, N., Kelley, D.E., and Perry, R.P. (1991). Delta, a transcription factor that binds to downstream elements in several polymerase II promoters, is a functionally versatile zinc finger protein. Proc. Natl. Acad. Sci. U.S.A. *88*, 9799–9803.

Hark, A.T., Schoenherr, C.J., Katz, D.J., Ingram, R.S., Levorse, J.M., and Tilghman, S.M. (2000). CTCF mediates methylation-sensitive enhancer-blocking activity at the H19/Igf2 locus. Nature *405*, 486–489.

Hendrich, B.D., Plenge, R.M., and Willard, H.F. (1997). Identification and characterization of the human XIST gene promoter: implications for models of X-chromosome inactivation. Nucleic Acids Res. *25*, 2661–2671.

Hirasawa, R., and Feil, R. (2010). Genomic imprinting and human disease. Essays Biochem. *48*, 187–200.

Hiromura, M., Choi, C.H., Sabourin, N.A., Jones, H., Bachvarov, D., and Usheva, A. (2003). YY1 is regulated by O-linked N-acetylglucosaminylation (O-glcNAcylation). J. Biol. Chem. *278*, 14046–14052.

Hongo, F., Garban, H., Huerta-Yepez, S., Vega, M., Jazirehi, A.R., Mizutani, Y., Miki, T., and Bonavida, B. (2005). Inhibition of the transcription factor Yin Yang 1 activity by S-nitrosation. Biochem. Biophys. Res. Commun. *336*, 692–701.

Horsthemke, B., and Wagstaff, J. (2008). Mechanisms of imprinting of the Prader–Willi/Angelman region. Am. J. Med. Genet. A *146A*, 2041–2052.

Hosler, B.A., LaRosa, G.J., Grippo, J.F., and Gudas, L.J. (1989). Expression of REX-1, a gene containing zinc finger motifs, is rapidly reduced by retinoic acid in F9 teratocarcinoma cells. Mol. Cell. Biol. *9*, 5623–5629.

Houbaviy, H.B., Usheva, A., Shenk, T., and Burley, S.K. (1996). Cocrystal structure of YY1 bound to the adeno-associated virus P5 initiator. Proc. Natl. Acad. Sci. U.S.A. *93*, 13577–13582.

Huang, W., Smaldino, P.J., Zhang, Q., Miller, L.D., Cao, P., Stadelman, K., Wan, M., Giri, B., Lei, M., Nagamine, Y., *et al.* (2012). Yin Yang 1 contains G-quadruplex structures in its promoter and 5′-UTR and its expression is modulated by G4 resolvase 1. Nucleic Acids Res. *40*, 1033–1049.

Huerta-Yepez, S., Vega, M., Escoto-Chavez, S.E., Murdock, B., Sakai, T., Baritaki, S., and Bonavida, B. (2009). Nitric oxide sensitizes tumor cells to TRAIL-induced apoptosis via inhibition of the DR5 transcription repressor Yin Yang 1. Nitric Oxide *20*, 39–52.

Hutter, B., Bieg, M., Helms, V., and Paulsen, M. (2010). Imprinted genes show unique patterns of sequence conservation. BMC Genomics *11*, 649.

Hyde-DeRuyscher, R.P., Jennings, E., and Shenk, T. (1995). DNA binding sites for the transcriptional activator/repressor YY1. Nucleic Acids Res. *23*, 4457–4465.

Ideraabdullah, F.Y., Vigneau, S., and Bartolomei, M.S. (2008). Genomic imprinting mechanisms in mammals. Mutat. Res. *647*, 77–85.

Jacobs, J.J., and van Lohuizen, M. (2002). Polycomb repression: from cellular memory to cellular proliferation and cancer. Biochim. Biophys. Acta *1602*, 151–161.

Jelinic, P., and Shaw, P. (2007). Loss of imprinting and cancer. J. Pathol. *211*, 261–268.

Jeon, Y., and Lee, J.T. (2011). YY1 tethers Xist RNA to the inactive X nucleation center. Cell *146*, 119–133.

Jeong, H.M., Lee, S.H., Yum, J., Yeo, C.Y., and Lee, K.Y. (2014). Smurf2 regulates the degradation of YY1. Biochim. Biophys. Acta *1843*, 2005–2011.

Kaludov, N.K., Bowman, T.L., Sikorski, E.M., and Hurt, M.M. (1996). Cell cycle-regulated binding of nuclear proteins to elements within a mouse H3.2 histone gene. Proc. Natl. Acad. Sci. U.S.A. *93*, 4465–4470.

Kassardjian, A., Rizkallah, R., Riman, S., Renfro, S.H., Alexander, K.E., and Hurt, M.M. (2012). The transcription factor YY1 is a novel substrate for Aurora B kinase at G2/M transition of the cell cycle. PLoS ONE 7, e50645.

Kennison, J.A., and Tamkun, J.W. (1988). Dosage-dependent modifiers of polycomb and antennapedia mutations in Drosophila. Proc. Natl. Acad. Sci. U.S.A. *85*, 8136–8140.

Kim, J. (2008). Multiple YY1 and CTCF binding sites in imprinting control regions. Epigenetics 3, 115–118.

Kim, J., and Kim, J.D. (2008). *In vivo* YY1 knockdown effects on genomic imprinting. Hum. Mol. Genet. *17*, 391–401.

Kim, J., Kollhoff, A., Bergmann, A., and Stubbs, L. (2003). Methylation-sensitive binding of transcription factor YY1 to an insulator sequence within the paternally expressed imprinted gene, Peg3. Hum. Mol. Genet. *12*, 233–245.

Kim, J.D., Hinz, A.K., Bergmann, A., Huang, J.M., Ovcharenko, I., Stubbs, L., and Kim, J. (2006). Identification of clustered YY1 binding sites in imprinting control regions. Genome Res. *16*, 901–911.

Kim, J.A., Kruhlak, M., Dotiwala, F., Nussenzweig, A., and Haber, J.E. (2007). Heterochromatin is refractory to gamma-H2AX modification in yeast and mammals. J. Cell Biol. *178*, 209–218.

Kim, J.D., Faulk, C., and Kim, J. (2007a). Retroposition and evolution of the DNA-binding motifs of YY1, YY2 and REX1. Nucleic Acids Res. 35, 3442–3452.

Kim, J.D., Hinz, A.K., Choo, J.H., Stubbs, L., and Kim, J. (2007b). YY1 as a controlling factor for the Peg3 and Gnas imprinted domains. Genomics 89, 262–269.

Kim, J.D., Kang, K., and Kim, J. (2009a). YY1's role in DNA methylation of Peg3 and Xist. Nucleic Acids Res. 37, 5656–5664.

Kim, J.D., Yu, S., and Kim, J. (2009b). YY1 is autoregulated through its own DNA-binding sites. BMC Mol. Biol. *10*, 85.

Kim, J.D., Kim, H., Ekram, M.B., Yu, S., Faulk, C., and Kim, J. (2011). Rex1/Zfp42 as an epigenetic regulator for genomic imprinting. Hum. Mol. Genet. *20*, 1353–1362.

Krippner-Heidenreich, A., Walsemann, G., Beyrouthy, M.J., Speckgens, S., Kraft, R., Thole, H., Talanian, R.V., Hurt, M.M., and Luscher, B. (2005). Caspase-dependent regulation and subcellular redistribution of the transcriptional modulator YY1 during apoptosis. Mol. Cell. Biol. *25*, 3704–3714.

Kurukuti, S., Tiwari, V.K., Tavoosidana, G., Pugacheva, E., Murrell, A., Zhao, Z., Lobanenkov, V., Reik, W., and Ohlsson, R. (2006). CTCF binding at the H19 imprinting control region mediates maternally inherited higher-order chromatin conformation to restrict enhancer access to Igf2. Proc. Natl. Acad. Sci. U.S.A. *103*, 10684–10689.

Kuzmichev, A., Nishioka, K., Erdjument-Bromage, H., Tempst, P., and Reinberg, D. (2002). Histone methyltransferase activity associated with a human multiprotein complex containing the Enhancer of Zeste protein. Genes Dev. *16*, 2893–2905.

Lee, J.S., Galvin, K.M., See, R.H., Eckner, R., Livingston, D., Moran, E., and Shi, Y. (1995). Relief of YY1 transcriptional repression by adenovirus E1A is mediated by E1A-associated protein p300. Genes Dev. *9*, 1188–1198.

Lee, J.T., Davidow, L.S., and Warshawsky, D. (1999). Tsix, a gene antisense to Xist at the X-inactivation centre. Nat. Genet. *21*, 400–404.

Lee, T.C., Shi, Y., and Schwartz, R.J. (1992). Displacement of BrdUrd-induced YY1 by serum response factor activates skeletal alpha-actin transcription in embryonic myoblasts. Proc. Natl. Acad. Sci. U.S.A. *89*, 9814–9818.

Lewis, E.B. (1978). A gene complex controlling segmentation in Drosophila. Nature *276*, 565–570.

Lieberthal, J.G., Kaminsky, M., Parkhurst, C.N., and Tanese, N. (2009). The role of YY1 in reduced HP1alpha gene expression in invasive human breast cancer cells. Breast Cancer Res. *11*, R42.

Lim, D.H., and Maher, E.R. (2010a). Genomic imprinting syndromes and cancer. Adv. Genet. *70*, 145–175.

Lim, D.H.K., and Maher, E.R. (2010b). Genomic imprinting syndromes and cancer. Adv. Genet. *70*, 145–175.

Liu, Q., Merkler, K.A., Zhang, X., and McLean, M.P. (2007). Prostaglandin F2alpha suppresses rat steroidogenic acute regulatory protein expression via induction of Yin Yang 1 protein and recruitment of histone deacetylase 1 protein. Endocrinology *148*, 5209–5219.

Lopez-Perrote, A., Alatwi, H.E., Torreira, E., Ismail, A., Ayora, S., Downs, J.A., and Llorca, O. (2014). Structure of Yin Yang 1 oligomers that cooperate with RuvBL1-RuvBL2 ATPases. J. Biol. Chem. *289*, 22614–22629.

Lucchesi, J.C., Kelly, W.G., and Parming, B. (2005). Chromatin remodeling in dosage compensation. Annu. Rev. Genet. 39, 615–651.

Luedi, P.P., Hartemink, A.J., and Jirtle, R.L. (2005). Genome-wide prediction of imprinted murine genes. Genome Res. *15*, 875–884.

Luedi, P.P., Dietrich, F.S., Weidman, J.R., Bosko, J.M., Jirtle, R.L., and Hartemink, A.J. (2007). Computational and experimental identification of novel human imprinted genes. Genome Res. *17*, 1723–1730.

Luke, M.P., Sui, G., Liu, H., and Shi, Y. (2006). Yin Yang 1 physically interacts with Hoxa11 and represses Hoxa11-dependent transcription. J. Biol. Chem. *281*, 33226–33232.

Lund, K., Adams, P.D., and Copland, M. (2014). EZH2 in normal and malignant hematopoiesis. Leukemia *28*, 44–49.

McCabe, M.T., and Creasy, C.L. (2014). EZH2 as a potential target in cancer therapy. Epigenomics *6*, 341–351.

Mager, J., Montgomery, N.D., de Villena, F.P., and Magnuson, T. (2003). Genome imprinting regulated by the mouse Polycomb group protein Eed. Nat. Genet. *33*, 502–507.

Marks, H., Chow, J.C., Denissov, S., Francoijs, K.J., Brockdorff, N., Heard, E., and Stunnenberg, H.G. (2009). High-resolution analysis of epigenetic changes associated with X inactivation. Genome Res. *19*, 1361–1373.

Molina, H., Horn, D.M., Tang, N., Mathivanan, S., and Pandey, A. (2007). Global proteomic profiling of phosphopeptides using electron transfer dissociation tandem mass spectrometry. Proc. Natl. Acad. Sci. U.S.A. *104*, 2199–2204.

Mongan, N.P., Martin, K.M., and Gudas, L.J. (2006). The putative human stem cell marker, Rex-1 (Zfp42): structural classification and expression in normal human epithelial and carcinoma cell cultures. Mol. Carcinog. *45*, 887–900.

Morison, I.M., Ramsay, J.P., and Spencer, H.G. (2005). A census of mammalian imprinting. Trends Genet. *21*, 457–465.

Moseley, S.C., Rizkallah, R., Tremblay, D.C., Anderson, B.R., Hurt, M.M., and Chadwick, B.P. (2012). YY1 associates with the macrosatellite DXZ4 on the inactive X-chromosome and binds with CTCF to a hypomethylated form in some male carcinomas. Nucleic Acids Res. *40*, 1596–1608.

Muller, J., Hart, C.M., Francis, N.J., Vargas, M.L., Sengupta, A., Wild, B., Miller, E.L., O'Connor, M.B., Kingston, R.E., and Simon, J.A. (2002). Histone methyltransferase activity of a Drosophila Polycomb group repressor complex. Cell *111*, 197–208.

Natesan, S., and Gilman, M.Z. (1993). DNA bending and orientation-dependent function of YY1 in the c-fos promoter. Genes Dev. *7*, 2497–2509.

Navarro, P., and Avner, P. (2010). An embryonic story: analysis of the gene regulative network controlling Xist expression in mouse embryonic stem cells. Bioessays *32*, 581–588.

Navarro, P., Page, D.R., Avner, P., and Rougeulle, C. (2006). Tsix-mediated epigenetic switch of a CTCF-flanked region of the Xist promoter determines the Xist transcription program. Genes Dev. *20*, 2787–2792.

Nicholson, S., Whitehouse, H., Naidoo, K., and Byers, R.J. (2011). Yin Yang 1 in human cancer. Crit. Rev. Oncog. *16*, 245–260.

de Nigris, F., Rossiello, R., Schiano, C., Arra, C., Williams-Ignarro, S., Barbieri, A., Lanza, A., Balestrieri, A., Giuliano, M.T., Ignarro, L.J., *et al.* (2008). Deletion of Yin Yang 1 protein in osteosarcoma cells on cell invasion and CXCR4/angiogenesis and metastasis. Cancer Res. *68*, 1797–1808.

Nousiainen, M., Sillje, H.H., Sauer, G., Nigg, E.A., and Korner, R. (2006). Phosphoproteome analysis of the human mitotic spindle. Proc. Natl. Acad. Sci. U.S.A. *103*, 5391–5396.

Oei, S.L., and Shi, Y. (2001). Poly(ADP-ribosyl)ation of transcription factor Yin Yang 1 under conditions of DNA damage. Biochem. Biophys. Res. Commun. *285*, 27–31.

Ohlsson, R., Bartkuhn, M., and Renkawitz, R. (2010). CTCF shapes chromatin by multiple mechanisms: the impact of 20 years of CTCF research on understanding the workings of chromatin. Chromosoma *119*, 351–360.

Okamoto, I., Otte, A.P., Allis, C.D., Reinberg, D., and Heard, E. (2004). Epigenetic dynamics of imprinted X inactivation during early mouse development. Science *303*, 644–649.

Ortiz, V., and de Pablo, J.J. (2011). Molecular origins of DNA flexibility: sequence effects on conformational and mechanical properties. Phys. Rev. Lett. *106*, 238107.

Park, K., and Atchison, M.L. (1991). Isolation of a candidate repressor/activator, NF-E1 (YY-1, delta), that binds to the immunoglobulin kappa 3′ enhancer and the immunoglobulin heavy-chain mu E1 site. Proc. Natl. Acad. Sci. U.S.A. *88*, 9804–9808.

Paro, R. (1990). Imprinting a determined state into the chromatin of Drosophila. Trends Genet. *6*, 416–421.

Patten, M., Wang, W., Aminololama-Shakeri, S., Burson, M., and Long, C.S. (2000). IL-1 beta increases abundance and activity of the negative transcriptional regulator yin yang-1 (YY1) in neonatal rat cardiac myocytes. J. Mol. Cell. Cardiol. *32*, 1341–1352.

Peters, A.H., Kubicek, S., Mechtler, K., O'Sullivan, R.J., Derijck, A.A., Perez-Burgos, L., Kohlmaier, A., Opravil, S., Tachibana, M., Shinkai, Y., *et al.* (2003). Partitioning and plasticity of repressive histone methylation states in mammalian chromatin. Mol. Cell *12*, 1577–1589.

Petkova, V., Romanowski, M.J., Sulijoadikusumo, I., Rohne, D., Kang, P., Shenk, T., and Usheva, A. (2001). Interaction between YY1 and the retinoblastoma protein. Regulation of cell cycle progression in differentiated cells. J. Biol. Chem. *276*, 7932–7936.

Phillips, J.E., and Corces, V.G. (2009). CTCF: master weaver of the genome. Cell *137*, 1194–1211.

Pirrotta, V. (1997a). Chromatin-silencing mechanisms in Drosophila maintain patterns of gene expression. Trends Genet. *13*, 314–318.

Pirrotta, V. (1997b). PcG complexes and chromatin silencing. Curr. Opin. Genet. Dev. *7*, 249–258.

Pirrotta, V. (1998). Polycombing the genome: PcG, trxG, and chromatin silencing. Cell 93, 333–336.

Pirrotta, V. (1999). Polycomb silencing and the maintenance of stable chromatin states. Results Probl. Cell. Differ. *25*, 205–228.

Pizzorno, M.C. (2001). Nuclear cathepsin B-like protease cleaves transcription factor YY1 in differentiated cells. Biochim. Biophys. Acta *1536*, 31–42.

Radulovic, V., de Haan, G., and Klauke, K. (2013). Polycomb-group proteins in hematopoietic stem cell regulation and hematopoietic neoplasms. Leukemia *27*, 523–533.

Ren, G., Zhang, G., Dong, Z., Liu, Z., Li, L., Feng, Y., Su, D., Zhang, Y., Huang, B., and Lu, J. (2009). Recruitment of HDAC4 by transcription factor YY1 represses HOXB13 to affect cell growth in AR-negative prostate cancers. Int. J. Biochem. Cell Biol. *41*, 1094–1101.

Rezai-Zadeh, N., Zhang, X., Namour, F., Fejer, G., Wen, Y.D., Yao, Y.L., Gyory, I., Wright, K., and Seto, E. (2003). Targeted recruitment of a histone H4-specific methyltransferase by the transcription factor YY1. Genes Dev. *17*, 1019–1029.

Riggs, K.J., Saleque, S., Wong, K.K., Merrell, K.T., Lee, J.S., Shi, Y., and Calame, K. (1993). Yin–yang 1 activates the c-myc promoter. Mol. Cell. Biol. *13*, 7487–7495.

Riman, S., Rizkallah, R., Kassardjian, A., Alexander, K.E., Luscher, B., and Hurt, M.M. (2012). Phosphorylation of the transcription factor YY1 by CK2alpha prevents cleavage by caspase 7 during apoptosis. Mol. Cell. Biol. *32*, 797–807.

Ringrose, L., and Paro, R. (2004). Epigenetic regulation of cellular memory by the Polycomb and Trithorax group proteins. Annu. Rev. Genet. *38*, 413–443.

Ringrose, L., and Paro, R. (2007). Polycomb/Trithorax response elements and epigenetic memory of cell identity. Development *134*, 223–232.

Rizkallah, R., and Hurt, M.M. (2009). Regulation of the transcription factor YY1 in mitosis through phosphorylation of its DNA-binding domain. Mol. Biol. Cell *20*, 4766–4776.

Rizkallah, R., Alexander, K.E., and Hurt, M.M. (2011a). Global mitotic phosphorylation of C2H2 zinc finger protein linker peptides. Cell Cycle *10*, 3327–3336.

Rizkallah, R., Alexander, K.E., Kassardjian, A., Luscher, B., and Hurt, M.M. (2011b). The transcription factor YY1 is a substrate for Polo-like kinase 1 at the G2/M transition of the cell cycle. PLoS ONE *6*, e15928.

Rizkallah, R., Batsomboon, P., Dudley, G.B., and Hurt, M.M. (2014). Identification of the oncogenic kinase TOPK/PBK as a master mitotic regulator of C2H2 zinc finger proteins. Oncotarget. Oncotarget. [Epub ahead of print].

Rodriguez-Jato, S., Nicholls, R.D., Driscoll, D.J., and Yang, T.P. (2005). Characterization of cis- and trans-acting elements in the imprinted human SNURF-SNRPN locus. Nucleic Acids Res. *33*, 4740–4753.

Rodriguez-Paredes, M., and Esteller, M. (2011). Cancer epigenetics reaches mainstream oncology. Nat. Med. *17*, 330–339.

Rogers, M.B., Hosler, B.A., and Gudas, L.J. (1991). Specific expression of a retinoic acid-regulated, zinc-finger gene, Rex-1, in preimplantation embryos, trophoblast and spermatocytes. Development (Cambridge, England) *113*, 815–824.

Safrany, G., and Perry, R.P. (1993). Characterization of the mouse gene that encodes the delta/YY1/NF-E1/UCRBP transcription factor. Proc. Natl. Acad. Sci. U.S.A. *90*, 5559–5563.

Satijn, D.P., and Otte, A.P. (1999). Polycomb group protein complexes: do different complexes regulate distinct target genes? Biochim. Biophys. Acta *1447*, 1–16.

Satijn, D.P., Hamer, K.M., den Blaauwen, J., and Otte, A.P. (2001). The polycomb group protein EED interacts with YY1, and both proteins induce neural tissue in *Xenopus* embryos. Mol. Cell. Biol. *21*, 1360–1369.

Schlisio, S., Halperin, T., Vidal, M., and Nevins, J.R. (2002). Interaction of YY1 with E2Fs, mediated by RYBP, provides a mechanism for specificity of E2F function. EMBO J. *21*, 5775–5786.

Schug, J., Schuller, W.-P., Kappen, C., Salbaum, J.M., Bucan, M., and Stoeckert, C.J., Jr. (2005). Promoter features related to tissue specificity as measured by Shannon entropy. Genome Biol. *6*, R33.

Schwartz, Y.B., and Pirrotta, V. (2013). A new world of Polycombs: unexpected partnerships and emerging functions. Nat. Rev. Genet. *14*, 853–864.

Sewalt, R.G., Gunster, M.J., van der Vlag, J., Satijn, D.P., and Otte, A.P. (1999). C-Terminal binding protein is a transcriptional repressor that interacts with a specific class of vertebrate Polycomb proteins. Mol. Cell. Biol. *19*, 777–787.

Shi, Y., Seto, E., Chang, L.S., and Shenk, T. (1991). Transcriptional repression by YY1, a human GLI-Kruppel-related protein, and relief of repression by adenovirus E1A protein. Cell *67*, 377–388.

Shi, Y., Lee, J.S., and Galvin, K.M. (1997). Everything you have ever wanted to know about Yin Yang 1. Biochim. Biophys. Acta *1332*, F49–66.

Shrivastava, A., and Calame, K. (1994). An analysis of genes regulated by the multi-functional transcriptional regulator Yin Yang-1. Nucleic Acids Res. *22*, 5151–5155.

Shrivastava, A., Saleque, S., Kalpana, G.V., Artandi, S., Goff, S.P., and Calame, K. (1993). Inhibition of transcriptional regulator Yin–Yang-1 by association with c-Myc. Science *262*, 1889–1892.

Shrivastava, A., Yu, J., Artandi, S., and Calame, K. (1996). YY1 and c-Myc associate *in vivo* in a manner that depends on c-Myc levels. Proc. Natl. Acad. Sci. U.S.A. *93*, 10638–10641.

Simon, J., Chiang, A., and Bender, W. (1992). Ten different Polycomb group genes are required for spatial control of the abdA and AbdB homeotic products. Development *114*, 493–505.

Simon, J.A., and Lange, C.A. (2008). Roles of the EZH2 histone methyltransferase in cancer epigenetics. Mutat. Res. *647*, 21–29.

Sing, A., Pannell, D., Karaiskakis, A., Sturgeon, K., Djabali, M., Ellis, J., Lipshitz, H.D., and Cordes, S.P. (2009). A vertebrate Polycomb response element governs segmentation of the posterior hindbrain. Cell *138*, 885–897.

Sparmann, A., and van Lohuizen, M. (2006). Polycomb silencers control cell fate, development and cancer. Nat. Rev. Cancer *6*, 846–856.

Srinivasan, L., and Atchison, M.L. (2004). YY1 DNA binding and PcG recruitment requires CtBP. Genes Dev. *18*, 2596–2601.

Strouboulis, J., Damjanovski, S., Vermaak, D., Meric, F., and Wolffe, A.P. (1999). Transcriptional repression by XPc1, a new Polycomb homolog in *Xenopus laevis* embryos, is independent of histone deacetylase. Mol. Cell. Biol. *19*, 3958–3968.

Struhl, G., and Brower, D. (1982). Early role of the esc+ gene product in the determination of segments in Drosophila. Cell *31*, 285–292.

Sui, G., Affar el, B., Shi, Y., Brignone, C., Wall, N.R., Yin, P., Donohoe, M., Luke, M.P., Calvo, D., and Grossman, S.R. (2004). Yin Yang 1 is a negative regulator of p53. Cell *117*, 859–872.

Takagi, N., and Sasaki, M. (1975). Preferential inactivation of the paternally derived X-chromosome in the extraembryonic membranes of the mouse. Nature *256*, 640–642.

Takasaki, N., Kurokawa, D., Nakayama, R., Nakayama, J., and Aizawa, S. (2007). Acetylated YY1 regulates Otx2 expression in anterior neuroectoderm at two cis-sites 90 kb apart. EMBO J. *26*, 1649–1659.

Tang, X., Milyavsky, M., Shats, I., Erez, N., Goldfinger, N., and Rotter, V. (2004). Activated p53 suppresses the histone methyltransferase EZH2 gene. Oncogene *23*, 5759–5769.

Thiaville, M.M., and Kim, J. (2011). Oncogenic potential of yin yang 1 mediated through control of imprinted genes. Crit. Rev. Oncog. *16*, 199–209.

Thomas, M.J., and Seto, E. (1999). Unlocking the mechanisms of transcription factor YY1: are chromatin modifying enzymes the key? Gene *236*, 197–208.

Tie, F., Prasad-Sinha, J., Birve, A., Rasmuson-Lestander, A., and Harte, P.J. (2003). A 1-megadalton ESC/E(Z) complex from *Drosophila* that contains polycomblike and RPD3. Mol. Cell. Biol. 23, 3352–3362.

Usheva, A., and Shenk, T. (1994). TATA-binding protein-independent initiation: YY1, TFIIB, and RNA polymerase II direct basal transcription on supercoiled template DNA. Cell 76, 1115–1121.

van der Vlag, J., and Otte, A.P. (1999). Transcriptional repression mediated by the human polycomb-group protein EED involves histone deacetylation. Nat. Genet. 23, 474–478.

Walowitz, J.L., Bradley, M.E., Chen, S., and Lee, T. (1998). Proteolytic regulation of the zinc finger transcription factor YY1, a repressor of muscle-restricted gene expression. J. Biol. Chem. *273*, 6656–6661.

Wang, A.M., Huang, T.T., Hsu, K.W., Huang, K.H., Fang, W.L., Yang, M.H., Lo, S.S., Chi, C.W., Lin, J.J., and Yeh, T.S. (2014). Yin Yang 1 is a target of microRNA-34 family and contributes to gastric carcinogenesis. Oncotarget 5, 5002–5016.

Wilkinson, F.H., Park, K., and Atchison, M.L. (2006). Polycomb recruitment to DNA *in vivo* by the YY1 REPO domain. Proc. Natl. Acad. Sci. U.S.A. *103*, 19296–19301.

Woo, C.J., Kharchenko, P.V., Daheron, L., Park, P.J., and Kingston, R.E. (2010). A region of the human HOXD cluster that confers polycomb-group responsiveness. Cell *140*, 99–110.

Wu, S., Shi, Y., Mulligan, P., Gay, F., Landry, J., Liu, H., Lu, J., Qi, H.H., Wang, W., Nickoloff, J.A., *et al.* (2007). A YY1–INO80 complex regulates genomic stability through homologous recombination-based repair. Nat. Struct. Mol. Biol. *14*, 1165–1172.

Wu, S., Hu, Y.C., Liu, H., and Shi, Y. (2009). Loss of YY1 impacts the heterochromatic state and meiotic double-strand breaks during mouse spermatogenesis. Mol. Cell. Biol. *29*, 6245–6256.

Xi, H., Yu, Y., Fu, Y., Foley, J., Halees, A., and Weng, Z. (2007). Analysis of overrepresented motifs in human core promoters reveals dual regulatory roles of YY1. Genome Res. *17*, 798–806.

Yang, W.M., Inouye, C., Zeng, Y., Bearss, D., and Seto, E. (1996). Transcriptional repression by YY1 is mediated by interaction with a mammalian homolog of the yeast global regulator RPD3. Proc. Natl. Acad. Sci. U.S.A. *93*, 12845–12850.

Yang, Y., Zhou, L., Lu, L., Wang, L., Li, X., Jiang, P., Chan, L.K., Zhang, T., Yu, J., Kwong, J., *et al.* (2013). A novel miR-193a-5p-YY1-APC regulatory axis in human endometrioid endometrial adenocarcinoma. Oncogene 32, 3432–3442.

Yant, S.R., Zhu, W., Millinoff, D., Slightom, J.L., Goodman, M., and Gumucio, D.L. (1995). High affinity YY1 binding motifs: identification of two core types (ACAT and CCAT) and distribution of potential binding sites within the human beta globin cluster. Nucleic Acids Res. 23, 4353–4362.

Yao, Y.L., Dupont, B.R., Ghosh, S., Fang, Y., Leach, R.J., and Seto, E. (1998). Cloning, chromosomal localization and promoter analysis of the human transcription factor YY1. Nucleic Acids Res. *26*, 3776–3783.

Yao, Y.L., Yang, W.M., and Seto, E. (2001). Regulation of transcription factor YY1 by acetylation and deacetylation. Mol. Cell. Biol. *21*, 5979–5991.

Zaravinos, A., and Spandidos, D.A. (2009). Yin Yang 1 as a prognostic factor. Cell Cycle *8*, 1305.

Zaravinos, A., and Spandidos, D.A. (2010). Yin yang 1 expression in human tumors. Cell Cycle *9*, 512–522.

Zhang, H., Niu, B., Hu, J.F., Ge, S., Wang, H., Li, T., Ling, J., Steelman, B.N., Qian, G., and Hoffman, A.R. (2011a). Interruption of intrachromosomal looping by CCCTC binding factor decoy proteins abrogates genomic imprinting of human insulin-like growth factor II. J. Cell Biol. *193*, 475–487.

Zhang, Q., Stovall, D.B., Inoue, K., and Sui, G. (2011b). The oncogenic role of Yin Yang 1. Crit. Rev. Oncog. *16*, 163–197.

SETting up the Epigenome Through the Histone Methyltransferase SETDB1

2

Zhuo Sun and Brian P. Chadwick

Abstract

SETDB1 is unusual among the SET-domain containing histone methyltransferases (HMTases) owing to the partitioning of its catalytic SET domain into two parts, hence its name 'SET Domain Bifurcated 1'. SETDB1 is one of several histone H3 lysine-9 HMTases, but is the only one that is essential in *Drosophila*. Through its ability to implement and maintain programmed silencing of genes that would otherwise interfere with differentiation and cell specialization, SETDB1 is indispensable for development from nematodes to mammals. In addition, SETDB1 is required to maintain pluripotency, in part through the silencing of endogenous retroviral elements that can act as promoters of chimeric transcripts or enhancers of surrounding genes when active. SETDB1 can be recruited to sites of action through a variety of proteins, among which targeting to KRAB-domain containing zinc finger proteins via the versatile adapter protein tripartite motif-containing 28 (TRIM28) is best characterized. Through this large and diverse family of DNA-binding proteins, TRIM28 guides SETDB1 to sites of action to shut down gene expression, which includes all silent alleles of known imprinted genes. SETDB1 is emerging as a key player in establishing and maintaining the epigenome. Owing to this pivotal role, levels of SETDB1 protein needs to be strictly regulated to avoid it contributing to disease and cancer progression.

SETting the stage

Eukaryotic genomes are packaged in the nucleus as chromatin. The fundamental repeating unit of chromatin is the nucleosome (Kornberg, 1974), which consists of approximately 146 base pairs (bp) of DNA wrapped almost twice around a central core of two molecules each of histones H2A, H2B, H3 and H4 (Luger *et al.*, 1997). While nucleosomes provide an efficient means to package chromosomes into the tight confines of the nucleus, they also restrict access to the underlying DNA and therefore permit a means to control who and what can see the DNA and when. At the level of the nucleosome, two distinct ways in which this can be achieved are through histone variants or via post-translational covalent modification of the core histones.

Histone variants are proteins that share structural similarity to one of the core histone and can be substituted into the nucleosome, bringing new physical attributes to the local chromatin environment (Maze *et al.*, 2014; Weber and Henikoff, 2014). Some variants differ substantially from their core histone counterpart, such as H2A Barr-body deficient (H2A-Bbd), which shares 48% amino acid identity with canonical H2A (Chadwick and Willard, 2001), or macroH2A that has an extensive carboxyl-terminal tail that makes up two-thirds of its mass (Pehrson and Fried, 1992). Others differ by only a few amino acids, but these subtle changes are sufficient to confer unique properties to the chromatin (Bush *et al.*, 2013; Lin *et al.*, 2013).

Alternatively, nucleosome function can be adjusted through regulated covalent modification of amino acid side chains. Most modifications occur to side chains of lysine, arginine, serine and threonine residues in the amino and carboxyl-terminal tails of histones (Tan *et al.*, 2011), and include a wide variety of modification types such as methylation, acetylation, phosphorylation and ubiquitination to name but a few (Kouzarides, 2007). Notably, some portions of the amino-terminal and carboxyl-terminal tails were not included in the atomic structure of the nucleosome (Luger *et al.*, 1997). Unlike the conserved and more rigid histone fold domain that makes up the central portion of each of the core histones (Arents *et al.*, 1991), the tails are more flexible and likely project outward which may assist with their accessibility for modification and in some cases the subsequent recruitment of effector proteins to execute the nature of the signal. The bulk of modifications occur to the amino-terminal tail of histones H3, and all lysine residues in this tail can be either acetylated or methylated (Guillemette *et al.*, 2011; Tan *et al.*, 2011), and the consequences of these modifications dictate chromatin-templated processes as proposed in the 'histone code' hypothesis (Strahl and Allis, 2000). Histone acetylation is primarily associated with euchromatin, which is nicely demonstrated by its near complete absence from the transcriptionally inert mammalian inactive X-chromosome (Xi) (Jeppesen and Turner, 1993). In contrast, histone methylation is associated with both euchromatin and heterochromatin, but its correlation with either state is dependent on the lysine residue in question as is shown in Fig. 2.1A. Methylated lysine 4 is almost exclusively associated with euchromatin (Boggs *et al.*, 2002), while methylated lysine 9 (Boggs *et al.*, 2002; Heard *et al.*, 2001; Mermoud *et al.*, 2002; Peters *et al.*, 2002) and 27 (Plath *et al.*, 2003; Silva *et al.*, 2003) are typically associated with heterochromatin.

Given that the same lysine residue can be either methylated or acetylated and that these two modifications can impart quite different outcomes on the local chromatin environment, the ability to transition between states when required to do so must exist. Indeed the dynamic reversible nature of histone acetylation through the actions of histone acetyltransferases (HATs) and histone

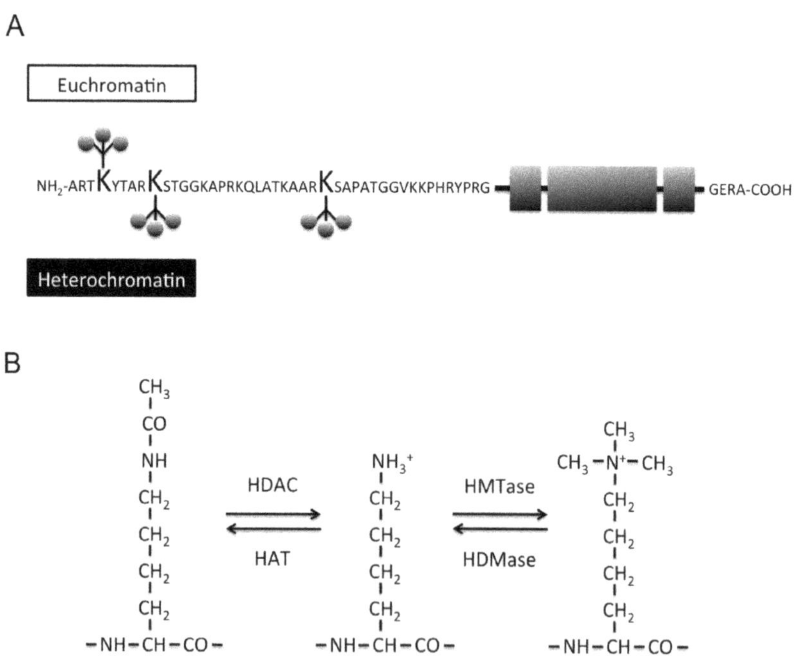

Figure 2.1 (A) Schematic representation of histone H3 showing the single letter amino acid code for the amino-terminal tail (left side) and carboxyl-terminal tail flanking the core histone fold domain (grey shaded boxes). Lysine residues 4, 9 and 27 are indicated as large letter K's with tri-methylation represented above for association with euchromatin or below with heterochromatin. (B) Transition between acetylated lysine (left side) through unmodified (middle) to tri-methylated (right side) with the enzyme type responsible for catalysing the transition labelled adjacent to the arrow indicating the reaction direction.

deacetylases (HDACs) was established early on (Kuo and Allis, 1998), but the identity of the enzymes responsible for histone methylation and demethylation remained unknown.

Histone methyltransferases

The breakthrough came at the turn of the century when Thomas Jenuwein and colleagues demonstrated that the mammalian homologues of *Drosophila* suppressor of variegation 3-9 (*Su(var)3-9*) possessed histone methyltransferase (HMTase) activity, specific for lysine 9 of H3 (Rea *et al.*, 2000). The authors went on to show that the activity was mediated through a domain shared with numerous other chromatin proteins through which it acquired its name (*Drosophila* SU(VAR)3-9, enhancer of zeste and trithorax: SET) (Jenuwein *et al.*, 1998), and that the *Schizosaccharomyces pombe* SET domain containing protein cryptic loci regulator 4 (Clr4) also possessed H3 HMTase activity (Rea *et al.*, 2000). After this seminal report, the race was on to determine if other SET domain containing proteins possessed lysine HMTase activity and to identify the target residue in histone tails. Importantly, lysine can be mono, di or trimethylated, and what appears to be only a subtle chemical difference can be interpreted quite differently in terms of function. For example, histone H3 trimethylated at lysine 4 (H3K4me3) is typically associated with active promoters, whereas H3K4me1 is a common feature of enhancer elements (Barski *et al.*, 2007). To date, enzymatic activity and specificity for a large number of SET-domain containing HMTases have been defined (Kouzarides, 2007). Like histone acetylation (Kuo and Allis, 1998), histone methylation is reversible and methyl marks can be removed by lysine-specific demethylase 1 (LSD1) (Metzger *et al.*, 2005; Shi *et al.*, 2004) that removes mono- and dimethyl lysine modifications or by members of the Jumonji C domain containing histone demethylases (Tsukada *et al.*, 2006) that can remove mono-, di- and trimethylation. Removal of methyl marks from lysine residues on histones has been reported for H3K4 and H3K9 (Klose *et al.*, 2006; Shi *et al.*, 2004; Tsukada *et al.*, 2006; Whetstine *et al.*, 2006; Yamane *et al.*, 2006), confirming that like histone acetylation and deacetylation, methylation is dynamic and enzymatically controlled, completing the puzzle of reversibility (Fig. 2.1B). Furthermore, the activity of these enzymes is modulated by interacting factors (Lee *et al.*, 2005) and electrostatic interactions due to combinatorial modifications on the histone tail (Forneris *et al.*, 2005). Finally, although LSD1 was initially described as a H3K4-demethylase (and hence a repressor), it can also activate transcription through demethylation of H3K9 (Garcia-Bassets *et al.*, 2007; Liang *et al.*, 2009; Metzger *et al.*, 2005; Wang *et al.*, 2007).

Lysine 9 HMTases, H3K9me and HP1

Interestingly, histones are not the only targets of HMTases, as these enzymes (Herz *et al.*, 2013) methylate numerous non-histone proteins. However, for the purpose of this chapter, the focus will be upon histone H3K9me. While in mammals two HMTases have been described that are responsible for methylation of H3K27 (Cao *et al.*, 2002; Czermin *et al.*, 2002; Kuzmichev *et al.*, 2002; Muller *et al.*, 2002; Shen *et al.*, 2008), seven catalyse methylation at H3K9, but their target regions in the genome differ (Table 2.1). As mentioned above, H3K9me is almost exclusively associated with heterochromatin (Litt *et al.*, 2001; Noma *et al.*, 2001). One way in which H3K9me is recognized and acted upon is that it provides a docking site for the chromodomain (CD) of heterochromatin protein 1 (HP1) (Aagaard *et al.*, 1999; Bannister *et al.*, 2001; Jacobs *et al.*, 2001; Lachner *et al.*, 2001; Nakayama *et al.*, 2001). HP1 was originally identified as a heterochromatin associated protein in *Drosophila* (James and Elgin, 1986), that subsequently emerged as a suppressor of position effect variegation in a genetic screen (Eissenberg *et al.*, 1990). HP1 is highly conserved, from *Schizosaccharomyces pombe* switching-deficient 6 (SWI6), to the three human homologues HP1α, HP1β and HP1γ (Saunders *et al.*, 1993; Singh *et al.*, 1991; Ye and Worman, 1996). In addition to the H3K9me recognizing CD, HP1 contains a chromoshadow domain (CSD) (Aasland and Stewart, 1995) that mediates HP1 dimerization and interaction with a wide variety of proteins via a PxVxL penta-peptide motif (Brasher *et al.*, 2000; Cowieson *et al.*, 2000). Although primarily

Table 2.1 Mammalian H3K9 HMTases

Name	Nomenclature assigned name[1]	Target region of activity	Reference
SETDB1	KMT1E	Euchromatin	Schultz *et al.* (2002), Yang *et al.* (2002)
SETDB2	KMT1F	Centromeres	Falandry *et al.* (2010)
SUV39H1	KMT1A	Constitutive heterochromatin	Rea *et al.* (2000)
SUVH39H2	KMT1B	Constitutive heterochromatin	O'Carroll *et al.* (2000)
G9a	KMT1C	Euchromatin	Tachibana *et al.* (2001)
GLP	KMT1D	Euchromatin	Ogawa *et al.* (2002)
RIZ	KMT8		Congdon *et al.* (2014)

[1]Nomenclature based on Allis *et al.* (2007).

associated with heterochromatin and gene silencing, HP1 is emerging as having important roles in many genomic functions including DNA repair, replication and euchromatin (Kwon and Workman, 2011). Collectively, these data indicate that HMTases are a group of 'writers' and HP1 is one of the many 'readers' (Musselman *et al.*, 2012) of the epigenetic code that work together to lock in programing of the genome.

SET domain bifurcated 1

Among the H3K9 HMTases, SET domain bifurcated 1 (SETDB1) and the related SETDB2 stand out. Unlike the other mammalian SET-domain containing HMTases, in these two proteins the SET domain contains a substantial insertion that splits it in two: 347 amino acids for SETDB1 (Harte *et al.*, 1999), and 218 amino acids for SETDB2 (Mabuchi *et al.*, 2001), indicating that the SET may actually be two functionally distinct domains (Harte *et al.*, 1999). Such an arrangement is not unique to mammals. Homologues of SETDB1 exist in a wide variety of organisms, including *Caenorhabditis elegans* (Andersen and Horvitz, 2007) and *Drosophila* (Stabell *et al.*, 2006), that share the same complement and order of domains as mammalian SETDB1, including bifurcation of the SET domain (Fig. 2.2).

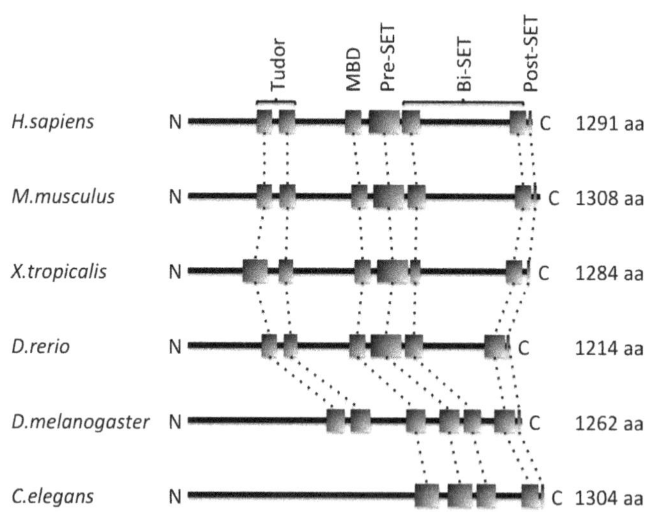

Figure 2.2 Conservation of the domain structure for SETDB1 from humans (*H. sapiens*) through to nematodes (*C. elegans*). The top row shows a schematic map of human SETDB1 with the location of domains indicated by shaded boxes and labelled above each. The protein is drawn from amino (N-left) to carboxyl (C-right) terminal and protein length indicated in amino acids (aa).

SETDB1, also known as ERG-associated protein with a SET domain (ESET), was mapped to chromosome 1q21 within a region that shows recurrent translocations in neoplasia (Harte *et al.*, 1999). Despite the insertion within the SET domain, both SETDB1 (Schultz *et al.*, 2002; Yang *et al.*, 2002) and SETDB2 (Falandry *et al.*, 2010) possess HMTase activity that was highly selective for lysine 9 of histone H3, with highest activity towards generation of trimethylated lysine 9 (H3K9me3). Interaction with H3 is mediated through the post-SET domain. Similar to the H3K9-HMTase G9a, but distinct from Suv39h1 (Nishioka *et al.*, 2002), SETDB1 can methylate H3 tails marked by mono- or dimethylated lysine 4 (H3K4me1/2), but it is unable to act at H3K4me3 *in vitro* (Binda *et al.*, 2010) and activity is reduced in the presence of H3 serine 10 phosphorylation or lysine 14 acetylation (Schultz *et al.*, 2002). Although, as described further below, SETDB1 associated H3K9me3 is observed at H3 marked by both H3K4me3 and H3K27me3 in embryonic stem cells (ESCs). Unlike Suv39h1/SUV39H1 and its homologue Suv39h2/SUV39H2 (Rea *et al.*, 2000) that direct HMTase activity at pericentric heterochromatin (Peters *et al.*, 2001; Rice *et al.*, 2003), SETDB1 localizes to euchromatic regions (Schultz *et al.*, 2002), similar to what is seen for the H3K9-HMTases G9a (Tachibana *et al.*, 2001) and G9a-like protein (GLP) (Tachibana *et al.*, 2005). Despite its distribution in euchromatic regions of the genome, SETDB1 overlaps with HP1, indicating that it likely is involved in gene silencing (Schultz *et al.*, 2002). In addition to the bifurcated SET domain, SETDB1 contains several other conserved domains, including an amino-terminal Tudor domain, a methyl-CpG-binding domain (MBD) and pre/post-SET domains (Fig. 2.2).

Tudor domain

The Tudor domain is named after a conserved motif identified in the *Drosophila tudor* gene that was isolated in a genetic screen for maternal-effect recessive lethality (Boswell and Mahowald, 1985). The Tudor domain consists of approximately 60 amino acids that form several anti-parallel β-sheets that aggregate to generate a barrel-like structure, and is found in many proteins involved in aspects of RNA metabolism, including splicing (Ponting, 1997) and small RNA pathways (Pek *et al.*, 2012). The Tudor domain mediates protein–protein interactions by recognizing and binding to methylated arginine in both histone and non-histone proteins (Lu and Wang, 2013), although more recently some Tudor-domain containing proteins form a chromodomain-like pocket that can recognize and bind methylated-lysine residues (Arita *et al.*, 2012; Bian *et al.*, 2011; Botuyan *et al.*, 2006; Huang *et al.*, 2006; Rothbart *et al.*, 2012; Sanders *et al.*, 2004; Vermeulen *et al.*, 2010; Wang *et al.*, 2011a; Yang *et al.*, 2012). As outlined below, SETDB1 interacts with several binding partners through this region of the protein such as Protein kinase B (Gao *et al.*, 2007). Interestingly, a conserved splice variant of SETDB1 has been reported in humans and mice that would generate a truncated form of the protein, that only contains the Tudor domains (Blackburn *et al.*, 2003). The functional significance of this isoform has yet to be determined.

Methyl-CpG-binding domain

The MBD is an approximately 70 amino acid conserved motif that recognizes and binds to methylated cytosine, and is found in a variety of chromatin proteins (Hendrich and Tweedie, 2003), most notable the methyl-CpG binding proteins MECP2 and MBD1-4 (Hendrich and Bird, 1998). Several amino acids within the MBD of SETDB1 differ from those that are important for binding activity (Ohki *et al.*, 2001), and therefore SETDB1 may not directly bind methylated DNA through this domain. However, the *Drosophila* SETDB1 homologue (dSETDB1) does bind methylated cytosine in the context CpA (Gou *et al.*, 2010), a form of DNA methylation that is more prominent in *Drosophila* than methylated CpG (Gowher *et al.*, 2000; Lyko *et al.*, 2000). In *Neurospora crassa*, the SET-domain containing HMTase defective in methylation-5 (DIM-5) (Tamaru and Selker, 2001) is required not only for establishing H3K9me3, but also for CpG-methylation (Tamaru *et al.*, 2003). Furthermore, in *Neurospora* the H3K9me3-binding protein HP1 is essential for DNA methylation (Freitag *et al.*, 2004), confirming the dependence of CpG-methylation on H3K9me3. However, in

Arabidopsis thaliana, CpNpG methylation, but not CpG methylation, is dependent on H3K9–HMTase activity (Jackson *et al.*, 2002; Malagnac *et al.*, 2002) and loss of CpG methylation results in concomitant loss of H3K9me (Tariq *et al.*, 2003). Therefore, the relationship between DNA methylation and H3K9me is complex, but the role of SETDB1 in this process is beginning to emerge as outlined in more detail below.

Pre- and post-SET domains

The pre-SET domain contains nine-invariant cysteine residues on the amino-terminal side of most, but not all SET-domain containing proteins. This cysteine-rich motif coordinates zinc ions, and provides a structural role for SET domain activity (Dillon *et al.*, 2005). The post-SET domain contains three-invariant cysteine residues located on the carboxyl-terminal side of the SET domain, and at least for SUV39H1 and DIM-5 this motif is essential for HMTase activity (Rea *et al.*, 2000; Zhang *et al.*, 2002), and is necessary for SETDB1 binding of the H3 amino-terminal tail (Binda *et al.*, 2010).

Like many chromatin modifiers and remodelling proteins, SETDB1 does not possess DNA binding activity or specificity and therefore in order to enable its H3K9–HMTase activity to impact chromatin structure, it needs to be directed to target sites. SETDB1 has been found to interact with a variety of proteins, including some that recruit SETDB1 to genomic loci.

SETDB1 interacting proteins

SETDB1 was originally independently identified through its interaction with two different proteins.

In the first report, Frank Rauscher III and colleagues performed a yeast 2-hybrid (Y2H) screen with the Plant homeodomain (PHD) finger and bromodomain of the protein tripartite motif containing 28 (TRIM28; also known as KRAB-associated protein-1 or KAP-1 and also as transcription intermediary factor 1-beta or TIF1B), to identify novel proteins that interact with this domain (Schultz *et al.*, 2002). TRIM28 is a multifunctional protein (Iyengar and Farnham, 2011) that can interact with proteins containing the Krüppel-associated box (KRAB) repression domain that is found in a variety of transcription factors, including a large number of zinc finger proteins (KRAB–ZFPs) (Friedman *et al.*, 1996; Mark *et al.*, 1999) and colocalizes with HP1 at interphase through the HP1 CSD (Ryan *et al.*, 1999). TRIM28 had previously been shown to recruit the nucleosome remodelling and deacetylase (NuRD) complex to silence target genes (Schultz *et al.*, 2001); a large multisubunit complex involved in repressing gene expression (Denslow and Wade, 2007). TRIM28 interacted with a region overlapping the Tudor domains of SETDB1 (Schultz *et al.*, 2002). In the second report, Yi Zhang and colleagues independently isolated SETDB1 through a Y2H screen as protein that interacted with the ETS-related gene (ERG) (Yang *et al.*, 2002); a member of the ETS-family transcription factors that are involved in numerous functions including cell growth and differentiation (Hollenhorst *et al.*, 2011).

Since then, numerous proteins have been shown to interact with SETDB1, with transcriptional repression and heterochromatin formation a unifying theme.

SIN3/HDAC co-repressor complex

Mammals code for two paralogues of yeast Sin3 called SIN3A and SIN3B (Ayer *et al.*, 1995), which are part of a multiprotein repressor complex that contains the histone deacetylase proteins HDAC1 and HDAC2 (Fleischer *et al.*, 2003). Immunoprecipitation of an epitope-tagged SETDB1 from human cells co-immunoprecipitated (CoIP) the SIN3/HDAC co-repressor complex (Yang *et al.*, 2003). Deletion analysis revealed a direct interaction between SETDB1 and SIN3 via the Tudor domains, and that removal of the carboxyl-terminal region of SETDB1 actually enhanced the interaction, suggesting that if the truncated splice-variant of SETDB1 produces a functional protein (Blackburn *et al.*, 2003), it may interact more readily with the SIN3/HDAC complex *in vivo* (Yang *et al.*, 2003).

MBD1 and MCAF1

DNA methylation inhibits transcription through the recruitment of methyl-CpG-binding proteins (Boyes and Bird, 1991). The methyl-CpG-binding domain 1 (MBD1) protein was isolated as part of a repressor complex (Cross *et al.*, 1997) and binds directly to methylated CpG (Ohki *et al.*, 2001). Precipitation of MECP2, a different methyl-CpG binding protein had shown CoIP of H3K9-HMTase activity, and recruited that activity to imprinted loci (Fuks *et al.*, 2003). Likewise, pull-down of MBD1 resulted in the CoIP of H3K9-HMTase activity (Sarraf and Stancheva, 2004), indicating that this property was not unique to MECP2. Using MBD1 as bait in a Y2H screen-identified SETDB1 and the large subunit (p150) of the chromatin assembly factor 1 (CAF1) as MBD1-interacting partners (Sarraf and Stancheva, 2004). MBD1 interacts via the second of its two CxxC motifs with a region of SETDB1 that encompasses the Tudor domains. MBD1 complexed with SETDB1 could be detected throughout the cell cycle. However, CAF1 only associated during the S-phase of the cell cycle. As its name suggests, CAF1 performs an important role in assembling histones onto replicating DNA (Kaufman *et al.*, 1995). Consistent with this, during S-phase, H3 and H4, but not H2A or H2B could be detected by CoIP with SETDB1. These data suggest a model in which as the replication fork passes during DNA replication, methylated DNA signals for MBD1:CAF1:SETDB1 to recognize the me-CpG, deposit a new $H3_2H4_2$-tetramer and methylate H3K9 to facilitate maintenance of repressed chromatin on the daughter chromosomes (Sarraf and Stancheva, 2004). Consistent with an anchoring role of MBD1 in this model, depletion of MBD1 levels through RNA interference (RNAi) reduced H3K9me3 levels. Chromatin assembly involving CAF1 can be replication coupled during S-phase, but also replication uncoupled during nuclear processes such as DNA repair or chromatin remodelling (Ransom *et al.*, 2010). Different CAF1 containing chaperone complexes exist to mediate these deposition pathways (Tagami *et al.*, 2004), including CAF1 associated with HP1α and HP1ϒ (Quivy *et al.*, 2004). Interestingly, a distinct CAF1 complex has been described that contains TRIM28, HP1α and H3K9-HMTase activity provided by SETDB1 (Loyola *et al.*, 2009). This complex does not appear to associate with nucleosomes, but rather free histones and mediates H3K9me1, an activity that is lost on depletion of SETDB1 by RNAi, suggesting that SETDB1 is responsible and that its ability to deposit mono-, di- or trimethyl signatures on H3 is an activity modulated by whom it interacts with. Furthermore, this complex colocalizes with the replication fork, as defined by the distribution of Proliferating Cell Nuclear Antigen (PCNA) (Prelich and Stillman, 1988), as pericentric heterochromatin is being replicated. Intriguingly, loss of Suv39h1/2 that is responsible for H3K9me3 at pericentric heterochromatin (Peters *et al.*, 2001), results in loss of H3K9me3 and gain of H3K9me1 at these genomic locations. This suggests that SETDB1 through the TRIM28–CAF1–HP1α complex provides H3K9me1 as a precursor at pericentric heterochromatin for Suv39h1/2 to bring to maturity.

MBD1 does not always interact directly with SETDB1, rather the two can be bridged through common association with the MBD1-containing chromatin-associated factor (MCAF) (Ichimura *et al.*, 2005; Wang *et al.*, 2003), that synergistically assist in maintaining silent chromatin as demonstrated by compromised repression of an X-linked transgene at the Xi when protein levels of each are reduced by RNAi (Minkovsky *et al.*, 2014). MBD1 recruits MCAF1 via its carboxyl transcriptional repression domain (TRD) (Fujita *et al.*, 2003). While *in vitro* both MCAF1 and MCAF2 can pull down SETDB1, *in vivo* SETDB1 preferentially associates with MCAF1 (Ichimura *et al.*, 2005). Interestingly, association of MCAF1 with SETDB1 has several effects. First, MCAF1 stimulates SETDB1 HMTase activity by increasing reaction turnover rate. Second, SETDB1 complexed with MCAF1 displays a modest increase in activity against oligo-nucleosomal H3 substrate, which is inhibited by histone H1. Finally, MCAF1 stimulates conversion of H3K9me2 to H3K9me3, which is associated with more potent repression of transcription (Wang *et al.*, 2003). These data further support the notion that factors associating with SETDB1 can alter its activity and substrate preference. Notably, MCAF1 is not part of the TRIM28–CAF1–HP1α–SETDB1 complex described above that is responsible for H3K9me1 (Loyola *et al.*, 2009).

The nature of the interactions between MBD1 and MCAF1 is more complex than stated here due to the post-translational modification sumoylation. Sumoylation involves the covalent attachment of small ubiquitin-related modifiers (SUMO) to target proteins, and occurs by an enzymatic pathway similar to ubiquitination (Johnson, 2004), resulting in altered DNA–DNA, protein–protein or protein–DNA interactions (Ulrich, 2014). MBD1 is sumoylated by protein Inhibitors of activated STAT 1 (PIAS1) and 3 (PIAS3) at specific lysine residues in its carboxyl terminus. While this does not impact the ability of MBD1 to bind methylated DNA, it was reported that modification by SUMO-1 blocks SETDB1 association and can de-repress target genes (Lyst *et al.*, 2006). Indeed, PIAS1 and PIAS3 were picked up as MBD1 interacting proteins in the same Y2H screen as SETDB1 and CAF1 (Sarraf and Stancheva, 2004). However, the impact of MBD1 sumoylation on MCAF1 and SETDB1 association is not so simple. In a separate report, sumoylation of MBD1 facilitated MCAF1 association, primarily through SUMO-2/3, and knockdown of SUMO-2/3 dissipated MBD1 heterochromatin through reduced H3K9me3 and HP1β and HP1γ colocalization (Uchimura *et al.*, 2006). Collectively, these data suggest the possibility that MBD1–SUMO-1 blocks direct SETDB1 recruitment, while MBD1–SUMO-2/3 recruits SETDB1 indirectly via MCAF1. Clearly Sumoylation has an important role in regulating SETDB1 targeting. GST-SUMO-2 affinity chromatography isolated SETDB1 (Rosendorff *et al.*, 2006). Furthermore, the transcription factor SP3, which can both activate or repress transcription (Suske, 1999), recruits SETDB1 (Stielow *et al.*, 2008) when modified by SUMO-1, which corresponds to the repressive form of SP3 (Ross *et al.*, 2002). As outlined below, Sumoylation also plays an important role in recruiting SETDB1 to TRIM28.

HP1

As outlined briefly earlier, mammals have three HP1 isoforms, HP1α, HP1β and HP1γ (Jones *et al.*, 2000), that are major features of pericentric and telomeric heterochromatin (Minc *et al.*, 1999; Wreggett *et al.*, 1994). HP1 is a classic modifier of position effect variegation in *Drosophila* (Eissenberg *et al.*, 1992; Locke *et al.*, 1988), a property that is also observed in mammals (Festenstein *et al.*, 1999). HP1 is a relatively small protein containing an amino-terminal CD (Paro and Hogness, 1991) and carboxyl-terminal CSD (Aasland and Stewart, 1995) that flank the central hinge domain (Smothers and Henikoff, 2001). While the CSD mediates protein–protein interactions (Lomberk *et al.*, 2006), the CD binds to H3K9me (Aagaard *et al.*, 1999; Bannister *et al.*, 2001; Jacobs *et al.*, 2001; Lachner *et al.*, 2001; Nakayama *et al.*, 2001), and can be seen enriched on portions of the mammalian Xi that are decorated by H3K9me3 (Chadwick and Willard, 2003; Cowell *et al.*, 2002). It is thought that initially HP1 makes weak association with H3K9me via the CD (Cheutin *et al.*, 2003; Jacobs *et al.*, 2001) which then interacts with the histone tail as a β-strand (Nielsen *et al.*, 2002) before the CD interacts with the histone fold domain of H3 (Nielsen *et al.*, 2001), locking in a stable interaction (Singh and Georgatos, 2002). These data all point to HP1 recruitment to heterochromatin as a step downstream of deposition of H3K9me, which is consistent with the inability of HP1 to associate with pericentric heterochromatin in the absence of H3K9me3 when Suv39h1/2 are lost (Peters *et al.*, 2001). Furthermore, in addition to the loss of H3K9me3 at pericentric regions in Suv39h1/2 null mice, H4K20me3 is lost, suggesting cross-talk between these two heterochromatin markers, with H4K20me3 dependent on H3K9me3 (Kourmouli *et al.*, 2004). It is likely that H4K20me3 is downstream of HP1 association, as HP1 is known to interact with the H4K20me3 HMTase Suv4-20h1/2 (Schotta *et al.*, 2004). Intriguingly, Kourmouli and colleagues demonstrated that anchoring HP1 back to pericentric DNA in cells lacking Suv39h1/2 was sufficient to restore H3K9me3 and H4K20me3 (Kourmouli *et al.*, 2005). HP1 was fused to the AT-hook of a high mobility group protein that bind AT-rich repeats at pericentric DNA (Reeves, 2001), targeting the chimeric HP1 protein back to pericentric DNA in the absence of H3K9me3. Restoration of H3K9me3 was achieved through HP1 recruiting SETDB1 (Kourmouli *et al.*, 2005), indicating that targeting of HP1 is sufficient to recruit SETDB1 to establish heterochromatin formation, consistent with the results of others using HP1-fusion proteins (Verschure *et al.*, 2005). This is not the only example of cooperation and coordination between H3K9-HMTases. Indeed, a subset of G9a, GLP, SETDB1 and SUV39H1 have been reported in a megacomplex and are targeted to satellite repeats (Fritsch *et al.*, 2010) and may assist in sister chromatid cohesion (Kim *et al.*, 2008).

DNA methyltransferases

Cytosine residues followed by guanine (CpG) can be methylated at the 5-position in mammalian genomes and is necessary for gene silencing, repression of repetitive elements and development (Li *et al.*, 1993). Several enzymes are responsible for DNA methylation and are known as the DNA methyltransferases (DNMT) and include the *de novo* methyltransferases DNMT3a and DNMT3b, the maintenance methyltransferase DNMT1 and the enzymatically inactive methyltransferase DNMT3L (Bestor, 2000). The mechanistic relationship between histone methylation and DNA methylation is beginning to be deciphered through interactions between HMTases and DNMTs (Cedar and Bergman, 2009). SETDB1 interacts via a region encompassing its Tudor domains directly with the PHD finger-like domain of DNMT3A (Li *et al.*, 2006b) and co-occupy methylated promoters of target genes. SETDB1 also interacts with DNMT3L (Kao *et al.*, 2014). Even though DNMT3L does not possess DNA methylation activity, it is expressed in prospermatogonia, oocytes and embryonic stem cells and is involved in establishing DNA methylation patterns during gametogenesis (Bourc'his and Bestor, 2004; Bourc'his *et al.*, 2001; Hata *et al.*, 2002). DNMT3L interacts with and stimulates the activity of DNMT3A/B and binds to unmethylated H3K4, directing *de novo* DNA methylation to the underlying sequence (Ooi *et al.*, 2007). In CoIP experiments, SETDB1 along with TRIM28, HDAC1 and DNMT3A were isolated alongside DNMT3L, but in DNMT3L null cells, association of SETDB1 and HDAC1 was reduced, whereas forced expression of DNMT3L in somatic cells was sufficient to recruit SETDB1 to target loci and introduce H3K9me3 (Kao *et al.*, 2014).

TRIM28 and KRAB–ZFPs

TRIM28 is a member of a large family of tripartite motif containing proteins (Reymond *et al.*, 2001), and recruits the NuRD HDAC complex to target genes (Schultz *et al.*, 2001). TRIM28 does not bind DNA directly, but interacts with domains within transcription factors for targeting. One of the best-characterized routes of TRIM28 targeting is as a corepressor for KRAB domain containing KRAB-ZFPs (Friedman *et al.*, 1996); one of the largest and most diverse (Dehal *et al.*, 2001; Krebs *et al.*, 2005) groups of transcription factors in mammals (Mark *et al.*, 1999), with greater than one third of all human ZFP containing a KRAB domain (Huntley *et al.*, 2006). As mentioned earlier, one of the routes through which SETDB1 was originally characterized as a repressive HMTase was through its association with TRIM28 (Schultz *et al.*, 2002). TRIM28 directly interacts with the KRAB domain (Abrink *et al.*, 2001; Peng *et al.*, 2000), and can be recruited to ectopic chimeric repressor proteins that are fused to the domain resulting in the establishment or mitotically stable heterochromatin (Ayyanathan *et al.*, 2003). TRIM28 also interacts directly with the CSD of HP1 (Ryan *et al.*, 1999), and mutation of the interaction site between HP1 and TRIM28 compromises acquisition of repressive histone modifications, and TRIM28 mediated silencing is attenuated when levels of TRIM28, HP1 or SETDB1 are reduced by RNAi (Sripathy *et al.*, 2006). TRIM28 contains a number of conserved domains including a PHD finger, a Bbromodomain and RING-finger (Fig. 2.3A). Bromodomains recognize acetylated lysine (Mujtaba *et al.*, 2007), whereas some PHD domains have been reported to bind to methylated lysine (Li *et al.*, 2006a; Shi *et al.*, 2006; Wysocka *et al.*, 2006). The TRIM28 PHD-finger is highly related to a RING finger (Capili *et al.*, 2001), and possesses E3 ligase activity, sumoylating the adjacent bromodomain (Ivanov *et al.*, 2007). Sumoylation is required for TRIM28-mediated silencing and is bound by SUMO-interacting motifs (SIM) in the NuRD subunit chromodomain helicase DNA binding protein 3 (CHD3) and with a SIM motif in the amino-terminal portion of SETDB1 (Fig. 2.3B).

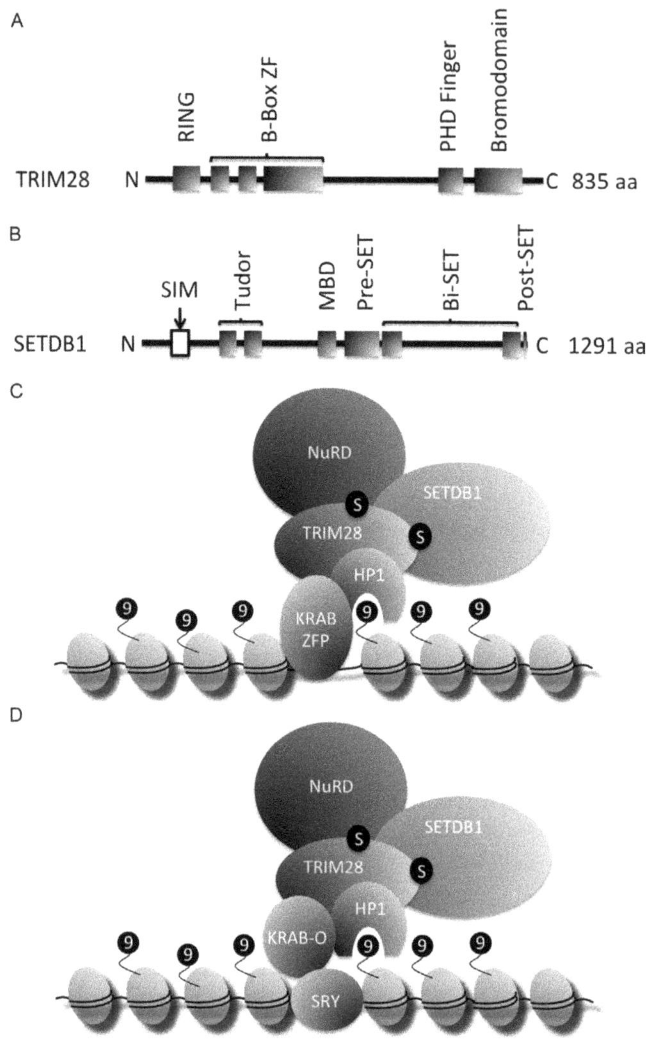

Figure 2.3 (A) Schematic representation of domain structure for TRIM28. The line is drawn from amino (N-left) to carboxyl (C-right) terminal and the protein length is indicated in amino acids (aa). The shaded boxes on the line represent the location and coverage of domains indicated. (B) Domain structure for SETDB1 with the location of the SUMO-interacting motif (SIM) represented by the open box. (C) Model depicting recruitment of TRIM28, SETDB1, NURD and HP1 to target sites by a ZRAB-domain containing zinc finger protein (KRAB-ZFP). The DNA is represented by the thin black-line that is wrapped around nucleosomes (spheres). Methylated lysine-9 of the amino-terminal tail of H3 is represented by the '9'-labelled black sphere at the end of the curved line. Association of SETDB1 and NURD with TRIM28 is mediated through their SIM-motifs and SUMO-modified sites on TRIM28 (black spheres labelled 'S'). (D) As for part-C, but with the addition of the KRAB-only protein (KRAB-O) acting as an adapter for SRY to recruit the TRIM28 repressor complex.

While TRIM28 is typically recruited to DNA targets through direct interaction with a DNA-tethered KRAB-ZFP (Fig. 2.3C), a notable exception is via KRAB-only (KRAB-O). KRAB-O is generated through alternative splicing of the Zfp748 gene. KRAB-O interacts with the sex-determining region Y (SRY) protein (Oh *et al.*, 2005), and recruits TRIM28–NURD–SETDB1–HP1 to SRY target genes (Fig. 2.3D) for testis differentiation during embryogenesis (Peng *et al.*, 2009).

Interestingly, the relationship between TRIM28 and KRAB–ZFPs extends beyond interactions in the proteome. Genome-wide analysis revealed that TRIM28 is frequently found in regions marked by H3K9me3 and that it is at the KRAB domain containing zinc finger genes (O'Geen *et al.*, 2007), which is most apparent on chromosome 19 where many KRAB-ZFP genes are arranged into several gene clusters (Grimwood *et al.*, 2004). Those KRAB–ZFP genes that are not bound by TRIM28 are actively transcribed, whereas those with TRIM28 association are expressed at very low levels. While TRIM28 is generally found at the promoter of target genes, for most KRAB–ZFP genes, TRIM28 is also enriched within the gene body (O'Geen *et al.*, 2007). The explanation for this is that the KRAB–ZFP ZNF274 recruits SETDB1 through TRIM28 to the 3′ end of zinc finger genes, and reduction of ZNF274 levels by RNAi only impacts association of TRIM28/SETDB1 at these sites (Frietze *et al.*, 2010). The association with the 3′ end of zinc finger genes is dependent upon the TRIM28 KRAB-binding domain and is stabilized by HP1 association (Iyengar *et al.*, 2011). The reason why the 3′ end of KRAB–ZFP is organized into H3K9me3 chromatin might be due to the fact that the coding sequence for multiple uninterrupted zinc finger motifs has the appearance of tandem repeat DNA. Organizing these sequences into heterochromatin might reduce the chance of inappropriate recombination between other family members and protect from genome instability. Indeed, KRAB–ZFP genes with more zinc fingers, and hence a more pronounced tandem repeat, have higher levels of H3K9me3 (Blahnik *et al.*, 2011).

SETDB1 in development

Gene targeting of H3K9-HMTases indicates the importance of H3K9me in development. Mice lacking Suv39h1/2 die ~8.5 days post coitum (dpc), and show cytogenetic abnormalities and compromised heterochromatin formation at pericentric chromosomal regions (Lehnertz *et al.*, 2003; Peters *et al.*, 2001). Likewise, loss of G9a or GLP is early embryonic lethal (Tachibana *et al.*, 2002; Tachibana *et al.*, 2005). Setdb1 expression begins at the blastocyst stage and is ubiquitous post-implantation. Not unsurprisingly, Setdb1 null mice are peri-implantation lethal, occurring ~3.5–5.5 dpc (Dodge *et al.*, 2004). Embryos are severely malformed, with embryonic tissues almost completely resorbed, while the trophectoderm-derived ectoplacental cone is relatively normal. Few blastocysts developed a normal morphology, but, of those that did, ESCs could not be derived from the inner cell mass (ICM) (Evans and Kaufman, 1981; Martin, 1981), pointing to a crucial role for Setdb1 in maintaining ESCs (Dodge *et al.*, 2004).

SETDB1, pluripotency and stem cells

Consistent with these observations, knock down of Setdb1 levels by RNAi in wild type mouse ESCs results in differentiation and the appearance of trophoblast-like cells that display expression of trophoblast associated genes (Yuan *et al.*, 2009). Furthermore, injection of GFP expressing mouse ESCs harbouring intact conditional Setdb1 alleles into four- to eight-cell stage embryos, show retention of GFP in the ICM, whereas induced deletion of Setdb1 prior to injection results in detection of GFP in the extraembryonic material, indicating that the cells migrated away from the ICM (Lohmann *et al.*, 2010).

The molecular basis for these observations is that Setdb1 associates with the POU-type homeodomain containing octamer-binding protein 4 (Oct4); a key gene in embryonic development and maintenance of stem cell pluripotency (Niwa *et al.*, 2000). Oct4 can CoIP with Setdb1 (Cho *et al.*, 2012) and Oct4-Setdb1 are found at, and silence trophoblast-associated genes (Yuan *et al.*, 2009).

Further evidence supporting an important role for Setdb1 in the maintenance of pluripotency and self renewal was obtained as part of a high-throughput RNAi screen to identify regulators for maintaining ESCs. Short-hairpin RNA constructs directed to Setdb1 resulted in the loss of ESC colony morphology and Oct4 expression, consistent with differentiation (Bilodeau *et al.*, 2009). Genome-wide analysis of Setdb1 distribution revealed that Setdb1 primarily associated with genes encoding developmental regulators that were silenced and marked by H3K9me3 (Bilodeau *et al.*, 2009). Previously, the promoters for a number of genes involved in development were reported to be marked by both H3K4me3 and H3K27me3 and that the genes were poised for activation on signalling. This chromatin configuration was termed 'bivalent' to account for the presence of both activating and repressive histone modifications (Bernstein *et al.*, 2006). Interestingly, a number of these genes are Setdb1 target sites and also are marked by H3K9me3 and are essentially 'trivalent' (Bilodeau *et al.*, 2009).

Mesenchymal stem cells (MSCs) are multipotent cells that have the ability to differentiate into a variety of cell lineages, including bone, cartilage, tendon, ligament and adipose tissue (Pittenger *et al.*, 1999). Setdb1 has been implicated as fulfilling an important role in MSC differentiation and control of downstream differentiation pathways (Lawson *et al.*, 2013a.b; Lee *et al.*, 2005; Musri *et al.*, 2010; Nifuji *et al.*, 2010; Takada *et al.*, 2007; Yang *et al.*, 2013), as well as mediating hormone differentiation and proliferation signals (Hong *et al.*, 2011), extending the importance of this protein beyond ESC maintenance.

SETDB1 and imprinting

Genomic imprinting refers to the situation whereby a subset of genes display parent of origin-specific transcription and as a consequence show strict monoallelic expression, an arrangement that is necessary for normal development (Barlow and Bartolomei, 2014). Previously it has been shown that the KRAB-ZFP Zfp57 is important for establishing and maintaining imprinting (Li *et al.*, 2008). Zfp57 recruits Trim28, Setdb1 and HP1ϒ to the CpG-methylated imprinting control region (ICR) (Quenneville *et al.*, 2011), while the architectural protein CCCTC-binding factor (Ctcf) associates with the unmethylated allele (Fig. 2.4) (Stedman *et al.*, 2008). The specificity of Zfp57 to imprinted genes is demonstrated by the fact that its deletion almost exclusively impacts

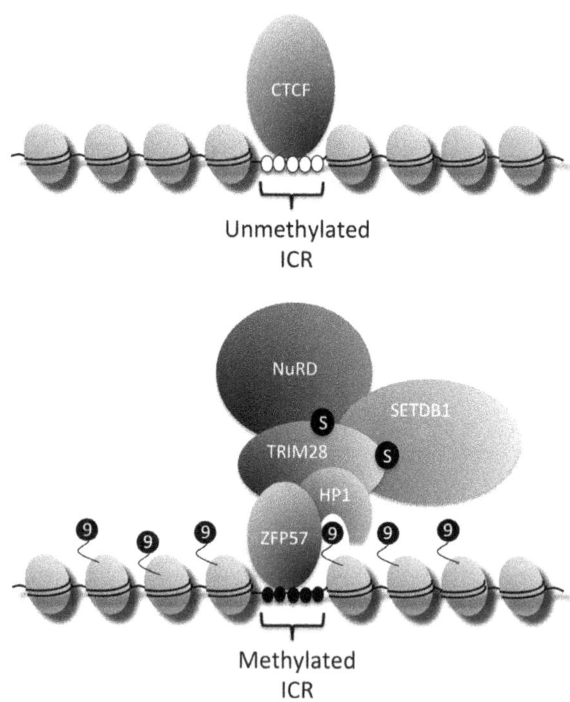

Figure 2.4 Model depicting the differential chromatin organization at the two alleles of imprinted genes. The top section shows the unmethylated imprinting control region (ICR) bound by CTCF. The unmethylated CpG residues at the CTCF binding site are represented by white-circles. The bottom section shows the methylated ICR allele with ZFP57 binding specifically to the methyl-CpG residues and recruiting the TRIM28 repressor complex that locks in a heterochromatin state through local modification of histone H3 tails by trimethylation through the activity of SETDB1.

the expression of imprinted genes (Li *et al.*, 2008), and genome-wide analysis of Zfp57 identified 91 genomic targets, which included all known ICRs. Zfp57 association with DNA is dependent on DNA methylation, because cells lacking Dnmt1, Dnmt3a and Dnmt3b lost Zfp57 association with ICRs, but Trim28 and Setdb1 remained associated with other non-imprinted targets as expected (Quenneville *et al.*, 2011), placing Zfp57 among the limited number of ZFPs that show sequence-specific recognition of methylated DNA (Sasai *et al.*, 2010). Consistent with these observations in mouse, the ZNF274 KRAB-ZFP specifically associates with and recruits SETDB1 to the silent maternal allele at the Prader–Willi syndrome locus at human chromosome 15q11–13 (Cruvinel *et al.*, 2014).

One of the biggest challenges facing epigenetic programming are the waves of demethylation that occur in the preimplantation embryo shortly after fertilization (Santos and Dean, 2004) and during early germ cell development in primordial germ cells (Hajkova *et al.*, 2002) as DNA methylation patterns are erased before being reset later in embryonic development or gametogenesis (Sasaki and Matsui, 2008). Central to this process are members of the ten-eleven translocation (TET) genes that catalyse the conversion of 5-methylcytosine (5mC) into 5-hydroxymethylcytosine (5hmC) (Tahiliani *et al.*, 2009), an intermediate in the demethylation pathway (He *et al.*, 2011; Inoue and Zhang, 2011) and direct the demethylation process (Hackett *et al.*, 2013; Iqbal *et al.*, 2011; Koh *et al.*, 2011; Vincent *et al.*, 2013; Williams *et al.*, 2011). The danger here is that important imprinted signals, in the form of DNA methylation are at risk of being caught up in this genome wide process. However, some sequences are resistant to demethylation (Hackett *et al.*, 2013; Hajkova *et al.*, 2002; Vincent *et al.*, 2013), including a subset of endogenous retroviruses (ERVs) (Guibert *et al.*, 2012; Hajkova *et al.*, 2002; Kobayashi *et al.*, 2012, 2013) and imprinted loci (Borgel *et al.*, 2010; Kobayashi *et al.*, 2012). Genome-wide 5mC profiles were compared between wild-type mouse ESCs and those lacking Dnmt3a and Dnmt3b, identifying thousands of sites with residual 5mC that were termed enriched residual methylation loci (ERML) (Leung *et al.*, 2014). A subset of ERML corresponded to ERV elements and imprinted loci and overlapped with peaks of H3K9me3. Knockdown of Dnmt1 or Setdb1 reactivated ERV expression, and was more pronounced if both were reduced simultaneously, indicating that both play an important role in maintaining ERML. Furthermore, reduction of Setdb1 correlated with reduced H3K9me3 and corresponding gains in 5hmC and Tet1 residency at ERML indicating that Setdb1 protects loci programmed to remain silent from demethylation waves during development (Leung *et al.*, 2014).

Lessons from *Drosophila* and *Caenorhabditis elegans*

As indicated in Fig. 2.2, homologues of SETDB1 exist in *Drosophila* (dSETDB1) (Stabell *et al.*, 2006) and *C. elegans* (cSETDB1) (Andersen and Horvitz, 2007).

In *C. elegans*, cSETDB1 was identified by two independent routes. In the first study, deletion alleles and RNAi were used to explore the role of all predicted lysine-specific HMTases in vulval development (Andersen and Horvitz, 2007). Development of the vulva provides a powerful system to explore cell fate determination, tissue remodelling and organogenesis assisted by the fact that it is dispensable for survival (Sternberg and Horvitz, 1991). Two HMTases altered cell fate decisions in vulva development: MET-1, a homologue of the H3K36 HMTase Set2, and MET-2, a homologue of SETDB1, indicating an important developmental role for H3K9me3 and SETDB1. In the second study, cSETDB1 was identified in a *N*-ethyl-*N*-nitrosourea (ENU) mutagenesis screen to identify female-sterile and lethal mutations (Clough *et al.*, 2007). PGC development in *C. elegans*, is accompanied by changes in histone acetylation and methylation (Schaner *et al.*, 2003), and SET domain proteins are required for germline development (Bender *et al.*, 2004; Feng *et al.*, 2010; Kelly and Fire, 1998). The ENU screen identified 13 mutants in egg development with weak to strong phenotypes. In one mutant, named *eggless* (*egg*), the levels of H3K9me2 were slightly elevated, but H3K9me3 was absent. Egg is cSETDB1, independently validating an important role for this protein in oogenesis (Clough *et al.*, 2007).

In *Drosophila,* dSETDB1 was identified through functional characterization of an annotated SET domain protein in FlyBase (Stabell *et al.,* 2006). dSETDB1 was detected after 3 hours of embryonic development through to adult flies, and was shown to be important for normal development through RNAi experiments. Like human SETDB1 (Schultz *et al.,* 2002), dSETDB1 localized to euchromatic regions of the genome based on polytene chromosome distribution, but was also detected at the chromocenter and was observed at the small fourth chromosome that is composed of interspersed bands of euchromatin and heterochromatin (Sun *et al.,* 2000). The relationship between dSETDB1 and the fourth chromosome was strengthened by the finding that mutation of dSETDB1 resulted in reduced expression of endogenous genes from the chromosome, as well as loss of H3K9me and HP1 association, and that dSETBD1 interacted with the chromosome 4-specific painting-of-fourth (POF) as determined by CoIP (Tzeng *et al.,* 2007). POF is required for up-regulation of genes on the fourth chromosome (Johansson *et al.,* 2007; Larsson *et al.,* 2001) and therefore, loss of dSETDB1 may impact the ability of POF to mediate this function. In contrast, dSETDB1 was required for silencing of a transgene inserted onto the fourth chromosome (Seum *et al.,* 2007b), indicating a role in maintaining endogenous genes active and ectopic genes silent on this chromosome, depending upon their location (Lundberg *et al.,* 2013). Loss of dSETDB1 reduces levels of mono-, di- and trimethylate H3K9, while overexpression increases all three (Seum *et al.,* 2007b; Tzeng *et al.,* 2007), confirming its role as an HMTase in *Drosophila,* and indicated that like mammalian and nematode SETDB1 (Dodge *et al.,* 2004), dSETDB1 was essential for viability (Brower-Toland *et al.,* 2009; Seum *et al.,* 2007b; Tzeng *et al.,* 2007; Wang *et al.,* 2011b; Yoon *et al.,* 2008). Several interesting parallels can be drawn between dSETDB1 and mammalian SETDB1. First, in *Drosophila* germline stem cells, dSETDB1 is the only H3K9me3 HMTase active and as such it is responsible for H3K9me3 and subsequent HP1 association with DAPI-dense pericentric heterochromatin, but later in germ cell development SU(VAR)3–9 takes over this role from dSETDB1 (Yoon *et al.,* 2008), similar to the Setdb1 priming histones for Suv39h1/2 in mammalian cells (Loyola *et al.,* 2009). Second, the *Drosophila* homologue of MCAF1, Windei, interacts with dSETDB1 and is essential for its proper localization and function (Koch *et al.,* 2009). One notable difference between dSETDB1 and mammalian SETDB1 is the ability of dSETDB1 to bind methylated DNA via the MBD domain. However, the binding in *Drosophila* is preferentially towards methylated CpA (Gou *et al.,* 2010), which is uncommon in mammals.

Notably, dSETDB1 is the only essential H3K9 HMTase in *Drosophila,* as loss of SU(VAR)3-9 (Tschiersch *et al.,* 1994) or G9a (Seum *et al.,* 2007a) is homozygous viable, further emphasizing the evolutionary importance of this protein in proper development.

SETDB1 and ERVs

ERV elements constitute 8% of the mouse genome (Mouse Genome Sequencing Consortium *et al.,* 2002) and approximately 10% of the human genome (Lander *et al.,* 2001), and can be divided into several classes based on their relationship to exogenous retroviruses (Nelson *et al.,* 2004; Stocking and Kozak, 2008). ERVs contain long terminal repeats (LTRs) that would have provided promoter function to their ancestral retrovirus, but now can provide promoter activity in their host genome (Cohen *et al.,* 2009), acting as tissue-specific enhancers and promoters (Buzdin *et al.,* 2006; Chuong *et al.,* 2013; Karimi *et al.,* 2011; Macfarlan *et al.,* 2011, 2012; Rebollo *et al.,* 2011). Indeed, many LTR elements function as *cis* regulatory elements during development through the recruitment of transcription factors (Bourque *et al.,* 2008; Kunarso *et al.,* 2010; Mey *et al.,* 2012; Schmidt *et al.,* 2012). Given that these elements can move in the genome, and that their complement, copy number and genomic locations differ between species, ERVs and other transposable elements are a major driving force in evolution (Chuong, 2013; Cordaux and Batzer, 2009; Jacques *et al.,* 2013; Stoye, 2012). However, as much as ERVs can provide new regulatory elements for genes or alter tissue or temporal expression patterns, they can also have detrimental impacts. Notably, ERVs can contribute to high-level inappropriate expression of nearby genes in cancer (Romanish *et al.*, 2010) or ageing tissue (Barbot *et al.*, 2002), and are a source of germline mutation (Maksakova *et al.*, 2006) and disease susceptibility (Kaer and Speek, 2013). As such, human and mouse ERVs are generally silenced through DNA methylation, but silencing in ESCs is largely independent of 5mC (Leung and Lorincz, 2012). Instead, in mouse ESCs, class I and class II ERVs involve silencing by H3K9me3 and H4K20me3 (Mikkelsen *et al.*, 2007) whereas Class III silencing involves LSD1 and the removal of H3K4me1/2 (Macfarlan *et al.*, 2011).

Silencing of mouse class I and class II ERVs is dependent on Setdb1 and Trim28, which lose association of H3K9me3 and are reactivated in the absence of either (Karimi *et al.*, 2011; Maksakova *et al.*, 2011; Matsui *et al.*, 2010; Rowe *et al.*, 2010, 2013). Mouse ESC triple-deleted for Dnmt1, Dnmt3a and Dnmt3b had a minimal effect on the association of Setdb1 and H3K9me3 at ERVs and lack of the H4K20me3 HMTases, Suv4-20h1/2 resulted in the loss of H4K20me3 at ERVs, but with no change in ERV expression or H3K9me3 levels (Karimi *et al.*, 2011; Matsui *et al.*, 2010). Furthermore, mouse ESCs lacking the other H3K9 HMTases (G9a, GLP, Suv39h1 and Suv39h2) did not show reactivation of class I or class II ERVs, and knockdown of individual HP1 isoforms or concurrent depletion of all three had a modest impact on ERV expression, although the slight spread of H4K20me3 into sequences flanking the ERV element was reduced (Maksakova *et al.*, 2011). Finally, deletion or knockdown of Trim28, which associates with the ERV LTR, resulted in loss of class I and class II H3K9me3 and reactivation of expression comparable to Setdb1 loss (Matsui *et al.*, 2010; Rowe *et al.*, 2010). Collectively, these data indicate that ERV silencing is dependent on Trim28 recruitment of Setdb1 to ERV elements in order to establish H3K9me3, which is sufficient to silence expression in the absence of DNA methylation, H4K20me3 or HP1. Furthermore, introduction of a catalytically inactive Setdb1 transgene into -/- Setdb1 cells was unable to repress ERVs or restore H3K9me3, supporting a key role for both in ERV silencing (Maksakova *et al.*, 2011).

Genome-wide analysis of Setdb1 binding sites in mouse ESCs revealed that 40% of Setdb1 binding sites overlap with, or are within 100 bp of an ERV (Karimi *et al.*, 2011). However, while ~21% of genes that are up-regulated in the absence of Setdb1 normally show Setdb1 binding in the promoter region, and only 13% of up-regulated genes are marked by H3K9me3 at the promoter in wild type cells, many up-regulated genes do not have either in their immediate promoter region. This then suggests that the expression of some genes are an indirect effect of Setdb1 loss, perhaps through the reactivation of a Setdb1-repressed transcription factor. Interestingly, in ESCs lacking Trim28 (Rowe *et al.*, 2013), most up-regulated genes coincided with peaks of H3K4me3/H3K27me3, indicating they are bivalent genes (Bernstein *et al.*, 2006). Given that many bivalent genes are also marked by H3K9me3 (Bilodeau *et al.*, 2009), these data would suggest that bivalent genes are silenced by H3K9me3 mediated through Trim28/Setdb1. Intriguingly, approximately 20% of those up-regulated genes that have promoter-proximal ERVs showed greater than a 10-fold increase in expression, suggesting that active ERV elements can impact local gene expression. In support of this notion, Rowe and colleagues showed that in mouse ESCs lacking Trim28, ERVs lost association of Trim28, Setdb1, H3K9me3 and H4K20me3, but many gained H3K4me1 and H3K27-acetylation (H3K27Ac) (Rowe *et al.*, 2013), which are generally accepted markers of active enhancer elements (Barski *et al.*, 2007; Bonn *et al.*, 2012; Consortium *et al.*, 2007; Creyghton *et al.*, 2010; Heintzman *et al.*, 2007; Koch *et al.*, 2007; Wang *et al.*, 2008; Zentner *et al.*, 2011). Of note, Trim28 recruits the NuRD complex (Schultz *et al.*, 2001) that deacetylates H3K27Ac (Reynolds *et al.*, 2012) and LSD1 (Macfarlan *et al.*, 2011, 2012) that demethylates H3K4me1 (Whyte *et al.*, 2012), preventing the establishment of an enhancer configuration in wild-type cells. Genome-wide analysis of TRIM28 distribution in human ESCs revealed 57,000 TRIM28 peaks, of which ~75% reside at ERVs and most coincide with peaks of H3K9me3 (Turelli *et al.*, 2014). Knockdown of TRIM28 levels mostly reactivated class I and II ERVs and, like in mouse ESCs, there was a gain of

H3K4me1 at 6% of former SETDB1/H3K9me3 peaks and H3K27Ac was gained at many class II ERVs, and that most up-regulated genes were close to former TRIM28/H3K9me3 marked class II ERVs (Turelli *et al.*, 2014).

Most striking is the fact that loss of Setdb1 resulted in the appearance of chimeric transcripts that consist of class I or class II ERV sequences at the 5′ end fused to exons of cellular genes (Karimi *et al.*, 2011). Most often this involved truncated ERV elements that lacked endogenous polyadenylation signals. Interestingly, chimeric ERV-cellular transcripts are common to Class III ERV elements, which are not all silenced in early development (Peaston *et al.*, 2004). LTR-driven transcripts have also been reported at the two-cell stage of development, and in a small fraction of pluripotent cell populations, suggesting a potentially important and naturally occurring role for this phenomenon (Macfarlan *et al.*, 2012). Although Setdb1/SETDB1 is essential for ERV silencing in ESCs, but not post-differentiation, the notable exception appears to be in brain, as some ERVs were obviously de-repressed in Setdb1 forebrain knockouts. The transcriptional start site for more than 25% of cellular genes up-regulated in these cells reside within 10kb of ERV LTRs, and like in ESCs, ERV-cellular chimeric transcripts were detected (Tan *et al.*, 2012).

Some ERVs are naturally expressed in ESCs and early development. Indeed, mouse, MuERV-L elements are expressed in preimplantation embryos (Peaston *et al.*, 2004) and fine-tune the genomic network of totipotent cells very early in development (Macfarlan *et al.*, 2011, 2012). While in humans, HERV-H is a pluripotency marker of human ESCs, and is an abundant ES-specific transcript (Santoni *et al.*, 2012) that is required for maintenance of ESC identity (Lu *et al.*, 2014). Reprogramming of somatic human or mouse cells to a pluripotent state through forced expression of transcription factors reactivates endogenous retroelements (Friedli *et al.*, 2014), a process made more efficient when Setdb1 levels are reduced (Sridharan *et al.*, 2013). Stem cell specific transcription factors have been reported at LTR elements in ESCs (Bourque *et al.*, 2008; Kunarso *et al.*, 2010) and many long non-coding RNAs in ESCs are derived from or contain transposable elements (Kelley and Rinn, 2012). Recently a novel class of non-annotated stem transcripts (NASTs) were described in human and mouse ESCs (Fort *et al.*, 2014). NASTs are expressed at low levels in ESCs and many correspond to reported long non-coding RNAs. Importantly, in humans NASTs are strongly associated with ERVK LTRs, and ERV1 in mouse, indicating that most are species specific. Histone modifications at NASTs are consistent with promoters and enhancers and are bound by NANOG, SOX2 and OCT4, whereas other non-transcribed ERVs are marked by H3K9me3 and are likely silenced by TRIM28/SETDB1. Through the use of chromatin interaction analysis by paired-end tag, LTRs were found to interact with target loci as much as 100kb away and mostly in *cis*, consistent with enhancers. Importantly, attempts to knock down NASTs by RNAi resulted in loss of ESC markers for a substantial number of NAST targets, indicating an important role for these non-coding RNAs in maintaining pluripotency (Fort *et al.*, 2014).

Of note, SETDB1 not only works to regulate ERV expression, but also recently was shown to have an important role in regulating pseudogenes (Guo *et al.*, 2014). While pseudogenes by definition have lost protein coding potential (Zheng *et al.*, 2007), many are actually transcribed (Pei *et al.*, 2012; Zheng *et al.*, 2007) and some regulate the parental gene from which they are derived (Pink *et al.*, 2011), highlighting another key role for SETDB1 in regulating gene expression and genomic integrity in complex genomes.

SETDB1 and disease

SETDB1 has been implicated in numerous health-related issues including regulating the transcription of integrated human immunodeficiency virus (Van Duyne *et al.*, 2008), facilitating promyelocytic leukaemia nuclear body mediated gene silencing (Cho *et al.*, 2011), proper heart development (Mysliwiec *et al.*, 2012) and autoimmune disease through pro-inflammatory cytokine regulation (Eames *et al.*, 2012). SETDB1 has also been connected to brain development/function and cancer.

SETDB1 and brain function

Setdb1 is expressed in the developing mouse brain, is essential for brain development (Tan *et al.*, 2012) and its overexpression results in an anti-depressant phenotype in transgenic mice (Jiang *et al.*, 2010). Neural progenitor cells (NPCs) give rise to various types of neurons through changes in gene expression. Only a small number of genes change in a forebrain specific conditional Setdb1 knockout. Most unregulated genes fall into two classes. The first class includes genes involved in ossification and gliogenesis, an important process for regeneration and repair in the central nervous system (Gallo and Deneen, 2014). The second class of genes includes those involved in spermatogenesis, meiosis and germ cell restricted processes and includes inappropriate activation of ERVs. Most genes in these two classes are normally bound by Setdb1, and its loss results in reduced H3K9me3 and concomitant gain of H3K9Ac (Tan *et al.*, 2012). Of those genes that are reduced in expression, most are involved in neurogenesis, consistent with disruption of this pathway.

SETDB1 is unregulated in the brain of Huntington's disease (HD) patients as well as in transgenic HD mice (Ryu *et al.*, 2006). The poly-glutamine in mutant Huntingtin interacts with CREB-binding protein (CBP), resulting in its subsequent degradation (Jiang *et al.*, 2003), blocking histone acetyltransferase activity (Nucifora *et al.*, 2001; Steffan *et al.*, 2001) which can be compensated for by overexpression of CBP or through HDAC inhibitors (Ferrante *et al.*, 2003). Loss of Cbp in mice was associated with increased Setdb1 expression and as a direct consequence, levels of H3K9me3 are elevated in mice lacking or showing reduced levels of Cbp (Lee *et al.*, 2008).

Interestingly, a rare potentially detrimental variant, Pro1067del, has been identified in *SETDB1* in patients with autism, adding SETDB1 to the list of MBD-containing proteins linked to this disorder (Cukier *et al.*, 2012). As outlined above, SETDB1 has an important role in normal brain development and function, and therefore variants of this protein are candidates for contributing to complex neurological and neurodevelopmental disorders.

SETDB1 in cancer

Mutations in the V-Raf murine sarcoma viral oncogene homologue B (BRAF) have been reported in many cancers, but is most commonly found in malignant melanoma (Davies *et al.*, 2002). SETDB1 is located at human chromosome 1q21 (Harte *et al.*, 1999), in a region that frequently shows copy number variation (CNV) in melanoma (Curtin *et al.*, 2005; Lin *et al.*, 2008). Increased copy number of SETDB1 on a BRAF mutant background accelerates melanoma formation causing a more aggressive and invasive cancer, confirming SETDB1 as a oncogene (Ceol *et al.*, 2011). Introduction of the common BRAF mutation (V600E) into cells induces senescence (Michaloglou *et al.*, 2005), but increasing SETDB1 levels (mimicking increased CNV) in these cells overcome this. Increased SETDB1 CNV has been reported in other forms of cancer, and depletion of SETDB1 levels resulted in reduced cancer growth in culture and in tumour induced nude mice, whereas overexpression of Setdb1 increased tumour invasiveness (Rodriguez-Paredes *et al.*, 2014). An increase in tumour invasion and metastasis is correlated with changes in microRNA (miR) expression (Pencheva and Tavazoie, 2013). Further evidence linking SETDB1 to cancer progression comes from experimental data indicating that miR-7 reduces SETDB1 levels through interaction with the 3′ untranslated region of SETDB1 messenger RNA and that miR-7 is down-regulated in breast cancer stem cells (Zhang *et al.*, 2014) permitting an increase in SETDB1 translation.

Concluding remarks

SETDB1 is central to establishing and maintaining heterochromatin throughout development. Therefore, it is of no surprise that its loss in nematodes through mammals is not tolerated, or that changes in protein levels impact cancer progression. Given how important this protein is in regulating gene expression and the epigenome, it is not unrealistic to consider that the connection between SETDB1 and disease susceptibility will continue to increase.

The multitude of KRAB-ZFPs that each possess methylation dependent, independent or perhaps methylation indifferent target sequence binding specificity, as well as developmental and cell-type specific expression patterns, provides enormous versatility to direct heterochromatin formation through the common recruitment of SETDB1 via TRIM28. KRAB-ZFP display extensive diversity between man and mouse, as do the location, copy number and type of ERV elements, which are not dormant, but can, and do, continue to move. Therefore SETDB1 can also be considered as an important player on the front lines, programming the evolving genome.

SETDB1 is also directed to target sites independently of TRIM28, for which far less is currently known. Furthermore, SETDB1 is subject to a variety of post-translational modifications, the functional consequences of which is also unknown. The surface has been but scratched, and the stage is SET for the exciting discoveries to come concerning this enigmatic protein.

Acknowledgements

This work was supported by the National Institutes of Health (GM073120 to B.P.C.). We apologize to those whose work relating to SETDB1 we do not cover in detail owing to space restrictions.

References

Aagaard, L., Laible, G., Selenko, P., Schmid, M., Dorn, R., Schotta, G., Kuhfittig, S., Wolf, A., Lebersorger, A., Singh, P.B., et al. (1999). Functional mammalian homologues of the Drosophila PEV-modifier Su(var)3-9 encode centromere-associated proteins which complex with the heterochromatin component M31. EMBO J. *18*, 1923–1938.

Aasland, R., and Stewart, A.F. (1995). The chromo shadow domain, a second chromo domain in heterochromatin-binding protein 1, HP1. Nucleic Acids Res. *23*, 3168–3173.

Abrink, M., Ortiz, J.A., Mark, C., Sanchez, C., Looman, C., Hellman, L., Chambon, P., and Losson, R. (2001). Conserved interaction between distinct Kruppel-associated box domains and the transcriptional intermediary factor 1 beta. Proc. Natl. Acad. Sci. U.S.A. *98*, 1422–1426.

Allis, C.D., Berger, S.L., Cote, J., Dent, S., Jenuwien, T., Kouzarides, T., Pillus, L., Reinberg, D., Shi, Y., Shiekhattar, R., et al. (2007). New nomenclature for chromatin-modifying enzymes. Cell *131*, 633–636.

Andersen, E.C., and Horvitz, H.R. (2007). Two *C. elegans* histone methyltransferases repress lin-3 EGF transcription to inhibit vulval development. Development *134*, 2991–2999.

Arents, G., Burlingame, R.W., Wang, B.C., Love, W.E., and Moudrianakis, E.N. (1991). The nucleosomal core histone octamer at 3.1 A resolution: a tripartite protein assembly and a left-handed superhelix. Proc. Natl. Acad. Sci. U.S.A. *88*, 10148–10152.

Arita, K., Isogai, S., Oda, T., Unoki, M., Sugita, K., Sekiyama, N., Kuwata, K., Hamamoto, R., Tochio, H., Sato, M., et al. (2012). Recognition of modification status on a histone H3 tail by linked histone reader modules of the epigenetic regulator UHRF1. Proc. Natl. Acad. Sci. U.S.A. *109*, 12950–12955.

Ayer, D.E., Lawrence, Q.A., and Eisenman, R.N. (1995). Mad-Max transcriptional repression is mediated by ternary complex formation with mammalian homologs of yeast repressor Sin3. Cell *80*, 767–776.

Ayyanathan, K., Lechner, M.S., Bell, P., Maul, G.G., Schultz, D.C., Yamada, Y., Tanaka, K., Torigoe, K., and Rauscher, F.J., 3rd (2003). Regulated recruitment of HP1 to a euchromatic gene induces mitotically heritable, epigenetic gene silencing: a mammalian cell culture model of gene variegation. Genes Dev. *17*, 1855–1869.

Bannister, A.J., Zegerman, P., Partridge, J.F., Miska, E.A., Thomas, J.O., Allshire, R.C., and Kouzarides, T. (2001). Selective recognition of methylated lysine 9 on histone H3 by the HP1 chromo domain. Nature *410*, 120–124.

Barbot, W., Dupressoir, A., Lazar, V., and Heidmann, T. (2002). Epigenetic regulation of an IAP retrotransposon in the aging mouse: progressive demethylation and de-silencing of the element by its repetitive induction. Nucleic Acids Res. *30*, 2365–2373.

Barlow, D.P., and Bartolomei, M.S. (2014). Genomic imprinting in mammals. Cold Spring Harbor Perspect. Biol. *6*, 1–20.

Barski, A., Cuddapah, S., Cui, K., Roh, T.Y., Schones, D.E., Wang, Z., Wei, G., Chepelev, I., and Zhao, K. (2007). High-resolution profiling of histone methylations in the human genome. Cell *129*, 823–837.

Bender, L.B., Cao, R., Zhang, Y., and Strome, S. (2004). The MES-2/MES-3/MES-6 complex and regulation of histone H3 methylation in *C. elegans*. Curr. Biol. *14*, 1639–1643.

Bernstein, B.E., Mikkelsen, T.S., Xie, X., Kamal, M., Huebert, D.J., Cuff, J., Fry, B., Meissner, A., Wernig, M., Plath, K., et al. (2006). A bivalent chromatin structure marks key developmental genes in embryonic stem cells. Cell *125*, 315–326.

Bestor, T.H. (2000). The DNA methyltransferases of mammals. Hum. Mol. Genet. *9*, 2395–2402.

Bian, C., Xu, C., Ruan, J., Lee, K.K., Burke, T.L., Tempel, W., Barsyte, D., Li, J., Wu, M., Zhou, B.O., et al. (2011). Sgf29 binds histone H3K4me2/3 and is required for SAGA complex recruitment and histone H3 acetylation. EMBO J. *30*, 2829–2842.

Bilodeau, S., Kagey, M.H., Frampton, G.M., Rahl, P.B., and Young, R.A. (2009). SetDB1 contributes to repression of genes encoding developmental regulators and maintenance of ES cell state. Genes Dev. *23*, 2484–2489.

Binda, O., LeRoy, G., Bua, D.J., Garcia, B.A., Gozani, O., and Richard, S. (2010). Trimethylation of histone H3 lysine 4 impairs methylation of histone H3 lysine

9: regulation of lysine methyltransferases by physical interaction with their substrates. Epigenetics 5, 767–775.

Blackburn, M.L., Chansky, H.A., Zielinska-Kwiatkowska, A., Matsui, Y., and Yang, L. (2003). Genomic structure and expression of the mouse ESET gene encoding an ERG-associated histone methyltransferase with a SET domain. Biochim. Biophys. Acta *1629*, 8–14.

Blahnik, K.R., Dou, L., Echipare, L., Iyengar, S., O'Geen, H., Sanchez, E., Zhao, Y., Marra, M.A., Hirst, M., Costello, J.F., et al. (2011). Characterization of the contradictory chromatin signatures at the 3′ exons of zinc finger genes. PLoS ONE 6, e17121.

Boggs, B.A., Cheung, P., Heard, E., Spector, D.L., Chinault, A.C., and Allis, C.D. (2002). Differentially methylated forms of histone H3 show unique association patterns with inactive human X-chromosomes. Nat. Genet. *30*, 73–76.

Bonn, S., Zinzen, R.P., Girardot, C., Gustafson, E.H., Perez-Gonzalez, A., Delhomme, N., Ghavi-Helm, Y., Wilczynski, B., Riddell, A., and Furlong, E.E. (2012). Tissue-specific analysis of chromatin state identifies temporal signatures of enhancer activity during embryonic development. Nat. Genet. *44*, 148–156.

Borgel, J., Guibert, S., Li, Y., Chiba, H., Schubeler, D., Sasaki, H., Forne, T., and Weber, M. (2010). Targets and dynamics of promoter DNA methylation during early mouse development. Nat. Genet. *42*, 1093–1100.

Boswell, R.E., and Mahowald, A.P. (1985). tudor, a gene required for assembly of the germ plasm in *Drosophila melanogaster*. Cell *43*, 97–104.

Botuyan, M.V., Lee, J., Ward, I.M., Kim, J.E., Thompson, J.R., Chen, J., and Mer, G. (2006). Structural basis for the methylation state-specific recognition of histone H4-K20 by 53BP1 and Crb2 in DNA repair. Cell *127*, 1361–1373.

Bourc'his, D., and Bestor, T.H. (2004). Meiotic catastrophe and retrotransposon reactivation in male germ cells lacking Dnmt3L. Nature *431*, 96–99.

Bourc'his, D., Xu, G.L., Lin, C.S., Bollman, B., and Bestor, T.H. (2001). Dnmt3L and the establishment of maternal genomic imprints. Science *294*, 2536–2539.

Bourque, G., Leong, B., Vega, V.B., Chen, X., Lee, Y.L., Srinivasan, K.G., Chew, J.L., Ruan, Y., Wei, C.L., Ng, H.H., et al. (2008). Evolution of the mammalian transcription factor binding repertoire via transposable elements. Genome Res. *18*, 1752–1762.

Boyes, J., and Bird, A. (1991). DNA methylation inhibits transcription indirectly via a methyl-CpG binding protein. Cell *64*, 1123–1134.

Brasher, S.V., Smith, B.O., Fogh, R.H., Nietlispach, D., Thiru, A., Nielsen, P.R., Broadhurst, R.W., Ball, L.J., Murzina, N.V., and Laue, E.D. (2000). The structure of mouse HP1 suggests a unique mode of single peptide recognition by the shadow chromo domain dimer. EMBO J. *19*, 1587–1597.

Brower-Toland, B., Riddle, N.C., Jiang, H., Huisinga, K.L., and Elgin, S.C. (2009). Multiple SET methyltransferases are required to maintain normal heterochromatin domains in the genome of Drosophila melanogaster. Genetics *181*, 1303–1319.

Bush, K.M., Yuen, B.T., Barrilleaux, B.L., Riggs, J.W., O'Geen, H., Cotterman, R.F., and Knoepfler, P.S. (2013). Endogenous mammalian histone H3.3 exhibits chromatin-related functions during development. Epigenetics Chromatin *6*, 7.

Buzdin, A., Kovalskaya-Alexandrova, E., Gogvadze, E., and Sverdlov, E. (2006). At least 50% of human-specific HERV-K (HML-2) long terminal repeats serve *in vivo* as active promoters for host nonrepetitive DNA transcription. J. Virol. *80*, 10752–10762.

Cao, R., Wang, L., Wang, H., Xia, L., Erdjument-Bromage, H., Tempst, P., Jones, R.S., and Zhang, Y. (2002). Role of histone H3 lysine 27 methylation in Polycomb-group silencing. Science *298*, 1039–1043.

Capili, A.D., Schultz, D.C., Rauscher, I.F., and Borden, K.L. (2001). Solution structure of the PHD domain from the KAP-1 corepressor: structural determinants for PHD, RING and LIM zinc-binding domains. EMBO J. *20*, 165–177.

Cedar, H., and Bergman, Y. (2009). Linking DNA methylation and histone modification: patterns and paradigms. Nat. Rev. Genet. *10*, 295–304.

Ceol, C.J., Houvras, Y., Jane-Valbuena, J., Bilodeau, S., Orlando, D.A., Battisti, V., Fritsch, L., Lin, W.M., Hollmann, T.J., Ferre, F., et al. (2011). The histone methyltransferase SETDB1 is recurrently amplified in melanoma and accelerates its onset. Nature *471*, 513–517.

Chadwick, B.P., and Willard, H.F. (2001). A novel chromatin protein, distantly related to histone H2A, is largely excluded from the inactive X-chromosome. J. Cell Biol. *152*, 375–384.

Chadwick, B.P., and Willard, H.F. (2003). Chromatin of the Barr body: histone and non-histone proteins associated with or excluded from the inactive X-chromosome. Hum. Mol. Genet. *12*, 2167–2178.

Cheutin, T., McNairn, A.J., Jenuwein, T., Gilbert, D.M., Singh, P.B., and Misteli, T. (2003). Maintenance of stable heterochromatin domains by dynamic HP1 binding. Science *299*, 721–725.

Cho, S., Park, J.S., and Kang, Y.K. (2011). Dual functions of histone-lysine N-methyltransferase Setdb1 protein at promyelocytic leukemia-nuclear body (PML-NB): maintaining PML-NB structure and regulating the expression of its associated genes. J. Biol. Chem. *286*, 41115–41124.

Cho, S., Park, J.S., Kwon, S., and Kang, Y.K. (2012). Dynamics of Setdb1 expression in early mouse development. Gene Expr. Patterns *12*, 213–218.

Chuong, E.B. (2013). Retroviruses facilitate the rapid evolution of the mammalian placenta. BioEssays *35*, 853–861.

Chuong, E.B., Rumi, M.A., Soares, M.J., and Baker, J.C. (2013). Endogenous retroviruses function as species-specific enhancer elements in the placenta. Nat. Genet. *45*, 325–329.

Clough, E., Moon, W., Wang, S., Smith, K., and Hazelrigg, T. (2007). Histone methylation is required for oogenesis in Drosophila. Development *134*, 157–165.

Cohen, C.J., Lock, W.M., and Mager, D.L. (2009). Endogenous retroviral LTRs as promoters for human genes: a critical assessment. Gene *448*, 105–114.

Congdon, L.M., Sims, J.K., Tuzon, C.T., and Rice, J.C. (2014). The PR-Set7 binding domain of Riz1 is required for the H4K20me1–H3K9me1 trans-tail 'histone code' and Riz1 tumor suppressor function. Nucleic Acids Res. *42*, 3580–3589.

Consortium, E.P., Birney, E., Stamatoyannopoulos, J.A., Dutta, A., Guigo, R., Gingeras, T.R., Margulies, E.H., Weng, Z., Snyder, M., Dermitzakis, E.T., et al. (2007). Identification and analysis of functional elements in 1% of the human genome by the ENCODE pilot project. Nature *447*, 799–816.

Cordaux, R., and Batzer, M.A. (2009). The impact of retrotransposons on human genome evolution. Nat. Rev. Genet. *10*, 691–703.

Cowell, I.G., Aucott, R., Mahadevaiah, S.K., Burgoyne, P.S., Huskisson, N., Bongiorni, S., Prantera, G., Fanti, L., Pimpinelli, S., Wu, R., et al. (2002). Heterochromatin, HP1 and methylation at lysine 9 of histone H3 in animals. Chromosoma *111*, 22–36.

Cowieson, N.P., Partridge, J.F., Allshire, R.C., and McLaughlin, P.J. (2000). Dimerisation of a chromo shadow domain and distinctions from the chromodomain as revealed by structural analysis. Curr. Biol. *10*, 517–525.

Creyghton, M.P., Cheng, A.W., Welstead, G.G., Kooistra, T., Carey, B.W., Steine, E.J., Hanna, J., Lodato, M.A., Frampton, G.M., Sharp, P.A., et al. (2010). Histone H3K27ac separates active from poised enhancers and predicts developmental state. Proc. Natl. Acad. Sci. U.S.A. *107*, 21931–21936.

Cross, S.H., Meehan, R.R., Nan, X., and Bird, A. (1997). A component of the transcriptional repressor MeCP1 shares a motif with DNA methyltransferase and HRX proteins. Nat. Genet. *16*, 256–259.

Cruvinel, E., Budinetz, T., Germain, N., Chamberlain, S., Lalande, M., and Martins-Taylor, K. (2014). Reactivation of maternal SNORD116 cluster via SETDB1 knockdown in Prader–Willi syndrome iPSCs. Hum. Mol. Genet. *23*, 4674–4685.

Cukier, H.N., Lee, J.M., Ma, D., Young, J.I., Mayo, V., Butler, B.L., Ramsook, S.S., Rantus, J.A., Abrams, A.J., Whitehead, P.L., et al. (2012). The expanding role of MBD genes in autism: identification of a MECP2 duplication and novel alterations in MBD5, MBD6, and SETDB1. Autism Res. *5*, 385–397.

Curtin, J.A., Fridlyand, J., Kageshita, T., Patel, H.N., Busam, K.J., Kutzner, H., Cho, K.H., Aiba, S., Brocker, E.B., LeBoit, P.E., et al. (2005). Distinct sets of genetic alterations in melanoma. N. Eng. J. Med. *353*, 2135–2147.

Czermin, B., Melfi, R., McCabe, D., Seitz, V., Imhof, A., and Pirrotta, V. (2002). Drosophila enhancer of Zeste/ESC complexes have a histone H3 methyltransferase activity that marks chromosomal Polycomb sites. Cell *111*, 185–196.

Davies, H., Bignell, G.R., Cox, C., Stephens, P., Edkins, S., Clegg, S., Teague, J., Woffendin, H., Garnett, M.J., Bottomley, W., et al. (2002). Mutations of the BRAF gene in human cancer. Nature *417*, 949–954.

Dehal, P., Predki, P., Olsen, A.S., Kobayashi, A., Folta, P., Lucas, S., Land, M., Terry, A., Ecale Zhou, C.L., Rash, S., et al. (2001). Human chromosome 19 and related regions in mouse: conservative and lineage-specific evolution. Science *293*, 104–111.

Denslow, S.A., and Wade, P.A. (2007). The human Mi-2/NuRD complex and gene regulation. Oncogene *26*, 5433–5438.

Dillon, S.C., Zhang, X., Trievel, R.C., and Cheng, X. (2005). The SET-domain protein superfamily: protein lysine methyltransferases. Genome Biol. *6*, 227.

Dodge, J.E., Kang, Y.K., Beppu, H., Lei, H., and Li, E. (2004). Histone H3-K9 methyltransferase ESET is essential for early development. Mol. Cell. Biol. *24*, 2478–2486.

Eames, H.L., Saliba, D.G., Krausgruber, T., Lanfrancotti, A., Ryzhakov, G., and Udalova, I.A. (2012). KAP1/TRIM28: an inhibitor of IRF5 function in inflammatory macrophages. Immunobiology *217*, 1315–1324.

Eissenberg, J.C., James, T.C., Foster-Hartnett, D.M., Hartnett, T., Ngan, V., and Elgin, S.C. (1990). Mutation in a heterochromatin-specific chromosomal protein is associated with suppression of position-effect variegation in *Drosophila melanogaster*. Proc. Natl. Acad. Sci. U.S.A. *87*, 9923–9927.

Eissenberg, J.C., Morris, G.D., Reuter, G., and Hartnett, T. (1992). The heterochromatin-associated protein HP-1 is an essential protein in Drosophila with dosage-dependent effects on position-effect variegation. Genetics *131*, 345–352.

Evans, M.J., and Kaufman, M.H. (1981). Establishment in culture of pluripotential cells from mouse embryos. Nature *292*, 154–156.

Falandry, C., Fourel, G., Galy, V., Ristriani, T., Horard, B., Bensimon, E., Salles, G., Gilson, E., and Magdinier, F. (2010). CLLD8/KMT1F is a lysine methyltransferase that is important for chromosome segregation. J. Biol. Chem. *285*, 20234–20241.

Feng, S., Jacobsen, S.E., and Reik, W. (2010). Epigenetic reprogramming in plant and animal development. Science *330*, 622–627.

Ferrante, R.J., Kubilus, J.K., Lee, J., Ryu, H., Beesen, A., Zucker, B., Smith, K., Kowall, N.W., Ratan, R.R., Luthi-Carter, R., et al. (2003). Histone deacetylase inhibition by sodium butyrate chemotherapy ameliorates the neurodegenerative phenotype in Huntington's disease mice. J. Neurosci. *23*, 9418–9427.

Festenstein, R., Sharghi-Namini, S., Fox, M., Roderick, K., Tolaini, M., Norton, T., Saveliev, A., Kioussis, D., and Singh, P. (1999). Heterochromatin protein 1 modifies mammalian PEV in a dose- and chromosomal-context-dependent manner. Nat. Genet. *23*, 457–461.

Fleischer, T.C., Yun, U.J., and Ayer, D.E. (2003). Identification and characterization of three new components of the mSin3A corepressor complex. Mol. Cell. Biol. *23*, 3456–3467.

Forneris, F., Binda, C., Vanoni, M.A., Battaglioli, E., and Mattevi, A. (2005). Human histone demethylase LSD1 reads the histone code. J. Biol. Chem. *280*, 41360–41365.

Fort, A., Hashimoto, K., Yamada, D., Salimullah, M., Keya, C.A., Saxena, A., Bonetti, A., Voineagu, I., Bertin, N.,

Kratz, A., et al. (2014). Deep transcriptome profiling of mammalian stem cells supports a regulatory role for retrotransposons in pluripotency maintenance. Nat. Genet. *46*, 558–566.

Freitag, M., Hickey, P.C., Khlafallah, T.K., Read, N.D., and Selker, E.U. (2004). HP1 is essential for DNA methylation in neurospora. Mol. Cell *13*, 427–434.

Friedli, M., Turelli, P., Kapopoulou, A., Rauwel, B., Castro-Diaz, N., Rowe, H.M., Ecco, G., Unzu, C., Planet, E., Lombardo, A., et al. (2014). Loss of transcriptional control over endogenous retroelements during reprogramming to pluripotency. Genome Res. *24*, 1251–1259.

Friedman, J.R., Fredericks, W.J., Jensen, D.E., Speicher, D.W., Huang, X.P., Neilson, E.G., and Rauscher, F.J., 3rd (1996). KAP-1, a novel corepressor for the highly conserved KRAB repression domain. Genes Dev. *10*, 2067–2078.

Frietze, S., O'Geen, H., Blahnik, K.R., Jin, V.X., and Farnham, P.J. (2010). ZNF274 recruits the histone methyltransferase SETDB1 to the 3′ ends of ZNF genes. PLoS ONE 5, e15082.

Fritsch, L., Robin, P., Mathieu, J.R., Souidi, M., Hinaux, H., Rougeulle, C., Harel-Bellan, A., Ameyar-Zazoua, M., and Ait-Si-Ali, S. (2010). A subset of the histone H3 lysine 9 methyltransferases Suv39h1, G9a, GLP, and SETDB1 participate in a multimeric complex. Mol. Cell 37, 46–56.

Fujita, N., Watanabe, S., Ichimura, T., Ohkuma, Y., Chiba, T., Saya, H., and Nakao, M. (2003). MCAF mediates MBD1-dependent transcriptional repression. Mol. Cell. Biol. 23, 2834–2843.

Fuks, F., Hurd, P.J., Wolf, D., Nan, X., Bird, A.P., and Kouzarides, T. (2003). The methyl-CpG-binding protein MeCP2 links DNA methylation to histone methylation. J. Biol. Chem. *278*, 4035–4040.

Gallo, V., and Deneen, B. (2014). Glial Development: The crossroads of regeneration and repair in the CNS. Neuron *83*, 283–308.

Gao, H., Yu, Z., Bi, D., Jiang, L., Cui, Y., Sun, J., and Ma, R. (2007). Akt/PKB interacts with the histone H3 methyltransferase SETDB1 and coordinates to silence gene expression. Mol. Cell. Biochem. *305*, 35–44.

Garcia-Bassets, I., Kwon, Y.S., Telese, F., Prefontaine, G.G., Hutt, K.R., Cheng, C.S., Ju, B.G., Ohgi, K.A., Wang, J., Escoubet-Lozach, L., et al. (2007). Histone methylation-dependent mechanisms impose ligand dependency for gene activation by nuclear receptors. Cell *128*, 505–518.

Gou, D., Rubalcava, M., Sauer, S., Mora-Bermudez, F., Erdjument-Bromage, H., Tempst, P., Kremmer, E., and Sauer, F. (2010). SETDB1 is involved in postembryonic DNA methylation and gene silencing in *Drosophila*. PLoS ONE 5, e10581.

Gowher, H., Leismann, O., and Jeltsch, A. (2000). DNA of *Drosophila melanogaster contains* 5-methylcytosine. EMBO J. *19*, 6918–6923.

Grimwood, J., Gordon, L.A., Olsen, A., Terry, A., Schmutz, J., Lamerdin, J., Hellsten, U., Goodstein, D., Couronne, O., Tran-Gyamfi, M., et al. (2004). The DNA sequence and biology of human chromosome 19. Nature *428*, 529–535.

Guibert, S., Forne, T., and Weber, M. (2012). Global profiling of DNA methylation erasure in mouse primordial germ cells. Genome Res. *22*, 633–641.

Guillemette, B., Drogaris, P., Lin, H.H., Armstrong, H., Hiragami-Hamada, K., Imhof, A., Bonneil, E., Thibault, P., Verreault, A., and Festenstein, R.J. (2011). H3 lysine 4 is acetylated at active gene promoters and is regulated by H3 lysine 4 methylation. PLoS Genet. 7, e1001354.

Guo, X., Lin, M., Rockowitz, S., Lachman, H.M., and Zheng, D. (2014). Characterization of human pseudogene-derived non-coding RNAs for functional potential. PLoS ONE 9, e93972.

Hackett, J.A., Sengupta, R., Zylicz, J.J., Murakami, K., Lee, C., Down, T.A., and Surani, M.A. (2013). Germline DNA demethylation dynamics and imprint erasure through 5-hydroxymethylcytosine. Science *339*, 448–452.

Hajkova, P., Erhardt, S., Lane, N., Haaf, T., El-Maarri, O., Reik, W., Walter, J., and Surani, M.A. (2002). Epigenetic reprogramming in mouse primordial germ cells. Mech. Dev. *117*, 15–23.

Harte, P.J., Wu, W., Carrasquillo, M.M., and Matera, A.G. (1999). Assignment of a novel bifurcated SET domain gene, SETDB1, to human chromosome band 1q21 by *in situ* hybridization and radiation hybrids. Cytogenet. Cell Genet. *84*, 83–86.

Hata, K., Okano, M., Lei, H., and Li, E. (2002). Dnmt3L cooperates with the Dnmt3 family of *de novo* DNA methyltransferases to establish maternal imprints in mice. Development *129*, 1983–1993.

He, Y.F., Li, B.Z., Li, Z., Liu, P., Wang, Y., Tang, Q., Ding, J., Jia, Y., Chen, Z., Li, L., et al. (2011). Tet-mediated formation of 5-carboxylcytosine and its excision by TDG in mammalian DNA. Science 333, 1303–1307.

Heard, E., Rougeulle, C., Arnaud, D., Avner, P., Allis, C.D., and Spector, D.L. (2001). Methylation of histone H3 at Lys-9 is an early mark on the X-chromosome during X inactivation. Cell *107*, 727–738.

Heintzman, N.D., Stuart, R.K., Hon, G., Fu, Y., Ching, C.W., Hawkins, R.D., Barrera, L.O., Van Calcar, S., Qu, C., Ching, K.A., et al. (2007). Distinct and predictive chromatin signatures of transcriptional promoters and enhancers in the human genome. Nat. Genet. *39*, 311–318.

Hendrich, B., and Bird, A. (1998). Identification and characterization of a family of mammalian methyl-CpG binding proteins. Mol. Cell. Biol. *18*, 6538–6547.

Hendrich, B., and Tweedie, S. (2003). The methyl-CpG binding domain and the evolving role of DNA methylation in animals. Trends Genet. *19*, 269–277.

Herz, H.M., Garruss, A., and Shilatifard, A. (2013). SET for life: biochemical activities and biological functions of SET domain-containing proteins. Trends Biochem. Sci. *38*, 621–639.

Hollenhorst, P.C., McIntosh, L.P., and Graves, B.J. (2011). Genomic and biochemical insights into the specificity of ETS transcription factors. Annu. Rev. Biochem. *80*, 437–471.

Hong, W., Li, J., Wang, B., Chen, L., Niu, W., Yao, Z., and Baniahmad, A. (2011). Epigenetic involvement of Alien/ESET complex in thyroid hormone-mediated repression of E2F1 gene expression and cell proliferation. Biochem. Biophys. Res. Commun. *415*, 650–655.

Huang, Y., Fang, J., Bedford, M.T., Zhang, Y., and Xu, R.M. (2006). Recognition of histone H3 lysine-4 methylation by the double tudor domain of JMJD2A. Science *312*, 748–751.

Huntley, S., Baggott, D.M., Hamilton, A.T., Tran-Gyamfi, M., Yang, S., Kim, J., Gordon, L., Branscomb, E., and Stubbs, L. (2006). A comprehensive catalog of human KRAB-associated zinc finger genes: insights into the evolutionary history of a large family of transcriptional repressors. Genome Res. *16*, 669–677.

Ichimura, T., Watanabe, S., Sakamoto, Y., Aoto, T., Fujita, N., and Nakao, M. (2005). Transcriptional repression and heterochromatin formation by MBD1 and MCAF/AM family proteins. J. Biol. Chem. *280*, 13928–13935.

Inoue, A., and Zhang, Y. (2011). Replication-dependent loss of 5-hydroxymethylcytosine in mouse preimplantation embryos. Science *334*, 194.

Iqbal, K., Jin, S.G., Pfeifer, G.P., and Szabo, P.E. (2011). Reprogramming of the paternal genome upon fertilization involves genome-wide oxidation of 5-methylcytosine. Proc. Natl. Acad. Sci. U.S.A. *108*, 3642–3647.

Ivanov, A.V., Peng, H., Yurchenko, V., Yap, K.L., Negorev, D.G., Schultz, D.C., Psulkowski, E., Fredericks, W.J., White, D.E., Maul, G.G., et al. (2007). PHD domain-mediated E3 ligase activity directs intramolecular sumoylation of an adjacent bromodomain required for gene silencing. Mol. Cell *28*, 823–837.

Iyengar, S., and Farnham, P.J. (2011). KAP1 protein: an enigmatic master regulator of the genome. J. Biol. Chem. *286*, 26267–26276.

Iyengar, S., Ivanov, A.V., Jin, V.X., Rauscher, F.J., 3rd, and Farnham, P.J. (2011). Functional analysis of KAP1 genomic recruitment. Mol. Cell. Biol. *31*, 1833–1847.

Jackson, J.P., Lindroth, A.M., Cao, X., and Jacobsen, S.E. (2002). Control of CpNpG DNA methylation by the KRYPTONITE histone H3 methyltransferase. Nature *416*, 556–560.

Jacobs, S.A., Taverna, S.D., Zhang, Y., Briggs, S.D., Li, J., Eissenberg, J.C., Allis, C.D., and Khorasanizadeh, S. (2001). Specificity of the HP1 chromo domain for the methylated N-terminus of histone H3. EMBO J. *20*, 5232–5241.

Jacques, P.E., Jeyakani, J., and Bourque, G. (2013). The majority of primate-specific regulatory sequences are derived from transposable elements. PLoS Genet. 9, e1003504.

James, T.C., and Elgin, S.C. (1986). Identification of a nonhistone chromosomal protein associated with heterochromatin in Drosophila melanogaster and its gene. Mol. Cell. Biol. *6*, 3862–3872.

Jenuwein, T., Laible, G., Dorn, R., and Reuter, G. (1998). SET domain proteins modulate chromatin domains in eu- and heterochromatin. Cell. Mol. Life Sci.54, 80–93.

Jeppesen, P., and Turner, B.M. (1993). The inactive X-chromosome in female mammals is distinguished by a lack of histone H4 acetylation, a cytogenetic marker for gene expression. Cell 74, 281–289.

Jiang, H., Nucifora, F.C., Jr., Ross, C.A., and DeFranco, D.B. (2003). Cell death triggered by polyglutamine-expanded huntingtin in a neuronal cell line is associated with degradation of CREB-binding protein. Hum. Mol. Genet. *12*, 1–12.

Jiang, Y., Jakovcevski, M., Bharadwaj, R., Connor, C., Schroeder, F.A., Lin, C.L., Straubhaar, J., Martin, G., and Akbarian, S. (2010). Setdb1 histone methyltransferase regulates mood-related behaviors and expression of the NMDA receptor subunit NR2B. J. Neurosci. *30*, 7152–7167.

Johansson, A.M., Stenberg, P., Bernhardsson, C., and Larsson, J. (2007). Painting of fourth and chromosome-wide regulation of the 4th chromosome in *Drosophila melanogaster*. EMBO J. *26*, 2307–2316.

Johnson, E.S. (2004). Protein modification by SUMO. Annu. Rev. Biochem. *73*, 355–382.

Jones, D.O., Cowell, I.G., and Singh, P.B. (2000). Mammalian chromodomain proteins: their role in genome organisation and expression. BioEssays *22*, 124–137.

Kaer, K., and Speek, M. (2013). Retroelements in human disease. Gene *518*, 231–241.

Kao, T.H., Liao, H.F., Wolf, D., Tai, K.Y., Chuang, C.Y., Lee, H.S., Kuo, H.C., Hata, K., Zhang, X., Cheng, X., et al. (2014). Ectopic DNMT3L triggers assembly of a repressive complex for retroviral silencing in somatic cells. J. Virol. *88*, 10680–10695.

Karimi, M.M., Goyal, P., Maksakova, I.A., Bilenky, M., Leung, D., Tang, J.X., Shinkai, Y., Mager, D.L., Jones, S., Hirst, M., et al. (2011). DNA methylation and SETDB1/H3K9me3 regulate predominantly distinct sets of genes, retroelements, and chimeric transcripts in mESCs. Cell Stem Cell *8*, 676–687.

Kaufman, P.D., Kobayashi, R., Kessler, N., and Stillman, B. (1995). The p150 and p60 subunits of chromatin assembly factor I: a molecular link between newly synthesized histones and DNA replication. Cell *81*, 1105–1114.

Kelley, D., and Rinn, J. (2012). Transposable elements reveal a stem cell-specific class of long noncoding RNAs. Genome Biol. 13, R107.

Kelly, W.G., and Fire, A. (1998). Chromatin silencing and the maintenance of a functional germline in Caenorhabditis elegans. Development *125*, 2451–2456.

Kim, B.J., Kang, K.M., Jung, S.Y., Choi, H.K., Seo, J.H., Chae, J.H., Cho, E.J., Youn, H.D., Qin, J., and Kim, S.T. (2008). Esco2 is a novel corepressor that associates with various chromatin modifying enzymes. Biochem. Biophys. Res. Commun. 372, 298–304.

Klose, R.J., Yamane, K., Bae, Y., Zhang, D., Erdjument-Bromage, H., Tempst, P., Wong, J., and Zhang, Y. (2006). The transcriptional repressor JHDM3A demethylates trimethyl histone H3 lysine 9 and lysine 36. Nature *442*, 312–316.

Kobayashi, H., Sakurai, T., Imai, M., Takahashi, N., Fukuda, A., Yayoi, O., Sato, S., Nakabayashi, K., Hata, K., Sotomaru, Y., et al. (2012). Contribution of intragenic DNA methylation in mouse gametic DNA methylomes to establish oocyte-specific heritable marks. PLoS Genet. 8, e1002440.

Kobayashi, H., Sakurai, T., Miura, F., Imai, M., Mochiduki, K., Yanagisawa, E., Sakashita, A., Wakai, T., Suzuki, Y., Ito, T., et al. (2013). High-resolution DNA methylome analysis of primordial germ cells identifies gender-specific reprogramming in mice. Genome Res. *23*, 616–627.

Koch, C.M., Andrews, R.M., Flicek, P., Dillon, S.C., Karaoz, U., Clelland, G.K., Wilcox, S., Beare, D.M., Fowler, J.C., Couttet, P., et al. (2007). The landscape of histone modifications across 1% of the human genome in five human cell lines. Genome Res. *17*, 691–707.

Koch, C.M., Honemann-Capito, M., Egger-Adam, D., and Wodarz, A. (2009). Windei, the Drosophila homolog of mAM/MCAF1, is an essential cofactor of the H3K9 methyl transferase dSETDB1/Eggless in germ line development. PLoS Genet. 5, e1000644.

Koh, K.P., Yabuuchi, A., Rao, S., Huang, Y., Cunniff, K., Nardone, J., Laiho, A., Tahiliani, M., Sommer, C.A., Mostoslavsky, G., et al. (2011). Tet1 and Tet2 regulate 5-hydroxymethylcytosine production and cell lineage specification in mouse embryonic stem cells. Cell Stem Cell *8*, 200–213.

Kornberg, R.D. (1974). Chromatin structure: a repeating unit of histones and DNA. Science *184*, 868–871.

Kourmouli, N., Jeppesen, P., Mahadevhaiah, S., Burgoyne, P., Wu, R., Gilbert, D.M., Bongiorni, S., Prantera, G., Fanti, L., Pimpinelli, S., et al. (2004). Heterochromatin and tri-methylated lysine 20 of histone H4 in animals. J. Cell Sci. *117*, 2491–2501.

Kourmouli, N., Sun, Y.M., van der Sar, S., Singh, P.B., and Brown, J.P. (2005). Epigenetic regulation of mammalian pericentric heterochromatin *in vivo* by HP1. Biochem. Biophys. Res. Commun. *337*, 901–907.

Kouzarides, T. (2007). SnapShot: Histone-modifying enzymes. Cell *131*, 822.

Krebs, C.J., Larkins, L.K., Khan, S.M., and Robins, D.M. (2005). Expansion and diversification of KRAB zinc-finger genes within a cluster including Regulator of sex-limitation 1 and 2. Genomics *85*, 752–761.

Kunarso, G., Chia, N.Y., Jeyakani, J., Hwang, C., Lu, X., Chan, Y.S., Ng, H.H., and Bourque, G. (2010). Transposable elements have rewired the core regulatory network of human embryonic stem cells. Nat. Genet. *42*, 631–634.

Kuo, M.H., and Allis, C.D. (1998). Roles of histone acetyltransferases and deacetylases in gene regulation. BioEssays *20*, 615–626.

Kuzmichev, A., Nishioka, K., Erdjument-Bromage, H., Tempst, P., and Reinberg, D. (2002). Histone methyltransferase activity associated with a human multiprotein complex containing the Enhancer of Zeste protein. Genes Dev. *16*, 2893–2905.

Kwon, S.H., and Workman, J.L. (2011). The changing faces of HP1: From heterochromatin formation and gene silencing to euchromatic gene expression: HP1 acts as a positive regulator of transcription. BioEssays *33*, 280–289.

Lachner, M., O'Carroll, D., Rea, S., Mechtler, K., and Jenuwein, T. (2001). Methylation of histone H3 lysine 9 creates a binding site for HP1 proteins. Nature *410*, 116–120.

Lander, E.S., Linton, L.M., Birren, B., Nusbaum, C., Zody, M.C., Baldwin, J., Devon, K., Dewar, K., Doyle, M., FitzHugh, W., et al. (2001). Initial sequencing and analysis of the human genome. Nature *409*, 860–921.

Larsson, J., Chen, J.D., Rasheva, V., Rasmuson-Lestander, A., and Pirrotta, V. (2001). Painting of fourth, a chromosome-specific protein in Drosophila. Proc. Natl. Acad. Sci. U.S.A. *98*, 6273–6278.

Lawson, K.A., Teteak, C.J., Gao, J., Li, N., Hacquebord, J., Ghatan, A., Zielinska-Kwiatkowska, A., Song, G., Chansky, H.A., and Yang, L. (2013a). ESET histone methyltransferase regulates osteoblastic differentiation of mesenchymal stem cells during postnatal bone development. FEBS Lett. *587*, 3961–3967.

Lawson, K.A., Teteak, C.J., Zou, J., Hacquebord, J., Ghatan, A., Zielinska-Kwiatkowska, A., Fernandes, R.J., Chansky, H.A., and Yang, L. (2013b). Mesenchyme-specific knockout of ESET histone methyltransferase causes ectopic hypertrophy and terminal differentiation of articular chondrocytes. J. Biol. Chem. *288*, 32119–32125.

Lee, J., Hagerty, S., Cormier, K.A., Kim, J., Kung, A.L., Ferrante, R.J., and Ryu, H. (2008). Monoallele deletion of CBP leads to pericentromeric heterochromatin condensation through ESET expression and histone H3 (K9) methylation. Hum. Mol. Genet. *17*, 1774–1782.

Lee, M.G., Wynder, C., Cooch, N., and Shiekhattar, R. (2005). An essential role for CoREST in nucleosomal histone 3 lysine 4 demethylation. Nature *437*, 432–435.

Lehnertz, B., Ueda, Y., Derijck, A.A., Braunschweig, U., Perez-Burgos, L., Kubicek, S., Chen, T., Li, E., Jenuwein, T., and Peters, A.H. (2003). Suv39h-mediated histone H3 lysine 9 methylation directs DNA methylation to major satellite repeats at pericentric heterochromatin. Curr. Biol. *13*, 1192–1200.

Leung, D.C., and Lorincz, M.C. (2012). Silencing of endogenous retroviruses: when and why do histone marks predominate? Trends Biochem. Sci. *37*, 127–133.

Leung, D., Du, T., Wagner, U., Xie, W., Lee, A.Y., Goyal, P., Li, Y., Szulwach, K.E., Jin, P., Lorincz, M.C., et al. (2014). Regulation of DNA methylation turnover at LTR retrotransposons and imprinted loci by the histone methyltransferase Setdb1. Proc. Natl. Acad. Sci. U.S.A. *111*, 6690–6695.

Li, E., Beard, C., Forster, A.C., Bestor, T.H., and Jaenisch, R. (1993). DNA methylation, genomic imprinting, and mammalian development. Cold Spring Harb. Symp. Quant. Biol. *58*, 297–305.

Li, H., Ilin, S., Wang, W., Duncan, E.M., Wysocka, J., Allis, C.D., and Patel, D.J. (2006a). Molecular basis for site-specific read-out of histone H3K4me3 by the BPTF PHD finger of NURF. Nature *442*, 91–95.

Li, H., Rauch, T., Chen, Z.X., Szabo, P.E., Riggs, A.D., and Pfeifer, G.P. (2006b). The histone methyltransferase SETDB1 and the DNA methyltransferase DNMT3A interact directly and localize to promoters silenced in cancer cells. J. Biol. Chem. *281*, 19489–19500.

Li, X., Ito, M., Zhou, F., Youngson, N., Zuo, X., Leder, P., and Ferguson-Smith, A.C. (2008). A maternal-zygotic effect gene, Zfp57, maintains both maternal and paternal imprints. Dev. Cell *15*, 547–557.

Liang, Y., Vogel, J.L., Narayanan, A., Peng, H., and Kristie, T.M. (2009). Inhibition of the histone demethylase LSD1 blocks alpha-herpesvirus lytic replication and reactivation from latency. Nat. Med. *15*, 1312–1317.

Lin, C.J., Conti, M., and Ramalho-Santos, M. (2013). Histone variant H3.3 maintains a decondensed chromatin state essential for mouse preimplantation development. Development *140*, 3624–3634.

Lin, W.M., Baker, A.C., Beroukhim, R., Winckler, W., Feng, W., Marmion, J.M., Laine, E., Greulich, H., Tseng, H., Gates, C., et al. (2008). Modeling genomic diversity and tumor dependency in malignant melanoma. Cancer Res. *68*, 664–673.

Litt, M.D., Simpson, M., Gaszner, M., Allis, C.D., and Felsenfeld, G. (2001). Correlation between histone lysine methylation and developmental changes at the chicken beta-globin locus. Science *293*, 2453–2455.

Locke, J., Kotarski, M.A., and Tartof, K.D. (1988). Dosage-dependent modifiers of position effect variegation in Drosophila and a mass action model that explains their effect. Genetics *120*, 181–198.

Lohmann, F., Loureiro, J., Su, H., Fang, Q., Lei, H., Lewis, T., Yang, Y., Labow, M., Li, E., Chen, T., et al. (2010). KMT1E mediated H3K9 methylation is required for the maintenance of embryonic stem cells by repressing trophectoderm differentiation. Stem Cells *28*, 201–212.

Lomberk, G., Wallrath, L., and Urrutia, R. (2006). The Heterochromatin Protein 1 family. Genome Biol. *7*, 228.

Loyola, A., Tagami, H., Bonaldi, T., Roche, D., Quivy, J.P., Imhof, A., Nakatani, Y., Dent, S.Y., and Almouzni, G. (2009). The HP1alpha–CAF1–SetDB1-containing complex provides H3K9me1 for Suv39-mediated K9me3 in pericentric heterochromatin. EMBO Rep. *10*, 769–775.

Lu, R., and Wang, G.G. (2013). Tudor: a versatile family of histone methylation 'readers'. Trends Biochem. Sci. *38*, 546–555.

Lu, X., Sachs, F., Ramsay, L., Jacques, P.E., Goke, J., Bourque, G., and Ng, H.H. (2014). The retrovirus HERVH is a long noncoding RNA required for human embryonic stem cell identity. Nat. Struct. Mol. Biol. *21*, 423–425.

Luger, K., Mader, A.W., Richmond, R.K., Sargent, D.F., and Richmond, T.J. (1997). Crystal structure of the nucleosome core particle at 2.8 A resolution. Nature *389*, 251–260.

Lundberg, L.E., Stenberg, P., and Larsson, J. (2013). HP1a, Su(var)3-9, SETDB1 and POF stimulate or repress gene expression depending on genomic position, gene length and expression pattern in Drosophila melanogaster. Nucleic Acids Res. *41*, 4481–4494.

Lyko, F., Ramsahoye, B.H., and Jaenisch, R. (2000). DNA methylation in Drosophila melanogaster. Nature *408*, 538–540.

Lyst, M.J., Nan, X., and Stancheva, I. (2006). Regulation of MBD1-mediated transcriptional repression by SUMO and PIAS proteins. EMBO J. *25*, 5317–5328.

Mabuchi, H., Fujii, H., Calin, G., Alder, H., Negrini, M., Rassenti, L., Kipps, T.J., Bullrich, F., and Croce, C.M. (2001). Cloning and characterization of CLLD6, CLLD7, and CLLD8, novel candidate genes for leukemogenesis at chromosome 13q14, a region commonly deleted in B-cell chronic lymphocytic leukemia. Cancer Res. *61*, 2870–2877.

Macfarlan, T.S., Gifford, W.D., Agarwal, S., Driscoll, S., Lettieri, K., Wang, J., Andrews, S.E., Franco, L., Rosenfeld, M.G., Ren, B., et al. (2011). Endogenous retroviruses and neighboring genes are coordinately repressed by LSD1/KDM1A. Genes Dev. *25*, 594–607.

Macfarlan, T.S., Gifford, W.D., Driscoll, S., Lettieri, K., Rowe, H.M., Bonanomi, D., Firth, A., Singer, O., Trono, D., and Pfaff, S.L. (2012). Embryonic stem cell potency fluctuates with endogenous retrovirus activity. Nature *487*, 57–63.

Maksakova, I.A., Romanish, M.T., Gagnier, L., Dunn, C.A., van de Lagemaat, L.N., and Mager, D.L. (2006). Retroviral elements and their hosts: insertional mutagenesis in the mouse germ line. PLoS Genet. 2, e2.

Maksakova, I.A., Goyal, P., Bullwinkel, J., Brown, J.P., Bilenky, M., Mager, D.L., Singh, P.B., and Lorincz, M.C. (2011). H3K9me3-binding proteins are dispensable for SETDB1/H3K9me3-dependent retroviral silencing. Epigenetics Chromatin *4*, 12.

Malagnac, F., Bartee, L., and Bender, J. (2002). An Arabidopsis SET domain protein required for maintenance but not establishment of DNA methylation. EMBO J. *21*, 6842–6852.

Mark, C., Abrink, M., and Hellman, L. (1999). Comparative analysis of KRAB zinc finger proteins in rodents and man: evidence for several evolutionarily distinct subfamilies of KRAB zinc finger genes. DNA Cell Biol. *18*, 381–396.

Martin, G.R. (1981). Isolation of a pluripotent cell line from early mouse embryos cultured in medium conditioned by teratocarcinoma stem cells. Proc. Natl. Acad. Sci. U.S.A. *78*, 7634–7638.

Matsui, T., Leung, D., Miyashita, H., Maksakova, I.A., Miyachi, H., Kimura, H., Tachibana, M., Lorincz, M.C., and Shinkai, Y. (2010). Proviral silencing in embryonic stem cells requires the histone methyltransferase ESET. Nature *464*, 927–931.

Maze, I., Noh, K.M., Soshnev, A.A., and Allis, C.D. (2014). Every amino acid matters: essential contributions of histone variants to mammalian development and disease. Nat. Rev. Genet. *15*, 259–271.

Mermoud, J.E., Popova, B., Peters, A.H., Jenuwein, T., and Brockdorff, N. (2002). Histone H3 lysine 9 methylation occurs rapidly at the onset of random X-chromosome inactivation. Curr. Biol. *12*, 247–251.

Metzger, E., Wissmann, M., Yin, N., Muller, J.M., Schneider, R., Peters, A.H., Gunther, T., Buettner, R., and Schule, R. (2005). LSD1 demethylates repressive histone marks to promote androgen-receptor-dependent transcription. Nature *437*, 436–439.

Mey, A., Acloque, H., Lerat, E., Gounel, S., Tribollet, V., Blanc, S., Curton, D., Birot, A.M., Nieto, M.A., and Samarut, J. (2012). The endogenous retrovirus ENS-1 provides active binding sites for transcription factors in embryonic stem cells that specify extra embryonic tissue. Retrovirology *9*, 21.

Michaloglou, C., Vredeveld, L.C., Soengas, M.S., Denoyelle, C., Kuilman, T., van der Horst, C.M., Majoor, D.M., Shay, J.W., Mooi, W.J., and Peeper, D.S. (2005). BRAFE600-associated senescence-like cell cycle arrest of human naevi. Nature *436*, 720–724.

Mikkelsen, T.S., Ku, M., Jaffe, D.B., Issac, B., Lieberman, E., Giannoukos, G., Alvarez, P., Brockman, W., Kim, T.K., Koche, R.P., et al. (2007). Genome-wide maps of chromatin state in pluripotent and lineage-committed cells. Nature *448*, 553–560.

Minc, E., Allory, Y., Worman, H.J., Courvalin, J.C., and Buendia, B. (1999). Localization and phosphorylation of HP1 proteins during the cell cycle in mammalian cells. Chromosoma *108*, 220–234.

Minkovsky, A., Sahakyan, A., Rankin-Gee, E., Bonora, G., Patel, S., and Plath, K. (2014). The Mbd1-Atf7ip-Setdb1 pathway contributes to the maintenance of X-chromosome inactivation. Epigenetics Chromatin *7*, 12.

Mouse Genome Sequencing Consortium, Waterston, R.H., Lindblad-Toh, K., Birney, E., Rogers, J., Abril, J.F., Agarwal, P., Agarwala, R., Ainscough, R., Alexandersson, M., et al. (2002). Initial sequencing and comparative analysis of the mouse genome. Nature *420*, 520–562.

Mujtaba, S., Zeng, L., and Zhou, M.M. (2007). Structure and acetyl-lysine recognition of the bromodomain. Oncogene *26*, 5521–5527.

Muller, J., Hart, C.M., Francis, N.J., Vargas, M.L., Sengupta, A., Wild, B., Miller, E.L., O'Connor, M.B., Kingston, R.E., and Simon, J.A. (2002). Histone methyltransferase activity of a Drosophila Polycomb group repressor complex. Cell *111*, 197–208.

Musri, M.M., Carmona, M.C., Hanzu, F.A., Kaliman, P., Gomis, R., and Parrizas, M. (2010). Histone demethylase LSD1 regulates adipogenesis. J. Biol. Chem. *285*, 30034–30041.

Musselman, C.A., Lalonde, M.E., Cote, J., and Kutateladze, T.G. (2012). Perceiving the epigenetic landscape through histone readers. Nat. Struct. Mol. Biol. *19*, 1218–1227.

Mysliwiec, M.R., Carlson, C.D., Tietjen, J., Hung, H., Ansari, A.Z., and Lee, Y. (2012). Jarid2 (Jumonji, AT rich interactive domain 2) regulates NOTCH1 expression via histone modification in the developing heart. J. Biol. Chem. *287*, 1235–1241.

Nakayama, J., Rice, J.C., Strahl, B.D., Allis, C.D., and Grewal, S.I. (2001). Role of histone H3 lysine 9 methylation in epigenetic control of heterochromatin assembly. Science *292*, 110–113.

Nelson, P.N., Hooley, P., Roden, D., Davari Ejtehadi, H., Rylance, P., Warren, P., Martin, J., Murray, P.G., and Molecular Immunology Research Group (2004). Human endogenous retroviruses: transposable elements with potential? Clin. Exp. Immunol. *138*, 1–9.

Nielsen, A.L., Oulad-Abdelghani, M., Ortiz, J.A., Remboutsika, E., Chambon, P., and Losson, R. (2001). Heterochromatin formation in mammalian cells: interaction between histones and HP1 proteins. Mol. Cell *7*, 729–739.

Nielsen, P.R., Nietlispach, D., Mott, H.R., Callaghan, J., Bannister, A., Kouzarides, T., Murzin, A.G., Murzina, N.V., and Laue, E.D. (2002). Structure of the HP1 chromodomain bound to histone H3 methylated at lysine 9. Nature *416*, 103–107.

Nifuji, A., Ideno, H., Ohyama, Y., Takanabe, R., Araki, R., Abe, M., Noda, M., and Shibuya, H. (2010). Nemo-like kinase (NLK) expression in osteoblastic cells and suppression of osteoblastic differentiation. Exp. Cell Res. *316*, 1127–1136.

Nishioka, K., Chuikov, S., Sarma, K., Erdjument-Bromage, H., Allis, C.D., Tempst, P., and Reinberg, D. (2002). Set9, a novel histone H3 methyltransferase that facilitates transcription by precluding histone tail modifications required for heterochromatin formation. Genes Dev. *16*, 479–489.

Niwa, H., Miyazaki, J., and Smith, A.G. (2000). Quantitative expression of Oct-3/4 defines differentiation, dedifferentiation or self-renewal of ES cells. Nat. Genet. *24*, 372–376.

Noma, K., Allis, C.D., and Grewal, S.I. (2001). Transitions in distinct histone H3 methylation patterns at the heterochromatin domain boundaries. Science *293*, 1150–1155.

Nucifora, F.C., Jr., Sasaki, M., Peters, M.F., Huang, H., Cooper, J.K., Yamada, M., Takahashi, H., Tsuji, S., Troncoso, J., Dawson, V.L., et al. (2001). Interference by huntingtin and atrophin-1 with cbp-mediated transcription leading to cellular toxicity. Science *291*, 2423–2428.

O'Carroll, D., Scherthan, H., Peters, A.H., Opravil, S., Haynes, A.R., Laible, G., Rea, S., Schmid, M., Lebersorger, A., Jerratsch, M., et al. (2000). Isolation and characterization of Suv39h2, a second histone H3 methyltransferase gene that displays testis-specific expression. Mol. Cell. Biol. *20*, 9423–9433.

O'Geen, H., Squazzo, S.L., Iyengar, S., Blahnik, K., Rinn, J.L., Chang, H.Y., Green, R., and Farnham, P.J. (2007). Genome-wide analysis of KAP1 binding suggests autoregulation of KRAB–ZNFs. PLoS Genet. 3, e89.

Ogawa, H., Ishiguro, K., Gaubatz, S., Livingston, D.M., and Nakatani, Y. (2002). A complex with chromatin modifiers that occupies E2F- and Myc-responsive genes in G0 cells. Science *296*, 1132–1136.

Oh, H.J., Li, Y., and Lau, Y.F. (2005). Sry associates with the heterochromatin protein 1 complex by interacting with a KRAB domain protein. Biol. Reprod. *72*, 407–415.

Ohki, I., Shimotake, N., Fujita, N., Jee, J., Ikegami, T., Nakao, M., and Shirakawa, M. (2001). Solution structure of the methyl-CpG binding domain of human MBD1 in complex with methylated DNA. Cell *105*, 487–497.

Ooi, S.K., Qiu, C., Bernstein, E., Li, K., Jia, D., Yang, Z., Erdjument-Bromage, H., Tempst, P., Lin, S.P., Allis, C.D., et al. (2007). DNMT3L connects unmethylated lysine 4 of histone H3 to de novo methylation of DNA. Nature *448*, 714–717.

Paro, R., and Hogness, D.S. (1991). The Polycomb protein shares a homologous domain with a heterochromatin-associated protein of Drosophila. Proc. Natl. Acad. Sci. U.S.A. *88*, 263–267.

Peaston, A.E., Evsikov, A.V., Graber, J.H., de Vries, W.N., Holbrook, A.E., Solter, D., and Knowles, B.B. (2004). Retrotransposons regulate host genes in mouse oocytes and preimplantation embryos. Dev. Cell 7, 597–606.

Pehrson, J.R., and Fried, V.A. (1992). MacroH2A, a core histone containing a large nonhistone region. Science *257*, 1398–1400.

Pei, B., Sisu, C., Frankish, A., Howald, C., Habegger, L., Mu, X.J., Harte, R., Balasubramanian, S., Tanzer, A., Diekhans, M., et al. (2012). The GENCODE pseudogene resource. Genome Biol. 13, R51.

Pek, J.W., Anand, A., and Kai, T. (2012). Tudor domain proteins in development. Development *139*, 2255–2266.

Pencheva, N., and Tavazoie, S.F. (2013). Control of metastatic progression by microRNA regulatory networks. Nat. Cell Biol. *15*, 546–554.

Peng, H., Begg, G.E., Schultz, D.C., Friedman, J.R., Jensen, D.E., Speicher, D.W., and Rauscher, F.J., 3rd (2000). Reconstitution of the KRAB-KAP-1 repressor complex: a model system for defining the molecular anatomy of RING-B box-coiled-coil domain-mediated protein–protein interactions. J. Mol. Biol. *295*, 1139–1162.

Peng, H., Ivanov, A.V., Oh, H.J., Lau, Y.F., and Rauscher, F.J., 3rd (2009). Epigenetic gene silencing by the SRY protein is mediated by a KRAB-O protein that recruits the KAP1 co-repressor machinery. J. Biol. Chem. *284*, 35670–35680.

Peters, A.H., O'Carroll, D., Scherthan, H., Mechtler, K., Sauer, S., Schofer, C., Weipoltshammer, K., Pagani, M., Lachner, M., Kohlmaier, A., et al. (2001). Loss of the Suv39h histone methyltransferases impairs mammalian heterochromatin and genome stability. Cell *107*, 323–337.

Peters, A.H., Mermoud, J.E., O'Carroll, D., Pagani, M., Schweizer, D., Brockdorff, N., and Jenuwein, T. (2002). Histone H3 lysine 9 methylation is an epigenetic imprint of facultative heterochromatin. Nat. Genet. *30*, 77–80.

Pink, R.C., Wicks, K., Caley, D.P., Punch, E.K., Jacobs, L., and Carter, D.R. (2011). Pseudogenes: pseudo-functional or key regulators in health and disease? RNA 17, 792–798.

Pittenger, M.F., Mackay, A.M., Beck, S.C., Jaiswal, R.K., Douglas, R., Mosca, J.D., Moorman, M.A., Simonetti, D.W., Craig, S., and Marshak, D.R. (1999). Multilineage potential of adult human mesenchymal stem cells. Science *284*, 143–147.

Plath, K., Fang, J., Mlynarczyk-Evans, S.K., Cao, R., Worringer, K.A., Wang, H., de la Cruz, C.C., Otte, A.P., Panning, B., and Zhang, Y. (2003). Role of histone H3 lysine 27 methylation in X inactivation. Science *300*, 131–135.

Ponting, C.P. (1997). Tudor domains in proteins that interact with RNA. Trends Biochem. Sci. 22, 51–52.

Prelich, G., and Stillman, B. (1988). Coordinated leading and lagging strand synthesis during SV40 DNA replication *in vitro* requires PCNA. Cell 53, 117–126.

Quenneville, S., Verde, G., Corsinotti, A., Kapopoulou, A., Jakobsson, J., Offner, S., Baglivo, I., Pedone, P.V., Grimaldi, G., Riccio, A., et al. (2011). In embryonic stem cells, ZFP57/KAP1 recognize a methylated hexanucleotide to affect chromatin and DNA methylation of imprinting control regions. Mol. Cell *44*, 361–372.

Quivy, J.P., Roche, D., Kirschner, D., Tagami, H., Nakatani, Y., and Almouzni, G. (2004). A CAF-1 dependent pool of HP1 during heterochromatin duplication. EMBO J. *23*, 3516–3526.

Ransom, M., Dennehey, B.K., and Tyler, J.K. (2010). Chaperoning histones during DNA replication and repair. Cell *140*, 183–195.

Rea, S., Eisenhaber, F., O'Carroll, D., Strahl, B.D., Sun, Z.W., Schmid, M., Opravil, S., Mechtler, K., Ponting, C.P., Allis, C.D., et al. (2000). Regulation of chromatin structure by site-specific histone H3 methyltransferases. Nature *406*, 593–599.

Rebollo, R., Karimi, M.M., Bilenky, M., Gagnier, L., Miceli-Royer, K., Zhang, Y., Goyal, P., Keane, T.M., Jones, S., Hirst, M., et al. (2011). Retrotransposon-induced heterochromatin spreading in the mouse revealed by insertional polymorphisms. PLoS Genet. 7, e1002301.

Reeves, R. (2001). Molecular biology of HMGA proteins: hubs of nuclear function. Gene *277*, 63–81.

Reymond, A., Meroni, G., Fantozzi, A., Merla, G., Cairo, S., Luzi, L., Riganelli, D., Zanaria, E., Messali, S., Cainarca, S., et al. (2001). The tripartite motif family identifies cell compartments. EMBO J. *20*, 2140–2151.

Reynolds, N., Salmon-Divon, M., Dvinge, H., Hynes-Allen, A., Balasooriya, G., Leaford, D., Behrens, A., Bertone, P., and Hendrich, B. (2012). NuRD-mediated deacetylation of H3K27 facilitates recruitment of Polycomb Repressive Complex 2 to direct gene repression. EMBO J. *31*, 593–605.

Rice, J.C., Briggs, S.D., Ueberheide, B., Barber, C.M., Shabanowitz, J., Hunt, D.F., Shinkai, Y., and Allis, C.D. (2003). Histone methyltransferases direct different degrees of methylation to define distinct chromatin domains. Mol. Cell *12*, 1591–1598.

Rodriguez-Paredes, M., Martinez de Paz, A., Simo-Riudalbas, L., Sayols, S., Moutinho, C., Moran, S., Villanueva, A., Vazquez-Cedeira, M., Lazo, P.A., Carneiro, F., et al. (2014). Gene amplification of the histone methyltransferase SETDB1 contributes to human lung tumorigenesis. Oncogene 33, 2807–2813.

Romanish, M.T., Cohen, C.J., and Mager, D.L. (2010). Potential mechanisms of endogenous retroviral-mediated genomic instability in human cancer. Semin. Cancer Biol. *20*, 246–253.

Rosendorff, A., Sakakibara, S., Lu, S., Kieff, E., Xuan, Y., DiBacco, A., Shi, Y., Shi, Y., and Gill, G. (2006). NXP-2 association with SUMO-2 depends on lysines required for transcriptional repression. Proc. Natl. Acad. Sci. U.S.A. *103*, 5308–5313.

Ross, S., Best, J.L., Zon, L.I., and Gill, G. (2002). SUMO-1 modification represses Sp3 transcriptional activation and modulates its subnuclear localization. Mol. Cell *10*, 831–842.

Rothbart, S.B., Krajewski, K., Nady, N., Tempel, W., Xue, S., Badeaux, A.I., Barsyte-Lovejoy, D., Martinez, J.Y., Bedford, M.T., Fuchs, S.M., et al. (2012). Association of UHRF1 with methylated H3K9 directs the maintenance of DNA methylation. Nat. Struct. Mol. Biol. *19*, 1155–1160.

Rowe, H.M., Jakobsson, J., Mesnard, D., Rougemont, J., Reynard, S., Aktas, T., Maillard, P.V., Layard-Liesching, H., Verp, S., Marquis, J., et al. (2010). KAP1 controls endogenous retroviruses in embryonic stem cells. Nature *463*, 237–240.

Rowe, H.M., Kapopoulou, A., Corsinotti, A., Fasching, L., Macfarlan, T.S., Tarabay, Y., Viville, S., Jakobsson, J., Pfaff, S.L., and Trono, D. (2013). TRIM28 repression of retrotransposon-based enhancers is necessary to preserve transcriptional dynamics in embryonic stem cells. Genome Res. *23*, 452–461.

Ryan, R.F., Schultz, D.C., Ayyanathan, K., Singh, P.B., Friedman, J.R., Fredericks, W.J., and Rauscher, F.J., 3rd (1999). KAP-1 corepressor protein interacts and colocalizes with heterochromatic and euchromatic HP1 proteins: a potential role for Kruppel-associated box-zinc finger proteins in heterochromatin-mediated gene silencing. Mol. Cell. Biol. *19*, 4366–4378.

Ryu, H., Lee, J., Hagerty, S.W., Soh, B.Y., McAlpin, S.E., Cormier, K.A., Smith, K.M., and Ferrante, R.J. (2006). ESET/SETDB1 gene expression and histone H3 (K9) trimethylation in Huntington's disease. Proc. Natl. Acad. Sci. U.S.A. *103*, 19176–19181.

Sanders, S.L., Portoso, M., Mata, J., Bahler, J., Allshire, R.C., and Kouzarides, T. (2004). Methylation of histone H4 lysine 20 controls recruitment of Crb2 to sites of DNA damage. Cell *119*, 603–614.

Santoni, F.A., Guerra, J., and Luban, J. (2012). HERV-H RNA is abundant in human embryonic stem cells and a precise marker for pluripotency. Retrovirology 9, 111.

Santos, F., and Dean, W. (2004). Epigenetic reprogramming during early development in mammals. Reproduction *127*, 643–651.

Sarraf, S.A., and Stancheva, I. (2004). Methyl-CpG binding protein MBD1 couples histone H3 methylation at lysine 9 by SETDB1 to DNA replication and chromatin assembly. Mol. Cell *15*, 595–605.

Sasai, N., Nakao, M., and Defossez, P.A. (2010). Sequence-specific recognition of methylated DNA by human zinc-finger proteins. Nucleic Acids Res. *38*, 5015–5022.

Sasaki, H., and Matsui, Y. (2008). Epigenetic events in mammalian germ-cell development: reprogramming and beyond. Nat. Rev. Genet. *9*, 129–140.

Saunders, W.S., Chue, C., Goebl, M., Craig, C., Clark, R.F., Powers, J.A., Eissenberg, J.C., Elgin, S.C., Rothfield, N.F., and Earnshaw, W.C. (1993). Molecular cloning of a human homologue of Drosophila heterochromatin protein HP1 using anti-centromere autoantibodies with anti-chromo specificity. J. Cell Sci. 104 (Pt 2), 573–582.

Schaner, C.E., Deshpande, G., Schedl, P.D., and Kelly, W.G. (2003). A conserved chromatin architecture marks and maintains the restricted germ cell lineage in worms and flies. Dev. Cell *5*, 747–757.

Schmidt, D., Schwalie, P.C., Wilson, M.D., Ballester, B., Goncalves, A., Kutter, C., Brown, G.D., Marshall, A., Flicek, P., and Odom, D.T. (2012). Waves of retrotransposon expansion remodel genome organization and CTCF binding in multiple mammalian lineages. Cell *148*, 335–348.

Schotta, G., Lachner, M., Sarma, K., Ebert, A., Sengupta, R., Reuter, G., Reinberg, D., and Jenuwein, T. (2004). A silencing pathway to induce H3-K9 and H4-K20 trimethylation at constitutive heterochromatin. Genes Dev. *18*, 1251–1262.

Schultz, D.C., Friedman, J.R., and Rauscher, F.J., 3rd (2001). Targeting histone deacetylase complexes via KRAB-zinc finger proteins: the PHD and bromodomains of KAP-1 form a cooperative unit that recruits a novel isoform of the Mi-2alpha subunit of NuRD. Genes Dev. *15*, 428–443.

Schultz, D.C., Ayyanathan, K., Negorev, D., Maul, G.G., and Rauscher, F.J., 3rd (2002). SETDB1: a novel KAP-1-associated histone H3, lysine 9-specific methyltransferase that contributes to HP1-mediated silencing of euchromatic genes by KRAB zinc-finger proteins. Genes Dev. *16*, 919–932.

Seum, C., Bontron, S., Reo, E., Delattre, M., and Spierer, P. (2007a). Drosophila G9a is a nonessential gene. Genetics *177*, 1955–1957.

Seum, C., Reo, E., Peng, H., Rauscher, F.J., 3rd, Spierer, P., and Bontron, S. (2007b). Drosophila SETDB1 is required for chromosome 4 silencing. PLoS Genet. 3, e76.

Shen, X., Liu, Y., Hsu, Y.J., Fujiwara, Y., Kim, J., Mao, X., Yuan, G.C., and Orkin, S.H. (2008). EZH1 mediates methylation on histone H3 lysine 27 and complements EZH2 in maintaining stem cell identity and executing pluripotency. Mol. Cell *32*, 491–502.

Shi, X., Hong, T., Walter, K.L., Ewalt, M., Michishita, E., Hung, T., Carney, D., Pena, P., Lan, F., Kaadige, M.R., et al. (2006). ING2 PHD domain links histone H3 lysine 4 methylation to active gene repression. Nature *442*, 96–99.

Shi, Y., Lan, F., Matson, C., Mulligan, P., Whetstine, J.R., Cole, P.A., Casero, R.A., and Shi, Y. (2004). Histone demethylation mediated by the nuclear amine oxidase homolog LSD1. Cell *119*, 941–953.

Silva, J., Mak, W., Zvetkova, I., Appanah, R., Nesterova, T.B., Webster, Z., Peters, A.H., Jenuwein, T., Otte, A.P., and Brockdorff, N. (2003). Establishment of histone h3 methylation on the inactive X-chromosome requires transient recruitment of Eed-Enx1 polycomb group complexes. Dev. Cell *4*, 481–495.

Singh, P.B., and Georgatos, S.D. (2002). HP1: facts, open questions, and speculation. J. Struct. Biol. *140*, 10–16.

Singh, P.B., Miller, J.R., Pearce, J., Kothary, R., Burton, R.D., Paro, R., James, T.C., and Gaunt, S.J. (1991). A sequence motif found in a Drosophila heterochromatin protein is conserved in animals and plants. Nucleic Acids Res. *19*, 789–794.

Smothers, J.F., and Henikoff, S. (2001). The hinge and chromo shadow domain impart distinct targeting of HP1-like proteins. Mol. Cell. Biol. *21*, 2555–2569.

Sridharan, R., Gonzales-Cope, M., Chronis, C., Bonora, G., McKee, R., Huang, C., Patel, S., Lopez, D., Mishra, N., Pellegrini, M., et al. (2013). Proteomic and genomic approaches reveal critical functions of H3K9 methylation and heterochromatin protein-1gamma in reprogramming to pluripotency. Nat. Cell Biol. *15*, 872–882.

Sripathy, S.P., Stevens, J., and Schultz, D.C. (2006). The KAP1 corepressor functions to coordinate the assembly of de novo HP1-demarcated microenvironments of heterochromatin required for KRAB zinc finger protein-mediated transcriptional repression. Mol. Cell. Biol. *26*, 8623–8638.

Stabell, M., Bjorkmo, M., Aalen, R.B., and Lambertsson, A. (2006). The Drosophila SET domain encoding gene dEset is essential for proper development. Hereditas *143*, 177–188.

Stedman, W., Kang, H., Lin, S., Kissil, J.L., Bartolomei, M.S., and Lieberman, P.M. (2008). Cohesins localize with CTCF at the KSHV latency control region and at cellular c-myc and H19/Igf2 insulators. EMBO J. *27*, 654–666.

Steffan, J.S., Bodai, L., Pallos, J., Poelman, M., McCampbell, A., Apostol, B.L., Kazantsev, A., Schmidt, E., Zhu, Y.Z., Greenwald, M., et al. (2001). Histone deacetylase inhibitors arrest polyglutamine-dependent neurodegeneration in Drosophila. Nature *413*, 739–743.

Sternberg, P.W., and Horvitz, H.R. (1991). Signal transduction during C. elegans vulval induction. Trends Genet. *7*, 366–371.

Stielow, B., Sapetschnig, A., Wink, C., Kruger, I., and Suske, G. (2008). SUMO-modified Sp3 represses transcription by provoking local heterochromatic gene silencing. EMBO Rep. *9*, 899–906.

Stocking, C., and Kozak, C.A. (2008). Murine endogenous retroviruses. Cell. Mol. Life Sci. *65*, 3383–3398.

Stoye, J.P. (2012). Studies of endogenous retroviruses reveal a continuing evolutionary saga. Nat. Rev. Microbiol. *10*, 395–406.

Strahl, B.D., and Allis, C.D. (2000). The language of covalent histone modifications. Nature *403*, 41–45.

Sun, F.L., Cuaycong, M.H., Craig, C.A., Wallrath, L.L., Locke, J., and Elgin, S.C. (2000). The fourth chromosome of Drosophila melanogaster: interspersed euchromatic and heterochromatic domains. Proc. Natl. Acad. Sci. U.S.A. *97*, 5340–5345.

Suske, G. (1999). The Sp-family of transcription factors. Gene *238*, 291–300.

Tachibana, M., Sugimoto, K., Fukushima, T., and Shinkai, Y. (2001). Set domain-containing protein, G9a, is a novel lysine-preferring mammalian histone methyltransferase with hyperactivity and specific selectivity to lysines 9 and 27 of histone H3. J. Biol. Chem. *276*, 25309–25317.

Tachibana, M., Sugimoto, K., Nozaki, M., Ueda, J., Ohta, T., Ohki, M., Fukuda, M., Takeda, N., Niida, H., Kato, H., et al. (2002). G9a histone methyltransferase plays a dominant role in euchromatic histone H3 lysine 9 methylation and is essential for early embryogenesis. Genes Dev. *16*, 1779–1791.

Tachibana, M., Ueda, J., Fukuda, M., Takeda, N., Ohta, T., Iwanari, H., Sakihama, T., Kodama, T., Hamakubo, T., and Shinkai, Y. (2005). Histone methyltransferases G9a and GLP form heteromeric complexes and are both crucial for methylation of euchromatin at H3-K9. Genes Dev. *19*, 815–826.

Tagami, H., Ray-Gallet, D., Almouzni, G., and Nakatani, Y. (2004). Histone H3.1 and H3.3 complexes mediate nucleosome assembly pathways dependent or independent of DNA synthesis. Cell *116*, 51–61.

Tahiliani, M., Koh, K.P., Shen, Y., Pastor, W.A., Bandukwala, H., Brudno, Y., Agarwal, S., Iyer, L.M., Liu, D.R., Aravind, L., et al. (2009). Conversion of 5-methylcytosine to 5-hydroxymethylcytosine in mammalian DNA by MLL partner TET1. Science *324*, 930–935.

Takada, I., Mihara, M., Suzawa, M., Ohtake, F., Kobayashi, S., Igarashi, M., Youn, M.Y., Takeyama, K., Nakamura, T., Mezaki, Y., et al. (2007). A histone lysine methyltransferase activated by non-canonical Wnt signalling suppresses PPAR-gamma transactivation. Nat. Cell Biol. *9*, 1273–1285.

Tamaru, H., and Selker, E.U. (2001). A histone H3 methyltransferase controls DNA methylation in Neurospora crassa. Nature *414*, 277–283.

Tamaru, H., Zhang, X., McMillen, D., Singh, P.B., Nakayama, J., Grewal, S.I., Allis, C.D., Cheng, X., and Selker, E.U. (2003). Trimethylated lysine 9 of histone H3 is a mark for DNA methylation in Neurospora crassa. Nat. Genet. *34*, 75–79.

Tan, M., Luo, H., Lee, S., Jin, F., Yang, J.S., Montellier, E., Buchou, T., Cheng, Z., Rousseaux, S., Rajagopal, N., et al. (2011). Identification of 67 histone marks and histone lysine crotonylation as a new type of histone modification. Cell *146*, 1016–1028.

Tan, S.L., Nishi, M., Ohtsuka, T., Matsui, T., Takemoto, K., Kamio-Miura, A., Aburatani, H., Shinkai, Y., and Kageyama, R. (2012). Essential roles of the histone methyltransferase ESET in the epigenetic control of neural progenitor cells during development. Development *139*, 3806–3816.

Tariq, M., Saze, H., Probst, A.V., Lichota, J., Habu, Y., and Paszkowski, J. (2003). Erasure of CpG methylation in Arabidopsis alters patterns of histone H3 methylation in heterochromatin. Proc. Natl. Acad. Sci. U.S.A. *100*, 8823–8827.

Tschiersch, B., Hofmann, A., Krauss, V., Dorn, R., Korge, G., and Reuter, G. (1994). The protein encoded by the Drosophila position-effect variegation suppressor gene Su(var)3-9 combines domains of antagonistic regulators of homeotic gene complexes. EMBO J. *13*, 3822–3831.

Tsukada, Y., Fang, J., Erdjument-Bromage, H., Warren, M.E., Borchers, C.H., Tempst, P., and Zhang, Y. (2006). Histone demethylation by a family of JmjC domain-containing proteins. Nature *439*, 811–816.

Turelli, P., Castro-Diaz, N., Marzetta, F., Kapopoulou, A., Raclot, C., Duc, J., Tieng, V., Quenneville, S., and Trono, D. (2014). Interplay of TRIM28 and DNA methylation in controlling human endogenous retroelements. Genome Res. *24*, 1260–1270.

Tzeng, T.Y., Lee, C.H., Chan, L.W., and Shen, C.K. (2007). Epigenetic regulation of the Drosophila chromosome 4 by the histone H3K9 methyltransferase dSETDB1. Proc. Natl. Acad. Sci. U.S.A. *104*, 12691–12696.

Uchimura, Y., Ichimura, T., Uwada, J., Tachibana, T., Sugahara, S., Nakao, M., and Saitoh, H. (2006). Involvement of SUMO modification in MBD1- and MCAF1-mediated heterochromatin formation. J. Biol. Chem. *281*, 23180–23190.

Ulrich, H.D. (2014). Two-way communications between ubiquitin-like modifiers and DNA. Nat. Struct. Mol. Biol. *21*, 317–324.

Van Duyne, R., Easley, R., Wu, W., Berro, R., Pedati, C., Klase, Z., Kehn-Hall, K., Flynn, E.K., Symer, D.E., and Kashanchi, F. (2008). Lysine methylation of HIV-1 Tat regulates transcriptional activity of the viral LTR. Retrovirology *5*, 40.

Vermeulen, M., Eberl, H.C., Matarese, F., Marks, H., Denissov, S., Butter, F., Lee, K.K., Olsen, J.V., Hyman, A.A., Stunnenberg, H.G., et al. (2010). Quantitative interaction proteomics and genome-wide profiling of epigenetic histone marks and their readers. Cell *142*, 967–980.

Verschure, P.J., van der Kraan, I., de Leeuw, W., van der Vlag, J., Carpenter, A.E., Belmont, A.S., and van Driel, R. (2005). *In vivo* HP1 targeting causes large-scale chromatin condensation and enhanced histone lysine methylation. Mol. Cell. Biol. *25*, 4552–4564.

Vincent, J.J., Huang, Y., Chen, P.Y., Feng, S., Calvopina, J.H., Nee, K., Lee, S.A., Le, T., Yoon, A.J., Faull, K., et al. (2013). Stage-specific roles for tet1 and tet2 in DNA demethylation in primordial germ cells. Cell stem cell *12*, 470–478.

Wang, H., An, W., Cao, R., Xia, L., Erdjument-Bromage, H., Chatton, B., Tempst, P., Roeder, R.G., and Zhang, Y. (2003). mAM facilitates conversion by ESET of dimethyl to trimethyl lysine 9 of histone H3 to cause transcriptional repression. Mol. Cell *12*, 475–487.

Wang, J., Scully, K., Zhu, X., Cai, L., Zhang, J., Prefontaine, G.G., Krones, A., Ohgi, K.A., Zhu, P., Garcia-Bassets, I., et al. (2007). Opposing LSD1 complexes function in developmental gene activation and repression programmes. Nature *446*, 882–887.

Wang, W., Chen, Z., Mao, Z., Zhang, H., Ding, X., Chen, S., Zhang, X., Xu, R., and Zhu, B. (2011a). Nucleolar protein Spindlin1 recognizes H3K4 methylation and stimulates the expression of rRNA genes. EMBO Rep. *12*, 1160–1166.

Wang, X., Pan, L., Wang, S., Zhou, J., McDowell, W., Park, J., Haug, J., Staehling, K., Tang, H., and Xie, T. (2011b). Histone H3K9 trimethylase Eggless controls germline stem cell maintenance and differentiation. PLoS Genet. 7, e1002426.

Wang, Z., Zang, C., Rosenfeld, J.A., Schones, D.E., Barski, A., Cuddapah, S., Cui, K., Roh, T.Y., Peng, W., Zhang, M.Q., et al. (2008). Combinatorial patterns of histone acetylations and methylations in the human genome. Nat. Genet. *40*, 897–903.

Weber, C.M., and Henikoff, S. (2014). Histone variants: dynamic punctuation in transcription. Genes Dev. *28*, 672–682.

Whetstine, J.R., Nottke, A., Lan, F., Huarte, M., Smolikov, S., Chen, Z., Spooner, E., Li, E., Zhang, G., Colaiacovo, M., et al. (2006). Reversal of histone lysine trimethylation by the JMJD2 family of histone demethylases. Cell *125*, 467–481.

Whyte, W.A., Bilodeau, S., Orlando, D.A., Hoke, H.A., Frampton, G.M., Foster, C.T., Cowley, S.M., and Young, R.A. (2012). Enhancer decommissioning by LSD1 during embryonic stem cell differentiation. Nature *482*, 221–225.

Williams, K., Christensen, J., Pedersen, M.T., Johansen, J.V., Cloos, P.A., Rappsilber, J., and Helin, K. (2011). TET1 and hydroxymethylcytosine in transcription and DNA methylation fidelity. Nature *473*, 343–348.

Wreggett, K.A., Hill, F., James, P.S., Hutchings, A., Butcher, G.W., and Singh, P.B. (1994). A mammalian homologue of Drosophila heterochromatin protein 1 (HP1) is a component of constitutive heterochromatin. Cytogenet. Cell Genet. *66*, 99–103.

Wysocka, J., Swigut, T., Xiao, H., Milne, T.A., Kwon, S.Y., Landry, J., Kauer, M., Tackett, A.J., Chait, B.T., Badenhorst, P., et al. (2006). A PHD finger of NURF couples histone H3 lysine 4 trimethylation with chromatin remodelling. Nature *442*, 86–90.

Yamane, K., Toumazou, C., Tsukada, Y., Erdjument-Bromage, H., Tempst, P., Wong, J., and Zhang, Y. (2006). JHDM2A, a JmjC-containing H3K9 demethylase, facilitates transcription activation by androgen receptor. Cell *125*, 483–495.

Yang, L., Xia, L., Wu, D.Y., Wang, H., Chansky, H.A., Schubach, W.H., Hickstein, D.D., and Zhang, Y. (2002). Molecular cloning of ESET, a novel histone H3-specific methyltransferase that interacts with ERG transcription factor. Oncogene *21*, 148–152.

Yang, L., Mei, Q., Zielinska-Kwiatkowska, A., Matsui, Y., Blackburn, M.L., Benedetti, D., Krumm, A.A., Taborsky, G.J., Jr., and Chansky, H.A. (2003). An ERG (ets-related gene)-associated histone methyltransferase interacts with histone deacetylases 1/2 and transcription co-repressors mSin3A/B. Biochem. J. *369*, 651–657.

Yang, L., Lawson, K.A., Teteak, C.J., Zou, J., Hacquebord, J., Patterson, D., Ghatan, A.C., Mei, Q., Zielinska-Kwiatkowska, A., Bain, S.D., et al. (2013). ESET histone methyltransferase is essential to hypertrophic differentiation of growth plate chondrocytes and formation of epiphyseal plates. Dev. Biol. *380*, 99–110.

Yang, N., Wang, W., Wang, Y., Wang, M., Zhao, Q., Rao, Z., Zhu, B., and Xu, R.M. (2012). Distinct mode of methylated lysine-4 of histone H3 recognition by tandem tudor-like domains of Spindlin1. Proc. Natl. Acad. Sci. U.S.A. *109*, 17954–17959.

Ye, Q., and Worman, H.J. (1996). Interaction between an integral protein of the nuclear envelope inner membrane and human chromodomain proteins homologous to Drosophila HP1. J. Biol. Chem. *271*, 14653–14656.

Yoon, J., Lee, K.S., Park, J.S., Yu, K., Paik, S.G., and Kang, Y.K. (2008). dSETDB1 and SU(VAR)3–9 sequentially function during germline-stem cell differentiation in *Drosophila melanogaster*. PLoS ONE 3, e2234.

Yuan, P., Han, J., Guo, G., Orlov, Y.L., Huss, M., Loh, Y.H., Yaw, L.P., Robson, P., Lim, B., and Ng, H.H. (2009). Eset partners with Oct4 to restrict extraembryonic trophoblast lineage potential in embryonic stem cells. Genes Dev. *23*, 2507–2520.

Zentner, G.E., Tesar, P.J., and Scacheri, P.C. (2011). Epigenetic signatures distinguish multiple classes of enhancers with distinct cellular functions. Genome Res. *21*, 1273–1283.

Zhang, H., Cai, K., Wang, J., Wang, X., Cheng, K., Shi, F., Jiang, L., Zhang, Y., and Dou, J. (2014). MiR-7, inhibited indirectly by LincRNA HOTAIR, directly inhibits SETDB1 and reverses the EMT of breast cancer stem cells by down-regulating the STAT3 pathway. Stem Cells *32*, 2858–2868.

Zhang, X., Tamaru, H., Khan, S.I., Horton, J.R., Keefe, L.J., Selker, E.U., and Cheng, X. (2002). Structure of the Neurospora SET domain protein DIM-5, a histone H3 lysine methyltransferase. Cell *111*, 117–127.

Zheng, D., Frankish, A., Baertsch, R., Kapranov, P., Reymond, A., Choo, S.W., Lu, Y., Denoeud, F., Antonarakis, S.E., Snyder, M., et al. (2007). Pseudogenes in the ENCODE regions: consensus annotation, analysis of transcription, and evolution. Genome Res. *17*, 839–851.

Sirtuin Deacetylases in Fungi: Connecting Metabolism to Lifecycle Progression, Stress Response and Genome Stability

3

Laura N. Rusche, Ashleigh S. Hanner, Justin M.H. Heltzel, Kristen M. Humphrey, Shivali Kapoor and Christopher B. Rupert

Abstract

Sirtuins are NAD$^+$-dependent deacylases that regulate biological processes such as the maintenance of genome stability, response to stress, adaptation to low or poor nutrients, and initiation of life cycle events such as mating and sporulation. Because sirtuins require NAD$^+$ for activity, changes in cellular metabolism that alter the NAD$^+$/NADH ratio or NAD$^+$ availability may influence the activity of sirtuins. Therefore, biological processes regulated by sirtuins could be linked to the metabolic state of the cell. Over the course of evolution, bringing new processes under the control of sirtuins could enable species to evolve novel responses to stresses that lower NAD$^+$ levels. Here, we explore this hypothesis by examining how sirtuins contribute to the biology of a variety of fungal species.

Properties of sirtuins

Enzymatic reactions catalysed by sirtuins

A defining feature of sirtuins is that they couple deacetylation of a lysine residue with the breakdown of NAD$^+$ (Landry *et al.*, 2000). NAD$^+$ (nicotinamide adenine dinucleotide) is a small molecule coenzyme which consists of nicotinamide mononucleotide and adenine mononucleotide joined by their respective phosphate groups. In the sirtuin catalysed reaction, NAD$^+$ is lysed to yield nicotinamide and ADP-ribose, and the acetyl group is concomitantly transferred to the ADP-ribose, generating a unique intermediate, O-acetyl-ADP ribose (Fig. 3.1) (Tanner *et al.*, 2000; Tanny and Moazed, 2001). A variation of this reaction is catalysed by sirtuins such as

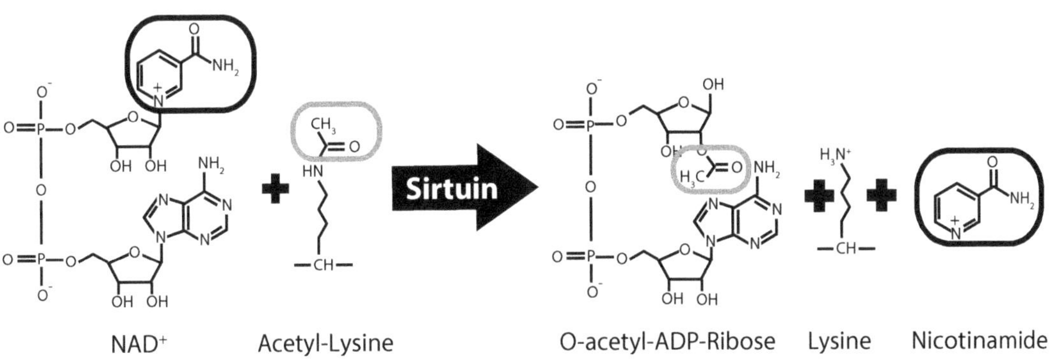

Figure 3.1 Deacetylation reaction catalysed by sirtuins. Deacetylation of a lysine residue is coupled to the cleavage of NAD$^+$, releasing nicotinamide and ADP-ribose. During the reaction, the acetyl group is transferred to ADP-ribose, generating the product O-acetyl-ADP-ribose. Dashed boxes indicate submolecular groupings that are preserved through the reaction.

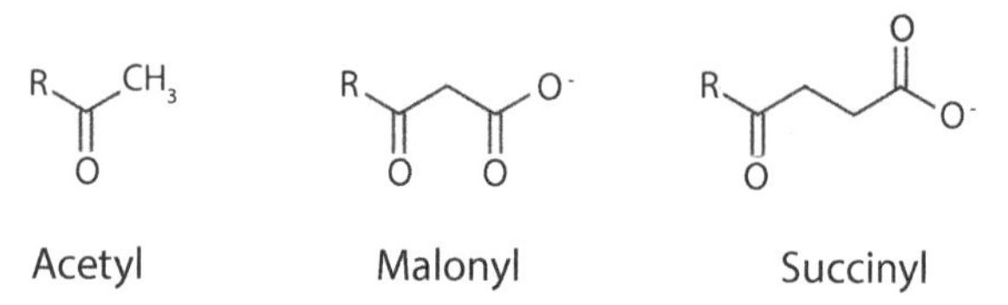

Figure 3.2 Acyl groups removed by sirtuins. Sirtuins catalyse the removal of acetyl- (left), malonyl- (centre), and succinyl- (right) groups from lysine side chains.

Sirt5 that remove a succinyl- or malonyl- moiety rather than an acetyl group (Du *et al.*, 2011). These functional groups are structurally related to an acetyl- group (Fig. 3.2), and the chemistry of the reaction is similar. Although succinylation and malonylation of lysines had not been detected until recently (Peng *et al.*, 2011; Zhang *et al.*, 2011), it is now thought that the abundant succinyl-CoA and malonyl-CoA molecules in cells contribute to the succinylation and malonylation of proteins. Succinyl-CoA is an intermediate of the citric acid cycle, and malonyl-CoA is synthesized in the mitochondrial membrane as a precursor of fatty acid synthesis. Thus, sirtuins are NAD^+-dependent deacylases.

It has also been proposed that some sirtuins transfer ADP-ribose from NAD^+ to a protein substrate in a mono-ADP-ribosylation reaction. This activity has been suggested for mammalian SirT4 (Haigis *et al.*, 2006) and SirT6 (Mao *et al.*, 2011), which have low deacetylase activity *in vitro*. However, ADP-ribosylation is not robust and may be a side reaction due to the susceptibility of NAD^+ in the active site to nucleophilic attack (Tanner *et al.*, 2000). In addition, the deacetylase activity of SirT6 can be stimulated by free fatty acids (Feldman *et al.*, 2013), suggesting that deacetylation may be more efficient *in vivo* than it appears *in vitro*.

Based on the deacylation reaction catalysed by sirtuins, there are three ways in which these proteins influence cellular processes. First, deacylation of a lysine changes the charge and shape of the residue, which could influence protein conformation and function. In addition, the two small molecule products of the reaction, nicotinamide and O-acetyl-ADP-ribose, could have secondary effects. In fact, nicotinamide inhibits sirtuins when present in sufficient concentrations (Landry *et al.*, 2000). Nicotinamide is ultimately recycled through the NAD^+ salvage pathway back into NAD^+. O-acetyl-ADP-ribose is a more distinct product than nicotinamide, and its potential as a second messenger has been much discussed. For example, it had been thought that O-acetyl-ADP-ribose generated by Sir2 in *S. cerevisiae* would interact with the heterochromatin protein Sir3, thereby inducing a conformational change favouring assembly of SIR heterochromatin (Liou *et al.*, 2005; Martino *et al.*, 2009). However, other evidence argues against a role for O-acetyl-ADP-ribose in promoting heterochromatin formation in *S. cerevisiae* (Chou *et al.*, 2008; Ehrentraut *et al.*, 2011). Another potential link with heterochromatin is that O-acetyl-ADP-ribose is bound by the macrodomain of one splice variant of macroH2A (Kustatscher *et al.*, 2005). This histone variant is associated with heterochromatin in mammalian cells (Chadwick *et al.*, 2001). However, the consequence of O-acetyl-ADP-ribose binding by macroH2A is unknown. Another potential function of O-acetyl-ADP-ribose is to influence metabolism. An overabundance of O-acetyl-ADP-ribose inhibits the electron transport chain and glycolysis, thereby decreasing reactive oxygen species (Tong *et al.*, 2009). Ultimately, O-acetyl-ADP-ribose can be degraded by nudix family hydrolases, such as Ysa1 in *S. cerevisiae* (Rafty *et al.*, 2002) or deacetylated by macro domain proteins such as macroD1, macroD2, or C6orf130 in humans (Chen *et al.*, 2011; Peterson *et al.*, 2011).

Structure of sirtuins

Structurally, sirtuins consist of a conserved catalytic core flanked by variable regions that can confer specificity for adaptor proteins or regulate catalysis. The catalytic core folds into two domains (Fig. 3.3) (Finnin *et al.*, 2001; Min *et al.*, 2001).

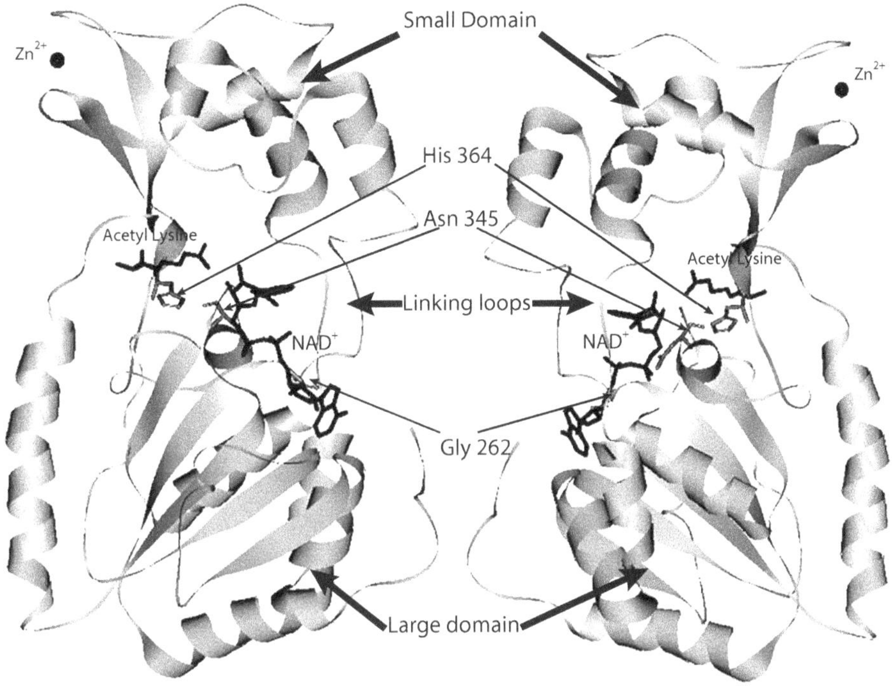

Figure 3.3 The conserved catalytic core of a representative sirtuin. The structure of ScHst2 bound to O-acetyl-ADP-ribose and nicotinamide was created from pdb file 1SZC using Viewerlite. The stabilizing zinc ion (black sphere) is found in the small domain. Key residues frequently mutated to generate catalytically inactive sirtuins are indicated (glycine 262, asparagine 345, and histidine 364).

The large domain comprises a classic Rossmann fold, known for binding NAD^+. The small domain is assembled from two insertions in the Rossmann fold and contains four conserved cysteine residues that bind zinc. Binding of substrates occurs in a cleft between the two domains. There are four linking loops which bridge the gap between the large and small domains and contribute to substrate binding and catalysis.

In the substrate cleft, NAD^+ and the peptide containing the acetylated lysine enter on opposite sides, with the nicotinamide group deep in the cleft next to the acetylated lysine (Avalos *et al.*, 2002; Zhao *et al.*, 2003b). Interactions between the peptide substrate and enzyme binding pocket are similar to that of beta strands, meaning that most of the contacts occur with the peptide backbone of the substrate. For Sirt5, which has desuccinylase and demalonylase activity, the binding pocket for the acylated lysine is larger than in deacetylases (Du *et al.*, 2011). When NAD^+ is bound, its conformation is distorted through bending of the glycosidic bond out of the plane of the nicotinamide ring (Avalos *et al.*, 2004). This high energy conformation of the nicotinamide moiety is further strained by rotating 150° from the most commonly observed position. This destabilization of the NAD^+ is believed to be the driving force for the nucleophilic attack during the reaction. Several residues in the NAD^+ binding pocket are commonly mutated to generate catalytically inactive sirtuins. These include an asparagine (345 in ScSir2), a histidine (364 in ScSir2), and a glycine (262 in ScSir2) (Fig. 3.3).

Zn^{2+} is a necessary cofactor for sirtuin function, but the ion does not bind in the catalytic site. Instead, Zn^{2+} stabilizes the small domain. This arrangement is unusual compared to other zinc dependent protein deacetylases (Finnin *et al.*, 1999). Removal of Zn^{2+} through either mutations of the cysteines or chelation of the Zn^{2+} ions abolishes Sir2 activity (Chakrabarty and Balaram, 2010; Sherman *et al.*, 1999).

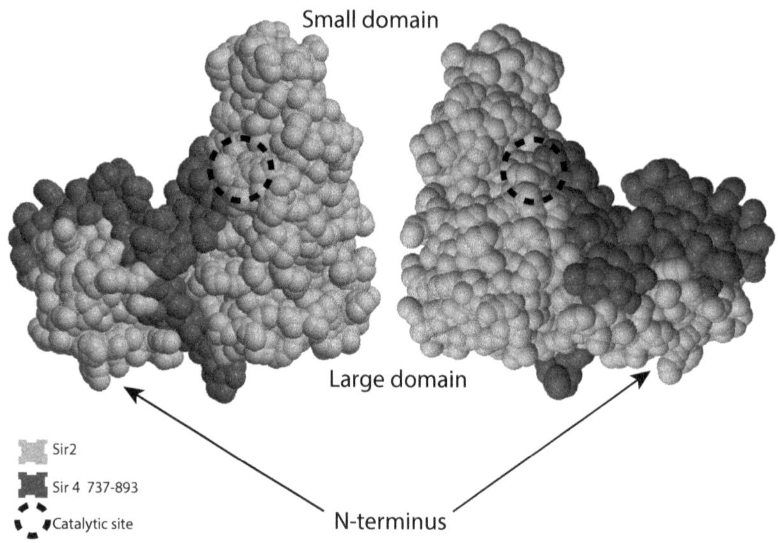

Figure 3.4 Interaction of ScSir2 with the adaptor protein ScSir4. The catalytic core and amino-terminal domain of ScSir2 (light grey) are shown in complex with residues 737–893 of ScSir4 (dark grey). This model was generated in viewerlite from pdb file 4IAO.

The structures of the variable amino and carboxyl terminal regions are less well characterized than that of the catalytic core. One case in which these domains have been studied is that of Sir2 from *S. cerevisiae*, which interacts with the heterochromatin forming protein Sir4 via its variable domains (Fig. 3.4). A co-crystal structure of Sir2 bound to a Sir4 fragment reveals a helical amino terminal domain of Sir2, which interacts extensively with Sir4 (Hsu *et al.*, 2013). In addition, an unstructured segment of Sir4 makes hydrophobic interactions with both the amino terminal domain and the catalytic core of Sir2. Importantly, this interaction stabilizes the active conformation of the catalytic domain, explaining how Sir4 stimulates the catalytic activity of Sir2.

Another example in which regions outside the core domain have been visualized is that of Hst2 from *S. cerevisiae*, for which the full length protein was crystallized (Zhao *et al.*, 2003a). In this case, the terminal helix folds back to make extensive contacts with the NAD^+ binding pocket, potentially blocking cofactor binding. Consistent with this idea, deleting this helix increases the binding affinity for NAD^+ (Zhao *et al.*, 2003a). Thus, it is proposed that Hst2 auto-regulates its activity through this terminal domain.

Specificity of sirtuins

Sirtuins are specific for particular lysine residues of proteins. For example, Sir2 from *S. cerevisiae* preferentially deacetylates lysine 16 over other lysines in the tail of histone H4 (Imai *et al.*, 2000). However, the basis for this sequence specificity is not fully understood. One observation is that targeted lysines are often in unstructured regions of proteins, such as histone tails (Khan and Lewis, 2005). In addition, structural analyses indicate that the N-1 and N+2 positions relative to the acetylated lysine are critical for determining specificity (Cosgrove *et al.*, 2006; Zhao *et al.*, 2004). For example, the N-1 position makes an interaction with a residue of the sirtuin that is highly variable, which could lead to specificity (Cosgrove *et al.*, 2006). In addition to structural studies, high-throughput analyses have revealed sequence preferences for mammalian SirT1 and SirT3 (Garske and Denu, 2006; Smith *et al.*, 2011).

In addition to targeting lysines in particular sequence contexts, sirtuins can be recruited to their substrates by adaptor proteins. These adaptor proteins usually associate with both the substrate and the sirtuin. Adaptor proteins may also enhance the catalytic activity of sirtuins. For example, *S. cerevisiae* Sir2 associates with the adaptor protein Sir4, which recruits Sir2 to appropriate genomic targets and enhances the deacetylase activity of Sir2.

Regulation of sirtuin catalysis

As regulatory proteins, sirtuins themselves need to be regulated such that the processes they govern can be switched on and off. The unusual requirement of sirtuins for NAD^+ implies that their activity is regulated by NAD^+ availability. Furthermore, since NAD^+ is a central cofactor in intermediary metabolism, the metabolic state of the cell could influence sirtuin activity. For example, NAD^+ is reduced to NADH as sugars are oxidized to CO_2 and H_2O through glycolysis and the citric acid cycle, and NADH is reoxidized to NAD^+ as ATP is generated through oxidative phosphorylation. Thus, it is proposed that as the metabolic flux shifts, changes in the NAD^+/NADH ratio and NAD^+ availability will influence the activity of sirtuins. Therefore, biological processes regulated by sirtuins could be linked to the metabolic state of the cell.

Another factor that may influence sirtuin activity is the subcellular distribution of NAD^+. Enzymes in the salvage pathway are localized in both the nucleus and cytoplasm of *S. cerevisiae* (Anderson *et al.*, 2002; Sandmeier *et al.*, 2002), indicating that NAD^+ may be generated independently in each compartment. In addition, the enzyme Tdh3 is proposed to provide NAD^+ directly to Sir2 (Ringel *et al.*, 2013). ScSir2 interacts physically with Tdh3 (glyceraldehyde 3-phosphate dehydrogenase), a glycolytic enzyme which reduces NAD^+ to NADH while converting glyceraldehyde-3-phosphate to 1,3-bis-phosphoglycerate. It is proposed that NAD^+ bound to Tdh3 can be utilized by ScSir2. In fact, Tdh3 associates with SIR heterochromatin and enhances SIR-dependent silencing (Ringel *et al.*, 2013).

In addition to NAD^+ availability, sirtuins are regulated by other mechanisms that restrict their expression to particular subcellular locations or phases of the cell cycle. For example, some sirtuins are localized to the nucleus or the mitochondria (Michishita *et al.*, 2005), and orthologues of the Hst2/SirT2 subfamily shuttle between the cytoplasm and nucleus (Durand-Dubief *et al.*, 2007; Vaquero *et al.*, 2006; Wilson *et al.*, 2006). In addition, the transcription of proteins of the Hst3/Hst4 subfamily is restricted to G2/M phase of the cell cycle (Celic *et al.*, 2006; Maas *et al.*, 2006), and these proteins are targeted by ubiquitylation for degradation in M/G1 phase (Delgoshaie *et al.*, 2014; Edenberg *et al.*, 2014).

Finally, some sirtuins have auto regulatory domains. For example, ScSir2 and ScHst2 form homotrimers. In the case of ScHst2, trimerization results in the amino terminal domain blocking the peptide binding site and decreasing peptide binding (Zhao *et al.*, 2003a). It is proposed that ScHst2 is activated by undergoing a conformational change that releases the amino terminal domain from the active site. In the case of ScSir2, homotrimerization is disrupted by the binding of the adaptor protein Sir4 (Cubizolles *et al.*, 2006), presumably activating Sir2. Curiously, mutations that disrupt homotrimerization reduce rDNA silencing, although the basis for this effect is unknown. A second form of autoregulation is documented for ScHst2, in which the carboxy-terminal domain folds back and interferes with NAD^+ binding (Zhao *et al.*, 2003a), as described above.

Biological functions of sirtuins

Although the enzymatic reactions catalysed by sirtuins are relatively conserved, the biological outcomes of these reactions are diverse and depend on the set of substrates targeted by a particular sirtuin. Examples of processes regulated by fungal sirtuins are response to stress, adaptation to low or poor nutrients, and initiation of life cycle events such as mating and sporulation. In addition, fungal sirtuins are critical in maintaining genome stability and modulating chromatin states through the cell cycle. It is important to note that sirtuins often repress the processes they regulate. Thus, under optimal conditions for sirtuin activity these processes are inhibited, whereas under stressful conditions that result in lowered NAD^+ levels sirtuin activity will be compromised, releasing the brake on regulated processes.

The most studied substrates of fungal sirtuins are histones. Histone octomers, comprised of two molecules each of H2A, H2B, H3, and H4, form a protein core around which 147 base pairs of DNA wraps to form a nucleosome. Nucleosomes pack relative to one another in various arrangements to form chromatin fibres that are more or less accessible to DNA metabolizing enzymes, such as DNA and RNA polymerases and repair proteins. Histones are rich in lysines, which are positively charged and make electrostatic interactions with the negatively charged DNA phosphodiester backbone. However, acetylated lysines are neutral, and consequently the acetylation state of histones has a profound impact on histone–DNA interactions and hence chromatin structure. Therefore, the deacetylation of histones by sirtuins increases histone–DNA interactions, thereby influencing gene expression, DNA replication and repair, and genome stability.

In addition to histones, Sir2 from *S. cerevisiae* is documented to deacetylate a few other proteins. For example, ScSir2 deacetylates the transcription factor Ifh1 (Downey *et al.*, 2013), which activates ribosomal protein genes. The unacetylated form of Ifh1 is active, and therefore deacetylation by ScSir2 maintains Ifh1 activity and ribosomal protein production. ScSir2 also deacetylates the enzyme Pck1 (phosphoenolpyruvate carboxy kinase) (Lin *et al.*, 2009), which is required for gluconeogenesis and growth on non-fermentable carbon sources. In this case, the acetylated form is active, and deacetylation by ScSir2 reduces activity and disfavours gluconeogenesis. Interestingly, in both these cases, ScSir2 activity promotes growth on rich nutrients and reduced ScSir2 activity is required to adapt to a poor nutrient environment.

One biological function of sirtuins that has garnered much attention is their ability to influence an organism's lifespan. In fact, the connection between Sir2 and ageing was first demonstrated in *S. cerevisiae* (Kaeberlein *et al.*, 1999). In budding yeast, replicative lifespan is measured by the number of times a mother cell buds. During cell division, there is an asymmetric segregation of toxic compounds, such as damaged proteins and extrachromosomal DNA. These compounds are largely retained in the mother cell due to their retrograde transport out of the bud. Consequently, mother cells accumulate damage and eventually senesce, whereas the daughter cells are born young with a full replicative lifespan, regardless of the age of the mother. ScSir2 influences replicative lifespan through its influence on recombination of the tandem rDNA genes (Kobayashi, 2008; Sinclair and Guarente, 1997). In addition, ScSir2 may impact the retrograde transport of toxic compounds out of the bud. ScSir2 is proposed to deacetylate the chaperonin CCT (Liu *et al.*, 2010), which folds actin. It is not clear that acetylation influences the activity of CCT, but there is evidence that in the absence of ScSir2 actin is less well folded. It is suggested that proper actin folding is required for the retrograde segregation of damaged proteins into mother cell. Sir2 has also been proposed to influence ageing in *C. albicans* (Fu *et al.*, 2008), as well as a number of animal species. Moreover, other sirtuins from *S. cerevisiae* have also been implicated in ageing (Hachinohe *et al.*, 2011; Lamming *et al.*, 2005).

Functions of sirtuins are evolving

A survey of sirtuins reveals substantial differences in the functions of orthologous sirtuins from different species. This observation raises the questions of how and why sirtuin functions shift over evolutionary time. The mechanism by which functions shift is likely related to changes in the set of deacetylated substrates. For sirtuins that deacetylate histones, their functions can also be changed by being directed to new genomic locations. Although the substrate specificity of sirtuins is determined in part by the sequence context of the lysine being deacylated, a more significant factor may be their association with adaptor proteins that direct sirtuins to particular protein substrates or genomic locations. Therefore, the set of adaptor proteins with which a sirtuin interacts is important in determining its biological function. Furthermore, if a sirtuin gains or loses interactions with targeting proteins, its substrates and hence its biological function will shift (Fig. 3.5). As to why the biological functions of sirtuins shift, we hypothesize that it may be evolutionarily advantageous for species to bring new cellular processes under the control of a sirtuin and thereby establish novel responses to stressful conditions that reduce the activity of NAD^+-dependent sirtuins.

Three processes that drive evolutionary changes in sirtuin function are: shifts in the set of deacetylated substrates, duplication and then subfunctionalization of sirtuin genes, and loss

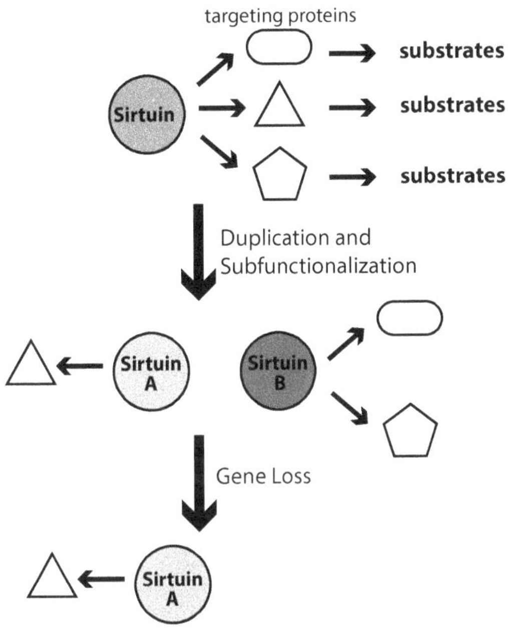

Figure 3.5 Step-wise changes in sirtuin functions. A multifunctional sirtuin interacts with several adaptor proteins that target the sirtuin to a variety of substrates. Fortuitous gene duplication provides the opportunity for subfunctionalization, such that each sirtuin paralogue interacts with a subset of the adaptor proteins. Loss of the gene encoding one paralogue reduces the range of sirtuin functions.

of sirtuin genes. Small shifts in substrates may fine tune the way in which a sirtuin regulates a particular biological process. For example, two orthologues of a sirtuin that represses gene expression may act upon different but overlapping gene sets in related species. If both gene sets are associated with the same biological processes, the biological function of the sirtuin is conserved. Nevertheless, because the regulated gene sets are not identical, each species will display a distinct response to the loss of sirtuin-mediated repression. In contrast, if the sirtuin evolves the ability to repress a new class of genes, a new biological process will come under the control of that sirtuin. Such an event could allow a species to evolve a novel response to conditions that compromise sirtuin function. An example of this process, described below, was discovered by comparing the genes repressed by Sir2 in *Kluyveromyces lactis* and its orthologue Hst1 in *S. cerevisiae*.

A second evolutionary process that influences sirtuin function is gene duplication followed by subfunctionalization. If a sirtuin has multiple biological functions, these functions may be partitioned between the paralogues after duplication. Most often, this subfunctionalization results in each paralogue deacetylating a distinct set of substrates. Mechanistically, subfunctionalization can be achieved by changes in the affinities of the two paralogues for adaptor proteins. For example, in *S. cerevisiae* the paralogues Sir2 and Hst1 arose through gene duplication. Both deacetylate histones, but they are recruited to different chromosomal locations through distinct adaptor proteins. In contrast, a non-duplicated Sir2 protein from *K. lactis* interacts with both sets of targeting proteins (Hickman and Rusche, 2009). After duplication, Sir2 and Hst1 acquired complementary inactivating mutations that reduced the affinity of each paralogue for a subset of interacting proteins (Froyd and Rusche, 2011). After subfunctionalization has occurred, the uncoupling of functions may enable one or both paralogues to evolve new properties that were not compatible with the multifunctional protein.

A third mechanism for evolutionary change is the loss of a sirtuin gene. In this case, the functions of that sirtuin are lost, unless another deacetylase has an overlapping function. This type of event appears to be relatively frequent in fungi, as few species have orthologues from all of the sirtuin subfamilies. Several examples of gene loss can be observed among budding yeasts. Most yeast in this lineage encode the mitochondrial sirtuin SirT5. However, those species, including *S. cerevisiae*, that descended from a whole genome duplication lack SirT5 (Byrne and Wolfe, 2005). Interestingly, the loss of SirT5 correlates with an increased reliance on fermentation and a reduced dependence on mitochondrial functions. A second example is the loss of the *SIR2* gene in the *Candida* lineage, which is described below for *Candida lusitaniae*.

Taken together, these three processes result in step-wise shifts in the biological functions of sirtuins (Fig. 3.5). First, a sirtuin acquires additional targets of deacetylation, perhaps by gaining an interaction with an adaptor protein. As this process is repeated, the sirtuin becomes multifunctional. Eventually, the gene encoding this sirtuin may become duplicated, allowing for subfunctionalization. Finally, one paralogue may be lost. Ultimately, the remaining paralogue, although still a deacetylase, may have a completely different set of targets and biological functions than the ancestral gene.

Subfamilies of sirtuins

Phylogenetic analyses reveal that sirtuins can be divided into subfamilies that diverged early in the eukaryotic lineage (Frye, 2000; Greiss and Gartner, 2009). The seven mammalian sirtuins represent these distinct subfamilies, although SirT2 and SirT3 are part of the same subfamily and SirT6 and SirT7 represent recently diverged subfamilies. Fungal sirtuins fall into these same subfamilies. However, Hst3 and Hst4 form an additional fungal-specific subfamily not found in animals. Although fungi encode orthologues from all the subfamilies, many fungal species possess a reduced set of sirtuins (Table 3.1), presumably due to gene loss. For this discussion, we have named the three subfamilies best characterized in fungi after their *S. cerevisiae* orthologues. The other three subfamilies are lacking in *S. cerevisiae* and are named for their mammalian orthologues.

Sir2/Hst1 subfamily

The Sir2/Hst1 subfamily of proteins is orthologous to mammalian SirT1, and these proteins regulate chromatin structure and gene expression through the deacetylation of histones and transcription factors. Sir2/Hst1 proteins are particularly implicated in deacetylating H4K16 and H3K9 residues in the histone tails (Fig. 3.6) (Alper *et al.*, 2013; Imai *et al.*, 2000). It is proposed that positively charged lysines in the histone tails interact with the negatively charged DNA backbone. This electrostatic interaction creates a condensed, transcriptionally inactive chromatin structure. In contrast, acetylated lysines are neutral and interact weakly with the DNA. This loosely packed chromatin is found at actively transcribed regions of DNA. Thus, the deacetylase activity of Sir2/Hst1 proteins promotes transcriptional repression.

Although deacetylation of histones is a conserved property of fungal Sir2/Hst1 family members, the genomic locations at which these proteins act are variable. Genomic regions that are targeted by Sir2/Hst1 orthologues include subtelomeric and pericentromeric regions, cryptic mating-type loci, the promoters of specific genes, and the tandemly repeated rDNA genes. At these loci, there are three main ways in which Sir2/Hst1 proteins act: formation of an extended heterochromatic domain, local repression of a specific promoter, and suppression of recombination.

Sir2/Hst1 proteins promote heterochromatin formation in subtelomeric and pericentromeric regions as well as at cryptic mating-type loci. This heterochromatin stabilizes telomeres and reinforces points of tension at centromeres. In addition, in many fungal species heterochromatin represses transcription of subtelomeric gene families encoding cell surface adhesins and transporters required for adaptation to different environments. Sir2/Hst1 proteins also silence the cryptic mating-type loci in yeasts, thus maintaining cell-type identity and haploid cell characteristics (Freeman-Cook *et al.*, 2005; Haber, 1998; Rine and Herskowitz, 1987). In *S. cerevisiae* and related species, heterochromatin formation occurs when Sir2 deacetylates H4K16 to promote the binding of the heterochromatin protein Sir3 to nucleosomes. In contrast, in *S. pombe* Sir2 deacetylates H3K9 in preparation for its methylation and recognition by heterochromatin proteins such as Swi6.

Sir2/Hst1 proteins also repress transcription through a local, promoter-specific mechanism. For example, in *S. cerevisiae* Hst1 is recruited to target genes through the DNA-binding Sum1 repressor (Xie *et al.*, 1999). The genes repressed by Sir2/Hst1 proteins include those that maintain NAD^+ homeostasis, promote life-cycle transitions such as mating and sporulation, and respond to environmental stresses.

Sir2/Hst1 proteins also suppress unequal mitotic and meiotic recombination at the tandem rDNA arrays (Gottlieb and Esposito, 1989; Smith and Boeke, 1997). In many fungi, the rDNA locus is comprised of 100–200 identical tandem repeats, each comprising two genes separated by non-transcribed spacers. The short gene, transcribed by polymerase III, encodes the 5.8S ribosomal RNA, and the long gene, transcribed by polymerase I, encodes the remaining

Table 3.1 Fungal sirtuins

Scer name	Sir2/Hst1	Hst2	Hst3/4			
Human name	SirT1	SirT2/3		SirT4	SirT5	SirT6/7
S. cerevisiae	Sir2 (YDL042C)	Hst2 (YPL015C)	Hst3 (YOR025W)			
	Hst1 (YOL068C)		Hst4 (YDR191W)			
C. glabrata	Sir2 (CAGL0K01463g)	Hst2 (CAGL0L08668g)	Hst3 (CAGL0H08239g)			
	Hst1 (CAGL0C05357g)		Hst4 (CAGL0F05621g)			
K. lactis	Sir2 (KLLA0F14663g)	Hst2 (KLLA0F11033g)	Hst3 (KLLA0A07172g)			
			Hst4 (KLLA0D18535g)			
C. albicans	Sir2 (orf19.1992)	Hst2 (orf19.2580)	Hst3 (orf19.1934)		orf19.2963	
	Hst1 (orf19.4761)					
C. lusitaniae	Hst1 (CLUG_01277)	CLUG_04582	CLUG_04799		CLUG_02628	
A. nidulans	sirA (AN10449)	AN7461	AN1226	hstA (AN11067)	AN1782	
		AN11873				
N. crassa	nst-1 (NCU04737.7)	nst-2 (NCU00523.7)	nst-3 (NCU03059.7)	nst-5 (NCU00203.7)	nst-6 (NCU05973.7)	nst-7 (NCU07624.7)
			nst-4 (NCU04859.7)			
S. pombe	sir2 (SPBC16D10.07c)	hst2 (SPCC132.02)	hst4 (SPAC1783.04c)			
C. neoformans	CNAG_04866	CNAG_07712	CNAG_00343	CNAG_03170		
			CNAG_02085			

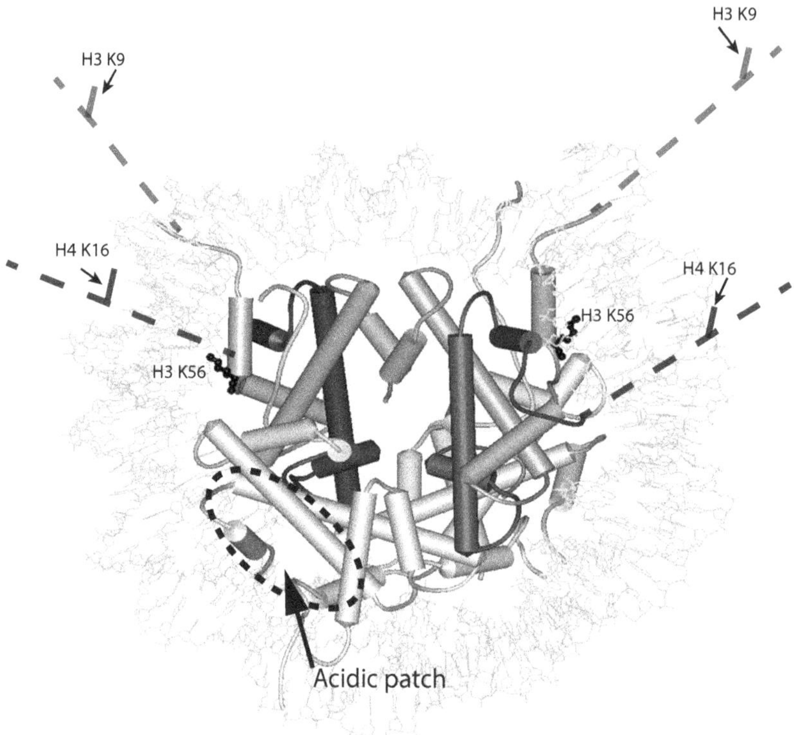

Figure 3.6 Positions of histone lysines targeted by sirtuins. Lysines H4K16 and H3K9 are deacetylated by Sir2/Hst1 proteins, lysine H4K16 is deacetylated by Hst2 proteins, and lysine H3K56 is deacetylated by Hst3/4 proteins. The dashed oval indicates an acidic patch on H2A proposed to interact with the H4 amino-terminal tail to stabilize internucleosome interactions. This model depicts a nucleosome composed of yeast histones bound to DNA and was generated from pdb file 1ID3.

ribosomal RNAs. Recombination between identical rDNA repeats on sister chromatids can result in unequal exchange, altering the copy numbers on the two chromatids. Alternatively, recombination between rDNA repeats on the same chromatid creates extrachromosomal rDNA circles. These circles accumulate in mother cells, and are thought to contribute to senescence. Sir2/Hst1 proteins can suppress recombination through the recruitment of cohesin, which keeps the sister chromatids aligned, thus minimizing unequal sister chromatid exchange and intrachromosomal recombination among repeats (Huang *et al.*, 2006). Sir2 stabilizes cohesin at the rDNA by repressing transcription from a bidirectional RNA polymerase II promoter whose activity would otherwise displace cohesin (Kobayashi and Ganley, 2005). The presence of Sir2 at the rDNA also leads to the repression of pol II transcribed reporter genes inserted into the tandem rDNA, a phenomena termed rDNA silencing. Suppression of recombination and rDNA silencing require the RENT complex (regulator of nucleolar silencing and telophase exit). This complex contains Sir2, Net1 and Cdc14 (Shou *et al.*, 1999; Straight *et al.*, 1999). Net1 recruits Sir2 to the non-transcribed spacer regions of the rDNA.

Although fungal Sir2/Hst1 proteins are largely described as histone deacetylases, they likely have other important substrates that are not yet identified. After all, the mammalian orthologue SirT1 has many documented non-histone substrates, and ScSir2 itself is known to deacetylate Ifh1 and Pck1, as described above.

The Sir2/Hst1 family has undergone several duplications in the budding yeast lineage, which have reshuffled the functions of these proteins (Fig. 3.7). One duplication coincided with a whole genome duplication that occurred in the *Saccharomyces* clade, giving rise to Sir2 and

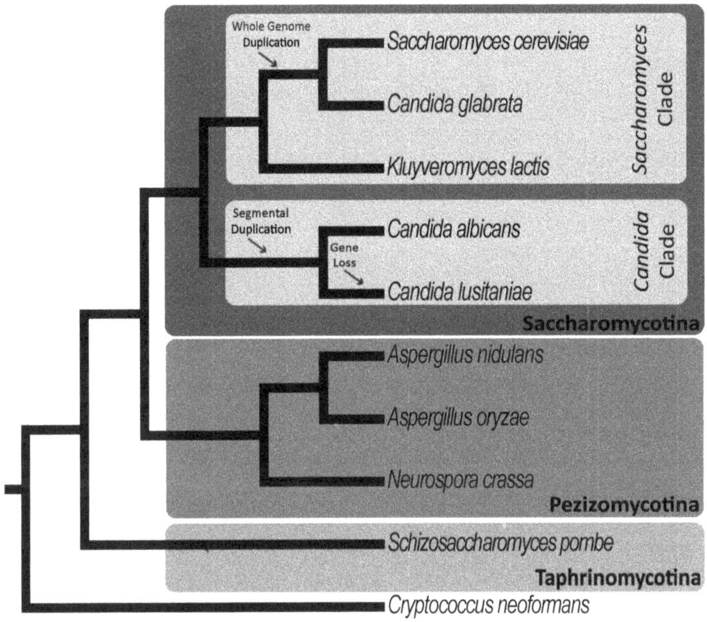

Figure 3.7 Fungal species discussed in this chapter. A cladogram represents the relative relationships of fungal species. The budding yeast clades *Saccharomyces* and *Candida* both belong to the subphylum Saccharomycotina, formerly known as Hemiascomycetes. The filamentous fungi belong to the subphylum Pezizomycotina, formerly known as Euascomycetes. The fission yeast belong to the subphylum Taphrinomycotina, formerly known as Archiascomycetes. All the species shown are members of the phylum Ascomycota, with the exception of *C. neoformans*, which is a Basidiomycete. Two duplications of *SIR2* are indicated, as well as the loss of one duplicate in the *C. lusitaniae* lineage.

Hst1 (Byrne and Wolfe, 2005). There was also a segmental duplication in the *Candida* clade. Unfortunately, the resulting paralogues are also named Sir2 and Hst1, despite the lack of orthology with the *Saccharomyces* sirtuins of the same name. In both cases, the functions of the ancestral protein were partitioned between the two paralogues through subfunctionalization. Subsequently, *SIR2* was lost in some branches of the *Candida* clade but not others (Maguire *et al.*, 2013). Thus, Sir2/Hst1 proteins have adapted and redistributed functions as genes have been duplicated and lost.

Hst2 subfamily

The Hst2 subfamily is orthologous to mammalian SirT2. Proteins in this subfamily are stable and have robust enzymatic activity *in vitro*. Consequently, these proteins have been widely used for structural and enzymatic studies. However, the *in vivo* functions of Hst2 orthologues in fungi have not been extensively examined, and hence the cellular substrates and biological consequences of deacetylation remain poorly understood.

One distinctive feature of Hst2 orthologues compared to other sirtuins is that they shuttle between the cytoplasm and the nucleus (Durand-Dubief *et al.*, 2007; Vaquero *et al.*, 2006; Wilson *et al.*, 2006). Nevertheless, most of the characterized functions of Hst2 proteins are nuclear, and H4K16 has been identified as a preferred substrate (Vaquero *et al.*, 2006). It should be noted that whereas mammalian SirT2 and fungal Hst2 proteins have similar subcellular localization patterns, the related SirT3 proteins are mitochondrial (Michishita *et al.*, 2005). SirT3 proteins are unique to animals and likely arose through a duplication of SirT2 (Greiss and Gartner, 2009). Thus, the ancestral SirT2/SirT3 may also have had mitochondrial functions, and an open question is whether any fungal Hst2 proteins share this trait and are mitochondrial.

Hst2 proteins deacetylate histones both genome-wide and at specific locations. The genome-wide deacetylation of H4K16 occurs specifically in M phase (Wilkins *et al.*, 2014). This deacetylation is proposed to promote the condensation of chromosomes by enabling the unacetylated, positively charged tail of histone H4, including K16, to interact with a negatively charged patch on histone H2A of a neighbouring nucleosome (Fig. 3.6). The recruitment of Hst2 to nucleosomes requires the phosphorylation of H3S10 (Wilkins *et al.*, 2014), which is catalysed specifically in M phase by the Aurora B kinase. Hst2 is also reported to repress particular genes in both *S. cerevisiae* (Halme *et al.*, 2004) and *S. pombe* (Durand-Dubief *et al.*, 2007), but it is not known how it is targeted to these genes.

Hst3/Hst4 subfamily

Sirtuins of the Hst3/Hst4 subfamily are found only in fungi and have evolved to deacetylate histone H3K56 throughout the genome (Celic *et al.*, 2006; Hachinohe *et al.*, 2011; Maas *et al.*, 2006; Thaminy *et al.*, 2007). Lysine 56 is located in the amino-terminal alpha helix (αN) of histone H3, which contacts DNA at the nucleosome entry and exit points (Fig. 3.6). It is proposed that acetylation of lysine 56 reduces the affinity of the αN helix for DNA, thereby destabilizing the nucleosome. H3K56 is acetylated by Rtt109 on newly synthesized histones prior to deposition in S phase (Han *et al.*, 2007; Tsubota *et al.*, 2007), and this lysine is deacetylated by Hst3 and Hst4 in G2/M phase after the histones are assembled into chromatin (Celic *et al.*, 2006; Maas *et al.*, 2006; Masumoto *et al.*, 2005). H3K56 acetylation is thought to promote chromatin assembly in the wake of DNA replication by increasing the affinity of histones for chaperones such as Rtt106 (Li *et al.*, 2008). In addition, H3K56 acetylation facilitates repair of replication-coupled DNA damage by stabilizing stalled replication forks and promoting sister chromatid recombination (Kadyrova *et al.*, 2013; Munoz-Galvan *et al.*, 2013). However, the continued presence of acetylation of H3K56 throughout the cell cycle results in spontaneous gross chromosomal rearrangements and other mutations (Kadyrova *et al.*, 2013). Consequently, loss of Hst3 and Hst4 results in sensitivity to DNA damage and genomic instability.

The expression of Hst3 and Hst4 fluctuates with the cell cycle. The genes are transcribed in G2 (Spellman *et al.*, 1998), and the proteins are degraded in M phase (Delgoshaie *et al.*, 2014; Edenberg *et al.*, 2014). Degradation occurs when degron sequences within the proteins are phosphorylated by cell cycle-regulated kinases and consequently attract ubiquitin ligases, resulting in polyubiquitylation and ultimately proteasome-mediated degradation. In addition, the degradation of Hst3 is accelerated by activation of the DNA damage checkpoint (Edenberg *et al.*, 2014), thereby causing H3K56 acetylation to persist and allow reassembly of nucleosomes after DNA repair has occurred.

The functions of Hst3 and Hst4 have been characterized primarily in *S. cerevisiae*. However, orthologues in *S. pombe* and *C. albicans* also deacetylate H3K56 and contribute to genome stability (Haldar and Kamakaka, 2008; Wurtele *et al.*, 2010), indicating that this function is conserved. It is interesting to note that many fungal species, including *S. pombe* and *C. albicans*, encode a single member of the Hst3/Hst4 subfamily, whereas some lineages encode two paralogues that apparently arose through gene duplications (Table 3.1). For example, in the *Saccharomyces* clade that includes *S. cerevisiae* and *K. lactis*, all the sequenced species encode both Hst3 and Hst4 (Byrne and Wolfe, 2005). This observation suggests that these two paralogues have distinct properties and that it is advantageous for yeast to retain both.

Other subfamilies

Most characterization of fungal sirtuins has been conducted in *S. cerevisiae* or related species, and therefore little is known about the functions of fungal sirtuins lacking orthologues in *S. cerevisiae*. However, the SirT4, SirT5, and SirT6/7 subfamilies are all represented in at least some fungi (Table 3.1), and it will be important to determine how the biological functions are partitioned among sirtuins in these species with a larger spectrum of sirtuins. For example, in mammals, one function of SirT6 is to establish heterochromatin at telomeres (Michishita *et al.*, 2008), which is similar to a function of Sir2 in some budding yeast. Similarly, SirT7 also has functions overlapping with ScSir2, as it acts on chromatin within the rDNA repeats and influences polymerase I transcription (Ford *et al., 2006; Grob et al., 2009). An important question is whether fungal SirT6/7 orthologues have similar functions, and if so how these functions overlap with those of Sir2/Hst1 orthologues in species that express both. Another example of functions that have yet to be investigated concerns the fungal SirT4 and SirT5 orthologues. In mammals, SirT4 and SirT5 are localized to mitochondria. SirT4 regulates β-oxidation of fatty acids, glutamine metabolism and ATP homeostasis (Haigis *et al.*, 2006; Ho *et al.*, 2013; Laurent *et al.*, 2013). SirT5 acts as a desuccinylase and demalonylase (Du *et al.*, 2011). An important question is whether the fungal orthologues are also mitochondrial, and if so, how they contribute to the regulation of fungal metabolism.

Contributions of sirtuins to fungal biology

Fungal species differ from one another in how they respond to low nutrients and other stresses and in the organization of their sexual cycles. We hypothesize that sirtuins may contribute to speciation by allowing organisms to evolve novel responses to stressful conditions that reduce NAD^+ availability. Bringing new cellular processes under the control of a sirtuin would enable organisms to activate that process when sirtuin activity drops. Support for this hypothesis can be found by examining the types of processes that are regulated by sirtuins in various fungal species.

Saccharomyces cerevisiae

S. cerevisiae, commonly called budding yeast, has been used by humans in winemaking, brewing and baking for thousands of years. It is also a popular experimental model organism and was the first eukaryote to have its genome sequenced. *S. cerevisiae* is the source of the first sirtuin described. *ScSIR2* (silent information regulator) was discovered based on its role in silencing the cryptic mating-type loci (Rine and Herskowitz, 1987). Subsequently, *S. cerevisiae* was found to encode four related genes called *HST1–4* (homologue of sir two) (Brachmann *et al.*, 1995; Derbyshire *et al.*, 1996).

ScSir2 acts at three loci, the cryptic mating-type loci (Rine and Herskowitz, 1987), the subtelomeres (Gottschling *et al.*, 1990), and the ribosomal DNA (rDNA) repeats (Gottlieb and Esposito, 1989; Smith and Boeke, 1997). At the cryptic mating-type loci and subtelomeres, ScSir2 generates heterochromatin as part of the SIR complex, containing Sir1–4 proteins. To generate heterochromatin, the Sir proteins are first recruited to the chromosome through DNA binding proteins. At the telomeres, multiple ScRap1 proteins bind the telomere repeat sequence and then recruit Sir proteins (Moretti *et al.*, 1994). At the cryptic mating-type loci, silencer sequences bind ScRap1 and the ORC (origin recognition complex), which in turn recruit Sir proteins (Triolo and Sternglanz, 1996). After recruitment, the next phase of heterochromatin assembly involves spreading of the SIR complex across the locus (Hoppe *et al.*, 2002; Rusche *et al.*, 2002). ScSir2 deacetylates nucleosomes adjacent to recruitment sites. ScSir3 then binds the newly deacetylated nucleosomes and recruits additional ScSir2. The cycle then repeats, with ScSir2 deacetylating additional nucleosomes. However, spreading is limited by boundary elements (Bi *et al.*, 1999; Donze *et al.*, 1999), competing acetyltransferases (Suka *et al.*, 2002), and the inherent instability of the SIR complex (Lynch and Rusche, 2009). Due to these limitations, protosilencers and looping of the chromatin fibre may be necessary to stabilize Sir proteins at the cryptic mating-type loci (Lebrun *et al.*, 2001; Rusche and Lynch, 2009). The loss of SIR heterochromatin leads to sterility and induction of subtelomeric genes, including adhesins and sugar transporters.

ScSir2 also acts at the rDNA, where its major function is to suppress recombination (Gottlieb and Esposito, 1989; Sinclair and Guarente, 1997), as described above. It is suggested that the accumulation of extrachromosomal rDNA circles in mother cells contributes to senescence by titrating critical chromatin proteins, such as DNA replication factors, away from chromosomes (Sinclair and Guarente, 1997). Alternatively, the instability of the rDNA locus and not the rDNA circles may lead to senescence (Kobayashi, 2008). In this model, genomic instability leads to replication stress, DNA repair foci, and chromatin reorganization that reinforce one another and delay progression of the cell cycle.

ScHst1 represses transcription in a promoter-specific fashion rather than by generating extended domains of heterochromatin. ScHst1 was first described as a repressor of middle-sporulation genes (Xie *et al.*, 1999). Subsequently, ScHst1 was found to repress over fifty genes involved in sporulation, NAD^+ biosynthesis and cell identity (Bedalov *et al.*, 2003; Li *et al.*, 2010; McCord *et al.*, 2003). ScHst1 is recruited to target promoters by the DNA binding protein Sum1 (McCord *et al.*, 2003; Rusche and Rine, 2001). ScHst1 is particularly sensitive to drops in NAD^+ concentration, and is proposed to respond by enabling the transcription of NAD^+ biosynthesis genes (Bedalov *et al.*, 2003).

Sir2 and Hst1 arose in the whole genome duplication that occurred about 100 million years ago in the *Saccharomyces* lineage. After duplication, Sir2 and Hst1 subfunctionalized by acquiring complementary inactivating mutations that reduced the affinities of the deacetylases for adaptor proteins (Froyd and Rusche, 2011; Hickman and Rusche, 2007; Hickman and Rusche, 2009; Mead *et al.*, 2007). Sir2 lost affinity for Sum1, whereas Hst1 lost affinity for Sir4 and Net1 (Froyd and Rusche, 2011). This finding is consistent with the Duplication Degeneration Complementation (DDC) model for subfunctionalization (Force *et al.*, 1999). The surfaces of Sir2 and Hst1 that interact with adaptor proteins are distinct, with the amino-terminal variable region of Sir2 interacting with Sir4 and the zinc-binding domain of Hst1 interacting with Sum (Cockell *et al.*, 2000; Hickman and Rusche, 2007; Mead *et al.*, 2007). Importantly, these same surfaces interact with the same adaptor proteins in a non-duplicated Sir2 from *K. lactis* (Froyd and Rusche, 2011), indicating that subfunctionalization was not accompanied by the acquisition of new interacting surfaces.

The remaining sirtuins from *S. cerevisiae*, Hst2–4, have been studied less extensively than ScSir2 and ScHst1. Moreover, the sirtuin subfamilies they represent have been studied primarily in *S. cerevisiae*. Therefore, the descriptions provided above for the Hst2 and Hst3/Hst4 subfamilies summarize the functions of these sirtuins in *S. cerevisiae*.

Candida glabrata

Candida glabrata is a pathogenic yeast that is one of the most frequent causes of human infections after *C. albicans*. Despite the name *Candida*, it is more closely related to *S. cerevisiae* than to *C. albicans* (Fig. 3.7). *C. glabrata* encodes five sirtuins, which are orthologous to the sirtuins in *S. cerevisiae* (Table 3.1). However, only CgSir2 and CgHst1 have been characterized experimentally. A distinctive property of *C. glabrata* is that it is auxotrophic for niacin, the precursor of NAD^+ (Li and Bao, 2007; Ma *et al.*, 2007). This requirement for environmental niacin arises because *C. glabrata* lacks the enzymes required for the biosynthesis of NAD^+ from tryptophan and instead relies solely on the salvage pathway. Consequently, when *C. glabrata* cells experience low environmental niacin, cellular NAD^+ levels fall and sirtuin activity is reduced.

Both CgSir2 and CgHst1 play a role in the pathogenicity of *C. glabrata*. These sirtuins repress genes that contribute to stress resistance and adaptation to a mammalian host, and these genes are induced when sirtuin activity is compromised in low niacin environments. In fact, common sites of *C. glabrata* infection, such as the genitourinary tract, are low in niacin, and consequently the deacetylase activities of CgSir2 and CgHst1 are reduced in these environments. Decreased CgHst1 activity results in the transcription of high-affinity niacin transporters (Ma *et al.*, 2009) and genes that contribute to resistance to antifungal drugs and hydrogen peroxide (Orta-Zavalza *et al.*, 2013). Decreased CgSir2 activity results in the loss of SIR-mediated subtelomeric silencing of adhesin genes of the EPA family (Castano *et al.*, 2005; Domergue *et al.*, 2005). These adhesins enable the yeast to attach to human epithelial tissue and to form biofilms. Thus, together CgHst1 and CgSir2 repress a set of genes that contribute to pathogenicity.

Kluyveromyces lactis

K. lactis is a budding yeast of the *Saccharomyces* clade that has important applications in biotechnology. For example, it is used to produce recombinant chymosin for cheese production. Two notable properties of *K. lactis* are that it is auxotrophic for niacin (Li and Bao, 2007) and that it diverged from *S. cerevisiae* and *C. glabrata* prior to the whole genome duplication (Fig. 3.7). Consequently, *K. lactis* has a single non-duplicated Sir2/Hst1 orthologue (Table 3.1). It also has orthologues of Hst2, Hst3, and Hst4, but their functions remain uncharacterized. Another important distinction is that mating occurs under different conditions in *K. lactis* and *S. cerevisiae*. Haploid *K. lactis* cells delay mating until nutrients become scarce (Booth *et al.*, 2010; Herman and Roman, 1966), whereas haploid *S. cerevisiae* cells mate in rich nutrient conditions.

The non-duplicated KlSir2 protein has functions similar to both ScSir2 and ScHst1 (Astrom *et al.*, 2000; Hickman and Rusche, 2009). KlSir2 acts with the SIR complex (KlSir4 and KlOrc1) to form heterochromatin at *HMLa* and subtelomeres, it acts with KlNet1 to influence chromatin at the rDNA repeats, and it acts with KlSum1 to represses genes in a promoter-specific fashion. Therefore, a multifunctional ancestral Sir2 must have subfunctionalized after the whole genome duplication, giving rise to the paralogues Sir2 and Hst1. However, there are some differences in the mechanisms by which heterochromatin forms in *K. lactis* compared to *S. cerevisiae*. For example, in *K. lactis* the *HMRa* locus is repressed by the Sum1–Sir2 complex rather than the SIR complex, and remarkably the Sum1–Sir2 complex appears to spread at this locus (Hickman and Rusche, 2009). In addition, at the *HMLa* locus, both the SIR and Sum1 complexes contribute to silencing (Hickman and Rusche, 2009). Thus, the Sum1–Sir2 complex has an expanded role in *K. lactis* compared to *S. cerevisiae*.

An ongoing study indicates that KlSir2 represses an overlapping but distinct set of genes compared to ScHst1 of CgHst1 (Humphrey, Hasan and Rusche, unpublished). KlSir2 represses genes involved in NAD^+ homeostasis, utilization of carbon sources other than glucose, and the sexual cycle, including mating and sporulation. We hypothesize that *K. lactis* has a tiered response to low NAD^+ levels, in which it first up-regulates the NAD^+ salvage pathway and adjusts its metabolic profile to utilize alternative carbon sources. If cellular NAD^+ levels remain low, cells then proceed through the sexual cycle to generate meiotic spores. Interestingly, the genes required for mating that are repressed by KlSir2 are not reported to be repressed by Hst1 or Sir2 in *S. cerevisiae*. This difference in gene expression corresponds with the alteration in the conditions under which mating occurs in the two species. Only in *K. lactis*, which mates in low nutrient conditions, is there a regulatory logic for Sir2 repressing mating genes in rich nutrient conditions.

Candida albicans

C. albicans is a diploid yeast of the *Candida* clade and is the most common cause of human yeast infections. It is a normal human commensal, found in the mouth and gastrointestinal tract. Overgrowth can result in infection of the mucous membranes or systemic infections, which can be fatal. An important property of *C. albicans* that contributes to its pathogenicity is its ability to switch among several morphological forms. Cells can grow either in budding yeast form, which can be disseminated, or filamentous form, which contributes to biofilm formation and tissue penetration. In addition, yeast cells can switch between white and opaque forms, which have different cell shapes and mating tendencies. These morphological forms are stable over multiple generations and are associated with distinct, metastable gene expression patterns. It is therefore possible that sirtuins promote the stability of particular morphological forms by maintaining repression of critical genes.

C. albicans encodes five sirtuins, representing four subfamilies (Table 3.1). As described above, a segmental duplication early in the *Candida* lineage generated two paralogues of the Sir2/Hst1 family. These two genes have unfortunately been named *SIR2* and *HST1*, despite the fact that they arose independently of the paralogues *SIR2* and *HST1* in *S. cerevisiae*. Thus, there is no reason to expect the functions of CaHst1 and ScHst1 or CaSir2 and ScSir2 to correspond.

An initial study suggested that *C. albicans* displays increased rates of phenotypic switching in the absence of CaSir2 (Perez-Martin *et al.*, 1999). However, the authors were unable to reproduce their results (personal communication), and the contribution of CaSir2 to stabilizing morphological forms has been largely disregarded in the field. However, CaHst1, CaHst2, and CaHst3 have been shown to influence the rate of switching between the white and opaque forms (Hnisz *et al.*, 2009; Stevenson and Liu, 2011). Perhaps because of this initial setback, the important question of which genomic loci are targeted by CaSir2 and its paralogue CaHst1 remains unanswered. It has been suggested based on expression of a reporter gene that CaSir2 acts at the ribosomal DNA (Fu *et al.*, 2008). However, no increase in rDNA circles occurs in the absence of CaSir2 (Fu *et al.*, 2008), inconsistent with CaSir2 repressing recombination in the rDNA repeats. CaSir2 or CaHst1 might also form heterochromatin at subtelomeres, as has been observed in other species. Indeed, CaSir2 is implicated in the expression of subtelomeric TLO genes and a URA3 reporter (Anderson et al., 2014). However, Rap1 in *C. albicans* is lacking the C-terminal domain (Yu *et al.*, 2010) that interacts with Sir proteins in *S. cerevisiae*. It is therefore unclear how CaSir2 or CaHst1 would be recruited to the telomeres in *C. albicans*. One observation is that loss of CaSir2 results in a decrease in lifespan, which has been attributed to increased oxidized proteins in the daughter cells (Fu *et al.*, 2008).

C. albicans possesses a single member of the Hst3/4 family, called CaHst3. Like its orthologues in *S. cerevisiae*, CaHst3 deacetylates H3K56 (Wurtele *et al.*, 2010). However, deletion of CaHst3 is lethal, indicating that high levels of H3K56 acetylation are toxic to *C. albicans*. Given this susceptibility and the fact that the Hst3/4 subfamily is unique to fungi, it is proposed that new antifungal drugs could be developed by targeting Hst3 (Wurtele *et al.*, 2010). Another function of CaHst3 that is specific to *C. albicans* is a role in the transition between white and opaque cells (Stevenson and Liu, 2011). The rate of switching from white to opaque cells is much higher in the absence of CaHst3, and it is proposed that increased H3K56 acetylation at the promoter of the master regulator *WOR1*, promotes its expression and hence the opaque state (Stevenson and Liu, 2013).

Candida lusitaniae

C. lusitaniae is a haploid yeast of the *Candida* clade that occasionally causes human infections, particularly in immunocompromised individuals. *C. lusitaniae* is notable for developing resistance to antifungal drugs, such as amphotericin B. *C. lusitaniae* encodes four sirtuins (Table 3.1), of which only the Sir2/Hst1 orthologue has been characterized. An important point is that *C. lusitaniae* is missing a Sir2/Hst1 orthologue compared to *C. albicans* (Table 3.1), apparently due to gene loss. The remaining Sir2/Hst1 protein is called ClHst1, based on its relationship to the *C. albicans* Hst1 (Maguire *et al.*, 2013).

Surprisingly, ClHst1 does not associate with subtelomeric regions or impact gene expression directly (Froyd *et al.*, 2013). Moreover, ClHst1 does not associate with centromeres (Kapoor and Rusche, unpublished), which might also be heterochromatic. Rather, ClHst1 is localized to the non-transcribed spacer regions of the rDNA repeats and preferentially deacetylates H3K9 (Froyd *et al.*, 2013). This observation reveals that *C. lusitaniae* is lacking known forms of heterochromatin and raises the question of how this organism maintains genome stability in the apparent absence of heterochromatin. It could be that *C. lusitaniae* harbours a unique alternative heterochromatic structure or that it employs alternative mechanisms to maintain genome stability.

The observation that ClHst1 is restricted to the rDNA repeats implies that maintenance of rDNA stability may be a more critical function of Sir2/Hst1 proteins than formation of heterochromatin or gene repression. It also suggests that the orthologous Hst1 protein in *C. albicans* may act at the rDNA. If so, CaSir2 may be responsible for generating heterochromatin at telomeres or centromeres.

Schizosaccharomyces pombe

S. pombe, or fission yeast, has served as a laboratory model organism since the 1950s. *S. pombe* cells are cylindrical, grow from the two tips, and divide by medial fission. Research on *S. pombe* has led to significant advances in the understanding of cell cycle control and chromosome biology in particular. *S. pombe* encodes only three sirtuins (Table 3.1), which are from the same subfamilies as the sirtuins in *S. cerevisiae*.

SpSir2 is primarily nuclear and is involved in the formation of heterochromatin at pericentromeric domains, subtelomeric regions and cryptic mating type loci (Alper *et al.*, 2013;

Buscaino *et al.*, 2013; Freeman-Cook *et al.*, 2005; Shankaranarayana *et al.*, 2003). The mechanism of heterochromatin formation in fission yeast differs from silencing by the SIR complex in *S. cerevisiae*. In *S. pombe*, methylation of H3K9 is a hallmark of heterochromatin, and this modification is specifically bound by heterochromatin proteins, such as Swi6. Importantly, a critical target of SpSir2 *in vivo* appears to be H3K9, and deacetylation of H3K9 is thought to promote the subsequent methylation of this lysine, which is then recognized by heterochromatin proteins (Shankaranarayana *et al.*, 2003). In addition, SpSir2 deacetylates H3K14, and this deacetylation is necessary for recognition of H3K9 by the methyltransferase Clr4 (Alper *et al.*, 2013). As in *S. cerevisiae*, SpSir2 also acts at the rDNA repeats (Wiren *et al.*, 2005), but this function remains poorly characterized. It is also striking that promoter specific repression by SpSir2 has not yet been documented suggesting that stabilization of the rDNA repeats and heterochromatin formation are the ancient functions of Sir2.

SpHst2 likely shares properties with ScHst2, since it also localizes both to the cytoplasm and nucleus (Durand-Dubief *et al.*, 2007). SpHst2 has been shown through genome-wide studies to represses genes involved in transport and membrane function (Durand-Dubief *et al.*, 2007). It also contributes modestly to heterochromatic silencing at the rDNA and centromeres.

SpHst4 is the only orthologue of *S. cerevisiae* Hst3 and Hst4. Like other subfamily members, it deacetylates H3K56ac and is involved in repair of DNA damage during S phase (Freeman-Cook *et al.*, 1999; Haldar and Kamakaka, 2008). It was also shown that the SpHst4 protein is initially decreased upon exposure to the DNA damaging agent H_2O_2, resulting in increased H3K56 acetylation to support repair (Chang *et al.*, 2011). However, SpHst4 is ultimately recruited to sites of repair to restore the deacetylation of H3K56. SpHst4 is concentrated in the nucleolus (Freeman-Cook *et al.*, 1999), but no function in the rDNA has been described. SpHst4 is reported to contribute to heterochromatic silencing at centromeres and telomeres but not at the mating-type loci (Freeman-Cook *et al.*, 1999). In addition, a role of SpHst4 in Tf2 5′ mRNA processing and repression of genes involved in amino acid biosynthesis and oxidoreductase activity was revealed in genome-wide studies (Durand-Dubief *et al.*, 2007).

Aspergillus species

The genus *Aspergillus* is comprised of several hundred species of moulds. Among these species, *Aspergillus nidulans* is an important laboratory organism, and *Aspergillus oryzae* is used to ferment soybeans and rice. Both species encode six sirtuins, representing all subfamilies except SirT6/7. *Aspergillus* species are well known for producing secondary metabolites in response to starvation and stress. These small, bioactive compounds are not essential for viability but modulate the mould's environment by influencing the behaviour of competing species and predators. Some secondary metabolites have been exploited by humans as medicines, cosmetics, or flavour enhancers, whereas others are toxic or carcinogenic. It is therefore of interest to understand how the production of secondary metabolites is regulated. One important observation is that biosynthetic genes for secondary metabolites are generally organized into clusters of genes acting in a common pathway. This organization leads to the hypothesis that the genes are co-regulated by a regional chromatin structure, and some studies have implicated sirtuins as repressors of biosynthetic genes.

In *A. nidulans*, SirA (Sir2/Hst1 subfamily) localizes to the promoters of genes involved in production of the secondary metabolites penicillin and sterigmatocystin, and these genes are induced in the absence of SirA (Shimizu *et al.*, 2012). Interestingly, penicillin and sterigmatocystin are produced during stationary phase when NAD^+ levels are lower (Shimizu *et al.*, 2012), leading to the model that changes in cellular NAD^+ concentrations influence SirA activity and hence secondary metabolite production. Another study in *A. nidulans* found that deletion of HstA (SirT4 subfamily) in combination with non-NAD^+-dependent deacetylases had a modest effect on expression of secondary metabolite producing genes (Shwab *et al.*, 2007). In *Aspergillus oryzae*, microarray analyses revealed that genes involved in secondary metabolite production and detoxification are induced in the absence of the sirtuin HstD (Hst3/4 subfamily) (Kawauchi *et al.*, 2013). Of particular note was the induction of LaeA, a global activator of secondary metabolite production. These results suggest that HstD acts as a promoter-specific repressor of a large number of genes, but chromatin IP assays are needed to determine whether HstD regulates these genes directly. Overall, these studies suggest that in *Aspergillus* species, coordinating secondary metabolite production with nutrient availability is achieved in part by bringing the process under the control of NAD^+-dependent sirtuins.

Neurospora crassa

Neurospora crassa is a filamentous fungus that has been widely used as a model organism. In the environment, it grows on plant matter after fires and on bread. The genome of *N. crassa* encodes seven sirtuins, which have been named Nst1 –Nst7 and represent all subfamilies (Table 3.1). Four of these sirtuins have been mutated, and the impact of these mutations on silencing of a subtelomeric reporter gene has been assessed (Smith *et al.*, 2008). Mutation of Nst3, which belongs to the Hst3/4 family, had the greatest impact on subtelomeric silencing, and a modest loss of silencing was also observed upon mutation of Nst1 (Sir2/Hst1 subfamily) or Nst5 (SirT4 subfamily). Thus, a role for sirtuin proteins at subtelomeres is conserved in *N. crassa*. Distinguishing the functions of the seven sirtuin deacetylases will require genome-wide studies to map the distributions of sirtuins along chromosomes and to identify genes whose transcription is regulated by sirtuins.

Conclusion

Research on fungal sirtuins reveals that these deacetylases often repress genes involved in responding to low nutrients or other stresses. Presumably, this repression is relieved in stressful conditions in part due to reduced NAD^+ availability. These observations are consistent with the hypothesis that sirtuins allow organisms to evolve novel responses to stresses. However, validation of this hypothesis will require additional comparative studies on the biological functions of sirtuins in a variety of species. In addition, it will be important to establish the conditions under which NAD^+ levels are reduced, and to correlate these reductions with sirtuin activity.

Fungal sirtuins are also critical in maintaining genome stability in the rDNA repeats and during chromosome replication and segregation. It is not clear whether any biological advantage is derived from utilizing NAD^+-dependent deacetylases for maintaining genome stability. Perhaps a transient period of genome rearrangement upon stress creates opportunities for novel mutations that confer stress resistance.

Finally, this survey of the literature reveals that few non-histone substrates of fungal sirtuins have been described. Nevertheless, it seems highly likely that such substrates exist, and it will be important to identify these substrates.

Acknowledgement

Research in the Rusche lab is supported by a grant from the National Science Foundation (MCB-1306862).

References

Alper, B.J., Job, G., Yadav, R.K., Shanker, S., Lowe, B.R., and Partridge, J.F. (2013). Sir2 is required for Clr4 to initiate centromeric heterochromatin assembly in fission yeast. EMBO J. 32, 2321–2335.

Anderson, R.M., Bitterman, K.J., Wood, J.G., Medvedik, O., Cohen, H., Lin, S.S., Manchester, J.K., Gordon, J.I., and Sinclair, D.A. (2002). Manipulation of a nuclear NAD+ salvage pathway delays aging without altering steady-state NAD+ levels. J. Biol. Chem. *277*, 18881–18890.

Anderson, M.Z., Gerstein, A.C., Wigen, L., Baller, J.A., and Berman, J. (2014). Silencing is noisy: population and cell level noise in telomere-adjacent genes is dependent on telomere position and sir2. PLoS Genet. *10*, e1004436.

Astrom, S.U., Kegel, A., Sjostrand, J.O., and Rine, J. (2000). *Kluyveromyces lactis* Sir2p regulates cation sensitivity and maintains a specialized chromatin structure at the cryptic alpha-locus. Genetics *156*, 81–91.

Avalos, J.L., Celic, I., Muhammad, S., Cosgrove, M.S., Boeke, J.D., and Wolberger, C. (2002). Structure of a Sir2 enzyme bound to an acetylated p53 peptide. Mol. Cell *10*, 523–535.

Avalos, J.L., Boeke, J.D., and Wolberger, C. (2004). Structural basis for the mechanism and regulation of Sir2 enzymes. Mol. Cell *13*, 639–648.

Bedalov, A., Hirao, M., Posakony, J., Nelson, M., and Simon, J.A. (2003). NAD^+-dependent deacetylase Hst1p controls biosynthesis and cellular NAD^+ levels in *Saccharomyces cerevisiae*. Mol. Cell. Biol. 23, 7044–7054.

Bi, X., Braunstein, M., Shei, G.J., and Broach, J.R. (1999). The yeast HML I silencer defines a heterochromatin domain boundary by directional establishment of silencing. Proc. Natl. Acad. Sci. U.S.A. *96*, 11934–11939.

Booth, L.N., Tuch, B.B., and Johnson, A.D. (2010). Intercalation of a new tier of transcription regulation into an ancient circuit. Nature *468*, 959–963.

Brachmann, C.B., Sherman, J.M., Devine, S.E., Cameron, E.E., Pillus, L., and Boeke, J.D. (1995). The *SIR2* gene family, conserved from bacteria to humans, functions in silencing, cell cycle progression, and chromosome stability. Genes Dev. *9*, 2888–2902.

Buscaino, A., Lejeune, E., Audergon, P., Hamilton, G., Pidoux, A., and Allshire, R.C. (2013). Distinct roles for Sir2 and RNAi in centromeric heterochromatin nucleation, spreading and maintenance. EMBO J. *32*, 1250–1264.

Byrne, K.P., and Wolfe, K.H. (2005). The Yeast Gene Order Browser: combining curated homology and syntenic context reveals gene fate in polyploid species. Genome Res. *15*, 1456–1461.

Castano, I., Pan, S.J., Zupancic, M., Hennequin, C., Dujon, B., and Cormack, B.P. (2005). Telomere length control and transcriptional regulation of subtelomeric adhesins in *Candida glabrata*. Mol Microbiol *55*, 1246–1258.

Celic, I., Masumoto, H., Griffith, W.P., Meluh, P., Cotter, R.J., Boeke, J.D., and Verreault, A. (2006). The sirtuins Hst3 and Hst4p preserve genome integrity by controlling histone H3 lysine 56 deacetylation. Current Biology *16*, 1280–1289.

Chadwick, B.P., Valley, C.M., and Willard, H.F. (2001). Histone variant macroH2A contains two distinct macrochromatin domains capable of directing macroH2A to the inactive X-chromosome. Nucleic Acids Res. *29*, 2699–2705.

Chakrabarty, S.P., and Balaram, H. (2010). Reversible binding of zinc in *Plasmodium falciparum* Sir2: structure and activity of the apoenzyme. Biochim. Biophys. Acta *1804*, 1743–1750.

Chang, D.Y., Shi, G., Durand-Dubief, M., Ekwall, K., and Lu, A.L. (2011). The role of MutY homolog (Myh1) in controlling the histone deacetylase Hst4 in the fission yeast *Schizosaccharomyces pombe*. J. Mol. Biol. *405*, 653–665.

Chen, D., Vollmar, M., Rossi, M.N., Phillips, C., Kraehenbuehl, R., Slade, D., Mehrotra, P.V., von Delft, F., Crosthwaite, S.K., Gileadi, O., *et al.* (2011). Identification of macrodomain proteins as novel O-acetyl-ADP-ribose deacetylases. J. Biol. Chem. *286*, 13261–13271.

Chou, C.C., Li, Y.C., and Gartenberg, M.R. (2008). Bypassing Sir2 and O-acetyl-ADP-ribose in transcriptional silencing. Mol. Cell *31*, 650–659.

Cockell, M.M., Perrod, S., and Gasser, S.M. (2000). Analysis of Sir2p domains required for rDNA and telomeric silencing in *Saccharomyces cerevisiae*. Genetics *154*, 1069–1083.

Cosgrove, M.S., Bever, K., Avalos, J.L., Muhammad, S., Zhang, X., and Wolberger, C. (2006). The structural basis of sirtuin substrate affinity. Biochemistry *45*, 7511–7521.

Cubizolles, F., Martino, F., Perrod, S., and Gasser, S.M. (2006). A homotrimer-heterotrimer switch in Sir2 structure differentiates rDNA and telomeric silencing. Mol. Cell *21*, 825–836.

Delgoshaie, N., Tang, X., Kanshin, E.D., Williams, E.C., Rudner, A.D., Thibault, P., Tyers, M., and Verreault, A. (2014). Regulation of the histone deacetylase Hst3 by cyclin-dependent kinases and the ubiquitin ligase SCFCdc4. J. Biol. Chem. *289*, 13186–13196.

Derbyshire, M.K., Weinstock, K.G., and Strathern, J.N. (1996). *HST1*, a new member of the *SIR2* family of genes. Yeast *12*, 631–640.

Domergue, R., Castano, I., De Las Penas, A., Zupancic, M., Lockatell, V., Hebel, J.R., Johnson, D., and Cormack, B.P. (2005). Nicotinic acid limitation regulates silencing of *Candida* adhesins during UTI. Science *308*, 866–870.

Donze, D., Adams, C.R., Rine, J., and Kamakaka, R.T. (1999). The boundaries of the silenced *HMR* domain in *Saccharomyces cerevisiae*. Genes Dev. *13*, 698–708.

Downey, M., Knight, B., Vashisht, A.A., Seller, C.A., Wohlschlegel, J.A., Shore, D., and Toczyski, D.P. (2013). Gcn5 and sirtuins regulate acetylation of the ribosomal protein transcription factor Ifh1. Curr. Biol. *23*, 1638–1648.

Du, J., Zhou, Y., Su, X., Yu, J.J., Khan, S., Jiang, H., Kim, J., Woo, J., Kim, J.H., Choi, B.H., *et al.* (2011). Sirt5 is a NAD-dependent protein lysine demalonylase and desuccinylase. Science *334*, 806–809.

Durand-Dubief, M., Sinha, I., Fagerstrom-Billai, F., Bonilla, C., Wright, A., Grunstein, M., and Ekwall, K. (2007). Specific functions for the fission yeast Sirtuins Hst2 and Hst4 in gene regulation and retrotransposon silencing. EMBO J. *26*, 2477–2488.

Edenberg, E.R., Vashisht, A.A., Topacio, B.R., Wohlschlegel, J.A., and Toczyski, D.P. (2014). Hst3 is turned over by a replication stress-responsive SCF(Cdc4) phospho-degron. Proc. Natl. Acad. Sci. U.S.A. *111*, 5962–5967.

Ehrentraut, S., Hassler, M., Oppikofer, M., Kueng, S., Weber, J.M., Mueller, J.W., Gasser, S.M., Ladurner, A.G., and Ehrenhofer-Murray, A.E. (2011). Structural basis for the role of the Sir3 AAA+ domain in silencing: interaction with Sir4 and unmethylated histone H3K79. Genes Dev. *25*, 1835–1846.

Feldman, J.L., Baeza, J., and Denu, J.M. (2013). Activation of the protein deacetylase SIRT6 by long-chain fatty acids and widespread deacylation by mammalian sirtuins. J. Biol. Chem. *288*, 31350–31356.

Finnin, M.S., Donigian, J.R., Cohen, A., Richon, V.M., Rifkind, R.A., Marks, P.A., Breslow, R., and Pavletich, N.P. (1999). Structures of a histone deacetylase homologue bound to the TSA and SAHA inhibitors. Nature *401*, 188–193.

Finnin, M.S., Donigian, J.R., and Pavletich, N.P. (2001). Structure of the histone deacetylase SIRT2. Nature Struct. Biol. *8*, 621–625.

Force, A., Lynch, M., Pickett, F.B., Amores, A., Yan, Y.L., and Postlethwait, J. (1999). Preservation of duplicate genes by complementary, degenerative mutations. Genetics *151*, 1531–1545.

Ford, E., Voit, R., Liszt, G., Magin, C., Grummt, I., and Guarente, L. (2006). Mammalian Sir2 homolog SIRT7 is an activator of RNA polymerase I transcription. Genes Dev. *20*, 1075–1080.

Freeman-Cook, L.L., Sherman, J.M., Brachmann, C.B., Allshire, R.C., Boeke, J.D., and Pillus, L. (1999). The *Schizosaccharomyces pombe hst4(+)* gene is a *SIR2* homologue with silencing and centromeric functions. Mol. Biol. Cell *10*, 3171–3186.

Freeman-Cook, L.L., Gomez, E.B., Spedale, E.J., Marlett, J., Forsburg, S.L., Pillus, L., and Laurenson, P. (2005). Conserved locus-specific silencing functions of *Schizosaccharomyces pombe sir2+*. Genetics *169*, 1243–1260.

Froyd, C.A., and Rusche, L.N. (2011). The duplicated deacetylases Sir2 and Hst1 subfunctionalized by acquiring complementary inactivating mutations. Mol. Cell. Biol. *31*, 3351–3365.

Froyd, C.A., Kapoor, S., Dietrich, F., and Rusche, L.N. (2013). The deacetylase Sir2 from the yeast *Clavispora lusitaniae* lacks the evolutionarily conserved capacity to generate subtelomeric heterochromatin. PLoS Genet. *9*, e1003935.

Frye, R.A. (2000). Phylogenetic classification of prokaryotic and eukaryotic Sir2-like proteins. Biochem. Biophys. Res. Commun. *273*, 793–798.

Fu, X.H., Meng, F.L., Hu, Y., and Zhou, J.Q. (2008). *Candida albicans*, a distinctive fungal model for cellular aging study. Aging Cell *7*, 746–757.

Garske, A.L., and Denu, J.M. (2006). SIRT1 top 40 hits: use of one-bead, one-compound acetyl-peptide libraries and quantum dots to probe deacetylase specificity. Biochemistry *45*, 94–101.

Gottlieb, S., and Esposito, R.E. (1989). A new role for a yeast transcriptional silencer gene, *SIR2*, in regulation of recombination in ribosomal DNA. Cell *56*, 771–776.

Gottschling, D.E., Aparicio, O.M., Billington, B.L., and Zakian, V.A. (1990). Position effect at *S. cerevisiae* telomeres: reversible repression of Pol II transcription. Cell *63*, 751–762.

Greiss, S., and Gartner, A. (2009). Sirtuin/Sir2 phylogeny, evolutionary considerations and structural conservation. Mol. Cells *28*, 407–415.

Grob, A., Roussel, P., Wright, J.E., McStay, B., Hernandez-Verdun, D., and Sirri, V. (2009). Involvement of SIRT7 in resumption of rDNA transcription at the exit from mitosis. J. Cell Sci. *122*, 489–498.

Haber, J.E. (1998). Mating-type gene switching in *Saccharomyces cerevisiae*. Annu. Rev. Genet. *32*, 561–599.

Hachinohe, M., Hanaoka, F., and Masumoto, H. (2011). Hst3 and Hst4 histone deacetylases regulate replicative lifespan by preventing genome instability in *Saccharomyces cerevisiae*. Genes Cells *16*, 467–477.

Haigis, M.C., Mostoslavsky, R., Haigis, K.M., Fahie, K., Christodoulou, D.C., Murphy, A.J., Valenzuela, D.M., Yancopoulos, G.D., Karow, M., Blander, G., *et al.* (2006). SIRT4 inhibits glutamate dehydrogenase and opposes the effects of calorie restriction in pancreatic beta cells. Cell *126*, 941–954.

Haldar, D., and Kamakaka, R.T. (2008). *Schizosaccharomyces pombe* Hst4 functions in DNA damage response by regulating histone H3 K56 acetylation. Eukaryot. Cell *7*, 800–813.

Halme, A., Bumgarner, S., Styles, C., and Fink, G.R. (2004). Genetic and epigenetic regulation of the *FLO* gene family generates cell-surface variation in yeast. Cell *116*, 405–415.

Han, J., Zhou, H., Li, Z., Xu, R.M., and Zhang, Z. (2007). The Rtt109-Vps75 histone acetyltransferase complex acetylates non-nucleosomal histone H3. J. Biol. Chem. *282*, 14158–14164.

Herman, A., and Roman, H. (1966). Allele specific determinants of homothallism in *Saccharomyces lactis*. Genetics *53*, 727–740.

Hickman, M.A., and Rusche, L.N. (2007). Substitution as a mechanism for genetic robustness: the duplicated deacetylases Hst1p and Sir2p in *Saccharomyces cerevisiae*. PLoS Genet. *3*, e126.

Hickman, M.A., and Rusche, L.N. (2009). The Sir2–Sum1 complex represses transcription using both promoter-specific and long-range mechanisms to regulate cell identity and sexual cycle in the yeast *Kluyveromyces lactis*. PLoS Genet. *5*, e1000710.

Hnisz, D., Schwarzmuller, T., and Kuchler, K. (2009). Transcriptional loops meet chromatin: a dual-layer network controls white-opaque switching in *Candida albicans*. Mol. Microbiol. *74*, 1–15.

Ho, L., Titus, A.S., Banerjee, K.K., George, S., Lin, W., Deota, S., Saha, A.K., Nakamura, K., Gut, P., Verdin, E., *et al.* (2013). SIRT4 regulates ATP homeostasis and mediates a retrograde signaling via AMPK. Aging *5*, 835–849.

Hoppe, G.J., Tanny, J.C., Rudner, A.D., Gerber, S.A., Danaie, S., Gygi, S.P., and Moazed, D. (2002). Steps in assembly of silent chromatin in yeast: Sir3-independent binding of a Sir2/Sir4 complex to silencers and role for Sir2-dependent deacetylation. Mol. Cell. Biol. *22*, 4167–4180.

Hsu, H.C., Wang, C.L., Wang, M., Yang, N., Chen, Z., Sternglanz, R., and Xu, R.M. (2013). Structural basis for allosteric stimulation of Sir2 activity by Sir4 binding. Genes Dev. *27*, 64–73.

Huang, J., Brito, I.L., Villen, J., Gygi, S.P., Amon, A., and Moazed, D. (2006). Inhibition of homologous recombination by a cohesin-associated clamp complex recruited to the rDNA recombination enhancer. Genes Dev. *20*, 2887–2901.

Imai, S., Armstrong, C.M., Kaeberlein, M., and Guarente, L. (2000). Transcriptional silencing and longevity protein Sir2 is an NAD-dependent histone deacetylase. Nature *403*, 795–800.

Kadyrova, L.Y., Mertz, T.M., Zhang, Y., Northam, M.R., Sheng, Z., Lobachev, K.S., Shcherbakova, P.V., and Kadyrov, F.A. (2013). A reversible histone H3 acetylation cooperates with mismatch repair and replicative polymerases in maintaining genome stability. PLoS Genet. *9*, e1003899.

Kaeberlein, M., McVey, M., and Guarente, L. (1999). The SIR2/3/4 complex and SIR2 alone promote longevity in *Saccharomyces cerevisiae* by two different mechanisms. Genes Dev. *13*, 2570–2580.

Kawauchi, M., Nishiura, M., and Iwashita, K. (2013). Fungus-specific sirtuin HstD coordinates secondary metabolism and development through control of LaeA. Eukaryot. Cell *12*, 1087–1096.

Khan, A.N., and Lewis, P.N. (2005). Unstructured conformations are a substrate requirement for the Sir2 family of NAD-dependent protein deacetylases. J. Biol. Chem. *280*, 36073–36078.

Kobayashi, T. (2008). A new role of the rDNA and nucleolus in the nucleus – rDNA instability maintains genome integrity. BioEssays *30*, 267–272.

Kobayashi, T., and Ganley, A.R. (2005). Recombination regulation by transcription-induced cohesin dissociation in rDNA repeats. Science *309*, 1581–1584.

Kustatscher, G., Hothorn, M., Pugieux, C., Scheffzek, K., and Ladurner, A.G. (2005). Splicing regulates NAD metabolite binding to histone macroH2A. Nat. Struct. Mol. Biol. *12*, 624–625.

Lamming, D.W., Latorre-Esteves, M., Medvedik, O., Wong, S.N., Tsang, F.A., Wang, C., Lin, S.J., and Sinclair, D.A. (2005). HST2 mediates SIR2-independent life-span extension by calorie restriction. Science *309*, 1861–1864.

Landry, J., Slama, J.T., and Sternglanz, R. (2000). Role of NAD(+) in the deacetylase activity of the SIR2-like proteins. Biochem. Biophys. Res. Commun. *278*, 685–690.

Laurent, G., German, N.J., Saha, A.K., de Boer, V.C., Davies, M., Koves, T.R., Dephoure, N., Fischer, F., Boanca, G., Vaitheesvaran, B., *et al.* (2013). SIRT4 coordinates the balance between lipid synthesis and catabolism by repressing malonyl CoA decarboxylase. Mol. Cell *50*, 686–698.

Lebrun, E., Revardel, E., Boscheron, C., Li, R., Gilson, E., and Fourel, G. (2001). Protosilencers in *Saccharomyces cerevisiae* subtelomeric regions. Genetics *158*, 167–176.

Li, M., Petteys, B.J., McClure, J.M., Valsakumar, V., Bekiranov, S., Frank, E.L., and Smith, J.S. (2010). Thiamine biosynthesis in *Saccharomyces cerevisiae* is regulated by the NAD+-dependent histone deacetylase Hst1. Mol. Cell. Biol. *30*, 3329–3341.

Li, Q., Zhou, H., Wurtele, H., Davies, B., Horazdovsky, B., Verreault, A., and Zhang, Z. (2008). Acetylation of histone H3 lysine 56 regulates replication-coupled nucleosome assembly. Cell *134*, 244–255.

Li, Y.F., and Bao, W.G. (2007). Why do some yeast species require niacin for growth? Different modes of NAD synthesis. FEMS Yeast Res. *7*, 657–664.

Lin, Y.Y., Lu, J.Y., Zhang, J., Walter, W., Dang, W., Wan, J., Tao, S.C., Qian, J., Zhao, Y., Boeke, J.D., *et al.* (2009). Protein acetylation microarray reveals that NuA4 controls key metabolic target regulating gluconeogenesis. Cell *136*, 1073–1084.

Liou, G.G., Tanny, J.C., Kruger, R.G., Walz, T., and Moazed, D. (2005). Assembly of the SIR complex and its regulation by O-acetyl-ADP-ribose, a product of NAD-dependent histone deacetylation. Cell *121*, 515–527.

Liu, B., Larsson, L., Caballero, A., Hao, X., Oling, D., Grantham, J., and Nystrom, T. (2010). The polarisome is required for segregation and retrograde transport of protein aggregates. Cell *140*, 257–267.

Lynch, P.J., and Rusche, L.N. (2009). A silencer promotes the assembly of silenced chromatin independently of recruitment. Mol. Cell. Biol. *29*, 43–56.

Ma, B., Pan, S.J., Zupancic, M.L., and Cormack, B.P. (2007). Assimilation of NAD(+) precursors in *Candida glabrata*. Mol. Microbiol. *66*, 14–25.

Ma, B., Pan, S.J., Domergue, R., Rigby, T., Whiteway, M., Johnson, D., and Cormack, B.P. (2009). High-affinity transporters for NAD+ precursors in *Candida glabrata* are regulated by Hst1 and induced in response to niacin limitation. Mol. Cell. Biol. *29*, 4067–4079.

Maas, N.L., Miller, K.M., DeFazio, L.G., and Toczyski, D.P. (2006). Cell cycle and checkpoint regulation of histone H3 K56 acetylation by Hst3 and Hst4. Mol. Cell *23*, 109–119.

McCord, R., Pierce, M., Xie, J., Wonkatal, S., Mickel, C., and Vershon, A.K. (2003). Rfm1, a novel tethering factor required to recruit the Hst1 histone deacetylase for repression of middle sporulation genes. Mol. Cell. Biol. *23*, 2009–2016.

Maguire, S.L., OhEigeartaigh, S.S., Byrne, K.P., Schroder, M.S., O'Gaora, P., Wolfe, K.H., and Butler, G. (2013). Comparative genome analysis and gene finding in *Candida* species using CGOB. Mol. Biol. Evol. *30*, 1281–1291.

Mao, Z., Hine, C., Tian, X., Van Meter, M., Au, M., Vaidya, A., Seluanov, A., and Gorbunova, V. (2011). SIRT6 promotes DNA repair under stress by activating PARP1. Science *332*, 1443–1446.

Martino, F., Kueng, S., Robinson, P., Tsai-Pflugfelder, M., van Leeuwen, F., Ziegler, M., Cubizolles, F., Cockell, M.M., Rhodes, D., and Gasser, S.M. (2009). Reconstitution of yeast silent chromatin: multiple contact sites and O-AADPR binding load SIR complexes onto nucleosomes *in vitro*. Mol. Cell *33*, 323–334.

Masumoto, H., Hawke, D., Kobayashi, R., and Verreault, A. (2005). A role for cell-cycle-regulated histone H3 lysine 56 acetylation in the DNA damage response. Nature *436*, 294–298.

Mead, J., McCord, R., Youngster, L., Sharma, M., Gartenberg, M.R., and Vershon, A.K. (2007). Swapping the gene-specific and regional silencing specificities of the Hst1 and Sir2 histone deacetylases. Mol. Cell. Biol. *27*, 2466–2475.

Michishita, E., Park, J.Y., Burneskis, J.M., Barrett, J.C., and Horikawa, I. (2005). Evolutionarily conserved and nonconserved cellular localizations and functions of human SIRT proteins. Mol. Biol. Cell *16*, 4623–4635.

Michishita, E., McCord, R.A., Berber, E., Kioi, M., Padilla-Nash, H., Damian, M., Cheung, P., Kusumoto, R., Kawahara, T.L.A., Barrett, J.C., *et al.* (2008). SIRT6 is a histone H3 lysine 9 deacetylase that modulates telomeric chromatin. Nature *452*, 492–496.

Min, J., Landry, J., Sternglanz, R., and Xu, R.M. (2001). Crystal structure of a SIR2 homolog–NAD complex. Cell *105*, 269–279.

Moretti, P., Freeman, K., Coodly, L., and Shore, D. (1994). Evidence that a complex of SIR proteins interacts with the silencer and telomere-binding protein RAP1. Genes Dev. *8*, 2257–2269.

Munoz-Galvan, S., Jimeno, S., Rothstein, R., and Aguilera, A. (2013). Histone H3K56 acetylation, Rad52, and non-DNA repair factors control double-strand break repair choice with the sister chromatid. PLoS Genet. *9*, e1003237.

Orta-Zavalza, E., Guerrero-Serrano, G., Gutierrez-Escobedo, G., Canas-Villamar, I., Juarez-Cepeda, J., Castano, I., and De Las Penas, A. (2013). Local silencing controls the oxidative stress response and the multidrug resistance in *Candida glabrata*. Mol. Microbiol. *88*, 1135–1148.

Peng, C., Lu, Z., Xie, Z., Cheng, Z., Chen, Y., Tan, M., Luo, H., Zhang, Y., He, W., Yang, K., *et al.* (2011). The first identification of lysine malonylation substrates and its regulatory enzyme. *Mol. Cell. Proteomics 10*, M111 012658.

Perez-Martin, J., Uria, J.A., and Johnson, A.D. (1999). Phenotypic switching in *Candida albicans* is controlled by a *SIR2* gene. EMBO J. *18*, 2580–2592.

Peterson, F.C., Chen, D., Lytle, B.L., Rossi, M.N., Ahel, I., Denu, J.M., and Volkman, B.F. (2011). Orphan macrodomain protein (human C6orf130) is an O-acyl-ADP-ribose deacylase: solution structure and catalytic properties. J. Biol. Chem. *286*, 35955–35965.

Rafty, L.A., Schmidt, M.T., Perraud, A.L., Scharenberg, A.M., and Denu, J.M. (2002). Analysis of O-acetyl-ADP-ribose as a target for Nudix ADP-ribose hydrolases. J. Biol. Chem. *277*, 47114–47122.

Rine, J., and Herskowitz, I. (1987). Four genes responsible for a position effect on expression from HML and HMR in Saccharomyces cerevisiae. Genetics *116*, 9–22.

Ringel, A.E., Ryznar, R., Picariello, H., Huang, K.L., Lazarus, A.G., and Holmes, S.G. (2013). Yeast Tdh3 (glyceraldehyde 3-phosphate dehydrogenase) is a Sir2-interacting factor that regulates transcriptional silencing and rDNA recombination. PLoS Genet. *9*, e1003871.

Rusche, L.N., and Rine, J. (2001). Conversion of a gene-specific repressor to a regional silencer. Genes Dev. *15*, 955–967.

Rusche, L.N., and Lynch, P.J. (2009). Assembling heterochromatin in the appropriate places: A boost is needed. J. Cell. Physiol. *219*, 525–528.

Rusche, L.N., Kirchmaier, A.L., and Rine, J. (2002). Ordered nucleation and spreading of silenced chromatin in *Saccharomyces cerevisiae*. Mol. Biol. Cell *13*, 2207–2222.

Sandmeier, J.J., Celic, I., Boeke, J.D., and Smith, J.S. (2002). Telomeric and rDNA silencing in *Saccharomyces cerevisiae* are dependent on a nuclear NAD(+) salvage pathway. Genetics *160*, 877–889.

Shankaranarayana, G.D., Motamedi, M.R., Moazed, D., and Grewal, S.I. (2003). Sir2 regulates histone H3 lysine 9 methylation and heterochromatin assembly in fission yeast. Curr. Biol. *13*, 1240–1246.

Sherman, J.M., Stone, E.M., Freeman-Cook, L.L., Brachmann, C.B., Boeke, J.D., and Pillus, L. (1999). The conserved core of a human SIR2 homologue functions in yeast silencing. Mol. Biol. Cell *10*, 3045–3059.

Shimizu, M., Masuo, S., Fujita, T., Doi, Y., Kamimura, Y., and Takaya, N. (2012). Hydrolase controls cellular NAD, sirtuin, and secondary metabolites. Mol. Cell. Biol. *32*, 3743–3755.

Shou, W.Y., Seol, J.H., Shevchenko, A., Baskerville, C., Moazed, D., Chen, Z.W.S., Jang, J., Shevchenko, A., Charbonneau, H., and Deshaies, R.J. (1999). Exit from mitosis is triggered by Tem1-dependent release of the protein phosphatase Cdc14 from nucleolar RENT complex. Cell *97*, 233–244.

Shwab, E.K., Bok, J.W., Tribus, M., Galehr, J., Graessle, S., and Keller, N.P. (2007). Histone deacetylase activity regulates chemical diversity in *Aspergillus*. Eukaryot. Cell *6*, 1656–1664.

Sinclair, D.A., and Guarente, L. (1997). Extrachromosomal rDNA circles – a cause of aging in yeast. Cell *91*, 1033–1042.

Smith, B.C., Settles, B., Hallows, W.C., Craven, M.W., and Denu, J.M. (2011). SIRT3 substrate specificity determined by peptide arrays and machine learning. ACS Chem. Biol. *6*, 146–157.

Smith, J.S., and Boeke, J.D. (1997). An unusual form of transcriptional silencing in yeast ribosomal DNA. Genes Dev. *11*, 241–254.

Smith, K.M., Kothe, G.O., Matsen, C.B., Khlafallah, T.K., Adhvaryu, K.K., Hemphill, M., Freitag, M., Motamedi, M.R., and Selker, E.U. (2008). The fungus *Neurospora crassa* displays telomeric silencing mediated by multiple sirtuins and by methylation of histone H3 lysine 9. Epigenetics Chromatin *1*, 5.

Spellman, P.T., Sherlock, G., Zhang, M.Q., Iyer, V.R., Anders, K., Eisen, M.B., Brown, P.O., Botstein, D., and Futcher, B. (1998). Comprehensive identification of cell cycle-regulated genes of the yeast *Saccharomyces cerevisiae* by microarray hybridization. Mol. Biol. Cell *9*, 3273–3297.

Stevenson, J.S., and Liu, H. (2011). Regulation of white and opaque cell-type formation in *Candida albicans* by Rtt109 and Hst3. Mol. Microbiol. *81*, 1078–1091.

Stevenson, J.S., and Liu, H. (2013). Nucleosome assembly factors CAF-1 and HIR modulate epigenetic switching frequencies in an H3K56 acetylation-associated manner in *Candida albicans*. Eukaryot. Cell *12*, 591–603.

Straight, A.F., Shou, W., Dowd, G.J., Turck, C.W., Deshaies, R.J., Johnson, A.D., and Moazed, D. (1999). Net1, a Sir2-associated nucleolar protein required for rDNA silencing and nucleolar integrity. Cell *97*, 245–256.

Suka, N., Luo, K., and Grunstein, M. (2002). Sir2p and Sas2p opposingly regulate acetylation of yeast histone H4 lysine16 and spreading of heterochromatin. Nat. Genet. *32*, 378–383.

Tanner, K.G., Landry, J., Sternglanz, R., and Denu, J.M. (2000). Silent information regulator 2 family of NAD-dependent histone/protein deacetylases generates a unique product, 1-O-acetyl-ADP-ribose. Proc. Natl. Acad. Sci. U.S.A. *97*, 14178–14182.

Tanny, J.C., and Moazed, D. (2001). Coupling of histone deacetylation to NAD breakdown by the yeast silencing protein Sir2: Evidence for acetyl transfer from substrate to an NAD breakdown product. Proc. Natl. Acad. Sci. U.S.A. *98*, 415–420.

Thaminy, S., Newcomb, B., Kim, J., Gatbonton, T., Foss, E., Simon, J., and Bedalov, A. (2007). Hst3 is regulated by Mec1-dependent proteolysis and controls the S phase checkpoint and sister chromatid cohesion by deacetylating histone H3 at lysine 56. J. Biol. Chem. *282*, 37805–37814.

Tong, L., Lee, S., and Denu, J.M. (2009). Hydrolase regulates NAD+ metabolites and modulates cellular redox. J. Biol. Chem. *284*, 11256–11266.

Triolo, T., and Sternglanz, R. (1996). Role of interactions between the origin recognition complex and SIR1 in transcriptional silencing. Nature *381*, 251–253.

Tsubota, T., Berndsen, C.E., Erkmann, J.A., Smith, C.L., Yang, L., Freitas, M.A., Denu, J.M., and Kaufman, P.D. (2007). Histone H3-K56 acetylation is catalyzed by histone chaperone-dependent complexes. Mol. Cell *25*, 703–712.

Vaquero, A., Scher, M.B., Lee, D.H., Sutton, A., Cheng, H.L., Alt, F.W., Serrano, L., Sternglanz, R., and Reinberg, D. (2006). SirT2 is a histone deacetylase with preference for histone H4 Lys 16 during mitosis. Genes Dev. *20*, 1256–1261.

Wilkins, B.J., Rall, N.A., Ostwal, Y., Kruitwagen, T., Hiragami-Hamada, K., Winkler, M., Barral, Y., Fischle, W., and Neumann, H. (2014). A cascade of histone modifications induces chromatin condensation in mitosis. Science *343*, 77–80.

Wilson, J.M., Le, V.Q., Zimmerman, C., Marmorstein, R., and Pillus, L. (2006). Nuclear export modulates the cytoplasmic Sir2 homologue Hst2. EMBO Rep. *7*, 1247–1251.

Wiren, M., Silverstein, R.A., Sinha, I., Walfridsson, J., Lee, H.M., Laurenson, P., Pillus, L., Robyr, D., Grunstein, M., and Ekwall, K. (2005). Genomewide analysis of nucleosome density histone acetylation and HDAC function in fission yeast. EMBO J. *24*, 2906–2918.

Wurtele, H., Tsao, S., Lepine, G., Mullick, A., Tremblay, J., Drogaris, P., Lee, E.H., Thibault, P., Verreault, A., and Raymond, M. (2010). Modulation of histone H3 lysine 56 acetylation as an antifungal therapeutic strategy. Nat. Med. *16*, 774–780.

Xie, J., Pierce, M., Gailus-Durner, V., Wagner, M., Winter, E., and Vershon, A.K. (1999). Sum1 and Hst1 repress middle sporulation-specific gene expression during mitosis in *Saccharomyces cerevisiae*. EMBO J. *18*, 6448–6454.

Yu, E.Y., Yen, W.F., Steinberg-Neifach, O., and Lue, N.F. (2010). Rap1 in *Candida albicans*: an unusual structural organization and a critical function in suppressing telomere recombination. Mol. Cell. Biol. *30*, 1254–1268.

Zhang, Z., Tan, M., Xie, Z., Dai, L., Chen, Y., and Zhao, Y. (2011). Identification of lysine succinylation as a new post-translational modification. Nat. Chem. Biol. *7*, 58–63.

Zhao, K., Chai, X., Clements, A., and Marmorstein, R. (2003a). Structure and autoregulation of the yeast Hst2 homolog of Sir2. Nature Struct. Biol. *10*, 864–871.

Zhao, K., Chai, X., and Marmorstein, R. (2003b). Structure of the yeast Hst2 protein deacetylase in ternary complex with 2′-O-acetyl ADP ribose and histone peptide. Structure *11*, 1403–1411.

Zhao, K., Chai, X., and Marmorstein, R. (2004). Structure and substrate binding properties of cobB, a Sir2 homolog protein deacetylase from *Escherichia coli*. J. Mol. Biol. *337*, 731–741.

Development-linked Differences in Cytosine 5-Hydroxymethylation in Mammalian DNA: Relationship to 5-Methylcytosine and Function

4

Melanie Ehrlich, Michelle Lacey, Guoqiang Zhang, Kenneth C. Ehrlich and Sriharsa Pradhan

Abstract

There has been a burst of research activity since the recent definitive determination of 5-hydroxymethylcytosine (5hmC) as a differentiation-linked base in mammalian DNA. Differences in 5hmC tissue-specificity are even more striking than those for 5-methylcytosine (5mC), the DNA base from which 5hmC is derived. There are strong associations of 5hmC with extended promoter regions but its distribution around transcription start sites is different depending on the genes' transcriptional activity. The levels of 5hmC in the body of genes and in weak or strong enhancer-type chromatin partly correlate with transcription levels. We summarize some of the frequently used methods for analysing 5hmC genomic profiles and recent evidence for correlations of genomic 5hmC with major histone modifications, differential splicing, differentiation, disease, and DNA demethylation. In addition, we compare DNA hydroxymethylation and DNA methylation in embryonic stem cells and the neural and skeletal muscle lineages as well as the distinct enrichment patterns of these two modifications along the genome. Because evidence indicates that 5hmC functions as both an intermediate in DNA demethylation and as a stable component of DNA with many different features from those of 5mC, 5hmC probably has a wide variety of roles in differentiation and sometimes in cell physiology.

Introduction

It is one of the ironies of the history of molecular biology research that the surprising discovery in 1952 of 5-hydroxymethylcytosine (5hmC) replacing all the cytosine in the DNA of the T-even bacteriophages (T2, T4, and T6) (Wyatt and Cohen, 1952) may have made the then-popular, tiny T4 genome appear less relevant to studies of vertebrate DNA. However, 57 years later, a report on human Purkinje and granule cell DNA (Kriaucionis and Heintz, 2009) and another on mouse embryonic stem cell (ESC) DNA and DNA modified *in vitro* by the human enzyme TET1 (Tahiliani *et al.*, 2009) provided unequivocal evidence that mammalian genomes contain highly significant, tissue-specific amounts of 5hmC. In T-even phage DNA, the 5hmC is further naturally modified by glucosylation in a reaction catalysed by phage DNA glucosyltransferases (Revel and Luria, 1970). One of these enzymes, T4 β-glucosyltransferase (β-GT) is now an extremely useful tool in analysing 5hmC in mammalian DNA even though human and mouse DNA do not have such glucosylation *in vivo*, as determined by chromatographic and/or mass spectrometric analyses (e.g. Bhattacharyya *et al.*, 2013; Chen *et al.*, 2013; Globisch *et al.*, 2011).

Another major difference between the 5hmC in mammalian and T-even phage DNAs is that this modified base in phage genomes fully replaces all C residues by complete conversion of the DNA substrate dCTP to 5hmdCTP in infected *E. coli* cells (Wyatt and Cohen, 1952). In contrast, 5hmC arises in mammalian DNA from enzymatic oxidation of a small fraction of genomic 5-methylcytosine (5mC) to 5hmC (Huang *et al.*, 2014; Tahiliani *et al.*, 2009). This reaction is catalysed by TET1, TET2, or TET3, which are oxoglutarate- and iron-dependent dioxygenases (Branco *et al.*, 2011). The 5-methylation of only a minor fraction of C residues and hydroxylation of usually a small fraction of 5mC residues in preferred locations in mammalian DNA in a cell context-dependent manner provides an additional layer of differential modification of mammalian DNA that is dependent upon differentiation, physiological changes, and pathogenic events (Almeida *et al.*, 2012b; Ehrlich and Lacey, 2013a,b; Nestor *et al.*, 2012; Pfeifer *et al.*, 2013; Smith and Meissner, 2013; Yu *et al.*, 2013a). After the formation of 5hmC from 5mC, the TET enzymes can further oxidize genomic 5hmC to 5-formylcytosine (5fC) and 5-carboxylcytosine (5caC), although the levels of these two products *in vivo* are much lower even than those of 5hmC in DNA (Song *et al.*, 2013; Wu and Zhang, 2011). The ability of TET proteins to regulate transcription independently of their catalytic domain and to stably associate with chromatin and their partly overlapping functionality (Chen *et al.*, 2012; Huang *et al.*, 2014; Kaas *et al.*, 2013; Wu *et al.*, 2011b) complicate interpretation of results from the experimental manipulation of TET levels to change the 5hmC content of DNA.

Despite the generation of 5hmC residues from 5mC residues in mammalian DNA (Ficz *et al.*, 2011; Tahiliani *et al.*, 2009), the functional correlates of 5mC and 5hmC are frequently different. These modified bases differ markedly in their associations with histone modifications and levels of gene expression (Coleman-Derr and Zilberman, 2012; Jin *et al.*, 2011b; Serandour *et al.*, 2012; Stroud *et al.*, 2011; Szulwach *et al.*, 2011; Yu *et al.*, 2013b) as well as with satellite DNA (Chen *et al.*, 2014). Another difference between 5hmC and 5mC residues in DNA is that some DNA-binding proteins have differential affinity for these two bases in a given DNA sequence (Hashimoto *et al.*, 2012; Otani *et al.*, 2013; Spruijt *et al.*, 2013). Even in naked DNA there are major differences between 5hmC- and 5mC-containing base pairs. As we had shown in 1975, 5-methylation of C residues in DNA greatly stabilizes the double helix (Ehrlich *et al.*, 1975). In contrast, Wanunu *et al.* (2011) recently demonstrated that 5-hydroxymethylation of genomic C residues destabilizes the double helix. In addition, 5hmC and 5mC differ in their hydrophobicity (Bianchi and Zangi, 2013; Wanunu *et al.*, 2011).

In this chapter, we discuss the relationship of DNA hydroxymethylation to human and rodent differentiation and contrast the functional implications of genomic hydroxymethylation with those of methylation on gene expression. Despite research on 5hmC in mammals being quite recent, it has seen explosive activity and so this survey necessarily considers only some of the findings and will be restricted to mammalian DNA although other vertebrates have been reported to have genomic 5hmC (Almeida *et al.*, 2012a). After a brief introduction to some of the methods for studying 5hmC in DNA, we will give an overview of differential DNA hydroxymethylation along genomic DNA and chromatin structures and among different cell and tissue types. Based upon these associations with normal differentiation, disease, and gene structure and with alterations in gene function upon experimental manipulation of genomic 5hmC, we discuss possible roles for 5hmC in the mammalian genome.

Analytic methods used to study 5hmC in vertebrate DNA

Genome-wide analysis of 5-hydroxymethylcytosine

A breakthrough in deciphering DNA epigenetic modifications was the inception of two different genome-wide technologies. The first was based on the use of large-scale DNA microarrays (Schena *et al.*, 1995) and the second involved the introduction of the next-generation sequencing workflow (ten Bosch and Grody, 2008; Tucker *et al.*, 2009). DNA microarrays were later modified for downstream application of chromatin-immunoprecipitated DNA (ChIP on chip) (Ren *et al.*, 2000). Antibodies specific for 5hmC (or for reaction products from it) were developed, enabling DNA immunoprecipitation of hydroxymethylated or methylated DNA followed by hybridization to oligonucleotide microarrays (hMeDIP-chip or MeDIP) (Thomson *et al.*, 2012; Wu *et al.*, 2011a) as well as immunocytochemical studies (Almeida *et al.*, 2012b; Chen *et al.*, 2014). However, the specificity and reactivity of the employed antibodies could be a major concern due to batch variability. Moreover, 5hmC genome-profiling methods that use immunoprecipitation or affinity capture are prone to a bias for detecting local sequences with a high density of 5hmC sites, which is a similar concern for 5mC genomic profiling with antibodies or methylated DNA binding protein pull-down.

The increased efficiency of high-throughput massive parallel sequencing of DNA led to use of these methods for DNA treated with bisulfite to distinguish 5mC residues (which appear as C after PCR) and unmodified cytosine residues (which appear as T after PCR). Such procedures were done with either complete coverage of the genome (bisulfite-seq or BS-seq) or partial, quite representative coverage (reduced representation bisulfite sequencing, RRBS) (Meissner *et al.*, 2008). However, bisulfite conversion does not discriminate between 5mC and 5hmC and discriminating these bases requires additional protocols. To date, two modified bisulfite-sequencing methods have been shown to detect 5hmC. The first one, Tet enzyme-assisted bisulfite sequencing (TAB-seq), utilizes a Tet enzyme and the second, oxidative bisulfite sequencing (oxBS-seq), depends on selective chemical oxidization of genomic DNA. Both methods provide base resolution mapping of genomic 5hmC by next-generation sequencing.

The TAB-seq method originally reported by Yu *et al.* (2012a,b) uses a comparative technique that compares parallel TAB-seq and standard BS-seq to allow mapping of both genomic 5mC and 5hmC. The DNA aliquot for the TAB-seq procedure is first glucosylated at 5hmC by T4 phage β-GT, then treated with excess Tet1 enzyme to oxidize all the 5mC to 5caC, bisulfite-converted to give U from C and 5caC, PCR-amplified (to give T at residues where there was U after conversion) and then sequenced. The glucosylation of 5hmC residues prevents their oxidation so that the original 5hmC residues will appear as C residues after PCR. Combining the results from BS-Seq (5mC and 5hmC appear as C after PCR) to those from TAB-seq (only 5hmC behaves as C after PCR) allows mapping of all the 5hmC and 5mC. Even though this comparative sequencing technique provides base resolution identification of both 5mC and 5hmC, there are still concerns about the efficiency of enzymatic conversion and the low abundance of 5hmC in the sample for accurate coverage. Any 5mC that is not completely converted to 5caC will be displayed as 5hmC in the TAB-Seq assay, so it is imperative that both glucosylation and TET-mediated conversion should be nearly 100% efficient.

The oxBS-seq method was first detailed by Booth *et al.* (2012, 2013). Like TAB-seq, the oxBS-seq method is performed in conjunction with standard BS-Seq to allow for the separate detection of 5mC and 5hmC throughout the genome. To detect 5hmC, the DNA sample is chemically oxidized with potassium perruthenate ($KRuO_4$), which will selectively oxidize 5hmC to 5fC. Upon subsequent bisulfite conversion, the 5fC will be chemically converted to U, which will be replaced by T after PCR amplification. In oxBS-seq, 5hmC and C appear as T upon PCR while 5mC appears as C after PCR. Both Tab-Seq and oxBS-Seq require much higher sequencing depth when performing genome-wide 5mC and 5hmC analyses than does BS-seq.

Recently, an additional single-base resolution method for profiling 5hmC, Aba-Seq, has been introduced by Sun *et al.* (2013). It requires substantially less sequencing depth than the previous methods because unlike them Aba-Seq inherently enriches for genomic DNA containing 5hmC and creates a positive display. Aba-Seq utilizes the AbaSI restriction enzyme, which efficiently cleaves glucosylated 5hmC. Genomic 5hmC is first glucosylated with β-GT. Then the DNA is digested with the AbaSI, ligated to biotin-tagged sequencing adapters, sheared to 300–500 bp, extracted with streptavidin beads, end-repaired/ligated to another sequencing adapter, amplified and sequenced. Next-generation sequencing of the DNA library reveals all the AbaSI restriction cut sites, thus providing 5hmC mapping at a very high degree of resolution in the genome. The sequencing read-number for each strand of DNA fragment created by AbaSI digestion strongly correlates with the amount of 5hmC on each strand. The finding that Aba-seq reproducibly revealed about six times more 5hmC sites in ESC DNA than did TAB-seq (Sun *et al.*, 2013) suggests that Aba-seq is much more sensitive and may be detecting a preponderance of low-frequency 5hmC sites that are missed by TAB-seq.

DNA hydroxymethylation in very early mammalian development: embryonic stem cells (ESC)

Overview of ESC and their modified C residues and DNA modifying enzymes

Although, immunocytochemical findings suggested that 5-hydroxymethylation plays a central role in demethylation occurring at the zygote stage (Wossidlo *et al.*, 2011), this conclusion was controversial because of the difficulties in studying such cells, especially given the potential for artefacts that can be associated with immunochemical techniques (Li and O'Neill, 2013). The latest studies suggest that the roles of DNA hydroxymethylation in early mammalian embryogenesis involve counteracting the repressive effects of 5mC and fostering demethylation during several of the earliest cell division cycles rather than in the zygote (Iqbal *et al.*, 2011; Smith *et al.*, 2012). Most studies of DNA hydroxymethylation very early in mammalian development have used murine ESC as models for the pluripotent embryoblast (later-stage inner cell mass, ICM) (Ficz *et al.*, 2013). ICM cells give rise to the embryo itself rather than supporting tissues. Several studies found that in ESC DNA about 0.3% of the cytosines are 5hmC and 5% of CCGG's (and probably approximately 5% of all CpGs) are 5hmCpG (Tahiliani *et al.*, 2009; Terragni *et al.*, 2012). Upon *in vitro* differentiation, ESC lose much of their genomic 5hmC (Kinney *et al.*, 2011) and DNA 5hmC levels decrease in concert with decreases in expression of the *Tet* genes (Ficz *et al.*, 2011; Koh *et al.*, 2011). The 5hmC in undifferentiated ESC DNA is predominantly at CpGs, with >95% or >99% of genomic 5hmC at CpG (Sun *et al.*, 2013; Yu *et al.*, 2012b).

About three times as much 5mC as 5hmC was found by Yu *et al.* (2012) in ESC DNA. In contrast to extremely low levels of non-CpG 5hmC, about 20% of the total 5mC is in non-CpG positions in these cells (Laurent *et al.*, 2010). In this regard, ESC are exceptional (but not unique; see neural cells below) in having a substantial amount of non-CpG 5mC. Probably relevant to the high global levels of non-CpG methylation in ESC cells is that, compared to 19 other cell types, these cells have high levels of expression of *DNMT3A/Dnmt3a* and especially of *DNMT3B/Dnmt3b*, the two DNA methyltransferase genes most associated with *de novo* DNA methylation (methylation of symmetrically unmethylated CpG dinucleotide pairs), as well as of *DNMT1*, which encodes the DNA methyltransferase most important for maintaining DNA methylation upon DNA replication (Ficz *et al.*, 2013; Tsumagari *et al.*, 2011). They also express *DNMT3L*, which encodes a cofactor that is essential for the activity of DNMT3A and DNMT3B (Ooi *et al.*, 2007). In addition, ESC have especially high levels of *TET1* RNA as well as lower levels of *TET2* and *TET3* RNA (Ficz *et al.*, 2013; Huang *et al.*, 2014; Kent *et al.*, 2014). When all three DNMT genes are mutationally inactivated, no 5hmC is detected (Ito *et al.*, 2010; Szwagierczak *et al.*, 2010). Therefore, it is likely that 5mC is the mandatory precursor to 5hmC in mammalian DNA.

Nanog a, gene associated with totipotency in ESC requires *Tet1* to maintain its high expression levels in these cells unless the *Dnmt* genes are knocked out (Ito *et al.*, 2010). Moreover, knockdown (kd) of *Tet1* in mouse ESC results in *de novo* DNA methylation in the tissue-differentially methylated region (T-DMR) proximal to *Nanog* and decreased maintenance of the totipotent state (Ito *et al.*, 2010). Differentiation in these mutants is abnormally skewed towards the trophectoderm lineage. During differentiation of normal ESC there is increased methylation of *Nanog* but no detectable 5hmC at the assayed site even in undifferentiated cells (Kinney *et al.*, 2011). These results suggest that 5hmC may form only transiently during *Tet1*-catalysed demethylation of DNA at critical genes like *Nanog* to maintain the gene's activity and, thereby, pluripotency in ESC. Consistent with this hypothesis, short-term, RNAi-mediated *Tet1* kd resulted in decreased global levels of 5hmC and concomitantly the loss of pluripotency in ESC (Freudenberg *et al.*, 2012). However, the loss of the non-enzymatic functions of Tet1 (as well as Tet2 and Tet3) in *Tet* kd cells complicates the interpretation of these results.

5hmC as an intermediate in DNA demethylation

Much evidence indicates that one of the functions of genomic 5hmC is to serve as an intermediate in DNA demethylation (Kohli and Zhang, 2013). This demethylation can be independent of DNA replication and can be mediated by repair of 5hmC residues or their further TET-catalysed oxidation products (Hashimoto *et al.*, 2012; Kohli and Zhang, 2013). In addition, 5hmC production can be an intermediate step in a DNA replication-dependent pathway for demethylation. A duplex DNA sequence with 5hmC in only one strand is not recognized well by DNMT1, the main maintenance DNA methyltransferase (Otani *et al.*, 2013). Such 5hmC-containing sequences were reported to be bound less well than 5mC sequences by UHRF1 (Otani *et al.*, 2013), a cofactor necessary for maintenance methylation. Therefore, upon cell cycling, 5hmC levels are diluted out. Although 5hmC is strongly implicated as an intermediate in DNA demethylation in early embryogenesis (Ficz *et al.*, 2013), the finding that it is maintained at considerable levels in the genome of undifferentiated ESC if *Tet1* and *Tet2* are not experimentally down-regulated (Ficz *et al.*, 2011; Kinney *et al.*, 2011; Koh *et al.*, 2011) strongly suggests additional roles for 5hmC in early embryogenesis. Moreover, the highly non-random distribution of this base along different DNA sequence elements and types of chromatin regions in ESC cells and other types of cells and tissues, as described below, is consistent with functions of 5hmC residues irrespective of DNA demethylation.

5hmC enrichment in gene bodies of ESC and relationship to gene expression and RNA splicing

As described above, the most representative analyses of the distribution of 5hmC among different types of DNA sequences come from base-resolution methylome analyses in which the entire genome is examined with the exception of highly conserved DNA repeats whose position in the reference genome cannot be assigned (Table 4.1, studies 7–9). These studies showed enrichment of 5hmC either in gene bodies in general (Booth *et al.*, 2012; Sun *et al.*, 2013) or in the gene bodies of weakly expressed genes (Yu *et al.*, 2012b). Exons are enriched in 5hmC. They are also enriched in 5mC, which probably mostly reflects the CpG-rich nature of coding exons, in general, due to codon restraints on DNA sequence (Saxonov *et al.*, 2006).

Huang *et al.* (2014) found that high levels of 5hmC in gene bodies were correlated with high levels of transcription (Table 4.1). Depletion of Tet1 or Tet2 (Tet1 or Tet2 kd) gave mostly losses of 5hmC and dysregulation of many genes. Surprisingly, there were also some gains in average 5mhC levels, especially at promoter regions in Tet2 kd mutants. There were no significant overall associations with the loss or gain of gene-body or promoter 5hmC and the direction of change of gene expression (increased vs. decreased RNA levels). This might be due to indirect cascades of effects of Tet1 or Tet2 loss on gene expression, especially in view of the stable binding of Tet proteins throughout the genome and their non-enzymatic gene regulatory functions that are independent of their roles in oxidation of 5mC residues (Chen *et al.*, 2012; Kaas *et al.*, 2013; Neri *et al.*, 2013; Wu *et al.*, 2011b). Several other genome-wide profiles of 5hmC in ESC also indicated that transcription of some large classes of genes is positively associated with gene-body hydroxymethylation (Stroud *et al.*, 2011; Yu *et al.*, 2012b). Neri *et al.* (2013)provided extensive evidence for an association of gene body 5hmC enrichment with steady-state RNA levels for the corresponding gene (Neri *et al.*, 2013). They showed that the levels of 5hmC throughout the gene body for weakly expressed, moderately expressed, and highly expressed genes were low, moderate, and high, respectively for ESC, as well as for mouse embryonic fibroblasts (MEF), brain and liver.

Intragenic 5hmC has also been implicated in regulating splicing (Table 4.1, study 10) as has 5mC. With respect to 5mC, methylome profiles revealed peaks of DNA methylation at exon–intron borders and/or within the exons (Anastasiadou *et al.*, 2011; Khare *et al.*, 2012; Maunakea *et al.*, 2013; Wen *et al.*, 2014). Studies of individual genes

Table 4.1 Enrichment of genomic 5hmC within gene or chromatin regions in embryonic stem cells[1]

Study no.	Cells or tissue	Method used to study 5hmC	Enrichment around the TSS?	Enrichment at enhancer chromatin?	Enrichment in gene bodies?	Association with transcription	Reference
1	Human ESC	β-GT, affinity purification, next-generation sequencing	Y, bifurcated peak with trough at TSS for genes with much txn; 5hmC peak at TSS for genes with little txn	Y	Y, especially for exons	Generally higher around TSS for moderately expressed genes	Szulwach *et al*. (2011)
2	Human ESC	hMeDIP-seq	Y, with txn dependence as for study 1	Y, especially at active enhancer chromatin	Y	More enriched in hESC-specific genes than for all genes	Stroud *et al*. (2011)
3	Mouse ESC	CMS-IP, or T4 β-GT, single-molecule seq.	Y	Y	Y, especially for exons	Genes with 5hmC at their promoter generally had lower txn than other genes but were more likely to be upregulated upon ESC differentiation to embryoid bodies	Pastor *et al*. (2011)
4	Mouse ESC	hMeDIP-seq	Y; seen especially in regions with an intermediate CpG density	ND	Y	ND	Williams *et al*. (2011)
5	Mouse ESC (WT or Tet1 kd)	hMeDIP-chip	Highest association with promoter was for genes with little or no txn	Y	Y	Intragenic enrichment in genes with high or moderately txn and in promoter regions of unexpressed genes	Wu *et al*. (2011)
6	Mouse ESC	hMeDIP-seq	Y	ND	Y, mostly for exons	Genes enriched for 5hmC in CpG-poor promoters had higher txn	Ficz *et al*. (2011)

7	Mouse ESC	oxBS-seq (RRBS protocol)	N for CGIs at promoters	Unclear	Y	Unclear	Booth *et al.* (2012)
8	Human and mouse ESC	TAB-seq	Y for TSS ± 1 kb	Y	Y	Weakly expressed genes had highest enrichment in intragenic regions for 5hmC	Yu *et al.* (2012)
9	Mouse ESC	Aba-seq	Y but only if no H3K4me3 enrichment	Yes for weak enhancers (H3K4me1 but not H3K27Ac)	Y for exons but not more than for 5mCpG	Unclear	Sun *et al.* (2013)
10	Mouse ESC (WT, Tet1 or Tet2 kd)	CMS-IP, next-generation sequencing	Y in WT; even more enrichment in Tet2 kd; N in Tet1 kd	Y	Y, especially at borders of exons with high txn; Tet1 or Tet2 kd gave altered splicing	Genes with highest txn had low 5hmC around TSS and high 5hmC in gene body but no simple correlation for altered txn levels in Tet1 or Tet2 kd vs. WT ESC.	Huang *et al.* (2014)
11	Mouse ESC vs. MEF, brain and liver	β-GT, affinity purification, next-generation sequencing	Y, high 5hmC surrounding a trough at the TSS for genes with moderate or high txn	Y for active enhancers in ESC; Y for poised enhancers in other samples	Y for genes with moderate or high txn in exons in all sample types	Genes with highest txn had low 5hmC very close to the TSS, higher 5hmC upstream and downstream of the TSS, and high 5hmC in gene body in all sample types	Neri *et al.* (2013), 5hmC only

β-GT, using bacteriophage T4 β-glucosyltransferase to glucosylate genomic 5hmC; purifcation, enrichment by an affinity method hMeDIP-seq, hMeDIP-chip, oxBS-seq, TAB-seq, and Aba-seq; explained in the text; CMS, bisulfite conversion of 5hmC to cytosine 5-methylenesulphonate followed by immunoprecipitation with an antibody to the modified base; ND, not described. TSS, transcription start site; Y, yes; N, no; txn, transcription as monitored by steady-state levels of RNA, although post-transcriptional processing can also affect RNA levels; WT, wild type; kd, knock-down; enhancer or enhancer chromatin, chromatin enriched in H3K4me1 (and H3K27Ac).

are consistent with these genome-wide findings (Tsumagari *et al.*, 2013b). It has been proposed that this differential DNA methylation helps identify alternative spliced exons and constitutive exons to be included in the mRNA, modulates binding of CTCF near alternative exons to affect splicing patterns, and potentiates a subfraction of exonic splicing enhancers (Oberdoerffer, 2012; Shukla *et al.*, 2011) (see 'Mechanistic insights into the associations of 5hmC and 5mC with chromatin structure, CTCF binding, gene expression and differentiation', below). Similarly, for 5hmC, enrichment in gene bodies was especially strong for highly transcribed exons, specifically from the region 200 bp into the exon to 100 bp into the adjacent intron sequence (Huang *et al.*, 2014). Moreover, both Tet1 and Tet2 kd cells displayed many changes in exclusion or inclusion of exons compared with the parental ESC. The differentially spliced exons affected by Tet1 kd were different from those affected by Tet2 kd (Huang *et al.*, 2014).

5hmC density at promoter regions of ESC and its relationship to gene expression

Many whole-genome studies of 5hmC in ESC indicate that its composite profile in gene regions contains one peak at the transcription start site (TSS) or high levels in the vicinity of the promoter but somewhat further away from the TSS (Table 4.1). CpG islands (CGIs), regions of unusually high CpG density compared with the rest of the genome (Saxonov *et al.*, 2006), overlap ~70% of mammalian gene promoters and are usually constitutively unmethylated or mostly unmethylated. Genome-wide, regions with a low or intermediate density of CpG in ESC and other cell types display higher percentages of 5hmC than the CpG-dense CGI (Booth *et al.*, 2012; Neri *et al.*, 2013). Although 5hmC is generally not stably associated with CGI promoters, low levels of 5hmC might play a pivotal role as a transient intermediate in demethylating inadvertently generated 5mC in normally unmethylated sequences (Williams *et al.*, 2011).

In interpreting conclusions about the relationship between 5hmC levels in promoter regions and transcriptional activity, the latter is usually defined by the steady-state levels of RNA, which is an imperfect surrogate for transcription rates. In these comparisons of DNA hydroxymethylation and transcription, the resolution and sensitivity of the 5hmC analysis, the distance of the studied DNA region from the TSS and whether the genes analysed have low, no, weak or moderate transcription need to be considered. Analyses of hydroxymethylomes from various human or mouse samples indicate that 5hmC is generally enriched about 0.5–2 kb upstream and downstream of the transcription start site (TSS) but is depleted closer to the TSS at moderately or highly transcribed genes (Neri *et al.*, 2013; Stroud *et al.*, 2011; Szulwach *et al.*, 2011). The increased levels of 5hmC at the outskirts of the promoter often appear as a bifurcated peak centred over the TSS (Table 4.1). Immediately surrounding the TSS, there are very low levels of 5hmC for the most highly transcribed genes or much reduced enrichment for the moderately transcribed genes. Untranscribed and especially weakly transcribed genes had the opposite 5hmC promoter profile. In the composite gene profile in these three studies, poorly or untranscribed genes had a peak of 5hmC at the TSS rather than twin peaks surrounding the TSS.

Tet1 stably binds throughout the genome and in ESC exhibits a peak of binding around the TSS of many promoters, especially those exhibiting repressive H3 lysine 27 trimethylation (H3K27me3) mark or the histone modification profile associated with poised promoters (H3K27me3 plus H3K4me3), which are referred to as bivalent promoters (Neri *et al.*, 2013). CXXC DNA-binding domains are implicated in recruiting Tet/TET proteins to CGIs either as part of the TET1/Tet1 or TET3/Tet3 polypeptides or as an accessory protein that binds to TET2/Tet2 (Jin *et al.*, 2014; Ko *et al.*, 2013). Moreover, there appears to be a second pathway for binding Tet1 to H3K27me3-rich CGI promoters in ESC but not in other tested cell types. This alternative pathway involves recruitment by the co-repressor Sin3A and the polycomb repressor PRC2 binding components Ezh2 and Suz12 (Neri *et al.*, 2013). Bivalent chromatin is much more prevalent in ESC than other cell types and is usually unmethylated in its DNA. Therefore, the targeting of a Tet enzyme for stable binding to CGI promoters is consistent with the hypothesized involvement of 5hmC as an intermediate in keeping unmethylated CpG island-overlapping promoters free of 5mC. However, the biological effects of 5hmC at promoter regions are clearly not limited to this role, as seen in the above-described stable introduction of 5hmC at a short distance upstream and downstream from the TSS preferentially in highly and moderately transcribed promoters.

5hmC density at enhancer-type chromatin in ESC and its relationship to gene expression

Most studies profiling 5hmC in the ESC genome have noted the enrichment of this modified base in enhancer-type chromatin (Table 4.1). Enhancers or putative enhancers are usually defined in these studies as chromatin regions with peaks of H3K4me1 and H3K27 acetylation (H3K27Ac), and empirical tests of actual enhancer activity of representative DNA regions with these characteristics in a given cell type usually confirm the enhancer designation (Ernst *et al.*, 2011). Peaks of H3K4me1 without appreciable H3K27Ac are often referred to as poised enhancers while peaks of H3K4me1 and H3K27Ac are called active enhancers (Yu *et al.*, 2012b). The finding that mouse ESC analysed by Aba-seq showed enrichment of 5hmC in poised enhancers rather than active enhancers could be due to the much greater sensitivity of Aba-seq for detecting 5hmC (Sun *et al.*, 2013; Yu *et al.*, 2012b).

There is also evidence for cell-type specificity as a determinant of whether 5hmC is associated with active versus poised enhancers. Neri *et al.* (2013) found that while ESC had a significant association of 5hmC with active enhancers, MEF, brain, and liver had a significant association only with poised enhancers. The correlation of 5hmC with poised enhancers might prime these cis-regulatory elements for rapid activation by demethylation of 5hmC residues during exposure to inducing conditions. Indeed, there is evidence for hydroxymethylation being an early event in formation of some types of enhancers (Serandour *et al.*, 2012). This priming could be due to the increased susceptibility of 5hmC residues versus 5mC residues to demethylation by active or passive means.

DNA hydroxymethylation changes during development: neurogenic, germline and adipogenic lineages

5hmC density at promoter regions, gene bodies and enhancer-type chromatin in non-ESC samples

Most of the 5-hydroxymethylome analyses to date on samples other than ESC involved mouse or human brain or neural progenitor cells. As for the ESC studies described above (Table 4.1) and the study comparing ESC to brain, liver, and MEF (Neri *et al.*, 2013), genomic 5hmC profiles of fetal and adult neural cells or brain had low levels of 5hmC centred around the TSS but enrichment a little further upstream and downstream in genes with moderate or high transcription (Table 4.2). The hydroxymethylome profiles of mouse primordial germ cells (PGC) gave similar results (Hackett *et al.*, 2013). Similarly, these investigations of non-ESC samples found generally higher 5hmC levels in gene bodies of highly transcribed genes, especially in coding exons (Table 4.2). In a study of human and mouse brain DNA, Wen *et al.* (2014) described a genome-wide enrichment for 5hmC at the 5′ splice site of internal exons that was irrespective of the genes' RNA levels. There was also an overall intragenic enrichment of 5hmC (in both exons and introns) that was positively associated with the genes' RNA levels. Similar to reports on ESC, 5hmC was generally found to be enriched in enhancer chromatin (Table 4.2).

The association of 5hmC with gene expression and differentiation

For genes containing substantial levels of 5hmC, the relative extent of enrichment in the promoter region versus the gene body (Table 4.2) usually depends on the cell or tissue type, the stage of differentiation, the transcription level and whether the genes were specific to the type of examined tissue or cell or were housekeeping genes (Hackett *et al.*, 2013; Hahn *et al.*, 2013; Kim *et al.*, 2013; Song *et al.*, 2011; Neri *et al.*, 2013). Tissue-specific

Table 4.2 Enrichment of genomic 5hmC within gene or chromatin regions in neural and other cells and tissues[1]

Study no.	Cells or tissue	Method used to study 5hmC	Enrichment around the TSS?	Enrichment at enhancer chromatin?	Enrichment in gene bodies?	Association with transcription	References
12	Adult and fetal mouse and human cerebellum	β-GT, affinity capture, NG seq	Y, bifurcated peak with trough at TSS esp. for genes with high or medium txn	ND	Y	~5400 genes had significantly more 5hmC in adult vs. P7 mouse brain and were enriched in neurodegeneration-related genes	Song *et al.* (2011), Wang *et al.* (2012)
13	Human & mouse brain (frontal cortex)	TAB-seq (& BS-seq)	Depleted in the TSS ± 2 kb relative to intragenic region for both fetal and 6-week mouse brain	Y, higher than for all intragenic regions in fetal brain but not adult brain	Y, for highly transcribed genes	Genes down-regulated from fetal to adult brain often had less 5hmC in the adult. More 5hmC and less non-CpG meth in the gene body was often in genes with high or moderate txn or neuron-associated genes	Lister *et al.* (2013)
14	Fetal mouse neurons and NPC	hMeDIP-chip	Y, in a bifurcated peak withTSS trough for many genes	N, depleted at p300 sites but enriched at adjacent DNA	Y, especially for genes with high txn in NPC or neurons	Top 25% of genes gaining expression from NPC to neurons had a corresponding enrichment of 5hmC in gene bodies	Hahn *et al.* (2013)
15	Mouse and human brain (frontal cortex)	β-GT-chip probes. for only 6 chroms.	ND	ND	Y, especially within 10 bp of the exon–intron border of constitutive exons	Only a weak association although high 5hmC was associated with the GO term 'synapse'	Khare *et al.* (2012)
16	Human brain (frontal lobe)	hMeDIP-chip, CGI and promoter probes	Y but microarrays were mostly for CGIs and promoters	ND	Y	5hmC enrichment at low-CpG density promoters and/or in the gene body were positively associated with more txn	Jin *et al.* (2011)
17	NPC from human ESC; neurons from NPC	hMeDIP-seq	Y for both 5hmC and 5mC, bifurcated peak aroundTSS in genes with much txn and just at TSS for oher genes	Y	Y, esp. in exons	ESC diffn. to NPC: genes often gained 5hmC near the TSS and 5hmC in the gene body; repressed genes often gained 5hmC only in the promoter region and lost it from the gene body	Kim *et al.* (2013)

18	WT & HD transgenic model mouse brain striatum & cortex	β-GT, affinity capture, NG seq	Depleted for HD and WT mice at the TSS	ND	Y, especially for coding exons, which had the highest enrichment of all tested regions	Large loss in global 5hmC in HD brain. qRT-PCR on ninegenes with HD vs. WT DhMRs showed correlation for all but one between their levels of 5hmC and expression in HD vs. WT	Wang *et al.* (2013)
19	Human breast, brain, liver, testes, colon and placenta	hMeDIP, selected genes on microarray; T4 β-GT, RE, PCR	Y	ND	The *HOXA* gene cluster was enriched in 5hmC over gene bodies in breast but not on brain	5hmC throughout the *HOXA* cluster in breast correlated with txn of most *HOXA* genes; *EGFR*, *VANGL1*, *H19 and IGF2;* no significant association between 5hmC andtxn	Nestor *et al.* (2012)
20	EC cell line, (NT2) undifferentiated or induced to differentiate	hMeDIP-seq	Examining *HOXA* genes, *Tet2* ko cells had less 5hmC and more 5mC in promoter and exon 1 of four genes	ND	Y for *HOXA* gene cluster	Upon differentiation of NT2 cells, 5hmC peaks increased in number and height in *HOXA1* through *HOXA6*), which had their txn induced by retinoic acid	Bocker *et al.* (2012)
21	Mouse primordial germ cells (PGC) at E10.5–13.5	hMeDIP-seq (and MeDIP-seq)	5hmC transiently increased as 5mC decreased at *Daz1* promoter during differentiation	ND	Y, from epiblast stem cells to E10.5 PGC, saw large increase in 5hmC and decrease in 5mC at internal exon borders	ND	Hackett *et al.* (2013)
22	P19 and 3T3 cells upon neural or adipocyte differentiation.=	hMeDIP-seq	ND	Y (H3K4me2 and H3K27Ac distal to promoter)	ND	Enrichment of 5hmC in activated genes but little or none for repressed genes upon differentiation. Activated genes enriched in expected GO terms	Serandour *et al.* (2012)
23	Human and mouse prefrontal cortex	TAB-seq	Y, high 5hmC surrounding a trough at the TSS for genes with moderate or high txn	Y, especially in poised enhancers and fetal/adult DhMRs	Y, both exons and introns; highest in exons of genes with high txn, especially at the 5¢ splice site	Expressed genes with 5hmC in gene gene body had a small but highly significant overall bias towards 5hmC in the sense strand and 5mC in antisense strand; no strand bias for unexpressed genes	Wen *et al.* (2014)

[1]Abbreviations are the same as for Table 4.1 with the addition of NPC, neural progenitor cells; EC cell line, embryonic carcinoma cell line; MeDIP, immunoprecipitation of methylated DNA; HD, Huntington's disease; chrom., chromosomes; DhMR, differentially hydroxymethylated region; GO, gene ontology.

differences in 5hmC levels in DNA are often rather stable but, in some cases, are transient. For example, during embryonic differentiation of mouse PGC, there are increases in exonic 5hmC from epiblast stem cells to E10.5, followed by progressive decreases on E11.5 and E12.5 and, subsequently, by a strong decrease on E13.5 (Hackett *et al.*, 2013). During this time period, the majority of exonic 5mC in PGC is lost. In this differentiation system, the genomic 5hmC is most likely an intermediate in the massive DNA demethylation associated with germ cell formation. As differentiation proceeds, there is eventually a decrease in ratios of 5hmC to total C in DNA due to dilution by DNA replication. However, in other developmental or physiological-response systems, there is evidence for a rapid loss of genomic 5hmC or rapid decreases in 5hmC in postmitotic cells (Klug *et al.*, 2013; Okashita *et al.*, 2014; Wheldon *et al.*, 2014). Such replication-independent (active) losses in genomic 5hmC can involve further TET oxidation of genomic 5hmC followed by DNA repair of the final oxidation products. Active DNA demethylation may occur as a result of alterations in cell physiology first increasing 5hmC levels in specific parts of the genome (Guo *et al.*, 2011a; Li *et al.*, 2014; Rudenko *et al.*, 2013).

Brain DNA has been the subject of many studies of 5hmC because it is so much richer in 5hmC than other tissue and cell types. In the DNA of Purkinje cells there is up to 40% as much 5hmC as 5mC (Kriaucionis and Heintz, 2009). By TAB-seq the hydroxymethylomes of fetal, juvenile, adult, and aged mouse or human brain (frontal cortex) (Lister *et al.*, 2013) or adult human brain and fetal brain (prefrontal cortex) (Wen *et al.*, 2014) were compared. In the former study, TAB-seq indicated about four times as much 5hmC in adult mouse brain as in fetal brain. In the latter study, a liquid chromatography-tandem mass spectrometry analysis indicated that there is almost six times as much 5hmC in adult human brain as in fetal brain. Despite the lower global levels of 5hmC in fetal brain, there was a minor fraction of 5hmCpG that was present in fetal but not adult brain (fetal > adult hmCGs). These fetal > adult hmCG sites generally are hypomethylated (loss of 5mC) in adult versus fetal tissue (Lister *et al.*, 2013). These findings suggest that many of the fetal > adult hmCGs are intermediates in developmentally determined DNA demethylation in the brain (Lister *et al.*, 2013). However, other intragenic regions enriched in 5hmC in fetal brain, especially in genes specifically associated with neurons or astrocytes, are still enriched in 5hmC in adult brain, indicating that some of the fetal 5hmC is long-lived 5hmC probably with its own functionality.

In adult prefrontal cortex 13% of all CpGs were reported to have higher average 5hmC levels than 5mC contents, and there was a strong overrepresentation of 5hmC in gene regions and intergenic regulatory elements (Wen *et al.*, 2014). In another study, it was shown that most of the changes in the hydroxymethylome occurred during the transition from fetus to young adult (Lister *et al.*, 2013), a time when synapse formation, learning and memory formation are at their peak. Similarly, maturing neurons in the cortical plate of the brain from EP15.5 mice have much higher levels of genomic 5hmC than neuronal progenitor cells (NPC) from fetuses (Hahn *et al.*, 2013). The increase in 5hmC is largely ascribable to increased gene-body methylation.

Genomic 5hmC is enriched in genes encoding tissue-specific proteins relevant to the studied cell type. For example, differentiation-linked alteration of DNA hydroxymethylation was found in relevant tissue-related genes in a study of the retinoic acid-induced differentiation of P19 embryonic carcinoma cells to neural cells and of insulin-induced conversion of 3T3 cells to adipocytes (Serandour *et al.*, 2012). The genes that had increased 5hmC upon differentiation of P19 and 3T3 cells were enriched in the following Gene Ontology terms: thalamus, future brain, neural ectoderm, and spinal cord (for P19 neural differentiation) and metabolism of lipids, adipose tissue and brown fat and also cellular response to insulin stimulus (for differentiation to adipocytes). Another example of tissue-relevant genes being enriched in 5hmC is the finding that age-related neurodegeneration-associated genes, including 15 of 23 genes linked to ataxia or Purkinje cell degeneration, are overrepresented among the ~5400 genes that have significantly more 5hmC in adult versus P7 mouse brain (Song *et al.*, 2011). In another study, upon differentiation of ESC to NPS and also of NPS to neurons, there were large numbers of DNA regions with increases and other regions with decreases in 5hmC (Kim *et al.*, 2013). Gains in 5hmC in the gene body were most overrepresented in genes involved in neurogenesis, ectoderm development, or signal transduction (e.g. Notch pathway genes).

The association of 5hmC with neurological disease

To test for a role of abnormalities in genomic 5hmC in a neurological disease, Wang *et al.* (2013) used a mice model of the autosomal dominant Huntington's disease (HD). The genome of these transgenic mice contains a disease-like expansion of a CAG repeat in the open reading frame of the gene encoding the Huntington protein. Concomitantly, the mice have many of the behavioural and neuropathological features of HD. Because neurons, the spinal cord, and various parts of the brain have by far the highest genomic 5hmC content of any studied cell or tissue types (Globisch *et al.*, 2011; Kriaucionis and Heintz, 2009; Munzel *et al.*, 2010; Terragni *et al.*, 2012) and 5hmC can be rapidly converted to C residues by replication-independent DNA repair even in postmitotic cells (Guo *et al.*, 2011b), changes in DNA hydroxymethylation might be a downstream consequence of having an HD allele. Indeed, by hydroxymethylome profiling of the striatum and the cortex of 3-month-old HD versus control mice, Wang *et al.* (2013) found that HD mice had about half as many 5mC-rich genomic peaks as did age-matched controls. This decrease might be due to the down-regulation of Tet genes that they observed in the striatum and cortex of the brains of HD versus control mice.

Importantly, these decreases in Tet and genomic 5hmC were seen in 3-month-old HD mice, which do not yet show the motor or pathological manifestations of HD (Wang *et al.*, 2013). This suggests that decreased DNA hydroxymethylation is an early event in the development of the disease. While much research is needed to understand the consequences of the strong decrease in 5hmC in HD brain, it is likely that some of the abnormal loss of 5hmC contributes to the pathology of HD. In addition, for a variety of other neurological diseases (including Rett syndrome, schizophrenia, fragile X-associated tremor/ataxia syndrome, and Alzheimer's disease) evidence is beginning to suggest the involvement of genomic 5hmC abnormalities (Cheng *et al.*, 2014).

Neurological diseases are not the only diseases manifesting pathologically linked changes in the hydroxymethylome. Many studies have shown abnormalities in genomic 5hmC in cancers, especially overall decreases and rearrangement in the location of the persisting 5hmC (Bhattacharyya *et al.*, 2013; Haffner *et al.*, 2011; Jin *et al.*, 2011a). These cancer-linked changes in the hydroxymethylome are reminiscent of cancer-linked DNA hypomethylation (Ehrlich, 2002), although generally greater in the per cent change.

5hmC in a variety of other tissue types

The strong tissue-specificity of global levels of 5hmC is well established (Globisch *et al.*, 2011; Terragni *et al.*, 2012, 2014). This includes differences in overall levels in different parts of the brain, although all studied brain regions have much higher levels of genomic 5hmC than did non-neural tissues (Munzel *et al.*, 2010). It is important to study the tissue-specificity of 5hmC in a range of tissues and cell types and not just in 5hmC-rich brain and ESC. Because of the expense and the work involved, examination of the tissue-specificity of hydroxymethylomes in a variety of samples has so far been done only rarely, with the notable exception of the above-cited study by Neri *et al.* (2013) in which single samples of ESC, MEF, brain and liver were analysed.

A limited 5hmC genomic profiling (hmeDIP and hybridization to DNA probes for a very small fraction of the genome on a microarray) was done on biological duplicates or triplicates of liver, brain, testes, colon, placenta and brain (Nestor *et al.*, 2012). Examining mostly the *HOXA* and *APO* gene clusters and the *HBB* and *CFTFR* genes, Nestor *et al.* (2012) demonstrated that the distribution of 5hmC was tissue-specific with little inter-individual variation. They quantified 5hmC and 5mC at selected CCGG sites with an assay that is similar to the above-described hydroxymethylome analysis with β-GT, except that only individual sites are tested by digestion in parallel with MspI and HpaII restriction endonucleases after *in vitro* glucosylation of 5hmC, followed by qPCR to quantify the extent of digestion (β-GT/RE/q-PCR assay). The levels of 5hmC at tested CCGG sites in 12 loci did not display an overall correlation with RNA levels of the corresponding genes determined by qRT-PCR. However, microarray analysis indicated some correlations, namely transcription of *HOXA* genes was positively associated with 5hmC levels in a comparison of breast versus brain.

Myogenic differential DNA methylation, 5hmC, and comparisons of 5hmC and 5mC levels

In our recent studies of differential methylation in the skeletal muscle lineage (myoblasts, Mb; myotubes, Mt, skeletal muscle tissue) by RRBS, we found evidence for massive DNA demethylation at later stages in development (Terragni *et al.*, 2014; Tsumagari *et al.*, 2013a,b). In these studies we compared skeletal muscle tissue with myogenic progenitor cells, myoblasts (Mb) and myotubes (Mt), their syncytial differentiation product. We first determined the set of CpG sites genome wide that were differentially methylated (DM) in the set of Mb plus Mt (MbMt) versus 16 types of non-muscle cell cultures using stringent criteria. The MbMt DM sites were in turn compared to DM sites from skeletal muscle tissue versus 14 types of normal non-muscle tissues (muscle DM sites). Although there are changes in DNA methylation that occur during cell culture, this type of comparison is valid because RRBS-derived methylomes from 82 human cell cultures or tissues showed that DM sites in cultured cells versus analogous cultured tissue occur mostly as unmethylated CpGs in the bodies of genes involved in controlling cellular proliferation (Varley *et al.*, 2013). In addition, in our study, MbMt DM sites were derived only by comparing myogenic to non-myogenic primary cell cultures (with the exception of lymphoblasts), and the muscle DM sites from a comparison of muscle to non-muscle tissues. Moreover, 30% of the MbMt-hypomethylated sites overlapped muscle hypomethylated DM sites (Tsumagari *et al.*, 2011). Demethylation occurring from the Mt to the skeletal muscle tissue stage was evidenced by our finding that 91% of genes with MbMt-hypermethylated sites did not have hypermethylated sites in skeletal muscle. Importantly, the genes exhibiting this loss of MbMt-hypermethylated sites were highly enriched in transcription factor genes, especially those involved in early development (*HOX* genes and T-box genes). This result is consistent with the hypothesis that the loss of DNA methylation after the Mt stage is related to differentiation-linked changes in gene expression. Of the small number of genes with significantly hypermethylated sites in skeletal muscle (89 genes), 83% were also MbMt hypermethylated. This finding suggests that most of the DNA hypermethylation found in muscle was acquired by the myogenic progenitor stage. In contrast, it appears that most of the DNA hypomethylation in muscle was acquired subsequently.

Besides the evidence for massive DNA demethylation in the skeletal muscle lineage after the Mt stage of development, another characteristic of Mb and Mt that led to our study of 5hmC in the muscle lineage was the especially high levels of *TET1* and *TET2* RNA in Mb and Mt. Therefore, we looked for 5hmC in the genome at selected CCGG sites that were also MbMt DM sites as well as a few sites in their vicinity (Terragni *et al.*, 2014; Tsumagari *et al.*, 2013a,b). Both 5mC and 5hmC levels were examined by the above-described β-GT/RE/qPCR assay. This assay not only quantifies both 5hmC and 5mC simultaneously by a simple procedure, but also it allows rapid determinations on many different samples types with biological replicates (two to three for each of eight sample types assayed in Table 4.3). Hydroxymethylation on only one strand at a CCGG duplex site will score positive for 5hmC in this assay (Kinney *et al.*, 2011). Therefore, the assay determines the percentage of hydroxymethylation on at least one strand of each studied CCGG duplex region. Most of the sites that we examined (Table 4.3) showed a high level of reproducibility between biological duplicates as contrasted with major tissue/cell type-specific differences, which is consistent with the findings of Nestor *et al.* (2012).

Although the majority of the studied CCGG sites were MbMt DM sites, we observed much

Table 4.3 Levels of 5hmC at assayed CCGG sites in gene

Location of examined site	Overlap CGI?	Average 5hmC content: 5hmC/(5hmC + 5mC + unmodified C)							
		Muscle	Myoblasts	Myotubes	Heart	Brain	Leucocytes	Fibroblasts	HUVECs
TBX1 (last) exon 9	Yes	0.20	0.00	NA	0.01	0.02	0.00	0.00	0.12
PAX3 −7.9 kb	No	0.23	0.00	NA	0.01	0.03	0.00	NA	0.00
EBF3 −4.2 kb	Yes	0.26	0.00	0.00	0.03	0.02	0.00	0.00	0.00
HOXB5 intron1	No	0.34	0.06	NA	0.04	0.02	0.02	0.03	0.02
HOXA5 exon 1	Yes	0.02	0.00	NA	0.00	0.01	0.00	0.00	0.00
HOXA7 upstream	No	0.04	0.00	NA	0.00	0.01	0.02	NA	0.00
HOXD4 intron 1	No	0.04	0.00	NA	0.02	0.03	0.00	0.03	0.06
PITX2 exon 2	Yes	0.04	0.01	NA	0.00	0.01	0.02	0.03	0.02
HOXC6 exon 2	Yes	0.04	0.01	NA	0.00	0.02	0.00	0.02	0.00
MYF6 +238 bp	No	0.02	0.00	0.03	0.02	0.18	0.00	0.00	0.00
MYOD1 −48 kb	No	0.05	0.00	NA	0.18	0.12	0.00	0.00	0.02
NOTCH1 intron 2, Mb enh.	No	0.37	0.00	0.01	0.81	0.30	0.03	0.00	0.03
NOTCH1 exon 32	Yes	0.78	0.00	0.02	0.67	0.54	0.00	0.01	0.03
DLL1 TTS +56 kb	No	0.43	NA	NA	0.28	0.47	0.00	NA	0.07
DLL1 TTS +97 kb, Mb enh.	No	0.63	0.00	0.05	0.30	0.34	0.00	0.00	0.32
JAG2 intron 5, Mb enh.	No	0.00	0.00	0.08	0.29	0.57	0.01	0.01	0.05
MYOD1 −25 kb, Mb enh.	No	0.01	0.25	0.30	0.18	0.40	0.00	0.01	0.02
NOTCH1 exon 20	Yes	0.08	0.01	0.00	0.16	0.28	0.00	0.01	0.05
MYOD1 −8.9 kb	No	0.22	0.00	NA	0.16	0.36	0.00	0.00	0.00
NOTCH2 exon 24	No	0.13	NA	NA	0.16	0.23	0.00	NA	0.02
MYOD1 −45.5 kb, Mb enh.	No	0.31	0.13	0.16	0.06	0.24	0.00	0.01	0.02
PITX3 exon 3	Yes	0.47	0.00	0.00	0.05	0.09	0.00	0.00	0.03

[1]5hmC (above) and 5mC (not shown) were quantified by the β-GT/RE/qPCR assay for two to three biological replicates of the human DNA samples [skeletal muscle, myoblasts, myotubes, heart, cerebellum, leucocytes, skin fibroblasts, and human umbilical vein endothelial cells (HUVECs)] for a tested site in indicated gene regions. The position of the assay site is shown relative to the transcription start site except in the case of the DLL1 sites, which are given relative to the transcription termination site. The cultured cell populations were primary cells. Mb enh, sites embedded in chromatin with histone modifications typical of enhancers as explained in the text. The samples are arranged in order of a hierarchical cluster analysis from individual values for each sample (using the Manhattan distance and then clustered with the Ward algorithm) with lines separating each cluster. This table is based on published (Terragni et al., 2014; Tsumagari et al., 2013a,b) and unpublished data.

more frequent 5hmC in skeletal muscle tissue, brain, and heart muscle than in Mb or Mt. Only 2 of the 22 studied CCGG sites (9%) displayed > 10% of their DNA bases as 5hmC in Mb or Mt (Table 4.3). In contrast, 12 of the sites (54%) had 10–78% 5hmC in skeletal muscle. The two sites enriched in 5hmC in Mb were both in enhancer regions far upstream of *MYOD1* (−25 and −45.5 kb relative to the TSS). MYOD1 is a myogenic transcription factor that is very highly and specifically expressed in Mb, Mt, and skeletal muscle (Blum and Dynlacht, 2013). These two sites enriched in 5hmC in Mb (averages of 25% and 13% 5hmC) were also enriched in 5hmC in Mt (averages of 30% and 16%, respectively). The enrichment of 5hmC at these two sites is in accord with the previously described overrepresentation of 5hmC in enhancers; however, three other sites that were embedded in enhancer-type chromatin in Mb (*NOTCH1* intron 2, *DLL1* +97 kb and *JAG2* intron 5) had no detectable 5hmC in Mb although two of them were very rich in 5hmC in skeletal muscle (Table 4.3).

Enrichment in 5hmC was observed in exons, introns and intergenic regions and in both CGI and non-CGI regions. Complex patterns of tissue/cell type-specificity for the levels of 5hmC were seen. At most studied sites associated with genes that are expressed in the skeletal muscle lineage, skeletal muscle displayed much higher levels of genomic 5hmC than did the myogenic progenitor cells. This is consistent with the finding that a variety of less differentiated cell types show much less immunohistochemically detectable 5hmC than do their more differentiated derivatives in the same tissue slide (Haffner *et al.*, 2011). An exception to this finding of more differentiation/less expression is for ESC which undergo a decrease in 5hmC content in their DNA upon differentiation (Ficz *et al.*, 2011).

As we previously reported (Terragni *et al.*, 2014; Tsumagari *et al.*, 2013a, b), many of the examined 5hmC-containing sites in skeletal muscle, heart, and brain had much more 5hmC than 5mC. This result is especially noteworthy because 5mC levels in mammalian DNA (including in 5hmC-rich brain DNA) are usually higher than 5hmC levels at a given CpG position that is positive for 5hmC, as seen in β-GT/RE/qPCR assays and hydroxymethylome/methylome comparisons (Lister *et al.*, 2013; Sun *et al.*, 2013; Terragni *et al.*, 2014). For example, by Tab-seq the median levels of 5hmC and 5mC in adult brain were 29% versus 60%, respectively (Wen *et al.*, 2014), and 21 versus 61%, respectively, in ESC (Yu *et al.*, 2012b).

With the 5hmC data shown in Table 4.3 and the accompanying 5mC data from each of these assays (not shown), we tested for the relationship between the levels of 5mC and the fraction of

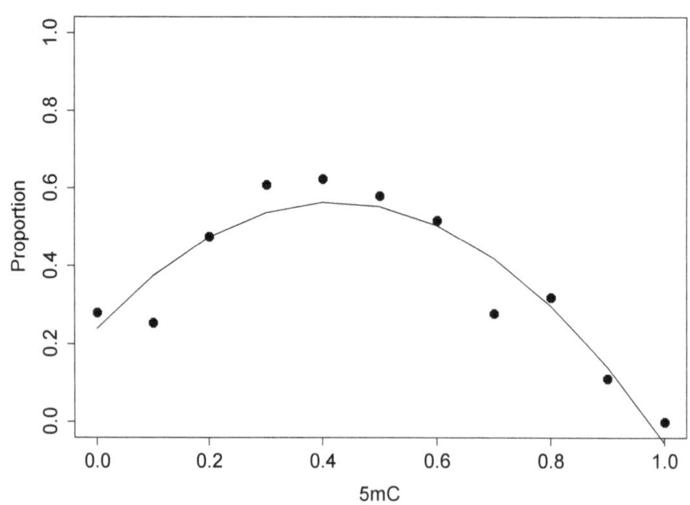

Figure 4.1 The proportion of sites with 5hmC equal to or greater than 5% of total C for specified binned levels of 5mC with a fitted quadratic curve. Adjusted $R^2 = 0.86$.

assayed CCGG sites that displayed >5% 5hmC. At the assayed CCGG sites, Fig. 4.1 shows there is a trend for intermediate 5mC levels to be associated with the highest proportions of sites containing appreciable 5hmC (>5% 5hmC). In addition, CCGG sites with moderately low levels of 5mC were more likely to be enriched in 5hmC than were CCGG sites with moderately high levels of m5C. This observation suggests that highly methylated CpGs are less of a target for hydroxymethylation. In addition, at many modified CpG sites, demethylation and re-methylation are probably in dynamic equilibrium, with the loss of 5mC being favoured over loss of 5hmC in post-mitotic nuclei of skeletal muscle, heart and brain.

5hmC versus 5mC: chromatin modifications, transcription factors, DNA strand biases and proposed functions

Associations of 5hmC and 5mC with histone modifications

In human and mouse ESC, 5hmC was more strongly associated with H3K4me1 (an enhancer-type modification) than with H3K4me3 (an active promoter-type modification) and more weakly with H3K27Ac (active enhancer- or promoter-type modification in conjunction with H3K4 methylation) (Sun *et al.*, 2013; Szulwach *et al.*, 2011). This finding is consistent with studies describing a stronger correlation of 5hmC with enhancers or poised enhancers than with the region immediately surrounding the TSS for moderately or highly transcribed genes (Tables 4.1 and 4.2). In contrast, 5mC is generally depleted in both the TSS-vicinity and enhancer chromatin. 5-Methylcytosine is clearly overrepresented in H3K36me3-marked regions (associated with transcription elongation) and in H3K9me3 (repressed) regions while 5hmC shows either little or no enrichment in such regions.

A moderately positive correlation between 5hmC and the polycomb repressive H3K27me3 mark was seen in a study of human ESC (Szulwach *et al.*, 2011). However, there may be cell type-specific differences in the dominant genome-wide relationships between 5hmC and polycomb repressive chromatin because in brain, H3K27me3-enriched regions were depleted in 5hmC although they contained higher than expected 5mC (Wen *et al.*, 2014). We (Tsumagari *et al.*, 2013a) and Kim *et al.* (2013) have proposed that gene regions which acquire both H3K27me3 and 5mC may have the need for double-layered protection against expression but this is on a cell population average. Recent findings indicate that individual chromatin molecules at a given genomic subregion is ESC or fibroplasts do not harbour both 5mC and H3K27me3 (Murphy *et al.*, 2013).

Hahn *et al.* (2013) compared whole-genome profiles of 5hmC and 5mC in DNA from fetal NPC and neurons by HMeDIP-chip (Table 4.2) and by an immunoprecipitation method for the methylome analysis. The biggest change in DNA epigenetics from NPC to neurons was the gain in gene-body 5hmC in genes that are activated during differentiation. For example, genes with the most loss of the repression-associated H3K27me3 mark in the gene body in the differentiated cells (top 15%) often gained 5hmC in the gene body and were prone to have a higher expression level than genes not losing H3K27me3 from the gene body. In contrast, there was little overall change in the methylome-derived, gene-subregion profile for 5mC upon differentiation *in vivo*. Moreover, in both neuron and NPC methylomes and hydroxymethylomes, the two types of cytosine modification were enriched at the 3′ end of genes but differed at the 5′ ends. Enrichment in 5hmC was seen in the several kb surrounding the TSS preferentially in genes that did not have gene body hydroxymethylation (and probably had only low transcription rates), but there was depletion of 5mC surrounding the TSS whether or not gene body hydroxymethylation was present.

Mechanistic insights into the associations of 5hmC and 5mC with chromatin structure, CTCF binding, gene expression and differentiation

The above-described depletion of 5mC around the TSS is probably largely due to the unmethylated CGIs overlapping most promoters [including the region immediately downstream of the TSS (Brenet *et al.*, 2011)]. At a minority of promoters, there are clusters of DNA methylated sites that can work in concert with differential binding of transcription factors and repressive histone modifications to down-modulate or largely repress gene expression (Della Ragione *et al.*, 2012; Lister *et al.*, 2013). In some cis-acting transcription regulatory sequences, differential methylation of an individual transcription factor binding site may alter gene expression directly by affecting binding of methylation-sensitive DNA-binding proteins (De Andres *et al.*, 2013; Pogribny *et al.*, 2000; Sengupta *et al.*, 1999). The ability of cytosine methylation to locally alter the extent of binding of sequence-specific DNA binding proteins to their cognate sites is being increasingly recognized (Hu *et al.*, 2013; Rodriguez *et al.*, 2010; Wang *et al.*, 2012).

Initiation of gene expression involves not only the regulation of specific binding of the RNA polymerase II (RNA PolII) complex to the promoter region, but also release of the complex so that it can begin movement down the gene (Chen *et al.*, 2011). Often the RNA PolII complex stalls within the first 0.2 kb of the TSS resulting in a lack of transcription of all but the very 5′ end of the gene. Tao and coworkers have shown that moderately high levels of cell type-specific CpG methylation at a CGI overlapping the TSS of *HoxC6* and also of *HoxC8* is permissive for binding of the RNA PolII complex in MEF (Tao *et al.*, 2011). However, this methylation was associated with stalling of the complex within several hundred base-pairs of the TSS. This methylation-associated stalling of RNA PolII, and not the inhibition of transcription initiation, can explain the repression at these genes in MEF.

As described above (Tables 4.1 and 4.2), 5hmC is depleted around the TSS of moderately and highly expressed genes in various tissues or cell types although it often displays high levels of binding approximately 0.5–2 kb on either side of the TSS of such active genes (Colquitt *et al.*, 2013; Hahn *et al.*, 2013; Neri *et al.*, 2013; Szulwach *et al.*, 2011). This pattern of 5hmC enrichment, frequently seen as a bifurcated peak, is similar to that described for the distribution of the active promoter- and enhancer-associated histone variant H2A.Z around the TSS of mouse lymphoid cells (Conerly *et al.*, 2010). The positioning of H2A.Z was found to be positively correlated with 5hmC in genome-wide profiles while 5mC is negatively correlated (Coleman-Derr and Zilberman, 2012; Yu *et al.*, 2013b).

Similar to the enrichment of 5hmC at the borders of promoters of actively transcribed genes, this base was also enriched at the borders of >1-kb regions identified solely by their low 5mC content (≤ 10%) (Jeong *et al.*, 2014). This association may partly reflect 5hmC deposition at the borders of unmethylated CGIs to prevent DNA methylation immediately outside an unmethylated CGI from aberrant spreading into the CGI (Jin *et al.*, 2014). Such spreading might be more likely during conditions of gene expression. In addition, 5hmC surrounding an active promoter might help directly in modulating gene expression levels, e.g. by being part of enhancer-like chromatin (Kim *et al.*, 2013; Neri *et al.*, 2013; Serandour *et al.*, 2012; Wen *et al.*, 2014) bordering a region of active promoter chromatin. In contrast to actively transcribed genes, untranscribed or poorly transcribed genes often have 5hmC in the immediate vicinity of the TSS in a given cell or tissue type (Hahn *et al.*, 2013; Neri *et al.*, 2013; Szulwach *et al.*, 2011). For ESC, but not other tested cell types, this enrichment of 5hmC at the TSS of weakly or untranscribed genes can be partly ascribed to the recruitment of Tet1 by the Polycomb repressive complex that generates repressive H3K27me3 at bivalent promoters (Neri *et al.*, 2013).

In addition to differential DNA modification controlling gene expression locally at promoters, large regions of DNA can be enriched in 5mC or 5hmC to coordinately control expression of adjacent genes. High levels of DNA methylation over such a large multigenic region can be associated with gene silencing, especially for clusters of functionally related genes (Tsumagari *et al.*, 2013a). In contrast, a multigenic region encompassing the anterior portion of the *HOXA* gene cluster became enriched in 5hmC upon differentiation of EC cells by exposure to retinoic acid as *TET2* and *TET3* RNA levels increased (Bocker *et al.*, 2012). Concomitant with the large increase in 5hmC in this region was the activation of transcription of these genes. When TET2 activity was knocked down there was less of an increase in 5hmC and less of a decrease in 5mC at several assayed sites and, coordinately, less expression of *HOXA* genes.

A part of the enrichment in 5hmC in multigenic regions and in gene bodies of actively transcribed genes is probably due to the association of 5hmC with enhancer activity in these regions (Chandra *et al.*, 2014; Lister *et al.*, 2013) as well as to a role in regulating RNA splicing (see below). Enhancers are usually depleted in 5mC but can be enriched in 5hmC (Table 4.2). How 5hmC is influencing enhancer function and also the inclusion or exclusion of exons during RNA splicing is only just starting to be elucidated. However, inverse relationships between DNA methylation and binding of CTCF, a transcription factor and insulator protein, are likely to be one mechanism by which differential modification of genomic C residues affects large-scale chromatin structure, promoter function, and enhancer function. CTCF is a ubiquitously expressed protein implicated in transcription activation, repression, imprinting, and insulation of enhancers from promoters. It also plays a major role in mediating intrachromosomal looping and interchromosomal contacts (Phillips and Corces, 2009). It is estimated that about 40% of the cell type-associated differences in CTCF binding to DNA are due to cell-type specific differential methylation. Increased methylation at CTCF DNA recognition sequences containing CpG dinucleotides inhibits the binding of CTCF (Wang *et al.*, 2012). CTCF binding to DNA can determine the location of intrachromosomal loops, which are able to mediate the ability of CTCF to serve as an insulator by disrupting enhancer/promoter interactions, including for intragenic enhancers (Yu *et al.*, 2013a).

In addition, CTCF can help regulate RNA splicing, e.g. by increasing the inclusion of weak upstream exons via mediating local pausing of RNA polymerase II (Wang *et al.*, 2012). Regulating pausing of transcription within the body of a gene can create a favourable conformation of the nascent RNA for a given set of alternative splice sites Methylation of a specific exon, in turn, can alter splicing patterns by inhibiting CTCF binding (Wang *et al.*, 2012). In contrast to 5mC depletion at CTCF binding sites detected by chromatin immunoprecipitation/next-generation DNA sequencing (ChIP-seq), 5hmC was found to be overrepresented at such regions when using low-resolution hydroxymethylome analysis (Ficz *et al.*, 2011; Stroud *et al.*, 2011). With single-base resolution identification of 5hmC, 5hmC was seen to form a bifurcated peak of binding centred around the CTCF site with a trough at the site and maxima about 150 bp (close to the length of one nucleosome) on each side surrounded by several yet lower peaks of binding with a similar periodicity (Sun *et al.*, 2013; Yu *et al.*, 2012b). These findings suggest that the postulated role of 5hmC in regulating alternative splicing of RNA or enhancer function could be mediated in part by nucleosome deposition surrounding CTCF sites.

As mentioned above, 5hmC and 5mC residues differ greatly in their contributions to the helix stability of DNA sequences enriched in one or the other of these modified cytosine residues (Ehrlich *et al.*, 1975; Wanunu *et al.*, 2011). However, it is probably the ability of some DNA-binding proteins to distinguish between the 5-methyl versus the 5-hydroxymethyl group of genomic cytosine residues that plays the largest role in differential effects of 5mC and stable 5hmC on gene expression and possibly other DNA functions. DNA-binding proteins may be able to distinguish DNA sequences rich in the hydrophilic hydroxymethyl moiety of 5hmC compared with sequences rich in the hydrophobic methyl moiety of 5mC. Steric effects could also alter protein binding since the hydroxymethyl group with its associated water molecules would be much larger than the methylated C in the binding cavity. Enrichment of stable 5hmC in a gene subregion in one cell type that is 5hmC-poor and 5mC-rich in another may just abrogate the effect of 5mC. Alternatively, the 5hmC may have its own functionality different from that of 5mC. In addition, as described above, the main function of 5hmC in some specific genome regions, cell types, and developmental contexts is probably to serve as an intermediate in DNA demethylation (Guo *et al.*, 2011b).

The persistence of hydroxymethylation of DNA cytosine residues introduced into DNA by TET1, TET2, or TET3 depends on whether the DNA subsequently replicates or is removed by DNA repair enzymes without further oxidation by these enzymes (Kohli and Zhang, 2013). Unlike 5mC, 5hmC is not recognized by a known maintenance system targeted to asymmetrically modified CpGs (Otani *et al.*, 2013) so that replication of hemi-hydroxymethylated DNA can lead to depletion of the modified base (Hackett *et al.*, 2013). Probably the substantial amount of genomic 5mC (~19–35%) in human and mouse ESC genomic and adult brain found in non-CpG contexts (Lister *et al.*, 2013; Wen *et al.*, 2014; Xie *et al.*, 2012; Yu *et al.*, 2012b), especially in CHH (H, any base except G), can also be diluted out upon DNA replication. However, in ESC cells the levels of the de novo DNMTs (DNMT3A and DNMT3B) and of TET enzymes are high and so can replenish non-CpG C methylation as well as 5hmC in DNA. In brain many of the cells are post-mitotic and so escape DNA replication-related dilution.

The prevalence of correlations between the epigenetics, gene expression and mammalian differentiation as well as candidate mechanisms for cause-and-effect relationships between changes in genomic cytosine modification and changes in gene expression argue that 5hmC or 5mC influences the initiation or maintenance of gene expression. However, some of the cell type-specific differences in the distribution of 5hmC along a given DNA region are likely to be side-effects of incomplete specificity of differentiation-linked targeting of DNA hydroxymethylation. Which differentiation- or physiology-linked changes in DNA epigenetics are regulatory will need to be determined on a case-by-case basis using methods mimicking as closely as possible normal *in vivo* epigenetics.

Associations of 5hmC and 5mC with transcription factors

Some transcription regulatory proteins bind preferentially to 5mC-rich DNA sequences regardless of the exact sequence although, in the context of chromatin, they may bind to certain other DNA sequences or chromatin regions even when they are not methylated (e.g. MeCP2) (Guy *et al.*, 2011). Other proteins bind preferentially to methylated DNA in a sequence-specific manner, at least for certain of their binding sites, as we first reported (RFX proteins/MDBP) (Huang *et al.*, 1984; Zhang *et al.*, 1993)) and has been shown recently to be true for many transcription factors evaluated in methylated DNA-binding screens (e.g. RFX5 and RFX1) (Hu *et al.*, 2013; Spruijt *et al.*, 2013). Furthermore, RFX5 and RFX1 bound poorly to their cognate sequences containing 5hmCpG or unmodified CpG instead of 5mCpG (Spruijt *et al.*, 2013). Yet other DNA-binding proteins bound specifically to certain sequences containing either 5mCpG or 5hmCpG (e.g. Kaiso) or just 5hmCpG (e.g. Uhrf2) (Spruijt *et al.*, 2013). In addition, some DNA-binding proteins require an unmethylated CpG (e.g. MLL1) (Birke *et al.*, 2002). However, interpretation of the results from studies of the relative binding of nuclear proteins to DNA is complicated by technical considerations about the different types of assays and the nature of the cells of origin of the proteins (possibly due to different protein modifications or concentrations of the DNA-binding protein or cofactors). In addition, binding can be greatly influenced by whether or not the DNA is in normal chromatin context (Hashimoto *et al.*, 2012; Mellen *et al.*, 2012; Spruijt *et al.*, 2013; Yildirim *et al.*, 2011). Besides affecting the binding of transcription factors, cytosine methylation or hydroxymethylation might have more direct effects on chromatin structure, including interacting with modified histones and altering nucleosome phasing or stability in certain chromatin contexts.

Associations of 5hmC and 5mC with DNA strands and base composition of adjacent sequences

Although 5hmC and 5mC are often (but not always, Table 4.3) depleted in CGIs and instead favour regions of low CpG content (Feldmann *et al.*, 2013; Sun *et al.*, 2013; Yu *et al.*, 2012b), they differ in sequence or strand preferences. Stroud *et al.* (2011) found that 5hmC-rich regions on a given strand of ESC DNA often had a base composition asymmetry such that the 5′ border of the region on one strand was G-rich and the 3′ border on the same strand was C-rich. Yu *et al.* (2012b), also studying ESC DNA, reported that there was frequent strand asymmetry at 5hmCpG dinucleotide pairs with one strand having an average of almost twice as much 5hmC as the opposite strand (average of 20% 5hmC on the strand with the higher 5hmC content and 11% on the opposite strand). In addition, they found that the strand with the higher 5hmC content was usually the strand with the higher local G content in the 100 bases surrounding around the 5hmC. In contrast, strand asymmetry for 5mC was found at only ~9% of methylated CpG dinucleotide pairs. With respect to the local sequence context of cytosine modification, 5mC is predominantly in CpG dinucleotides with the prominent exceptions of ESC (before differentiation) and the neural lineage, where non-CpG 5mC is implicated in contributing to the regulation of gene expression (Lister *et al.*, 2013; Varley *et al.*, 2013; Wen *et al.*, 2014; Yu *et al.*, 2012b). In ESC and neural cells, 5hmC is almost always bordered by a G residue at the 3′ side unlike 5mC, for which a substantial, albeit minor, fraction is at non-CpG sites.

Two other important features of genomic 5hmC were recently discovered. In a study of human and mouse samples, Wen *et al.* (2014) demonstrated that brain, neurons and astrocytes have a small, but highly significant, bias in 5hmC levels towards the sense strand versus the antisense strand (average bias, ~0.7% 5hmC) in the gene bodies of highly transcribed genes. This bias was seen throughout the gene bodies of actively transcribed genes but not in intergenic regions nor in untranscribed genes. The bias was apparent, but weaker, in moderately transcribed genes. The opposite result was found for DNA methylation, because there is a small but significant strand bias in 5mC for the antisense strand, rather than the sense strand, of highly transcribed genes (average bias, ~ 1.0% 5mC). These strand biases were observed for constitutive as well as cell type-specific genes. The biases suggest regulation of transcription elongation by DNA methylation and hydroxymethylation in a subset of genes distinguished by certain undefined DNA sequences or chromatin structures within the gene body. In addition to this relationship between transcription and hydroxymethylation of DNA, from a study of CGI genome-wide methylation, evidence was provided by Otani *et al.* (2013) that Dnmt3a or Dnmt3b is needed for hydroxymethylation of DNA in ESC. Therefore, hydroxymethylation of CGI in ESC seems to be dependent on newly introduced DNA methylation. This conclusion is consistent with a role for 5hmC in keeping levels of 5mC at CGI low through its function as an intermediate in DNA demethylation.

DNA hydroxymethylation and methylation: cellular heterogeneity and dynamics

The findings that 5hmC enrichment usually involves only moderate-to-low levels of genomic hydroxymethylation at a given CpG site and that there are usually higher average 5mC levels than 5hmC levels at such 5hmC-enriched sites need to be considered to understand the biological functions of DNA hydroxymethylation. For example, in ESC analysed by oxBS-seq and BS-seq (using the RRBS protocol), the highest levels of 5hmC were seen among C residues that had 25–50% methylation but most of those sites had only about 8–12% hydroxymethylation (Booth *et al.*, 2013). Similarly, in ESC or brain tissue, the scales used for modified C levels in figures comparing genome-wide 5hmC and 5mC often differ by a factor of 2 to 5 (Booth *et al.*, 2013; Lister *et al.*, 2013; Wen *et al.*, 2014). This is done to compensate for the lower levels of 5hmC than of 5mC, even in 5hmC-enriched subregions of the genome; however, there are prominent exceptions at individual genes (Table 4.3, CCGG sites in *NOTCH1* exon 32 in skeletal muscle and in *NOTCH1* intron 2 in heart). An additional source of 5mC at 5hmC-enriched sites can be asymmetric hydroxymethylation and methylation (5′-5hmCpG-3′/3¢-Gp5mC-5′) (Yu *et al.*, 2012b).

The frequent mixture of 5mC, 5hmC, and unmodified C at a given site in steady-state populations of either cycling or postmitotic cells suggests the dynamic nature of C modification *in vivo*. The heterogeneity in 5hmC levels at a given DNA site can be a type of biologically useful variation, especially if clusters of sites (Chen and Riggs, 2005) differ in 5hmC levels. An example of a category of genes exhibiting tissue-specific 5hmC enrichment that may help control gene expression is the Notch genes, which play critical roles in cell fate decisions. Within or near assayed sites in several Notch genes there is strong, but partial, enrichment of 5hmC in skeletal muscle and brain (Hahn *et al.*, 2013; Terragni *et al.*, 2014).

Heterogeneity of genomic C modification among individual nuclei in these two very dynamic and largely postmitotic tissues and the reversibility of C modification by active processes might contribute functional diversity needed by these tissues in long-lived physiological responses to the environment. The highly negative side of this natural, partly stochastic, variability in DNA C modification and the mechanisms for converting unmodified C to modified C or vice versa is that the modification of CpGs is a breeding ground for cancer-associated epigenetic changes (Ehrlich and Lacey, 2013a; Jin *et al.*, 2011a; Liu *et al.*, 2013).

Conclusions

Much evidence indicates that 5hmC in mammalian DNA acts sometimes as an intermediate in DNA demethylation by active or passive (DNA replication-dependent) methods, and other times has independent functionality. The role of 5hmC in DNA demethylation may often involve transient production of this modified base to help keep certain methylation-prone regions mostly unmethylated (Feldmann *et al.*, 2013), e.g. at the edges of unmethylated CpG islands (Jin *et al.*, 2014). This modified base is implicated in transcription up-regulation (especially in gene bodies, active enhancers, and surrounding the TSS, but not in its immediate vicinity), down-regulation (at and close to the TSS), and dynamic modulation (poised enhancers). In addition, much evidence suggests that 5hmC is involved in regulating RNA splicing occurring co-transcriptionally. 5-Hydroxymethylcytosine residues in DNA may be more abundant and influential than indicated by their steady-state levels because they may often be only transiently present in certain genomic locations.

The effects of genomic 5hmC on transcription might be mediated by negating interactions of transcription factors with sites previously rich in 5mC, by binding selectively to certain other transcription factors, by altering directly the stability of DNA or the conformation of chromatin or, indirectly, by favouring differential histone modification. Genomic 5hmC clearly has a close relationship to differentiation and differential control of gene expression that is cell type related and development stage related. This base is generally more prevalent in slowly cycling or non-cycling cell types consistent with its propensity to be diluted out by DNA replication because asymmetrical 5hmCpG base pairs (5′-5hmCpG-3′/3¢-GpC-5′ and 5hmCpG/3¢-Gp5mC-5′) are not well recognized for maintenance DNA methylation as are hemimethylated base pairs (5′-5mCpG-3′/3’-GpC-5′). Moreover, as an intermediate in DNA demethylation, 5hmC may be especially important in dynamic tissues such as brain and skeletal muscle in helping to regulate gene expression in response to physiological changes.

Acknowledgements

Supported in part by grants from the National Institutes of Health to ME (NS04885) and to SP 4R44GM096723-02 and internal funding for basic research at New England Biolabs.

References

Almeida, R.D., Loose, M., Sottile, V., Matsa, E., Denning, C., Young, L., Johnson, A.D., Gering, M., and Ruzov, A. (2012a). 5-hydroxymethyl-cytosine enrichment of non-committed cells is not a universal feature of vertebrate development. Epigenetics 7, 383–389.

Almeida, R.D., Sottile, V., Loose, M., De Sousa, P.A., Johnson, A.D., and Ruzov, A. (2012b). Semi-quantitative immunohistochemical detection of 5-hydroxymethyl-cytosine reveals conservation of its tissue distribution between amphibians and mammals. Epigenetics 7, 137–140.

Anastasiadou, C., Malousi, A., Maglaveras, N., and Kouidou, S. (2011). Human epigenome data reveal increased CpG methylation in alternatively spliced sites and putative exonic splicing enhancers. DNA Cell Biol. *30*, 267–275.

Bhattacharyya, S., Yu, Y., Suzuki, M., Campbell, N., Mazdo, J., Vasanthakumar, A., Bhagat, T.D., Nischal, S., Christopeit, M., Parekh, S., *et al.* (2013). Genome-wide hydroxymethylation tested using the HELP-GT assay shows redistribution in cancer. Nucleic Acids Res. *41*, e157.

Bianchi, C., and Zangi, R. (2013). UHRF1 discriminates against binding to fully methylated CpG-sites by steric repulsion. Biophys. Chem. *171*, 38–45.

Birke, M., Schreiner, S., Garcia-Cuellar, M.P., Mahr, K., Titgemeyer, F., and Slany, R.K. (2002). The MT domain of the proto-oncoprotein MLL binds to CpG-containing DNA and discriminates against methylation. Nucleic Acids Res. *30*, 958–965.

Blum, R., and Dynlacht, B.D. (2013). The role of MyoD1 and histone modifications in the activation of muscle enhancers. Epigenetics *8*, 778–784.

Bocker, M.T., Tuorto, F., Raddatz, G., Musch, T., Yang, F.C., Xu, M., Lyko, F., and Breiling, A. (2012).

Hydroxylation of 5-methylcytosine by TET2 maintains the active state of the mammalian HOXA cluster. Nat. Commun. *3*, 818.

Booth, M.J., Branco, M.R., Ficz, G., Oxley, D., Krueger, F., Reik, W., and Balasubramanian, S. (2012). Quantitative sequencing of 5-methylcytosine and 5-hydroxymethylcytosine at single-base resolution. Science *336*, 934–937.

Booth, M.J., Ost, T.W., Beraldi, D., Bell, N.M., Branco, M.R., Reik, W., and Balasubramanian, S. (2013). Oxidative bisulfite sequencing of 5-methylcytosine and 5-hydroxymethylcytosine. Nat. Protoc. *8*, 1841–1851.

ten Bosch, J.R., and Grody, W.W. (2008). Keeping up with the next generation: massively parallel sequencing in clinical diagnostics. J. Mol. Diagn. *10*, 484–492.

Branco, M.R., Ficz, G., and Reik, W. (2011). Uncovering the role of 5-hydroxymethylcytosine in the epigenome. Nat. Rev. Genet. *13*, 7–13.

Brenet, F., Moh, M., Funk, P., Feierstein, E., Viale, A.J., Socci, N.D., and Scandura, J.M. (2011). DNA methylation of the first exon is tightly linked to transcriptional silencing. PLoS ONE *6*, e14524.

Chandra, S., Baribault, C., Lacey, M., and Ehrlich, M. (2014). Myogenic differential methylation: diverse associations with chromatin structure. Biology (Basel) *3*, 426–451.

Chen, M.L., Shen, F., Huang, W., Qi, J.H., Wang, Y., Feng, Y.Q., Liu, S.M., and Yuan, B.F. (2013). Quantification of 5-methylcytosine and 5-hydroxymethylcytosine in genomic DNA from hepatocellular carcinoma tissues by capillary hydrophilic-interaction liquid chromatography/quadrupole TOF mass spectrometry. Clin. Chem. *59*, 824–832.

Chen, Q., Chen, Y., Bian, C., Fujiki, R., and Yu, X. (2012). TET2 promotes histone O-GlcNAcylation during gene transcription. Nature *493*, 561–564.

Chen, Y., Jorgensen, M., Kolde, R., Zhao, X., Parker, B., Valen, E., Wen, J., and Sandelin, A. (2011). Prediction of RNA Polymerase II recruitment, elongation and stalling from histone modification data. BMC Genomics *12*, 544.

Chen, Y., Damayanti, N.P., Irudayaraj, J., Dunn, K., and Zhou, F.C. (2014). Diversity of two forms of DNA methylation in the brain. Front. Genet. *5*, 46.

Chen, Z.X., and Riggs, A.D. (2005). Maintenance and regulation of DNA methylation patterns in mammals. Biochem. Cell Biol. *83*, 438–448.

Cheng, Y., Bernstein, A., Chen, D., and Jin, P. (2014). 5-Hydroxymethylcytosine: A new player in brain disorders? Exp. Neurol. pii: S0014-4886(14)00150-2 (epub ahead of print).

Coleman-Derr, D., and Zilberman, D. (2012). DNA methylation, H2A.Z, and the regulation of constitutive expression. Cold Spring Harb. Symp. Quant. Biol. 77, 147–154.

Colquitt, B.M., Allen, W.E., Barnea, G., and Lomvardas, S. (2013). Alteration of genic 5-hydroxymethylcytosine patterning in olfactory neurons correlates with changes in gene expression and cell identity. Proc. Natl. Acad. Sci. U.S.A. *110*, 14682–14687.

Conerly, M.L., Teves, S.S., Diolaiti, D., Ulrich, M., Eisenman, R.N., and Henikoff, S. (2010). Changes in H2A.Z occupancy and DNA methylation during B-cell lymphomagenesis. Genome Res. *20*, 1383–1390.

De Andres, M.C., Imagawa, K., Hashimoto, K., Gonzalez, A., Roach, H.I., Goldring, M.B., and Oreffo, R.O. (2013). Loss of methylation in CpG sites in the NF-kappaB enhancer elements of iNOS is responsible for gene induction in human articular chondrocytes. Arthritis Rheum. *65*, 732–742.

Della Ragione, F., Filosa, S., Scalabri, F., and D'Esposito, M. (2012). MeCP2 as a genome-wide modulator: the renewal of an old story. Front. Genet. *3*, 181.

Ehrlich, M. (2002). DNA methylation in cancer: too much, but also too little. Oncogene *21*, 5400–5413.

Ehrlich, M., and Lacey, M. (2013a). DNA hypomethylation and hemimethylation in cancer. Adv. Exp. Med. Biol. *754*, 31–56.

Ehrlich, M., and Lacey, M. (2013b). DNA methylation and differentiation: silencing, upregulation and modulation of gene expression. Epigenomics *5*, 553–568.

Ehrlich, M., Ehrlich, K., and Mayo, J.A. (1975). Unusual properties of the DNA from Xanthomonas phage XP-12 in which 5-methylcytosine completely replaces cytosine. Biochim. Biophys. Acta *395*, 109–119.

Ernst, J., Kheradpour, P., Mikkelsen, T.S., Shoresh, N., Ward, L.D., Epstein, C.B., Zhang, X., Wang, L., Issner, R., Coyne, M., *et al.* (2011). Mapping and analysis of chromatin state dynamics in nine human cell types. Nature *473*, 43–49.

Feldmann, A., Ivanek, R., Murr, R., Gaidatzis, D., Burger, L., and Schubeler, D. (2013). Transcription factor occupancy can mediate active turnover of DNA methylation at regulatory regions. PLoS Genet. *9*, e1003994.

Ficz, G., Branco, M.R., Seisenberger, S., Santos, F., Krueger, F., Hore, T.A., Marques, C.J., Andrews, S., and Reik, W. (2011). Dynamic regulation of 5-hydroxymethylcytosine in mouse ES cells and during differentiation. Nature *473*, 398–402.

Ficz, G., Hore, T.A., Santos, F., Lee, H.J., Dean, W., Arand, J., Krueger, F., Oxley, D., Paul, Y.L., Walter, J., *et al.* (2013). FGF signaling inhibition in ESCs drives rapid genome-wide demethylation to the epigenetic ground state of pluripotency. Cell Stem Cell *13*, 351–359.

Freudenberg, J.M., Ghosh, S., Lackford, B.L., Yellaboina, S., Zheng, X., Li, R., Cuddapah, S., Wade, P.A., Hu, G., and Jothi, R. (2012). Acute depletion of Tet1-dependent 5-hydroxymethylcytosine levels impairs LIF/Stat3 signaling and results in loss of embryonic stem cell identity. Nucleic Acids Res. *40*, 3364–3377.

Globisch, D., Munzel, M., Muller, M., Michalakis, S., Wagner, M., Koch, S., Bruckl, T., Biel, M., and Carell, T. (2011). Tissue distribution of 5-hydroxymethylcytosine and search for active demethylation intermediates. PLoS ONE *5*, e15367.

Guo, J.U., Ma, D.K., Mo, H., Ball, M.P., Jang, M.H., Bonaguidi, M.A., Balazer, J.A., Eaves, H.L., Xie, B., Ford, E., *et al.* (2011a). Neuronal activity modifies the DNA methylation landscape in the adult brain. Nature Neurosci. *14*, 1345–1351.

Guo, J.U., Su, Y., Zhong, C., Ming, G.L., and Song, H. (2011b). Emerging roles of TET proteins and 5-hydroxymethylcytosines in active DNA demethylation and beyond. Cell Cycle *10*, 2662–2668.

Guy, J., Cheval, H., Selfridge, J., and Bird, A. (2011). The role of MeCP2 in the brain. Annu. Rev. Cell Dev. Biol. *27*, 631–652.

Hackett, J.A., Sengupta, R., Zylicz, J.J., Murakami, K., Lee, C., Down, T.A., and Surani, M.A. (2013). Germline DNA demethylation dynamics and imprint erasure through 5-hydroxymethylcytosine. Science *339*, 448–452.

Haffner, M.C., Chaux, A., Meeker, A.K., Esopi, D.M., Gerber, J., Pellakuru, L.G., Toubaji, A., Argani, P., Iacobuzio-Donahue, C., Nelson, W.G., *et al.* (2011). Global 5-hydroxymethylcytosine content is significantly reduced in tissue stem/progenitor cell compartments and in human cancers. Oncotarget *2*, 627–637.

Hahn, M.A., Qiu, R., Wu, X., Li, A.X., Zhang, H., Wang, J., Jui, J., Jin, S.G., Jiang, Y., Pfeifer, G.P., *et al.* (2013). Dynamics of 5-hydroxymethylcytosine and chromatin marks in mammalian neurogenesis. Cell Rep. *3*, 291–300.

Hashimoto, H., Liu, Y., Upadhyay, A.K., Chang, Y., Howerton, S.B., Vertino, P.M., Zhang, X., and Cheng, X. (2012). Recognition and potential mechanisms for replication and erasure of cytosine hydroxymethylation. Nucleic Acids Res. *40*, 4841–4849.

Hu, S., Wan, J., Su, Y., Song, Q., Zeng, Y., Nguyen, H.N., Shin, J., Cox, E., Rho, H.S., Woodard, C., *et al.* (2013). DNA methylation presents distinct binding sites for human transcription factors. Elife *2*, e00726.

Huang, L.-H., Wang, R., Gama-Sosa, M.A., Shenoy, S., and Ehrlich, M. (1984). A protein from human placental nuclei binds preferentially to 5-methylcytosine-rich DNA. Nature *308*, 293–295.

Huang, Y., Chavez, L., Chang, X., Wang, X., Pastor, W.A., Kang, J., Zepeda-Martinez, J.A., Pape, U.J., Jacobsen, S.E., Peters, B., *et al.* (2014). Distinct roles of the methylcytosine oxidases Tet1 and Tet2 in mouse embryonic stem cells. Proc. Natl. Acad. Sci. U.S.A. *111*, 1361–1366.

Iqbal, K., Jin, S.G., Pfeifer, G.P., and Szabo, P.E. (2011). Reprogramming of the paternal genome upon fertilization involves genome-wide oxidation of 5-methylcytosine. Proc. Natl. Acad. Sci. U.S.A. *108*, 3642–3647.

Ito, S., D'Alessio, A.C., Taranova, O.V., Hong, K., Sowers, L.C., and Zhang, Y. (2010). Role of Tet proteins in 5mC to 5hmC conversion, ES-cell self-renewal and inner cell mass specification. Nature *466*, 1129–1133.

Jeong, M., Sun, D., Luo, M., Huang, Y., Challen, G.A., Rodriguez, B., Zhang, X., Chavez, L., Wang, H., Hannah, R., *et al.* (2014). Large conserved domains of low DNA methylation maintained by Dnmt3a. Nat. Genet. *46*, 17–23.

Jin, C., Lu, Y., Jelinek, J., Liang, S., Estecio, M.R., Barton, M.C., and Issa, J.P. (2014). TET1 is a maintenance DNA demethylase that prevents methylation spreading in differentiated cells. Nucleic Acids Res. *42*, 6956–6971.

Jin, S.G., Jiang, Y., Qiu, R., Rauch, T.A., Wang, Y., Schackert, G., Krex, D., Lu, Q., and Pfeifer, G.P. (2011a). 5-Hydroxymethylcytosine is strongly depleted in human cancers but its levels do not correlate with IDH1 mutations. Cancer Res. *71*, 7360–7365.

Jin, S.G., Wu, X., Li, A.X., and Pfeifer, G.P. (2011b). Genomic mapping of 5-hydroxymethylcytosine in the human brain. Nucleic Acids Res. *39*, 5015–5024.

Kaas, G.A., Zhong, C., Eason, D.E., Ross, D.L., Vachhani, R.V., Ming, G.L., King, J.R., Song, H., and Sweatt, J.D. (2013). TET1 controls CNS 5-methylcytosine hydroxylation, active DNA demethylation, gene transcription, and memory formation. Neuron *79*, 1086–1093.

Kent, W.J., Sugnet, C.W., Furey, T.S., Roskin, K.M., Pringle, T.H., Zahler, A., M., and Haussler, D. (2014). UCSC Genome Browser. http://genome.ucsc.edu

Khare, T., Pai, S., Koncevicius, K., Pal, M., Kriukiene, E., Liutkeviciute, Z., Irimia, M., Jia, P., Ptak, C., Xia, M., *et al.* (2012). 5-hmC in the brain is abundant in synaptic genes and shows differences at the exon–intron boundary. Nat. Struct. Mol. Biol. *19*, 1037–1043.

Kim, M., Park, Y.K., Kang, T.W., Lee, S.H., Rhee, Y.H., Park, J.L., Kim, H.J., Lee, D., Lee, D., Kim, S.Y., *et al.* (2013). Dynamic changes in DNA methylation and hydroxymethylation when hES cells undergo differentiation toward a neuronal lineage. Hum. Mol. Genet. *23*, 657–667.

Kinney, S.M., Chin, H.G., Vaisvila, R., Bitinaite, J., Zheng, Y., Esteve, P.O., Feng, S., Stroud, H., Jacobsen, S.E., and Pradhan, S. (2011). Tissue-specific distribution and dynamic changes of 5-hydroxymethylcytosine in mammalian genomes. J. Biol. Chem. *286*, 24685–24693.

Klug, M., Schmidhofer, S., Gebhard, C., Andreesen, R., and Rehli, M. (2013). 5-Hydroxymethylcytosine is an essential intermediate of active DNA demethylation processes in primary human monocytes. Genome Biol. *14*, R46.

Ko, M., An, J., Bandukwala, H.S., Chavez, L., Aijo, T., Pastor, W.A., Segal, M.F., Li, H., Koh, K.P., Lahdesmaki, H., *et al.* (2013). Modulation of TET2 expression and 5-methylcytosine oxidation by the CXXC domain protein IDAX. Nature *497*, 122–126.

Koh, K.P., Yabuuchi, A., Rao, S., Huang, Y., Cunniff, K., Nardone, J., Laiho, A., Tahiliani, M., Sommer, C.A., Mostoslavsky, G., *et al.* (2011). Tet1 and Tet2 regulate 5-hydroxymethylcytosine production and cell lineage specification in mouse embryonic stem cells. Cell Stem Cell *8*, 200–213.

Kohli, R.M., and Zhang, Y. (2013). TET enzymes, TDG and the dynamics of DNA demethylation. Nature *502*, 472–479.

Kriaucionis, S., and Heintz, N. (2009). The nuclear DNA base 5-hydroxymethylcytosine is present in Purkinje neurons and the brain. Science *324*, 929–930.

Laurent, L., Wong, E., Li, G., Huynh, T., Tsirigos, A., Ong, C.T., Low, H.M., Kin Sung, K.W., Rigoutsos, I., Loring, J., *et al.* (2010). Dynamic changes in the human methylome during differentiation. Genome Res. *20*, 320–331.

Li, X., Wei, W., Zhao, Q.Y., Widagdo, J., Baker-Andresen, D., Flavell, C.R., D'Alessio, A., Zhang, Y., and Bredy, T.W. (2014). Neocortical Tet3-mediated accumulation of 5-hydroxymethylcytosine promotes rapid behavioral adaptation. Proc. Natl. Acad. Sci. U.S.A. *111*, 7120–7125.

Li, Y., and O'Neill, C. (2013). 5′-Methylcytosine and 5′-hydroxymethylcytosine: Each provide epigenetic information to the mouse zygote. PLoS ONE *8*, e63689.

Lister, R., Mukamel, E.A., Nery, J.R., Urich, M., Puddifoot, C.A., Johnson, N.D., Lucero, J., Huang, Y., Dwork, A.J., Schultz, M.D., *et al.* (2013). Global epigenomic reconfiguration during mammalian brain development. Science *341*, 1237905.

Liu, C., Liu, L., Chen, X., Shen, J., Shan, J., Xu, Y., Yang, Z., Wu, L., Xia, F., Bie, P., *et al.* (2013). Decrease of 5-hydroxymethylcytosine is associated with progression of hepatocellular carcinoma through downregulation of TET1. PLoS ONE *8*, e62828.

Maunakea, A.K., Chepelev, I., Cui, K., and Zhao, K. (2013). Intragenic DNA methylation modulates alternative splicing by recruiting MeCP2 to promote exon recognition. Cell Res. *23*, 1256–1269.

Meissner, A., Mikkelsen, T.S., Gu, H., Wernig, M., Hanna, J., Sivachenko, A., Zhang, X., Bernstein, B.E., Nusbaum, C., Jaffe, D.B., *et al.* (2008). Genome-scale DNA methylation maps of pluripotent and differentiated cells. Nature *454*, 766–770.

Mellen, M., Ayata, P., Dewell, S., Kriaucionis, S., and Heintz, N. (2012). MeCP2 binds to 5hmC enriched within active genes and accessible chromatin in the nervous system. Cell *151*, 1417–1430.

Munzel, M., Globisch, D., Bruckl, T., Wagner, M., Welzmiller, V., Michalakis, S., Muller, M., Biel, M., and Carell, T. (2010). Quantification of the sixth DNA base hydroxymethylcytosine in the brain. Angew. Chem. Int. Ed. Engl. *49*, 5375–5377.

Murphy, P.J., Cipriany, B.R., Wallin, C.B., Ju, C.Y., Szeto, K., Hagarman, J.A., Benitez, J.J., Craighead, H.G., and Soloway, P.D. (2013). Single-molecule analysis of combinational epigenomic states in normal and tumor cells. Proc. Natl. Acad. Sci. U.S.A. *110*, 7772–7777.

Neri, F., Incarnato, D., Krepelova, A., Rapelli, S., Pagnani, A., Zecchina, R., Parlato, C., and Oliviero, S. (2013). Genome-wide analysis identifies a functional association of Tet1 and Polycomb repressive complex 2 in mouse embryonic stem cells. Genome Biol. *14*, R91.

Nestor, C.E., Ottaviano, R., Reddington, J., Sproul, D., Reinhardt, D., Dunican, D., Katz, E., Dixon, J.M., Harrison, D.J., and Meehan, R.R. (2012). Tissue type is a major modifier of the 5-hydroxymethylcytosine content of human genes. Genome Res. *22*, 467–477.

Oberdoerffer, S. (2012). A conserved role for intragenic DNA methylation in alternative pre-mRNA splicing. Transcription *3*, 106–109.

Okashita, N., Kumaki, Y., Ebi, K., Nishi, M., Okamoto, Y., Nakayama, M., Hashimoto, S., Nakamura, T., Sugasawa, K., Kojima, N., *et al.* (2014). PRDM14 promotes active DNA demethylation through the ten-eleven translocation (TET)-mediated base excision repair pathway in embryonic stem cells. Development *141*, 269–280.

Ooi, S.K., Qiu, C., Bernstein, E., Li, K., Jia, D., Yang, Z., Erdjument-Bromage, H., Tempst, P., Lin, S.P., Allis, C.D., *et al.* (2007). DNMT3L connects unmethylated lysine 4 of histone H3 to de novo methylation of DNA. Nature *448*, 714–717.

Otani, J., Kimura, H., Sharif, J., Endo, T.A., Mishima, Y., Kawakami, T., Koseki, H., Shirakawa, M., Suetake, I., and Tajima, S. (2013). Cell cycle-dependent turnover of 5-hydroxymethyl cytosine in mouse embryonic stem cells. PLoS ONE *8*, e82961.

Pfeifer, G.P., Kadam, S., and Jin, S.G. (2013). 5-hydroxymethylcytosine and its potential roles in development and cancer. Epigenetics Chromatin *6*, 10.

Phillips, J.E., and Corces, V.G. (2009). CTCF: master weaver of the genome. Cell *137*, 1194–1211.

Pogribny, I.P., Pogribna, M., Christman, J.K., and James, S.J. (2000). Single-site methylation within the p53 promoter region reduces gene expression in a reporter gene construct: possible *in vivo* relevance during tumorigenesis. Cancer Res. *60*, 588–594.

Ren, B., Robert, F., Wyrick, J.J., Aparicio, O., Jennings, E.G., Simon, I., Zeitlinger, J., Schreiber, J., Hannett, N., Kanin, E., *et al.* (2000). Genome-wide location and function of DNA binding proteins. Science *290*, 2306–2309.

Revel, H.R., and Luria, S.E. (1970). DNA-glucosylation in T-even phage: genetic determination and role in phagehost interaction. Annu. Rev. Genet. *4*, 177–192.

Rodriguez, C., Borgel, J., Court, F., Cathala, G., Forne, T., and Piette, J. (2010). CTCF is a DNA methylation-sensitive positive regulator of the INK/ARF locus. Biochem. Biophys. Res. Commun. *392*, 129–134.

Rudenko, A., Dawlaty, M.M., Seo, J., Cheng, A.W., Meng, J., Le, T., Faull, K.F., Jaenisch, R., and Tsai, L.H. (2013). Tet1 is critical for neuronal activity-regulated gene expression and memory extinction. Neuron *79*, 1109–1122.

Saxonov, S., Berg, P., and Brutlag, D.L. (2006). A genome-wide analysis of CpG dinucleotides in the human genome distinguishes two distinct classes of promoters. Proc. Natl. Acad. Sci. U.S.A. *103*, 1412–1417.

Schena, M., Shalon, D., Davis, R.W., and Brown, P.O. (1995). Quantitative monitoring of gene expression patterns with a complementary DNA microarray. Science *270*, 467–470.

Sengupta, P.K., Ehrlich, M., and Smith, B.D. (1999). A methylation-responsive MDBP/RFX site is in the first exon of the collagen alpha2(I) promoter. J. Biol. Chem. *274*, 36649–36655.

Serandour, A.A., Avner, S., Oger, F., Bizot, M., Percevault, F., Lucchetti-Miganeh, C., Palierne, G., Gheeraert, C., Barloy-Hubler, F., Peron, C.L., *et al.* (2012). Dynamic hydroxymethylation of deoxyribonucleic acid marks differentiation-associated enhancers. Nucleic Acids Res. *40*, 8255–8265.

Shukla, S., Kavak, E., Gregory, M., Imashimizu, M., Shutinoski, B., Kashlev, M., Oberdoerffer, P., Sandberg, R., and Oberdoerffer, S. (2011). CTCF-promoted

RNA polymerase II pausing links DNA methylation to splicing. Nature *479*, 74–79.

Smith, Z.D., and Meissner, A. (2013). DNA methylation: roles in mammalian development. Nat. Rev. Genet. *14*, 204–220.

Smith, Z.D., Chan, M.M., Mikkelsen, T.S., Gu, H., Gnirke, A., Regev, A., and Meissner, A. (2012). A unique regulatory phase of DNA methylation in the early mammalian embryo. Nature *484*, 339–344.

Song, C.X., Szulwach, K.E., Fu, Y., Dai, Q., Yi, C., Li, X., Li, Y., Chen, C.H., Zhang, W., Jian, X., *et al.* (2011). Selective chemical labeling reveals the genome-wide distribution of 5-hydroxymethylcytosine. Nat. Biotechnol. *29*, 68–72.

Song, C.X., Szulwach, K.E., Dai, Q., Fu, Y., Mao, S.Q., Lin, L., Street, C., Li, Y., Poidevin, M., Wu, H., *et al.* (2013). Genome-wide profiling of 5-formylcytosine reveals its roles in epigenetic priming. Cell *153*, 678–691.

Spruijt, C.G., Gnerlich, F., Smits, A.H., Pfaffeneder, T., Jansen, P.W., Bauer, C., Munzel, M., Wagner, M., Muller, M., Khan, F., *et al.* (2013). Dynamic readers for 5-(hydroxy)methylcytosine and its oxidized derivatives. Cell *152*, 1146–1159.

Stroud, H., Feng, S., Morey Kinney, S., Pradhan, S., and Jacobsen, S.E. (2011). 5-Hydroxymethylcytosine is associated with enhancers and gene bodies in human embryonic stem cells. Genome Biol. *12*, R54.

Sun, Z., Terragni, J., Borgaro, J.G., Liu, Y., Yu, L., Guan, S., Wang, H., Sun, D., Cheng, X., Zhu, Z., *et al.* (2013). High-resolution enzymatic mapping of genomic 5-hydroxymethylcytosine in mouse embryonic stem cells. Cell Rep. *3*, 567–576.

Szulwach, K.E., Li, X., Li, Y., Song, C.X., Han, J.W., Kim, S., Namburi, S., Hermetz, K., Kim, J.J., Rudd, M.K., *et al.* (2011). Integrating 5-hydroxymethylcytosine into the epigenomic landscape of human embryonic stem cells. PLoS Genet. *7*, e1002154.

Szwagierczak, A., Bultmann, S., Schmidt, C.S., Spada, F., and Leonhardt, H. (2010). Sensitive enzymatic quantification of 5-hydroxymethylcytosine in genomic DNA. Nucleic Acids Res. *38*, e181.

Tahiliani, M., Koh, K.P., Shen, Y., Pastor, W.A., Bandukwala, H., Brudno, Y., Agarwal, S., Iyer, L.M., Liu, D.R., Aravind, L., *et al.* (2009). Conversion of 5-methylcytosine to 5-hydroxymethylcytosine in mammalian DNA by MLL partner TET1. Science *324*, 930–935.

Tao, Y., Xi, S., Briones, V., and Muegge, K. (2011). Lsh mediated RNA polymerase II stalling at HoxC6 and HoxC8 involves DNA methylation. PLoS ONE *5*, e9163.

Terragni, J., Bitinaite, J., Zheng, Y., and Pradhan, S. (2012). Biochemical characterization of recombinant beta-glucosyltransferase and analysis of global 5-hydroxymethylcytosine in unique genomes. Biochemistry *51*, 1009–1019.

Terragni, J., Zhang, G., Sun, Z., Pradhan, S., Song, L., Crawford, G.E., Lacey, M., and Ehrlich, M. (2014). Notch signaling genes: Myogenic DNA hypomethylation and 5-hydroxymethylcytosine. Epigenetics *9*, 842–850.

Thomson, J.P., Lempiainen, H., Hackett, J.A., Nestor, C.E., Muller, A., Bolognani, F., Oakeley, E.J., Schubeler, D., Terranova, R., Reinhardt, D., *et al.* (2012). Non-genotoxic carcinogen exposure induces defined changes in the 5-hydroxymethylome. Genome Biol. *13*, R93.

Tsumagari, K., Chang, S.-C., Lacey, M., Baribault, C., Chittur, S.V., Sowden, J., Tawil, R., Crawford, G.E., and Ehrlich, M. (2011). Gene expression during normal and FSHD myogenesis. BMC Med. Genomics *4*, 67.

Tsumagari, K., Baribault, C., Terragni, J., Chandra, S., Renshaw, C., Sun, Z., Song, L., Crawford, G.E., Pradhan, S., Lacey, M., *et al.* (2013a). DNA methylation and differentiation: HOX genes in muscle cells. Epigen. Chromatin *6*, 25.

Tsumagari, K., Baribault, C., Terragni, J., Varley, K.E., Gertz, J., Pradhan, S., Baddoo, M., Crain, C.M., Song, L., Crawford, G.E., *et al.* (2013b). Early de novo DNA methylation and prolonged demethylation in the muscle lineage. Epigenetics *8*, 317–332.

Tucker, T., Marra, M., and Friedman, J.M. (2009). Massively parallel sequencing: the next big thing in genetic medicine. Am. J. Hum. Genet. *85*, 142–154.

Varley, K.E., Gertz, J., Bowling, K.M., Parker, S.L., Reddy, T.E., Pauli-Behn, F., Cross, M.K., Williams, B.A., Stamatoyannopoulos, J.A., Crawford, G.E., *et al.* (2013). Dynamic DNA methylation across diverse human cell lines and tissues. Genome Res. *23*, 555–567.

Wang, F., Yang, Y., Lin, X., Wang, J.Q., Wu, Y.S., Xie, W., Wang, D., Zhu, S., Liao, Y.Q., Sun, Q., *et al.* (2013). Genome-wide loss of 5-hmC is a novel epigenetic feature of Huntington's disease. Hum. Mol. Genet. *22*, 3641–3653.

Wang, H., Maurano, M.T., Qu, H., Varley, K.E., Gertz, J., Pauli, F., Lee, K., Canfield, T., Weaver, M., Sandstrom, R., *et al.* (2012). Widespread plasticity in CTCF occupancy linked to DNA methylation. Genome Res. *22*, 1680–1688.

Wanunu, M., Cohen-Karni, D., Johnson, R.R., Fields, L., Benner, J., Peterman, N., Zheng, Y., Klein, M.L., and Drndic, M. (2011). Discrimination of methylcytosine from hydroxymethylcytosine in DNA molecules. J. Am. Chem. Soc. *133*, 486–492.

Wen, L., Li, X., Yan, L., Tan, Y., Li, R., Zhao, Y., Wang, Y., Xie, J., Zhang, Y., Song, C., *et al.* (2014). Whole-genome analysis of 5-hydroxymethylcytosine and 5-methylcytosine at base resolution in the human brain. Genome Biol. *15*, R49.

Wheldon, L.M., Abakir, A., Ferjentsik, Z., Dudnakova, T., Strohbuecker, S., Christie, D., Dai, N., Guan, S., Foster, J.M., Correa, I.R., Jr., *et al.* (2014). Transient accumulation of 5-carboxylcytosine indicates involvement of active demethylation in lineage specification of neural stem cells. Cell Rep. *7*, 1353–1361.

Williams, K., Christensen, J., and Helin, K. (2011). DNA methylation: TET proteins-guardians of CpG islands? EMBO Rep. *13*, 28–35.

Wossidlo, M., Nakamura, T., Lepikhov, K., Marques, C.J., Zakhartchenko, V., Boiani, M., Arand, J., Nakano, T., Reik, W., and Walter, J. (2011).

5-Hydroxymethylcytosine in the mammalian zygote is linked with epigenetic reprogramming. Nat. Commun. *2*, 241.

Wu, H., and Zhang, Y. (2011). Mechanisms and functions of Tet protein-mediated 5-methylcytosine oxidation. Genes Dev. *25*, 2436–2452.

Wu, H., D'Alessio, A.C., Ito, S., Wang, Z., Cui, K., Zhao, K., Sun, Y.E., and Zhang, Y. (2011a). Genome-wide analysis of 5-hydroxymethylcytosine distribution reveals its dual function in transcriptional regulation in mouse embryonic stem cells. Genes Dev. *25*, 679–684.

Wu, H., D'Alessio, A.C., Ito, S., Xia, K., Wang, Z., Cui, K., Zhao, K., Sun, Y.E., and Zhang, Y. (2011b). Dual functions of Tet1 in transcriptional regulation in mouse embryonic stem cells. Nature *473*, 389–393.

Wyatt, G.R., and Cohen, S.S. (1952). A new pyrimidine base from bacteriophage nucleic acids. Nature *170*, 1072–1073.

Xie, W., Barr, C.L., Kim, A., Yue, F., Lee, A.Y., Eubanks, J., Dempster, E.L., and Ren, B. (2012). Base-resolution analyses of sequence and parent-of-origin dependent DNA methylation in the mouse genome. Cell *148*, 816–831.

Yildirim, O., Li, R., Hung, J.H., Chen, P.B., Dong, X., Ee, L.S., Weng, Z., Rando, O.J., and Fazzio, T.G. (2011). Mbd3/NURD complex regulates expression of 5-hydroxymethylcytosine marked genes in embryonic stem cells. Cell *147*, 1498–1510.

Yu, D.H., Ware, C., Waterland, R.A., Zhang, J., Chen, M.H., Gadkari, M., Kunde-Ramamoorthy, G., Nosavanh, L.M., and Shen, L. (2013a). Developmentally programmed 3′ CpG island methylation confers tissue- and cell-type specific transcriptional activation. Mol. Cell. Biol. *33*, 1845–1858.

Yu, M., Hon, G.C., Szulwach, K.E., Song, C.X., Jin, P., Ren, B., and He, C. (2012a). Tet-assisted bisulfite sequencing of 5-hydroxymethylcytosine. Nat. Protoc. *7*, 2159–2170.

Yu, M., Hon, G.C., Szulwach, K.E., Song, C.X., Zhang, L., Kim, A., Li, X., Dai, Q., Shen, Y., Park, B., *et al.* (2012b). Base-resolution analysis of 5-hydroxymethylcytosine in the mammalian genome. Cell *149*, 1368–1380.

Yu, P., Xiao, S., Xin, X., Song, C.X., Huang, W., McDee, D., Tanaka, T., Wang, T., He, C., and Zhong, S. (2013b). Spatiotemporal clustering of the epigenome reveals rules of dynamic gene regulation. Genome Res. *23*, 352–364.

Zhang, X.Y., Jabrane-Ferrat, N., Asiedu, C.K., Samac, S., Peterlin, B.M., and Ehrlich, M. (1993). The major histocompatibility complex class II promoter-binding protein RFX (NF-X) is a methylated DNA-binding protein. Mol. Cell. Biol. *13*, 6810–6818.

The Identification of Mammalian Proteins Involved in Epigenetics

5

Luke Isbel, Harry Oey and Emma Whitelaw

Abstract

Mutagenesis screening in *Drosophila* has identified genes involved in epigenetic processes, and in most cases the mammalian homologues have been found to have similar functions. However, there are some gene-silencing mechanisms that are present in mammals but absent in *Drosophila*, such as DNA methylation. So mutagenesis screens have also been carried out in the mouse and some novel genes have been found. Random mutagenesis screens in the mouse require a massive effort but they do facilitate the study of the phenotypic consequences of haploinsufficiency, and this is providing us with clear associations between epigenetic dysfunction and human disease. *In vitro* RNAi screening in cell lines provides a high-throughput alternative but faces some difficulties, in particular off-target effects. Biochemical approaches such as affinity pull-down and mass spectrometry have provided researchers with a means to identify binding partners. Here we review the efforts made to identify the mammalian complexes necessary for epigenetic events. While the set of proteins found to be involved is now extensive, the challenge is to understand how they work together to achieve the highly regulated patterns of gene expression seen in the adult organism.

Lessons from position effect variegation in *Drosophila* and yeast

Mutagenesis screening for modifiers of PEV in *Drosophila*

It is well established in *Drosophila* that euchromatic, transcriptionally active loci that have integrated into or adjacent to heterochromatic regions can become silenced in a variegated manner, termed position effect variegation (PEV). Such silencing is a classic example of epigenetic regulation of gene expression. A well-studied example is the *white* (*w*) locus, responsible for eye pigmentation. It exhibits a mosaic pattern of expression after translocation to pericentromeric heterochromatin (Muller, 1930). Early work showed that variegating alleles have reduced levels of mRNA in a subset of cells (Rushlow *et al.*, 1984) and the affected locus acquires a dense chromatin banding pattern (Henikoff, 1981; Zhimulev *et al.*, 1986).

Mutagenesis screening for modifiers of PEV using one of these variegating alleles, the w^{m4} strain, has identified loci that contribute to the process (Locke *et al.*, 1988; Schotta *et al.*, 2003). Mapping these mutations has identified factors that interact with and/or modify chromatin. Genes that contribute to the silencing are termed suppressors of variegation, *Su(var)*s, and mutations in these genes result in an increase in the proportion of cells expressing *w*. Genes that have the reverse function, i.e. in their wild type form they activate transcription, are termed enhancers of variegation, *E(var)*s, and mutations in them result in a decrease in the proportion of cells expressing *w*. For example, *Su(var)*2-5 (Sinclair *et al.*, 1983) encodes the protein heterochromatin protein 1 (HP1), which was independently identified as a non-histone heterochromatin-associated protein (James and Elgin, 1986). Other proteins identified in these screens include factors affecting N-terminal histone tail modifications, chromatin remodellers involved in repositioning nucleosomes, DNA binding factors, histone variants, DNA replication machinery, regulators of higher order chromatin structure, and others, as reviewed (Eissenberg and Reuter, 2009; Fodor *et al.*, 2010; Schotta *et al.*, 2003).

The complex behaviour of modifiers of variegation

In the case of w^{m4}, a translocation event has repositioned the *w* gene to pericentromeric heterochromatin. Interestingly, some Su(var) proteins, HP1 and Su(var)201, identified to function at w^{m4}, are unable to modify silencing if the *w* locus has moved to telomeric heterochromatic regions (Wallrath and Elgin, 1995). This is consistent with the notion of 'position effects', i.e. that different factors are involved in epigenetic regulation of one particular gene at different chromosomal locations (Wilson *et al.*, 1990).

In addition, variegation can sometimes be produced by *trans* effects. For example, when a block of heterochromatin is inserted into $brown^{Dominant}$ (bw^D), the allele itself becomes null but the wild type allele then displays variegation (Henikoff *et al.*, 1992). A screen for modifiers of variegation at this locus was carried out and, while most modifiers identified were also able to modify PEV at w^{m4} (33/37 tested modifiers), none did so at *w* at a telomeric position (0/34 tested modifiers) (Talbert *et al.*, 1994). Although homologous pairing in somatic cells is not common in mammals, it has been reported but the underlying mechanisms by which the epigenetic state is communicated across the two alleles remains unclear (Krueger *et al.*, 2012; Yoon *et al.*, 2002).

Another example of a locus that can variegate after chromosomal rearrangement is *yellow* (*y*) (Le *et al.*, 1995). The protein Bonus was found to act as a Su(var) at *yellow* (*y*) but an E(var) at *white* (*w*), suggesting different functions at different loci (Beckstead *et al.*, 2005). Bonus is the fly homologue of the mammalian transcriptional intermediary factor 1 family (TIF1) proteins, members of the Trim (Tripartite motif-containing) family. TIF1 proteins have been shown to act as either activators or repressors (Iyengar *et al.*, 2011), probably as a result of forming complexes with active (Rooney and Calame, 2001) or repressive (Schultz *et al.*, 2002) cofactors (see Fig. 5.1).

Some active genes are found embedded in heterochromatic regions (Hoskins *et al.*, 2002). Surprisingly, these genes can undergo mosaic silencing after being translocated into euchromatic regions. For example, the *light* (*lt*) gene, responsible for pigmentation in the eye, and five adjacent loci, normally located near centromeric heterochromatin, are variably silenced upon rearrangement into euchromatic regions (Wakimoto and Hearn, 1990). Fourteen Su(var)s of w^{m4} were tested for their ability to modify variegation of lt^x; eight had no effect and six, including HP1, acted in the opposite direction from expected, i.e. as E(var) s (Hearn *et al.*, 1991). There are other examples of genes located in the pericentromeric heterochromatin of chromosome 3 that show decreased expression in the absence of HP1 (Schulze *et al.*, 2005). This fascinating finding suggests that genes located in heterochromatic regions have acquired

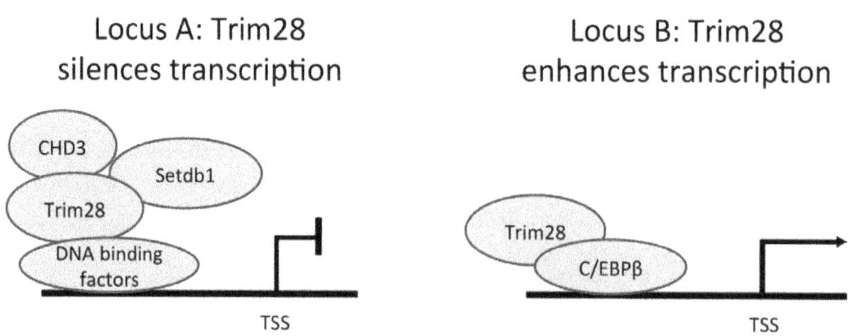

Figure 5.1 Trim28 can be found binding distally to genes that are responsive to its loss, in both transcriptionally negative (locus A) and positive (locus B) directions. As Trim28 has no DNA binding ability or catalytic domains it must localize to the genome through DNA binding factors and modify chromatin through the requirement of chromatin remodelling co-factors.

the ability to use proteins normally associated with repression to remain transcriptionally active.

Further evidence of dual functions of some modifiers comes from a recent study of genome-wide transcriptional changes in Drosophila carrying mutations in the classic repressors, HP1 and Su(var)3-9. Decreased expression of some genes located in centromeric heterochromatin was observed and additional experiments suggested that this was a direct consequence of reduced levels of HP1 and Su(var)3-9 at the affected loci (Lundberg *et al.*, 2013). In other words, at these loci HP1 and Su(var)3-9 appear to act directly as transcriptional co-activators.

Conserved domains in modifiers of PEV

The identification of functional domains in *Drosophila* modifiers of PEV can be used to search for mammalian proteins with conserved function. For example, mutation mapping identified a particular section of HP1 that conferred the ability to bind chromatin (Platero *et al.*, 1995) and similar regions were found in other proteins with overlapping function (Paro and Hogness, 1991). This region has become known as the chromo domain. The mammalian homologues, called HP1β and HP1γ, were identified by screening mouse cDNA libraries for sequence homology to the *Drosophila* chromo domain (Singh *et al.*, 1991). Indeed, the function is so similar that the mammalian homologue of *Drosophila* Su(var)3-9, SUV39H1, can modify PEV in *Drosophila* (Aagaard *et al.*, 1999).

Many modifiers of variegation have more than one type of functional domain. For example, Su(var)3-9 contains a chromo domain and a SET domain (suppressor of variegation, Enhancer of zeste and Trithorax), both of which are conserved across species in proteins that regulate chromatin (Jenuwein *et al.*, 1998; Tschiersch *et al.*, 1994). In this way, families of proteins can be organized by domains, for example SET domains, chromodomains, bromodomain, plant homeodomain (PHD), as reviewed by Musselman *et al.* (2012).

Mutagenesis screens for modifiers of PEV in yeast

Mutagenesis screens for modifiers of PEV have been carried out in fission yeast *Schizosaccharomyces pombe* (Ekwall *et al.*, 1999). Genes placed close to the mating-type (MAT) locus, centromeric repeats or telomeres, exhibit stochastic silencing and can be screened for modifiers. Some overlap has been found between the factors identified in the *Drosophila* and yeast screens. For example, the genes *Swi6* and *Clr4*, identified as repressors in the yeast screen, are homologues of the *HP1* and *SuVar3-9* genes, respectively, in *Drosophila* (Allshire *et al.*, 1995; Ekwall and Ruusala, 1994; Ivanova *et al.*, 1998; Lorentz *et al.*, 1994).

Finding modifiers of PEV in mammals

Position effect variegation in mammals

As in *Drosophila*, there are genes in the genome of mammals that variegate due to stochastic gene silencing. These are called metastable epialleles and include both endogenous single copy genes, such as *Agouti viable yellow* (A^{vy}) (see Fig. 5.2) and *axin fused* ($Axin^{Fu}$), and transgene arrays that have been inserted into the genome (Rakyan *et al.*, 2002).

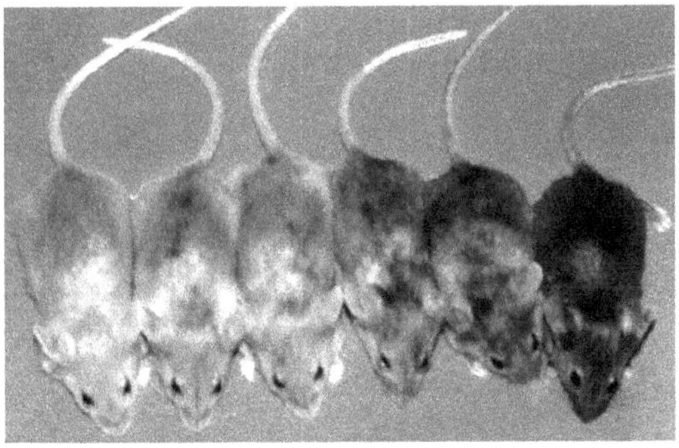

Figure 5.2 The coat colour spectrum in isogenic A^{vy} mice demonstrates the phenotypic outcome of stochastic gene silencing at a locus under the control of an upstream promoter. Low DNA methylation at the upstream promoter is associated with constitutive expression of the *agouti* gene (left). Intermediate methylation states are seen in mottled mice and no expression occurs when the upstream promoter is completely methylated in pseudoagouti mice (right). A colour version of this figure is available in the plate section at the back of the book.

Modifiers of epigenetic function in mammals

It is well established that transgenes inserted into the mouse genome are often associated with variegated expression (McGowan *et al.*, 1989) as a result of gene silencing (Robertson *et al.*, 1996). By using transgenes designed to be expressed in blood cells, simple quantitative technologies can be used to measure silencing at the single cell level. Promoters that come on exclusively in erythroid cells (Robertson *et al.*, 1996) or in T-cells (Festenstein *et al.*, 1996) have been used.

Fluorescence activating cell sorting (FACS) analysis of T-cell populations from mice with CD2 transgenes driven by T-cell-specific regulatory elements, showed 'mosaic' expression, when integrated into heterochromatin (Festenstein *et al.*, 1999). Molecular analysis of this CD2 transgene identified epigenetic marks associated with both repression and activation of transcription, e.g. histone marks representative of pericentromeric heterochromatin, and DNase1 hypersensitivity (Hiragami-Hamada *et al.*, 2009). These mice were crossed to others carrying a transgene that overexpressed HP1. Offspring inheriting both transgenes showed increased numbers of T-cells positive for CD2 transgene expression (Festenstein *et al.*, 1999). This result was contrary to expectation at the time because HP1 had been identified as a Su(var) in *Drosophila*; however, we now know that HP1 can work in some contexts as an activator (see above). This and other similar studies (Garrick *et al.*, 1998; Gaudet *et al.*, 2004) suggested that one should be able to identify novel modifiers of variegation in the mouse using random mutagenesis.

Screening for modifiers of epigenetic function in mammals

Random mutagenesis screens can be carried out in mice to find, in an unbiased manner, genes involved in a particular process or phenotypic trait (Hrabe de Angelis *et al.*, 2000; Nolan *et al.*, 2000). A large mutagenesis screen for modifiers of PEV has been carried out in the mouse and has identified mutations that act as dominant *Modifiers of Murine Metastable Epialleles* (*MommeDs*) (Blewitt *et al.*, 2005). Screening was carried out in mice carrying a transgene containing ~10 copies of a GFP reporter linked to α-globin promoter/enhancer elements. This transgene displays variegation in erythrocytes (Preis *et al.*, 2003). Male mice homozygous for the transgene were treated with the mutagen *N*-ethyl-*N*-nitrosourea (ENU), which preferentially affects spermatogonial stem cells. Treated males were crossed to females that were also homozygous for the transgene. Peripheral blood from F1 offspring (> 2500 individuals) was screened for a change in the percentage of erythrocytes expressing the transgene using flow cytometry.

To date, over 50 *MommeD* lines, both E(var)s and Su(var)s, have been produced and the underlying mutation has been found in the majority of these (Daxinger *et al.*, 2013; Harten *et al.*, 2014). More recently, candidate mutations have been identified for some additional lines and an updated summary is shown in Table 5.1. In total, the underlying mutation has been found for 40 *MommeDs* and they map back to 24 unique genes. Given that ~80 genes were identified in the *Drosophila* screens for modifiers of PEV, it is unlikely that the *MommeD* screen is saturated. Genetic screens are defined as having reached saturation when no new genes or gene mutations can be found (Pollock and Larkin, 2004). With respect to the *MommeD* screen, further work is likely to find more genes but it would require considerable resources.

Known modifier genes identified in the *MommeD* screen

For many of the *Momme* genes (Table 5.1), *Drosophila* homologues had already been identified in the PEV screens (Fodor *et al.*, 2010) and most of these have well-established roles in regulating epigenetic events during development. They represent broad categories of epigenetic control mechanisms: DNA methylation, histone modification and chromatin remodelling.

DNA methyltransferases and their partners

DNA methylation does not occur to any significant extent in flies and it was hoped that the mouse screen would identify novel proteins

Table 5.1 A summary of the genes identified in the *MommeD* screen is shown with the nature of *Momme* alleles, and their role as either an *E(var)* or *Su(var)* in the screen

Momme gene			Protein complex	Known or predicted function
DNA methylation				
Dnmt1	MommeD2 null	Su(var)	DNMT1–PCNA complex	Maintenance DNA methyltransferase
	MommeD32 null			
Dnmt3b	MommeD14 hypomorph	Su(var)	DNMT3B complex	*De novo* methyltransferase
	MommeD55*			
Uhrf1	MommeD40 null	Su(var)	DNMT1–PCNA complex	Binds hemi-methylated DNA at replication forks
Chromatin remodelling				
Baz1b	MommeD10 hypomorph	E(var)	WICH and WINAC chromatin remodelling complexes	Nucleosome remodelling
	MommeD16 hypomorph			
Pbrm1	MommeD27 hypomorph	Su(var)	SWI/SNF complex	Nucleosome remodelling
Smarca4	MommeD39 null	Su(var)	SWI/SNF complex, WINAC complex and NCOR1 complex	Nucleosome remodelling
Smarca5	MommeD4 hypomorph	E(var)	NURF complex and WICH complex	Nucleosome remodelling
	MommeD35 hypomorph			
	MommeD37 null			
Smarcc1	MommeD19 null	Su(var)	SWI/SNF complex and NCOR1 complex	Nucleosome remodelling
Histone modification				
Brd1	MommeD42 null	E(var)	MOZ/MORF acetyltransferase complex	Histone 3 binding activity
	MommeD54*			
Dot1l	MommeD52*	Su(var)	EAP complex	H3K79 methyltransferase
Hdac1	MommeD5 Null	E(var)	Sin3A, NuRD and CoREST complexes	Histone deacetylase
Nsd1	MommeD50*	E(var)	Unknown	H3K36 methyltransferase
Setdb1	MommeD13 null	Su(var)	MBD1–MCAF1–SETDB1 complex	H3K9 methyltransferase
	MommeD17 hypomorph			
Su-var39h1	MommeD33 null	Su(var)	SUVAR39H1/HP1 complex	H3K9 methyltransferase
Trim28	MommeD9 null	E(var)	NCOR1 complex	Transcriptional cofactor
	MommeD31 null			
Epigenetic role yet to be determined				
D14Abb1e	MommeD6*	Su(var)	Unknown	Unknown
	MommeD20 null			
	MommeD51*			
Eif3h	MommeD12 null	E(var)	Eif3 complex	mRNA translation
	MommeD38 null			

Table 5.1 continued

Momme gene			Protein complex	Known or predicted function
Klf1	MommeD11 MommeD45	Su(var)	Transcription factor	Erythroid transcriptional activator
Morc3	MommeD21 MommeD41	Su(var)	unknown	H3K4me3 binding activity
Rif1	MommeD18 null	Su(var)	Bloom complex, H3K9 methylation complex	DNA damage response, telomere transcriptional repressor
Rlf	MommeD8 hypomorph	E(var)	Unknown	unknown
	MommeD28 null			
	MommeD34 null			
Smchd1	MommeD1 null	Su(var)	Unknown	Transcriptional repressor, maintenance of X-inactivation
	MommeD22 null			
	MommeD23 null			
	MommeD36 null			
Wiz	MommeD30 null	E(var)	Wiz–G9a–Glp complex	Unknown

involved in this pathway. The DNA-methylating enzymes DNA (cytosine-5-)-methyltransferase 1 (*Dnmt1*) and DNA (cytosine-5-)-methyltransferase 3 Beta (*Dnmt3b*) were both identified in the screen and emerged as Su(var)s, as expected. They are required for global DNA methylation in developing embryos (Li *et al.*, 1992; Okano *et al.*, 1999) and their roles in repression is well established (Bird, 2002). Dnmt1 is known to complex with components of the DNA replication fork and believed to be involved in maintenance of DNA methylation during replication (Chuang *et al.*, 1997). Dnmt3b plays a role in establishing DNA methylation in the germ line (Velasco *et al.*, 2010) and at imprinted loci (Kato *et al.*, 2007).

Ubiquitin-like with PHD and ring finger domains 1 (Uhrf1) was identified in the screen as a Su(var). Uhrf1 is required for global DNA methylation in the embryo (Sharif *et al.*, 2007) and is necessary for DNMT1 localization to chromatin (Bostick *et al.*, 2007). Uhrf1 binds hemi-methylated DNA through a specific SET and ring finger associated domain (SRA) domain (Avvakumov *et al.*, 2008) and to trimethylated lysine 9 on histone 3 (H3K9me3) via a tandem Tudor domain (Nady *et al.*, 2011), supporting an interplay between the two epigenetic systems; DNA methylation and histone modification (Rothbart *et al.*, 2012).

Occasionally, ENU mutagenesis produces an allele of a gene with altered function but not null status. For example, the *MommeD14* contains a mutation in *Dnmt3b* that behaves as a hypomorphic allele. Homozygous mutants are defective in DNA methylation, display weight loss and reduced viability but they remain fertile (Youngson *et al.*, 2013). This provided a rare opportunity to examine potential transgenerational effects of reduced DNA methylation, which could not be tested in the non-viable *Dnmt3b* null mice (Okano *et al.*, 1999). The notion that altered epigenetic state might be able to be inherited across generations, providing a novel form of evolution, has been of interest to the broader scientific community for some years (Daxinger and Whitelaw, 2012). Interestingly, no successive worsening of the *MommeD14* phenotypes was detected over multiple generations (Youngson *et al.*, 2013).

Histone modifiers

The screen identified a number of factors that interact with H3K9me3, including *SET domain, Bifurcated 1* (*Setdb1*) (Schultz *et al.*, 2002), *Suvar39H1* (Chin *et al.*, 2006) and *Trim28* (Matsui *et al.*, 2010). They are known to either bind or modify this mark, which is generally associated with transcriptional repression.

Somewhat surprisingly, these were not all identified as Su(var)s; *Trim 28* was found as an E(Var). Similarly, *Histone deacetylase 1* (*Hdac1*), a histone deacetylase, normally associated with transcriptional repression was identified as an E(var). Interestingly, Hdac1 has also been shown to interact with Trim28 and as such might indirectly bind the H3K9me3 mark (Schultz *et al.*, 2001).

The gene *DOT1-like histone H3K79 methyltransferase* (*Dot1l*), was found in the screen as a Su(var). Dot1L is responsible for methylating H3K79 (Min *et al.*, 2003), a mark associated with active transcription in both mouse and human cells (Steger *et al.*, 2008; Wang *et al.*, 2008). Dot1l is a core member of the ENL-associated proteins (EAP) complex that functions in transcriptional elongation, linking H3K79 methylation with active transcription (Mueller *et al.*, 2009). The finding of *Dot1l* as a Su(var) in the screen is interesting because there is some evidence in yeast that Dot1l is involved in silencing of a reporter placed near heterochromatin but not in silencing endogenous loci in these regions (Takahashi *et al.*, 2011).

So, as in the *Drosophila* screens for modifiers of PEV, it appears that the proteins involved in influencing stochastic transcriptional states at variegating loci have complex functions that are dependent on context.

Chromatin remodellers

Chromatin remodellers were also identified. These include *SWI/SNF related, matrix associated, Actin dependent regulator of chromatin, subfamily A, member 5* (*Smarca5*), *SWI/SNF related, matrix associated, actin dependent regulator of chromatin, subfamily A, Member 4* (*Smarca4*), *bromodomain adjacent to zinc finger domain 1B* (*Baz1b*), *SWI/snf related, matrix associated, actin dependent regulator of chromatin, subfamily C, member 1* (*Smarcc1*) and *Polybromo 1* (*Pbrm1*). Chromatin remodelling factors are required for the correct reestablishment of nucleosome density following replication of heterochromatin, as reviewed (Mermoud *et al.*, 2011), and have specific roles in developmental processes as reviewed (Ho and Crabtree, 2010). Chromatin remodellers are well recognized in the literature as having the capacity to act in various protein complexes as both suppressors and enhancers of transcription (Clapier and Cairns, 2009) and, consistent with this, they are found in the screen as both E(var)s and Su(var)s (Table 5.1).

Hypomorphic alleles of *Smarca5* (*MommeD4*) and *Baz1b* (*MommeD10*) were produced as a result of an amino acid changes in the respective proteins (Ashe *et al.*, 2008; Chong *et al.*, 2007). These have provided opportunities to study the effects of reduced activity of certain proteins in viable whole animals and have been well described in the cited works. In particular, *MommeD10* homozygous mutants were noted to have abnormalities that overlapped with the human disease Williams syndrome, providing the first evidence that loss of function of this gene in humans might explain some of the phenotypes in these individuals (Ashe *et al.*, 2008).

Momme genes not previously linked to epigenetic regulation of transcription

Structural maintenance of chromosome hinge domain 1 (*Smchd1*)

A mutation in *Smchd1* was identified in the *Momme* screen. Smchd1 contains a SMC hinge domain present in proteins needed for chromosome condensation and cohesion but its function was otherwise unknown. Identification in the screen suggested the involvement of cohesion-like proteins in epigenetic silencing (Blewitt *et al.*, 2008). Study of the mutant mice showed a role for Smchd1 in X-inactivation. Female embryos homozygous for the *Smchd1* null mutation died around mid-gestation, while male homozygotes survived for longer. In female homozygotes, a failure in silencing of genes on the inactive X-chromosome was accompanied by DNA hypomethylation (Blewitt *et al.*, 2008).

Smchd1 was also found to regulate multiple clusters of imprinted genes (Gendrel *et al.*, 2013; Mould *et al.*, 2013). In both humans and mice, clustered protocadherin-beta genes,(Pcdh-β), which are known to display stochastic and combinatorial expression (Yokota *et al.*, 2011), are deregulated in expression following the reduction of Smchd1 levels (Gendrel *et al.*, 2013; Massah *et al.*, 2014; Mould *et al.*, 2013). The underlying mechanism by which Smchd1 represses transcription at these specific set of loci is still not understood. There is no clear homologue of *Smchd1* in *Drosophila* and so it is possible that this gene is part of a mammalian-specific mechanism. Recently, mutations in human *Smchd1* were found to cause deregulation of the *DUX4* gene in the telomeric D4Z4 macrosatellite repeat array, resulting in Facioscapulohumeral dystrophy type 2 (FSHD2) (see Chapter 11).

Widely interspaced zinc finger (*Wiz*) and rearranged L-*myc* fusion (*Rlf*)

Two genes identified in the screen code for proteins containing multiple zinc fingers with wide spacing between each domain, *Wiz* and *Rlf* (Daxinger *et al.*, 2013), about which little is known. Both genes act as *E(var)s* in the screen. Most proteins containing multiple zinc fingers have them organized in closely spaced tandem repeats and the significance of the wide spacing in Wiz and Rlf remains unclear. Classic zinc fingers can bind to a range of macromolecules, including DNA, RNA and proteins (Liew *et al.*, 2005) but whether widely spaced fingers behave in a similar way is unclear.

The mutation in *Wiz* is a single base deletion resulting in a null allele (Daxinger *et al.*, 2013). Although *Wiz* was identified in the *Momme* screen as an *E(var)*, previous biochemical studies have shown that it is part of a protein complex with histone H3 lysine 9 (H3K9) methyltransferases, G9a and GLP, usually associated with transcriptional repression (Ueda *et al.*, 2006).

Rlf also acts as an *E(var)* in the screen and even less is known about its role in the cell. One unusual finding with respect to mice lacking Rlf is that significant increases in DNA methylation are detected at the transgene, suggesting a role in preventing methylation at specific loci. This is somewhat reminiscent of the behaviour of a group of transcription factors called pioneers (Sérandour *et al.*, 2011).

The *Drosophila* Suvar(3)7 protein (Reuter *et al.*, 1990) has widely spaced zinc fingers and has been shown to bind DNA (Cleard and Spierer, 2001) and HP1 (Cleard *et al.*, 1997). It is possible that Wiz and Rlf, which possess no recognizable catalytic domains, mediate molecular interactions in a similar way, linking effector proteins to DNA targets.

Rap1-interacting factor (*Rif1*)

Rif1 was identified as a Su(var) in the *Momme* screen; the mutation results in a null allele and the effect of Rif1 haploinsufficiency on the reporter transgene was validated using a gene trap allele (Daxinger *et al.*, 2013). The protein was initially found in yeast as a binding partner of Ras-related protein 1a, Rap1, known to bind telomeres. Rif1 is required for telomere length maintenance and transcriptional silencing in yeast (Hardy *et al.*, 1992). The *Drosophila* Rif1 protein localizes to heterochromatin, specifically histone H3K9me3. However, unlike yeast Rif1, the *Drosophila* protein was not found to regulate transcription (Sreesankar *et al.*, 2012).

Human Rif1 has been shown in differentiated cell lines to be associated with telomeres in the context of DNA damage (Silverman *et al.*, 2004; Xu and Blackburn, 2004) and was found at stalled replication forks (Xu *et al.*, 2010). In primary human cell lines, Rif1 accumulates at pericentromeric heterochromatin during DNA replication (Buonomo *et al.*, 2009).

Since our finding of Rif1 in the screen, it has been shown to be required for the maintenance of H3K9me3 levels at telomeric, subtelomeric and pericentric heterochromatic loci, probably through a physical interaction with histone methyltransferase complex proteins – Trim28, Suv39h1, HP1, G9a and Glp (Dan *et al.*, 2014). It is likely that Rif1 is involved in a similar way in the early establishment of heterochromatin at the transgene reporter, hence its identification in the screen.

Eukaryotic translation initiation factor 3 (*Eif3h*) and DNA segment, chromosome 14, Abbott 1 expressed (*D14Abb1e*)

Two genes were identified that have not previously been linked with transcriptional regulation, an E(var) allele of *Eif3h* (Daxinger et al., 2012) and a Su(var) allele of *D14Abb1e* (Harten *et al.*, 2014). Eif3h is a subunit of the eukaryotic translation initiation factor 3 (Eif3) complex (Masutani *et al.*, 2007), known to regulate 40S ribosome translation dynamics (Hinnebusch, 2006). Its identification in the screen suggests a role in regulating expression of the reporter locus but whether this is a result of transcriptional changes at the reporter locus or translational changes is unknown (Daxinger *et al.*, 2012).

D14Abb1e, also known as *Fam208a*, is another gene identified in the screen with no history of involvement in transcription. D14Abb1e is located in the nucleus and mice haploinsufficient for the protein display an increase in the percentage of erythrocytes expressing the reporter (Harten *et al.*, 2014). D14Abb1e has one recognizable protein domain of unknown function and, recently, protein binding experiments have identified it as a putative interacting factor with H3K9me3 (Eberl *et al.*, 2013).

Other types of genetic screens for factors involved in epigenetic function

Mice carrying targeted knockouts of genes with potential epigenetic function

Large-scale projects to generate knockout mice have recently been made feasible by the ability to create transgenic mice from ES cells carrying targeted mutations. Large collections of ES cell knock-out lines now exist for both protein-coding genes (Skarnes *et al.*, 2011) and regulatory elements, such as miRNAs (Park *et al.*, 2012). Recent efforts have been made to create mice from these mutant ES cells and to screen their offspring for phenotypic defects, thereby carrying out a reasonably unbiased screen for phenotypes without having to identify the underlying mutation because that is already known (White *et al.*, 2013). From 900 mutant lines available from the KOMP/EUCOMM resource (www.sanger.ac.uk/mouseportal/), 489 were screened for viability and 250 mutations were screened for adult phenotypes. Among those selected for screening as adults, 25 were picked because there was some prior evidence of a role for the gene in chromatin. In many cases the resultant mutants are the first characterized in whole animals. Initial phenotypes have been reported and should aid in understanding the role of these chromatin proteins in the biology of the mouse. Examples include integrator complex subunit 12 (*Int12*) and MIS18-binding protein 1 (*Mis18bp1*). Int12 is a part of an integrator-RNA Pol II complex (Baillat *et al.*, 2005), and Mis18bp1 is recruited in a cell cycle-specific manner to centromeric chromatin by CENP-C (Dambacher *et al.*, 2012).

RNA interference (RNAi) screens for genes involved in epigenetic regulation of transcription

Cell lines offer a high-throughput method of screening for modifiers of specific molecular functions, including epigenetic processes. Libraries of inhibiting RNA constructs (i.e. siRNAs and shRNAs) of known chromatin modifiers can be produced. These are restricted to proteins with known epigenetic function, but candidate lists can be expanded based upon the presence of conserved domains. While novel gene discovery is somewhat limited in this approach, new epigenetic functions in the context of a different cell type can be found.

An RNAi library of 200 predicted epigenetic regulators was screened for their ability to alter an epigenetically silent GFP reporter in Hela cells (Poleshko *et al.*, 2010). Of the 200 candidates tested, 15 were required for silencing, including histone modifiers, chromatin remodellers, DNA methylation and Polycomb group protein-related factors, are represented and some overlap can be seen with *Momme* genes (*Hdac1*, *Trim28* and *Pbrm1*).

Another screen was conducted in the U2OS human cell line to find proteins required for activation of Herpes simplex virus genes upon entry into the host nucleus. shRNAs against 372 candidate epigenetic regulators were tested (Oh *et al.*, 2014). Of the 17 factors identified as regulators of viral expression, several were components of either the SWI/SNF, SIN3A complex or the Mediator complex. Again, some overlap can be seen with the factors found in the *Momme* screen (*Baz1b*).

RNAi screens for chromatin-associated genes involved in maintaining an undifferentiated state

RNAi screens have also been used to identify genes necessary for maintaining an undifferentiated state. ES cells were screened with a RNAi library of 1008 genes with known chromatin function or with domains common to chromatin regulators (Fazzio *et al.*, 2008). Sixty-eight factors were identified that contribute to ES cell function; many were known to play important roles in mRNA synthesis and DNA replication, and a large number of proteins known to participate in chromatin remodelling complexes. A similar screen was carried out to identify factors necessary for regulation of differentiation of epidermal cells, 332 genes with known or predicted epigenetic function were knocked down using siRNA constructs, and cells were assessed on their ability to differentiate under multiple conditions using epidermal differentiation markers (Mulder *et al.*, 2012). By clustering genes based on their ability to enable or inhibit differentiation, known protein complexes were identified, including the NURF, MORF, LSD1, BRD4, NuRD and SWI/SNF complexes. Furthermore, epigenomic analysis (i.e. genome-wide ChIPseq data of chromatin bound factors and genome-wide DNA methylation data) combined with transcriptional analysis (i.e. microarray) of differentiating cells showed genetic targets that are regulated.

An RNAi screen to study the role of chromatin proteins in cancer cell-line models

A library of 615 siRNAs that targeted genes with gene ontology (GO) annotations related to epigenetics was used to screen for effects on cultured prostate cancer cells (Bjorkman *et al.*, 2012). Arrays of cells carrying these siRNAs were tested for their ability to divide, survive and express androgen receptor (AR), the gene for which is linked to the development of prostate cancer. Of these, 231 were shown to affect at least one of the phenotypes tested, and interestingly, genes with particular protein domains, for example, PHD-fingers and JmjC domains, behaved in related ways. A similar screen used a shRNA library of chromatin-associated proteins to identify effects on an acute myeloid leukaemia cell line (Zuber *et al.*, 2011). The bromodomain-containing 4 (Brd4) protein, which binds acetylated histones, was found to be required for cellular proliferation and may be a potential therapeutic target in acute myeloid leukaemia.

Yeast two-hybrid screens can identify novel epigenetic regulators

Yeast two-hybrid screening for partners of chromatin-associated proteins

Yeast two-hybrid screening is a method of detecting protein partners by expressing a 'bait' protein and a library of cDNAs, which code for potential 'prey' proteins. If a protein produced from the cDNA library interacts with the bait, a detectable response is produced, i.e. a change in reporter gene transcription. Individual positive clones can be recovered and sequenced. Core components of heterochromatin and euchromatin, as well as novel partners, have been identified in this way. For example, using Su(var)3-7 as a bait, HP1was identified as a partner (Delattre *et al.*, 2000), and in a similar screen, HP1 was identified as a partner of Trim24 (Tif1alpha) (LeDouarin *et al.*, 1996). Further dissection can be carried out to identify the specific interacting domains by *in vitro* mutagenesis.

Interestingly, some surprises have also emerged from these screens; the chromatin assembly factor, chromatin assembly factor 1 (Caf-1), (Murzina *et al.*, 1999), the inner centromere protein antigens 135/155 kDa (INCENP) protein, involved in chromosome segregation (Ainsztein *et al.*, 1998) and origin recognition complex, subunit 1 (Orc1), a factor in the origin of replication complex (Pak *et al.*, 1997), all interact with HP1. A screen for novel partners of DNMT3b identified centromeric protein C (CENP-C), a centromeric protein required for mitosis and chromosome segregation (Gopalakrishnan *et al.*, 2009). The identification of these classic epigenetic regulators as binding partners of proteins that function during the regulation of replication, supports the notion that transcriptional regulation and mitosis are functionally coupled (Zaidi *et al.*, 2010).

A number of mammalian epigenetic regulators were first identified via yeast two-hybrid screening. The cofactor protein, Tat-1 interacting protein 60 kDa (Tip60), was identified in a screen for factors that bind the HIV-transactivator protein Tat1 (Kamine *et al.*, 1996). Subsequently, it was discovered that Tip60 contains a conserved domain sufficient for histone acetyltransferase activity (Yamamoto and Horikoshi, 1997). Similarly, lysine demethylase 5A (KDM5A) also known as retinoblastoma binding protein 2 (RBP2) was initially identified as a binding partner of the cell cycle regulator, retinoblastoma protein (Defeo-Jones *et al.*, 1991). This information helped others to identify the conserved JmjC domain found in KDM5A and two other proteins, shown to be necessary for their function as histone lysine demethylases (Tsukada *et al.*, 2006).

Using yeast one-hybrid assays to dissect mechanisms of transcriptional control

Variations on the yeast two-hybrid assay exist and have been used to identify protein interactions that contribute to epigenetic function. A yeast one-hybrid system was developed to identify proteins that interact with a bait CpG island that directs transcription of a reporter. This CpG island can be methylated by a bacterial CpG methyltransferase. A cDNA library was screened for proteins that activate transcription of the reporter by binding the CpG island in a methylation-dependent manner (Feng *et al.*, 2004). This system identified known methylated CpG-binding proteins, including methyl-CpG-binding domain protein 2 (MBD2), Methyl-CpG-binding domain protein 4 (MBD4) and zinc finger and BTB domain containing 33 (Kaiso) but no novel interactors were found (Feng *et al.*, 2010).

Affinity purification of proteins involved in epigenetics

Immunoprecipitation can be used to identify protein complexes

Specific antibodies can be used to pull down protein partners. The identity of these partners can then be determined by either testing for candidates using antibodies and Western blotting or in an unbiased way by mass spectroscopy. These methods can also been used to identify the changing membership of protein complexes in different cellular contexts (Malovannaya *et al.*, 2010). In this way, a number of complexes involved in epigenetic regulation have been identified, such as the MLL1–WDR5 complex (Dou *et al.*, 2005), the DNMT3B complex (Geiman *et al.*, 2004), the DNMT1–PCNA complex (Estève *et al.*, 2006), the polycomb repressive complex 1 (Levine *et al.*, 2002) and 2 (Kuzmichev *et al.*, 2002), the SWI/SNF complexes (Sif *et al.*, 2001) and others.

Pull-down approaches can be used to identify novel histone partners

Pull-down and identification methods can be applied to histone proteins, allowing the identification of partners associated with particular histone modifications. Initial approaches used the peptide sequence of the N-terminal tail of histones with specific modifications (Bannister *et al.*, 2001; Lachner *et al.*, 2001). Similarly, biotin-tagged N-terminal peptides from Histone 3, di- or tri-methylated at lysine 4, H3K4me2/3, was used to pull down members of the MLL complex (Wysocka *et al.*, 2005).

Quantitative methods to identify novel histone binding factors

Chromatin pull-down and mass spectrometry methodology became quantitative by incorporating SILAC (stable isotope labelling by amino acids in culture), which directly compares samples by metabolic tagging. Initial approaches used artificial peptides to the modified or unmodified N-terminal tails (Vermeulen *et al.*, 2010) or the modified or unmodified mononucleosomes (Bartke *et al.*, 2010) as baits. The studies focussed on specific modifications associated with active or repressive transcription, including H3K4me3 and H3K9me3 and H3K27me3, respectively. In addition, mononucleosome histone methylation was coupled with CpG methylation, with the aim of investigating epigenetic 'crosstalk' (Bartke *et al.*, 2010). Both studies identified proteins known to be associated with specific epigenetic marks, as well as novel proteins. The protein LRWD1 was validated as a binding partner of repressive H3K9me3 by Western blotting.

LRWD1 was found to bind the origin recognition complex (ORC) which in yeast acts as a replication initiator and had previously been linked to transcriptional silencing (Bell *et al.*, 1995). A role for LRWD1 in silencing mammalian repetitive elements was recently demonstrated (Chan and Zhang, 2012).

Using similar methodology, a bait peptide representing the N-terminal sequence of histone H3K4me3 was found to pull down the MORC family CW-type zinc finger 3 (Morc3) protein (Li *et al.*, 2012). Morc3 was previously identified in a screen for proteins associated with the nuclear matrix (Kimura *et al.*, 2002) and the gene has also been identified as a *Su(var)* in the *Momme* screen (Table 5.1). A combination approach, using both modified peptides and recombinant nucleosome arrays, identified factors binding the histone H3K4me3 (active) and H3K9me3 (repressive) modifications (Nikolov *et al.*, 2011), including FANCF and SMCHD1.

A recent report has demonstrated the ability of this approach to examine partners of nucleosomes constituted with the histone H2A or its variants H2A.Z or H2A.X (Fujimoto *et al.*, 2012). Nuclear extracts were incubated with nucleosomes made exclusively from one or other of the three variants and new proteins were identified. A similar approach was used to identify flag-tagged H2A.Z binding partners, among which ubiquitin specific peptidase 39 (USP39), a proteins required for regulation of mitosis (Van Leuken *et al.*, 2008), and PWWP domain containing 2A (PWWP2A), a protein about which little is known, were validated (Draker *et al.*, 2012).

Chromatin binding proteins identified in mouse tissues as opposed to cell lines

Recent advances in the field of mass spectroscopy have permitted the label-free analysis of protein interactions from tissues, as an alternative to SILAC methodology (Luber *et al.*, 2010). This approach was applied to active and repressive histone marks, using extract purified from a number of different mouse tissues. This allowed the identification of histone interactors in a more natural cellular environment (Eberl *et al.*, 2013). Interestingly, some novel tissue specific chromatin binding proteins were identified, including zinc finger protein 462 (ZNF462) binding H3K9me3 in the brain and kidney, and spermiogenesis-specific transcript on the Y 1 (SSTY1) and spermiogenesis-specific transcript on the Y 2 (SSTY2) binding H3K4me3 in the testis, both previously been shown to be required for sperm development (Touré *et al.*, 2004).

Immunoprecipitation can be used to identify genome-wide occupancy of specific proteins

Chromatin immunoprecipitation followed by sequencing, ChIP-seq can be used to map the binding sites of proteins that interact with DNA. Early experiments of this type examined the patterns of modified histones across the genome and the power of the methodology was quickly apparent. For example, it was found that H3K4me3 marks the promoters of actively transcribed genes (Guenther *et al.*, 2007), while H3K9me3 marks inactive loci such as repetitive elements (Barski *et al.*, 2007). These experiments were extended to identify DNA binding sites of non-histone proteins, such as transcription factors and the RNA polymerase machinery.

Using these methods, the complexity of some protein complexes can be unravelled. For example, global mapping of the binding regions for several components of the SWI/SNF chromatin remodelling complex revealed that configurations of different subunits varies with high and low transcriptional states (Euskirchen *et al.*, 2011).

Affinity purification of DNA elements, telomeres and repetitive loci and single loci

Proteins that bind DNA in a methylation-dependent manner

Tagged oligonucleotides can be used as probes and pull-down and identification methods, as described above, can be used to identify proteins bound to them. Methylated versus unmethylated CpG oligonucleotide arrays have been used to identify DNA methylation binding protein, using SILAC methodology. Several proteins were identified as methylation-dependent interactors, among them was actin-like 8 (ACTL8), which had not previously been described as having a role in epigenetics (Mittler *et al.*, 2009). This approach is complicated by the specific sequence of the probe used because proteins can be both sequence-specific and methylation-state-dependent binders. For example, the Notch signalling transcription factor recombination signal binding protein for immunoglobulin kappa J region (RBP-J) has been found to be both CpG methylation and sequence specific (Bartels *et al.*, 2011).

Affinity purification of replication fork associated proteins

Recent mass spectroscopy techniques have been described that enable the purification and identification of proteins that bind newly replicated DNA (Sirbu *et al.*, 2011). zinc finger protein 24 (ZNF24) and WIZ were both found at newly replicated DNA and knockdown of both led to a decrease in replication fork speed (Lopez-Contreras *et al.*, 2013). While Wiz is known to bind histone modifying enzymes (Ueda *et al.*, 2006) and was identified in the *Momme* screen (Daxinger *et al.*, 2013), the molecular role of ZNF24 is less well described. It has been reported to be part of the JMJD3 histone demethylase complex (Chen *et al.*, 2012). Other reports of epigenetic modifier proteins at the replication fork highlight the importance of the rapid reestablishment of epigenetic state of the chromatin following mitosis (Aranda *et al.*, 2014). Proteins identified include family with sequence similarity 111, member A (FAM111A), family with sequence similarity 178, member A (FAM178A) and ATPase family, AAA domain containing 2B (ATAD2B) (Alabert *et al.*, 2014). ATAD2B was previously found to interact with histone acetylation marks via a bromodomain (Filippakopoulos *et al.*, 2012).

Proteins that bind to telomeres

Probes with complementary sequence to a specific locus can be used to pull down endogenous binding factors. Probes hybridizing to telomeric sequence were used to find both known and novel telomere-interacting proteins (Déjardin and Kingston, 2009). Telomeric sequences are thought of as transcriptionally silent in mammals; all three HP1 proteins were identified at telomeres and the protein BRCA1-interacting protein C-terminal helicase 1 (FANCJ) was also identified as enriched, previously known only as a helicase able to bind to BRCA1 (Cantor *et al.*, 2001). More recently, FANCJ has been identified as a suppressor of heterochromatin spreading via its actions at the replication fork (Schwab *et al.*, 2013).

Proteins that bind to single loci

Recently, a similar approach was used to identify factors whose occupancy at the γ-globin locus resulted in the suppression of gene expression that occurs during embryonic development (Pourfarzad *et al.*, 2013). A number of proteins that bind the γ-globin locus, zinc finger DNA-binding protein 89 (ZBP-89), cell division cycle 5-like (CDC5L), APEX nuclease (multifunctional DNA repair enzyme) 1 (APEX1) and transcription factor EB (TFEB), were found necessary for repression after knockdown using shRNAs. The authors validated several γ-globin locus binding proteins from a screen of more than 200 proteins. It is likely that this technology will be applied more broadly to mammalian loci to identify proteins regulating transcription under various conditions. This has been successfully demonstrated recently in yeast (Byrum *et al.*, 2012).

Concluding remarks

Screening in *Drosophila*, yeast and the mouse for modifiers of epigenetic processes has uncovered the identity of a surprisingly large number of proteins. Some of these are species specific, illustrating the importance of carrying out mutagenesis screens in a number of different model organisms. Many turn out to be homologues of known proteins. Emerging approaches, such as RNAi screens and protein enrichment coupled to mass spectrometry, have identified many more. The ultimate value of such techniques might lie in their ability to provide information about the cellular context; the same locus will be bound by different complexes of proteins in different cell types. A number of the factors have been shown to be required for activation of transcription at one locus and repression at another. However, while we might be close to identifying all the factors involved in these processes, we are far from understanding how they work together to ensure proper reprogramming during early development and the maintenance of these states over rounds of cellular division. We await the development of new technologies to enable us to understand the linear order of event

References

Aagaard, L., Laible, G., Selenko, P., Schmid, M., Dorn, R., Schotta, G., Kuhfittig, S., Wolf, A., Lebersorger, A., Singh, P.B., *et al.* (1999). Functional mammalian homologues of the Drosophila PEV-modifier Su(var)3-9 encode centromere-associated proteins which complex with the heterochromatin component M31. EMBO J. *18*, 1923–1938.

Ainsztein, A.M., Kandels-Lewis, S.E., Mackay, A.M., and Earnshaw, W.C. (1998). INCENP centromere and spindle targeting: identification of essential conserved motifs and involvement of heterochromatin protein HP1. J. Cell Biol. *143*, 1763–1774.

Alabert, C., Bukowski-Wills, J.-C., Lee, S.-B., Kustatscher, G., Nakamura, K., de Lima Alves, F., Menard, P., Mejlvang, J., Rappsilber, J., and Groth, A. (2014). Nascent chromatin capture proteomics determines chromatin dynamics during DNA replication and identifies unknown fork components. Nat. Cell Biol. *16*, 281–293.

Allshire, R.C., Nimmo, E.R., Ekwall, K., Javerzat, J.P., and Cranston, G. (1995). Mutations derepressing silent centromeric domains in fission yeast disrupt chromosome segregation. Genes Dev. *9*, 218–233.

Aranda, S., Rutishauser, D., and Ernfors, P. (2014). Identification of a large protein network involved in epigenetic transmission in replicating DNA of embryonic stem cells. Nucleic Acids Res. *42*, 6972–6986.

Ashe, A., Morgan, D., Whitelaw, N., Bruxner, T., Vickaryous, N., Cox, L., Butterfield, N., Wicking, C., Blewitt, M., Wilkins, S., *et al.* (2008). A genome-wide screen for modifiers of transgene variegation identifies genes with critical roles in development. Genome Biol. *9*, R182.1–182.16.

Avvakumov, G.V., Walker, J.R., Xue, S., Li, Y., Duan, S., Bronner, C., Arrowsmith, C.H., and Dhe-Paganon, S. (2008). Structural basis for recognition of hemi-methylated DNA by the SRA domain of human UHRF1. Nature *455*, 822–825.

Baillat, D., Hakimi, M.-A., Näär, A.M., Shilatifard, A., Cooch, N., and Shiekhattar, R. (2005). Integrator, a multiprotein mediator of small nuclear RNA processing, associates with the C-terminal repeat of RNA polymerase II. Cell *123*, 265–276.

Bannister, A.J., Zegerman, P., Partridge, J.F., Miska, E.A., Thomas, J.O., Allshire, R.C., and Kouzarides, T. (2001). Selective recognition of methylated lysine 9 on histone H3 by the HP1 chromo domain. Nature *410*, 120–124.

Barski, A., Cuddapah, S., Cui, K., Roh, T.-Y., Schones, D.E., Wang, Z., Wei, G., Chepelev, I., and Zhao, K. (2007). High-resolution profiling of histone methylations in the human genome. Cell *129*, 823–837.

Bartels, S.J.J., Spruijt, C.G., Brinkman, A.B., Jansen, P.W.T.C., Vermeulen, M., and Stunnenberg, H.G. (2011). A SILAC-based screen for methyl-CpG binding proteins identifies RBP-J as a DNA methylation and sequence-specific binding protein. PLoS ONE *6*, e25884.

Bartke, T., Vermeulen, M., Xhemalce, B., Robson, S.C., Mann, M., and Kouzarides, T. (2010). Nucleosome-interacting proteins regulated by DNA and histone methylation. Cell *143*, 470–484.

Beckstead, R.B., Ner, S.S., Hales, K.G., Grigliatti, T.A., Baker, B.S., and Bellen, H.J. (2005). Bonus, a Drosophila TIF1 homolog, is a chromatin-associated protein that acts as a modifier of position-effect variegation. Genetics *169*, 783–794.

Bell, S.P., Mitchell, J., Leber, J., Kobayashi, R., and Stillman, B. (1995). The multidomain structure of Orc1 p reveals similarity to regulators of DNA replication and transcriptional silencing. Cell *83*, 563–568.

Bird, A. (2002). DNA methylation patterns and epigenetic memory. Genes Dev. *16*, 6–21.

Bjorkman, M., Ostling, P., Harma, V., Virtanen, J., Mpindi, J.P., Rantala, J., Mirtti, T., Vesterinen, T., Lundin, M., Sankila, A., *et al.* (2012). Systematic knockdown of epigenetic enzymes identifies a novel histone demethylase PHF8 overexpressed in prostate cancer with an impact on cell proliferation, migration and invasion. Oncogene *31*, 3444–3456.

Blewitt, M.E., Gendrel, A.-V., Pang, Z., Sparrow, D.B., Whitelaw, N., Craig, J.M., Apedaile, A., Hilton, D.J., Dunwoodie, S.L., Brockdorff, N., *et al.* (2008). SmcHD1, containing a structural-maintenance-of-chromosomes hinge domain, has a critical role in X inactivation. Nat. Genet. *40*, 663–669.

Bostick, M., Kim, J.K., Estève, P.-O., Clark, A., Pradhan, S., and Jacobsen, S.E. (2007). UHRF1 plays a role in maintaining DNA methylation in mammalian cells. Science *317*, 1760–1764.

Buonomo, S.B.C., Wu, Y., Ferguson, D., and de Lange, T. (2009). Mammalian Rif1 contributes to replication stress survival and homology-directed repair. J. Cell Biol. *187*, 385–398.

Byrum, S.D., Raman, A., Taverna, S.D., and Tackett, A.J. (2012). ChAP-MS: a method for identification of proteins and histone post-translational modifications at a single genomic locus. Cell Rep. *2*, 198–205.

Cantor, S.B., Bell, D.W., Ganesan, S., Kass, E.M., Drapkin, R., Grossman, S., Wahrer, D.C.R., Sgroi, D.C., Lane, W.S., Haber, D.A., *et al.* (2001). BACH1, a novel helicase-like protein, interacts directly with BRCA1 and contributes to its DNA repair function. Cell *105*, 149–160.

Chan, K.M., and Zhang, Z. (2012). Leucine-rich repeat and WD repeat-containing protein 1 is recruited to pericentric heterochromatin by trimethylated lysine 9 of histone H3 and maintains heterochromatin silencing. J. Biol. Chem. *287*, 15024–15033.

Chen, S., Ma, J., Wu, F., Xiong, L.-j., Ma, H., Xu, W., Lv, R., Li, X., Villen, J., Gygi, S.P., *et al.* (2012). The histone H3 Lys 27 demethylase JMJD3 regulates gene expression by impacting transcriptional elongation. Genes Dev. *26*, 1364–1375.

Chin, H.G., Patnaik, D., Estève, P.-O., Jacobsen, S.E., and Pradhan, S. (2006). Catalytic properties and kinetic mechanism of human recombinant lys-9 histone H3 methyltransferase SUV39H1: participation of the chromodomain in enzymatic catalysis. Biochemistry *45*, 3272–3284.

Chong, S., Vickaryous, N., Ashe, A., Zamudio, N., Youngson, N., Hemley, S., Stopka, T., Skoultchi, A., Matthews, J., Scott, H.S., *et al.* (2007). Modifiers of epigenetic reprogramming show paternal effects in the mouse. Nat. Genet. *39*, 614–622.

Chuang, L.S.-H., Ian, H.-I., Koh, T.-W., Ng, H.-H., Xu, G., and Li, B.F.L. (1997). Human DNA–(cytosine-5) methyltransferase–PCNA complex as a target for p21WAF1. Science *277*, 1996–2000.

Clapier, C.R., and Cairns, B.R. (2009). The biology of chromatin remodeling complexes. Annu. Rev. Biochem. *78*, 273–304.

Cleard, F., and Spierer, P. (2001). Position-effect variegation in Drosophila: the modifier Su(var)3–7 is a modular DNA-binding protein. EMBO Rep. *2*, 1095–1100.

Cleard, F., Delattre, M., and Spierer, P. (1997). SU(VAR)3–7, a Drosophila heterochromatin-associated protein and companion of HP1 in the genomic silencing of position-effect variegation. EMBO J. *16*, 5280–5288.

Dambacher, S., Deng, W., Hahn, M., Sadic, D., Fröhlich, J., Nuber, A., Hoischen, C., Diekmann, S., Leonhardt, H., and Schotta, G. (2012). CENP-C facilitates the recruitment of M18BP1 to centromeric chromatin. Nucleus 3, 101–110.

Dan, J., Liu, Y., Liu, N., Chiourea, M., Okuka, M., Wu, T., Ye, X., Mou, C., Wang, L., Wang, L., *et al.* (2014). Rif1 maintains telomere length homeostasis of ESCs by mediating heterochromatin silencing. Dev. Cell *29*, 7–19.

Daxinger, L., and Whitelaw, E. (2012). Understanding transgenerational epigenetic inheritance via the gametes in mammals. Nat. Rev. Genet. *13*, 153–162.

Daxinger, L., Oey, H., Apedaile, A., Sutton, J., Ashe, A., and Whitelaw, E. (2012). A forward genetic screen identifies eukaryotic translation initiation factor 3, subunit H (eIF3h), as an enhancer of variegation in the mouse. G3 (Bethesda) *2*, 1393–1396.

Daxinger, L., Harten, S., Oey, H., Epp, T., Isbel, L., Huang, E., Whitelaw, N., Apedaile, A., Sorolla, A., Yong, J., *et al.* (2013). An ENU mutagenesis screen identifies novel and known genes involved in epigenetic processes in the mouse. Genome Biol. *14*, R96.

Defeo-Jones, D., Huang, P.S., Jones, R.E., Haskell, K.M., Vuocolo, G.A., Hanobik, M.G., Huber, H.E., and Oliff, A. (1991). Cloning of cDNAs for cellular proteins that bind to the retinoblastoma gene product. Nature *352*, 251–254.

Déjardin, J., and Kingston, R.E. (2009). Purification of proteins associated with specific genomic loci. Cell *136*, 175–186.

Delattre, M., Spierer, A., Tonka, C.H., and Spierer, P. (2000). The genomic silencing of position-effect variegation in *Drosophila melanogaster*: interaction between the heterochromatin-associated proteins Su(var)3-7 and HP1. J. Cell Sci. *113*, 4253–4261.

Dou, Y., Milne, T.A., Tackett, A.J., Smith, E.R., Fukuda, A., Wysocka, J., Allis, C.D., Chait, B.T., Hess, J.L., and Roeder, R.G. (2005). Physical association and coordinate function of the H3 K4 methyltransferase MLL1 and the H4 K16 acetyltransferase MOF. Cell *121*, 873–885.

Draker, R., Ng, M.K., Sarcinella, E., Ignatchenko, V., Kislinger, T., and Cheung, P. (2012). A combination of H2A.Z and H4 acetylation recruits Brd2 to chromatin during transcriptional activation. PLoS Genet. *8*, e1003047.

Eberl, H.C., Spruijt, Cornelia, G., Kelstrup, Christian, D., Vermeulen, M., and Mann, M. (2013). A map of general and specialized chromatin readers in mouse tissues generated by label-free interaction proteomics. Mol. Cell *49*, 368–378.

Eissenberg, J.C., and Reuter, G. (2009). Cellular mechanism for targeting heterochromatin formation in Drosophila. In International Review of Cell and Molecular Biology, W.J. Kwang, ed. (Academic Press, San Diego), pp. 1–47.

Ekwall, K., and Ruusala, T. (1994). Mutations in rik1, clr2, clr3 and clr4 genes asymmetrically derepress the silent mating-type loci in fission yeast. Genetics *136*, 53–64.

Ekwall, K., Cranston, G., and Allshire, R.C. (1999). Fission yeast mutants that alleviate transcriptional silencing in centromeric flanking repeats and disrupt chromosome segregation. Genetics *153*, 1153–1169.

Estève, P.-O., Chin, H.G., Smallwood, A., Feehery, G.R., Gangisetty, O., Karpf, A.R., Carey, M.F., and Pradhan, S. (2006). Direct interaction between DNMT1 and G9a coordinates DNA and histone methylation during replication. Genes Dev. *20*, 3089–3103.

Euskirchen, G.M., Auerbach, R.K., Davidov, E., Gianoulis, T.A., Zhong, G., Rozowsky, J., Bhardwaj, N., Gerstein, M.B., and Snyder, M. (2011). Diverse roles and interactions of the SWI/SNF chromatin remodeling complex revealed using global approaches. PLoS Genet. *7*, e1002008.

Fazzio, T.G., Huff, J.T., and Panning, B. (2008). An RNAi screen of chromatin proteins identifies Tip60-p400 as a regulator of embryonic stem cell identity. Cell *134*, 162–174.

Feng, S.-Y., Ota, K., Yamada, Y., Sawabu, N., and Ito, T. (2004). A yeast one-hybrid system to detect methylation-dependent DNA–protein interactions. Biochem. Biophys. Res. Commun. *313*, 922–925.

Feng, S.-Y., Ota, K., and Ito, T. (2010). A yeast one-hybrid system to screen for methylated DNA-binding proteins. Nucleic Acids Res. *38*, e189.

Festenstein, R., Tolaini, M., Corbella, P., Mamalaki, C., Parrington, J., Fox, M., Miliou, A., Jones, M., and Kioussis, D. (1996). Locus control region function and heterochromatin-induced position effect variegation. Science *271*, 1123–1125.

Festenstein, R., Sharghi-Namini, S., Fox, M., Roderick, K., Tolaini, M., Norton, T., Saveliev, A., Kioussis, D., and Singh, P. (1999). Heterochromatin protein 1 modifies mammalian PEV in a dose- and chromosomal-context-dependent manner. Nat. Genet. *23*, 457–461.

Filippakopoulos, P., Picaud, S., Mangos, M., Keates, T., Lambert, J.-P., Barsyte-Lovejoy, D., Felletar, I., Volkmer, R., Müller, S., Pawson, T., *et al.* (2012). Histone recognition and large-scale structural analysis of the human bromodomain family. Cell *149*, 214–231.

Fodor, B.D., Shukeir, N., Reuter, G., and Jenuwein, T. (2010). Mammalian Su(var) genes in chromatin control. Annu. Rev. Cell Dev. Biol. *26*, 471–501.

Fujimoto, S., Seebart, C., Guastafierro, T., Prenni, J., Caiafa, P., and Zlatanova, J. (2012). Proteome analysis of protein partners to nucleosomes containing canonical H2A or the variant histones H2A.Z or H2A.X. Biol. Chem. *393*, 47.

Garrick, D., Fiering, S., Martin, D.I.K., and Whitelaw, E. (1998). Repeat-induced gene silencing in mammals. Nat. Genet. *18*, 56–59.

Gaudet, F., Rideout, W.M., Meissner, A., Dausman, J., Leonhardt, H., and Jaenisch, R. (2004). Dnmt1 expression in pre- and postimplantation embryogenesis and the maintenance of IAP silencing. Mol. Cell. Biol. *24*, 1640–1648.

Geiman, T.M., Sankpal, U.T., Robertson, A.K., Chen, Y., Mazumdar, M., Heale, J.T., Schmiesing, J.A., Kim, W., Yokomori, K., Zhao, Y., *et al.* (2004). Isolation and characterization of a novel DNA methyltransferase complex linking DNMT3B with components of the mitotic chromosome condensation machinery. Nucleic Acids Res. *32*, 2716–2729.

Gendrel, A.-V., Tang, Y.A., Suzuki, M., Godwin, J., Nesterova, T.B., Greally, J.M., Heard, E., and Brockdorff, N. (2013). Epigenetic functions of Smchd1 repress gene clusters on the inactive X-chromosome and on autosomes. Mol. Cell. Biol. *33*, 3150–3165.

Gopalakrishnan, S., Sullivan, B.A., Trazzi, S., Della Valle, G., and Robertson, K.D. (2009). DNMT3B interacts with constitutive centromere protein CENP-C to modulate DNA methylation and the histone code at centromeric regions. Hum. Mol. Genet. *18*, 3178–3193.

Guenther, M.G., Levine, S.S., Boyer, L.A., Jaenisch, R., and Young, R.A. (2007). A chromatin landmark and transcription initiation at most promoters in human cells. Cell *130*, 77–88.

Hardy, C.F., Sussel, L., and Shore, D. (1992). A RAP1-interacting protein involved in transcriptional silencing and telomere length regulation. Genes Dev. *6*, 801–814.

Harten, S., Bruxner, T., Bharti, V., Blewitt, M., Nguyen, T.-M.-T., Whitelaw, E., and Epp, T. (2014). The first mouse mutants of D14Abb1e (Fam208a) show that it is critical for early development. Mamm. Genome *25*, 293–303.

Hearn, M.G., Hedrick, A., Grigliatti, T.A., and Wakimoto, B.T. (1991). The effect of modifiers of position-effect variegation on the variegation of heterochromatic genes of *Drosophila melanogaster*. Genetics *128*, 785–797.

Henikoff, S. (1981). Position-effect variegation and chromosome structure of a heat shock puff in *Drosophila*. Chromosoma *83*, 381–393.

Henikoff, S., Loughney, K., and Dreesen, T.D. (1992). The enigma of dominant position-effect variegation in *Drosophila*. The Chromosome Tenth John Innes Symposium, September 7–10, 1992, 193–206.

Hinnebusch, A.G. (2006). eIF3: a versatile scaffold for translation initiation complexes. Trends Biochem. Sci. *31*, 553–562.

Hiragami-Hamada, K., Xie, S., Saveliev, A., Uribe-Lewis, S., Pombo, A., and Festenstein, R. (2009). The molecular basis for stability of heterochromatin-mediated silencing in mammals. Epigenetics Chromatin *2*, 14.

Ho, L., and Crabtree, G.R. (2010). Chromatin remodelling during development. Nature *463*, 474–484.

Hoskins, R., Smith, C., Carlson, J., Carvalho, A.B., Halpern, A., Kaminker, J., Kennedy, C., Mungall, C., Sullivan, B., Sutton, G., *et al.* (2002). Heterochromatic sequences in a Drosophila whole-genome shotgun assembly. Genome Biol. *3*, research0085.0081–0085.0016.

Hrabe de Angelis, M., Flaswinkel, H., Fuchs, H., Rathkolb, B., Soewarto, D., Marschall, S., Heffner, S., Pargent, W., Wuensch, K., Jung, M., *et al.* (2000). Genome-wide, large-scale production of mutant mice by ENU mutagenesis. Nat. Genet. *25*, 444–447.

Ivanova, A.V., Bonaduce, M.J., Ivanov, S.V., and Klar, A.J.S. (1998). The chromo and SET domains of the Clr4 protein are essential for silencing in fission yeast. Nat. Genet. *19*, 192–195.

Iyengar, S., Ivanov, A.V., Jin, V.X., Rauscher, F.J., and Farnham, P.J. (2011). Functional analysis of KAP1 genomic recruitment. Mol. Cell. Biol. *31*, 1833–1847.

James, T.C., and Elgin, S.C. (1986). Identification of a nonhistone chromosomal protein associated with heterochromatin in *Drosophila melanogaster* and its gene. Mol. Cell. Biol. *6*, 3862–3872.

Jenuwein, T., Laible, G., Dorn, R., and Reuter, G. (1998). SET domain proteins modulate chromatin domains in eu- and heterochromatin. Cell. Mol. Life Sci. *54*, 80–93.

Kamine, J., Elangovan, B., Subramanian, T., Coleman, D., and Chinnadurai, G. (1996). Identification of a cellular protein that specifically interacts with the essential cysteine region of the HIV-1 Tat transactivator. Virology *216*, 357–366.

Kato, Y., Kaneda, M., Hata, K., Kumaki, K., Hisano, M., Kohara, Y., Okano, M., Li, E., Nozaki, M., and Sasaki, H. (2007). Role of the Dnmt3 family in de novo methylation of imprinted and repetitive sequences during male germ cell development in the mouse. Hum. Mol. Genet. *16*, 2272–2280.

Kimura, Y., Sakai, F., Nakano, O., Kisaki, O., Sugimoto, H., Sawamura, T., Sadano, H., and Osumi, T. (2002). The newly identified human nuclear protein NXP-2 possesses three distinct domains, the nuclear matrix-binding, RNA-binding, and coiled-coil domains. J. Biol. Chem. *277*, 20611–20617.

Krueger, C., King, M.R., Krueger, F., Branco, M.R., Osborne, C.S., Niakan, K.K., Higgins, M.J., and Reik, W. (2012). Pairing of homologous regions in the mouse genome is associated with transcription but not imprinting status. PLoS ONE 7, e38983.

Kuzmichev, A., Nishioka, K., Erdjument-Bromage, H., Tempst, P., and Reinberg, D. (2002). Histone methyltransferase activity associated with a human multiprotein complex containing the Enhancer of Zeste protein. Genes Dev. *16*, 2893–2905.

Lachner, M., O'Carroll, D., Rea, S., Mechtler, K., and Jenuwein, T. (2001). Methylation of histone H3 lysine 9 creates a binding site for HP1 proteins. Nature *410*, 116–120.

Le, M.H., Duricka, D., and Karpen, G.H. (1995). Islands of complex DNA are widespread in *Drosophila* centric heterochromatin. Genetics *141*, 283–303.

LeDouarin, B., Nielsen, A.L., Garnier, J.M., Ichinose, H., Jeanmougin, F., Losson, R., and Chambon, P. (1996). A possible involvement of TIF1 alpha and TIF1 beta in the epigenetic control of transcription by nuclear receptors. EMBO J. *15*, 6701–6715.

Levine, S.S., Weiss, A., Erdjument-Bromage, H., Shao, Z., Tempst, P., and Kingston, R.E. (2002). The core of the polycomb repressive complex is compositionally and functionally conserved in flies and humans. Mol. Cell. Biol. *22*, 6070–6078.

Li, E., Bestor, T.H., and Jaenisch, R. (1992). Targeted mutation of the DNA methyltransferase gene results in embryonic lethality. Cell *69*, 915–926.

Li, X., Foley, E.A., Molloy, K.R., Li, Y., Chait, B.T., and Kapoor, T.M. (2012). Quantitative chemical proteomics approach to identify post-translational modification-mediated protein–protein interactions. J. Am. Chem. Soc. *134*, 1982–1985.

Liew, C.K., Simpson, R.J.Y., Kwan, A.H.Y., Crofts, L.A., Loughlin, F.E., Matthews, J.M., Crossley, M., and Mackay, J.P. (2005). Zinc fingers as protein recognition motifs: Structural basis for the GATA-1/Friend of GATA interaction. Proc. Natl. Acad. Sci. U.S.A. *102*, 583–588.

Locke, J., Kotarski, M.A., and Tartof, K.D. (1988). Dosage-dependent modifiers of position effect variegation in Drosophila and a mass action model that explains their effect. Genetics *120*, 181–198.

Lopez-Contreras, A.J., Ruppen, I., Nieto-Soler, M., Murga, M., Rodriguez-Acebes, S., Remeseiro, S., Rodrigo-Perez, S., Rojas, A.M., Mendez, J., Muñoz, J., *et al.* (2013). A Proteomic characterization of factors enriched at nascent DNA molecules. Cell Rep. 3, 1105–1116.

Lorentz, A., Ostermann, K., Fleck, O., and Schmidt, H. (1994). Switching gene swi6, involved in repression of silent mating-type loci in fission yeast, encodes a homologue of chromatin-associated proteins from *Drosophila* and mammals. Gene *143*, 139–143.

Luber, C.A., Cox, J., Lauterbach, H., Fancke, B., Selbach, M., Tschopp, J., Akira, S., Wiegand, M., Hochrein, H., O'Keeffe, M., *et al.* (2010). Quantitative proteomics reveals subset-specific viral recognition in dendritic cells. Immunity 32, 279–289.

Lundberg, L.E., Stenberg, P., and Larsson, J. (2013). HP1a, Su(var)3-9, SETDB1 and POF stimulate or repress gene expression depending on genomic position, gene length and expression pattern in Drosophila melanogaster. Nucleic Acids Res. *41*, 4481–4494.

McGowan, R., Campbell, R., Peterson, A., and Sapienza, C. (1989). Cellular mosaicism in the methylation and expression of hemizygous loci in the mouse. Genes Dev. *3*, 1669–1676.

Malovannaya, A., Li, Y., Bulynko, Y., Jung, S.Y., Wang, Y., Lanz, R.B., O'Malley, B.W., and Qin, J. (2010). Streamlined analysis schema for high-throughput identification of endogenous protein complexes. Proc. Natl Acad. Sci. U.S.A. *107*, 2431–2436.

Massah, S., Hollebakken, R., Labrecque, M.P., Kolybaba, A.M., Beischlag, T.V., and Prefontaine, G.G. (2014). Epigenetic characterization of the growth hormone gene identifies SmcHD1 as a regulator of autosomal gene clusters. PLoS ONE *9*, e97535.

Masutani, M., Sonenberg, N., Yokoyama, S., and Imataka, H. (2007). Reconstitution reveals the functional core of mammalian eIF3. EMBO J. *26*, 3373–3383.

Matsui, T., Leung, D., Miyashita, H., Maksakova, I.A., Miyachi, H., Kimura, H., Tachibana, M., Lorincz, M.C., and Shinkai, Y. (2010). Proviral silencing in embryonic stem cells requires the histone methyltransferase ESET. Nature *464*, 927–931.

Mermoud, J.E., Rowbotham, S.P., and Varga-Weisz, P.D. (2011). Keeping chromatin quiet: How nucleosome remodeling restores heterochromatin after replication. Cell Cycle *10*, 4017–4025.

Min, J., Feng, Q., Li, Z., Zhang, Y., and Xu, R.-M. (2003). Structure of the catalytic domain of human DOT1L, a non-SET domain nucleosomal histone methyltransferase. Cell *112*, 711–723.

Mittler, G., Butter, F., and Mann, M. (2009). A SILAC-based DNA protein interaction screen that identifies candidate binding proteins to functional DNA elements. Genome Res. *19*, 284–293.

Mould, A., Pang, Z., Pakusch, M., Tonks, I., Stark, M., Carrie, D., Mukhopadhyay, P., Seidel, A., Ellis, J., Deakin, J., *et al.* (2013). Smchd1 regulates a subset of autosomal genes subject to monoallelic expression in addition to being critical for X inactivation. Epigenetics Chromatin *6*, 19.

Mueller, D., García-Cuéllar, M.-P., Bach, C., Buhl, S., Maethner, E., and Slany, R.K. (2009). Misguided transcriptional elongation causes mixed lineage leukemia. PLoS Biol. 7, e1000249.

Mulder, K.W., Wang, X., Escriu, C., Ito, Y., Schwarz, R.F., Gillis, J., Sirokmany, G., Donati, G., Uribe-Lewis, S., Pavlidis, P., *et al.* (2012). Diverse epigenetic strategies interact to control epidermal differentiation. Nat. Cell Biol. *14*, 753–763.

Muller, H.J. (1930). Types of visible variations induced by X-rays in *Drosophila*. J. Genet. *22*, 299–334.

Murzina, N., Verreault, A., Laue, E., and Stillman, B. (1999). Heterochromatin dynamics in mouse cells: interaction between chromatin assembly factor 1 and HP1 proteins. Mol. Cell *4*, 529–540.

Musselman, C.A., Lalonde, M.-E., Cote, J., and Kutateladze, T.G. (2012). Perceiving the epigenetic landscape through histone readers. Nat. Struct. Mol. Biol. *19*, 1218–1227.

Nady, N., Lemak, A., Walker, J.R., Avvakumov, G.V., Kareta, M.S., Achour, M., Xue, S., Duan, S., Allali-Hassani, A., Zuo, X., *et al.* (2011). Recognition of multivalent histone states associated with heterochromatin by UHRF1 protein. J. Biol. Chem. *286*, 24300–24311.

Nikolov, M., Stützer, A., Mosch, K., Krasauskas, A., Soeroes, S., Stark, H., Urlaub, H., and Fischle, W. (2011). Chromatin affinity purification and quantitative mass spectrometry defining the interactome of histone modification patterns. Mol. Cell. Proteomics *10*, M.110.005371.1.–17.

Nolan, P.M., Peters, J., Strivens, M., Rogers, D., Hagan, J., Spurr, N., Gray, I.C., Vizor, L., Brooker, D., Whitehill, E., *et al.* (2000). A systematic, genome-wide, phenotype-driven mutagenesis programme for gene function studies in the mouse. Nat. Genet. *25*, 440–443.

Oh, H.S., Bryant, K.F., Nieland, T.J.F., Mazumder, A., Bagul, M., Bathe, M., Root, D.E., and Knipe, D.M. (2014). A targeted RNA interference screen reveals novel epigenetic factors that regulate herpesviral gene expression. MBio *5*, e01086-13.

Okano, M., Bell, D.W., Haber, D.A., and Li, E. (1999). DNA methyltransferases Dnmt3a and Dnmt3b are essential for *de novo* methylation and mammalian development. Cell *99*, 247–257.

Pak, D.T., Pflumm, M., Chesnokov, I., Huang, D.W., Kellum, R., Marr, J., Romanowski, P., and Botchan, M.R. (1997). Association of the origin recognition complex with heterochromatin and HP1 in higher eukaryotes. Cell *91*, 311–323.

Park, C.Y., Jeker, L.T., Carver-Moore, K., Oh, A., Liu, H.J., Cameron, R., Richards, H., Li, Z., Adler, D., Yoshinaga, Y., *et al.* (2012). A resource for the conditional ablation of microRNAs in the mouse. Cell Reports *1*, 385–391.

Paro, R., and Hogness, D.S. (1991). The polycomb protein shares a homologous domain with a heterochromatin-associated protein of drosophila. Proc. Natl. Acad. Sci. U.S.A. *88*, 263–267.

Platero, J.S., Hartnett, T., and Eissenberg, J.C. (1995). Functional-analysis of the chromo domain of HP1. EMBO J. *14*, 3977–3986.

Poleshko, A., Einarson, M.B., Shalginskikh, N., Zhang, R., Adams, P.D., Skalka, A.M., and Katz, R.A. (2010). Identification of a functional network of human epigenetic silencing factors. J. Biol. Chem. *285*, 422–433.

Pollock, D.D., and Larkin, J.C. (2004). Estimating the degree of saturation in mutant screens. Genetics *168*, 489–502.

Pourfarzad, F., Aghajanirefah, A., de Boer, E., Ten Have, S., Bryn van Dijk, T., Kheradmandkia, S., Stadhouders, R., Thongjuea, S., Soler, E., Gillemans, N., *et al.* (2013). Locus-specific proteomics by TChP: targeted chromatin purification. Cell Rep. *4*, 589–600.

Preis, J.I., Downes, M., Oates, N.A., Rasko, J.E.J., and Whitelaw, E. (2003). Sensitive flow cytometric analysis reveals a novel type of parent-of-origin effect in the mouse genome. Curr. Biol. *13*, 955–959.

Rakyan, V.K., Blewitt, M.E., Druker, R., Preis, J.I., and Whitelaw, E. (2002). Metastable epialleles in mammals. Trends Genet. *18*, 348–351.

Reuter, G., Giarre, M., Farah, J., Gausz, J., Spierer, A., and Spierer, P. (1990). Dependence of position-effect variegation in Drosophila on dose of a gene encoding an unusual zinc-finger protein. Nature *344*, 219–223.

Robertson, G., Garrick, D., Wilson, M., Martin, D.I.K., and Whitelaw, E. (1996). Age-dependent silencing of globin transgenes in the mouse. Nucleic Acids Res. *24*, 1465–1471.

Rooney, J.W., and Calame, K.L. (2001). TIF1β functions as a coactivator for C/EBPβ and is required for induced differentiation in the myelomonocytic cell line U937. Genes Dev. *15*, 3023–3038.

Rothbart, S.B., Krajewski, K., Nady, N., Tempel, W., Xue, S., Badeaux, A.I., Barsyte-Lovejoy, D., Martinez, J.Y., Bedford, M.T., Fuchs, S.M., *et al.* (2012). Association of UHRF1 with methylated H3K9 directs the maintenance of DNA methylation. Nat. Struct. Mol. Biol. *19*, 1155–1160.

Rushlow, C.A., Bender, W., and Chovnick, A. (1984). Studies on the mechanism of heterochromatic position effect at the rosy locus of *Drosophila melanogaster*. Genetics *108*, 603–615.

Schotta, G., Ebert, A., Dorn, R., and Reuter, G. (2003). Position-effect variegation and the genetic dissection of chromatin regulation in *Drosophila*. Semin Cell Dev. Biol. *14*, 67–75.

Schultz, D.C., Friedman, J.R., and Rauscher, F.J. (2001). Targeting histone deacetylase complexes via KRAB-zinc finger proteins: the PHD and bromodomains of KAP-1 form a cooperative unit that recruits a novel isoform of the Mi-2α subunit of NuRD. Genes Dev. *15*, 428–443.

Schultz, D.C., Ayyanathan, K., Negorev, D., Maul, G.G., and Rauscher, F.J. (2002). SETDB1: a novel KAP-1-associated histone H3, lysine 9-specific methyltransferase that contributes to HP1-mediated silencing of euchromatic genes by KRAB zinc-finger proteins. Genes Dev. *16*, 919–932.

Schulze, S.R., Sinclair, D.A.R., Fitzpatrick, K.A., and Honda, B.M. (2005). A genetic and molecular characterization of two proximal heterochromatic genes on chromosome 3 of *Drosophila melanogaster*. Genetics *169*, 2165–2177.

Sérandour, A.A., Avner, S., Percevault, F., Demay, F., Bizot, M., Lucchetti-Miganeh, C., Barloy-Hubler, F., Brown, M., Lupien, M., Métivier, R., *et al.* (2011). Epigenetic switch involved in activation of pioneer factor FOXA1-dependent enhancers. Genome Res. *21*, 555–565.

Sharif, J., Muto, M., Takebayashi, S.-i., Suetake, I., Iwamatsu, A., Endo, T.A., Shinga, J., Mizutani-Koseki, Y., Toyoda, T., Okamura, K., *et al.* (2007). The SRA protein Np95 mediates epigenetic inheritance by recruiting Dnmt1 to methylated DNA. Nature *450*, 908–912.

Sif, S., Saurin, A.J., Imbalzano, A.N., and Kingston, R.E. (2001). Purification and characterization of mSin3A-containing Brg1 and hBrm chromatin remodeling complexes. Genes Dev. *15*, 603–618.

Silverman, J., Takai, H., Buonomo, S.B.C., Eisenhaber, F., and de Lange, T. (2004). Human Rif1, ortholog of a yeast telomeric protein, is regulated by ATM and 53BP1 and functions in the S-phase checkpoint. Genes Dev. *18*, 2108–2119.

Sinclair, D.R., Mottus, R., and Grigliatti, T. (1983). Genes which suppress position-effect variegation in *Drosophila melanogaster* are clustered. Mol. Gen. Genet. *191*, 326–333.

Singh, P.B., Miller, J.R., Pearce, J., Kothary, R., Burton, R.D., Paro, R., James, T.C., and Gaunt, S.J. (1991). A sequence motif found in a *Drosophila* heterochromatin protein is conserved in animals and plants. Nucleic Acids Res. *19*, 789–794.

Sirbu, B.M., Couch, F.B., Feigerle, J.T., Bhaskara, S., Hiebert, S.W., and Cortez, D. (2011). Analysis of protein dynamics at active, stalled, and collapsed replication forks. Genes Dev. *25*, 1320–1327.

Skarnes, W.C., Rosen, B., West, A.P., Koutsourakis, M., Bushell, W., Iyer, V., Mujica, A.O., Thomas, M., Harrow, J., Cox, T., *et al.* (2011). A conditional knockout resource for the genome-wide study of mouse gene function. Nature *474*, 337–342.

Sreesankar, E., Senthilkumar, R., Bharathi, V., Mishra, R., and Mishra, K. (2012). Functional diversification of yeast telomere associated protein, Rif1, in higher eukaryotes. BMC Genomics *13*, 255.

Steger, D.J., Lefterova, M.I., Ying, L., Stonestrom, A.J., Schupp, M., Zhuo, D., Vakoc, A.L., Kim, J.-E., Chen, J., Lazar, M.A., *et al.* (2008). DOT1L/KMT4 recruitment and H3K79 methylation are ubiquitously coupled with gene transcription in mammalian cells. Mol. Cell. Biol. *28*, 2825–2839.

Takahashi, Y.-H., Schulze, J.M., Jackson, J., Hentrich, T., Seidel, C., Jaspersen, S.L., Kobor, M.S., and Shilatifard, A. (2011). Dot1 and histone H3K79 methylation in natural telomeric and HM silencing. Mol. Cell *42*, 118–126.

Talbert, P.B., LeCiel, C.D., and Henikoff, S. (1994). Modification of the Drosophila heterochromatic mutation brown[Dominant] by linkage alterations. Genetics *136*, 559–571.

Touré, A., Szot, M., Mahadevaiah, S.K., Rattigan, Á., Ojarikre, O.A., and Burgoyne, P.S. (2004). a new deletion of the mouse Y chromosome long arm associated with the loss of Ssty expression, abnormal sperm development and sterility. Genetics *166*, 901–912.

Tschiersch, B., Hofmann, A., Krauss, V., Dorn, R., Korge, G., and Reuter, G. (1994). The protein encoded by the Drosophila position-effect variegation suppressor gene Su(var)3-9 combines domains of antagonistic regulators of homeotic gene complexes. EMBO J. *13*, 3822–3831.

Tsukada, Y.-i., Fang, J., Erdjument-Bromage, H., Warren, M.E., Borchers, C.H., Tempst, P., and Zhang, Y. (2006). Histone demethylation by a family of JmjC domain-containing proteins. Nature *439*, 811–816.

Ueda, J., Tachibana, M., Ikura, T., and Shinkai, Y. (2006). Zinc finger protein Wiz links G9a/GLP histone methyltransferases to the co-repressor molecule CtBP. J. Biol. Chem. *281*, 20120–20128.

Van Leuken, R.J., Luna-Vargas, M.P., Sixma, T.K., Wolthuis, R.M.F., and Medema, R.H. (2008). Usp39 is essential for mitotic spindle checkpoint integrity and controls mRNA-levels of Aurora B. Cell Cycle *7*, 2710–2719.

Velasco, G., Hubé, F., Rollin, J., Neuillet, D., Philippe, C., Bouzinba-Segard, H., Galvani, A., Viegas-Péquignot, E., and Francastel, C. (2010). Dnmt3b recruitment through E2F6 transcriptional repressor mediates germ-line gene silencing in murine somatic tissues. Proc. Natl Acad. Sci. U.S.A. *107*, 9281–9286.

Vermeulen, M., Eberl, H.C., Matarese, F., Marks, H., Denissov, S., Butter, F., Lee, K.K., Olsen, J.V., Hyman, A.A., Stunnenberg, H.G., *et al.* (2010). Quantitative interaction proteomics and genome-wide profiling of epigenetic histone marks and their readers. Cell *142*, 967–980.

Wakimoto, B.T., and Hearn, M.G. (1990). The effects of chromosome rearrangements on the expression of heterochromatic genes in chromosome 2L of *Drosophila melanogaster*. Genetics *125*, 141–154.

Wallrath, L.L., and Elgin, S.C. (1995). Position effect variegation in Drosophila is associated with an altered chromatin structure. Genes Dev. *9*, 1263–1277.

Wang, Z., Zang, C., Rosenfeld, J.A., Schones, D.E., Barski, A., Cuddapah, S., Cui, K., Roh, T.-Y., Peng, W., Zhang, M.Q., *et al.* (2008). Combinatorial patterns of histone acetylations and methylations in the human genome. Nat. Genet. *40*, 897–903.

White, J.K., Gerdin, A.-K., Karp, N.A., Ryder, E., Buljan, M., Bussell, J.N., Salisbury, J., Clare, S., Ingham, N.J., Podrini, C., *et al.* (2013). Genome-wide generation and systematic phenotyping of knockout mice reveals new roles for many genes. Cell *154*, 452–464.

Wilson, C., Bellen, H.J., and Gehring, W.J. (1990). position effects on eukaryotic gene expression. Annu. Rev. Cell Biol. *6*, 679–714.

Wysocka, J., Swigut, T., Milne, T.A., Dou, Y., Zhang, X., Burlingame, A.L., Roeder, R.G., Brivanlou, A.H., and Allis, C.D. (2005). WDR5 associates with histone H3 methylated at K4 and is essential for H3 K4 methylation and vertebrate development. Cell *121*, 859–872.

Xu, D., Muniandy, P., Leo, E., Yin, J., Thangavel, S., Shen, X., Ii, M., Agama, K., Guo, R., Fox, D., *et al.* (2010). Rif1 provides a new DNA-binding interface for the Bloom syndrome complex to maintain normal replication. EMBO J. *29*, 3140–3155.

Xu, L., and Blackburn, E.H. (2004). Human Rif1 protein binds aberrant telomeres and aligns along anaphase midzone microtubules. J. Cell Biol. *167*, 819–830.

Yamamoto, T., and Horikoshi, M. (1997). Novel substrate specificity of the histone acetyltransferase activity of HIV-1-Tat interactive protein Tip60. J. Biol. Chem. *272*, 30595–30598.

Yokota, S., Hirayama, T., Hirano, K., Kaneko, R., Toyoda, S., Kawamura, Y., Hirabayashi, M., Hirabayashi, T., and Yagi, T. (2011). Identification of the cluster control region for the protocadherin-β genes located beyond the protocadherin-γ cluster. J. Biol. Chem. *286*, 31885–31895.

Yoon, B.J., Herman, H., Sikora, A., Smith, L.T., Plass, C., and Soloway, P.D. (2002). Regulation of DNA methylation of Rasgrf1. Nat. Genet. *30*, 92–96.

Youngson, N., Epp, T., Roberts, A., Daxinger, L., Ashe, A., Huang, E., Lester, K., Harten, S., Kay, G., Cox, T., *et al.* (2013). No evidence for cumulative effects in a Dnmt3b hypomorph across multiple generations. Mamm. Genome *24*, 206–217.

Zaidi, S.K., Young, D.W., Montecino, M.A., Lian, J.B., van Wijnen, A.J., Stein, J.L., and Stein, G.S. (2010). Mitotic bookmarking of genes: a novel dimension to epigenetic control. Nat. Rev. Genet. *11*, 583–589.

Zhimulev, I.F., Belyaeva, E.S., Fomina, O.V., Protopopov, M.O., and Bolshakov, V.N. (1986). Cytogenetic and molecular aspects of position effect variegation in *Drosophila melanogaster*. Chromosoma *94*, 492–504.

Zuber, J., Shi, J., Wang, E., Rappaport, A.R., Herrmann, H., Sison, E.A., Magoon, D., Qi, J., Blatt, K., Wunderlich, M., *et al.* (2011). RNAi screen identifies Brd4 as a therapeutic target in acute myeloid leukaemia. Nature *478*, 524–528.

Chromatin-mediated Response to Stimuli

6

Daniel L. Vera, Lauren A. Cole, Benjamin Hoffman and Jonathan H. Dennis

Abstract

In 1950, Edgar and Eileen Stedman hypothesized for the first time that 'one of [histones'] physiological functions is to act as gene suppressors' (Stedman, 1950). This hypothesis implied that histones may block access to the underlying DNA sequence rendering it functionally unavailable. Sixty-four years later, the expectations of this statement remain unfulfilled. Indeed, cells with differing physiologies display unique gene expression patterns, specific DNA methylation patterns, and different histone post translational modifications. The scientific community had anticipated that cells with different physiologies would also have strikingly different nucleosome distributions; however this has not been the case. Cells with different physiologies have remarkably similar nucleosome distributions. Recently, it has been observed that nucleosome distributions do indeed change in a widespread manner in response to stimulus. This event had not been previously observed, as nucleosome redistributions are transient, returning to their basal positions later in the response. Additionally, a new set of nucleosome mapping experiments have led to the observation that nucleosomes exhibit differential sensitivity to nuclease. These results indicate that these fragile nucleosomes may be targets for regulatory factor binding. The discovery of the widespread but transient nature of nucleosome remodelling combined with the observation that basal nucleosome positions exhibit differential sensitivity to nuclease, has allowed for the development of a new model of a role for chromatin in genome regulation.

Introduction to chromatin structure

The eukaryotic nuclear genome is organized into nucleosomes, 147-base-pair segments of DNA that are wrapped 1.65 times around histone octamers (Luger *et al.*, 1997). Nucleosomes appear to play important roles in regulating DNA-templated events such as transcription, replication, repair, and recombination by regulating access of proteins to DNA (Venter *et al.*, 1994; Stunkel *et al.*, 1997; Lomvardas and Thanos, 2002; Lieb and Clarke, 2005; Lam *et al.*, 2008). The application of microarray and next-generation sequencing to nucleosome mapping has allowed genome-scale characterizations of basal nucleosome positions in a variety of model organisms (Johnson *et al.*, 2006; Dennis *et al.*, 2007; Mavrich *et al.*, 2008; Schones *et al.*, 2008; Yuan and Liu, 2008; Chodavarapu *et al.*, 2010). Despite technological advances in nucleosome mapping, there has not been a concomitant increase in our understanding of the dynamic changes in nucleosome positioning.

Determinates of nucleosome position

Two schools of thought dominate the discussion regarding major determinates of nucleosome distribution in the genome: (i) statistical positioning of nucleosomes and (ii) genetically determined nucleosome positioning.

The 'statistical positioning' theory of nucleosome organization, proposed by Kornberg more than 25 years ago (Kornberg, 1981; Kornberg and Lorch, 1999) suggests that nucleosomes are generally stochastically positioned along the genome and are distributed between boundaries that comprise nucleosome-free regions, such as those known to be found at the promoters of active or poised genes. Moreover according to the statistical positioning theory, short regions of DNA, where regulatory factors bind in place of canonical nucleosomes, dynamically punctuate the repetitive nucleosomal structure. Therefore, in this model, nucleosomes are well positioned adjacent to non-nucleosomal boundaries with positioning decreasing as distance from the non-nucleosomal barrier increase. Whether a particular genomic position is occupied by a nucleosome may therefore vary from cell to cell within a population of cells and between different cell types. However, it is expected that the vast majority of the genome at any given time is covered by nucleosomes.

The other model describing nucleosome positioning, genetically determined nucleosome positioning, posits that genomic DNA contains a positioning code that facilitates nucleosome spacing (Kiyama and Trifonov, 2002; Kaplan *et al.*, 2009). In this model, the DNA sequence is the primary determinant for directing nucleosome position and it is this signal, rather than the non-nucleosomal boundaries of statistical positioning theory, that facilitates chromatin regulatory functions. Evidence for both statistical positioning and a nucleosome positioning code may be found in the literature. It is likely that both modes of nucleosome positioning play a role in the regulation of chromatin structure for transcription, replication, recombination and repair. Indeed, the hierarchical model for genome response presented in this chapter resolves a role for both models of nucleosome positioning. In order to appreciate the current understanding on determinants of nucleosome position in genome response it is important to understand the classic chromatin structure work that provided the foundation for this perspective.

History of dynamic chromatin structure

A series of seminal experiments undertaken in the 1970s and 1980s established a link between chromatin structure, nucleosome remodelling and gene expression.

In 1976, a classic experiment by Weintraub and Groudine (1976) reported a relationship between the sensitivity of chromatin to nuclease and tissue specific gene expression potential. Pancreatic deoxyribonuclease I (DNase I) was used to partially digest adult and embryonic chicken erythrocyte chromatin, which was subsequently probed with globin cDNA. Hybridization rates are inversely related to the extent of digestion, indeed relatively low amount of cDNA probe binding to its target gene indicate high sensitivity to the nuclease used of the DNA in which that gene is found. Highly digested globin genes were found in tissues that actively express these genes; yet the globin DNA largely escaping digestion – and which thus extensively hybridized with the globin cDNA probe – was taken from cells in which globin is not expressed. These results led to the understanding that actively transcribed genes must exhibit greater sensitivity to nuclease than those that are not expressed. Similar studies followed and highlighted other examples of tissue specific differential nuclease sensitivity. Many sensitive sites are present within the 5′ promoter region of genes that function in transcriptional regulation; as such, it has been theorized that they represent locations where structural changes enable transcription factor binding (Latchman, 1997, 2001; Stadler *et al.*, 2011).

The first promoter shown to exhibit nucleosomal reorganization following induction of transcription was of the *Saccharomyces cerevisiae* gene PHO5. Deletion mutants revealed that the promoter contained four upstream activating sequences (UAS), one of which was found within a micrococcal nuclease (MNase)/DNase I hypersensitivity site and was thus necessarily free of nucleosomes (Almer and Hörz, 1986; Almer *et al.*, 1986). PHO5 – which encodes a phosphatase – is transcriptionally inactive when the growth medium contains high concentration of inorganic phosphate; and, conversely, transcriptionally active when phosphate levels are low (Han *et al.*, 1988). Performing digests on cells cultured under repressive conditions revealed that two nucleosomes flanked the hypersensitive site on either side, while two other nucleosomes were found within separate UAS. In total, six nucleosomes were consistently located at the described position within the repressed promoter. However, digests performed on induced cells showed the active PHO5 promoter to be absent of nucleosomes (Almer and Hörz, 1986; Almer *et al.*, 1986). Furthermore, high rates of transcription have been observed in cells grown in repressive media and subjected to inhibition of histone H4, the latter prevents nucleosome formation, further implicating the regulatory role of nucleosomal architecture. This experiment was bolstered by the observation that PHO5 expression is independent of UAS in nucleosome-depleted cells (Han *et al.*, 1988).

Other studies expanded the focus to yeast genes besides PHO5. Bernstein, *et al.* (2004) examined over 4300 intergenic regions, measuring histone H3 and H2B content – an indication of nucleosome occupancy – as well as transcription levels of their associated downstream genes; the negative association between nucleosome presence and gene expression showed that the relationship was not limited to PHO5. Genomic regions in which nucleosome depletion was observed were noted to contain an overrepresentation of transcription factor binding motifs, such as that for Rap1; said depletion could be reversed through the removal of the Rap1 binding sites or via treatment of cells with rapamycin (Bernstein *et al.*, 2004). Yuan reiterated this finding in reporting that over 87% of the transcription factor binding sites examined were free of nucleosomes. Additionally, a greater proportion of delocalized nucleosomes were found in highly transcribed genes, while some 65% of observed nucleosomes exhibited consistent positioning (Yuan *et al.*, 2005). Similarly, access to the promoter of the *S. cerevisiae* gene HO is under the regulation of chromatin remodelling that exhibits dynamism across the period of the cell cycle (Takahata *et al.*, 2011).

Thus the regulatory function of chromatin remodelling during gene expression was described in yeast. Similar experiments identified a role for functional nucleosome remodelling in higher eukaryotes. The long terminal repeat (LTR) of the mouse mammalian tumour virus (MMTV) is under the control of a glucocorticoid receptor (GR)-activated promoter; binding of GR to the promoter induces delocalization of a nucleosome occupying the GR response element and modification of a transcription factor binding site. Coupling this information to the data that the various yeast studies had yielded, Perlmann and Wrange (1988, 1991) injected *Xenopus laevis* oocytes with plasmids separately encoding the MMTV LTR, GR and competitor DNA, then incubated the cells in medium containing glucocorticoid. This treatment resulted in reduced nucleosome formation and increased MMTV LTR transcription (Perlmann and Wrange, 1988, 1991). Subsequent work revealed that, upon hormone-induced remodelling of what was termed 'nucleosome B' – leading to establishment of a chromatin conformation conducive to transcription – a DNaseI hypersensitivity site is exposed. The same results were obtained when treating Xenopus oocytes with glucocorticoid agonists. Treatment with antagonists, however, caused nucleosomes to shift from the precisely arrayed configuration that enables transcription, returning them to the organization characteristic of the inactive promoter (Perlmann and Wrange, 1988; Belikov *et al.*, 2000, 2001).

Recent advances in mapping nucleosome dynamics

The seminal experiments that identified regulatory nucleosomal organizations were followed by genome-wide studies of nucleosome occupancy. Recent advances in nucleosome mapping technologies (e.g. microarrays, next-generation parallel sequencing) have allowed the measurement of nucleosome distributions on a genome-wide scale in several physiological contexts (Yuan *et al.*, 2005; Albert *et al.*, 2007; Dennis *et al.*, 2007; Ozsolak *et al.*, 2007; Schones *et al.*, 2008). These experiments affirmed a canonical transcription start site architecture consisting of a nucleosome free region flanked by statistically positioned nucleosomes. More recently, a nucleosome mapping study of seven different lymphoblastoid cells revealed strikingly consistent nucleosome positioning among all cell types tested (Gaffney *et al.*, 2012). This consistency in nucleosome positioning among disparate cell types has been truly remarkable (Fig. 6.1).

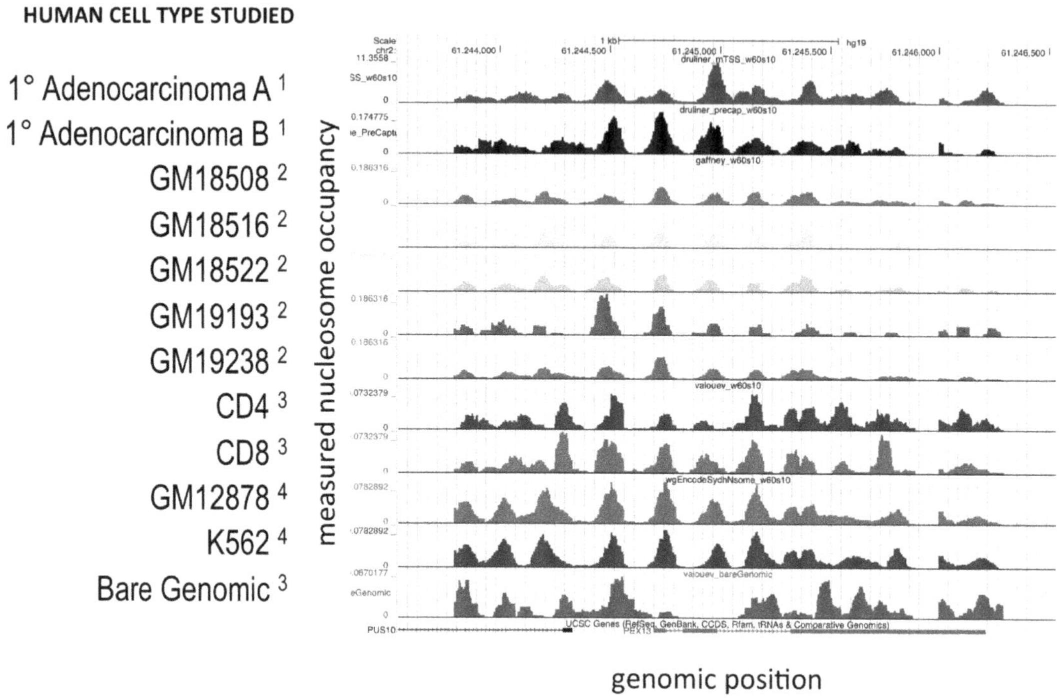

Figure 6.1 Nucleosome positions are strikingly similar among disparate cell types. This UCSC browser shot shows nucleosome positions at the PUS10 transcription start site from multiple disparate cell types. Peaks represent well-positioned nucleosomes, and these peaks are consistent among disparate cell types including: primary human lung cancer tumours, EBV transformed human B-lymphocytes, human cancer cell lines. (1) Druliner *et al.* (unpublished), (2) Gaffney *et al.* (2012), (3) Valouev *et al.* (2011) and (4) Kundaje *et al.* (2012).

Genome-scale nucleosome mapping and dynamics

Much of the work characterizing genome-wide nucleosome occupancy has focused primarily on basal nucleosome positions and not on nucleosome dynamics (Yuan *et al.*, 2005; Lee *et al.*, 2007; Valouev *et al.*, 2008, 2011; Gaffney *et al.*, 2012; Kundaje *et al.*, 2012). Many studies that have attempted to identify discrete nucleosome repositioning events have not provided strong evidence for such discrete changes at individual loci (Schones *et al.*, 2008). In addition, there have been hundreds of published analyses of data produced by the Encyclopedia of DNA Elements (ENCODE) Consortium focusing on transcription factor binding and histone post-translational modifications, however no publication using ENCODE data has highlighted differences in nucleosome occupancy between the two cell lines from which nucleosome positions have been mapped. While global differences in nucleosome occupancy have been observed in aggregate profiles of hundreds of loci, only recently has there been evidence for a widespread nucleosome redistributions and genomic regulation at individual loci (Druliner *et al.*, 2013; Sexton *et al.*, 2014a,b).

Studies of nucleosome distribution have shown that disparate cell types have similar nucleosome distribution. High quality data sets have now begun to shed more light on nucleosome occupancy across the genome or surrounding transcription start sites, but still view nucleosomes as mostly static entities observed at a single point in time (Schones *et al.*, 2008; Valouev *et al.*, 2008; Gaffney *et al.*, 2012). A new approach to nucleosome mapping experiments allowed for the identification of widespread nucleosome remodelling events. Recent studies have mapped the distribution of nucleosomes at multiple time points during a genomic response in order to generate a detailed picture of nucleosome dynamics in response to a given stimulus. High temporal resolution maps of nucleosome distribution have captured widespread changes throughout the genome in response to a stimulus. These changes are transient with nucleosomes ultimately returning to their basal positions. This phenomenon has been observed in two systems: reactivation of Kaposi's sarcoma-associated herpes virus (KSHV) and the progression of lung and colon adenocarcinoma (LAC and CRC).

KSHV lytic reactivation is well studied, and in this case serves as common immune stimulus to which a large set of genes respond. Nucleosome distribution was measured at 472 immunity-related genes at 6 hours, 8 hours, 24 hours, and 48 hours post KSHV reactivation. The greatest change in nucleosome distribution occurred at the 24 hour time point, with more than half of genes studied showing significant translational repositioning of nucleosomes (Fig. 6.2). Nearly all loci with altered nucleosome distribution returned to their basal nucleosome distributions at 48 hours, which is also the point of maximal viral reactivation (Sexton *et al.*, 2014a).

A similar observation was made for the progression of both lung adenocarcinoma (LAC) and colorectal cancer (CRC). Histologically defined increasingly advanced patient tumour samples, as well as their matched normal tissue, serve as sequential time points in this case. Across patient samples, nucleosome distribution changes occur most frequently in early tumours: grade 1 in LAC and stages 2 and 3 in CRC when compared with normal adjacent tissue. More advanced tumour samples (LAC grades 2 and 3, CRC stage 4) do not show changes, and in fact have a higher correlation with the normal tissue samples (Druliner *et al.*, 2013). These studies support a new hypothesis that nucleosome redistributions are widespread and that the nucleosomes at the transcription start site deviate from their basal distributions only fleetingly. These observations represent a major step forward in our understanding of the role of nucleosome position in genome response, and suggest that nucleosome redistributions are not rare events, as prior literature would indicate, but common to a variety of genomic responses.

The role of nucleosome position and sensitivity in genome regulation

A recently discovered feature of chromatin structure is the presence of individual nucleosomal

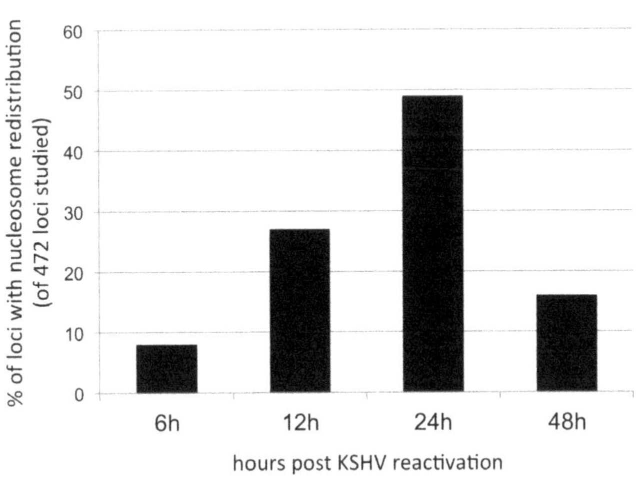

Figure 6.2 Nucleosome redistributions are widespread and transient. This bar chart shows the number of loci displaying translational repositioning of nucleosomes at 6, 12, 24, and 48 hours post reactivation of the Kaposi's sarcoma-associated herpesvirus. The greatest number of nucleosome distribution changes occurred at the 24 hours, and the nucleosome redistributions transiently return to the basal state nucleosome distribution (0 hours) at the 48 hours.

footprints that are hypersensitive to cleavage by MNase. Unlike DNase I-hypersensitive sites, MNase hypersensitive nucleosomes are characterized by a pair of hypersensitive cut sites surrounding nucleosome-sized footprints. By mapping nucleosomes with different concentrations of MNase, several groups have identified thousands of such nucleosomal footprints in the yeast genome (Weiner *et al.*, 2010; Henikoff *et al.*, 2011; Xi *et al.*, 2011). Specifically, a large proportion of the yeast genome displays nucleosomal footprints under light-digest conditions, but a small fraction of these footprints show significantly less signal under heavy-digest conditions. The differences in signal with different digestion conditions have been attributed to an altered conformation of nucleosomal DNA at these regions in the yeast genome that render these footprints hypersensitive to MNase cleavage (Henikoff *et al.*, 2011). These regions partially overlap with transcription factor binding sites and H2A.Z-containing nucleosomes (Xi *et al.*, 2011), and are associated with highly transcribed genes (Weiner *et al.*, 2010). The potential significance of these nucleosomes is not clear, but available evidence suggests that these regions are targets for transcription factor binding (Xi *et al.*, 2011).

A hierarchical model for genome response

When the classic dynamic nucleosome experiments are viewed through the prism of the more recent discoveries, it is interesting to develop a model that explains all of the results in one context (Fig. 6.3). The recent discovery of the widespread and transient nature of genome response leads to the proposal of a 'genomic transient intermediate state.' This 'genomic transient intermediate state' is induced by a widespread chromatin remodelling event in which many loci exhibit translational repositioning of their nucleosomes. In this transient intermediate a superset of genomic loci are made available for regulatory licensing. Gene loci in the superset in a physiology or cell type with the appropriate regulatory machinery will be licensed. Those without the regulatory machinery will not be affected. This transient reorganization maximizes the potential for multiple specific concerted responses with a limited number of genomic architectures. This model is quite practical when one considers that there must be a restriction on the number of architectures that the 3-metre human genome can adopt in the 5-mm nucleus of the cell.

Not all of the effects of genome response, however, are short lived. While nucleosome

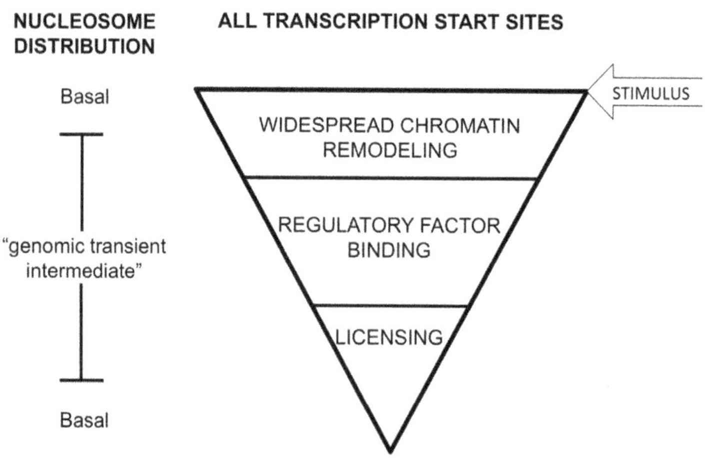

Figure 6.3 A hierarchical model for genome response. In this model, stimulus initiates a widespread chromatin remodelling event. This widespread remodelling event creates the 'genomic transient intermediate.' The 'genomic transient intermediate' potentiates regulatory factor binding and ultimately the licensing of individual loci. The remodelled nucleosomes return to their original basal positions, but individual nucleosomes display a new sensitivity to nuclease.

redistributions are transient, it appears that the sensitivity of nucleosomes to nuclease is a longer-lived feature of chromatin structure that has regulatory potential. In this hierarchical model for genome response, the different biochemical properties of the nuclease sensitive nucleosomes represent the cell-type specific regulatory potential of the chromatin. The sensitive nucleosomes are likely associated with multiple different features, including transcription factor binding sites, conserved non-coding sequences, gene expression levels, and tissue-specific gene expression. The identification of these sensitive nucleosomes solves the conundrum that nucleosome distribution is largely similar between disparate cell types. Even though the nucleosome distribution is similar in different cell types, the sensitivity of the individual nucleosomes is cell type specific and plays a role in the potential of the cell.

Conclusions

It will be important to test this model in multiple cellular contexts. If the 'genomic transient intermediate' can be identified in multiple cellular responses then we have an important tool with which to study the mechanisms underlying genome response. Highly time resolved studies before during and after the 'genomic transient intermediate' will go far to delineate the order of recruitment of chromatin modifying activities and their biochemical consequence. Likewise, measurement of the nuclease sensitivity of individual nucleosomes before and after the 'genomic transient intermediate' will give insights into the potential of the nuclear response for a given cell type. The fundamental and distinct observations that (1) nucleosome redistributions are widespread and transient and (2) individual nucleosomes show differential sensitivity to nuclease are important discoveries. These observations and their interpretation in the hierarchical model of genome response pave the way to transform nucleosome-mapping studies.

References

Albert, I., Mavrich, T.N., Tomsho, L.P., Qi, J., Zanton, S.J., Schuster, S.C., and Pugh, B.F. (2007). Translational and rotational settings of H2A.Z nucleosomes across the *Saccharomyces cerevisiae* genome. Nature *446*, 572–576.

Almer, A., and Hörz, W. (1986). Nuclease hypersensitive regions with adjacent positioned nucleosomes mark the gene boundaries of the PHO5/PHO3 locus in yeast. EMBO J. *5*, 2681–2687.

Almer, A., Rudolph, H., Hinnen, A., and Hörz, W. (1986). Removal of positioned nucleosomes from the yeast PHO5 promoter upon PHO5 induction releases additional upstream activating DNA elements. EMBO J. *5*, 2689–2696.

Belikov, S., Gelius, B., Almouzni, G., and Wrange, O. (2000). Hormone activation induces nucleosome positioning *in vivo*. EMBO J. *19*, 1023–1033.

Belikov, S., Gelius, B., and Wrange, O. (2001). Hormone-induced nucleosome positioning in the MMTV promoter is reversible. EMBO J. *20*, 2802–2811.

Bernstein, B.E., Liu, C.L., Humphrey, E.L., Perlstein, E.O., and Schreiber, S.L. (2004). Global nucleosome occupancy in yeast. Genome Biol. *5*, R62, 1–11.

Chodavarapu, R.K., Feng, S., Bernatavichute, Y.V., Chen, P.Y., Stroud, H., Yu, Y., Hetzel, J.A., Kuo, F., Kim, J., Cokus, S.J., *et al.* (2010). Relationship between nucleosome positioning and DNA methylation. Nature *466*, 388–392.

Dennis, J.H., Fan, H.Y., Reynolds, S.M., Yuan, G., Meldrim, J.C., Richter, D.J., Peterson, D.G., Rando, O.J., Noble, W.S., and Kingston, R.E. (2007). Independent and complementary methods for large-scale structural analysis of mammalian chromatin. Genome Res. *17*, 928–939.

Druliner, B.R., Fincher, J.A., Sexton, B.S., Vera, D.L., Roche, M., Lyle, S., and Dennis, J.H. (2013). Chromatin patterns associated with lung adenocarcinoma progression. Cell Cycle *12*, 1536–1543.

Gaffney, D.J., McVicker, G., Pai, A.A., Fondufe-Mittendorf, Y.N., Lewellen, N., Michelini, K., Widom, J., Gilad, Y., and Pritchard, J.K. (2012). Controls of nucleosome positioning in the human genome. PLoS Genet. *8*, e1003036.

Han, M., Kim, U.J., Kayne, P., and Grunstein, M. (1988). Depletion of histone H4 and nucleosomes activates the PHO5 gene in *Saccharomyces cerevisiae*. EMBO J. *7*, 2221–2228.

Henikoff, J.G., Belsky, J.A., Krassovsky, K., MacAlpine, D.M., and Henikoff, S. (2011). Epigenome characterization at single base-pair resolution. Proc. Natl. Acad. Sci. U.S.A. *108*, 18318–18323.

Johnson, S.M., Tan, F.J., McCullough, H.L., Riordan, D.P., and Fire, A.Z. (2006). Flexibility and constraint in the nucleosome core landscape of *Caenorhabditis elegans* chromatin. Genome Res. *16*, 1505–1516.

Kaplan, N., Moore, I.K., Fondufe-Mittendorf, Y., Gossett, A.J., Tillo, D., Field, Y., LeProust, E.M., Hughes, T.R., Lieb, J.D., Widom, J., *et al.* (2009). The DNA-encoded nucleosome organization of a eukaryotic genome. Nature *458*, 362–366.

Kiyama, R., and Trifonov, E.N. (2002). What positions nucleosomes? – A model. FEBS Lett. *523*, 7–11.

Kornberg, R. (1981). The location of nucleosomes in chromatin: specific or statistical. Nature *292*, 579–580.

Kornberg, R.D., and Lorch, Y. (1999). Twenty-five years of the nucleosome, fundamental particle of the eukaryote chromosome. Cell *98*, 285–294.

Kundaje, A., Kyriazopoulou-Panagiotopoulou, S., Libbrecht, M., Smith, C.L., Raha, D., Winters, E.E., Johnson, S.M., Snyder, M., Batzoglou, S., *et al.* (2012). Ubiquitous heterogeneity and asymmetry of the chromatin environment at regulatory elements. Genome Res. *22*, 1735–1747.

Lam, F.H., Steger, D.J., and O'Shea, E.K. (2008). Chromatin decouples promoter threshold from dynamic range. Nature *453*, 246–250.

Latchman, D.S. (1997). Landmarks in gene regulation (London: Portland Press).

Latchman, D.S. (2001). Transcription factors: bound to activate or repress. Trends Biochem. Sci. *26*, 211–213.

Lee, W., Tillo, D., Bray, N., Morse, R.H., Davis, R.W., Hughes, T.R., and Nislow, C. (2007). A high-resolution atlas of nucleosome occupancy in yeast. Nat. Genet. *39*, 1235–1244.

Lieb, J.D., and Clarke, N.D. (2005). Control of transcription through intragenic patterns of nucleosome composition. Cell *123*, 1187–1190.

Lomvardas, S., and Thanos, D. (2002). Opening chromatin. Mol. Cell *9*, 209–211.

Luger, K., Mader, A.W., Richmond, R.K., Sargent, D.F., and Richmond, T.J. (1997). Crystal structure of the nucleosome core particle at 2.8 A resolution. Nature *389*, 251–260.

Mavrich, T.N., Jiang, C., Ioshikhes, I.P., Li, X., Venters, B.J., Zanton, S.J., Tomsho, L.P., Qi, J., Glaser, R.L., Schuster, S.C., *et al.* (2008). Nucleosome organization in the Drosophila genome. Nature *453*, 358–362.

Ozsolak, F., Song, J.S., Liu, X.S., and Fisher, D.E. (2007). High-throughput mapping of the chromatin structure of human promoters. Nat. Biotechnol. *25*, 244–248.

Perlmann, T., and Wrange, O. (1988). Specific glucocorticoid receptor binding to DNA reconstituted in a nucleosome. EMBO J. *7*, 3073–3079.

Perlmann, T., and Wrange, O. (1991). Inhibition of chromatin assembly in Xenopus oocytes correlates with derepression of the mouse mammary tumor virus promoter. Mol. Cell. Biol. *11*, 5259–5265.

Schones, D.E., Cui, K., Cuddapah, S., Roh, T.Y., Barski, A., Wang, Z., Wei, G., and Zhao, K. (2008). Dynamic regulation of nucleosome positioning in the human genome. Cell *132*, 887–898.

Sexton, B.S., Avey, D., Druliner, B.R., Fincher, J.A., Vera, D.L., Grau, D.J., Borowsky, M.L., Gupta, S., Girimurugan, S.B., Chicken, E., *et al.* (2014a). The spring-loaded genome: nucleosome redistributions are widespread, transient, and DNA-directed. Genome Res. *24*, 251–259.

Sexton, B.S., Druliner, B.R., Avey, D., Zhu, F., and Dennis, J.H. (2014b). Changes in nucleosome occupancy occur in a chromosome specific manner. Genom. Data *2*, 114–116.

Stadler, M.B., Murr, R., Burger, L., Ivanek, R., Lienert, F., Scholer, A., van Nimwegen, E., Wirbelauer, C., Oakeley, E.J., Gaidatzis, D., *et al.* (2011). DNA-binding factors shape the mouse methylome at distal regulatory regions. Nature *480*, 490–495.

Stedman, E. (1950). Cell specificity of histones. Nature *166*, 780–781.

Stunkel, W., Kober, I., and Seifart, K.H. (1997). A nucleosome positioned in the distal promoter region activates transcription of the human U6 gene. Mol. Cell. Biol. *17*, 4397–4405.

Takahata, S., Yu, Y., and Stillman, D.J. (2011). Repressive chromatin affects factor binding at yeast HO (homothallic switching) promoter. J. Biol. Chem. *286*, 34809–34819.

Valouev, A., Ichikawa, J., Tonthat, T., Stuart, J., Ranade, S., Peckham, H., Zeng, K., Malek, J.A., Costa, G., McKernan, K., *et al.* (2008). A high-resolution, nucleosome position map of *C. elegans* reveals a lack of universal sequence-dictated positioning. Genome Res. *18*, 1051–1063.

Valouev, A., Johnson, S.M., Boyd, S.D., Smith, C.L., Fire, A.Z., and Sidow, A. (2011). Determinants of nucleosome organization in primary human cells. Nature *474*, 516–520.

Venter, U., Svaren, J., Schmitz, J., Schmid, A., and Horz, W. (1994). A nucleosome precludes binding of the transcription factor Pho4 *in vivo* to a critical target site in the PHO5 promoter. EMBO J. *13*, 4848–4855.

Weiner, A., Hughes, A., Yassour, M., Rando, O.J., and Friedman, N. (2010). High-resolution nucleosome mapping reveals transcription-dependent promoter packaging. Genome Res. *20*, 90–100.

Weintraub, H., and Groudine, M. (1976). Chromosomal subunits in active genes have an altered conformation. Science *193*, 848–856.

Xi, Y., Yao, J., Chen, R., Li, W., and He, X. (2011). Nucleosome fragility reveals novel functional states of chromatin and poises genes for activation. Genome Res. *21*, 718–724.

Yuan, G.C., and Liu, J.S. (2008). Genomic sequence is highly predictive of local nucleosome depletion. PLoS Comput. Biol. *4*, e13.

Yuan, G.C., Liu, Y.J., Dion, M.F., Slack, M.D., Wu, L.F., Altschuler, S.J., and Rando, O.J. (2005). Genome-scale identification of nucleosome positions in *S. cerevisiae*. Science *309*, 626–630.

The Epigenetics of Centromere Function

7

Justyne E. Ross*, Shannon M. McNulty* and Beth A. Sullivan

Abstract

The centromere is a complex chromosomal locus where the kinetochore is formed and microtubules attach during cell division. Centromere specification in eukaryotes largely depends on sequence-independent (epigenetic) mechanisms. In this chapter, we discuss current understanding of the epigenetic basis for centromere structure and function. In addition to endogenous centromeres in model organisms and humans, we review studies of atypical centromeres (neocentromeres, artificial chromosomes, dicentric chromosomes, engineered chromosomes) that have contributed to the current molecular understanding of centromere function. In particular, we focus on the behaviour and regulation of the centromeric histone variant CENP-A, chromatin modifications, and transcription in centromere establishment and maintenance. Over the past three decades, our view of centromere biology has expanded significantly from the initial characterization of centromere proteins to a more mature understanding of the DNA, RNA, and protein components that constitute this complex chromosomal locus.

Introduction

The centromere is an essential chromosomal locus of unique chromatin composition located at the primary constriction of eukaryotic chromosomes. It is required to segregate chromosomes in mitosis and meiosis and is important for maintaining sister chromatid (mitosis) or homologue (meiosis) attachments prior to anaphase. Each chromosome must contain one, and usually only one, centromere. Chromosomes without centromeres are lost during cell division and those with more than one centromere often undergo anaphase lag and/or chromosome breakage. Defective centromere function is associated with aneuploidy, birth defects, infertility, cancer, and ageing. Studying centromere structure, mechanics, and maintenance is important for understanding human health and disease.

Because the centromere has such a pivotal role in chromosome inheritance, it would seem logical that the genetic and/or genomic components of centromeres would be identical across organisms. This, however, is not the case. Centromeres are generally located in or near regions of repetitive DNA, but not only do centromeric DNA sequences differ between organisms, they are often dissimilar among chromosomes within the same organism (Fig. 7.1). Instead, many of the protein components of centromeres, such as the constitutive centromere-associated network (CCAN) and the kinetochore–microtubule network (KMN), are shared among organisms and chromosomes (McAinsh and Meraldi, 2011). These proteins are present at atypical centromeric sites, but are absent from centromeres that lack function. Altogether, this evidence points to epigenetic, or *sequence-independent*, mechanisms that regulate centromere identity and function.

This chapter covers the epigenetic nature of centromeres. We discuss research in various organisms that studied variant centromeres and naturally occurring and engineered centromeres to yield insight into how histone modifications,

*These authors contributed equally to this chapter

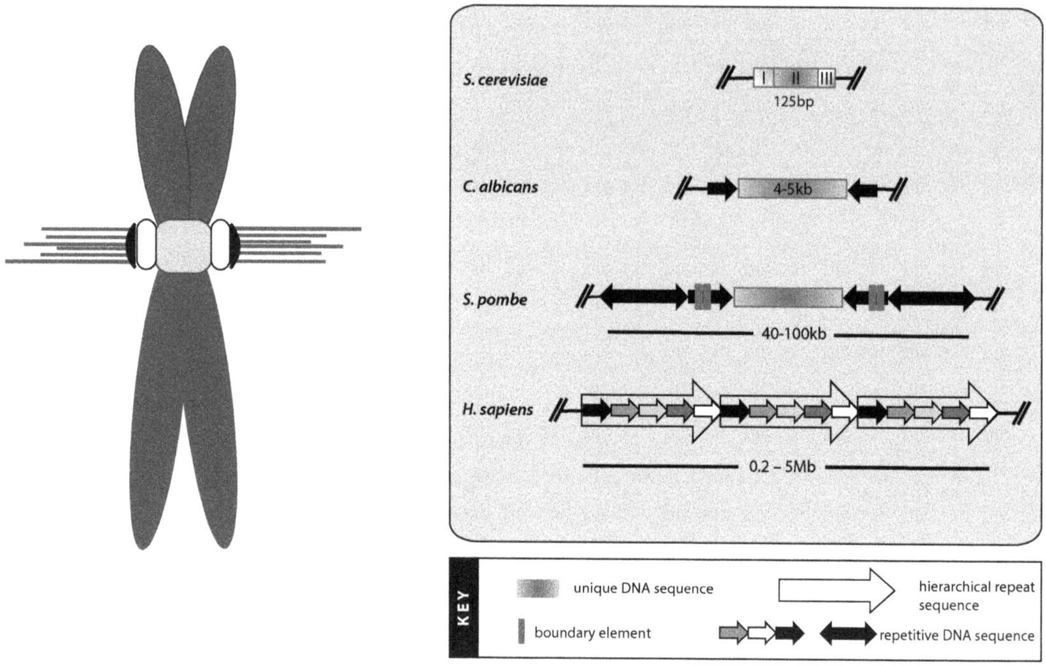

Figure 7.1 Eukaryotic centromeres lack DNA sequence identity. The centromere is a region on the chromosome that is essential for chromosome segregation. It is the site of nucleoprotein kinetochore complex and spindle microtubule attachment. Despite this essential functional role in chromosome inheritance, the DNA sequences at centromeres differ among organisms and chromosomes. With the exception of the budding yeast *S. cerevisiae* that is genetically and functionally defined by DNA sequence, there is little DNA homology among centromeres of larger eukaryotes. *C. albicans* and *S. pombe* centromeres have a unique central core region that is flanked by repetitive DNA, while human centromeres are comprised entirely of homogenous arrays of repetitive alpha satellite DNA.

protein cascades, and cell cycle regulation of protein recruitment confer centromere specification and function.

Centromeric chromatin and the histone variant CENP-A

As centromeres in model organisms and among chromosomes within the same organism were discovered, it became clear that no common sequence element was strictly linked to centromere identity. However, it was quickly realized that centromeres are epigenetically defined by the presence of centromeric protein A (CENP-A), a H3 histone variant. CENP-A was discovered in 1985 when blotting of serum isolated from CREST (calcinosis, Raynaud's phenomenon, oesophageal dysmotility, sclerodactyly, telangiectasia) syndrome patients revealed three recurrent bands. These antigens were identified as centromere components by immunofluorescence labelling of mitotic cells (Earnshaw and Migeon, 1985). The 17 kDa species was designated CENP-A while the other two bands were identified as CENP-B and CENP-C. Biochemical studies showed that CENP-A co-purified with nucleosome core particles and histones, providing the first suggestion that CENP-A functioned as a centromere-specific core histone and contributed to the fundamental chromatin nucleoprotein complex (Palmer *et al.*, 1987). Continued investigation confirmed that CENP-A is a core histone with some sequences highly similar to regions of histone H3 (Palmer *et al.*, 1991). Sequence analysis of CENP-A and histone H3 found 62% identity between the C-terminal histone fold domains, whereas the major differences were concentrated in the N-terminus (Sullivan *et al.*, 1994b). Since the initial discovery of CENP-A in human samples, CENP-A homologues have been identified in many eukaryotic species: HCP-3 in *Caenorhabditis elegans*, CID in *Drosophila melanogaster*, Cse4 in *Saccharomyces cerevisiae*, Cnp1 in *Schizosaccharomyces pombe*, and CenH3 in *Arabidopsis thaliana* (Blower and Karpen, 2001; Buchwitz *et al.*, 1999; Stoler *et al.*, 1995; Takahashi *et al.*, 2000; Talbert *et al.*, 2002).

Characterization of CENP-A and centromeric nucleosomes

Understanding structural and functional features of CENP-A provides insight into the epigenetic regulation of centromeres and chromatin in general. Many unique structural and dynamic characteristics have been identified in CENP-A and CENP-A-containing nucleosomes. Using an epitope-tagged derivative of human CENP-A, centromere targeting activity was found to be dependent on the presence of the histone fold domain of CENP-A (Sullivan *et al.*, 1994b). Located within the histone fold domain is the CENP-A targeting domain (CATD) composed of loop 1 and α2 helix (Black *et al.*, 2004). Additional unique features were identified using deuterium exchange/MS coupled with hydrodynamic measures; CENP-A and H4 form sub-nucleosomal tetramers that are more compact and conformationally more rigid than the corresponding H3/H4 tetramers (Black *et al.*, 2004). The increase in conformational rigidity of centromeric nucleosomes was determined to be caused by the CATD; substitution of the CATD into H3 created the same rigid nucleosome observed when CENP-A is present (Black *et al.*, 2007). Thus, the CATD confers both targeting information and conformational rigidity. Furthermore, direct support for centromere identity being maintained by a unique nucleosome structure came from creating nucleosomes consisting of modified H3 containing the CATD. In both human cells and yeast, CENP-A depletion was lethal but could be rescued by H3+CATD nucleosomes (Conde e Silva *et al.*, 2007).

These studies highlighted the functional importance of the CATD. In 2010 the crystal structure of the $(\text{CENP-A-H4})_2$ tetramer revealed three distinguishing properties conferred by the CATD: (1) a rotated CENP-A–CENP-A interface; (2) a bulge in loop1 of CENP-A with the opposite charge as that on H3; and (3) hydrophobic interactions at the CENP-A–H4 interface (Sekulic *et al.*, 2010). These unique features may serve to physically mark the centromere location. These characteristics, specific to CENP-A, have led to the proposal of a two-step mechanism for epigenetic specification of centromere identity and function (Fachinetti *et al.*, 2013). Gene targeting was utilized in human cells and fission yeast to determine these two steps: (1) centromere position is replicated and maintained by CATD-containing chromatin (2) either the amino- or carboxy-terminal tail of CENP-A is required for nucleation of kinetochore assembly (Fachinetti *et al.*, 2013).

Further characterization of CENP-A has extended to examining the make-up of CENP-A-containing nucleosomes. Despite much research, the exact composition of the CENP-A nucleosomal structure has not been conclusively determined. A more conservative model suggests that CENP-A-containing nucleosomes at the centromere form octamers, like those of canonical histones. The crystal structure of human CENP-A reconstituted *in vivo* has demonstrated that the nucleosome is a homotypic octamer containing two copies each of histones H1, H2, CENP-A, and H4 (Tachiwana *et al.*, 2011). This octameric structure has been identified *in vitro* in several species, either with or without histone chaperones (Black *et al.*, 2004; Camahort *et al.*, 2009; Dechassa *et al.*, 2011; Kingston *et al.*, 2011; Sekulic *et al.*, 2010; Yoda *et al.*, 2000). These CENP-A octamers formed *in vitro* induce left-handed negative supercoiling to DNA, similar to canonical nucleosomes (Barnhart *et al.*, 2011; Conde e Silva *et al.*, 2007; Panchenko *et al.*, 2011; Tachiwana and Kurumizaka, 2011; Yoda *et al.*, 2000).

However, an alternative tetrameric structure, also referred to as a hemisome, has also been suggested and supported by studies in multiple organisms (Dalal *et al.*, 2007; Furuyama *et al.*, 2013). Nucleosome cross-linking followed by immunoprecipitation and atomic force microscopy revealed that the height of CID-containing chromatin from *Drosophila* is half the height of canonical H3 nucleosomes (Dalal *et al.*, 2007). Additionally the electrophoretic behaviour of the CID-nucleosomal particles was consistent with the presence of only one copy of each histone.

Similar observations have been made in human cells regarding particle heights and presence of core histones (Dimitriadis *et al.*, 2010). Budding yeast centromeres are reported to consist of a single Cse4-containing nucleosome, wrapping only half the length of DNA, compared to a canonical nucleosome, once in a right-handed manner (Henikoff and Henikoff, 2012; Krassovsky *et al.*, 2012). These results supported earlier extrachromosomal plasmid work, showing that centromeric DNA is wrapped in a right-handed manner, causing less negative supercoiling or even positive supercoiling (Furuyama and Henikoff, 2009). Octamer-to-tetramer conversion models have also been proposed, as potential resolutions to the discrepancies in these observations. In these models, octameric CENP-A nucleosomes are split into tetrameric hemisomes during S phase to allow for the equal inheritance of the epigenetic mark to the daughter strands (Allshire and Karpen, 2008; Probst *et al.*, 2009). Two studies provide evidence for a cell cycle-coupled structural transition of CENP-A, in yeast and humans, in which atomic force microscopy revealed nucleosomes that changed in size depending on cell cycle timing (Bui *et al.*, 2012; Shivaraju *et al.*, 2011). However, further work is required to reach a consensus on the nature of CENP-A nucleosome as octameric, tetrameric, or adjustable.

Epigenetic state of centromeric chromatin

CENP-A serves as the defining epigenetic marker of the centromere but it does not function alone. The organization of centromeric chromatin, conserved in flies and humans, consists of blocks of CID/CENP-A interspersed with blocks of H3-containing nucleosomes (Blower *et al.*, 2002). Despite interspersion of CID/CENP-A and H3 blocks, at metaphase CID/CENP-A appears as a 3D structure lacking histone H3. It was proposed that this structure functions to present centromeric chromatin to the poleward face of the condensing mitotic chromosome (Blower *et al.*, 2002). An alternative model for the 3D structure of centromeric chromatin has been suggested based on super-resolution microscopy and field emission scanning electron microscopy. In the alternative model centromeric chromatin folding generates planar sinusoidal layers, forming a boustrophedon, with exposure of both CENP-A and H3 nucleosomes at the surface (Ribeiro *et al.*, 2010; Schroeder-Reiter *et al.*, 2012). In either model, it is apparent that the chromatin organization at the centromere has multiple layers of complexity contributing to the unique structure that results from the presence of CENP-A nucleosomes.

Even though centromeric chromatin is generally embedded in heterochromatin, it displays a histone modification pattern that is distinct from both euchromatin and heterochromatin (Fig. 7.2). H3 subdomains present within mammalian centromeric chromatin are enriched for the euchromatic modification H3K4me2 but lack the H3K9 and H4 acetylation generally found in euchromatin (Sullivan and Karpen, 2004). They also do not contain the heterochromatin-associated modifications of H3K9me2 or H3K9me3 (Greaves *et al.*, 2007; Sullivan and Karpen, 2004). More recent work has shown that centromeric chromatin is enriched for H3K4me1, H3K36me2, H3K36me3, and H4K20me1 (Bergmann *et al.*, 2011; Hori *et al.*, 2014). The centromeric chromatin is typically surrounded by pericentric heterochromatin, containing modifications of H3K9me2, H3K9me3, and H3K27me3 (Partridge *et al.*, 2000; Sullivan and Karpen, 2004). The presence of pericentric heterochromatin has been found to be necessary for the proper distribution of CENP-A (Smith *et al.*, 2011). In *Neurospora crassa*, mutation of dim-5, an H3K9 methyltransferase responsible for the large majority of H3K9me3, resulted in altered distribution of CENP-A, leading to the proposal that centromere proteins require DIM-5 and H3K9me3 for spreading. These post-translational modifications both in the centromeric and pericentric chromatin further shape the unique landscape of the centromere.

The CENP-A nucleosome itself is also modified, including CENP-A and the H4 histone. Post-translational modifications of CENP-A, such as trimethylation of Gly1 and phosphorylation of Ser16 and Ser18, further influence the conformation of centromeric chromatin (Bailey *et al.*, 2013; Bui *et al.*, 2012). The double phosphorylation allows for formation of a salt-bridged secondary

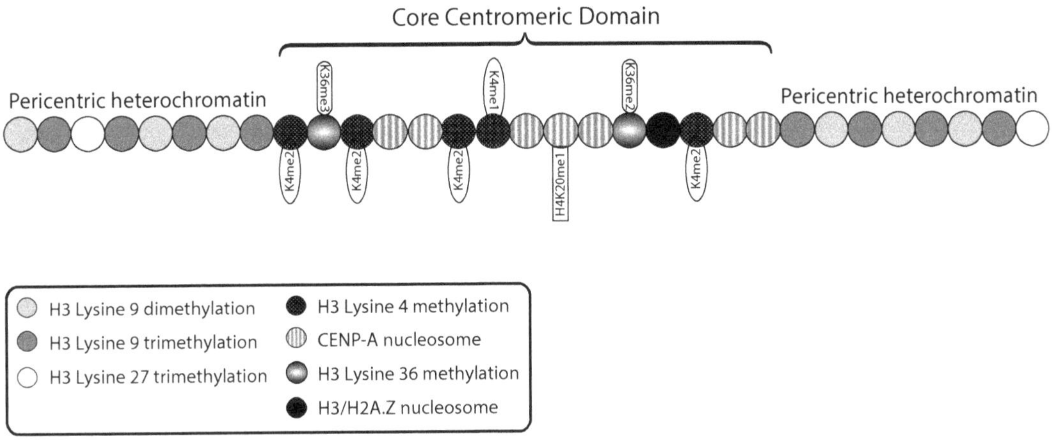

Figure 7.2 The unique chromatin landscape of the centromere. The eukaryotic centromere is composed of a core domain surrounding by pericentric heterochromatin. The core domain contains interspersed H3- and CENP-A-containing nucleosome subdomains. H3 subdomains are marked by euchromatic modifications such as H3K4me2 and H3K36me3. The flanking pericentric chromatin contains heterochromatic modifications such as H3K9me3 and H3K27me3.

structure, altering the physical properties of the chromatin at the centromere. The phosphorylation of CENP-A is differentially regulated in G2/M phase (Zeitlin *et al.*, 2001). CENP-A is phosphorylated at the motif it shares with H3; however, H3 and CENP-A phosphorylation are separate events. Additionally, in chicken cells, it was found that H4K20me1 within the CENP-A nucleosome is essential for kinetochore assembly (Hori *et al.*, 2014).

Post-translational modifications are not the only epigenetic modifications observed at the centromere. The pericentric heterochromatic region surrounding CENP-A chromatin also contains DNA methylation. DNMT3B (DNA methyltransferase 3B) modulates both DNA methylation and histone modifications at the centromere (Gopalakrishnan *et al.*, 2009). In addition, other histone variants are present at centromeres; H2A.Z is found within the H3 subdomains of centromeric chromatin, where it contributes to the 3D structure of the centromere (Greaves *et al.*, 2007). Another mechanism by which the epigenetic code is extended at the centromere is through the binding of the other centromere proteins, such as the CENP-T/W/S/X complex. The CENP-T/W/S/X complex forms a nucleosome-like structure at the centromere (Nishino *et al.*, 2012). The centromeric nucleosome-like CENP-T/W/S/X binds ~100 bp of otherwise nucleosome-free DNA as a dimer and induces positive DNA supercoils (Takeuchi *et al.*, 2014). CENP-T/W/S/X may function to switch centromeric chromatin to a mitosis-competent state rich in H3 and CENP-T/W/S/X, rather than a replication-competent state that is rich in CENP-A (Dornblut *et al.*, 2014). DNA methylation, histone modifications, and the presence of nucleosomes with unique compositions all function to define the epigenetic state of the centromere. Finally, the CENP-A loading protein Holliday junction recognition protein (HJURP; see more below) is also phosphorylated to accommodate CENP-A incorporation into chromatin (Muller *et al.*, 2014). Thus, epigenetic mechanisms acting on histones and chromatin-associated proteins are important for establishing and maintaining functional centromeres.

Centromere assembly and propagation

Centromere identity must be established and maintained to ensure genomic stability; faithful propagation of the centromere is essential and requires transmission of the unique epigenetic features found at the centromere, including the incorporation of CENP-A. The process of centromere transmission has been extensively linked to cell cycle progression. In both *Drosophila* and human cells, centromeric DNA is replicated during mid-to-late S phase (Shelby *et al.*, 1997, 2000; Sullivan and Karpen, 2001). However, during the S-phase replication of centromeric DNA there is no new CENP-A deposition, and existing CENP-A molecules are semi-conservatively distributed to both daughter strands (Jansen *et al.*, 2007). Instead, during S phase, H3.1 and H3.3 are deposited at centromeres, where H3.3 acts as a placeholder until new CENP-A is later added (Dunleavy *et al.*, 2012). New CENP-A is synthesized in G2 and deposited early in the next G1 phase and only after exit from mitosis (Hemmerich *et al.*, 2008; Jansen *et al.*, 2007; Schuh *et al.*, 2007; Shelby *et al.*, 1997, 2000). While the exact timing and the proteins involved differ to some degree, cell cycle-restricted deposition of CENP-A is a common mechanism across many organisms. CID is incorporated at *Drosophila* centromeres during anaphase and in *S. pombe* CENP-A^{Cnp1} deposition occurs in S phase (Pearson *et al.*, 2004; Schuh *et al.*, 2007). The uncoupling of CENP-A synthesis and deposition from DNA synthesis indicates that there is a dedicated mechanism for epigenetic transmission of centromeres. A three-step mechanism has been described for this process: priming of the centromere, loading of CENP-A, and maintenance of newly incorporated CENP-A (Fujita *et al.*, 2007).

Priming of the centromere for CENP-A deposition involves the Mis18 licensing complex, consisting of KNL-2 and Mis18α/β. This complex has been shown to be required for CENP-A localization to centromeres in many organisms, including *S. pombe*, mammals, worms, and other eukaryotes (Fujita *et al.*, 2007; Hayashi *et al.*, 2004; Maddox *et al.*, 2007). This is the first event known in the path to CENP-A deposition and, in humans, occurs following anaphase when the licensing complex is recruited to the centromere (Fujita *et al.*, 2007; Maddox *et al.*, 2007). The centromere localization of Mis18BP1 is controlled by phosphorylation states in a cell cycle-dependent manner (Silva *et al.*, 2012). Inhibition of Cdk1 and 2, responsible for controlling the phosphorylation state of Mis18BP1, results in CENP-A assembly throughout the cell cycle. This suggests that the CENP-A assembly machinery is prepared for activity throughout the cell cycle but is kept inactive by Cdk activity during S, G2, and M phases. Only when the licensing complex is recruited can the centromere propagation processes begin, however the exact priming reaction is as yet undetermined; no physical contact is known to be made between the licensing complex and CENP-A itself. One possibility is that the licensing complex alters the state of histone acetylation at the centromere (Fujita *et al.*, 2007; Hayashi *et al.*, 2004).

Loading of CENP-A occurs in complex with histone H4 and requires the CENP-A specific chaperone protein HJURP (Dunleavy *et al.*, 2009). The N-terminus of HJURP contains a domain that interacts with CENP-A and directs its deposition through the CATD (Foltz *et al.*, 2009). CENP-A specific chaperone proteins have also been identified in *Drosophila*, CAL1, and fission yeast, Scm3 (Dunleavy *et al.*, 2007). HJURP is enriched at human centromeres in late telophase and early G1 (Dunleavy *et al.*, 2009; Foltz *et al.*, 2009; Silva and Jansen, 2009). Similarly, Scm3 in fission yeast localizes to the centromere during only a portion of the cell cycle, not including mitosis (Pidoux *et al.*, 2009). However in budding yeast, Scm3 remains stably incorporated at centromeric sites throughout the cell cycle (Shivaraju *et al.*, 2011). In humans, HJURP-mediated G1 assembly of CENP-A represents a unique cell cycle-restricted chromatin assembly that does not involve other H3 variants (Bodor *et al.*, 2013). While the CENP-A/HJURP interaction sites have been identified, there is not yet a clear picture of how CENP-A is preferentially selected away from the bulk H3 histones. It is possible that phosphorylation of HJURP, as well as its binding to centromeric DNA, concentrates CENP-A to centromeric regions at the proper time in the cell cycle (Muller *et al.*, 2014).

Other proteins that have been found to be important for CENP-A deposition include the centromere proteins CENP-H and CENP-N. The CENP-H-containing complex assists in centromere deposition of CENP-A in collaboration with FACT (facilitates chromatin transcription) and CHD1 (chromodomain helicase DNA binding protein 1) (Okada *et al.*, 2009). Knockout mutants of the CENP-H-I complex cannot efficiently incorporate newly synthesized CENP-A (Okada *et al.*, 2006). CENP-N was the first protein found to selectively bind CENP-A, not H3, and depletion of CENP-N by siRNA causes centromere assembly defects resulting in reduced assembly of newly synthesized CENP-A (Carroll *et al.*, 2009). This suggests that CENP-N may function to interpret the information encoded within CENP-A nucleosomes and mediate centromere propagation.

Once CENP-A has been deposited it must mature in order to be maintained. The ATP-dependent remodeller RSF complex has been proposed to function in maturation of CENP-A following deposition in G1 to help complete nucleosome assembly (Perpelescu *et al.*, 2009). While the exact role of the RSF complex at the centromere requires further investigation, it may function similarly to ATP-dependent remodelling events that occur in conventional nucleosomes (Hargreaves and Crabtree, 2011). Later in G1, MgcRacGAP, a GTPase-activating protein, has also been implicated as functioning to stabilize newly incorporated CENP-A (Lagana *et al.*, 2010). MgcRacGAP is proposed to create a GTPase cycle with ECT2, a guanine nucleotide exchange factor, and the GTPase Cdc42, that maintains newly incorporated CENP-A. Maintenance of CENP-A is necessary so that it may serve as the epigenetic mark for the next round of centromere propagation. The mechanisms of epigenetic inheritance found at the centromere provide important implications and insights into widespread epigenetic transmission via nucleosome propagation. Defining molecular pathways for building functional centromeres is also necessary for understanding how variant centromeres are formed, how centromeres are turned off (see below), and how centromere function may be purposely regulated.

Dicentric chromosomes and centromere inactivation

In addition to the studies already discussed that support epigenetic components of centromere identity, variant or atypical centromeres represent strong evidence for epigenetic mechanisms of centromere function. These include dicentric chromosomes, the products of genome rearrangement that place two centromeres in proximity on the same chromosome (Fig. 7.3A). Generally, each chromosome contains a single region of centromeric DNA or centromeric chromatin where the kinetochore assembles and microtubules attach. However, dicentrics frequently form during repair of double strand breaks that results in the joining of two different chromosomes or the fusion of sister chromatids. The former mechanism leads to non-homologous dicentrics, while the latter produces isodicentric chromosomes.

Dicentrics were first described in maize by the cytogeneticist Barbara McClintock. She observed them to be inherently unstable, undergoing multiple rounds of anaphase breakage, bridge formation and fusion of broken ends (i.e. the BFB cycle), leading to significant genome reshuffling or deletion (McClintock, 1939, 1941) (Fig. 7.3B). Dicentric chromosomes in *Drosophila* undergo similar mis-segregation and breakage (Morgan, 1915; Novitski, 1955). Although early studies did not recover the products of dicentric segregation, suggesting that *Drosophila* are less tolerant of dicentric chromosomes than other organisms, more recent studies have been able to capture the fate of dicentrics during cell division (Ahmad and Golic, 1998). In the budding yeast *S. cerevisiae*, dicentric chromosomes are unstable in both meiosis and mitosis, typically undergoing breakage–fusion–bridge cycles and other types of instability that rearrange the genome (Haber *et al.*, 1984; Hill and Bloom, 1989; Koshland *et al.*, 1987; Pennaneach and Kolodner, 2009). These findings in multiple organisms have led to the general view that most, if not all, dicentrics are unstable.

Functional states of structurally dicentric chromosomes

Human dicentric X isochromosomes were first identified in the 1960s. Because they were present in the cells of adult women, representing hundreds of thousands of cell divisions, they were proposed to be stable by inactivation of one centromere (De la Chapelle *et al.*, 1966; Ockey *et al.*, 1966). Centromere inactivation was described as a phenomenon by which one centromere of a dicentric is functionally suppressed (Fig. 7.3). Although

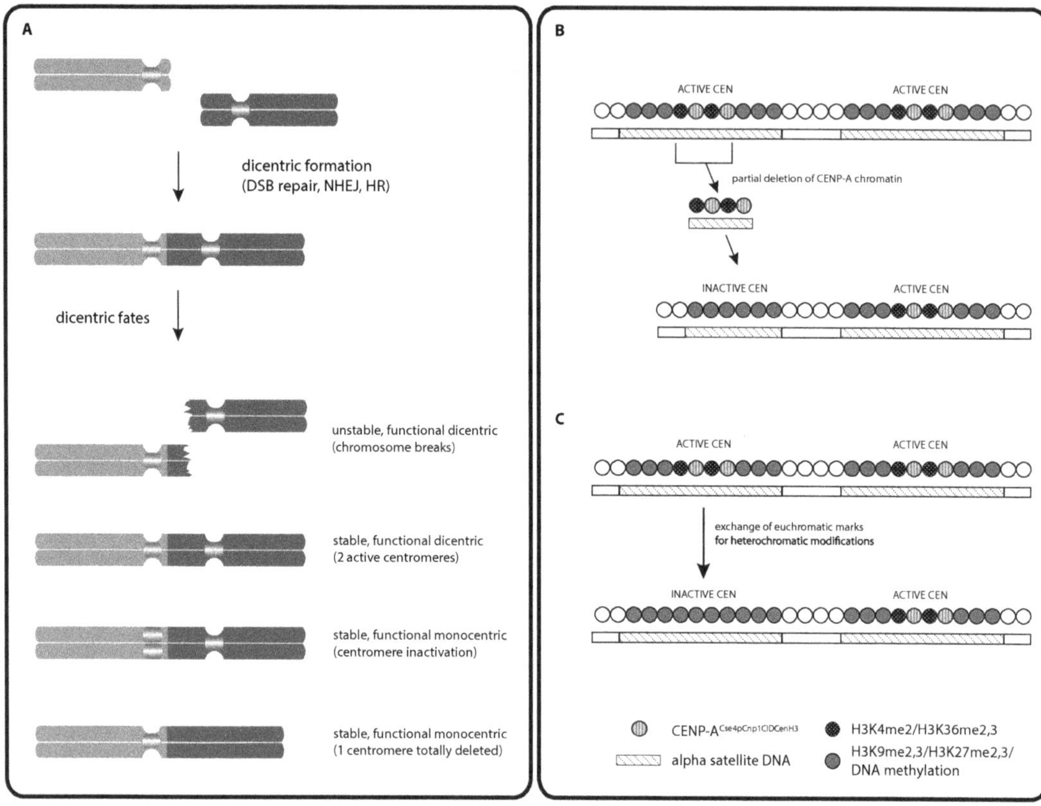

Figure 7.3 Dicentric chromosome formation and fate. (A) Chromosomes with two centromere regions, called dicentrics, are frequently formed due to DNA damage repair or non-homologous recombination between chromosomes. Studies in various organisms have shown that after formation, a dicentric chromosome can undergo one of several fates. If both centromeres remain active and segregate towards opposite spindle poles, the dicentric may break. Alternatively, if the centromeres are closely located, they may both remain functional and cooperate during cell division. In humans, most dicentric chromosomes undergo centromere inactivation, the process by which one centromere is functionally suppressed. In yeasts, centromere inactivation can occur if one centromere is completely deleted. (B) Centromere inactivation in larger eukaryotes can occur by several known mechanisms. Part of the centromeric DNA array, specifically the portion that contains CENP-A chromatin, can be deleted, resulting in an inactive centromere that is reduced in size. (C) In the absence of DNA changes (mutation or deletion), centromere inactivation appears to occurs through chromatin remodelling that replaces euchromatin-associated epigenetic markers with heterochromatic histone modifications.

originally thought to occur predominantly in humans, more recently it has been observed in model organisms, including maize (Han *et al.*, 2006). However, not all dicentrics undergo centromere inactivation. Some remain functionally dicentric (i.e. two active centromeres), either temporarily or permanently (Higgins *et al.*, 2005; Stimpson *et al.*, 2010; Sullivan and Willard, 1998). This suggests that, when first formed, dicentric chromosomes can experience one of several outcomes (Fig. 7.3).

Centromere cooperation and centromere inactivation: multiple ways to stabilize a dicentric chromosome

The use of engineered dicentric chromosomes in various organisms has shed some answers on outstanding questions of dicentric fate and centromere inactivation. In budding yeast, dicentrics can be stable and retain two active centromeres if the distance between the two centromeres is minimal (Hill and Bloom, 1989; Koshland *et al.*, 1987) (Fig. 7.3C). This is also true for human dicentrics, including dicentric Xs and Robertsonian translocations. In these cases, inter-centromeric distances of less than 20Mb are associated with functionally dicentric chromosomes (Higgins *et al.*, 1999; Page and Shaffer, 1998; Sullivan and Schwartz, 1995; Sullivan and Willard, 1998). Such functionally dicentric chromosomes are thought to be stable because either the two centromeres are so close that they 'cooperate' or the limited distance between them constrains the dicentric on the spindle and prevents orientation of centromeres towards opposite poles.

Most of the time, at least in humans, dicentric chromosomes undergo centromere inactivation (Fig. 7.3). Owing to the excellent cytology of human chromosomes, it is known that inactive centromeres appear morphologically different than active centromeres. Inactive centromeres do not have a constricted appearance, and instead visually appear more similar in compaction to the chromosomes arms. In addition, inactive centromeres lack key centromere proteins, such as CENP-A, CENP-C and CENP-E (Earnshaw and Migeon, 1985; Sato *et al.*, 2012; Sullivan *et al.*, 1994a). The fundamental question is how centromere inactivation occurs. Does it involve epigenetic or genomic mechanisms, or both? Clues from model organisms suggest both can be involved. In budding yeast, dicentrics can be stabilized and/or undergo inactivation of one centromere if (1) centromeric DNA sequences are mutated (genomic), (2) one centromere of the dicentric is completely deleted (genomic) or (3) transcription is driven through one centromere (epigenetic) (Fig. 7.3). Such mechanisms also appear to be present in larger eukaryotes. In *S. pombe* deletion of one of the centromeres of the dicentric occurs in about one-third of engineered dicentrics (Sato *et al.*, 2012). Removal of the region of CENP-A chromatin that is assembled on a portion of alpha satellite DNA is believed to underlie the partial centromeric deletions observed on induced human dicentric chromosomes (Stimpson *et al.*, 2010, 2012) (Fig. 7.3B).

Despite these observations, epigenetic mechanisms seem to be the predominant modes of centromere inactivation in larger eukaryotes, including most plants, *S. pombe*, and humans (Han *et al.*, 2006; Sato *et al.*, 2012; Stimpson *et al.*, 2010; Sullivan and Schwartz, 1995; Zhang *et al.*, 2010). The model for an epigenetic mechanism of centromere inactivation is largely based on the fact that changes (deletion, mutation) to the underlying DNA of the inactive centromere have not been detected (Higgins *et al.*, 2005). Thus, chromatin remodelling is thought to be involved in centromere inactivation (Fig. 7.3). Studies in fission yeast, plants (*Arabidopsis* and maize), and mammals have confirmed that inactive centromeres do not contain CENP-A and are depleted for euchromatic histone modifications (i.e. H3K9ac, H3K14ac, H3K4me2, H3K36me2) and other modifications, such as phosphorylated H3 (PH3S10) that are found at active centromeres (Han *et al.*, 2009; Maloney *et al.*, 2012; Sato *et al.*, 2012). Instead, inactive centromeres are enriched for heterochromatin-associated epigenetic marks (H3K9me2, 3 and H3K27me2, 3) (Sato *et al.*, 2012; Zhang *et al.*, 2010) (Fig. 7.3C). Epigenetic DNA changes, such as DNA methylation, have also been observed at inactive centromeres, suggesting that centromere inactivation involves the creation of more compact or distinctly packaged chromatin (Fu *et al.*, 2012; Koo *et al.*, 2011).

Centromere inactivation: a permanent or reversible functional state?

Epigenetic changes to the genome are often reversible, depending on developmental or cell cycle cues. Since centromere inactivation appears to be largely sequence-independent, an open question is whether inactive centromeres are permanently inactivated or are dormant and can be reactivated under the proper conditions or signals. This has been investigated using both engineered isodicentrics in humans and those created by ectopically inserting alpha satellite DNA into non-centromeric chromosomal sites. Isodicentrics containing two functional centromeres underwent centromere inactivation in the presence of Trichostatin A (TSA), an inhibitor of histone deacetylases (Higgins *et al.*, 2005). In contrast, inactive ectopic alpha satellite arrays were reactivated in the presence of TSA (Nakano *et al.*, 2003). These conflicting results failed to clarify if active or repressive chromatin is strictly associated with centromere inactivation, although they have supported the conclusion that many inactive centromeres are suppressed and not permanently inactivated. It is notable that high levels of local transcription were observed to counteract centromere inactivation (Nakano *et al.*, 2003; Nakashima *et al.*, 2005), suggesting that RNA-related pathways, or the convergence of epigenetic and transcriptional mechanisms, play a role in centromere function and inactivation (see 'RNA, transcription, and centromere function').

Artificial chromosomes

Artificial chromosomes are a valuable tool for the study of complex chromatin and are utilized for understanding the epigenetic aspects of centromere biology. These chromosomes are genetically engineered and contain endogenous DNA, synthetic DNA, or a combination of both. In order to be classified as a true artificial chromosome, the construct must be mitotically stable, exhibiting the ability to replicate and segregate faithfully with the endogenous chromosomes over a number of generations in the absence of selection (Harrington *et al.*, 1997).

Sequence-based elements of artificial chromosomes

Initially, the creation of artificial chromosomes in eukaryotic cells aimed to shed light on the basic genomic requirements of functional chromosomes. The construction of artificial chromosomes in *S. cerevisiae*, first completed over three decades ago, identified three elements necessary for mitotically stable yeast chromosomes: telomeres, origins of replication, and a centromere. The creation of these yeast artificial chromosomes (YACs) depended on inclusion of DNA sequences constituting the centromere (Clarke and Carbon, 1980; Murray and Szostak, 1983; Struhl *et al.*, 1979). The point centromere of budding yeast was first discovered as a DNA segment of a plasmid containing sequences known to be located close to the yeast centromere of chromosome 3 (Clarke and Carbon, 1980). This segment was observed to have a stabilizing effect when present on a plasmid, conferring mitotic stability. Further characterization of this sequence, and similar sequences isolated from other chromosomes, revealed the 125 base pair consensus sequence constituting all budding yeast centromeres (Clarke and Carbon, 1985) (Fig. 7.1). In contrast to this relatively simple centromere, the identification of restriction fragments of the *S. pombe* genome that conferred mitotic stability to artificial chromosomes revealed a much longer 40–100 kb partially repetitive sequence of fission yeast centromeres (Polizzi and Clarke, 1991). When first created in 1989, fission yeast artificial chromosomes opened the door for the study of the role of specific repetitive sequences observed in *S. pombe* and larger eukaryotes, as well as the creation of artificial chromosomes in other organisms (Hahnenberger *et al.*, 1989). The *S. pombe* experiments were important for pushing the field forward, since the highly repetitive nature and complexity of mammalian centromeres impeded the construction of *de novo* artificial chromosomes in humans until the late 1990s (Harrington *et al.*, 1997). Since then, two generations of human artificial chromosomes have been created, enabling both the study of *de novo* centromere formation and the direct alteration of the epigenetic factors involved in centromere specification.

Pioneering studies leading to the creation of human artificial chromosomes (HACs)

Telomere-directed chromosomal truncation represents a method originally used to create engineered human chromosomes (Farr *et al.*, 1991). In this approach, endogenous chromosomes containing an integrated telomeric repeat that causes terminal deletions are progressively shortened (Itzhaki *et al.*, 1992). The reliance of this method on chromosome engineering via recombination typically required the transfer of the chromosome to a homologous recombination-proficient cell line, such as chicken DT40 cells (Kuroiwa *et al.*, 1998). Known as the 'top-down' assembly method, the approach can create stable minichromosomes distinct from the endogenous chromosome, but is unable to produce a *de novo* chromosome or centromere. The first *de novo* human artificial chromosomes were constructed in *tour de force* efforts by two labs in the late 1990s. In this approach, HT1080 fibrosarcoma cells were transfected with a combination of genomic DNA, telomeric sequences, and synthetic centromeric alpha satellite arrays derived from chromosomes 17, 21 or Y (Harrington *et al.*, 1997; Ikeno *et al.*, 1998). The *de novo* artificial chromosomes formed by this approach, also referred to as the 'bottom-up' assembly method, were determined to be mitotically stable, as the microchromosome remained present in the majority of cells over six months after transfection in the absence of selection (Harrington *et al.*, 1997). Additionally, the *de novo* centromeres were found to be functional, as indicated by the presence of both the inner kinetochore protein CENP-C and outer kinetochore protein CENP-E, marks of an active kinetochore.

The first HACs were linear, a conformation requiring telomeric sequence, and supported the observation in *S. cerevisiae* that telomeres are required to form functional chromosomes. However, HACs developed in later experiments were efficiently formed in the absence of telomeric repeat DNA using circular DNA as the assembly construct (Ebersole *et al.*, 2000; Grimes *et al.*, 2002a). Similar to linear HACs, these circular chromosomes did not acquire endogenous telomere sequences and were stably maintained for hundreds of generations. Circular HACs were also found to be generally comparable in size to their linear counterparts, ranging from 1 to 10 Mb (reviewed in Larin and Mejia, 2002).

Several approaches have been taken to build the repetitive alpha satellite DNA inserts in the HAC assembly vectors. Single copies of the 171 base pair monomer of alpha satellite DNA were initially assembled in the same head-to-tail pattern as observed at endogenous centromeres to create synthetic centromeric sequence (Harrington *et al.*, 1997), however, more recent efforts have used transformation-associated recombination (TAR) cloning to isolate and expand centromeric DNA from the endogenous chromosome of choice (Ebersole *et al.*, 2005; Kouprina *et al.*, 2003). Regardless of the origin of the alpha satellite sequence included in the HAC, the minimum length of alpha satellite array required for formation of a functional *de novo* centromere is thought to be approximately 100 kb (Yang *et al.*, 2000) and must contain CENP-B boxes, a 17 base pair consensus sequence essential for CENP-B protein binding (Ohzeki *et al.*, 2002). As a result, Y chromosome centromeric sequence, which lacks this crucial sequence, is unable to form HACs, explaining early observations that Y chromosome sequence was an inefficient substrate for *de novo* HAC formation (Grimes *et al.*, 2002a; Harrington *et al.*, 1997; Mejia *et al.*, 2002). These studies also highlighted the possibility that the chromosome Y alpha satellite DNA maintains an active centromere through a currently unidentified functionally redundant mechanism that does not require CENP-B recognition and assembly.

Chromatin elements of HACs

Distinct regions of HACs have been identified based on their unique patterns of histone modifications and found to mimic those found in endogenous centromeres (Lam *et al.*, 2006; Moralli *et al.*, 2013). Just as at normal human centromeres, on HACs, CENP-A is interspersed with histone H3, specifically histone H3 dimethylated at lysine 4 (H3K4me2), to form the core centromeric domain (Lam *et al.*, 2006; Moralli *et al.*, 2013) (Fig. 7.4). The current view of the centromeric chromatin environment of HACs is based on findings at endogenous human chromosomes, defined by histone modifications in the core that exhibit markers of permissive chromatin, and flanked by the pericentromere domain that is characterized by the presence of H3K9me2 and H3K9me3, markers of facultative heterochromatin (Sullivan and Karpen, 2004; and reviewed in Bergmann *et al.*, 2012b). The overall structure of the HAC core centromeric domain has also been studied and found to resemble that of the endogenous centromere (Grimes *et al.*, 2004; Lam *et al.*, 2006). The assembly method of synthetic alpha satellite arrays introduces non-centromeric sequence, such as vector sequences and selectable markers, which may adopt a chromatin state that differs from nearby centromeric sequence. Comparison of endogenous and *de novo* centromeres has shown that despite the presence of non-centromeric sequence, the centromeric chromatin domain of *de novo* centromeres, defined by the presence of CENP-A and interspersed histone H3, is continuous (Lam *et al.*, 2006; Moralli *et al.*, 2013).

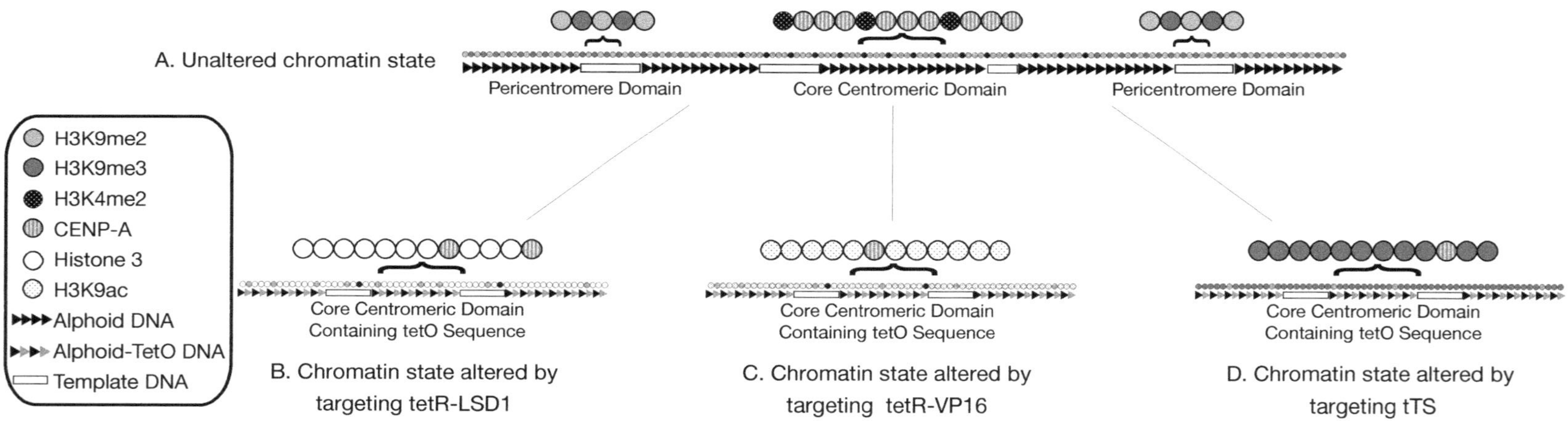

Figure 7.4 Human artificial chromosomes: tools to study chromatin properties at functional centromeres. (A) The chromatin state of unaltered HACs is characterized by the presence of both CENP-A and H3K4me2 in the core domain and H3K9me2 and H3K9me3 in the pericentromere domain. (B) Targeting of a tetR–LSD1 fusion protein to tetO sequences integrated into the alpha satellite (alphoid) DNA causes loss of both H3K4me2 and CENP-A, leading to the gradual inactivation of the centromere. (C) Targeting of a tetR–VP16 fusion protein to the tetO sequences increases transcription through the region and leads to the incorporation of H3K9ac, loss of CENP-A, and HAC destabilization. (D) Targeting of a tetR–tTS fusion protein to the tetO sequences causes H3K9me3 incorporation, inducing a more heterochromatic state and leading to CENP-A loss and centromere inactivation.

Creating conditional centromeres on HACs

Many recent studies have taken advantage of using HACs as a unique conditional centromere to directly alter the epigenetic landscape of the centromere. This approach leaves the centromeres of endogenous chromosomes unchanged and does not affect overall viability of the mammalian cells (Bergmann *et al.*, 2011; Cardinale *et al.*, 2009; Nakano *et al.*, 2008). To create these conditional HACs, tet operator (tetO) sequences were incorporated into the alpha satellite arrays of assembly vectors, replacing the CENP-B box in every other monomer. These structures represent the development of the second generation of HACs, known as alphoid^{tetO}-HACs. A variety of proteins can be fused to the tet repressor (tetR) protein, which binds to tetO sequences, and effectively tethers a protein of choice to the tetO array and thus, to the centromere (Fig. 7.4B–D). Multiple studies have used this system to track HACs during mitosis, to tether chromatin modifiers to the centromere of HACs, and to explore essential characteristics of active centromeres.

Beyond introducing the alphoid^{tetO}-HAC, Nakano and colleagues used the system to manipulate the chromatin state of centromeres to shed light on the incompatibility of heterochromatin and centromere function (Cardinale *et al.*, 2009; Nakano *et al.*, 2008). In these studies, transcriptional activators and repressors were tethered to the centromere to effectively open or close the chromatin on the HAC. Two different proteins, tTA and tTS, the tetracycline transcriptional activator and the tetracycline transcriptional repressor, respectively, were targeted to alphoid^{tetO}-HACs. The presence of a transcriptional activator (tTA) at the centromere caused kinetochore inactivation and led to a partial loss of HAC stability. Tethering a transcriptional repressor (tTS) to the centromere caused a much more dramatic loss of HAC mitotic stability, denoted by depletion of CENP-A, disruption of kinetochore activity, and an increase in H3K9me3 (Fig. 7.4D). These findings suggested that the presence of heterochromatin is incompatible with an active centromere and may indicate that some level of centromeric transcription is required for proper function (see 'Dicentric chromosomes and centromere inactivation' and 'RNA, transcription, and centromere function'). These studies also introduced the possibility of using a conditional centromere to remove or 'cure' the cell of the HAC for gene therapy purposes, since tethering transcriptional repressors can inactivate the kinetochore and cause mitotic instability that will rid the cell of the HAC after a limited number of generations (Kouprina *et al.*, 2014).

Using HACs to study epigenetic characteristics of centromeric chromatin

Additional studies using the tetO/tetR tethering system have investigated the histone H3 chromatin modifications, specifically acetylation and methylation, which underlie active centromeres and kinetochores. Targeting lysine-specific demethylase (LSD1) to the tetO array of HACs in HeLa 1C7 cells caused a loss of H3K4me2 (Bergmann *et al.*, 2011) (Fig. 7.4B). In addition to a number of consequences related to the loss of transcription through the centromere (see 'RNA, transcription, and centromere function'), depletion of H3K4me2 causes loss of CENP-A from the kinetochore and destabilization of the HAC. This inability to incorporate newly synthesized CENP-A as a result of the loss of H3K4me2 implicates this chromatin modification as a regulator of CENP-A incorporation, either directly or indirectly.

The tetO system has also been used to tether histone acetyltransferases (HATs), enzymes that modify the chromatin state by acetylating histone lysine residues, to the *de novo* centromere of HACs (Ohzeki *et al.*, 2012). Historically, HT1080 cells have been a preferred cell line for HAC formation due to their use in the formation of the first *de novo* HACs, as well as their well-characterized ability to form mitotically stable HACs. In contrast, it has been observed that stable HAC formation in HeLa cells and other popular human cell lines is inefficient (Ohzeki *et al.*, 2012). In these cells, CENP-A initially assembles at the alpha satellite DNA to form a *de novo* centromere but is progressively depleted over successive generations. Capitalizing on this observation, several groups have used HeLa cells as an experimental system to test the stability of CENP-A over time.

For instance, when HAT–tetR fusion proteins are targeted to the alphoid^{tetO}-HAC of HeLa cells, CENP-A is stably assembled on these regions, leading to the establishment of a mitotically stable HAC with a functional kinetochore (Ohzeki *et al.*, 2012). Tethered HATs were discovered to catalyse the acetylation of histone H3 to form H3K9ac, possibly preventing the establishment of H3K9me and mimicking the chromatin state that is present within the HT1080 line and promotes such efficient formation of *de novo* HACs in these cells. Tethering of HATs to an ectopic tetO-alpha satellite array integration site located on a chromosome arm can induce CENP-A assembly, as well as the formation of the inner and outer kinetochore (Ohzeki *et al.*, 2012). Studies like this not only revealed the importance of histone modifications typically associated with euchromatin (H3K9ac) in active centromere maintenance, but also provided a method to produce HACs in previously intractable cell lines.

The alphoid^{tetO}-HAC studies in HeLa cells also showed that increasing the levels of H3K9ac and decreasing the levels of H3K9me3 promoted efficient HAC formation (Ohzeki *et al.*, 2012), indicating the importance of specific histone modifications, or at least the necessity for a preferred chromatin state at the time of HAC assembly and/or centromere assembly. Further evidence for the role of H3K9ac in maintaining active centromeres in the HAC-tetO/tetR system was discovered when two activation domains, NF-κB p65 and herpes virus VP16, that induce low and high levels of transcription, respectively, were fused to tetR and tethered to the alphoid^{tetO}-HACs (Bergmann *et al.*, 2012a) (Fig. 7.4C). Both activation domains increased the proportion of H3K9ac to similar levels. However, while the tethering of p65 to the centromere had little effect on centromere function or stability, strongly increasing the transcription through the region with VP16 both removed CENP-A present at the centromere and prevented the future incorporation of the histone variant, resulting in HAC destabilization (see 'RNA, transcription, and centromere function'). This study further emphasized the role of chromatin modifications, specifically at H3K9, that impact CENP-A deposition and maintenance.

Gene expression and silencing on HACs

The formation of HACs also presented a promising new construct for gene expression purposes. HAC vectors provide a number of advantages, including reduced immunogenicity compared to viral vectors, large capacity for transgenes and associated regulatory elements, and stable maintenance of the gene-containing artificial chromosome in a single copy (reviewed in Kazuki and Oshimura, 2011). Initial experiments that expressed a gene from a HAC utilized a circular HAC containing the human *HPRT* gene that was introduced into *HPRT*-deficient HT1080 cells (Grimes *et al.*, 2002b; Mejia *et al.*, 2001). Stably maintained circular HACs were observed and stable expression of the *HPRT* transgene was confirmed by both northern blots and phenotypic selection. More recent HAC studies have shown that insertion of other genes, such as the transgene encoding Factor IX, can be properly expressed, suggesting these HACs could be used to correct the gene deficiency in the appropriate target cells (Breman *et al.*, 2008). These studies have demonstrated that human artificial chromosomes are capable of maintaining and expressing a transgene, and are a promising avenue for disease therapy.

A caveat of the prior studies is that expression of genes inserted on HAC vectors has not been well regulated. New advances in the creation of transgene-containing HACs for gene expression have aimed at increasing control during the formation process by dictating the location and copy number of the inserted gene. A modification of the previously described alphoid^{tetO}-HAC system introduced a *loxP* cassette into the alpha satellite array construct (Iida *et al.*, 2010). The *loxP* site can then be used to recombine the HAC with a *loxP*-flanked transgene of choice to introduce the gene into the exact location desired and, more importantly, in a single copy. In proof-of-principle experiments, the *EGFP* gene (20 kb in length) was inserted into a HAC in Chinese hamster ovary cells and its expression was monitored over a 12-week period (Iida *et al.*, 2010). Subsequent studies successfully inserted two different transgenes, *VHL* (25 kb in length) and *NBS1* (55 kb in length), into HACs in the same cell line and also observed stable expression (Kim *et al.*,

2011). Investigators were able to insert sizable (> 50 kb) transgenes and surrounding sequences, demonstrating the ability of the HAC system to deliver and express large genes, as well as their upstream and downstream elements.

Using HACs to study other epigenetic phenomenon

Silencing of genes located proximal to the centromere, a phenomenon known as position effect variegation (PEV) and which was originally described in *Drosophila* by Morgan and Bridges in the early 20th century (Morgan *et al.*, 1915), has been documented in many organisms (Allshire *et al.*, 1994; Karpen and Allshire, 1997; Wallrath and Elgin, 1995). This presents a concern when considering the use of HACs as gene expression vectors, since nearby heterochromatin domains flanking the centromere might silence the target gene (Lee *et al.*, 2013). A study investigating long-term expression of a gene incorporated into an alphoid^{tetO}-HAC observed a decline in expression of the transgene (*EGFP*) (Lee *et al.*, 2013). To determine if this finding was related to heterochromatin spreading, a number of insulator sequences were tested for activity as barriers to prevent the potential silencing of the transgene. All three insulators tested, tDNA, cHS4, derived from the chicken β-globin locus, and gamma satellite DNA, derived from human chromosome 8, were able to prevent the silencing of the gene and loss of expression noted in the absence of an insulator sequence (Kim *et al.*, 2009; Lee *et al.*, 2013). Future studies utilizing HACs as transgene vectors may require the inclusion of these sequences in order to ensure the efficiency of this new vector.

Neocentromeres and ectopic/induced centromeres

Neocentromeres are functional centromeres that form at ectopic sites and allow kinetochores to assemble on previously non-centromeric (unique sequence) DNA. The identification of neocentromeres has emphasized the role of epigenetics in centromere specification. The study of neocentromeres has provided insight into the formation of endogenous centromeres, and has allowed for distinction between features of centromeres that depend on centromeric DNA versus those that are epigenetically specified.

Neocentromeres occur in a variety of organisms

Neocentromeres have been described in eukaryotes ranging from fungi to mammals. These fully functional ectopic centromeres are quite different from the facultative 'classical' neocentromeres originally described in maize (Rhoades, 1942). The first description of a fully functional centromere that had arisen where no centromere previously existed came in 1993 with the discovery of a marker chromosome, designated mardel(10), that contained an ectopic centromere lacking detectable alpha satellite DNA (Voullaire *et al.*, 1993). Despite the complete absence of alpha satellite DNA, the neocentromere was able to form a primary constriction and assemble a functional kinetochore that was stable in mitosis. Cytogenetic and molecular studies showed that mardel(10) was formed through a *de novo* rearrangement of human chromosome 10, resulting in a ring chromosome containing the original centromere and a linear chromosome that lost the original centromere. The acentric marker chromosome was rescued by the formation of a new centromere at the cytogenetic band 10q25, which had not undergone any rearrangement or sequence change. Since the initial discovery of the human neocentromere, more than 100 neocentromeres have been identified in clinical samples (Alonso *et al.*, 2010; Klein *et al.*, 2012; Marshall *et al.*, 2008a).

Types of neocentromeres

In general, clinically discovered neocentromeres were found to exist on marker chromosomes that formed when an acentric chromosomal fragment gained a centromere at a novel location. There is considerable variability in the chromosomal origin and location of neocentromeres. They have been reported to originate on 21 of the 22 autosomes and both sex chromosomes (Liehr *et al.*, 2010; Marshall *et al.*, 2008a). The formation of a neocentromere is generally associated with a chromosomal rearrangement, such as an inverted duplication (class I) or an interstitial deletion (class II) (Fig. 7.5). Approximately 75%

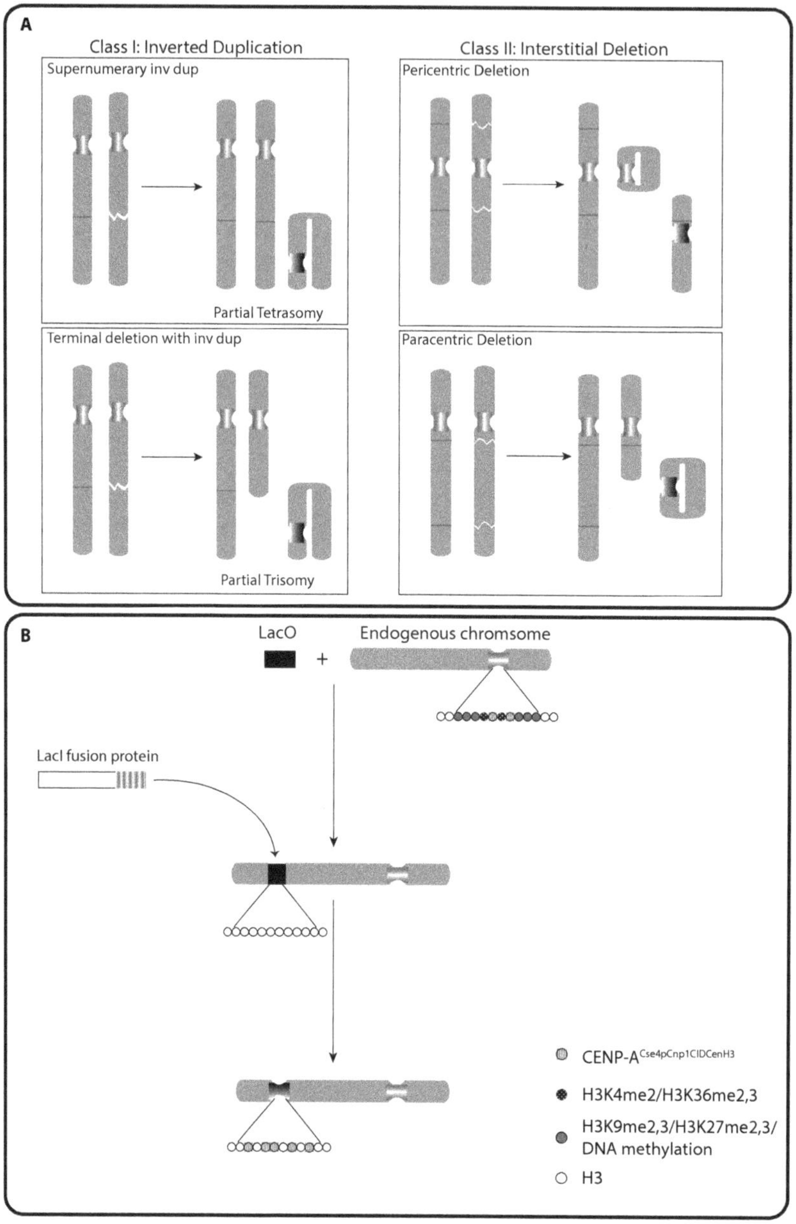

Figure 7.5 Natural and induced formation of neocentromeres (A) Naturally occurring neocentromeres form in humans via mechanisms of Class I inverted duplication and Class II interstitial deletion. Each type of rearrangement results in an acentric marker chromosome that is retained by gaining a neocentromere. Class I rearrangements result in unbalanced karyotypes, while Class II rearrangements remain balanced. (B) Neocentromeres can be induced in a variety of organisms utilizing protein-tethering strategies. For instance, a LacO array can be integrated into a non-centromeric location of an endogenous chromosome. LacI fusion proteins can then be tethered to the LacO array where they direct the formation of an ectopic centromere. CENP-C-, CENP-T- and HJURP–LacI fusion proteins have all been shown to induce ectopic centromeres when targeted to non-centromeric locations.

of all neocentromeres are class I neocentric chromosomal rearrangements in which an inverted duplication of the distal part of a chromosome results in an unbalanced karyotype. The second primary type of neocentric chromosomal rearrangements, class II, in which interstitial deletions result in the formation of a ring chromosome and a linear marker chromosome, produces a balanced karyotype. Class II rearrangements generally have no phenotypic consequences, and as such many may not be detected through clinical screening. However, at least three examples of class II rearrangements have been identified in phenotypically normal individuals: twice in which the rearrangement was discovered in the offspring of an individual (Chueh *et al.*, 2005; Wandall *et al.*, 1998) and once discovered only after multiple miscarriages (Knegt *et al.*, 2003). In addition to class I and II cases, there are rare instances of neocentromere formation on a non-rearranged chromosome with the original centromere location remaining intact but inactivated (Amor *et al.*, 2004; Bukvic *et al.*, 1996; Rivera *et al.*, 1996; Tyler-Smith *et al.*, 1999; Ventura *et al.*, 2004). In at least one case, a partial deletion of the pericentromeric alpha satellite repeats led to the formation of a neocentromere at 8p22 despite remaining alpha satellite at the normal location (Burnside *et al.*, 2011). Overall, clinically identified neocentromeres are rare, reportedly occurring in 1 in every 70,000 to 200,000 live births (Marshall *et al.*, 2008a).

Genomic characteristics of neocentromeres

Regardless of the type of neocentromere, establishment occurs almost exclusively on single-copy DNA. The exact locations are found in highly variable DNA regions, in either gene deserts or at regions that include actively transcribed genes (Alonso *et al.*, 2010; Marshall *et al.*, 2008a). Analysis of a 10q25 neocentromere, formed on single-copy DNA, and its inactive progenitor uncovered identical primary nucleotide sequence, further supporting an epigenetic mechanism of neocentromerization (Barry *et al.*, 2000). Hotspots of neocentromere formation have been identified on human chromosomes in regions of 3q, 8p, 13q, and 15q, accounting for approximately half of all reported cases. However, in all investigated cases, detailed analysis of the exact location of the neocentromeres in these so-called 'hotspots' has revealed distinct locations for the different neocentromere regions (Alonso *et al.*, 2003; Cardone *et al.*, 2006; Hasson *et al.*, 2011; Marshall *et al.*, 2008b; Warburton *et al.*, 2000). Mapping of neocentromeric DNA has failed to identify any significant deviation from the genome average for sequence motifs or repetitive elements (Alonso *et al.*, 2003; Cardone *et al.*, 2006; Chueh *et al.*, 2005; Lo *et al.*, 2001a,b). One trend that did emerge was that the AT content of neocentromeric DNA was increased, ranging from 59.9% to 66.1%, compared with the genome average of 59%. Additionally, neocentromeric sequences were found to be associated with an increase in the density of long interspersed nuclear elements (LINEs) and a decrease in short interspersed nuclear elements (SINEs) (Alonso *et al.*, 2007). These discoveries at neocentromeres may be informative for identifying important genomic or epigenetic cues that are also present within native centromeric DNA.

The biological phenomenon of neocentromere formation is not limited to humans or even primates. Neocentromeres have been identified in *Drosophila* (Maggert and Karpen, 2000, 2001; Williams *et al.*, 1998), maize (Topp *et al.*, 2009), barley (Nasuda *et al.*, 2005), rice (Gong *et al.*, 2009), wheat, *S. pombe* (Ishii *et al.*, 2008), and *Candida albicans* (Ketel *et al.*, 2009; Thakur and Sanyal, 2013). Model organisms have an advantage in neocentromere studies, because of the ease of chromosome manipulation. These studies have provided further insight into formation and function of both neocentromeres and endogenous centromeres. Model organisms have been particularly valuable for their ability to produce neocentromeres to test factors that are required for new centromere formation (see below).

Properties of centromeric domains on neocentromeres

Similar to endogenous centromeres, neocentromeric chromatin is defined by the presence of CENP-A. However, unlike endogenous human centromeres, neocentromeres in humans afford the possibility to map the organization of centromeric chromatin domains onto the underlying DNA, without the complications of highly homologous repetitive DNA. Thirteen human neocentromeres have been mapped by ChIP-chip, revealing CENP-A regions that range in size from ~54 to 450 kb (Alonso *et al.*, 2003, 2007; Capozzi et al., 2008; Saffery et al., 2003). In contrast to endogenous centromeres, which are defined by interspersed blocks of CENP-A and canonical H3 nucleosome, neocentromeres were found to have individual nucleosomes of CENP-A interspersed with H3 nucleosomes (Alonso et al., 2007). Thus, while there may be some differences in the centromeric chromatin organization at endogenous versus ectopic centromeres, it is clear that CENP-A is a key element that defines the centromeric domain at both endogenous centromeres and neocentromeres.

Neocentromeres associate with all known centromere/kinetochore proteins, except CENP-B, a protein that specifically binds to alpha satellite DNA (Craig et al., 2003a; Saffery et al., 2000). Immunostaining has demonstrated that CENP-A and several other centromeric proteins co-localize on neocentromeres. Co-localization of CENP-C with CENP-A and CENP-H/CENP-A has been shown by ChIP-on-chip for several neocentromeres (Alonso et al., 2007; Cardone et al., 2006). It is notable that on the first-described neocentromere, mardel(10), CENP-H did not colocalize with CENP-A, and instead was located over 1 Mb away from the CENP-A domain (Saffery et al., 2003). This suggests that centromere and kinetochore domains may be flexible among neocentromeres. However, analysis of two neocentromeres, derived from 10q25 and 20p12, demonstrated that for over 20 kinetochore-associated proteins, the pattern of protein binding was indistinguishable from that of the endogenous centromeres, except for the absence of CENP-B (Saffery et al., 2000). These findings confirmed earlier electron-microscopy studies of neocentromeres, showing microtubule-associated kinetochores exhibited sizes and morphologies that were identical to endogenous centromeres (Wandall et al., 1998). In addition to perfectly mimicking kinetochore structure, neocentromeres also reproduce the increase of scaffold attachment regions seen in endogenous centromeres (Saffery et al., 2003; Sumer et al., 2003). The lack of repetitive DNA allowed for further study of the enhanced scaffold attachment region on neocentromeres; the extent of the enhanced scaffold attachment region was found to be much larger than the CENP-A domain. The proportion was not consistent between the two neocentromeres analysed, with the CENP-A region occupying 10% of the scaffold attachment region in mardel(10) but 25% in invdup(20). These studies demonstrated that there are common features shared by neocentromeres and endogenous centromeres, making neocentromeres an appropriate model to study normal centromere function. Likewise, the differences, namely unique DNA sequence, provide opportunities for added discoveries not afforded due to the more complex genomics of endogenous centromeres.

Pericentromeric DNA exhibits characteristics of heterochromatin, and neocentromeres provide the unique opportunity to separate genetic and epigenetic properties of centromeres. Proximal heterochromatic regions, identified by ChIP for histone H3K9me and heterochromatin protein 1 (HP1), are not required for either neocentromere formation or function in chromosome segregation. HP1 showed only slight enrichment on mardel(10) as compared to the progenitor chromosome 10 (Saffery et al., 2003). Neocentromeres with little or no detectable heterochromatin do exhibit slight cohesion defects, indicating that even though heterochromatin is not necessary for centromere function, it may ensure mitotic stability of neocentric chromosomes (Alonso et al., 2010).

Additional epigenetic characteristics of neocentromeres have been studied using TSA to induce partial histone hyperacetylation. This treatment was able to shift the position of CENP-A and reduced the density of CENP-A per unit length (Craig et al., 2003b). Despite these changes, no measurable deleterious effects on mitosis or neocentromere function were detected, indicating flexibility of centromeric chromatin in response to epigenetic triggers. In support of this conclusion, removal of TSA caused the CENP-A domain to return to its original site, suggesting a strong memory for the original neocentromeric chromatin site. Another histone modifier, SUV39H1, a histone methyltransferase that accumulates at the centromere during prometaphase and dissociates at the metaphase to anaphase transition, was present at a neocentromere (Aagaard et al., 2000). These results reveal the importance of epigenetic mechanisms in centromere maintenance.

Neocentromeres and nuclear processes

Neocentromeres also provide a platform to study important nuclear and cell cycle processes, such as gene expression and replication. Neocentromeres in model organisms emphasize this best. Fungal centromeres of *S. cerevisiae, C. albicans,* and *S. pombe* replicate early (Kim *et al.*, 2003; Koren *et al.*, 2010; Raghuraman *et al.*, 2001). In *C. albicans,* neocentromere formation dramatically shifted the replication-timing pattern on the chromosome such that the neocentromere region now contained the earliest and most efficient origin, resembling an endogenous centromere (Koren *et al.*, 2010). In humans, the region of the chromosome where the mardel(10) neocentromere was derived replicated during mid-S phase prior to neocentromere formation, but shifted to mid-to-late S-phase replication after neocentromere formation. Replication timing of the 450 kb neocentromere region containing the CENP-A domain was in mid-to-late S before and after neocentromere formation. However, the timing in the surrounding area shifted such that the replication timing of the entire neocentromere region aligned with the other centromeres (Lo *et al.*, 2001a). Shifting of replication timing has also been demonstrated for neocentromeres in DT40 chicken cells (Shang *et al.*, 2013). These studies established that the impact of centromere presence on replication timing is independent of DNA sequence and suggest that epigenetic cues are involved.

Transcription has also been reported at neocentromeres, despite some exhibiting epigenetic marks of transcriptional silencing (Chueh *et al.*, 2009; Shang *et al.*, 2013). The repetitive nature of endogenous centromeres introduces difficulty in determining the origin of transcripts. However, using a fully sequenced human neocentromere, it was found that despite overall hypermethylation of the neocentromeric DNA, within the centromeric domain, there were areas of DNA hypomethylation that corresponded to sites of active transcription (Wong *et al.*, 2006). Active transcription has also been identified at other human neocentromeres, as well as those in *S. pombe* and DT40 chicken cells (Chueh *et al.*, 2009; Ogiyama *et al.*, 2013; Shang *et al.*, 2013).

Stability of neocentromeres

Neocentromeres are fully functional, but there are often differences in centromeric behaviour and chromosome stability of centromeres formed at ectopic versus native locations. For instance, human neocentromeres can have defects in the localization of aurora B kinase, an essential regulator of kinetochore-microtubule attachments (Bassett *et al.*, 2010). Even though epigenetic centromere specification directs the accumulation of aurora B at neocentromeres, there can be insufficient levels of aurora B on some neocentromeres to properly correct mitotic errors. In *Drosophila,* neocentromeres that lack heterochromatin are also less stable, even though they all contain comparable amounts of CENP-A^{CID} (Blower and Karpen, 2001; Maggert and Karpen, 2001; Williams *et al.*, 1998). It has also been demonstrated in *Drosophila* that CENP-A^{CID} mislocalization leads to neocentromerization on ectopic sites. Many of these neocentromeres exhibit significant mitotic segregation defects, aneuploidy, and a delay in progression from metaphase to anaphase (Heun *et al.*, 2006). It is possible that neocentromere formation goes through a transition period, in which neocentromeres can either mature and acquire more efficient centromere function or, conversely, lose function altogether (Scott and Sullivan, 2014). This model is support by studies in multiple organisms. For instance, in maize, chromosome stability increases in later generations, suggesting that neocentromeres can become more functional as they are propagated or inherited (Topp *et al.*, 2009). Other neocentromeres that are less functional may be eliminated via gene conversion in favour of the original endogenous centromere (Thakur and Sanyal, 2013). The maize, *Drosophila, S. pombe* and *C. albicans* studies suggest that initial CENP-A binding and neocentromere initiation may occur at many locations and that stable neocentromere formation requires a location where more CENP-A can incorporate in order to assemble and maintain a neocentromere of substantive size or chromatin organization.

Induction of neocentromeres

Recent experimental induction of neocentromeres has provided increasing insight into the biology of centromere assembly. The location of neocentromere formation appears to be non-random. In *S. pombe*, cre-lox excision of endogenous centromeres leads to the formation of neocentromeres at subtelomeric regions, but not at internal locations (Ishii *et al.*, 2008). In fact, formation is favoured at regions, such as telomeres, that are depleted of H2A.Z (Ogiyama *et al.*, 2013). However, these regions may need to have a more heterochromatic structure, since formation of subtelomeric neocentromeres is decreased by the inactivation of genes involved in RNAi dependent heterochromatin formation (Folco *et al.*, 2008). Neocentromere formation in *Drosophila* also appears to be favoured at regions in which there is heterochromatin (Olszak *et al.*, 2011).

In several *C. albicans* engineering experiments, centromeric DNA was replaced with a URA3 gene, and the formation of neocentromeres was observed at multiple locations along the chromosome (Ketel *et al.*, 2009). Neocentromeres arose in both transcriptionally active and intergenic, non-expressed regions, similar to the apparently random locations observed in human cases. In a separate study from a different group, systematic deletion of variably sized *Candida* centromere/CENP-A regions on different chromosomes suggested that neocentromeres favour formation near the site once occupied by the endogenous centromere (Thakur and Sanyal, 2013). Local neocentromere formation near endogenous centromeres also occurs in chicken cells. Highly efficient formation of neocentromeres in DT40 cells occurred when a 127 kb region including the 35 kb CENP-A domain of the chicken Z chromosome centromere was conditionally deleted. In fact, up to 90% of neocentromeres formed adjacent to the excised centromere (Shang *et al.*, 2013). In both the yeast and chicken studies, the neocentromeres were structurally and functionally equivalent to endogenous centromeres, although they did not exhibit the same level of enrichment for histone modifications H3K9me3, H3K4me2, and H3K36me3 or exhibit early replication timing like endogenous centromeres.

In *Drosophila*, neocentromeres that form proximal to active centromeres have also been described (Maggert and Karpen, 2001; Williams *et al.*, 1998). However, the mechanism of formation may be important for understanding neocentromere biology. When chromosome breakage or deletion is used to induce neocentromere formation, it is possible that they preferentially arise in proximity to the original, endogenous centromere (Ketel *et al.*, 2009; Maggert and Karpen, 2001; Shang *et al.*, 2013; Thakur and Sanyal, 2013). Other strategies to produce neocentromeres that do not require chromosome breakage suggest other modes of neocentromere formation. For instance, overexpression of CENP-A^{CID} in *Drosophila* was shown to produce ectopic centromeres/neocentromeres that exhibited kinetochore activity and successfully engaged microtubules at mitosis (Heun *et al.*, 2006). Neocentromere assembly appeared to favour the boundaries between heterochromatin and euchromatin, suggesting chromatin environments that especially promote CENP-A incorporation or maintenance (Olszak *et al.*, 2011). In human cells, similar overexpression studies have been performed. However, despite a high degree of CENP-A incorporation in chromosome arms, overexpression did not cause neocentromere formation as in *Drosophila* (Van Hooser *et al.*, 2001).

Several systems have been developed to target or tether proteins to specific genomic sites and induce an ectopic kinetochore at non-centromeric DNA. Replacement of the DNA-binding regions of CENP-C and CENP-T with alternate targeting domains has been shown to recruit these proteins to an ectopic locus, resulting in CENP-A-independent kinetochore assembly (Gascoigne *et al.*, 2011). Likewise, tethering of CENP-A^{CID} or CAL1 using a LacI/LacO system is sufficient for ectopic kinetochore assembly in *Drosophila* (Chen *et al.*, 2014; Mendiburo *et al.*, 2011). The neocentromere/ectopic kinetochore is maintained even when LacI–centromere fusion protein is shut off. In human cells, LacI-HJURP, the CENP-A chaperone protein, can recruit CENP-A to a LacO array at a non-centromeric locus on chromosome 1 (Barnhart *et al.*, 2011). The ectopically targeted CENP-A chromatin at the LacO array can direct the recruitment of other centromere proteins and result in the assembly of a functional centromere. Finally, just as in these other organisms, a chicken chromosome-engineering system that targets centromere and kinetochore proteins (HJURP, CENP-C, CENP-I, or CENP-C) to a non-centromeric site after removal of the endogenous centromere also produces neocentromeres/ectopic kinetochores (Hori *et al.*, 2013). These powerful chromosome-engineering systems coordinately have expanded our knowledge of the epigenetic assembly and maintenance of endogenous centromeres, neo-centromeres, and ectopic/*de novo* centromeres.

Models for the mechanistic basis of neocentromere formation in eukaryotes

Collectively, these studies highlight several models for how non-centromeric DNA might acquire the epigenetic features of a centromere. One model suggests centromere activity spreads from existing centromeric regions to neighbouring DNA (Murphy and Karpen, 1998; Williams *et al.*, 1998). This 'activation model' is supported by some *Drosophila* data (Maggert and Karpen, 2001; Williams *et al.*, 1998), and data from *Candida* and chicken cells. However, this mechanism does not appear to be a major pathway of neocentromere formation in humans, since neocentromeres are usually located far away from endogenous centromeres.

A second model – the lateral inhibition model – suggests that there are multiple sites along a chromosome that are capable of initiating centromere activity but are normally repressed in *cis* by the presence of a more dominant centromere or by chromatin surveillance mechanisms. These processes must coordinate proper deposition of CENP-A with removal of ectopic CENP-A in order to protect against neocentromere formation. Because only a small proportion (<5%) of CENP-A in human cells is actually incorporated into centromeres (Bodor *et al.*, 2014), the surveillance and ectopic CENP-A removal mechanisms must be particularly robust. It might involve cell cycle regulation of centromeric deposition in G1, modification of the CENP-A loading factor HJURP, and/or HJURP's affinity for specific DNA sequences (Jansen *et al.*, 2007; Muller *et al.*, 2014).

A third model that has yet to be tested is that the formation of a functional neocentromere precedes the loss of the original centromere. In this case, a functionally dicentric chromosome would be created that is predisposed to rearrangement. It may become stabilized by breakage and deletion of the neocentromere region that then persists in the karyotype.

RNA, transcription, and centromere function

RNA sits at the intersection of genomics and epigenetics in chromosome function. The roles of long non-coding RNAs (lncRNA) in epigenetic processes such as imprinting and X-chromosome inactivation have been well established. lncRNAs are involved in regulation of chromosomal functions, chromatin organization, and gene expression in the absence of changes to the underlying DNA sequence. Transcription of centromeric sequences to produce various RNA species has emerged as one way to ensure centromere assembly and maintenance through recruitment of histone modifying enzymes and proteins that read and interpret epigenetic signatures.

Implication of non-coding RNAs in centromere function

The first studies investigating transcription within and proximal to the centromere were completed in the model organism *S. cerevisiae* and defined the classical view of the incompatibility of transcription and centromere maintenance (Chlebowicz-Sledziewska and Sledziewski, 1985; Hill and Bloom, 1987). In a key experiment, the strong, inducible *GAL1* promoter was placed upstream of the *CEN* sequence and utilized to induce transcription through the centromere of budding yeast, resulting in chromosome instability and loss (Hill and Bloom, 1987). This observation led to the model that a functional centromere was incompatible with transcription and, therefore, must be heterochromatic in nature. While strong transcription through the centromere of both yeast and metazoan chromosomes does have a disruptive effect on centromere maintenance (Hill and Bloom, 1987; Nakano *et al.*, 2008), the centromeres of these organisms also appear to require, or at least tolerate, transcription, in order to remain functional (Bergmann *et al.*, 2012a, 2011; Ohkuni and Kitagawa, 2011).

These findings have led to the proposal that active centromeres depend on the establishment of a 'balance of transcriptional activity' (Bergmann *et al.*, 2012a).

Supporting the functional studies of directly altered transcription levels, careful analyses of the chromatin composition of centromeres argued for transcription at the centromere. Despite the perception that eukaryotic centromeres are generally considered to be heterochromatic and transcriptionally silent, an in-depth study of the chromatin environment of the centromeres in *Drosophila* and humans revealed the presence of euchromatic H3K4me2 marks, a hallmark of transcriptionally permissive DNA (Lam *et al.*, 2006; Sullivan and Karpen, 2004). These studies, combined with experiments utilizing human neocentromeres (Saffery *et al.*, 2003) (see 'Neocentromeres and ectopic/induced centromeres'), demonstrated that centromeric chromatin is accessible to transcriptional machinery. Additional studies in yeast, plants, and metazoans have uncovered the great diversity of transcripts emanating from and associated with the centromere and are beginning to shed light on their wide range of functions.

Discovery and function of centromeric RNAs

Pericentromeric transcripts

Initial endeavours to characterize the centromeric sequence of *S. pombe* have sought to identify centromeric RNA transcripts in this species. RNA at the centromere of wild type fission yeast was not detected in these early experiments, further propagating the idea that centromeres are heterochromatic and inaccessible to transcription (Fishel *et al.*, 1988). Centromeric RNAs in fission yeast were not discovered until 2002, when a mutant, *swi6*, was found to accumulate double-stranded RNA transcripts derived from the pericentromeric region in high levels (Volpe *et al.*, 2003), leading to the first demonstration that pericentromeric RNA transcription contributes to centromere function.

Despite the similarity of the repeats found in the *S. pombe* central core and innermost repeats and the flanking outer repeat region of the pericentromere, the two regions are functionally distinct (Partridge *et al.*, 2000). The deposition of CENP-A^{Cnp1} occurs only in the kinetochore domain (Partridge *et al.*, 2000; Takahashi *et al.*, 2000). RNA-interference (RNAi) has been implicated as the mechanism for initiating and limiting the deposition of CENP-A^{Cnp1} in this confined region of the fission yeast centromere, as well as establishing the heterochromatin environment of the pericentromeric region (Fig. 7.6). In this pathway, small interfering RNAs (siRNAs) transcribed from the outer repeat region by RNA polymerase II (RNAPII) (Kato *et al.*, 2005) contribute to chromatin modification by promoting the formation of heterochromatin in the pericentromere region, characterized by the presence of H3K9me2 (Volpe *et al.*, 2003). The formation of heterochromatin is required for the initial deposition of CENP-A^{Cnp1} within the central core domain. Interruption of the RNAi pathway (by deletion of *chp1* or *dcr1*, the genes encoding two key components of the RNAi pathway, Chp1 and Dicer) or heterochromatin formation (by the deletion of *swi6*, the gene encoding a component of pericentromeric heterochromatin, the yeast HP1 homologue) disrupts the assembly of CENP-A^{Cnp1} (Folco *et al.*, 2008; Volpe *et al.*, 2002, 2003). The small RNAs emanating from the outer repeat region of the centromere play a role in localizing the RNA-induced transcriptional silencing complex (RITS complex) to the heterochromatin, where it acts as an effector domain to induce silencing (Verdel *et al.*, 2004). Once heterochromatin is established, CENP-A^{Cnp1} deposition can occur (Folco *et al.*, 2008). After CENP-A is assembled, however, RNAi-dependent heterochromatin formation is no longer required for the stable maintenance of CENP-A^{Cnp1} (Folco *et al.*, 2008).

The RNAi pathway and pericentromeric transcription have been observed to play a key role in determining the chromatin state of the centromere in a number of other organisms as well, including *Arabidopsis* (Chen *et al.*, 2008), *Drosophila* (Pal-Bhadra *et al.*, 2004) and multiple vertebrate species (Fukagawa *et al.*, 2004; Kanellopoulou *et al.*, 2005). The budding yeast *S. cerevisiae* is a clear exception, since RNAi pathway proteins, and consequently siRNAs, are absent from this species (Aravind *et al.*, 2000; Nakayashiki *et al.*, 2006). Although there does not

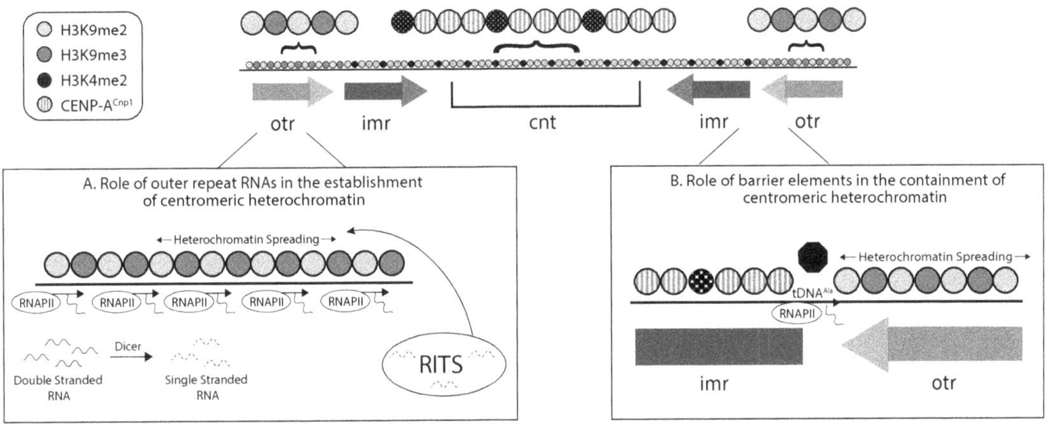

Figure 7.6 RNA and transcription at eukaryotic centromeres. (A) The centromere of *S. pombe* is characterized by the presence of CENP-A^{Cnp1} and H3K4me2 in the central core (cnt) and innermost repeat region (imr) and H3K9me2 and H3K9me3 in the outer repeat region (otr). RNAPII transcribes non-coding double stranded RNAs from the otr, which are processed by Dicer to produce single stranded siRNAs. These RNAs then localize the RITS complex to heterochromatin in order to further the spread of heterochromatin. (B) Heterochromatin spread is limited by the presence of the gene encoding tRNA^{Ala}. Located between the otr and imr, tDNA^{Ala} must also be actively transcribed to preserve the chromatin state of the central core.

appear to be a conserved centromeric sequence, RNAi may represent an epigenetic mechanism of active centromere specification that is, to some degree, present across many eukaryotes. A key difference, however, has been noted between the function of the RNAi pathway in *S. pombe* and higher eukaryotes. Similarly to *S. pombe*, RNAi in higher eukaryotes is involved in heterochromatin formation at the pericentromere, however, it does not appear to play a role in CENP-A deposition (Fukagawa *et al.*, 2004; Kim *et al.*, 2012).

In mammals, the characterization of larger, non-siRNA centromeric RNAs originating from the pericentromeric regions has helped provide a better understanding of the role of the process of transcription in the maintenance of the centromeric chromatin environment, particularly as it relates to the cell cycle (Bouzinba-Segard *et al.*, 2006; Guenatri *et al.*, 2004; Lu and Gilbert, 2007). One study in mice determined that the stage of the cell cycle was related to the size of the RNA transcribed from the pericentromeric major satellite region by RNAPII, with smaller transcripts being more abundant during mitosis and larger transcripts dominating from late in G1 phase to mid S phase (Lu and Gilbert, 2007). The relationship observed between the differences in RNA transcript length and the stage of the cell cycle led the authors to suggest that transcription at the pericentromere region may control heterochromatin formation, which must be reassembled after genome replication.

Central domain/CENP-A core domain transcripts

Transcription within the central domain of centromeres has also been investigated. In addition to transcribing pericentromeric DNA in *S. pombe*, RNAPII has also been implicated in the transcription of a second set of non-coding RNAs originating from the CENP-A^{Cnp1} chromatin domain, likely at a much lower level than observed at the pericentromere (Choi *et al.*, 2011). This transcription may be a mechanism for ensuring that a balance between histone H3 and CENP-A is maintained in the centromere, as two factors involved in RNAPII transcription, FACT and Clr6 complex II, have been found to prevent over-incorporation of CENP-A (Choi *et al.*, 2012).

A combination of *in vitro* and *in vivo* experiments in maize have revealed the ability of long RNA transcripts to stabilize the assembly of the CENP-C protein at the inner kinetochore (Du *et al.*, 2010). Biochemical experiments demonstrated that CENP-C binds more effectively to centromeric DNA in the presence of single-stranded RNA greater than 40 nucleotides in length derived from CentC, a marker of the maize centromere core domain, and suggest that these transcripts play a larger role in the targeting and deposition of new CENP-A^{CENH3} (Du *et al.*, 2010).

Further research in humans has identified transcription at the kinetochore region to produce transcripts with a key role in kinetochore assembly during mitosis. Human alpha satellite RNA transcripts have been shown to accumulate in the nucleolus during interphase, where it is likely involved in sequestering the inner centromere protein CENP-C and the inner kinetochore protein INCENP, as RNase treatment at this stage disrupts the localization of these two proteins to the nucleolus (Wong *et al.*, 2007). Once the cell enters metaphase, CENP-C and INCENP become localized to the centromere for kinetochore assembly. In addition to being transcribed during interphase, human alpha satellite RNA transcription during mitosis has also been discovered to stabilize CENP-C at the centromere (Chan *et al.*, 2012).

Transposable element transcripts

Transposable elements (TEs), DNA sequences that have the ability to move throughout the genome either autonomously or with the assistance of an activator element, constitute a portion of the centromere in humans, *Drosophila*, and plants (reviewed in Dawe, 2003). Specifically, in humans, the heterochromatic pericentromere region includes LINE and SINE retrotransposons. Plant genomes also contain substantial amounts of TEs. The majority of retrotransposons at plant centromeres can be traced to a subset of the viral family *Chromoviridae* (Neumann *et al.*, 2011) and are referred to as CRR elements in rice and CRM elements in maize (Jiang *et al.*, 2003; Neumann *et al.*, 2011). Studies of these elements in rice and maize have revealed transcripts that emanate from the CRR and CRM elements, as well as the satellite repeats (Topp *et al.*, 2004; Yan *et al.*, 2005). In maize, these transcripts subsequently associate with CENP-A^{CENH3} (Topp *et al.*, 2004). These transcripts are single-stranded and range from 40 to 200 nucleotides in length, indicating that they are not siRNAs and are likely unrelated to the RNAi processing that is involved in heterochromatin formation. Instead, the CRM transcripts might function to promote CENP-A^{CENH3} deposition and maintenance in plants (Topp *et al.*, 2004). Similarly, centromeric TE transcripts derived from KERV-1 *gag* and KERV-1 LTR, kangaroo-specific retroviral elements, have been identified in the tammar wallaby (Bulazel *et al.*, 2007; O'Neill *et al.*, 1998). These small RNAs directly establish the chromatin environment of the tammar wallaby centromere, as demonstrated by RNase treatment that reduced CENP-B localization and H3K9me3 foci within the pericentromeric heterochromatin domain (Carone *et al.*, 2009).

Transcription and centromere architecture

The complex structure of the *S. pombe* centromere, composed of a core domain dominated by CENP-A^{CNP1} chromatin and flanked by a heterochromatic pericentromere region, requires distinct barriers between the disparate regions to maintain their separate identities. The observed ability of heterochromatic DNA to spread in fission yeast (Allshire *et al.*, 1994) indicated that a specific mechanism prevents pericentromeric heterochromatin from spreading into the core domain. Recent studies showed that an insulator or, more specifically, a chromatin barrier, maintains these two types of chromatin in close proximity and is key in defining the architecture of the *S. pombe* centromere (Scott *et al.*, 2006). Utilizing the expression of a *ura4*$^{+}$ reporter gene integrated into a variety of positions within centromere 1 as a readout for heterochromatin spread, the barrier sequence was narrowed to a 480 bp region located in the inner repeat region. This region includes the alanine transfer RNA (tRNA^{Ala}) gene, and deletion of this gene resulted in the spread of heterochromatin to the core domain, indicating the role of this sequence in maintaining a boundary between CENP-A^{Cnp1} chromatin and euchromatin (Fig. 7.6). Follow-up studies determined that this region requires transcription factor IIIC and RNAPII to function as a barrier region, indicating that the presence of the barrier sequence is not sufficient to prevent heterochromatin spreading (Scott *et al.*, 2007). The transcription of this gene is also required, a characteristic of this particular barrier that has not been observed in other barrier sequences (Scott *et al.*, 2007). The presence and transcription of this sequence appear to protect the integrity of the core centromere chromatin and, thus, the activity of the kinetochore. Although similar barrier sequences within the centromere have yet to be identified in larger eukaryotes, this work provides a framework for the investigation of the mechanisms that link transcription and centromere architecture and function in other organisms.

Transcription on engineered chromosomes

Human artificial chromosomes (HACs) are a useful tool for probing the complex relationship among transcription, chromatin environment, and centromere and kinetochore maintenance. In particular, the development of alphoid^{tetO}-HACs (Nakano *et al.*, 2008) has provided a new strategy to directly manipulate levels of transcription at the centromere and to monitor effects on the epigenetic marks of the centromere (see 'Artificial chromosomes'). Collectively, studies utilizing this system have shown that the level of transcription at the centromere can be altered by targeting tetracycline transcriptional activators or repressors (Nakano *et al.*, 2008), activation domains NF-κB p65 and herpes virus VP16 (Bergmann *et al.*, 2012a), or the transcriptional corepressor KAP1 (Cardinale *et al.*, 2009). Each of these approaches results in changes to chromatin modifications, including increased H3K9me3, reduced H3K4me2, and loss of CENP-A that is linked to long-term consequences for the chromosome, such as kinetochore inactivation and chromosome instability. In other tethering experiments, alphoid^{tetO}-HACs have been used to alter the chromatin environment and observe the effect on transcription. Targeting lysine-specific demethylase to HACs to catalyse the loss of H3K4me2 at the core centromere region can deplete CENP-A from the kinetochore and block loading of new CENP-A, thereby resulting in destabilization of the HAC (Bergmann *et al.*, 2011). These studies have helped to confirm many of the transcription-related phenomena observed in endogenous centromeres, while taking advantage of the flexibility and ease of manipulation of HACs.

Concluding remarks

Over the past several decades, significant progress has been made in understanding the importance of epigenetic factors in centromere identity and function. Cell cycle-regulated incorporation of CENP-A, identification of the CENP-A loading factor, characterization of the centromeric licensing complex MIS18BP1, identification of RNA-dependent mechanisms of centromere domain assembly, and the histone code of centromeres are among the notable advances in our knowledge of this complex chromosomal locus. However, these findings have raised additional questions about properties of eukaryotic centromeres. Future studies will be important for understanding how centromeric epigenetic marks are regulated and inherited and for identifying which comes first – the chromatin landscape that is amenable for centromere assembly or CENP-A recruitment that initiates an epigenetic cascade of protein recruitment. In addition, the use of genome and chromosome engineering methods, such as HACs and CRISPR, will prove useful for determining genomic and epigenetic components that are required for centromere assembly and maintenance. Finally, the role of RNAs in centromere function is an area of biology that has only been minimally explored. Considering that substantial roles for non-coding RNAs in genome stability and chromatin biology are being increasingly appreciated, it is expected that a function for RNA and transcription in centromere identity and structure will be forthcoming. Another key area of focus will be identifying the mechanisms that restrict a centromere to a specific genome location and prevent unwanted centromeres from appearing at undesirable genomic sites. Such studies will be important for understanding the fundamental mechanisms that ensure genome stability and chromosome inheritance. More comprehensive knowledge of centromere biology will be equally important for pinpointing how defective centromeres affect human health and disease and for developing treatments that target mis-regulated centromeres or that use centromeres to drive gene therapy vectors.

References

Aagaard, L., Schmid, M., Warburton, P., and Jenuwein, T. (2000). Mitotic phosphorylation of SUV39H1, a novel component of active centromeres, coincides with transient accumulation at mammalian centromeres. J. Cell Sci. *113 (Pt 5)*, 817–829.

Ahmad, K., and Golic, K.G. (1998). The transmission of fragmented chromosomes in *Drosophila melanogaster*. Genetics *148*, 775–792.

Allshire, R.C., and Karpen, G.H. (2008). Epigenetic regulation of centromeric chromatin: old dogs, new tricks? Nat. Rev. Genet. *9*, 923–937.

Allshire, R.C., Javerzat, J.-P., Redhead, N.J., and Cranston, G. (1994). Position effect variegation at fission yeast centromeres. Cell *76*, 157–169.

Alonso, A., Mahmood, R., Li, S., Cheung, F., Yoda, K., and Warburton, P.E. (2003). Genomic microarray analysis reveals distinct locations for the CENP-A binding domains in three human chromosome 13q32 neocentromeres. Hum. Mol. Genet. *12*, 2711–2721.

Alonso, A., Fritz, B., Hasson, D., Abrusan, G., Cheung, F., Yoda, K., Radlwimmer, B., Ladurner, A.G., and Warburton, P.E. (2007). Co-localization of CENP-C and CENP-H to discontinuous domains of CENP-A chromatin at human neocentromeres. Genome Biol. *8*, R148.

Alonso, A., Hasson, D., Cheung, F., and Warburton, P.E. (2010). A paucity of heterochromatin at functional human neocentromeres. Epigenetics Chromatin *3*, 6.

Amor, D.J., Bentley, K., Ryan, J., Perry, J., Wong, L., Slater, H., and Choo, K.H. (2004). Human centromere repositioning "in progress". Proc. Natl. Acad. Sci. U.S.A. *101*, 6542–6547.

Aravind, L., Watanabe, H., Lipman, D.J., and Koonin, E.V. (2000). Lineage-specific loss and divergence of functionally linked genes in eukaryotes. Proc. Natl. Acad. Sci. U.S.A. *97*, 11319–11324.

Bailey, A.O., Panchenko, T., Sathyan, K.M., Petkowski, J.J., Pai, P.J., Bai, D.L., Russell, D.H., Macara, I.G., Shabanowitz, J., Hunt, D.F., *et al.* (2013). Post-translational modification of CENP-A influences the conformation of centromeric chromatin. Proc. Natl. Acad. Sci. U.S.A. *110*, 11827–11832.

Barnhart, M.C., Kuich, P.H., Stellfox, M.E., Ward, J.A., Bassett, E.A., Black, B.E., and Foltz, D.R. (2011). HJURP is a CENP-A chromatin assembly factor sufficient to form a functional *de novo* kinetochore. J. Cell Biol. *194*, 229–243.

Barry, A.E., Bateman, M., Howman, E.V., Cancilla, M.R., Tainton, K.M., Irvine, D.V., Saffery, R., and Choo, K.H. (2000). The 10q25 neocentromere and its inactive progenitor have identical primary nucleotide sequence: further evidence for epigenetic modification. Genome Res. *10*, 832–838.

Bassett, E.A., Wood, S., Salimian, K.J., Ajith, S., Foltz, D.R., and Black, B.E. (2010). Epigenetic centromere specification directs aurora B accumulation but is insufficient to efficiently correct mitotic errors. J. Cell Biol. *190*, 177–185.

Bergmann, J.H., Rodriguez, M.G., Martins, N.M., Kimura, H., Kelly, D.A., Masumoto, H., Larionov, V., Jansen, L.E., and Earnshaw, W.C. (2011). Epigenetic engineering shows H3K4me2 is required for HJURP targeting and CENP-A assembly on a synthetic human kinetochore. EMBO J. *30*, 328–340.

Bergmann, J.H., Jakubsche, J.N., Martins, N.M., Kagansky, A., Nakano, M., Kimura, H., Kelly, D.A., Turner, B.M., Masumoto, H., Larionov, V., *et al.* (2012a). Epigenetic engineering: histone H3K9 acetylation is compatible with kinetochore structure and function. J. Cell Sci. *125*, 411–421.

Bergmann, J.H., Martins, N.M., Larionov, V., Masumoto, H., and Earnshaw, W.C. (2012b). HACking the centromere chromatin code: insights from human artificial chromosomes. Chromosome Res. *20*, 505–519.

Black, B.E., Foltz, D.R., Chakravarthy, S., Luger, K., Woods, V.L. Jr., and Cleveland, D.W. (2004). Structural determinants for generating centromeric chromatin. Nature *430*, 578–582.

Black, B.E., Brock, M.A., Bedard, S., Woods, V.L. Jr., and Cleveland, D.W. (2007). An epigenetic mark generated by the incorporation of CENP-A into centromeric nucleosomes. Proc. Natl. Acad. Sci. U.S.A. *104*, 5008–5013.

Blower, M.D., and Karpen, G.H. (2001). The role of Drosophila CID in kinetochore formation, cell-cycle progression and heterochromatin interactions. Nat. Cell Biol. *3*, 730–739.

Blower, M.D., Sullivan, B.A., and Karpen, G.H. (2002). Conserved organization of centromeric chromatin in flies and humans. Dev. Cell *2*, 319–330.

Bodor, D.L., Valente, L.P., Mata, J.F., Black, B.E., and Jansen, L.E. (2013). Assembly in G1 phase and long-term stability are unique intrinsic features of CENP-A nucleosomes. Mol. Biol. Cell *24*, 923–932.

Bodor, D.L., Mata, J.F., Sergeev, M., David, A.F., Salimian, K.J., Panchenko, T., Cleveland, D.W., Black, B.E., Shah, J.V., and Jansen, L.E. (2014). The quantitative architecture of centromeric chromatin. Elife *3*, e02137.

Bouzinba-Segard, H., Guais, A., and Francastel, C. (2006). Accumulation of small murine minor satellite transcripts leads to impaired centromeric architecture and function. Proc. Natl. Acad. Sci. U.S.A. *103*, 8709–8714.

Breman, A.M., Steiner, C.M., Slee, R.B., and Grimes, B.R. (2008). Input DNA ratio determines copy number of the 33 kb Factor IX gene on de novo human artificial chromosomes. Mol. Ther. *16*, 315–323.

Buchwitz, B.J., Ahmad, K., Moore, L.L., Roth, M.B., and Henikoff, S. (1999). A histone-H3-like protein in C. elegans. Nature *401*, 547–548.

Bui, M., Dimitriadis, E.K., Hoischen, C., An, E., Quenet, D., Giebe, S., Nita-Lazar, A., Diekmann, S., and Dalal, Y. (2012). Cell-cycle-dependent structural transitions in the human CENP-A nucleosome *in vivo*. Cell *150*, 317–326.

Bukvic, N., Susca, F., Gentile, M., Tangari, E., Ianniruberto, A., and Guanti, G. (1996). An unusual dicentric Y chromosome with a functional centromere with no detectable alpha-satellite. Hum. Genet. *97*, 453–456.

Bulazel, K.V., Ferreri, G.C., Eldridge, M.D., and O'Neill, R.J. (2007). Species-specific shifts in centromere sequence composition are coincident with breakpoint reuse in karyotypically divergent lineages. Genome Biol. *8*, R170.

Burnside, R.D., Ibrahim, J., Flora, C., Schwartz, S., Tepperberg, J.H., Papenhausen, P.R., and Warburton, P.E. (2011). Interstitial deletion of proximal 8q including part of the centromere from unbalanced segregation of a paternal deletion/marker karyotype with neocentromere formation at 8p22. Cytogenet Genome Res. *132*, 227–232.

Camahort, R., Shivaraju, M., Mattingly, M., Li, B., Nakanishi, S., Zhu, D., Shilatifard, A., Workman, J.L., and Gerton, J.L. (2009). Cse4 is part of an octameric nucleosome in budding yeast. Mol. Cell 35, 794–805.

Capozzi, O., Purgato, S., Verdun di Cantogno, L., Grosso, E., Ciccone, R., Zuffardi, O., Della Valle, G., and Rocchi, M. (2008). Evolutionary and clinical neocentromeres: two faces of the same coin? Chromosoma *117*, 339–344.

Cardinale, S., Bergmann, J.H., Kelly, D., Nakano, M., Valdivia, M.M., Kimura, H., Masumoto, H., Larionov, V., and Earnshaw, W.C. (2009). Hierarchical inactivation of a synthetic human kinetochore by a chromatin modifier. Mol. Biol. Cell *20*, 4194–4204.

Cardone, M.F., Alonso, A., Pazienza, M., Ventura, M., Montemurro, G., Carbone, L., de Jong, P.J., Stanyon, R., D'Addabbo, P., Archidiacono, N., *et al.* (2006). Independent centromere formation in a capricious, gene-free domain of chromosome 13q21 in Old World monkeys and pigs. Genome Biol. *7*, R91.

Carone, D.M., Longo, M.S., Ferreri, G.C., Hall, L., Harris, M., Shook, N., Bulazel, K.V., Carone, B.R., Obergfell, C., O'Neill, M.J., *et al.* (2009). A new class of retroviral and satellite encoded small RNAs emanates from mammalian centromeres. Chromosoma *118*, 113–125.

Carroll, C.W., Silva, M.C., Godek, K.M., Jansen, L.E., and Straight, A.F. (2009). Centromere assembly requires the direct recognition of CENP-A nucleosomes by CENP-N. Nat. Cell Biol. *11*, 896–902.

Chan, F.L., Marshall, O.J., Saffery, R., Kim, B.W., Earle, E., Choo, K.H., and Wong, L.H. (2012). Active transcription and essential role of RNA polymerase II at the centromere during mitosis. Proc. Natl. Acad. Sci. U.S.A. *109*, 1979–1984.

Chen, C.C., Dechassa, M.L., Bettini, E., Ledoux, M.B., Belisario, C., Heun, P., Luger, K., and Mellone, B.G. (2014). CAL1 is the Drosophila CENP-A assembly factor. J. Cell Biol. *204*, 313–329.

Chen, M., Ha, M., Lackey, E., Wang, J., and Chen, Z.J. (2008). RNAi of met1 reduces DNA methylation and induces genome-specific changes in gene expression and centromeric small RNA accumulation in Arabidopsis allopolyploids. Genetics *178*, 1845–1858.

Chlebowicz-Sledziewska, E., and Sledziewski, A.Z. (1985). Construction of multicopy yeast plasmids with regulated centromere function. Gene 39, 25–31.

Choi, E.S., Stralfors, A., Castillo, A.G., Durand-Dubief, M., Ekwall, K., and Allshire, R.C. (2011). Identification of noncoding transcripts from within CENP-A chromatin at fission yeast centromeres. J. Biol. Chem. *286*, 23600–23607.

Choi, E.S., Stralfors, A., Catania, S., Castillo, A.G., Svensson, J.P., Pidoux, A.L., Ekwall, K., and Allshire, R.C. (2012). Factors that promote H3 chromatin integrity during transcription prevent promiscuous deposition of CENP-A(Cnp1) in fission yeast. PLoS Genet. *8*, e1002985.

Chueh, A.C., Wong, L.H., Wong, N., and Choo, K.H. (2005). Variable and hierarchical size distribution of L1-retroelement-enriched CENP-A clusters within a functional human neocentromere. Hum. Mol. Genet. *14*, 85–93.

Chueh, A.C., Northrop, E.L., Brettingham-Moore, K.H., Choo, K.H., and Wong, L.H. (2009). LINE retrotransposon RNA is an essential structural and functional epigenetic component of a core neocentromeric chromatin. PLoS Genet. *5*, e1000354.

Clarke, L., and Carbon, J. (1980). Isolation of a yeast centromere and construction of functional small circular chromosomes. *287*, 504–509.

Clarke, L., and Carbon, J. (1985). The structure and function of yeast centromeres. Annu. Rev. Genet. *19*, 29–55.

Conde e Silva, N., Black, B.E., Sivolob, A., Filipski, J., Cleveland, D.W., and Prunell, A. (2007). CENP-A-containing nucleosomes: easier disassembly versus exclusive centromeric localization. J. Mol. Biol. *370*, 555–573.

Craig, J.M., Earle, E., Canham, P., Wong, L.H., Anderson, M., and Choo, K.H. (2003a). Analysis of mammalian proteins involved in chromatin modification reveals new metaphase centromeric proteins and distinct chromosomal distribution patterns. Hum. Mol. Genet. *12*, 3109–3121.

Craig, J.M., Wong, L.H., Lo, A.W., Earle, E., and Choo, K.H. (2003b). Centromeric chromatin pliability and memory at a human neocentromere. EMBO J. *22*, 2495–2504.

Dalal, Y., Wang, H., Lindsay, S., and Henikoff, S. (2007). Tetrameric structure of centromeric nucleosomes in interphase Drosophila cells. PLoS Biol. *5*, e218.

Dawe, R.K. (2003). RNA interference, transposons, and the centromere. Plant Cell *15*, 297–301.

De la Chapelle, A., Wennstrom, J., Hortling, H., and Ockey, C.H. (1966). Isochromosome-X in man. I. Hereditas *54*, 260–276.

Dechassa, M.L., Wyns, K., Li, M., Hall, M.A., Wang, M.D., and Luger, K. (2011). Structure and Scm3-mediated assembly of budding yeast centromeric nucleosomes. Nat. Commun. *2*, 313.

Dimitriadis, E.K., Weber, C., Gill, R.K., Diekmann, S., and Dalal, Y. (2010). Tetrameric organization of vertebrate centromeric nucleosomes. Proc. Natl. Acad. Sci. U.S.A. *107*, 20317–20322.

Dornblut, C., Quinn, N., Monajambashi, S., Prendergast, L., van Vuuren, C., Munch, S., Deng, W., Leonhardt, H., Cardoso, M.C., Hoischen, C., *et al.* (2014). A CENP-S/X complex assembles at the centromere in S and G2 phases of the human cell cycle. Open Biol. *4*, 130229.

Du, Y., Topp, C.N., and Dawe, R.K. (2010). DNA binding of centromere protein C (CENPC) is stabilized by single-stranded RNA. PLoS Genet. *6*, e1000835.

Dunleavy, E.M., Pidoux, A.L., Monet, M., Bonilla, C., Richardson, W., Hamilton, G.L., Ekwall, K., McLaughlin, P.J., and Allshire, R.C. (2007). A NASP (N1/N2)-related protein, Sim3, binds CENP-A and is required for its deposition at fission yeast centromeres. Mol. Cell *28*, 1029–1044.

Dunleavy, E.M., Roche, D., Tagami, H., Lacoste, N., Ray-Gallet, D., Nakamura, Y., Daigo, Y., Nakatani, Y., and Almouzni-Pettinotti, G. (2009). HJURP is a cell-cycle-dependent maintenance and deposition factor of CENP-A at centromeres. Cell *137*, 485–497.

Dunleavy, E.M., Beier, N.L., Gorgescu, W., Tang, J., Costes, S.V., and Karpen, G.H. (2012). The cell cycle timing of centromeric chromatin assembly in Drosophila meiosis is distinct from mitosis yet requires CAL1 and CENP-C. PLoS Biol. *10*, e1001460.

Earnshaw, W.C., and Migeon, B.R. (1985). Three related centromere proteins are absent from the inactive centromere of a stable isodicentric chromosome. Chromosoma 92, 290–296.

Ebersole, T., Okamoto, Y., Noskov, V.N., Kouprina, N., Kim, J.H., Leem, S.H., Barrett, J.C., Masumoto, H., and Larionov, V. (2005). Rapid generation of long synthetic tandem repeats and its application for analysis in human artificial chromosome formation. Nucleic Acids Res. *33*, e130.

Ebersole, T.A., Ross, A., Clark, E., McGill, N., Schindelhauer, D., Cooke, H., and Grimes, B. (2000). Mammalian artificial chromosome formation from circular alphoid input DNA does not require telomere repeats. Hum. Mol. Genet. *9*, 1623–1631.

Fachinetti, D., Diego Folco, H., Nechemia-Arbely, Y., Valente, L.P., Nguyen, K., Wong, A.J., Zhu, Q., Holland, A.J., Desai, A., Jansen, L.E., *et al.* (2013). A two-step mechanism for epigenetic specification of centromere identity and function. Nat. Cell Biol. *15*, 1056–1066.

Fishel, B., Amstutz, H., Baum, M., Carbon, J., and Clarke, L. (1988). Structural organization and functional analysis of centromeric DNA in the fission yeast *Schizosaccharomyces pombe*. Mol. Cell. Biol. *8*, 754-763.

Folco, H.D., Pidoux, A.L., Urano, T., and Allshire, R.C. (2008). Heterochromatin and RNAi are required to establish CENP-A chromatin at centromeres. Science *319*, 94–97.

Foltz, D.R., Jansen, L.E., Bailey, A.O., Yates, J.R., 3rd, Bassett, E.A., Wood, S., Black, B.E., and Cleveland, D.W. (2009). Centromere-specific assembly of CENP-a nucleosomes is mediated by HJURP. Cell *137*, 472–484.

Fu, S., Gao, Z., Birchler, J., and Han, F. (2012). Dicentric chromosome formation and epigenetics of centromere formation in plants. J. Genet. Genomics *39*, 125–130.

Fujita, Y., Hayashi, T., Kiyomitsu, T., Toyoda, Y., Kokubu, A., Obuse, C., and Yanagida, M. (2007). Priming of centromere for CENP-A recruitment by human hMis18alpha, hMis18beta, and M18BP1. Dev. Cell *12*, 17–30.

Fukagawa, T., Nogami, M., Yoshikawa, M., Ikeno, M., Okazaki, T., Takami, Y., Nakayama, T., and Oshimura, M. (2004). Dicer is essential for formation of the heterochromatin structure in vertebrate cells. Nat. Cell Biol. *6*, 784–791.

Furuyama, T., and Henikoff, S. (2009). Centromeric nucleosomes induce positive DNA supercoils. Cell *138*, 104–113.

Furuyama, T., Codomo, C.A., and Henikoff, S. (2013). Reconstitution of hemisomes on budding yeast centromeric DNA. Nucleic Acids Res. *41*, 5769–5783.

Gascoigne, K.E., Takeuchi, K., Suzuki, A., Hori, T., Fukagawa, T., and Cheeseman, I.M. (2011). Induced ectopic kinetochore assembly bypasses the requirement for CENP-A nucleosomes. Cell *145*, 410–422.

Gong, Z., Yu, H., Huang, J., Yi, C., and Gu, M. (2009). Unstable transmission of rice chromosomes without functional centromeric repeats in asexual propagation. Chromosome Res. *17*, 863–872.

Gopalakrishnan, S., Sullivan, B.A., Trazzi, S., Della Valle, G., and Robertson, K.D. (2009). DNMT3B interacts with constitutive centromere protein CENP-C to modulate DNA methylation and the histone code at centromeric regions. Hum. Mol. Genet. *18*, 3178–3193.

Greaves, I.K., Rangasamy, D., Ridgway, P., and Tremethick, D.J. (2007). H2A.Z contributes to the unique 3D structure of the centromere. Proc. Natl. Acad. Sci. U.S.A. *104*, 525–530.

Grimes, B.R., Rhoades, A.A., and Willard, H.F. (2002a). Alpha-satellite DNA and vector composition influence rates of human artificial chromosome formation. Mol. Ther. *5*, 798–805.

Grimes, B.R., Warburton, P.E., and Farr, C.J. (2002b). Chromosome engineering: prospects for gene therapy. Gene Ther. *9*, 713–718.

Grimes, B.R., Babcock, J., Rudd, M.K., Chadwick, B., and Willard, H.F. (2004). Assembly and characterization of heterochromatin and euchromatin on human artificial chromosomes. Genome Biol. *5*, R89.81–R89.14.

Guenatri, M., Bailly, D., Maison, C., and Almouzni, G. (2004). Mouse centric and pericentric satellite repeats form distinct functional heterochromatin. J. Cell Biol. *166*, 493–505.

Haber, J.E., Thorburn, P.C., and Rogers, D. (1984). Meiotic and mitotic behavior of dicentric chromosomes in *Saccharomyces cerevisiae*. Genetics *106*, 185–205.

Hahnenberger, K.M., Baum, M.P., Polizzi, C.M., Carbon, J., and Clarke, L. (1989). Construction of functional artificial minichromosomes in the fission yeast *Schizosaccharomyces pombe*. Proc. Natl. Acad. Sci. U.S.A. *86*, 577–581.

Han, F., Lamb, J.C., and Birchler, J.A. (2006). High frequency of centromere inactivation resulting in stable dicentric chromosomes of maize. Proc. Natl. Acad. Sci. U.S.A. *103*, 3238–3243.

Han, F., Gao, Z., and Birchler, J.A. (2009). Reactivation of an inactive centromere reveals epigenetic and structural components for centromere specification in maize. Plant Cell *21*, 1929–1939.

Hargreaves, D.C., and Crabtree, G.R. (2011). ATP-dependent chromatin remodeling: genetics, genomics and mechanisms. Cell Res. *21*, 396–420.

Harrington, J.J., Bokkelen, G.V., Mays, R.W., Gustashaw, K., and Willard, H.F. (1997). Formation of de novo centromeres and construction of first-generation human artificial microchromosomes. Nat. Genet. *15*, 345–355.

Hasson, D., Alonso, A., Cheung, F., Tepperberg, J.H., Papenhausen, P.R., Engelen, J.J., and Warburton, P.E. (2011). Formation of novel CENP-A domains on tandem repetitive DNA and across chromosome breakpoints on human chromosome 8q21 neocentromeres. Chromosoma *120*, 621–632.

Hayashi, T., Fujita, Y., Iwasaki, O., Adachi, Y., Takahashi, K., and Yanagida, M. (2004). Mis16 and Mis18 are required for CENP-A loading and histone deacetylation at centromeres. Cell *118*, 715–729.

Hemmerich, P., Weidtkamp-Peters, S., Hoischen, C., Schmiedeberg, L., Erliandri, I., and Diekmann, S. (2008). Dynamics of inner kinetochore assembly and maintenance in living cells. J. Cell Biol. *180*, 1101–1114.

Henikoff, S., and Henikoff, J.G. (2012). "Point" centromeres of *Saccharomyces* harbor single centromere-specific nucleosomes. Genetics *190*, 1575–1577.

Heun, P., Erhardt, S., Blower, M.D., Weiss, S., Skora, A.D., and Karpen, G.H. (2006). Mislocalization of the Drosophila centromere-specific histone CID promotes formation of functional ectopic kinetochores. Dev. Cell *10*, 303–315.

Higgins, A.W., Schueler, M.G., and Willard, H.F. (1999). Chromosome engineering: generation of mono- and dicentric isochromosomes in a somatic cell hybrid system. Chromosoma *108*, 256–265.

Higgins, A.W., Gustashaw, K.M., and Willard, H.F. (2005). Engineered human dicentric chromosomes show centromere plasticity. Chromosome Res. *13*, 745–762.

Hill, A., and Bloom, K. (1987). Genetic manipulation of centromere function. Mol. Cell. Biol. *7*, 2397–2405.

Hill, A., and Bloom, K. (1989). Acquisition and processing of a conditional dicentric chromosome in *Saccharomyces cerevisiae*. Mol. Cell. Biol. *9*, 1368–1370.

Hori, T., Shang, W.H., Takeuchi, K., and Fukagawa, T. (2013). The CCAN recruits CENP-A to the centromere and forms the structural core for kinetochore assembly. J. Cell Biol. *200*, 45–60.

Hori, T., Shang, W.H., Toyoda, A., Misu, S., Monma, N., Ikeo, K., Molina, O., Vargiu, G., Fujiyama, A., Kimura, H., *et al.* (2014). Histone H4 Lys 20 Monomethylation of the CENP-A Nucleosome is essential for kinetochore assembly. Dev. Cell *29*, 740–749.

Iida, Y., Kim, J.H., Kazuki, Y., Hoshiya, H., Takiguchi, M., Hayashi, M., Erliandri, I., Lee, H.S., Samoshkin, A., Masumoto, H., *et al.* (2010). Human artificial chromosome with a conditional centromere for gene delivery and gene expression. DNA Res. *17*, 293–301.

Ikeno, M., Grimes, B., Okazaki, T., Nakano, M., Saitoh, K., Hoshino, H., McGill, N.I., Cooke, H., and Masumoto, H. (1998). Construction of YAC-based mammalian artificial chromosomes. Nat. Biotechnol. *16*, 431–439.

Ishii, K., Ogiyama, Y., Chikashige, Y., Soejima, S., Masuda, F., Kakuma, T., Hiraoka, Y., and Takahashi, K. (2008). Heterochromatin integrity affects chromosome reorganization after centromere dysfunction. Science *321*, 1088–1091.

Jansen, L.E., Black, B.E., Foltz, D.R., and Cleveland, D.W. (2007). Propagation of centromeric chromatin requires exit from mitosis. J. Cell Biol. *176*, 795–805.

Jiang, J., Birchler, J.A., Parrott, W.A., and Kelly Dawe, R. (2003). A molecular view of plant centromeres. Trends Plant Sci. *8*, 570–575.

Kanellopoulou, C., Muljo, S.A., Kung, A.L., Ganesan, S., Drapkin, R., Jenuwein, T., Livingston, D.M., and Rajewsky, K. (2005). Dicer-deficient mouse embryonic stem cells are defective in differentiation and centromeric silencing. Genes Dev. *19*, 489–501.

Karpen, G.H., and Allshire, R.C. (1997). The case for epigenetic effects on centromere identity and function. Trends Genet. *13*, 489–496.

Kato, H., Goto, D.B., Martienssen, R.A., Urano, T., Furukawa, K., and Murakami, Y. (2005). RNA polymerase II is required for RNAi-dependent heterochromatin assembly. Science *309*, 467–469.

Kazuki, Y., and Oshimura, M. (2011). Human artificial chromosomes for gene delivery and the development of animal models. Mol. Ther. *19*, 1591–1601.

Ketel, C., Wang, H.S., McClellan, M., Bouchonville, K., Selmecki, A., Lahav, T., Gerami-Nejad, M., and Berman, J. (2009). Neocentromeres form efficiently at multiple possible loci in *Candida albicans*. PLoS Genet. *5*, e1000400.

Kim, I.S., Lee, M., Park, K.C., Jeon, Y., Park, J.H., Hwang, E.J., Jeon, T.I., Ko, S., Lee, H., Baek, S.H., *et al.* (2012). Roles of Mis18alpha in epigenetic regulation of centromeric chromatin and CENP-A loading. Mol. Cell *46*, 260–273.

Kim, J.H., Ebersole, T., Kouprina, N., Noskov, V.N., Ohzeki, J., Masumoto, H., Mravinac, B., Sullivan, B.A., Pavlicek, A., Dovat, S., *et al.* (2009). Human gamma-satellite DNA maintains open chromatin structure and protects a transgene from epigenetic silencing. Genome Res. *19*, 533–544.

Kim, J.H., Kononenko, A., Erliandri, I., Kim, T.A., Nakano, M., Iida, Y., Barrett, J.C., Oshimura, M., Masumoto, H., Earnshaw, W.C., *et al.* (2011). Human artificial chromosome (HAC) vector with a conditional centromere for correction of genetic deficiencies in human cells. Proc. Natl. Acad. Sci. U.S.A. *108*, 20048–20053.

Kim, S.M., Dubey, D.D., and Huberman, J.A. (2003). Early replicating heterochromatin. Genes Dev. *17*, 330–335.

Kingston, I.J., Yung, J.S., and Singleton, M.R. (2011). Biophysical characterization of the centromere-specific nucleosome from budding yeast. J. Biol. Chem. *286*, 4021–4026.

Klein, E., Rocchi, M., Ovens-Raeder, A., Kosyakova, N., Weise, A., Ziegler, M., Meins, M., Morlot, S., Fischer, W., Volleth, M., *et al.* (2012). Five novel locations of neocentromeres in human: 18q22.1,

Xq27.1 approximately 27.2, Acro p13, Acro p12, and heterochromatin of unknown origin. Cytogenet. Genome Res. *136*, 163–166.

Knegt, A.C., Li, S., Engelen, J.J., Bijlsma, E.K., and Warburton, P.E. (2003). Prenatal diagnosis of a karyotypically normal pregnancy in a mother with a supernumerary neocentric 13q21 *13q22 chromosome and balancing reciprocal deletion. Prenat. Diagn. *23*, 215–220.

Koo, D.H., Han, F., Birchler, J.A., and Jiang, J. (2011). Distinct DNA methylation patterns associated with active and inactive centromeres of the maize B chromosome. Genome Res. *21*, 908–914.

Koren, A., Tsai, H.J., Tirosh, I., Burrack, L.S., Barkai, N., and Berman, J. (2010). Epigenetically inherited centromere and neocentromere DNA replicates earliest in S-phase. PLoS Genet. *6*, e1001068.

Koshland, D., Rutledge, L., Fitzgerald-Hayes, M., and Hartwell, L.H. (1987). A genetic analysis of dicentric minichromosomes in *Saccharomyces cerevisiae*. Cell *48*, 801–812.

Kouprina, N., Ebersole, T., Koriabine, M., Pak, E., Rogozin, I.B., Katoh, M., Oshimura, M., Ogi, K., Peredelchuk, M., Solomon, G., *et al.* (2003). Cloning of human centromeres by transformation-associated recombination in yeast and generation of functional human artificial chromosomes. Nucleic Acids Res. *31*, 922–934.

Kouprina, N., Tomilin, A.N., Masumoto, H., Earnshaw, W.C., and Larionov, V. (2014). Human artificial chromosome-based gene delivery vectors for biomedicine and biotechnology. Expert Opin. Drug Deliv. *11*, 517–535.

Krassovsky, K., Henikoff, J.G., and Henikoff, S. (2012). Tripartite organization of centromeric chromatin in budding yeast. Proc. Natl. Acad. Sci. U.S.A. *109*, 243–248.

Lagana, A., Dorn, J.F., De Rop, V., Ladouceur, A.M., Maddox, A.S., and Maddox, P.S. (2010). A small GTPase molecular switch regulates epigenetic centromere maintenance by stabilizing newly incorporated CENP-A. Nat. Cell Biol. *12*, 1186–1193.

Lam, A.L., Boivin, C.D., Bonney, C.F., Rudd, M.K., and Sullivan, B.A. (2006). Human centromeric chromatin is a dynamic chromosomal domain that can spread over noncentromeric DNA. Proc. Natl. Acad. Sci. U.S.A. *103*, 4186–4191.

Larin, Z., and Mejia, J.E. (2002). Advances in human artificial chromosome technology. Trends Genet. *18*, 313–319.

Lee, N.C., Kononenko, A.V., Lee, H.S., Tolkunova, E.N., Liskovykh, M.A., Masumoto, H., Earnshaw, W.C., Tomilin, A.N., Larionov, V., and Kouprina, N. (2013). Protecting a transgene expression from the HAC-based vector by different chromatin insulators. Cell. Mol. Life Sci. *70*, 3723–3737.

Liehr, T., Kosyakova, N., Weise, A., Ziegler, M., and Raabe-Meyer, G. (2010). First case of a neocentromere formation in an otherwise normal chromosome 7. Cytogenet. Genome Res. *128*, 189–191.

Lo, A.W., Craig, J.M., Saffery, R., Kalitsis, P., Irvine, D.V., Earle, E., Magliano, D.J., and Choo, K.H. (2001a). A 330 kb CENP-A binding domain and altered replication timing at a human neocentromere. EMBO J. *20*, 2087–2096.

Lo, A.W., Magliano, D.J., Sibson, M.C., Kalitsis, P., Craig, J.M., and Choo, K.H. (2001b). A novel chromatin immunoprecipitation and array (CIA) analysis identifies a 460-kb CENP-A-binding neocentromere DNA. Genome Res. *11*, 448–457.

Lu, J., and Gilbert, D.M. (2007). Proliferation-dependent and cell cycle regulated transcription of mouse pericentric heterochromatin. J. Cell Biol. *179*, 411–421.

Maddox, P.S., Hyndman, F., Monen, J., Oegema, K., and Desai, A. (2007). Functional genomics identifies a Myb domain-containing protein family required for assembly of CENP-A chromatin. J. Cell Biol. *176*, 757–763.

Maggert, K.A., and Karpen, G.H. (2000). Acquisition and metastability of centromere identity and function: sequence analysis of a human neocentromere. Genome Res. *10*, 725–728.

Maggert, K.A., and Karpen, G.H. (2001). The activation of a neocentromere in Drosophila requires proximity to an endogenous centromere. Genetics *158*, 1615–1628.

Maloney, K.A., Sullivan, L.L., Matheny, J.E., Strome, E.D., Merrett, S.L., Ferris, A., and Sullivan, B.A. (2012). Functional epialleles at an endogenous human centromere. Proc. Natl. Acad. Sci. U.S.A. *109*, 13704–13709.

Marshall, O.J., Chueh, A.C., Wong, L.H., and Choo, K.H. (2008a). Neocentromeres: new insights into centromere structure, disease development, and karyotype evolution. Am. J. Hum. Genet. *82*, 261–282.

Marshall, O.J., Marshall, A.T., and Choo, K.H. (2008b). Three-dimensional localization of CENP-A suggests a complex higher order structure of centromeric chromatin. J. Cell Biol. *183*, 1193–1202.

McAinsh, A.D., and Meraldi, P. (2011). The CCAN complex: linking centromere specification to control of kinetochore-microtubule dynamics. Semin. Cell Dev. Biol. *22*, 946–952.

McClintock, B. (1939). The behaviour of successive nuclear divisions of a chromosome broken at meiosis. Proc. Natl. Acad. Sci. U.S.A. *25*, 405–416.

McClintock, B. (1941). The stability of broken ends of chromosomes in *Zea mays*. Genetics *26*, 234–282.

Mejia, J.E., Willmott, A., Levy, E., Earnshaw, W.C., and Larin, Z. (2001). Functional complementation of a genetic deficiency with human artificial chromosomes. Am. J. Hum. Genet. *69*, 315–326.

Mejia, J.E., Alazami, A., Willmott, A., Marschall, P., Levy, E., Earnshaw, W.C., and Larin, Z. (2002). Efficiency of de novo centromere formation in human artificial chromosomes. Genomics *79*, 297–304.

Mendiburo, M.J., Padeken, J., Fulop, S., Schepers, A., and Heun, P. (2011). Drosophila CENH3 is sufficient for centromere formation. Science *334*, 686–690.

Moralli, D., Jefferson, A., Volpi, E.V., and Monaco, Z.L. (2013). Comparative study of artificial chromosome centromeres in human and murine cells. Eur. J. Hum. Genet. *21*, 948–956.

Morgan, T.H., Sturtevant, A.H., Muller, H.J., and Bridges, C.B. (1915). The Mechanisms of Mendelian Heredity (Henry Holt, New York, NY, USA).

Muller, S., Montes de Oca, R., Lacoste, N., Dingli, F., Loew, D., and Almouzni, G. (2014). Phosphorylation and DNA binding of HJURP determine its centromeric recruitment and function in CenH3(CENP-A) loading. Cell Rep. *8*, 190–203.

Murphy, T.D., and Karpen, G.H. (1998). Centromeres take flight: alpha satellite and the quest for the human centromere. Cell *93*, 317–320.

Murray, A.W., and Szostak, J.W. (1983). Construction of artificial chromosomes in yeast. Nature *305*, 189–193.

Nakano, M., Okamoto, Y., Ohzeki, J., and Masumoto, H. (2003). Epigenetic assembly of centromeric chromatin at ectopic alpha-satellite sites on human chromosomes. J. Cell Sci. *116*, 4021–4034.

Nakano, M., Cardinale, S., Noskov, V.N., Gassmann, R., Vagnarelli, P., Kandels-Lewis, S., Larionov, V., Earnshaw, W.C., and Masumoto, H. (2008). Inactivation of a human kinetochore by specific targeting of chromatin modifiers. Dev. Cell *14*, 507–522.

Nakashima, H., Nakano, M., Ohnishi, R., Hiraoka, Y., Kaneda, Y., Sugino, A., and Masumoto, H. (2005). Assembly of additional heterochromatin distinct from centromere-kinetochore chromatin is required for de novo formation of human artificial chromosome. J. Cell Sci. *118*, 5885–5898.

Nakayashiki, H., Kadotani, N., and Mayama, S. (2006). Evolution and diversification of RNA silencing proteins in fungi. J. Mol. Evol. *63*, 127–135.

Nasuda, S., Hudakova, S., Schubert, I., Houben, A., and Endo, T.R. (2005). Stable barley chromosomes without centromeric repeats. Proc. Natl. Acad. Sci. U.S.A. *102*, 9842–9847.

Neumann, P., Navrátilová, A., Koblížková, A., Kejnovský, E., Hřibová, E., Hobza, R., Widmer, A., Doležel, J., and Macas, J. (2011). Plant centromeric retrotransposons: a structural and cytogenetic perspective. Mob. DNA 2, 4.

Nishino, T., Takeuchi, K., Gascoigne, K.E., Suzuki, A., Hori, T., Oyama, T., Morikawa, K., Cheeseman, I.M., and Fukagawa, T. (2012). CENP-T-W-S-X forms a unique centromeric chromatin structure with a histone-like fold. Cell *148*, 487–501.

Novitski, E. (1955). Genetic measures of centromere activity in Drosophila melanogaster. J. Cell. Physiol. *45*(Suppl. 2), 151–169.

O'Neill, R.J., O'Neill, M.J., and Graves, J.A. (1998). Undermethylation associated with retroelement activation and chromosome remodelling in an interspecific mammalian hybrid. Nature *393*, 68–72.

Ockey, C.H., Wennstrom, J., and De la Chapelle, A. (1966). Isochromosome-X in man. II. Hereditas *54*, 277–292.

Ogiyama, Y., Ohno, Y., Kubota, Y., and Ishii, K. (2013). Epigenetically induced paucity of histone H2A.Z stabilizes fission-yeast ectopic centromeres. Nat. Struct. Mol. Biol. *20*, 1397–1406.

Ohkuni, K., and Kitagawa, K. (2011). Endogenous transcription at the centromere facilitates centromere activity in budding yeast. Curr. Biol. *21*, 1695–1703.

Ohzeki, J., Bergmann, J.H., Kouprina, N., Noskov, V.N., Nakano, M., Kimura, H., Earnshaw, W.C., Larionov, V., and Masumoto, H. (2012). Breaking the HAC barrier: histone H3K9 acetyl/methyl balance regulates CENP-A assembly. EMBO J. *31*, 2391–2402.

Okada, M., Cheeseman, I.M., Hori, T., Okawa, K., McLeod, I.X., Yates, J.R. 3rd, Desai, A., and Fukagawa, T. (2006). The CENP-H-I complex is required for the efficient incorporation of newly synthesized CENP-A into centromeres. Nat. Cell Biol. *8*, 446–457.

Okada, M., Okawa, K., Isobe, T., and Fukagawa, T. (2009). CENP-H-containing complex facilitates centromere deposition of CENP-A in cooperation with FACT and CHD1. Mol. Biol. Cell *20*, 3986–3995.

Olszak, A.M., van Essen, D., Pereira, A.J., Diehl, S., Manke, T., Maiato, H., Saccani, S., and Heun, P. (2011). Heterochromatin boundaries are hotspots for de novo kinetochore formation. Nat. Cell Biol. *13*, 799–808.

Page, S.L., and Shaffer, L.G. (1998). Chromosome stability is maintained by short intercentromeric distance in functionally dicentric human Robertsonian translocations. Chromosome Res. *6*, 115–122.

Pal-Bhadra, M., Leibovitch, B.A., Gandhi, S.G., Chikka, M.R., Bhadra, U., Birchler, J.A., and Elgin, S.C. (2004). Heterochromatic silencing and HP1 localization in Drosophila are dependent on the RNAi machinery. Science *303*, 669–672.

Palmer, D.K., O'Day, K., Wener, M.H., Andrews, B.S., and Margolis, R.L. (1987). A 17-kD centromere protein (CENP-A) copurifies with nucleosome core particles and with histones. J. Cell Biol. *104*, 805–815.

Palmer, D.K., O'Day, K., Trong, H.L., Charbonneau, H., and Margolis, R.L. (1991). Purification of the centromere-specific protein CENP-A and demonstration that it is a distinctive histone. Proc. Natl. Acad. Sci. U.S.A. *88*, 3734–3738.

Panchenko, T., Sorensen, T.C., Woodcock, C.L., Kan, Z.Y., Wood, S., Resch, M.G., Luger, K., Englander, S.W., Hansen, J.C., and Black, B.E. (2011). Replacement of histone H3 with CENP-A directs global nucleosome array condensation and loosening of nucleosome superhelical termini. Proc. Natl. Acad. Sci. U.S.A. *108*, 16588–16593.

Partridge, J.F., Borgstrom, B., and Allshire, R.C. (2000). Distinct protein interaction domains and protein spreading in a complex centromere. Genes Dev. *14*, 783–791.

Pearson, C.G., Yeh, E., Gardner, M., Odde, D., Salmon, E.D., and Bloom, K. (2004). Stable kinetochore-microtubule attachment constrains centromere positioning in metaphase. Curr. Biol. *14*, 1962–1967.

Pennaneach, V., and Kolodner, R.D. (2009). Stabilization of dicentric translocations through secondary rearrangements mediated by multiple mechanisms in *S. cerevisiae*. PLoS ONE *4*, e6389.

Perpelescu, M., Nozaki, N., Obuse, C., Yang, H., and Yoda, K. (2009). Active establishment of centromeric

CENP-A chromatin by RSF complex. J. Cell Biol. *185*, 397–407.

Pidoux, A.L., Choi, E.S., Abbott, J.K., Liu, X., Kagansky, A., Castillo, A.G., Hamilton, G.L., Richardson, W., Rappsilber, J., He, X., *et al.* (2009). Fission yeast Scm3: A CENP-A receptor required for integrity of subkinetochore chromatin. Mol. Cell 33, 299–311.

Polizzi, C., and Clarke, L. (1991). The chromatin structure of centromeres from fission yeast: differentiation of the central core that correlates with function. J. Cell Biol. *112*, 191–201.

Probst, A.V., Dunleavy, E., and Almouzni, G. (2009). Epigenetic inheritance during the cell cycle. Nat. Rev. Mol. Cell. Biol. *10*, 192–206.

Raghuraman, M.K., Winzeler, E.A., Collingwood, D., Hunt, S., Wodicka, L., Conway, A., Lockhart, D.J., Davis, R.W., Brewer, B.J., and Fangman, W.L. (2001). Replication dynamics of the yeast genome. Science *294*, 115–121.

Rhoades, M.M. (1942). Preferential segregation in maize. Genetics 27, 395–407.

Ribeiro, S.A., Vagnarelli, P., Dong, Y., Hori, T., McEwen, B.F., Fukagawa, T., Flors, C., and Earnshaw, W.C. (2010). A super-resolution map of the vertebrate kinetochore. Proc. Natl. Acad. Sci. U.S.A. *107*, 10484–10489.

Rivera, H., Vassquez, A.I., Ayala-Madrigal, M.L., Ramirez-Duenas, M.L., and Davalos, I.P. (1996). Alphoidless centromere of a familial unstable inverted Y chromosome. Ann. Genet. 39, 236–239.

Saffery, R., Irvine, D.V., Griffiths, B., Kalitsis, P., Wordeman, L., and Choo, K.H. (2000). Human centromeres and neocentromeres show identical distribution patterns of >20 functionally important kinetochore-associated proteins. Hum. Mol. Genet. *9*, 175–185.

Saffery, R., Sumer, H., Hassan, S., Wong, L.H., Craig, J.M., Todokoro, K., Anderson, M., Stafford, A., and Choo, K.H.A. (2003). Transcription within a functional human centromere. Mol. Cell *12*, 509–516.

Sato, H., Masuda, F., Takayama, Y., Takahashi, K., and Saitoh, S. (2012). Epigenetic inactivation and subsequent heterochromatinization of a centromere stabilize dicentric chromosomes. Curr. Biol. *22*, 658–667.

Schroeder-Reiter, E., Sanei, M., Houben, A., and Wanner, G. (2012). Current SEM techniques for de- and re-construction of centromeres to determine 3D CENH3 distribution in barley mitotic chromosomes. J. Microsc. *246*, 96–106.

Schuh, M., Lehner, C.F., and Heidmann, S. (2007). Incorporation of Drosophila CID/CENP-A and CENP-C into centromeres during early embryonic anaphase. Curr. Biol. *17*, 237–243.

Scott, K.C., and Sullivan, B.A. (2014). Neocentromeres: a place for everything and everything in its place. Trends Genet. *30*, 66–74.

Scott, K.C., Merrett, S.L., and Willard, H.F. (2006). A heterochromatin barrier partitions the fission yeast centromere into discrete chromatin domains. Curr. Biol. *16*, 119–129.

Scott, K.C., White, C.V., and Willard, H.F. (2007). An RNA polymerase III-dependent heterochromatin barrier at fission yeast centromere 1. PLoS ONE 2, e1099.

Sekulic, N., Bassett, E.A., Rogers, D.J., and Black, B.E. (2010). The structure of (CENP-A-H4)(2) reveals physical features that mark centromeres. Nature *467*, 347–351.

Shang, W.H., Hori, T., Martins, N.M., Toyoda, A., Misu, S., Monma, N., Hiratani, I., Maeshima, K., Ikeo, K., Fujiyama, A., *et al.* (2013). Chromosome engineering allows the efficient isolation of vertebrate neocentromeres. Dev. Cell *24*, 635–648.

Shelby, R.D., Vafa, O., and Sullivan, K.F. (1997). Assembly of CENP-A into centromeric chromatin requires a cooperative array of nucleosomal DNA contact sites. J. Cell Biol. *136*, 501–513.

Shelby, R.D., Monier, K., and Sullivan, K.F. (2000). Chromatin assembly at kinetochores is uncoupled from DNA replication. J. Cell Biol. *151*, 1113–1118.

Shivaraju, M., Camahort, R., Mattingly, M., and Gerton, J.L. (2011). Scm3 is a centromeric nucleosome assembly factor. J. Biol. Chem. *286*, 12016–12023.

Silva, M.C., and Jansen, L.E. (2009). At the right place at the right time: novel CENP-A binding proteins shed light on centromere assembly. Chromosoma *118*, 567–574.

Silva, M.C., Bodor, D.L., Stellfox, M.E., Martins, N.M., Hochegger, H., Foltz, D.R., and Jansen, L.E. (2012). Cdk activity couples epigenetic centromere inheritance to cell cycle progression. Dev. Cell *22*, 52–63.

Smith, K.M., Phatale, P.A., Sullivan, C.M., Pomraning, K.R., and Freitag, M. (2011). Heterochromatin is required for normal distribution of *Neurospora crassa* CenH3. Mol. Cell. Biol. *31*, 2528–2542.

Stimpson, K.M., Song, I.Y., Jauch, A., Holtgreve-Grez, H., Hayden, K.E., Bridger, J.M., and Sullivan, B.A. (2010). Telomere disruption results in non-random formation of de novo dicentric chromosomes involving acrocentric human chromosomes. PLoS Genet. *6*, e1001061.

Stimpson, K.M., Matheny, J.E., and Sullivan, B.A. (2012). Dicentric chromosomes: unique models to study centromere function and inactivation. Chromosome Res. *20*, 595–605.

Stoler, S., Keith, K.C., Curnick, K.E., and Fitzgerald-Hayes, M. (1995). A mutation in CSE4, an essential gene encoding a novel chromatin-associated protein in yeast, causes chromosome nondisjunction and cell cycle arrest at mitosis. Genes Dev. *9*, 573–586.

Struhl, K., Stinchcomb, D.T., Scherer, S., and Davis, R.W. (1979). High-frequency transformation of yeast: autonomous replication of hybrid DNA molecules. Proc. Natl. Acad. Sci. U.S.A. *76*, 1035–1039

Sullivan, B., and Karpen, G. (2001). Centromere identity in Drosophila is not determined *in vivo* by replication timing. J. Cell Biol. *154*, 683–690.

Sullivan, B.A., and Schwartz, S. (1995). Identification of centromeric antigens in dicentric Robertsonian translocations: CENP-C and CENP-E are necessary components of functional centromeres. Hum. Mol. Genet. *4*, 2189–2197.

Sullivan, B.A., and Willard, H.F. (1998). Stable dicentric X-chromosomes with two functional centromeres. Nat. Genet. *20*, 227–228.

Sullivan, B.A., and Karpen, G.H. (2004). Centromeric chromatin exhibits a histone modification pattern that is distinct from both euchromatin and heterochromatin. Nat. Struct. Mol. Biol. *11*, 1076–1083.

Sullivan, B.A., Wolff, D.J., and Schwartz, S. (1994a). Analysis of centromeric activity in Robertsonian translocations: implications for a functional acrocentric hierarchy. Chromosoma *103*, 459–467.

Sullivan, K.F., Hechenberger, M., and Masri, K. (1994b). Human CENP-A contains a histone H3 related histone fold domain that is required for targeting to the centromere. J. Cell Biol. *127*, 581–592.

Sumer, H., Craig, J.M., Sibson, M., and Choo, K.H. (2003). A rapid method of genomic array analysis of scaffold/matrix attachment regions (S/MARs) identifies a 2.5-Mb region of enhanced scaffold/matrix attachment at a human neocentromere. Genome Res. *13*, 1737–1743.

Tachiwana, H., and Kurumizaka, H. (2011). Structure of the CENP-A nucleosome and its implications for centromeric chromatin architecture. Genes Genet. Syst. *86*, 357–364.

Tachiwana, H., Kagawa, W., Shiga, T., Osakabe, A., Miya, Y., Saito, K., Hayashi-Takanaka, Y., Oda, T., Sato, M., Park, S.Y., *et al.* (2011). Crystal structure of the human centromeric nucleosome containing CENP-A. Nature *476*, 232–235.

Takahashi, K., Chen, E.S., and Yanagida, M. (2000). Requirement of Mis6 centromere connector for localizing a CENP-A-like protein in fission yeast. Science *288*, 2215–2219.

Takeuchi, K., Nishino, T., Mayanagi, K., Horikoshi, N., Osakabe, A., Tachiwana, H., Hori, T., Kurumizaka, H., and Fukagawa, T. (2014). The centromeric nucleosome-like CENP-T-W-S-X complex induces positive supercoils into DNA. Nucleic Acids Res. *42*, 1644–1655.

Talbert, P.B., Masuelli, R., Tyagi, A.P., Comai, L., and Henikoff, S. (2002). Centromeric localization and adaptive evolution of an Arabidopsis histone H3 variant. Plant Cell *14*, 1053–1066.

Thakur, J., and Sanyal, K. (2013). Efficient neocentromere formation is suppressed by gene conversion to maintain centromere function at native physical chromosomal loci in *Candida albicans*. Genome Res. *23*, 638–652.

Topp, C.N., Zhong, C.X., and Dawe, R.K. (2004). Centromere-encoded RNAs are integral components of the maize kinetochore. Proc. Natl. Acad. Sci. U.S.A. *101*, 15986–15991.

Topp, C.N., Okagaki, R.J., Melo, J.R., Kynast, R.G., Phillips, R.L., and Dawe, R.K. (2009). Identification of a maize neocentromere in an oat-maize addition line. Cytogenet. Genome Res. *124*, 228–238.

Tyler-Smith, C., Gimelli, G., Giglio, S., Floridia, G., Pandya, A., Terzoli, G., Warburton, P.E., Earnshaw, W.C., and Zuffardi, O. (1999). Transmission of a fully functional human neocentromere through three generations. Am. J. Hum. Genet. *64*, 1440–1444.

Van Hooser, A.A., Ouspenski, I.I., Gregson, H.C., Starr, D.A., Yen, T.J., Goldberg, M.L., Yokomori, K., Earnshaw, W.C., Sullivan, K.F., and Brinkley, B.R. (2001). Specification of kinetochore-forming chromatin by the histone H3 variant CENP-A. J. Cell Sci. *114*, 3529–3542.

Ventura, M., Weigl, S., Carbone, L., Cardone, M.F., Misceo, D., Teti, M., D'Addabbo, P., Wandall, A., Bjorck, E., de Jong, P.J., *et al.* (2004). Recurrent sites for new centromere seeding. Genome Res. *14*, 1696–1703.

Verdel, A., Jia, S., Gerber, S., Sugiyama, T., Gygi, S., Grewal, S.I., and Moazed, D. (2004). RNAi-mediated targeting of heterochromatin by the RITS complex. Science *303*, 672–676.

Volpe, T., Schramke, V., Hamilton, G.L., White, S.A., Teng, G., Martienssen, R.A., and Allshire, R.C. (2003). RNA interference is required for normal centromere function in fission yeast. Chromosome Res. *11*, 137–146.

Volpe, T.A., Kidner, C., Hall, I.M., Teng, G., Grewal, S.I., and Martienssen, R.A. (2002). Regulation of heterochromatic silencing and histone H3 lysine-9 methylation by RNAi. Science *297*, 1833–1837.

Voullaire, L.E., Slater, H.R., Petrovic, V., and Choo, K.H. (1993). A functional marker centromere with no detectable alpha-satellite, satellite III, or CENP-B protein: activation of a latent centromere? Am. J. Hum. Genet. *52*, 1153–1163.

Wallrath, L.L., and Elgin, S.C. (1995). Position effect variegation in Drosophila is associated with an altered chromatin structure. Genes Dev. *9*, 1263–1277.

Wandall, A., Tranebjaerg, L., and Tommerup, N. (1998). A neocentromere on human chromosome 3 without detectable alpha-satellite DNA forms morphologically normal kinetochores. Chromosoma *107*, 359–365.

Warburton, P.E., Dolled, M., Mahmood, R., Alonso, A., Li, S., Naritomi, K., Tohma, T., Nagai, T., Hasegawa, T., Ohashi, H., *et al.* (2000). Molecular cytogenetic analysis of eight inversion duplications of human chromosome 13q that each contain a neocentromere. Am. J. Hum. Genet. *66*, 1794–1806.

Williams, B.C., Murphy, T.D., Goldberg, M.L., and Karpen, G.H. (1998). Neocentromere activity of structurally acentric mini-chromosomes in Drosophila. Nat. Genet. *18*, 30–37.

Wong, L.H., Brettingham-Moore, K.H., Chan, L., Quach, J.M., Anderson, M.A., Northrop, E.L., Hannan, R., Saffery, R., Shaw, M.L., Williams, E., *et al.* (2007). Centromere RNA is a key component for the assembly of nucleoproteins at the nucleolus and centromere. Genome Res. *17*, 1146–1160.

Wong, N.C., Wong, L.H., Quach, J.M., Canham, P., Craig, J.M., Song, J.Z., Clark, S.J., and Choo, K.H. (2006). Permissive transcriptional activity at the centromere through pockets of DNA hypomethylation. PLoS Genet. *2*, e17.

Yan, H., Jin, W., Nagaki, K., Tian, S., Ouyang, S., Buell, C.R., Talbert, P.B., Henikoff, S., and Jiang, J. (2005). Transcription and histone modifications in the recombination-free region spanning a rice centromere. Plant Cell *17*, 3227–3238.

Yoda, K., Ando, S., Morishita, S., Houmura, K., Hashimoto, K., Takeyasu, K., and Okazaki, T. (2000). Human centromere protein A (CENP-A) can replace histone H3 in nucleosome reconstitution *in vitro*. Proc. Natl. Acad. Sci. U.S.A. *97*, 7266–7271.

Zeitlin, S.G., Barber, C.M., Allis, C.D., and Sullivan, K.F. (2001). Differential regulation of CENP-A and histone H3 phosphorylation in G2/M. J. Cell Sci. *114*, 653–661.

Zhang, W., Friebe, B., Gill, B.S., and Jiang, J. (2010). Centromere inactivation and epigenetic modifications of a plant chromosome with three functional centromeres. Chromosoma *119*, 553–563.

Dosage Compensation in Frogs and Toads

8

John H. Malone

Abstract

Dosage compensation is a mechanism that facilitates changes in expression relative to DNA abundance. Intensive study of worms, flies, and human sex chromosomes during the last several decades has shown that intrinsic effects, non-coding RNA, macromolecular complexes, or a combination act to manipulate expression. The strength of studying these model organisms comes from detailed genetic knowledge and an array of molecular tools, however with the advent of whole genome sequencing, it becomes possible to understand dosage compensation in a wider evolutionary context. In this chapter, I provide an overview of dosage compensation and what is known about dosage compensation in frogs and toads. Frogs and toads represent the basal tetrapods and are useful models for understanding gene dosage during early development, a window that is difficult to study in vertebrates. While few studies exist, there are curious examples of dosage compensation in frogs. Applying genomic tools in frogs and toads will provide a better understanding for the role of compensation in solving problems of abnormal gene dosage, and create new opportunities for understanding the role of dosage compensation in evolution, genome function, and early development.

Introduction to gene dosage and dosage compensation

Gene dosage is a DNA abundance metric. In most cases, vertebrate organisms inherit a copy of genes from their mother and a copy of the same genes from their father. These gene copies are joined during fertilization, creating a gene dosage that theoretically must be balanced in every cell. Changes in gene dose by either the loss or gain of copies leave cells to face the challenge of gene dosage imbalance.

DNA dosage has been recognized as important for genome function since the early days of genetics and cell biology (Boveri, 1902; Goldschmidt, 1920; Bridges, 1925; Blakeslee, 1934), but only recently has it been possible to accurately assess gene dosage genome-wide. Past molecular work has focused on understanding changes in the quality of genetic information (SNPs, frame-shifts, nonsense, and other types of information changing mutations). Polymerase chain reaction and DNA sequencing made it easier to identify qualitative changes in DNA sequence but difficult to detect quantitative changes in DNA abundance. The invention of comparative genomic hybridization microarrays and, more recently, high-throughput sequencing have made it possible to assess deletions and duplications genome-wide and these developments have shown that variation in gene dosage is more extensive than variation at the nucleotide polymorphism level (Zhang *et al.*, 2009; Stankiewicz and Lupski, 2010; Carvalho *et al.*, 2011).

How each copy of a gene is expressed is controlled by complex genetic and epigenetic mechanisms, but the RNA and protein level output should be at a state that satisfies phenotypic function. Abnormally high or low levels can change phenotype. Having three copies of chromosome 21 associates with Down syndrome, a single copy of the X-chromosome in females is associated with Turner syndrome, and copy number variants

(CNVs) on a variety of autosomes are associated with behavioural disease states, including autism, schizophrenia and Alzheimer's disease. In some cases, when a region of the genome is reduced or amplified in copy number this results in mirror phenotypes (i.e. obesity vs. underweightness), highlighting how slight changes in dosage can control the range of phenotypic extremes for a trait (Grafodatskaya *et al.*, 2010; Jacquemont *et al.*, 2011; Golzio *et al.*, 2012; Valsesia *et al.*, 2013; Chung *et al.*, 2014; Pinto *et al.*, 2014).

While these examples illustrate that altering gene dosage can be deleterious, there are examples of adaptive changes that associate with variation in gene dosage (Schrider and Hahn, 2010; Katju and Bergthorsson, 2013). Copy number variation in the salivary amylase gene (*AMY1*) of humans correlates with protein level and varies with diet. Individuals from populations with high-starch diets have more *AMY1* copies than from populations with low-starch diets (Perry *et al.*, 2007; Schrider and Hahn, 2010; Katju and Bergthorsson, 2013; Schrider *et al.*, 2013; Mokhtar *et al.*, 2014; Skinner *et al.*, 2014). Protein levels contribute to more efficient digestion of high-starch foods implicating selection on gene dosage. Low copy number of *AMY1* is correlated with increased obesity risk, suggesting more complicated effects of copy number variation and trade-offs to human health (Falchi *et al.*, 2014). Copy number variation for *AMY2B*, an amylase gene in dogs, is correlated with amylase activity, varies extensively among breeds, and together suggests that digestibility of starch is associated with variation in gene dosage (Arendt *et al.*, 2014). Copy number variation is fixed between species of birds, correlated with immune function and loss of sexual dimorphism in a quail species, and copy number variation varies extensively between populations of humans and *Drosophila* (Emerson *et al.*, 2008; Conrad *et al.*, 2010; Schrider *et al.*, 2013; Mokhtar *et al.*, 2014).

Variation in gene dosage creates the possibility that levels of RNA and protein could be altered and changes in levels can explain changes in phenotypes. If we assume some relationship between the number of copies of a gene (gene dosage) and the level of RNA and proteins, then changes in gene dosage may change levels of RNA and protein. Lower abundance of structural proteins necessary for cell function could be deleterious. Lower amounts of transcription factors and other regulatory proteins could change the amounts of regulated gene products. Insufficient transcription factor concentration could lower transcription rate, leading to lower abundances of RNA and protein products. Increased abundance of *cis* regulatory elements could disrupt transcriptional output, leading to additional cascades of gene networks and creating widespread genomic imbalance (Birchler and Veitia, 2012; Veitia *et al.*, 2013).

While it is intuitively appealing that a change in gene dosage should change RNA and protein levels, and there are clear examples of this relationship [e.g. salivary amylase gene (Perry *et al.*, 2007)], complex non-linear biochemical mechanisms predict the opposite. Transcription involves protein/DNA interactions that can be modelled as enzyme/substrate reactions similar to Michaelis–Menten dynamics. These dynamics of enzyme and substrate do not produce linear curves. Feedforward and feedback control, network dosage effects, (Chen *et al.*, 2014b; Song *et al.*, 2014), and non-linear interactions between gene dosage and gene expression produce non-straightforward predictions of the relationship between gene dosage and gene expression (Veitia *et al.*, 2013). These biochemical and network level interactions create the possibility of adjustments to gene expression, and are examples of dosage compensation.

What is dosage compensation?

Dosage compensation is a change in expression or functional output of genes that is not correlated with DNA abundance. Dosage compensation has the potential to rescue organisms from the deleterious effects of gene dose variation by restoring expression to sufficient or even normal phenotypic function. Dosage compensation could be viewed as an adaptive response; however, the degree of compensation is relevant: if expression is adjusted more than normal function, an overdose problem could lead to deleterious phenotypes. Gene network interactions, dose response networks, manipulation of chromatin, post-transcriptional control by non-coding RNAs, and changes in three-dimensional structure of the genome are all epigenetic mechanisms of genome function. It is possible that all of these mechanisms can play a role in dosage compensation.

Dosage compensation is usually regarded as a mechanism specific for sex chromosomes: mammalian X-inactivation is one of the best-known dosage compensation mechanisms. However, changes in gene dosage can occur anywhere in the genome and the response is often compensated (Johansson *et al.*, 2007; Stenberg *et al.*, 2009; Zhang *et al.*, 2010; Stenberg and Larsson, 2011; Malone *et al.*, 2012). In dosage manipulations in maize and *Drosophila*, mutants often have similar levels of expression as wild-type (Birchler, 1979, 1981; Birchler and Newton, 1981; Birchler *et al.*, 1990). In engineered segmental aneuploidies of *Drosophila*, the expression of genes in the aneuploid state does not have a linear relationship with gene dosage (Gupta *et al.*, 2006; Stenberg *et al.*, 2009; McAnally and Yampolsky, 2010; Stenberg and Larsson, 2011). In an aneuploid *Drosophila* cell line, buffering of gene expression occurred for abnormal gene dosage in a background with significant genome-wide aneuploidy (Zhang *et al.*, 2010). Genes on trisomy 21 do not have as much expression as would be expected (Aït Yahya-Graison *et al.*, 2007; Prandini *et al.*, 2007; Jiang *et al.*, 2013). In a series of autosomal hemizygous fruit flies, mediation of the dosage compensation response was due to network interactions that facilitated gene-specific responses (Malone *et al.*, 2012). These examples highlight that dosage compensation is taxonomically widespread, but also widespread across scales ranging from single genes to entire chromosomes.

Genetic sex determination, sex chromosomes and dosage compensation

A problem of gene dosage that can be viewed as a consequence of genetic sex determination is the evolution of heteromorphic sex chromosomes. Early cytogenetic analysis of chromosomes by Hermann Henking in true bugs (*Pyrrhocoris*) revealed a peculiar chromosome with potential gene dosage problems and named the strange chromosome 'X' to reflect its unknown nature (Henking, 1891). Careful cytogenetic work by Nellie Stevens revealed that sex chromosomes, which contain information relevant for control of female or male phenotypes, are inherited uniparentally (Stevens, 1905). Two common modes of sex determination involve male heterogamety (XY) and female heterogamety (ZW). In XY sex determination, some fish, frogs, reptiles, all mammals and *Drosophila* have either a dominant allele on the Y chromosome, or a counting mechanism for the X-chromosome to trigger male development. In ZW sex determination, other fish, frogs, reptiles, all birds and lepidopterans, a dominant allele on the W chromosome, or dosage dependent system on the Z controls femaleness (Smith *et al.*, 2009; Yoshimoto and Ito, 2011; Sahara *et al.*, 2012; Graves, 2014; Kiuchi *et al.*, 2014).

Population genetic theory predicts that, as a consequence of uniparental inheritance of the sex chromosomes and the loss of recombination, deleterious mutations accumulate and drive the gradual degradation of sex chromosomes over time (Charlesworth, 1991, 1996; Vicoso and Charlesworth, 2006; Bachtrog, 2013). Thus, as sex chromosomes mature and genes are lost from one chromosome (Y or W) a 2-fold reduction in gene dosage occurs for thousands of genes in the genome (Bachtrog, 2006; Vicoso and Bachtrog, 2009). Because loss of gene copies changes gene dosage and decreased dosage can be lethal, the loss of gene dosage creates deleterious phenotypes (Davoli and de Lange, 2011; Davoli *et al.*, 2013; Chen *et al.*, 2014b). As sex chromosomes evolve, initially one sex will suffer the consequence of reduced gene dosage. Theoretically, sex chromosome evolution, a consequence of genetic sex determination, must be balanced by mechanisms that prevent loss of gene expression for genes that become one copy in one sex relative to two copies in the opposite sex (Straub and Becker, 2007; Veitia *et al.*, 2008; Birchler and Veitia, 2014; Conant *et al.*, 2014). Dosage compensation is a mechanism favoured by natural selection to maintain fitness due to loss of gene dosage from sex chromosome evolution (Muller, 1932).

How are sex chromosomes dosage compensated?

In vertebrates, a non-coding RNA, *Xist*, and its marsupial homologue, *Rsx*, are the most well-characterized dosage compensation system (Namekawa *et al.*, 2007; Yen *et al.*, 2007; Deakin *et al.*, 2009; Grant *et al.*, 2012; Engreitz *et al.*, 2013). In mammals, expression of the non-coding RNA is restricted to females, and in each cell *Xist* randomly selects a copy of the X-chromosome for silencing. *Xist* binds to the X-chromosome, recruiting heterochromatin and factors associated with gene silencing. This process causes a randomly selected single copy of the entire X-chromosome to become silenced in every cell of female eutherian mammals (Yen *et al.*, 2007; Sin and Namekawa, 2013; Deng *et al.*, 2014). Because a copy of the X-chromosome is silenced, inactivating a chromosome in females creates a dosage problem similar to males. Ohno (1967) proposed that as the Y chromosome loses genes, selection acts to increase expression on the single copy X by fixing regulatory changes that increase the expression level of the X. This creates an overdose problem because females have two copies of the X-chromosome, and combined with increased expression, creates a problem. The solution is that females inactivate an X-chromosome to avoid overdosing on gene expression (Forsdyke, 2012; but see Pessia *et al.*, 2014).

In *Drosophila*, a macromolecular complex composed of *msl* genes, *mof*, *maless*, *JIL-1*, and two non-coding RNAs (*roX1*, *roX2*) is expressed and assembled exclusively in males and binds to the X-chromosome (Sass *et al.*, 2003; Taipale and Akhtar, 2005; Rodriguez *et al.*, 2007; Straub *et al.*, 2008; Zhang *et al.*, 2010). The dosage compensation complex is hypothesized to change the chromatin configuration for genes on the X-chromosome in males, contributing to a 2-fold increase in expression. However, the dosage compensation complex is not entirely responsible for this effect. Instead, an intrinsic compensation system related to network dynamics and genome balance better explains the majority of the increased expression (Birchler *et al.*, 2011; Philip and Stenberg, 2013; Sun *et al.*, 2013a,b,c; Birchler and Hall, 2014). Transposable element evolution on the X-chromosome in flies contributes to the recruitment of the dosage compensation complex, suggesting the need for regulatory transformation to achieve action of the dosage compensation complex and intrinsic compensation system (Chuong and Feschotte, 2013; Ellison and Bachtrog, 2013). Changes in RNA turnover contribute to the compensation response for genes on the X-chromosome, and chromatin state of the X-chromosome differs between female and male *Drosophila* (Zhang and Oliver, 2010; Larschan *et al.*, 2011). The mechanisms of compensation in *Drosophila* remain controversial, but in flies there is manipulation to increase expression for genes on the X-chromosome in males, whereas in mammals, there is inactivation of an X-chromosome in females. In these two types of XY organisms, there are opposite mechanisms that have evolved to adjust gene dosage to balance expression as a result of sex chromosome evolution (Birchler, 2009).

How to detect the presence or absence of dosage compensation?

Given the theoretical and empirical support for dosage compensation, it is useful to consider how we can detect and quantify compensation, a critical step preceding the search for mechanisms of dosage compensation. Distributional analysis of gene expression throughout the genome is one way to detect dosage compensation. RNA from cells or tissues are extracted, then transcriptome techniques such as microarrays or deep sequencing are used to estimate levels of gene expression for each gene in the genome (Malone and Oliver, 2011; Malcom and Malone, 2014). Because of difficulty in accurately assessing isoform expression level, most studies have used gene level estimates of expression, where expression of individual or subsets of exons from gene models are pooled to obtain an estimate of expression. New studies that characterize alternative splicing and abundance of transcript isoforms will provide directions for understanding compensation at the splicing and isoform level (Sturgill *et al.*, 2013; Smith *et al.*, 2014).

Because a change in gene dosage at the chromosome level can involve thousands of genes, measuring the ratio of expression between wild-type and the dose variant state provides a readout for the presence or absence of dosage compensation. When dose changes are involved between the sexes, this analysis is not straightforward because of the presence of genes with sex-biased expression. Genes on a one-copy sex chromosome that are expressed more than the same genes when in two-copy reflect a different form of regulation that involves the function of sex than genes that are expressed the same between sexes.

Expression levels calculated from fluorescent intensity on microarrays or sequencing output and then compared between samples that vary in dosage provides a way to detect the presence or absence of dosage compensation. As detailed above, sex chromosomes can vary in dosage between the sexes. In *Drosophila*, females have two copies of the X-chromosome and males have one copy, creating a dosage problem in males. If there was no dosage compensation, then comparing the level of expression for each gene on the X-chromosome would show two times more expression in females compared to males. In principle, dosage compensation is best assayed as a chromosome-wide phenomenon because genes that differ in expression owing to sex bias can mislead the inference of dosage compensation. If only a few genes were studied, then one or two genes that had expression with sex-bias (higher in one sex relative to the other regardless of dosage) would tend to miss the chromosome-wide phenomena and create false inferences for the presence or absence of dosage compensation. There is good evidence for dosage compensation of individual loci in segmental aneuploids, but the response is varied and analysing a few genes does not provide as much power to control against false inferences (Stenberg *et al.*, 2009; McAnally and Yampolsky, 2010; Malone *et al.*, 2012; Birchler and Hall, 2014).

For genes on the sex chromosome, comparing between the sexes is a direct way to test for dosage compensation, but more relevant to the cell is comparing expression for genes on the sex chromosome to the expression level of the autosomes. As changes in dosage occur, this affects gene products that interact with autosomes. Dosage changes within the cell create perturbations by altering stoichiometry of macromolecular complexes that regulate the function of other genes. If not corrected, misassembled complexes that fail to regulate other genes upset genome balance by creating cascades of perturbations in gene networks (Birchler and Veitia, 2012; Malone *et al.*, 2012; Veitia *et al.*, 2013; Kemmeren *et al.*, 2014). Insufficient gene product within the cell becomes biologically relevant, whereas comparisons between sexes are not meaningful to the biology of the cell.

Comparing expression values for individuals that differ in copy number from samples that do not have these genotypes allows calculation of a ratio of gene expression. If the ratio = 1, then dosage compensation occurs, because the level of expression is the same even though there is a change in gene dosage. If the ratio is not equal to 1, then dosage compensation does not occur on average and this would suggest tolerance for changes in gene dosage. One danger with comparing the expression level for genes on the sex chromosome between the sexes are genes with sex-biased expression. These types of genes change the inference of dosage compensation because a ratio greater than or equal to 1 between the sexes could be mostly due to genes with sex-biased expression. For this type of pattern, it is reasonable to hypothesize that as sex chromosomes evolve there would have to be reorganization of the regulation of genes on the sex chromosome to deal with the gene dosage problem.

Some caveats are necessary when quantifying gene expression genome-wide. Knowing the detectable level of expression becomes critical for properly analysing distributions, because including genes that have no expression, or expression levels at the lower threshold of detection (i.e. effectively including many zeros in the analysis), will lower the mean of the distribution and different analytical approaches create problems for inferring dosage compensation (Xiong *et al.*, 2010; Deng *et al.*, 2011; Jiang *et al.*, 2011; Birchler, 2012; Jue *et al.*, 2013).

Techniques for measuring gene expression for a limited set of genes have also been used to assess dosage compensation. At the gene level, RNA fluorescent *in situ* hybridization probes have been hybridized and fluorescence levels detected as a way to compare expression for a few genes between genotypes. Quantitative real-time PCR is useful for measuring gene expression for a limited set of genes. New technologies like Nanostrings (Golubeva *et al.*, 2013) and digital droplet PCR (Mazaika and Homsy, 2014) provide techniques for comparing expression levels for genes that vary in gene dosage. These techniques are effective but limited in scope since only a few genes can be compared, rather than measuring expression throughout the genome.

Most studies of dosage compensation have focused on quantifying levels of RNA. There is a need to examine dosage compensation at the protein level. Quantitative proteomic techniques and methods that provide a protein abundance index [i.e. SILAC, Riboprofiling (Ong *et al.*, 2002; Ingolia *et al.*, 2009; Sury *et al.*, 2010; Artieri and Fraser, 2014; Dephoure *et al.*, 2014; McManus *et al.*, 2014)] allow better detection of small fold changes in expression and more accurate measurements of protein levels, compared to label free methods. Comparing gene dosage to RNA and protein levels will determine if compensation can occur post-transcription and post-translationally, creating a more complete view of the impact of changes in gene dosage on cellular function.

Variation in the extent of dosage compensation

Heteromorphic ZW sex chromosomes are present in some fish, reptiles, and most birds, yet there is a different pattern for dosage compensation in these organisms. In general, there is incomplete dosage compensation for genes on the Z chromosome in fish, birds, and snakes (Itoh *et al.*, 2007, 2010; Mank and Ellegren, 2009; Wolf and Bryk, 2011; Naurin *et al.*, 2012; Vicoso *et al.*, 2013; Chen *et al.*, 2014a). Additionally, liver flukes and some Lepidopterans, both with ZW sex determination and heteromorphic sex chromosomes, lack complete dosage compensation across the Z chromosome (Vicoso and Bachtrog, 2011; Harrison *et al.*, 2012). Measurements from microarrays and high-throughput sequencing show higher expression on average in males compared to females.

There are some exceptions to the lack of dosage compensation in organisms with ZW sex determination. In the silkmoth, *Bombyx mori*, a Lepidopteran with heteromorphic ZW sex chromosomes, initial analyses suggested that like other ZW species, there was no complete dosage compensation (Zha *et al.*, 2009) but re-analyses on the same data suggest the opposite, though the pattern of expression for Z-linked genes was lower than expression of autosomal genes (Walters and Hardcastle, 2011). In the gypsy moth, another heteromorphic system, RNA-Seq analyses showed complete dosage compensation (Smith *et al.*, 2014). In chicken, RNA-fluorescence *in situ* hybridization in fibroblast cells showed that 11 genes on the chicken Z were inactivated (Livernois *et al.*, 2013) and suggested that inactivation is a mechanism of dosage compensation in birds.

While it is tantalizing to suggest that something about ZW sex determination provides a tolerance to a change in gene dosage compared to XY sex determination, the examples above illustrate that the relationship is not so simple and there is a need to understand more about the origin and evolutionary history of dosage compensation in more lineages and more sex determination systems.

Outstanding challenges

The overview of dosage compensation in the preceding sections highlights that the mechanisms effecting the compensatory changes are mostly epigenetic controls. RNAs and macromolecular complexes change the chromatin context of dosage imbalanced chromosomes and chromosome segments to effectively re-balance the amount of gene product expressed. The phenomenon of dosage compensation is widespread among eukaryotes, and multiple mechanisms appear to have independently evolved in different invertebrate and vertebrate lineages.

Despite advances in identifying and understanding some of the mechanisms of dosage compensation, several important questions remain. When did dosage compensation arise in vertebrate lineages? How often has sex determination switched between male- and female-heterogametic, or how often are autosomes co-opted as new sex chromosomes? Given the presence of switching, is the evolution of ageing sex chromosome content systematic (e.g. parallel loss of genes) or haphazard, and what are the consequences for compensation? Are some compensation mechanisms more likely to arise before others as dosage imbalance increases with sex chromosome age? Are the genes on dosage imbalanced but compensated sex chromosomes particularly dose sensitive, while those on imbalanced but uncompensated sex chromosomes dose tolerant? Are there novel dosage compensation mechanisms that have yet to be described? Very little is known about dosage compensation through development; how dynamic is dosage compensation within an individual's lifetime? Addressing these questions will require moving beyond the well-studied classical systems.

Frogs and toads as models for dosage compensation

Plotting what is known about dosage compensation among vertebrate lineages we observe that coelacanths (*Latimeria*) and the basal tetrapods – amphibians – are completely untested (Fig. 8.1). The lack of dosage compensation data for *Latimeria*, frogs, toads, and salamanders suggests two hypotheses regarding the evolutionary origin (and/or loss) of compensation mechanisms:

- Hypothesis 1. Dosage compensation arose just once, after the divergence of mammals and reptiles.
- Hypothesis 2. Dosage compensation arose at the base of tetrapods and was lost in the reptile lineage.

These two hypotheses can be tested, but the biology of these groups presents certain challenges. First, the coelacanth genome is of moderate size, but the organism is very hard to study (Bruton and Stobbs, 1991), and whether they even have genetic sex determination is unknown (but see Forconi *et al.*, 2013). Caudates (salamanders) are easier to study, but the smallest genomes are ~15 Gb (Sun *et al.*, 2012) and repeat-heavy, making sequencing very expensive and complicated. Frogs and toads, in contrast, fill the missing gap: the genomes of several species are smaller than the human genome; frogs and toads are relatively simple to study in nature and in the laboratory; and although there have been no exhaustive assessments of dosage compensation

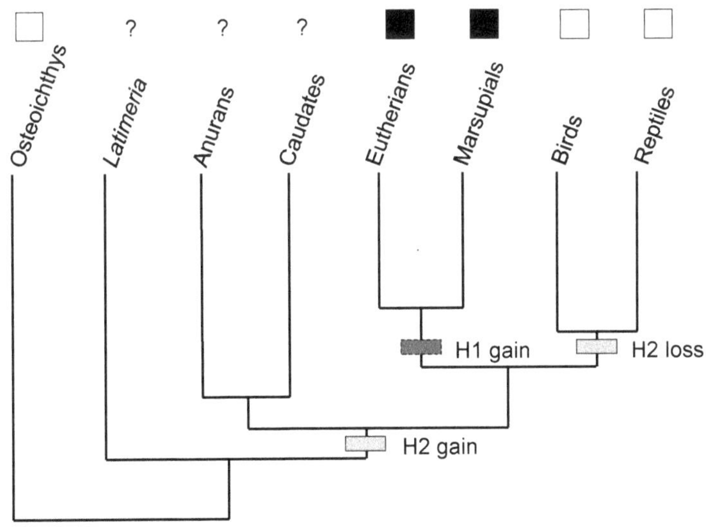

Figure 8.1 Phylogeny of vertebrates showing the distribution of global dosage compensation mechanisms. Black boxes indicate the presence, white boxes indicate the absence, and question marks indicate no knowledge of global dosage compensation. Evidence from coelacanth and amphibians would inform our understanding of when in vertebrate evolution dosage compensation evolved. Hypotheses for gain and loss of dosage compensation represented by H1 gain, H2 gain, and H2 loss, as discussed in the main text.

in frogs to date, there are tantalizing results indicating, at a minimum, autosomal dosage compensation.

Frog biology

Many attributes of frogs make them ideal organisms for studying the mechanisms and evolution of gene dosage. First and foremost, frogs and toads have a variety of genetic sex determination patterns. Sex determination system changes both within and between frog lineages (Hillis and Green, 1990; Miura *et al.*, 1998; Ezaz *et al.*, 2006; Quinn *et al.*, 2011; Sarre *et al.*, 2011; Livernois *et al.*, 2012), creating an interesting opportunity to study how sex determination system and sex chromosome evolution relates to dosage compensation. Frogs and toads also have a variety of sex chromosome maturation states. In some lineages, the sex chromosomes are radically divergent (heteromorphic) and in others, the sex chromosomes are similar (homomorphic), providing an opportunity to understand expression changes before and after aneuploidy and gene dosage problems evolve (Humphrey, 1942, 1945; Ponse, 1942; Hayes, 1998; Schmid and Steinlein, 2001; Schmid *et al.*, 2010, 2012). In addition to the typical pattern of diminutive Y (W) relative to X (Z), some frogs such as *Eleutherodactylus oxyrhynchus* and *Pristimantis euphronides* are characterized by gigantic uniparental (Y or W) sex chromosomes (Schmid *et al.*, 2010).

Frogs have evolved sex chromosomes independently many times. Miura (2007) analysed linkage data and suggested a total of 17 different male sex-determining loci among five different chromosomes in ranid frogs. An analysis of *Xenopus laevis* has shown that *DM-W*, the female-determining gene in *X. laevis*, does not exist in *X. tropicalis* and that the W chromosomes of *X. laevis* is not the same as the W chromosome in *Rana rugosa* (Ogata *et al.*, 2008; Uno *et al.*, 2008; Bewick *et al.*, 2011). These different evolutionary origins of sex chromosomes and sex determination genes create a natural experiment to examine the contribution of gene dosage variation to genome function (Malcom *et al.*, 2014).

The African clawed frog (*Xenopus*) has contributed significantly to our understanding of vertebrate development, endocrinology, cell cycle, stem cell biology, and cell reprogramming (Kashiwagi *et al.*, 2010). Development is easy to study in frogs because the eggs are large, fertilization is external, and genetic manipulations can be done in the embryo. Ploidy in frogs can also be manipulated in the laboratory using temperature or pressure treatment (Gates, 1923; Gurdon, 1959; Tompkins, 1978; Reinschmidt *et al.*, 1979; Kawamura, 1984; Tompkins and Reinschmidt, 1991; Kondo and Kashiwagi, 2004; Tucker and Gerhardt, 2012). Phenotypic sex can be switched experimentally, allowing one to separate the contribution of genes with sex-biased expression, chromosomal differences, or sex hierarchy, from the sex chromosome environment (Chang and Witschi, 1956; Mikamo and Witschi, 1963, 1964; Malone and Michalak, 2008). Many species can be easily raised in the laboratory allowing for crosses to be produced (Malone *et al.*, 2007; Malone and Fontenot, 2008). There is extreme variation in the level of B and supernumerary chromosomes and translocations of sex chromosomes among populations of the same species of frogs and toads (Green, 2004; Camacho *et al.*, 2011). Last, because of extensive homology between frog and human genomes (Hellsten *et al.*, 2010), studying gene dosage function in frogs can help inform understanding of genome function in mammals. These observations of the varied aspects of anuran biology suggest that frogs and toads can prove extremely useful for studies of gene dose.

Dosage compensation in sex chromosomes of frogs and toads

We know very little about dosage compensation of sex chromosomes in frogs: no studies have quantified the expression of genes on frog heteromorphic sex chromosomes genome-wide. The barrier to progress in frogs has been lack of genomic tools and appropriate study systems. The classic developmental biology model frog, *Xenopus*, has homomorphic sex chromosomes and the sex determination gene has only been identified recently (Yoshimoto *et al.*, 2008; Okada *et al.*, 2009; Yoshimoto and Ito, 2011) in *X. laevis*. The majority of frogs examined for karyotypes show that most frogs, like *Xenopus*, have homomorphic sex chromosomes in which gene dosage imbalance is thought to be less extensive. There are, however, frogs with heteromorphic sex chromosomes, and these sex chromosome systems appear to have evolved several times independently (Miura, 2007; Schmid *et al.*, 2012; Brelsford *et al.*, 2013). These species that vary in the morphology, and presumably gene content, of sex chromosomes provide a natural experiment to examine the presence or absence of dosage compensation, and doing so will enrich knowledge of the origin and diversification of dosage compensation. Furthermore, sequencing and annotating gene content and structural variation for closely related species with homomorphic sex chromosomes will allow comparisons of chromatin content and structural variation between the two gene dosage states, fulfilling an important model for epigenetic control and response to changes in gene dosage.

Despite the lack of genome-wide assays of gene dosage and compensation in frogs, work on gene expression from sex-linked genes that vary in dose in several species is informative. *Buergeria buergeri*, a Japanese rhacophorid frog with morphologically undifferentiated ZW sex chromosomes, illustrates a lack of similarity with mammalian dosage compensation. In *Buergeria*, the only major difference between the Z and W chromosome is the presence of a nucleolus organizer region (NOR) on the Z chromosome and absence of NOR on W. Therefore, males have two nucleoli per nucleus because of two Z chromosomes, whereas females have one (ZW) (Ohta, 1986, 1987; Schmid *et al.*, 1993). The NOR contains loci essential for function of the ribosome, which are necessary for translation. The amphibian NOR includes between 45–80 nucleolar units with copies of the 5.8S, 18S and 28S ribosomal genes clustered together on the chromosome (Schmid *et al.*, 2012). Variable gene dosage for this critical component of cell function has been linked to genetic diseases such as Down syndrome (Jackson-Cook *et al.*, 1985; Demirtas, 2009), and sickly phenotypes in *Drosophila* (Marygold *et al.*, 2007).

In addition to direct assays of gene expression there are other characteristics indicative of dosage compensation. Delayed replication timing of the inactive X-chromosome is a feature of dosage compensation in mammals; all autosome pairs have similar replication timing. If inactivation is a mechanism to balance expression in *Buergeria*, then a change in replication timing would be consistent with dosage compensation in mammals. Similarly, the presence of a Barr body structure in frogs would also be evidence of congruent mechanisms of dosage compensation between frogs and mammals.

Studies of replication timing using 5-bromodeoxyuridine (BrdU) labelling of chromosomes show no difference in replication rate in *Buergeria*, *Hyla*, and *Gastrotheca* (Schmid and Steinlein, 2003). Banding analyses, analyses of nucleoli in somatic interphase nuclei, and silver staining of transcriptionally active NORs suggest that no inactivation of Z chromosomes occur in the homogametic sex of *Buergeria* (Schmid *et al.*, 1993; Hayes, 1998). Sex reversal experiments, coupled with mating of sex-reversed frogs, showed that nucleoli-less individuals in WW genotypes die in early development whereas hetero- or homozygous NOR genotypes develop normally. Gene dosage for NOR is clearly variable, but it is not clear whether there is dosage compensation for expression levels of the NORs or tolerance for the 2-fold difference in dosage. Either scenario is interesting because of further unanswered questions that are raised. If there is dosage compensation, by what mechanism does compensation occur? In contrast, if dosage compensation is absent, then the cost of transcribing unnecessary rRNAs should be high in males; why has compensation not evolved?

An outstanding question about variable NORs is whether they could contribute as triggers of sex determination in a dose dependent manner. Given the importance of the ribosome to cell function, it seems unlikely however in experimental ploidy work from *Buergeria*, ZWW become female and ZZZ became male. In ZZW frogs however, about half became female and the other half became male. Simple two-copy NOR does not always lead to male development suggesting interesting interactions between compensation and W dominance. Expression profiling of NORs in these triploid frogs would be very useful (Ohta *et al.*, 1999).

Another morphologically undifferentiated sex chromosome system that has been studied is *Hyla femoralis*, which exhibits XY sex determination (Schmid and Steinlein, 2003). Like *Buergeria*, but phylogenetically unrelated, the NOR is sex-linked but, because of XY sex determination, the NOR is localized to the X-chromosome. In *H. femoralis*, females have two NORs and males have one, creating an asymmetry in gene dosage for NOR loci between the sexes. Silver staining of transcriptionally active NORs in *H. femoralis* demonstrate that females have double the dosage and theoretically double the expression level of NOR loci compared to males. This suggests a tolerance mechanism, but it is also possible that at the resolution of silver straining it is not possible to detect a compensation response in males, which have only one copy of NOR. Finally, no evidence of a Barr body has been observed, and no delayed replication patterns have been seen for X-chromosomes in *H. femoralis* providing an additional line of evidence that an inactivating mechanism similar to mammals does not operate in an XY frog with morphologically indistinguishable sex chromosomes (Schmid and Steinlein, 2003). As with *Buergeria*, the observations with *H. femoralis* beg the questions whether there could be dosage compensation, and if not, why not for something as energetically expensive as rRNA transcription.

No evidence of dosage compensation has been found in frogs with heteromorphic sex chromosomes. *Gastrotheca riobambae* has heteromorphic XY chromosomes and the NOR is X-linked (Schmid *et al.*, 1983). Silver staining of NORs shows two active NORs in females (XX) and one in males (X), suggesting no inactivation of an X-chromosome of females and consistent with the absence of dosage compensation in *Hyla* and *Buergeria* (Schmid *et al.*, 2012). Taken together, silver staining and BrdU studies do not provide evidence for inactivation of sex chromosomes as observed in mammals.

An interesting possibility for frog sex chromosome dosage compensation comes from the recent discovery of a novel class of microRNAs termed siteRNAs (Harding *et al.*, 2014). In the course of profiling small RNAs in *X. tropicalis* larvae, Harding and colleagues (2014) discovered a novel class of transposable element-derived small RNA that localizes to introns of protein-coding genes during early embryonic development. This localization causes a suppression of gene expression by deposition of the repressive histones H3K9me3 and H4K20me3 to inhibit transcription. If the activity of the siteRNAs varies between sexes in species with heteromorphic sex chromosomes, or if it can be directed to cope with genes that vary in dosage off the sex chromosomes, then the RNAs offer a novel mechanism for compensating gene dosage differences. The mere fact that exploring a 'non-traditional' model species turned up a new class of regulatory RNA suggests that further expanding the repertoire of species should bode well for additional discoveries.

Evidence of autosomal dosage compensation in frogs

One of the more intriguing observations regarding dosage compensation in frogs comes in the form of autosomal dosage compensation. Quantitative PCR assays of NOR gene expression have not been performed in sex-linked NORs, however polymorphic NOR loci have been examined when autosomal in *Xenopus laevis*. The Oxford mutant of *X. laevis* is polymorphic for the NOR and the NOR is located on the autosome in *Xenopus*. Hetero and homozygotes develop normally, but heterozygous mutants have one copy of the NOR whereas homozygotes have two. Crossing heterozygotes produces individuals without NORs and these anucleolate mutants die in the early swimming tadpole stage (Wallace, 1960; Rafferty and Sherwin, 1969).

Brown and Gurdon (1964) measured RNA synthesis by radioactive labelling anucleate, hetero, and homozygous NORs in *X. laevis*. Heterozygous mutant frogs had half the total nucleolar copies compared to wild type. The rates of ribosome synthesis in heterozygous mutants were comparable to diploids. Interesting, null mutants survive until swimming tadpole stages and show normal differentiation of all main cell types despite the inability to synthesize new ribosomal RNA. The 28S and 18S ribosomal RNA were expressed as in heterozygote mutants relative to wild type, suggesting the presence of autosomal dosage compensation at the level of synthesis (Brown and Gurdon, 1966; Wallace and Birnstiel, 1966). The heterozygote also had a nucleolus twice the surface area of wild-type nucleoli which might reflect this increased transcriptional capacity (Barr and Esper, 1963; Brown and Gurdon, 1964; Esper and Barr, 1964; Rafferty and Sherwin, 1969).

Compensation of the NOR in frogs, contradicts the lack of compensation found on sex chromosomes. While no evidence of inactivation of sex chromosomes can be observed from replication banding or karyotype studies, there are many species with sex-linked NORs. It has not been shown conclusively whether expression of sex-linked NORs are dosage compensated but given the compensation found for autosomal NORs in *Xenopus* it is interesting to speculate about links between the mechanisms of compensation found on autosomal NOR and compensation systems found on the sex chromosomes in frogs. Have redundancy mechanisms evolved in frogs to maintain expression of ribosome stoichiometry? If so, can these be the same mechanisms that allow heteromorphic sex chromosomes to evolve in frogs?

Amplification of ribosomal DNA occurs in *Xenopus* and many other organisms (Brown and Dawid, 1968). Amplification is also a response to loss of rDNA dosage. In XX versus XO *Drosophila*, compensation by differential rDNA replication occurs between XX and XO flies to produce the same rDNA content (Tartof, 1971; Hawley and Marcus, 1989; Lyckegaard and Clark, 1991). Amplification in heterozygote NOR mutants would explain dosage compensation of gene expression in frogs and could explain why the NOR is so frequently sex-linked in frogs.

Summary and future directions

Dosage compensation is an epigenetic phenomenon that is at the core of genome dynamics and function. Changes in gene dosage can have important effects in the progression of cancer, and also be involved with adaptive changes in gene function. Frogs provide a much-needed research resource both in terms of their evolutionary position at the base of tetrapods and as a developmental biology model for the study of gene dosage.

While most frogs have homomorphic sex chromosomes there are several lineages with heteromorphic sex chromosomes, and likely many more remain to be discovered. Given the rapid decline of frog and toad populations around the world, there is an urgent need to catalogue these sex chromosomes before these lineages become extinct (Mendelson *et al.*, 2006; Collins, 2010; Hayes *et al.*, 2010). Several frog species have comparable levels of aneuploidy as the sex chromosomes of mammals, however there is no indication of X or Z chromosome inactivation. It remains an open question whether dosage compensation of heteromorphic sex chromosomes exists in frogs, and measuring the transcriptome in a species with dramatic sex chromosomes is an important objective to understand the evolutionary history of dosage compensation of sex chromosomes. With current high-throughput sequencing technology, it is straightforward to quantify levels of transcripts in female and male non-model species or species that do not have a genome reference in order to understand differences in expression for genes that are sex-linked (Malcom and Malone, 2014). A bigger challenge will be to identify the genomic location of these sequences. The only frog genome reference available is for *Xenopus tropicalis* and this species has homomorphic sex chromosomes (Hellsten *et al.*, 2010). Sequencing the genome of females and males separately can enable identification of sex-linkage by comparing the ratios of DNA-Seq coverage between the sexes. For sequences that are sex-linked and in a one-copy state, DNA-Seq coverage should be 2-fold different between the sexes (Zhou and Bachtrog, 2012; Vicoso *et al.*, 2013). Coupling DNA and RNA sequencing will enable rapid progress on the sex chromosomes of frogs and allow comparison of these sequences to sex chromosomes of other vertebrates and test for the presence or absence of dosage compensation. Comparisons with closely related homomorphic lineages will allow investigating and understanding the expression state for genes before and after they become aneuploid. Comparative genomic approaches combined with experimental techniques to manipulate ploidy levels, sex, and new genome editing techniques, will allow experimental flexibility in understanding dosage during development. Experiments in frogs will provide understanding of the role of gene dosage sensitivity in genome function, and given the high degree of synteny between amphibian and mammalian genomes, inform development of a dosage sensitivity map of the human genomes.

Acknowledgements

I thank Jacob Malcom, Randy Kudra and Jessica Strein for providing comments that greatly improved the manuscript.

References

Aït Yahya-Graison, E., Aubert, J., Dauphinot, L., Rivals, I., Prieur, M., Golfier, G., Rossier, J., Personnaz, L., Creau, N., Bléhaut, H., *et al.* (2007). Classification of human chromosome 21 gene-expression variations in Down syndrome: impact on disease phenotypes. Am. J. Hum. Genet. *81*, 475–491.

Arendt, M., Fall, T., Lindblad-Toh, K., and Axelsson, E. (2014). Amylase activity is associated with AMY2B copy numbers in dog: implications for dog domestication, diet and diabetes. Anim. Genet. *45*, 716–722

Artieri, C.G., and Fraser, H.B. (2014). Evolution at two levels of gene expression in yeast. Genome Res. *24*, 411–421.

Bachtrog, D. (2006). A dynamic view of sex chromosome evolution. Curr. Opin. Genet. Dev. *16*, 578–585.

Bachtrog, D. (2013). Y-chromosome evolution: emerging insights into processes of Y-chromosome degeneration. Nat. Rev. Genet. *14*, 113–124.

Barr H.J., and Esper, H. (1963). Nucleolar size in cells of *Xenopus laevis* in relation to nucleolar competition. Exp. Cell Res. *31*, 211–214.

Bewick, A.J., Anderson, D.W., and Evans, B.J. (2011). Evolution of the closely related, sex-related genes DM-W and DMRT1 in African clawed frogs (Xenopus). Evolution *65*, 698–712.

Birchler, J.A. (1979). A study of enzyme activities in a dosage series of the long arm of chromosome one in maize. Genetics *92*, 1211–1229.

Birchler, J.A. (1981). The genetic basis of dosage compensation of alcohol dehydrogenase-1 in maize. Genetics *97*, 625–637.

Birchler, J.A. (2009). Compensation contemplation. In SeX-chromosomes: Genetics, Abnormalities, and Disorders, C.N. Weingarten and S.E. Jefferson, eds. (New York: Nova Publishers), pp. 207–224.

Birchler, J.A. (2012). Claims and counterclaims of X-chromosome compensation. Nat. Struct. Mol. Biol. *19*, 3–5.

Birchler, J.A., and Hall, T. (2014). Facts and artifacts in studies of gene expression in aneuploids and sex chromosomes. Chromosoma *23*, 459–469.

Birchler, J.A., and Newton, K.J. (1981). Modulation of protein levels in chromosomal dosage series of maize: the biochemical basis of aneuploid syndromes. Genetics *99*, 247–266.

Birchler, J.A., and Veitia, R.A. (2012). Gene balance hypothesis: connecting issues of dosage sensitivity across biological disciplines. Proc. Natl. Acad. Sci. U.S.A. *109*, 14746–14753.

Birchler, J.A., and Veitia, R.A. (2014). The Gene Balance Hypothesis: dosage effects in plants. Methods Mol. Biol. *1112*, 25–32.

Birchler, J., Sun, L., Fernandez, H., Donohue, R., Xie, W., and Sanyal, A. (2011). Re-evaluation of the function of the male specific lethal complex in Drosophila. J. Genet. Genomics *38*, 327–332.

Birchler, J.A., Hiebert, J.C., and Paigen, K. (1990). Analysis of autosomal dosage compensation involving the alcohol dehydrogenase locus in *Drosophila melanogaster*. Genetics *124*, 679–686.

Blakeslee, A.F. (1934). New Jimson weeds from old chromosomes. J. Hered. *25*, 80–108.

Boveri, T. (1902). Über mehrpolige Mitosen als Mittel zur Analyse des Zellkerns. Verh. Phys.-Med. Ges. *35*, 67–90.

Brelsford, A., Stöck, M., Betto-Colliard, C., Dubey, S., Dufresnes, C., Jourdan-Pineau, H., Rodrigues, N., Savary, R., Sermier, R., and Perrin, N. (2013). Homologous sex chromosomes in three deeply divergent anuran species. Evolution *67*, 2434–2440.

Bridges, C.B. (1925). Sex in relation to chromosomes and genes. Am. Nat. *59*, 127–137.

Brown, D.D., and Dawid, I.B. (1968). Specific gene amplification in oocytes. Oocyte nuclei contain extrachromosomal replicas of the genes for ribosomal RNA. Science *160*, 272–280.

Brown, D.D., and Gurdon, J.B. (1964). Absence of ribosomal RNA synthesis in the anucleolate mutant of *Xenopus laevis*. Proc. Natl. Acad. Sci. U.S.A. *51*, 139–146.

Brown, D.D., and Gurdon, J.B. (1966). Size distribution and stability of DNA-like RNA synthesized during development of anucleolate embryos of Xenopus laevis. J. Mol. Biol. *19*, 399–422.

Bruton, M.N., and Stobbs, R.E. (1991). The ecology and conservation of the coelacanth Latimeria chalumnae. In The Biology of Latimeria Chalumnae and Evolution of Coelacanths, J.A. Musick, M.N. Bruton, and E.K. Balon, eds. (Springer, Netherlands), pp. 313–340.

Camacho, J.P.M., Schmid, M., and Cabrero, J. (2011). B chromosomes and sex in animals. Sex Dev. *5*, 155–166.

Carvalho, C.M.B., Zhang, F., and Lupski, J.R. (2011). Structural variation of the human genome: mechanisms, assays, and role in male infertility. Syst. Biol. Reprod. Med. *57*, 3–16.

Chang, C.Y., and Witschi, E. (1956). Genic control and hormonal reversal of sex differentiation in Xenopus. Proc. Soc. Exp. Biol. Med. *93*, 140–144.

Charlesworth, B. (1991). The evolution of sex chromosomes. Science *251*, 1030–1033.

Charlesworth, B. (1996). The evolution of chromosomal sex determination and dosage compensation. Curr. Biol. *6*, 149–162.

Chen, S., Zhang, G., Shao, C., Huang, Q., Liu, G., Zhang, P., Song, W., An, N., Chalopin, D., Volff, J.-N., *et al.* (2014a). Whole-genome sequence of a flatfish provides insights into ZW sex chromosome evolution and adaptation to a benthic lifestyle. Nat. Genet. *46*, 253–260.

Chen, Z.-X., Golovnina, K., Sultana, H., Kumar, S., and Oliver, B. (2014b). Transcriptional effects of gene dose reduction. Biol. Sex Differ. *5*, 5.

Chung, B.H.-Y., Tao, V.Q., and Tso, W.W.-Y. (2014). Copy number variation and autism: new insights and clinical implications. J. Formos. Med. Assoc. *113*, 400–408.

Chuong, E.B., and Feschotte, C. (2013). Evolution. Transposons up the dosage. Science *342*, 812–813.

Collins, J.P. (2010). Amphibian decline and extinction: what we know and what we need to learn. Dis. Aquat. Organ. *92*, 93–99.

Conant, G.C., Birchler, J.A., and Pires, J.C. (2014). Dosage, duplication, and diploidization: clarifying the interplay of multiple models for duplicate gene evolution over time. Curr. Opin. Plant Biol. *19C*, 91–98.

Conrad, D.F., Pinto, D., Redon, R., Feuk, L., Gokcumen, O., Zhang, Y., Aerts, J., Andrews, T.D., Barnes, C., Campbell, P., *et al.* (2010). Origins and functional impact of copy number variation in the human genome. Nature *464*, 704–712.

Davoli, T., and de Lange, T. (2011). The causes and consequences of polyploidy in normal development and cancer. Annu. Rev. Cell Dev. Biol. *27*, 585–610.

Davoli, T., Xu, A.W., Mengwasser, K.E., Sack, L.M., Yoon, J.C., Park, P.J., and Elledge, S.J. (2013). Cumulative haploinsufficiency and triplosensitivity drive aneuploidy patterns and shape the cancer genome. Cell *155*, 948–962.

Deakin, J.E., Chaumeil, J., Hore, T.A., and Marshall Graves, J.A. (2009). Unravelling the evolutionary origins of X-chromosome inactivation in mammals: insights from marsupials and monotremes. Chromosome Res. *17*, 671–685.

Demirtas, H. (2009). AgNOR status in Down's syndrome infants and a plausible phenotype formation hypothesis. Micron *40*, 511–518.

Deng, X., Hiatt, J.B., Nguyen, D.K., Ercan, S., Sturgill, D., Hillier, L.W., Schlesinger, F., Davis, C.A., Reinke, V.J., Gingeras, T.R., *et al.* (2011). Evidence for compensatory upregulation of expressed X-linked genes in mammals, *Caenorhabditis elegans and Drosophila melanogaster*. Nat. Genet. *43*, 1179–1185.

Deng, X., Berletch, J.B., Nguyen, D.K., and Disteche, C.M. (2014). X-chromosome regulation: diverse patterns in development, tissues and disease. Nat. Rev. Genet. *15*, 367–378.

Dephoure, N., Hwang, S., O'Sullivan, C., Dodgson, S.E., Gygi, S.P., Amon, A., and Torres, E.M. (2014). Quantitative proteomic analysis reveals post-translational responses to aneuploidy in yeast. Elife 3, e03023.

Ellison, C.E., and Bachtrog, D. (2013). Dosage compensation via transposable element mediated rewiring of a regulatory network. Science *342*, 846–850.

Emerson, J.J., Cardoso-Moreira, M., Borevitz, J.O., and Long, M. (2008). Natural selection shapes genome-wide patterns of copy-number polymorphism in *Drosophila melanogaster*. Science *320*, 1629–1631.

Engreitz, J.M., Pandya-Jones, A., McDonel, P., Shishkin, A., Sirokman, K., Surka, C., Kadri, S., Xing, J., Goren, A., Lander, E.S., *et al.* (2013). The Xist lncRNA exploits three-dimensional genome architecture to spread across the X-chromosome. Science *341*, 1237973.

Esper, H., and Barr, H.J. (1964). A study of the developmental cytology of a mutation affecting nucleoli in *Xenopus* embryos. Dev. Biol. *10*, 105–121.

Ezaz, T., Stiglec, R., Veyrunes, F., and Marshall Graves, J.A. (2006). Relationships between vertebrate ZW and XY sex chromosome systems. Curr. Biol. *16*, R736–43.

Falchi, M., El-Sayed Moustafa, J.S., Takousis, P., Pesce, F., Bonnefond, A., Andersson-Assarsson, J.C., Sudmant, P.H., Dorajoo, R., Al-Shafai, M.N., Bottolo, L., *et al.* (2014). Low copy number of the salivary amylase gene predisposes to obesity. Nat. Genet. *46*, 492–497.

Forconi, M., Canapa, A., Barucca, M., Biscotti, M.A., Capriglione, T., Buonocore, F., Fausto, A.M., Makapedua, D.M., Pallavicini, A., Gerdol, M., *et al.* (2013). Characterization of sex determination and sex differentiation genes in *Latimeria*. PLoS ONE *8*, e56006.

Forsdyke, D.R. (2012). Ohno's hypothesis and Muller's paradox: sex chromosome dosage compensation may serve collective gene functions. Bioessays *34*, 930–933.

Gates, R.R. (1923). Polyploidy. Br. J. Exp. Biol. *1*, 153–182.

Goldschmidt, R.B. (1920). Untersuchungen uber intersexualitat. Z. Indukt. Abstamm. Vererbungsl. 23, 1–199.

Golubeva, Y., Salcedo, R., Mueller, C., Liotta, L.A., and Espina, V. (2013). Laser capture microdissection for protein and NanoString RNA analysis. Methods Mol. Biol. *931*, 213–257.

Golzio, C., Willer, J., Talkowski, M.E., Oh, E.C., Taniguchi, Y., Jacquemont, S., Reymond, A., Sun, M., Sawa, A., Gusella, J.F., *et al.* (2012). KCTD13 is a major driver of mirrored neuroanatomical phenotypes of the 16p11.2 copy number variant. Nature *485*, 363–367.

Grafodatskaya, D., Chung, B., Szatmari, P., and Weksberg, R. (2010). Autism spectrum disorders and epigenetics. J. Am. Acad. Child Adolesc. Psychiatry *49*, 794–809.

Grant, J., Mahadevaiah, S.K., Khil, P., Sangrithi, M.N., Royo, H., Duckworth, J., McCarrey, J.R., VandeBerg, J.L., Renfree, M.B., Taylor, W., *et al.* (2012). Rsx is a metatherian RNA with Xist-like properties in X-chromosome inactivation. Nature *487*, 254–258.

Graves, J.A.M. (2014). Avian sex, sex chromosomes, and dosage compensation in the age of genomics. Chromosome Res. *22*, 45–57.

Green, D.M. (2004). Structure and evolution of B chromosomes in amphibians. Cytogenet. Genome Res. *106*, 235–242.

Gupta, V., Parisi, M., Sturgill, D., Nuttall, R., Doctolero, M., Dudko, O.K., Malley, J.D., Eastman, P.S., and Oliver, B. (2006). Global analysis of X-chromosome dosage compensation. J. Biol. *5*, 3.

Gurdon, J.M. (1959). Tetraploid frogs. J. Exp. Zool. *141*, 519–543.

Harding, J.L., Horswell, S., Heliot, C., Armisen, J., Zimmerman, L.B., Luscombe, N.M., Miska, E.A., and Hill, C.S. (2014). Small RNA profiling of Xenopus embryos reveals novel miRNAs and a new class of small RNAs derived from intronic transposable elements. Genome Res. *24*, 96–106.

Harrison, P.W., Mank, J.E., and Wedell, N. (2012). Incomplete sex chromosome dosage compensation in the Indian meal moth, Plodia interpunctella, based on de novo transcriptome assembly. Genome Biol. Evol. *4*, 1118–1126.

Hawley, R.S., and Marcus, C.H. (1989). Recombinational controls of rDNA redundancy in *Drosophila*. Annu. Rev. Genet. *23*, 87–120.

Hayes, T.B. (1998). Sex determination and primary sex differentiation in amphibians: genetic and developmental mechanisms. J. Exp. Zool. *281*, 373–399.

Hayes, T.B., Falso, P., Gallipeau, S., and Stice, M. (2010). The cause of global amphibian declines: a developmental endocrinologist's perspective. J. Exp. Biol. *213*, 921–933.

Hellsten, U., Harland, R.M., Gilchrist, M.J., Hendrix, D., Jurka, J., Kapitonov, V., Ovcharenko, I., Putnam, N.H., Shu, S., Taher, L., *et al.* (2010). The genome of the Western clawed frog *Xenopus tropicalis*. Science *328*, 633–636.

Henking, H. (1891). Untersuchungen uber die ersten Entwicklungsvorigoenge in den Eiern der Insektern II Uber spermatogenese und Beziehung zur Eientwicklung bei Pyrrhocoris apterus. Z. Wiss. Zool. *51*, 685–736.

Hillis, D.M., and Green, D.M. (1990). Evolutionary changes in the heterogametic sex in the phylogenetic history of amphibians. J. Evol. Biol. 3, 49–64.

Humphrey, R.R. (1942). Sex of the offspring fathered by two Ambystoma females experimental converted into males. Anat. Rec. *82*, 469.

Humphrey, R.R. (1945). Sex determination in ambystomid salamanders: a study of the progeny of females experimentally converted into males. Am. J. Anat. 76, 33–36.

Ingolia, N.T., Ghaemmaghami, S., Newman, J.R.S., and Weissman, J.S. (2009). Genome-wide analysis *in vivo* of translation with nucleotide resolution using ribosome profiling. Science *324*, 218–223.

Itoh, Y., Melamed, E., Yang, X., Kampf, K., Wang, S., Yehya, N., Van Nas, A., Replogle, K., Band, M.R., Clayton, D.F., *et al.* (2007). Dosage compensation is less effective in birds than in mammals. J. Biol. *6*, 2.

Itoh, Y., Replogle, K., Kim, Y.-H., Wade, J., Clayton, D.F., and Arnold, A.P. (2010). Sex bias and dosage compensation in the zebra finch versus chicken genomes: general and specialized patterns among birds. Genome Res. *20*, 512–518.

Jackson-Cook, C.K., Flannery, D.B., Corey, L.A., Nance, W.E., and Brown, J.A. (1985). Nucleolar organizer region variants as a risk factor for Down syndrome. Am. J. Hum. Genet. *37*, 1049–1061.

Jacquemont, S., Reymond, A., Zufferey, F., Harewood, L., Walters, R.G., Kutalik, Z., Martinet, D., Shen, Y., Valsesia, A., Beckmann, N.D., *et al.* (2011). Mirror extreme BMI phenotypes associated with gene dosage at the chromosome 16p11.2 locus. Nature *478*, 97–102.

Jiang, J., Jing, Y., Cost, G.J., Chiang, J.-C., Kolpa, H.J., Cotton, A.M., Carone, D.M., Carone, B.R., Shivak, D.A., Guschin, D.Y., *et al.* (2013). Translating dosage compensation to trisomy 21. Nature *500*, 296–300.

Jiang, L., Schlesinger, F., Davis, C.A., Zhang, Y., Li, R., Salit, M., Gingeras, T.R., and Oliver, B. (2011). Synthetic spike-in standards for RNA-seq experiments. Genome Res. *21*, 1543–1551.

Johansson, A.-M., Stenberg, P., Bernhardsson, C., and Larsson, J. (2007). Painting of fourth and chromosome-wide regulation of the 4th chromosome in Drosophila melanogaster. EMBO J. *26*, 2307–2316.

Jue, N.K., Murphy, M.B., Kasowitz, S.D., Qureshi, S.M., Obergfell, C.J., Elsisi, S., Foley, R.J., O'Neill, R.J., and O'Neill, M.J. (2013). Determination of dosage compensation of the mammalian X-chromosome by RNA-seq is dependent on analytical approach. BMC Genomics *14*, 150.

Kashiwagi, K., Kashiwagi, A., Kurabayashi, A., Hanada, H., Nakajima, K., Okada, M., Takase, M., and Yaoita, Y. (2010). Xenopus tropicalis: an ideal experimental animal in amphibia. Exp. Anim. *59*, 395–405.

Katju, V., and Bergthorsson, U. (2013). Copy-number changes in evolution: rates, fitness effects and adaptive significance. Front. Genet. *4*, 273.

Kawamura, T. (1984). Polyploidy in amphibians. Zoolog. Sci. *1*, 1–15.

Kemmeren, P., Sameith, K., van de Pasch, L.A.L., Benschop, J.J., Lenstra, T.L., Margaritis, T., O'Duibhir, E., Apweiler, E., van Wageningen, S., Ko, C.W., *et al.* (2014). Large-scale genetic perturbations reveal regulatory networks and an abundance of gene-specific repressors. Cell *157*, 740–752.

Kiuchi, T., Koga, H., Kawamoto, M., Shoji, K., Sakai, H., Arai, Y., Ishihara, G., Kawaoka, S., Sugano, S., Shimada, T., *et al.* (2014). A single female-specific piRNA is the primary determiner of sex in the silkworm. Nature *509*, 633–636.

Kondo, Y., and Kashiwagi, A. (2004). Experimentally induced autotetraploidy and allotetraploidy in two japanese pond frogs. J. Herpetol. *38*, 381–392.

Larschan, E., Bishop, E.P., Kharchenko, P.V., Core, L.J., Lis, J.T., Park, P.J., and Kuroda, M.I. (2011). X-chromosome dosage compensation via enhanced transcriptional elongation in Drosophila. Nature *471*, 115–118.

Livernois, A.M., Graves, J.A.M., and Waters, P.D. (2012). The origin and evolution of vertebrate sex chromosomes and dosage compensation. Heredity *108*, 50–58.

Livernois, A.M., Waters, S.A., Deakin, J.E., Marshall Graves, J.A., and Waters, P.D. (2013). Independent evolution of transcriptional inactivation on sex chromosomes in birds and mammals. PLoS Genet. *9*, e1003635.

Lyckegaard, E.M., and Clark, A.G. (1991). Evolution of ribosomal RNA gene copy number on the sex chromosomes of *Drosophila melanogaster*. Mol. Biol. Evol. *8*, 458–474.

McAnally, A.A., and Yampolsky, L.Y. (2010). Widespread transcriptional autosomal dosage compensation in Drosophila correlates with gene expression level. Genome Biol. Evol. *2*, 44–52.

McManus, C.J., May, G.E., Spealman, P., and Shteyman, A. (2014). Ribosome profiling reveals post-transcriptional buffering of divergent gene expression in yeast. Genome Res. *24*, 422–430.

Malcom, J.W., and Malone, J.H. (2014). Analysis of deep sequencing data: insights and challenges. Comprehensive Analytical Chemistry *63*, 325–354.

Malcom, J.W., Kudra, R.S., and Malone, J.H. (2014). The sex chromosomes of frogs: variability and tolerance offer clues to genome evolution and function. J. Genomics *2*, 68–76.

Malone, J.H., and Fontenot, B.E. (2008). Patterns of reproductive isolation in toads. PLoS ONE *3*, e3900.

Malone, J.H., and Michalak, P. (2008). Physiological sex predicts hybrid sterility regardless of genotype. Science *319*, 59.

Malone, J.H., and Oliver, B. (2011). Microarrays, deep sequencing and the true measure of the transcriptome. BMC Biol. *9*, 34.

Malone, J.H., Chrzanowski, T.H., and Michalak, P. (2007). Sterility and gene expression in hybrid males of Xenopus laevis and X. muelleri. PLoS ONE *2*, e781.

Malone, J.H., Cho, D.-Y., Mattiuzzo, N.R., Artieri, C.G., Jiang, L., Dale, R.K., Smith, H.E., McDaniel, J., Munro, S., Salit, M., *et al.* (2012). Mediation of Drosophila autosomal dosage effects and compensation by network interactions. Genome Biol. *13*, r28.

Mank, J.E., and Ellegren, H. (2009). All dosage compensation is local: gene-by-gene regulation of sex-biased expression on the chicken Z chromosome. Heredity *102*, 312–320.

Marygold, S.J., Roote, J., Reuter, G., Lambertsson, A., Ashburner, M., Millburn, G.H., Harrison, P.M., Yu, Z., Kenmochi, N., Kaufman, T.C., *et al.* (2007). The ribosomal protein genes and Minute loci of *Drosophila melanogaster*. Genome Biol. *8*, R216.

Mazaika, E., and Homsy, J. (2014). Digital Droplet PCR: CNV analysis and other applications. Curr. Protoc. Hum. Genet. *82*, 7.24.1–7.24.13.

Mendelson, J.R., Lips, K.R., Gagliardo, R.W., Rabb, G.B., Collins, J.P., Diffendorfer, J.E., Daszak, P., Ibanez, D.R., Zippel, K.C., Lawson, D.P., *et al.* (2006). Confronting amphibian declines and extinctions. Science *313*, 48–48.

Mikamo, K., and Witschi, E. (1963). Functional sex-reversal in genetic females of *Xenopus laevis*, induced by implanted testes. Genetics *48*, 1411–1421.

Mikamo, K., and Witschi, E. (1964). Masculinization and breeding of the WW *Xenopus*. Experientia *20*, 622–623.

Miura, I. (2007). An evolutionary witness: the frog rana rugosa underwent change of heterogametic sex from XY male to ZW female. Sex Dev. *1*, 323–331.

Miura, I., Ohtani, H., Nakamura, M., Ichikawa, Y., and Saitoh, K. (1998). The origin and differentiation of the heteromorphic sex chromosomes Z, W, X, and Y in the frog *Rana rugosa*, inferred from the sequences of a sex-linked gene, ADP/ATP translocase. Mol. Biol. Evol. *15*, 1612–1619.

Mokhtar, S.S., Marshall, C.R., Phipps, M.E., Thiruvahindrapuram, B., Lionel, A.C., Scherer, S.W., and Peng, H.B. (2014). Novel population specific autosomal copy number variation and its functional analysis amongst Negritos from Peninsular Malaysia. PLoS ONE *9*, e100371.

Muller, H.J. (1932). Further studies on the nature and causes of gene mutations. Proc. Sixth Int. Congress Genetics *1*, 213–255.

Namekawa, S.H., VandeBerg, J.L., McCarrey, J.R., and Lee, J.T. (2007). SeX-chromosome silencing in the marsupial male germ line. Proc. Natl. Acad. Sci. U.S.A. *104*, 9730–9735.

Naurin, S., Hasselquist, D., Bensch, S., and Hansson, B. (2012). Sex-biased gene expression on the avian Z chromosome: highly expressed genes show higher male-biased expression. PLoS ONE *7*, e46854.

Ogata, M., Hasegawa, Y., Ohtani, H., Mineyama, M., and Miura, I. (2008). The ZZ/ZW sex-determining mechanism originated twice and independently during evolution of the frog, Rana rugosa. Heredity *100*, 92–99.

Ohno, S. (1967). Sex chromosomes and sex-linked genes. Monographs on endocrinology, Vol. 1 (Springer Verlag, Berlin, Germany).

Ohta, S. (1986). Sex determining mechanism in *Buergeria buergeri* (SCHLEGAL): I. Heterozygosity of chromosome pair no. 7 in the female. Sci. Rep. Lab. Amphib. Biol. Hiroshima Univ. *8*, 29–43.

Ohta, S. (1987). Sex determining mechanism in *Buergeria buergeri* (SCHLEGEL): II. The effects of sex hormones on the differentiation of gonads and the offspring of sex-reversed females. Sci. Rep. Lab. Amphibian Biol. Hiroshima Univ. *9*, 213–238.

Ohta, S., Sumida, M., and Nishioka, M. (1999). Sex-determining mechanism in *Buergeria buergeri* (Anura, Rhacophoridae). III. Does the ZZW triploid frog become female or male? J. Exp. Zool. *283*, 295–306.

Okada, E., Yoshimoto, S., Ikeda, N., Kanda, H., Tamura, K., Shiba, T., Takamatsu, N., and Ito, M. (2009). Xenopus W-linked DM-W induces Foxl2 and Cyp19 expression during ovary formation. Sex Dev. *3*, 38–42.

Ong, S.-E., Blagoev, B., Kratchmarova, I., Kristensen, D.B., Steen, H., Pandey, A., and Mann, M. (2002). Stable isotope labeling by amino acids in cell culture, SILAC, as a simple and accurate approach to expression proteomics. Mol. Cell. Proteomics *1*, 376–386.

Perry, G.H., Dominy, N.J., Claw, K.G., Lee, A.S., Fiegler, H., Redon, R., Werner, J., Villanea, F.A., Mountain, J.L., Misra, R., *et al.* (2007). Diet and the evolution of human amylase gene copy number variation. Nat. Genet. *39*, 1256–1260.

Pessia, E., Engelstädter, J., and Marais, G.A.B. (2014). The evolution of X-chromosome inactivation in mammals: the demise of Ohno's hypothesis? Cell. Mol. Life Sci. *71*, 1383–1394.

Philip, P., and Stenberg, P. (2013). Male X-linked genes in *Drosophila melanogaster* are compensated independently of the male-specific lethal complex. Epigenetics Chromatin *6*, 35.

Pinto, D., Delaby, E., Merico, D., Barbosa, M., Merikangas, A., Klei, L., Thiruvahindrapuram, B., Xu, X., Ziman, R., Wang, Z., *et al.* (2014). Convergence of genes and cellular pathways dysregulated in autism spectrum disorders. Am. J. Hum. Genet. *94*, 677–694.

Ponse, K. (1942). Sur la digametie du crapaud hermaphrodite. Rev. Suisse Zool. *49*, 185–189.

Prandini, P., Deutsch, S., Lyle, R., Gagnebin, M., Delucinge Vivier, C., Delorenzi, M., Gehrig, C., Descombes, P., Sherman, S., Dagna Bricarelli, F., *et al.* (2007). Natural gene-expression variation in Down syndrome modulates the outcome of gene-dosage imbalance. Am. J. Hum. Genet. *81*, 252–263.

Quinn, A.E., Sarre, S.D., Ezaz, T., Marshall Graves, J.A., and Georges, A. (2011). Evolutionary transitions between mechanisms of sex determination in vertebrates. Biol. Lett. *7*, 443–448.

Rafferty, K.A., and Sherwin, R.W. (1969). The length of secondary constrictions in normal individuals and in a nucleolar mutant of *Xenopus laevis*. Cytogenetics *8*, 427–438.

Reinschmidt, D.C., Simon, S.J., Volpe, E.P., and Tompkins, R. (1979). Production of tetraploid and homozygous diploid amphibians: suppression of first cleavage. J. Exp. Zool. *210*, 137–143.

Rodriguez, M.A., Vermaak, D., Bayes, J.J., and Malik, H.S. (2007). Species-specific positive selection of the male-specific lethal complex that participates in dosage compensation in Drosophila. Proc. Natl. Acad. Sci. U.S.A. *104*, 15412–15417.

Sahara, K., Yoshido, A., and Traut, W. (2012). SeX-chromosome evolution in moths and butterflies. Chromosome Res. *20*, 83–94.

Sarre, S.D., Ezaz, T., and Georges, A. (2011). Transitions between sex-determining systems in reptiles and amphibians. Annu. Rev. Genomics Hum. Genet. *12*, 391–406.

Sass, G.L., Pannuti, A., and Lucchesi, J.C. (2003). Male-specific lethal complex of Drosophila targets activated regions of the X-chromosome for chromatin remodeling. Proc. Natl. Acad. Sci. U.S.A. *100*, 8287–8291.

Schmid, M., and Steinlein, C. (2001). SeX-chromosomes, sex-linked genes, and sex determination in the vertebrate class amphibia. EXS *2001*, 143–176.

Schmid, M., and Steinlein, C. (2003). Chromosome banding in Amphibia. XXIX. The primitive XY/XX sex chromosomes of Hyla femoralis (Anura, Hylidae). Cytogenet. Genome Res. *101*, 74–79.

Schmid, M., Sims, S.H., Haaf, T., and Macgregor, H.C. (1983). Chromosome banding in Amphibia. X. 18S and 28S ribosomal genes, nucleolus organizers and nucleoli in *Gastrotheca riobambae*. Chromosoma *94*, 139–145.

Schmid, M., Ohta, S., Steinlein, C., and Guttenbach, M. (1993). Chromosome banding in Amphibia. XIX. Primitive ZW/ZZ sex chromosomes in Buergeria buergeri (Anura, Rhacophoridae). Cytogenetic Cell Genetics *62*, 238–246.

Schmid, M., Steinlein, C., Bogart, J.P., Feichtinger, W., León, P., La Marca, E., Díaz, L.M., Sanz, A., Chen, S.-H., and Hedges, S.B. (2010). The chromosomes of terraranan frogs. Insights into vertebrate cytogenetics. Cytogenet. Genome Res. *130–131*, 1–568.

Schmid, M., Steinlein, C., Bogart, J.P., Feichtinger, W., Haaf, T., Nanda, I., del Pino, E.M., Duellman, W.E., and Hedges, S.B. (2012). The hemiphractid frogs. Phylogeny, embryology, life history, and cytogenetics. Cytogenet. Genome Res. *138*, 69–384.

Schrider, D.R., and Hahn, M.W. (2010). Gene copy-number polymorphism in nature. Proc. Biol. Sci. *277*, 3213–3221.

Schrider, D.R., Navarro, F.C.P., Galante, P.A.F., Parmigiani, R.B., Camargo, A.A., Hahn, M.W., and de Souza, S.J. (2013). Gene copy-number polymorphism caused by retrotransposition in humans. PLoS Genet. *9*, e1003242.

Sin, H.-S., and Namekawa, S.H. (2013). The great escape: active genes on inactive sex chromosomes and their evolutionary implications. Epigenetics *8*, 887–892.

Skinner, B.M., Al Mutery, A., Smith, D., Völker, M., Hojjat, N., Raja, S., Trim, S., Houde, P., Boecklen, W.J., and Griffin, D.K. (2014). Global patterns of apparent copy number variation in birds revealed by cross-species comparative genomic hybridization. Chromosome Res. *22*, 59–70.

Smith, C.A., Roeszler, K.N., Ohnesorg, T., Cummins, D.M., Farlie, P.G., Doran, T.J., and Sinclair, A.H. (2009). The avian Z-linked gene DMRT1 is required for male sex determination in the chicken. Nature *461*, 267–271.

Smith, G., Chen, Y.-R., Blissard, G.W., and Briscoe, A.D. (2014). Complete dosage compensation and sex-biased gene expression in the moth Manduca sexta. Genome Biol. Evol. *6*, 526–537.

Song, R., Liu, P., and Acar, M. (2014). Network-dosage compensation topologies as recurrent network motifs in natural gene networks. BMC Syst. Biol. *8*, 69.

Stankiewicz, P., and Lupski, J.R. (2010). Structural variation in the human genome and its role in disease. Annu. Rev. Med. *61*, 437–455.

Stenberg, P., and Larsson, J. (2011). Buffering and the evolution of chromosome-wide gene regulation. Chromosoma *120*, 213–225.

Stenberg, P., Lundberg, L.E., Johansson, A.-M., Rydén, P., Svensson, M.J., and Larsson, J. (2009). Buffering of segmental and chromosomal aneuploidies in *Drosophila melanogaster*. PLoS Genet. *5*, e1000465.

Stevens, N.M. (1905). Studies in spermatogenesis with especial reference to the "accessory chromosome" (Carnegie Institution of Washington, Publication No. 36, Washington, DC).

Straub, T., and Becker, P.B. (2007). Dosage compensation: the beginning and end of generalization. Nat. Rev. Genet. *8*, 47–57.

Straub, T., Grimaud, C., Gilfillan, G.D., Mitterweger, A., and Becker, P.B. (2008). The chromosomal high-affinity binding sites for the Drosophila dosage compensation complex. PLoS Genet. *4*, e1000302.

Sturgill, D., Malone, J.H., Sun, X., Smith, H.E., Rabinow, L., Samson, M.-L., and Oliver, B. (2013). Design of RNA splicing analysis null models for post hoc filtering of Drosophila head RNA-Seq data with the splicing analysis kit (Spanki). BMC Bioinformatics *14*, 320.

Sun, C., López Arriaza, J.R., and Mueller, R.L. (2012). Slow DNA loss in the gigantic genomes of salamanders. Genome Biol. Evol. *4*, 1340–1348.

Sun, L., Johnson, A.F., Li, J., Lambdin, A.S., Cheng, J., and Birchler, J.A. (2013a). Differential effect of aneuploidy on the X-chromosome and genes with sex-biased expression in Drosophila. Proc. Natl. Acad. Sci. U.S.A. *110*, 16514–16519.

Sun, L., Johnson, A.F., Donohue, R.C., Li, J., Cheng, J., and Birchler, J.A. (2013b). Dosage compensation and inverse effects in triple X metafemales of Drosophila. Proc. Natl. Acad. Sci. U.S.A. *110*, 7383–7388.

Sun, L., Fernandez, H.R., Donohue, R.C., Li, J., Cheng, J., and Birchler, J.A. (2013c). Male-specific lethal complex in Drosophila counteracts histone acetylation and does not mediate dosage compensation. Proc. Natl. Acad. Sci. U.S.A. *110*, E808–17.

Sury, M.D., Chen, J.-X., and Selbach, M. (2010). The SILAC fly allows for accurate protein quantification *in vivo*. Mol. Cell. Proteomics *9*, 2173–2183.

Taipale, M., and Akhtar, A. (2005). Chromatin mechanisms in Drosophila dosage compensation. Prog. Mol. Subcell. Biol. *38*, 123–149.

Tartof, K.D. (1971). Increasing the multiplicity of ribosomal RNA genes in *Drosophila melanogaster*. Science *171*, 294–297.

Tompkins, R. (1978). Triploid and gynogenetic diploid *Xenopus laevis*. J. Exp. Zool. *203*, 251–256.

Tompkins, R., and Reinschmidt (1991). Experimentally induced homozygosity in Xenopus laevis. Methods Cell Biol. *36*, 35–44.

Tucker, M.A., and Gerhardt, H.C. (2012). Parallel changes in mate-attracting calls and female preferences in autotriploid tree frogs. Proc. Biol. Sci. *279*, 1583–1587.

Uno, Y., Nishida, C., Oshima, Y., Yokoyama, S., Miura, I., Matsuda, Y., and Nakamura, M. (2008). Comparative chromosome mapping of sex-linked genes and identification of seX-chromosomal rearrangements in the Japanese wrinkled frog (Rana rugosa, Ranidae) with ZW and XY sex chromosome systems. Chromosome Res. *16*, 637–647.

Valsesia, A., Macé, A., Jacquemont, S., Beckmann, J.S., and Kutalik, Z. (2013). The growing importance of CNVs: new insights for detection and clinical interpretation. Front. Genet. *4*, 92.

Veitia, R.A., Bottani, S., and Birchler, J.A. (2008). Cellular reactions to gene dosage imbalance: genomic, transcriptomic and proteomic effects. Trends Genet. *24*, 390–397.

Veitia, R.A., Bottani, S., and Birchler, J.A. (2013). Gene dosage effects: nonlinearities, genetic interactions, and dosage compensation. Trends Genet. *29*, 385–393.

Vicoso, B., and Bachtrog, D. (2009). Progress and prospects toward our understanding of the evolution of dosage compensation. Chromosome Res. *17*, 585–602.

Vicoso, B., and Bachtrog, D. (2011). Lack of global dosage compensation in Schistosoma mansoni, a female-heterogametic parasite. Genome Biol. Evol. *3*, 230–235.

Vicoso, B., and Charlesworth, B. (2006). Evolution on the X-chromosome: unusual patterns and processes. Nat. Rev. Genet. *7*, 645–653.

Vicoso, B., Emerson, J.J., Zektser, Y., Mahajan, S., and Bachtrog, D. (2013). Comparative sex chromosome genomics in snakes: differentiation, evolutionary strata, and lack of global dosage compensation. PLoS Biol. *11*, e1001643.

Wallace, H. (1960). The development of anucleolate embryos of *Xenopus laevis*. J. Embryol. Exp. Morph. *8*, 405–413.

Wallace, H., and Birnstiel, M.L. (1966). Ribosomal cistrons and the nucleolar organizer. Biochim. Biophys. Acta *114*, 296–310.

Walters, J.R., and Hardcastle, T.J. (2011). Getting a full dose? Reconsidering sex chromosome dosage compensation in the silkworm, *Bombyx mori*. Genome Biol. Evol. *3*, 491–504.

Wolf, J.B., and Bryk, J. (2011). General lack of global dosage compensation in ZZ/ZW systems? Broadening the perspective with RNA-seq. BMC Genomics *12*, 91.

Xiong, Y., Chen, X., Chen, Z., Wang, X., Shi, S., Wang, X., Zhang, J., and He, X. (2010). RNA sequencing shows no dosage compensation of the active X-chromosome. Nat. Genet. *42*, 1043–1047.

Yen, Z.C., Meyer, I.M., Karalic, S., and Brown, C.J. (2007). A cross-species comparison of X-chromosome inactivation in Eutheria. Genomics *90*, 453–463.

Yoshimoto, S., and Ito, M. (2011). A ZZ/ZW-type sex determination in *Xenopus laevis*. FEBS J. *278*, 1020–1026.

Yoshimoto, S., Okada, E., Umemoto, H., Tamura, K., Uno, Y., Nishida-Umehara, C., Matsuda, Y., Takamatsu, N., Shiba, T., and Ito, M. (2008). A W-linked DM-domain gene, DM-W, participates in primary ovary development in *Xenopus laevis*. Proc. Natl. Acad. Sci. U.S.A. *105*, 2469–2474.

Zha, X., Xia, Q., Duan, J., Wang, C., He, N., and Xiang, Z. (2009). Dosage analysis of Z chromosome genes using microarray in silkworm, *Bombyx mori*. Insect Biochem. Mol. Biol. *39*, 315–321.

Zhang, F., Gu, W., Hurles, M.E., and Lupski, J.R. (2009). Copy number variation in human health, disease, and evolution. Annu. Rev. Genomics Hum. Genet. *10*, 451–481.

Zhang, Y., and Oliver, B. (2010). An evolutionary consequence of dosage compensation on *Drosophila melanogaster* female X-chromatin structure? BMC Genomics *11*, 6.

Zhang, Y., Malone, J.H., Powell, S.K., Periwal, V., Spana, E., Macalpine, D.M., and Oliver, B. (2010). Expression in aneuploid Drosophila S2 cells. PLoS Biol. *8*, e1000320.

Zhou, Q., and Bachtrog, D. (2012). Sex-specific adaptation drives early sex chromosome evolution in Drosophila. Science 337, 341–345.

Ingenious Genes: The Diverse Roles of Long Non-coding RNA in Regulatory Processes

9

Emily M. Darrow and Brian P. Chadwick

Abstract

The general principles of the Central Dogma describe the flow of genetic information from the transcription of DNA into RNA that is subsequently translated into protein. While there has long been an appreciation for those transcriptional units that do not complete this course and function at the level of RNA, the breadth and extent of molecules that fall into this category is undoubtedly far greater than could have been imagined in the early 1990s when the first long non-coding RNAs (lncRNA) were discovered. Experimental evidence firmly associates lncRNAs with epigenetic regulation of gene expression. In this chapter, we provide a broad overview of the many types and functions of lncRNA and briefly cover the various ways in which these RNA molecules impact and regulate protein-coding genes that do conform to the Central Dogma.

Introduction

One of the surprising discoveries that arose from early genome sequencing efforts and the completion of the Human Genome Project was the degree of transcription that extends far beyond the body of protein coding genes (PCG) (Lander *et al.*, 2001; Venter *et al.*, 2001). While only approximately 1% of the genome is coding, approximately 70% is transcribed into RNA (Carninci *et al.*, 2005; Cheng *et al.*, 2005; Djebali *et al.*, 2012; Kampa *et al.*, 2004; Kapranov *et al.*, 2002; Rinn *et al.*, 2003). The purpose of several types of the small non-coding RNAs (microRNA, small interfering RNA, and PIWI-interacting RNA) is well known (Morris and Mattick, 2014), but long non-coding RNAs (lncRNAs), transcripts generally accepted to be greater than 200 nucleotides (nts) (Kapranov *et al.*, 2007) without any obvious coding potential are less well understood and the majority remain uncharacterized (Mercer *et al.*, 2009). Advances in RNA sequencing technologies (Core *et al.*, 2008; Mercer *et al.*, 2012; Mortazavi *et al.*, 2008) and combinatorial approaches such as those using chromatin signatures to identify transcriptional units (Guttman *et al.*, 2009; Yamashita *et al.*, 2011) have brought to light the large number lncRNAs present throughout the eukaryotic genome. To date, the identification of transcripts has far outstripped detailed mechanistic explanations as to how individual lncRNAs exert their influence and as the list of functions and protein interacting partners rapidly increases, the focus of the field is shifting from identifying and broadly characterizing these transcripts to understanding and predicting the roles of individual lncRNA in both health and disease. Of those lncRNAs for which functional attributes have been assigned, these occupy diverse roles in the epigenome ranging from the well-known silencing activities of the lncRNA X-inactive specific transcript (Xist/XIST) on the inactive X-chromosome (Brockdorff *et al.*, 1992; Brown *et al.*, 1991, 1992) or the H19 lncRNA involved in imprinting (Brannan *et al.*, 1990), to a more recently proposed role for lncRNA in protein synthesis (Carrieri *et al.*, 2012; van Heesch *et al.*, 2014).

The initial definition of lncRNAs is slowly resolving into subcategories based on genomic location and putative regulatory roles, but despite this, the diversity and often multiplicity of lncRNA functions continue to blur categorical lines (Cabili *et al.*, 2011). The intrinsic ability of RNA to base pair with other nucleic acids, form secondary structures, interact with proteins, and transcriptionally mark specific loci make them ideal regulatory molecules. Deciphering the epigenetic contribution of lncRNAs requires disentangling their influence from the often co-occurring changes in histone modifications, DNA methylation, and chromatin organization that are associated with epigenetic regulation.

Over the past decade our appreciation of lncRNA has grown exponentially, as has the number of related publications, and these versatile molecules have emerged as a vital link in our understanding of epigenetic regulation. Here, we discuss the discovery, characterization, and biological roles of lncRNA in addition to highlighting the mechanisms through which these molecules carry out their function.

Identifying and defining lncRNAs

Characteristics and identification

lncRNA transcripts range from 200 nts to greater that 100,000 nts (Korostowski *et al.*, 2012; Vallot *et al.*, 2013) and like messenger RNA (mRNA), they are often spliced, polyadenylated, have a 5′–7-methylguanosine cap, and are transcribed by RNA polymerase II (Pol II) (Djebali *et al.*, 2012). However, a substantial number of lncRNAs are either not polyadenylated (Livyatan *et al.*, 2013), or are present in both polyA+ and polyA– RNA fractions (Yang *et al.*, 2011). There are also those that undergo non-canonical processing such as small nucleolar lncRNAs and the lncRNA metastasis associated lung adenocarcinoma transcript 1 (MALAT1), which is processed into small cytoplasmic RNAs with tRNA-like structures (Wilusz *et al.*, 2008). While the majority of lncRNAs appear to be non-coding (Banfai *et al.*, 2012), some do associate with ribosomes (van Heesch *et al.*, 2014) and a subset of these give rise to small peptides (Guttman *et al.*, 2013; Ingolia *et al.*, 2011).

Cataloguing the complete non-coding RNA transcriptome is complicated by the genomic diversity of the transcripts as well as their low abundance, tissue specificity, and increased alternative splicing relative to PCGs, and as such the number and variety of non-coding transcripts is undoubtedly set to increase substantially over the next decade (Cabili *et al.*, 2011; Derrien *et al.*, 2012; Mercer *et al.*, 2012; Nakaya *et al.*, 2007).

High throughput sequencing of RNA (RNA-Seq) and variations thereof, are currently the methods of choice for identifying novel transcripts and measuring genome-wide transcription (Djebali *et al.*, 2012; Washietl *et al.*, 2012). One variation, global run-on sequencing (GRO-Seq), is particularly useful for low abundant transcripts (Core *et al.*, 2008) or for those produced in response to stimuli, such as in association with the hormone signalling (Hah *et al.*, 2013) or response of the immune system (Kaikkonen *et al.*, 2013). In this procedure, transcripts still attached to actively engaging RNA Pol II complexes are isolated, reverse transcribed, sequenced, and then mapped back to the genome to identify and quantify newly transcribed, rather than steady state RNA.

Chromatin immunoprecipitation followed by high throughput sequencing (ChIP-Seq) of various covalent histone modifications revealed that active PCGs have a distinct chromatin signature (Barski *et al.*, 2007; Heintzman *et al.*, 2007; Mikkelsen *et al.*, 2007). This signature in its most condensed form denotes transcribed genes by the presence of high levels of histone H3 tri-methylated at lysine 4 (H3K4me3) within the promoter region and histone H3 tri-methylated at lysine 36 (H3K36me3) throughout the transcribed gene body. Most lncRNA display the same signature as PCGs and by excluding annotated PCGs, Guttman and colleagues catalogued a large number of long intergenic non-coding RNAs (lincRNAs), many of which are conserved between human and mouse (Guttman *et al.*, 2009, 2010). The presence of elevated levels of H3K4me3 used in conjunction with RNA-Seq and other transcription-based approaches is often used to validate putative lncRNAs (Kutter *et al.*, 2012; Sati *et al.*, 2012).

It is becoming clear that most genes produce many more transcript isoforms than originally anticipated, including splice variants (Kapranov *et al.*, 2005; Mercer *et al.*, 2012; Trapnell *et al.*, 2010). This is particularly evident for lncRNAs, which may reflect the fact that they are not under the constraints that PCGs are, which act as an intermediary in protein synthesis. The majority of lncRNAs have many isoforms and often transcripts overlap or converge with other lncRNA or PCGs (Fig. 9.1), further complicating accurate annotation (Derrien *et al.*, 2012; He *et al.*, 2008; Sigova *et al.*, 2013). The definition of a lncRNA is broad and only restricted by an apparent lack of or limited coding potential, and a length of at least 200 nts. Consequently, the RNAs within this group feature a wide range of characteristics and potential functions, many of which are not mutually exclusive and serve to further complicate a standardized system of lncRNA classification. As the further functional validation of individual lncRNAs proceeds, it is likely that the current subcategories of lncRNAs based on genomic location or biological activity will continue to become more refined in definition.

Genomics of lncRNA

lncRNAs can be transcribed from the same or opposite strand relative to associated PCGs (Fig. 9.1), and the extent of the transcriptional unit can vary substantially. Transcripts that have significant overlap, especially those encoded on the same strand as the PCG, are difficult to characterize without using methods such as 5′ or 3′ rapid amplification of complementary DNA ends (RACE) to define the limits of transcription in order to distinguish between it and the PCG mRNA, or in some cases with other overlapping lncRNAs. Studies attempting to create large catalogues of lncRNAs expressed at various stages of development (Ulitsky *et al.*, 2011) or those expressed in specific tissues often focus on lincRNAs, primarily because lincRNAs share chromatin signatures with PCGs, such as H3K4me3 and H43K36me3, but are spatially separate, making validation and characterization far less complex (Cabili *et al.*, 2011; Khaitovich *et al.*, 2006; Khalil *et al.*, 2009; Washietl *et al.*, 2014).

Natural antisense transcripts

LncRNA that overlap with a PCG and are transcribed from the opposite DNA strand are referred to as natural antisense transcripts (NATs) (Arthanari *et al.*, 2014; He *et al.*, 2008; Katayama *et al.*, 2005). NATs are involved in both repression and activation of their protein-coding partners. While the functional effects of lincRNAs can operate in *cis* or *trans*, NATs typically only regulate the gene with which they overlap (Arthanari *et al.*, 2014; H. Wang *et al.*, 2014; Yu *et al.*, 2008).

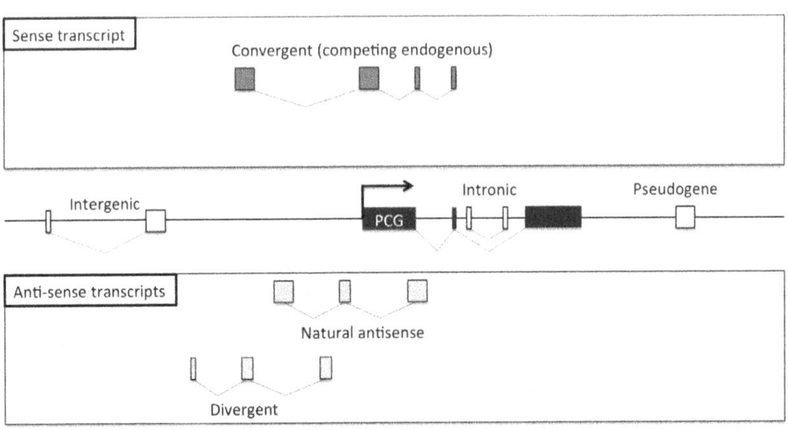

Figure 9.1 Image to illustrate the various types of lncRNA. In the middle is a schematic map of a theoretical protein-coding gene (PCG). The right-facing hooked arrow represents the promoter and the three black blocks are the exons. Upstream is an example of an Intergenic lncRNA, downstream is a Pseudogene and embedded in intron-2 of the PCG is an intronic lncRNA. These can be sense or antisense relative to the PCG. In the box above is an example of a Convergent lncRNA that originates from the sense strand relative to the PCG. In the box below are examples of lncRNAs that originate specifically from the antisense strand relative to the PCG.

Pseudogenes

Importantly, despite their inability to code for protein, many pseudogenes are transcribed (Pei *et al.*, 2012; Zheng *et al.*, 2007), and we are moving away from the long-held view that these simply represent genomic fossils (Han *et al.*, 2011; Poliseno *et al.*, 2010; Rapicavoli *et al.*, 2013). Pseudogenes are categorized by their method of formation. Processed pseudogenes are produced by retrotransposition of mature mRNA from a functional PCG, and therefore typically lack introns and regulatory sequences when compared to their gene or origin, and depending on the time interval since insertion, they accumulate mutations and diverge from their parental gene. Unitary pseudogenes, are created by mutations that render a PCG non-functional, and duplicated pseudogenes are a result of gene duplication followed by the acquisition of mutations that render them incapable of coding for functional protein. Several pseudogene-derived ncRNAs are able to regulate their parental genes through the recruitment of chromatin modifiers and competitive binding of miRNAs (Poliseno *et al.*, 2010), supporting the notion that pseudogenes may provide a source for lncRNA genesis.

Conservation

Early scepticism regarding the functionality and significance of lncRNAs (Struhl, 2007) was due to a lack of sequence conservation between mouse and human homologues, or complete absence of the orthologous loci (Babak *et al.*, 2005). Sequence conservation of PCGs is highest in the regions that code for amino acids critical to protein function. If lncRNAs are viewed in a similar fashion, whereby sequences of critical function are conserved, such as regions necessary for secondary structure formations or interaction with specific proteins, then it is not unrealistic to expect at least some portions of lncRNAs to retain sequence similarity, such as the various domains of the human and mouse XIST/Xist lncRNA (Brown *et al.*, 1992; Brockdorff *et al.*, 1992). However, conservation of an important secondary structure for function, does not necessarily translate to a requirement for conserved primary nucleotide sequence. Therefore, although some lncRNAs may be subject to evolutionary pressures that reflect their function, such restraints might not be reflected in DNA sequence conservation (Khaitovich *et al.*, 2006; Kutter *et al.*, 2012; Pang *et al.*, 2006; Ponjavic *et al.*, 2007). Indeed, several studies have noted that lncRNAs lacking DNA sequence conservation between distant species are often found within syntenic regions of the genome. Furthermore, the genes flanking the lncRNA are enriched for those involved in the immune response and development, suggesting that the importance of these molecules lies in their genomic location (Ulitsky *et al.*, 2011; Washietl *et al.*, 2014).

Function or fiction?

Are some lncRNAs simply a by-product of aberrant RNA Pol II activity or are they all bonafide functional molecules? Following the initial identification of widespread intergenic transcription, this was the main question surrounding lncRNA. Although lncRNAs have distinct patterns of tissue specificity and are differentially expressed in many scenarios, studies that reduced expression levels through genome engineering or knockdown with RNA interference (RNAi), firmly linked their importance to genome function. RNAi utilizing short hairpin RNA (shRNA) or small interfering RNAs (siRNA) has proved a cost-effective way to reduce the levels of a large number of lncRNA transcripts and probe their function in biological processes. Several studies have used this approach to identify transcripts with roles in development and differentiation.

Knockdown of 147 lincRNA genes expressed in mouse embryonic stem cells (mESCs), demonstrated the importance of specific lincRNA during embryogenesis (Guttman *et al.*, 2011). Reducing the levels of the majority of the lincRNAs resulted in changes in gene expression on par with that observed when reducing levels of mESC regulatory proteins. Furthermore, 26 of these lincRNAs caused a decrease in pluripotency markers such as Sox2 and Klf4 in addition to altered mESC cell morphology. These same pluripotency markers were also found at many lincRNA promoters, suggesting an active role for pluripotency transcription factors in the transcriptional regulation of lncRNAs. To determine if these lncRNA had an impact on specific differentiation pathways, they monitored the expression of lineage-specific transcription factors and found that indeed, reduced expression of some lincRNA led to the up-regulation of specific gene markers (e.g. Sox17 in the endoderm), while not affecting a luciferase reporter gene driven by the endogenous promoter of *Nanog*, a well-known marker of pluripotency. Together these data suggest that while some lncRNA are important for maintaining pluripotency, loss of their expression alone is not sufficient to trigger differentiation, indicating that other factors are required.

Mechanisms and functions of lncRNAs

Mechanisms of regulation

While many lncRNAs have documented roles and preferential expression in numerous biological processes and contribute to disease, uncovering the importance of their expression and the mechanisms through which lncRNAs exact their influence is an ongoing process. Like proteins, lncRNAs are an end-point molecule and as such their intrinsic abilities to base pair with RNA and DNA, distinguish between alleles, direct protein localization, and form secondary structure provide numerous functional opportunities that extend the repertoire of epigenetic tools. By behaving as molecular sponges (Cesana *et al.*, 2011; Johnsson *et al.*, 2013; K. Wang *et al.*, 2014a,b), scaffolds for protein complexes (Simon *et al.*, 2013; Tsai *et al.*, 2010; Wu *et al.*, 2013; Yoon *et al.*, 2013; Zhao *et al.*, 2008, 2010), and guides that direct the localization of protein complexes without DNA binding specificity (Beckedorff *et al.*, 2013; Wu *et al.*, 2013), lncRNAs are able to influence gene regulation at nearly every step of the process. It is through these mechanisms that lncRNA are able to influence transcriptional regulation, mRNA maturation and transport, mRNA stability, genomic organization, and protein synthesis as detailed below.

Decoys and sponges

By serving as sponges or decoys, some lncRNA interfere with molecular signals and miRNA targeting attempts. For example, the glucocorticoid receptor (GR) is a hormone–dependent transcription factor that, upon agonist binding, travels to the nucleus whereupon it stimulates expression of glucocorticoid-response genes by binding to the glucocorticoid response elements (GRE) within gene promoters, recruiting the transcriptional machinery. *Gas5* is a lncRNA that prevents GR binding to GREs by associating with the DNA-binding domain of the GR, acting as a molecular decoy (Kino *et al.*, 2010).

Competing endogenous RNAs (ceRNAs) (Fig. 9.1) are lncRNAs with miRNA binding sites that sequester miRNAs away from the intended target, thereby preventing target gene repression (Salmena *et al.*, 2011). Pseudogenes PTENP1 and CD44 behave as ceRNAs and prevent repression of their respective parental gene, PTEN and CD44 (Jeyapalan *et al.*, 2011; Poliseno *et al.*, 2010). Another example of a lncRNA behaving as a molecular decoy occurs during mammalian myogenesis. Linc-MD1 is expressed during myoblast differentiation and is localized to the cytoplasm where it competitively binds miR-135 and miRNA-133 which target the mRNAs myocyte-specific enhancer factor 2C (Mef2c) and the mastermind-like-1 transcription factor (MamL1). Consequently, the presence of Linc-MD1 reduces the ability of the miRNAs to bind to their intended targets, permitting the translation of Mef2c and MamL1, transcription factors that have important roles in muscle differentiation (Cesana *et al.*, 2011).

Guides, tethers, and scaffolds

Epigenetic regulation requires the direction of appropriate regulatory factors to specific locations at specified times; however, the means by which a complex without sequence-specific DNA binding abilities (like the Polycomb repressor complex 2 – PRC2), is able to act at specific loci is not fully understood (Moran *et al.*, 2012). An emerging mechanism involves lncRNA directing chromatin modifying complex to specific targets (Goff and Rinn, 2013). The large number of lncRNAs capable of binding and targeting PRC2 to specific loci in *cis* and *trans* suggests the regulatory capacity of these transcripts is far reaching (Khalil *et al.*, 2009; Zhao *et al.*, 2010).

Xist lncRNA is necessary for X-chromosome inactivation (Brockdorff *et al.*, 1992; Brown *et al.*,

1991), the mammalian form of dosage compensation (Lyon, 1961). Early in female embryonic development, Xist is expressed from and associates in cis with the X chosen to become the inactive X-chromosome (Clemson *et al.*, 1996). Xist recruits PRC2 to sites throughout the length of the X-chromosome, establishing the repressive modification histone H3 tri-methylated at lysine 27 (H3K27me3), triggering a cascade of epigenetic changes that lock in chromosome-wide gene silencing (Simon *et al.*, 2013; Zhao *et al.*, 2008).

The HOX gene clusters are developmentally regulated and decorated by both activating (H3K4me3) and repressive (H3K27me3) chromatin modifications (Rinn *et al.*, 2007), referred to as a bivalent domain (Bernstein *et al.*, 2006; Mikkelsen *et al.*, 2007). HOTAIR is a lncRNAs transcribed from the HOXC locus that maintains the repression of HOXD genes prior to differentiation. HOTAIR coordinates histone modifications by acting as a scaffold for the repressive chromatin complexes PRC2 and lysine-specific demethylase 1 (LSD1)/corepressor for element-1-silencing transcription (CoREST) (Tsai *et al.*, 2010). Establishment of H3K27me3 by enhancer of zeste (Ezh2), the catalytic subunit of PRC2 (Margueron *et al.*, 2008), and removal of the euchromatin associated mark histone H3 di-methylated at lysine 4 (H3K4me2) by LSD1 are responsible for repression. Each complex interacts with a specific region of HOTAIR at either the 5′ or 3′ end of the transcript. The distinct RNA secondary structures predicted at each of the binding domains may facilitate complex-specific recognition and agrees with previously proposed mechanism by which a complex lacking a sequence-specific DNA binding subunit could target specific loci. Disrupting these sequences prevents the RNA–protein interactions in vitro and reducing the levels of HOTAIR leads to the de-repression of HOXD genes as well as other genomic targets (Tsai *et al.*, 2010).

Transcriptional regulation

Repression and activation of gene expression occurs *in* cis and *in* trans. Cis-acting lncRNAs activity is confined to the location of its transcription where it interacts with nearby genes and those on the same chromosome, whereas trans-acting lncRNAs typically function on a larger scale to control expression of many genes within transcriptional networks.

HOTTIP (HOXA transcript at the distal tip) is a lincRNA that behaves in cis to activate expression within the developmentally regulated HOXA gene cluster (Wang *et al.*, 2011). HOTTIP directly binds WD repeat domain 5 (WDR5), a subunit of the histone methyltransferase (HMTase) complex Mixed Lineage Leukaemia (MLL), and, through chromatin looping, brings the activating HMTase complex within close proximity to the 5′ HOXA genes resulting in the deposition of H3K4me3. HOTTIP expression, long-range chromosomal interactions, and HOXA gene expression were present at distal and/or posterior anatomical sites during development which points to this lncRNA as a temporal and spatially specific regulator (Wang *et al.*, 2011).

The act of transcription, rather than the transcript itself can contribute to gene regulation through a mechanism known as transcriptional interference (Shearwin *et al.*, 2005; Sneppen *et al.*, 2005). The SER3 gene in *Saccharomyces cerevisiae* encodes a protein involved in serine biosynthesis and its repression in rich medium prevents unnecessary serine synthesis. Transcription of the lncRNA, SRG1, across the promoter of SER3 interferes with the binding of Gal4, a transcription factor required for SER expression (Martens *et al.*, 2004).

Although considered a separate class of non-coding RNA, enhancer derived RNAs (eRNAs) have functional overlap with other lncRNAs through the use of similar mechanisms (Darrow and Chadwick, 2013; De Santa *et al.*, 2010; Hah *et al.*, 2011; Kim *et al.*, 2010; Li *et al.*, 2013). eRNAs can associate with structural complexes such as Mediator (Lai *et al.*, 2013) and Cohesin (Li *et al.*, 2013) to facilitate long-range interactions necessary for transcriptional activation or positive modulation of target genes (Schmidt *et al.*, 2010). A number of similarities between the two classes mottle their definition, but the determining factor is the origin of transcription and chromatin signature. LincRNAs are transcribed from intergenic regions void of underlying enhancer activity and have H3K36me3 throughout transcriptional unit, where as eRNAs originate from within enhancer sequences characterized by coactivator binding and are typically characterized by a high histone H3 mono-methylated at lysine 4 (H3K4me1) to H3K4me3 ratio (De Santa *et al.*, 2010; Koch *et al.*, 2011; Pekowska *et al.*, 2011). Many eRNAs are bidirectionally transcribed from a promoter within the enhancer region, although some only show unidirectional transcription (De Santa *et al.*, 2010; Hah *et al.*, 2011; Kim *et al.*, 2010; Lam *et al.*, 2013). Importantly, in several cases, eRNA itself, not the act of transcription has been shown as the critical component for the long-range interactions between promoters and enhancers (Hah *et al.*, 2011; Lai *et al.*, 2013; Lam *et al.*, 2013; Li *et al.*, 2013). Although transcription from active enhancers is common and functional roles of transcripts have been reported, the natural association of RNA Pol II with enhancers still raises the question of what proportion of eRNAs are involved in enhancer directed gene activation (Struhl, 2007).

mRNA maturation and transport

Prior to exiting the nucleus, most pre-mRNAs are spliced to produce mature mRNA. Nuclear speckles are sub-nuclear domains enriched with splicing factors such as the serine/arginine-rich (SR) proteins that are responsible for recognizing exon-splicing enhancers and the recruitment of other splicing factors to form the spliceosome. MALAT1 is enriched at nuclear speckles during interphase. The 5′ end of MALAT1 is capable of binding several SR proteins and localizing them to the nuclear speckles as well as controlling the distribution and ratio of phosphorylated to dephosphorylated SR proteins within the nucleus, thereby influencing alternative splicing of many pre-mRNAs (Tripathi *et al.*, 2010). The exact mechanism of its control over the levels of active SR proteins is not understood; however, it is clear that MALAT1 mediates alternative splicing of many mRNAs though its association with SR proteins (Tripathi *et al.*, 2010). Although it is not vital during mouse development (Zhang *et al.*, 2012) or for viability of a lung carcinoma cell line (Eissmann *et al.*, 2012), it does contribute, in certain tissues, to the formation of paraspeckles (Nakagawa *et al.*, 2012).

mRNA stability

BACE1-AS is an antisense transcript that originates within the gene for beta-secretase-1 (BASE1), a protein expressed in the brain and required for normal neurological functions. Decreases in BASE1 levels cause neurological abnormalities, and increased levels have been detected in the brains of patients with Alzheimer's disease. The orthologous sense–antisense transcriptional unit present in mice suggests a conserved function in mammals. BACE1-AS is another example of a lncRNA that influences gene expression through more than one mechanism (Faghihi *et al.*, 2008, 2010). Cell stress causes an up-regulation of the antisense transcript, which concordantly regulates BACE1 transcription; however, the mechanism of this co-regulation is unclear. The complementarity between BACE1-AS and BACE1 enables the formation of a RNA duplex capable of stabilizing the transcript and protecting the overlapping regions from degradation in an RNase protection assay (Faghihi *et al.*, 2008). A subsequent study improved upon the model and found that BACE1-AS inhibits miRNA-induced BACE1 mRNA degradation within the cytoplasm by competing with miR-485-5p for binding of the mRNA (Faghihi *et al.*, 2010). Transcription and translation of BACE1 is tightly controlled in healthy individuals, whereas in Alzheimer's patients expression of both in patients is higher than those of controls individuals.

LncRNAs can also function to promote degradation of mRNA. Some mRNAs have secondary structures within their 3′ untranslated regions (UTRs) that are recognized by the Staufen double-stranded RNA binding protein STAU1. STAU1 binding recruits UPF1, a component of nonsense-mediated decay, and leads to the degradation of select mRNAs. Some STAU1 targets without the necessary 3′ UTR secondary structures rely on association with the lncRNA 1/2-sbsRNA (half-STAU1 binding site). Partial base pairing between Alu elements in the 3′ UTR of some mRNA and 1/2-sbsRNA create STAU1 binding sites sufficient to direct the transcripts down the pathway of STAU1-medited mRNA decay (Gong and Maquat, 2011).

LncRNA are regulated in similar manners to PCG and their involvement in a number of diseases makes them potential therapeutic targets (Spizzo *et al.*, 2012; Wahlestedt, 2013; Wheeler *et al.*, 2012). One of the original studies linking antisense ncRNA transcripts to gene repression, identified p15AS, an antisense transcript to the tumour suppressor p15 as the cause of local heterochromatin formation and DNA methylation upon stem cell differentiation (Yu *et al.*, 2008). Interestingly siRNA and miRNA targeting the tumour suppressor p21, E-cadherin, or vascular endothelial growth factor (VEGF) resulted in higher, rather than lower expression (Li *et al.*, 2006). In the case of p21, a tumour suppressor gene, the siRNA targeted an antisense transcript responsible for moderating the expression of the p21 mRNA. While mRNA and antisense transcripts were expressed at comparable levels in untreated cells, post-transcriptional repression of the antisense transcript with siRNA resulted in increased mRNA expression as well as a reduction of the low levels of H3K27me3 and Argonaute 1 at the sense promoter. This presents a model whereby inappropriate antisense transcription could be a feature of epigenetically silenced tumour suppressor genes in cancer and a potential avenue for RNAi-mediated therapeutic strategies (Morris *et al.*, 2008).

Translational regulation

Several lncRNAs are involved specifically in regulating translation through association with ribosome complexes (van Heesch *et al.*, 2014). AS Uchl1 is an antisense transcript in mice that overlaps the first two exons of ubiquitin carboxy-terminal hydrolase L1 (Uchl1) gene, and extends past the 5′ end before it splices into two repetitive elements (Carrieri *et al.*, 2012). This sense–antisense relationship is conserved among mammals. Base pairing between the 5′ complementary sense and antisense transcripts in the cytoplasm increases Uchl1 translation by recruiting ribosomes. The 5′ end of AS Uchl1, which is complementary to Uchl1, and the short interspersed nuclear element B2 (SINEB2) repeat element at the 3′ end are required for translational up-regulation. While the exact mechanism underlying ribosomal recruitment is unclear, the same study identified another sense–antisense SINEB2-containing transcriptional unit that underwent translational regulation, suggesting that this may be a means of regulating translation.

Dual roles

LincRNA-p21 behaves as a transcriptional and translational repressor (Huarte *et al.*, 2010). The translation of the jun B proto-oncogene (JUNB) and the catenin (cadherin-associated protein), beta 1 (CTNNB1) mRNA, is repressed through partial base pairing and recruitment of RCK, a DEAD box helicase, which leads to reduced polysome sizes. HuR controls the lincRNA-p21 stability and consequently its effects on JUNB and CTNNB1 by binding the lncRNA and recruiting the miRNA let-7, which leads to degradation by the RNA-induced silencing complex (RISC) (Yoon *et al.*, 2012). LincRNA-21 is one of several lincRNAs regulated by the tumour suppressor protein p53. LincRNA transcription is induced by p53 binding at its promoter. LincRNA-p21 binds hnRNP-K (heterogeneous nuclear riboprotein K) and guides it to specific locations where it transcriptionally represses p53 target genes to induce apoptosis (Huarte *et al.*, 2010).

Regulating the regulators

While the functions of many lncRNA are known, the mechanisms responsible for regulating their expression are less explored. Although their purposes are quite different, the same mechanisms responsible for regulating PCGs have been found to operate on lncRNAs. This is unsurprising as often the main difference between the two types of RNA is coding potential.

Transcription factor binding

There is some evidence to suggest that while the mechanism are the same, certain lncRNA promoters may be more sensitive to TF binding than neighbouring PCGs (Zheng *et al.*, 2014). Whether this is a consequence of increased TF recruitment by lncRNA promoters or more efficient conversion of TF binding signals, remains to be seen.

Several indirect roles for miRNA in lncRNA regulation exist. A recent study found an indirect role for miRNA in the lncRNA regulation by way of targeting cMyc, a regulator of pluripotency in ES cells (Zheng *et al.*, 2014). Loss of Dicer, a RNase III enzyme necessary for maturation of miRNA, unexpectedly resulted in down-regulation of approximately 17% of the detected lncRNAs in mESC. Rather than having an affect on the degradation of lncRNA transcripts, which would indicate direct miRNA involvement, they found reduced rates of transcription and decreased H3K36me3 in the gene body, indicating that the down-regulation was taking place at the level of transcription, rather than transcript stability. cMyc, a transcription factor involved in lncRNA regulation, was enriched specifically at the promoter regions of many down-regulated lncRNA and, additionally, cMyc expression was reduced 10-fold in the Dicer KO. The importance of cMyc expression in regulating lncRNA was confirmed when enforced expression of cMyc in Dicer −/− cells was able to rescue the expression of most of the lncRNAs that were originally down-regulated. Dicer indirectly affects lncRNA regulation by preventing maturation of miRNAs that regulate cMyc. Many of cMyc-regulated lncRNAs were among those with roles in maintaining pluripotency and directing differentiation (Guttman *et al.*, 2011).

Concluding remarks

Clearly, while our appreciation and understanding of the many roles of lncRNA in nuclear processes has increased dramatically over the past decade, there is still much to be learned and discovered about this versatile class of RNA. lncRNAs are closely aligned with gene regulation and epigenetic processes, and as they garner more attention due to their connection with serious health issues, such as diabetes and obesity, new functional assays and improved analytical tools will drive the field onward and further enlighten us about this enigmatic class of molecule.

References

Arthanari, Y., Heintzen, C., Griffiths-Jones, S., and Crosthwaite, S.K. (2014). Natural antisense transcripts and long non-coding RNA in Neurospora crassa. PLoS ONE 9, e91353.

Babak, T., Blencowe, B.J., and Hughes, T.R. (2005). A systematic search for new mammalian noncoding RNAs indicates little conserved intergenic transcription. BMC Genomics 6, 104.

Banfai, B., Jia, H., Khatun, J., Wood, E., Risk, B., Gundling, W.E. Jr., Kundaje, A., Gunawardena, H.P., Yu, Y., Xie, L., *et al.* (2012). Long noncoding RNAs are rarely translated in two human cell lines. Genome Res. *22*, 1646–1657.

Barski, A., Cuddapah, S., Cui, K., Roh, T.Y., Schones, D.E., Wang, Z., Wei, G., Chepelev, I., and Zhao, K. (2007). High-resolution profiling of histone methylations in the human genome. Cell *129*, 823–837.

Beckedorff, F.C., Ayupe, A.C., Crocci-Souza, R., Amaral, M.S., Nakaya, H.I., Soltys, D.T., Menck, C.F., Reis, E.M., and Verjovski-Almeida, S. (2013). The intronic long noncoding RNA ANRASSF1 recruits PRC2 to the RASSF1A promoter, reducing the expression of RASSF1A and increasing cell proliferation. PLoS Genet. *9*, e1003705.

Bernstein, B.E., Mikkelsen, T.S., Xie, X., Kamal, M., Huebert, D.J., Cuff, J., Fry, B., Meissner, A., Wernig, M., Plath, K., *et al.* (2006). A bivalent chromatin structure marks key developmental genes in embryonic stem cells. Cell *125*, 315–326.

Brannan, C.I., Dees, E.C., Ingram, R.S., and Tilghman, S.M. (1990). The product of the H19 gene may function as an RNA. Mol. Cell. Biol. *10*, 28–36.

Brockdorff, N., Ashworth, A., Kay, G.F., Mccabe, V.M., Norris, D.P., Cooper, P.J., Swift, S., and Rastan, S. (1992). The product of the mouse Xist gene is a 15 kb inactive X-specific transcript containing no conserved ORF and located in the nucleus. Cell *71*, 515–526.

Brown, C.J., Ballabio, A., Rupert, J.L., Lafreniere, R.G., Grompe, M., Tonlorenzi, R., and Willard, H.F. (1991). A gene from the region of the human X-inactivation center is expressed exclusively from the inactive X-chromosome. Nature *349*, 38–44.

Brown, C.J., Hendrich, B.D., Rupert, J.L., Lafreniere, R.G., Xing, Y., Lawrence, J., and Willard, H.F. (1992). The human Xist gene – analysis of a 17 kb inactive X-specific RNA that contains conserved repeats and is highly localized within the nucleus. Cell *71*, 527–542.

Cabili, M.N., Trapnell, C., Goff, L., Koziol, M., Tazon-Vega, B., Regev, A., and Rinn, J.L. (2011). Integrative annotation of human large intergenic noncoding RNAs reveals global properties and specific subclasses. Gene Dev. *25*, 1915–1927.

Carninci, P., Kasukawa, T., Katayama, S., Gough, J., Frith, M.C., Maeda, N., Oyama, R., Ravasi, T., Lenhard, B., Wells, C., *et al.* (2005). The transcriptional landscape of the mammalian genome. Science *309*, 1559–1563.

Carrieri, C., Cimatti, L., Biagioli, M., Beugnet, A., Zucchelli, S., Fedele, S., Pesce, E., Ferrer, I., Collavin, L., Santoro, C., *et al.* (2012). Long non-coding antisense RNA controls Uchl1 translation through an embedded SINEB2 repeat. Nature *491*, 454–457.

Cesana, M., Cacchiarelli, D., Legnini, I., Santini, T., Sthandier, O., Chinappi, M., Tramontano, A., and Bozzoni, I. (2011). A long noncoding RNA controls muscle differentiation by functioning as a competing endogenous RNA. Cell *147*, 358–369.

Cheng, J., Kapranov, P., Drenkow, J., Dike, S., Brubaker, S., Patel, S., Long, J., Stern, D., Tammana, H., Helt, G., *et al.* (2005). Transcriptional maps of 10 human chromosomes at 5-nucleotide resolution. Science *308*, 1149–1154.

Clemson, C.M., McNeil, J.A., Willard, H.F., and Lawrence, J.B. (1996). XIST RNA paints the inactive X-chromosome at interphase: Evidence for a novel RNA involved in nuclear chromosome structure. J. Cell Biol. *132*, 259–275.

Core, L.J., Waterfall, J.J., and Lis, J.T. (2008). Nascent RNA sequencing reveals widespread pausing and divergent initiation at human promoters. Science *322*, 1845–1848.

Darrow, E.M., and Chadwick, B.P. (2013). Boosting transcription by transcription: enhancer-associated transcripts. Chromosome Res. *21*, 713–724.

Derrien, T., Johnson, R., Bussotti, G., Tanzer, A., Djebali, S., Tilgner, H., Guernec, G., Martin, D., Merkel, A., Knowles, D.G., *et al.* (2012). The GENCODE v7 catalog of human long noncoding RNAs: analysis of their gene structure, evolution, and expression. Genome Res. *22*, 1775–1789.

De Santa, F., Barozzi, I., Mietton, F., Ghisletti, S., Polletti, S., Tusi, B.K., Muller, H., Ragoussis, J., Wei, C.L., and Natoli, G. (2010). A large fraction of extragenic RNA pol II transcription sites overlap enhancers. PLoS Biol. *8*, e1000384.

Djebali, S., Davis, C.A., Merkel, A., Dobin, A., Lassmann, T., Mortazavi, A., Tanzer, A., Lagarde, J., Lin, W., Schlesinger, F., *et al.* (2012). Landscape of transcription in human cells. Nature *489*, 101–108.

Eissmann, M., Gutschner, T., Hammerle, M., Gunther, S., Caudron-Herger, M., Gross, M., Schirmacher, P., Rippe, K., Braun, T., Zornig, M., *et al.* (2012). Loss of the abundant nuclear non-coding RNA MALAT1 is compatible with life and development. RNA Biol. *9*, 1076–1087.

Faghihi, M.A., Modarresi, F., Khalil, A.M., Wood, D.E., Sahagan, B.G., Morgan, T.E., Finch, C.E., Laurent, G.S., Kenny, P.J., and Wahlestedt, C. (2008). Expression of a noncoding RNA is elevated in Alzheimer's disease and drives rapid feed-forward regulation of beta-secretase. Nat. Med. *14*, 723–730.

Faghihi, M.A., Zhang, M., Huang, J., Modarresi, F., Van der Brug, M.P., Nalls, M.A., Cookson, M.R., St-Laurent, G., and Wahlestedt, C. (2010). Evidence for natural antisense transcript-mediated inhibition of microRNA function. Genome Biol. *11*, 1–13.

Goff, L.A., and Rinn, J.L. (2013). Poly-combing the genome for RNA. Nat. Struct. Mol. Biol. *20*, 1344–1346.

Gong, C.G., and Maquat, L.E. (2011). lncRNAs transactivate STAU1-mediated mRNA decay by duplexing with 3' UTRs via Alu elements. Nature *470*, 284–288.

Guttman, M., Amit, I., Garber, M., French, C., Lin, M.F., Feldser, D., Huarte, M., Zuk, O., Carey, B.W., Cassady, J.P., *et al.* (2009). Chromatin signature reveals over a thousand highly conserved large non-coding RNAs in mammals. Nature *458*, 223–227.

Guttman, M., Garber, M., Levin, J.Z., Donaghey, J., Robinson, J., Adiconis, X., Fan, L., Koziol, M.J., Gnirke, A., Nusbaum, C., *et al.* (2010). *Ab initio* reconstruction of cell type-specific transcriptomes in mouse reveals the conserved multi-exonic structure of lincRNAs. Nat. Biotechnol. *28*, 503–510.

Guttman, M., Donaghey, J., Carey, B.W., Garber, M., Grenier, J.K., Munson, G., Young, G., Lucas, A.B., Ach, R., Bruhn, L., *et al.* (2011). lincRNAs act in the circuitry controlling pluripotency and differentiation. Nature *477*, 295–300.

Guttman, M., Russell, P., Ingolia, N.T., Weissman, J.S., and Lander, E.S. (2013). ribosome profiling provides evidence that large noncoding RNAs do not encode proteins. Cell *154*, 240–251.

Hah, N., Danko, C.G., Core, L., Waterfall, J.J., Siepel, A., Lis, J.T., and Kraus, W.L. (2011). A rapid, extensive, and transient transcriptional response to estrogen signaling in breast cancer cells. Cell *145*, 622–634.

Hah, N., Murakami, S., Nagari, A., Danko, C.G., and Kraus, W.L. (2013). Enhancer transcripts mark active estrogen receptor binding sites. Genome Res. *23*, 1210–1223.

Han, Y.J., Ma, S.F., Yourek, G., Park, Y.D., and Garcia, J.G. (2011). A transcribed pseudogene of MYLK promotes cell proliferation. FASEB J. *25*, 2305–2312.

He, Y.P., Vogelstein, B., Velculescu, V.E., Papadopoulos, N., and Kinzler, K.W. (2008). The antisense transcriptomes of human cells. Science *322*, 1855–1857.

van Heesch, S., van Iterson, M., Jacobi, J., Boymans, S., Essers, P.B., de Bruijn, E., Hao, W., Macinnes, A.W., Cuppen, E., and Simonis, M. (2014). Extensive localization of long noncoding RNAs to the cytosol and mono- and polyribosomal complexes. Genome Biol. *15*, R6.

Heintzman, N.D., Stuart, R.K., Hon, G., Fu, Y., Ching, C.W., Hawkins, R.D., Barrera, L.O., Van Calcar, S., Qu, C., Ching, K.A., *et al.* (2007). Distinct and predictive chromatin signatures of transcriptional promoters and enhancers in the human genome. Nat. Genet. *39*, 311–318.

Huarte, M., Guttman, M., Feldser, D., Garber, M., Koziol, M.J., Kenzelmann-Broz, D., Khalil, A.M., Zuk, O., Amit, I., Rabani, M., *et al.* (2010). A large intergenic noncoding RNA induced by p53 mediates global gene repression in the p53 response. Cell *142*, 409–419.

Ingolia, N.T., Lareau, L.F., and Weissman, J.S. (2011). Ribosome profiling of mouse embryonic stem cells reveals the complexity and dynamics of mammalian proteomes. Cell *147*, 789–802.

Jeyapalan, Z., Deng, Z., Shatseva, T., Fang, L., He, C., and Yang, B.B. (2011). Expression of CD44 3'-untranslated region regulates endogenous microRNA functions in tumorigenesis and angiogenesis. Nucleic Acids Res. *39*, 3026–3041.

Johnsson, P., Ackley, A., Vidarsdottir, L., Lui, W.O., Corcoran, M., Grander, D., and Morris, K.V. (2013). A pseudogene long-noncoding-RNA network regulates PTEN transcription and translation in human cells. Nat. Struct. Mol. Biol. *20*, 440–446.

Kaikkonen, M.U., Spann, N.J., Heinz, S., Romanoski, C.E., Allison, K.A., Stender, J.D., Chun, H.B., Tough, D.F., Prinjha, R.K., Benner, C., *et al.* (2013). Remodeling of the enhancer landscape during macrophage activation is coupled to enhancer transcription. Mol. Cell *51*, 310–325.

Kampa, D., Cheng, J., Kapranov, P., Yamanaka, M., Brubaker, S., Cawley, S., Drenkow, J., Piccolboni, A., Bekiranov, S., Helt, G., *et al.* (2004). Novel RNAs identified from an in-depth analysis of the transcriptome of human chromosomes 21 and 22. Genome Res. *14*, 331–342.

Kapranov, P., Cawley, S.E., Drenkow, J., Bekiranov, S., Strausberg, R.L., Fodor, S.P., and Gingeras, T.R. (2002). Large-scale transcriptional activity in chromosomes 21 and 22. Science *296*, 916–919.

Kapranov, P., Drenkow, J., Cheng, J., Long, J., Helt, G., Dike, S., and Gingeras, T.R. (2005). Examples of the complex architecture of the human transcriptome revealed by RACE and high-density tiling arrays. Genome Res. *15*, 987–997.

Kapranov, P., Cheng, J., Dike, S., Nix, D.A., Duttagupta, R., Willingham, A.T., Stadler, P.F., Hertel, J., Hackermuller, J., Hofacker, I.L., *et al.* (2007). RNA maps reveal new RNA classes and a possible function for pervasive transcription. Science *316*, 1484–1488.

Katayama, S., Tomaru, Y., Kasukawa, T., Waki, K., Nakanishi, M., Nakamura, M., Nishida, H., Yap, C.C., Suzuki, M., Kawai, J., *et al.* (2005). Antisense transcription in the mammalian transcriptome. Science *309*, 1564–1566.

Khaitovich, P., Kelso, J., Franz, H., Visagie, J., Giger, T., Joerchel, S., Petzold, E., Green, R.E., Lachmann, M., and Paabo, S. (2006). Functionality of intergenic transcription: an evolutionary comparison. PLoS Genet. *2*, e171.

Khalil, A.M., Guttman, M., Huarte, M., Garber, M., Raj, A., Morales, D.R., Thomas, K., Presser, A., Bernstein, B.E., van Oudenaarden, A., *et al.* (2009). Many human large intergenic noncoding RNAs associate with chromatin-modifying complexes and affect gene expression. Proc. Natl. Acad. Sci. U.S.A. *106*, 11667–11672.

Kim, T.K., Hemberg, M., Gray, J.M., Costa, A.M., Bear, D.M., Wu, J., Harmin, D.A., Laptewicz, M., Barbara-Haley, K., Kuersten, S., *et al.* (2010). Widespread transcription at neuronal activity-regulated enhancers. Nature *465*, 182–187.

Kino, T., Hurt, D.E., Ichijo, T., Nader, N., and Chrousos, G.P. (2010). Noncoding RNA Gas5 is a growth arrest- and starvation-associated repressor of the glucocorticoid receptor. Sci. Signal 3, 1–15.

Koch, F., Fenouil, R., Gut, M., Cauchy, P., Albert, T.K., Zacarias-Cabeza, J., Spicuglia, S., de la Chapelle, A.L., Heidemann, M., Hintermair, C., *et al.* (2011). Transcription initiation platforms and GTF recruitment at tissue-specific enhancers and promoters. Nat. Struct. Mol. Biol. *18*, 956–963.

Korostowski, L., Sedlak, N., and Engel, N. (2012). The Kcnq1ot1 long non-coding RNA affects chromatin conformation and expression of Kcnq1, but does not regulate its imprinting in the developing heart. PLoS Genet. *8*, e1002956.

Kutter, C., Watt, S., Stefflova, K., Wilson, M.D., Goncalves, A., Ponting, C.P., Odom, D.T., and Marques, A.C. (2012). Rapid turnover of long noncoding RNAs and the evolution of gene expression. PLoS Genet. *8*, 1–15.

Lai, F., Orom, U.A., Cesaroni, M., Beringer, M., Taatjes, D.J., Blobel, G.A., and Shiekhattar, R. (2013). Activating RNAs associate with mediator to enhance chromatin architecture and transcription. Nature *494*, 497–501.

Lam, M.T., Cho, H., Lesch, H.P., Gosselin, D., Heinz, S., Tanaka-Oishi, Y., Benner, C., Kaikkonen, M.U., Kim, A.S., Kosaka, M., *et al.* (2013). Rev-Erbs repress macrophage gene expression by inhibiting enhancer-directed transcription. Nature *498*, 511–515.

Lander, E.S., Linton, L.M., Birren, B., Nusbaum, C., Zody, M.C., Baldwin, J., Devon, K., Dewar, K., Doyle, M., FitzHugh, W., *et al.* (2001). Initial sequencing and analysis of the human genome. Nature *409*, 860–921.

Li, L.C., Okino, S.T., Zhao, H., Pookot, D., Place, R.F., Urakami, S., Enokida, H., and Dahiya, R. (2006). Small dsRNAs induce transcriptional activation in human cells. Proc. Natl. Acad. Sci. U.S.A. *103*, 17337–17342.

Li, W., Notani, D., Ma, Q., Tanasa, B., Nunez, E., Chen, A.Y., Merkurjev, D., Zhang, J., Ohgi, K., Song, X., *et al.* (2013). Functional roles of enhancer RNAs for oestrogen-dependent transcriptional activation. Nature *498*, 516–520.

Livyatan, I., Harikumar, A., Nissim-Rafinia, M., Duttagupta, R., Gingeras, T.R., and Meshorer, E. (2013). Non-polyadenylated transcription in embryonic stem cells reveals novel non-coding RNA related to pluripotency and differentiation. Nucleic Acids Res. *41*, 6300–6315.

Lyon, M.F. (1961). Gene action in the X-chromosome of the mouse (Mus musculus L.). Nature *190*, 372–373.

Margueron, R., Li, G.H., Sarma, K., Blais, A., Zavadil, J., Woodcock, C.L., Dyniacht, B.D., and Reinberg, D. (2008). Ezh1 and Ezh2 maintain repressive chromatin through different mechanisms. Mol. Cell 32, 503–518.

Martens, J.A., Laprade, L., and Winston, F. (2004). Intergenic transcription is required to repress the Saccharomyces cerevisiae SER3 gene. Nature *429*, 571–574.

Mercer, T.R., Dinger, M.E., and Mattick, J.S. (2009). Long non-coding RNAs: insights into functions. Nat. Rev. Genet. *10*, 155–159.

Mercer, T.R., Gerhardt, D.J., Dinger, M.E., Crawford, J., Trapnell, C., Jeddeloh, J.A., Mattick, J.S., and Rinn, J.L. (2012). Targeted RNA sequencing reveals the deep complexity of the human transcriptome. Nat. Biotechnol. *30*, 99–104.

Mikkelsen, T.S., Ku, M.C., Jaffe, D.B., Issac, B., Lieberman, E., Giannoukos, G., Alvarez, P., Brockman, W., Kim, T.K., Koche, R.P., *et al.* (2007). Genome-wide maps of chromatin state in pluripotent and lineage-committed cells. Nature *448*, 553-U552.

Moran, V.A., Perera, R.J., and Khalil, A.M. (2012). Emerging functional and mechanistic paradigms of mammalian long non-coding RNAs. Nucleic Acids Res. *40*, 6391–6400.

Morris, K.V., and Mattick, J.S. (2014). The rise of regulatory RNA. Nat. Rev. Genet. *15*, 423–437.

Morris, K.V., Santoso, S., Turner, A.M., Pastori, C., and Hawkins, P.G. (2008). Bidirectional transcription directs both transcriptional gene activation and suppression in human cells. PLoS Genet. *4*, 1–9.

Mortazavi, A., Williams, B.A., McCue, K., Schaeffer, L., and Wold, B. (2008). Mapping and quantifying mammalian transcriptomes by RNA-Seq. Nat. Methods *5*, 621–628.

Nakagawa, S., Naganuma, T., Shioi, G., and Hirose, T. (2011). Paraspeckles are subpopulation-specific nuclear bodies that are not essential in mice. J. Cell Biol. *193*, 31–39.

Nakaya, H.I., Amaral, P.P., Louro, R., Lopes, A., Fachel, A.A., Moreira, Y.B., El-Jundi, T.A., da Silva, A.M., Reis, E.M., and Verjovski-Almeida, S. (2007). Genome mapping and expression analyses of human intronic noncoding RNAs reveal tissue-specific patterns and enrichment in genes related to regulation of transcription. Genome Biol. *8*, 1–25.

Pang, K.C., Frith, M.C., and Mattick, J.S. (2006). Rapid evolution of noncoding RNAs: lack of conservation does not mean lack of function. Trends Genet. *22*, 1–5.

Pei, B., Sisu, C., Frankish, A., Howald, C., Habegger, L., Mu, X.J., Harte, R., Balasubramanian, S., Tanzer, A., Diekhans, M., *et al.* (2012). The GENCODE pseudogene resource. Genome Biol. *13*, R51.

Pekowska, A., Benoukraf, T., Zacarias-Cabeza, J., Belhocine, M., Koch, F., Holota, H., Imbert, J., Andrau, J.-C., Ferrier, P., and Spicuglia, S. (2011). H3K4 tri-methylation provides an epigenetic signature of active enhancers. EMBO J. *30*, 4198–4210.

Poliseno, L., Salmena, L., Zhang, J., Carver, B., Haveman, W.J., and Pandolfi, P.P. (2010). A coding-independent function of gene and pseudogene mRNAs regulates tumour biology. Nature *465*, 1033–1038.

Ponjavic, J., Ponting, C.P., and Lunter, G. (2007). Functionality or transcriptional noise? Evidence for selection within long noncoding RNAs. Genome Res. *17*, 556–565.

Rapicavoli, N.A., Qu, K., Zhang, J.J., Mikhail, M., Laberge, R.M., and Chang, H.Y. (2013). A mammalian pseudogene lncRNA at the interface of inflammation and anti-inflammatory therapeutics. Elife *2*, 1–16.

Rinn, J.L., Euskirchen, G., Bertone, P., Martone, R., Luscombe, N.M., Hartman, S., Harrison, P.M., Nelson, F.K., Miller, P., Gerstein, M., *et al.* (2003). The transcriptional activity of human chromosome 22. Genes Dev. *17*, 529–540.

Rinn, J.L., Kertesz, M., Wang, J.K., Squazzo, S.L., Xu, X., Brugmann, S.A., Goodnough, L.H., Helms, J.A., Farnham, P.J., Segal, E., *et al.* (2007). Functional demarcation of active and silent chromatin domains in human HOX loci by noncoding RNAs. Cell *129*, 1311–1323.

Salmena, L., Poliseno, L., Tay, Y., Kats, L., and Pandolfi, P.P. (2011). A ceRNA hypothesis: the Rosetta Stone of a hidden RNA language? Cell *146*, 353–358.

Sati, S., Ghosh, S., Jain, V., Scaria, V., and Sengupta, S. (2012). Genome-wide analysis reveals distinct patterns of epigenetic features in long non-coding RNA loci. Nucleic Acids Res. *40*, 10018–10031.

Schmidt, D., Schwalie, P.C., Ross-Innes, C.S., Hurtado, A., Brown, G.D., Carroll, J.S., Flicek, P., and Odom, D.T. (2010). A CTCF-independent role for cohesin in tissue-specific transcription. Genome Res. *20*, 578–588.

Shearwin, K.E., Callen, B.P., and Egan, J.B. (2005). Transcriptional interference – a crash course. Trends Genet. *21*, 339–345.

Sigova, A.A., Mullen, A.C., Molinie, B., Gupta, S., Orlando, D.A., Guenther, M.G., Almada, A.E., Lin, C., Sharp, P.A., Giallourakis, C.C., *et al.* (2013). Divergent transcription of long noncoding RNA/mRNA gene pairs in embryonic stem cells. Proc. Natl. Acad. Sci. U.S.A. *110*, 2876–2881.

Simon, M.D., Pinter, S.F., Fang, R., Sarma, K., Rutenberg-Schoenberg, M., Bowman, S.K., Kesner, B.A., Maier, V.K., Kingston, R.E., and Lee, J.T. (2013). High-resolution Xist binding maps reveal two-step spreading during X-chromosome inactivation. Nature *504*, 465–459.

Sneppen, K., Dodd, I.B., Shearwin, K.E., Palmer, A.C., Schubert, R.A., Callen, B.P., and Egan, J.B. (2005). A mathematical model for transcriptional interference by RNA polymerase traffic in *Escherichia coli*. J. Mol. Biol. *346*, 399–409.

Spizzo, R., Almeida, M.I., Colombatti, A., and Calin, G.A. (2012). Long non-coding RNAs and cancer: a new frontier of translational research? Oncogene *31*, 4577–4587.

Struhl, K. (2007). Transcriptional noise and the fidelity of initiation by RNA polymerase II. Nat. Struct. Mol. Biol. *14*, 103–105.

Trapnell, C., Williams, B.A., Pertea, G., Mortazavi, A., Kwan, G., van Baren, M.J., Salzberg, S.L., Wold, B.J., and Pachter, L. (2010). Transcript assembly and quantification by RNA-Seq reveals unannotated transcripts and isoform switching during cell differentiation. Nat. Biotechnol. *28*, 511–515.

Tripathi, V., Ellis, J.D., Shen, Z., Song, D.Y., Pan, Q., Watt, A.T., Freier, S.M., Bennett, C.F., Sharma, A., Bubulya, P.A., *et al.* (2010). The nuclear-retained noncoding RNA MALAT1 regulates alternative splicing by modulating SR splicing factor phosphorylation. Mol. Cell *39*, 925–938.

Tsai, M.C., Manor, O., Wan, Y., Mosammaparast, N., Wang, J.K., Lan, F., Shi, Y., Segal, E., and Chang, H.Y. (2010). Long noncoding RNA as modular scaffold of histone modification complexes. Science *329*, 689–693.

Ulitsky, I., Shkumatava, A., Jan, C.H., Sive, H., and Bartel, D.P. (2011). Conserved function of lincRNAs in vertebrate embryonic development despite rapid sequence evolution. Cell *147*, 1537–1550.

Vallot, C., Huret, C., Lesecque, Y., Resch, A., Oudrhiri, N., Bennaceur-Griscelli, A., Duret, L., and Rougeulle, C. (2013). XACT, a long noncoding transcript coating the active X-chromosome in human pluripotent cells. Nat. Genet. *45*, 239–241.

Venter, J.C., Adams, M.D., Myers, E.W., Li, P.W., Mural, R.J., Sutton, G.G., Smith, H.O., Yandell, M., Evans, C.A., Holt, R.A., *et al.* (2001). The sequence of the human genome. Science *291*, 1304–1351.

Wahlestedt, C. (2013). Targeting long non-coding RNA to therapeutically upregulate gene expression. Nat. Rev. Drug Discov. *12*, 433–446.

Wang, H., Chung, P.J., Liu, J., Jang, I.C., Kean, M.J., Xu, J., and Chua, N.H. (2014). Genome-wide identification of long noncoding natural antisense transcripts and their responses to light in Arabidopsis. Genome Res. *24*, 444–453.

Wang, K., Liu, F., Zhou, L.Y., Long, B., Yuan, S.M., Wang, Y., Liu, C.Y., Sun, T., Zhang, X.J., and Li, P.F. (2014a). The long noncoding RNA CHRF regulates cardiac hypertrophy by targeting miR-489. Circ. Res. *114*, 1377–1388.

Wang, K., Long, B., Zhou, L.Y., Liu, F., Zhou, Q.Y., Liu, C.Y., Fan, Y.Y., and Li, P.F. (2014b). CARL lncRNA inhibits anoxia-induced mitochondrial fission and apoptosis in cardiomyocytes by impairing miR-539-dependent PHB2 downregulation. Nat. Commun. *5*, 3596.

Wang, K.C., Yang, Y.W., Liu, B., Sanyal, A., Corces-Zimmerman, R., Chen, Y., Lajoie, B.R., Protacio, A., Flynn, R.A., Gupta, R.A., *et al.* (2011). A long noncoding RNA maintains active chromatin to coordinate homeotic gene expression. Nature *472*, 120–124.

Washietl, S., Will, S., Hendrix, D.A., Goff, L.A., Rinn, J.L., Berger, B., and Kellis, M. (2012). Computational analysis of noncoding RNAs. Wires RNA 3, 759–778.

Washietl, S., Kellis, M., and Garber, M. (2014). Evolutionary dynamics and tissue specificity of human long noncoding RNAs in six mammals. Genome Res. *24*, 616–628.

Wheeler, T.M., Leger, A.J., Pandey, S.K., MacLeod, A.R., Nakamori, M., Cheng, S.H., Wentworth, B.M., Bennett, C.F., and Thornton, C.A. (2012). Targeting nuclear RNA for *in vivo* correction of myotonic dystrophy. Nature *488*, 111–115.

Wilusz, J.E., Freier, S.M., and Spector, D.L. (2008). 3° End processing of a long nuclear-retained noncoding RNA yields a tRNA-like cytoplasmic RNA. Cell *135*, 919–932.

Wu, L., Murat, P., Matak-Vinkovic, D., Murrell, A., and Balasubramanian, S. (2013). Binding interactions between long noncoding RNA HOTAIR and PRC2 proteins. Biochemistry *52*, 9519–9527.

Yamashita, R., Sathira, N.P., Kanai, A., Tanimoto, K., Arauchi, T., Tanaka, Y., Hashimoto, S.-i., Sugano, S., Nakai, K., and Suzuki, Y. (2011). Genome-wide characterization of transcriptional start sites in humans by integrative transcriptome analysis. Genome Res. *21*, 775–789.

Yang, L., Duff, M.O., Graveley, B.R., Carmichael, G.G., and Chen, L.L. (2011). Genomewide characterization of non-polyadenylated RNAs. Genome Biol.*12*, R16.

Yoon, J.H., Abdelmohsen, K., Srikantan, S., Yang, X.L., Martindale, J.L., De, S., Huarte, M., Zhan, M., Becker, K.G., and Gorospe, M. (2012). LincRNA-p21 Suppresses Target mRNA Translation. Mol. Cell *47*, 648–655.

Yoon, J.H., Abdelmohsen, K., Kim, J., Yang, X.L., Martindale, J.L., Tominaga-Yamanaka, K., White, E.J., Orjalo, A.V., Rinn, J.L., Kreft, S.G., *et al.* (2013). Scaffold function of long non-coding RNA HOTAIR in protein ubiquitination. Nat. Commun. *4*, 1–14.

Yu, W.Q., Gius, D., Onyango, P., Muldoon-Jacobs, K., Karp, J., Feinberg, A.P., and Cui, H.M. (2008). Epigenetic silencing of tumour suppressor gene p15 by its antisense RNA. Nature *451*, 202-U210.

Zhang, B., Arun, G., Mao, Y.S., Lazar, Z., Hung, G.N., Bhattacharjee, G., Xiao, X.K., Booth, C.J., Wu, J., Zhang, C.L., *et al.* (2012). The lncRNA Malat1 is dispensable for mouse development but its transcription plays a cis-regulatory role in the adult. Cell Rep 2, 111–123.

Zhao, J., Sun, B.K., Erwin, J.A., Song, J.J., and Lee, J.T. (2008). Polycomb proteins targeted by a short repeat RNA to the mouse X-chromosome. Science *322*, 750–756.

Zhao, J., Ohsumi, T.K., Kung, J.T., Ogawa, Y., Grau, D.J., Sarma, K., Song, J.J., Kingston, R.E., Borowsky, M., and Lee, J.T. (2010). Genome-wide identification of polycomb-associated RNAs by RIP-seq. Mol. Cell *40*, 939–953.

Zheng, D., Frankish, A., Baertsch, R., Kapranov, P., Reymond, A., Choo, S.W., Lu, Y., Denoeud, F., Antonarakis, S.E., Snyder, M., *et al.* (2007). Pseudogenes in the ENCODE regions: consensus annotation, analysis of transcription, and evolution. Genome Res. *17*, 839–851.

Zheng, G.X., Do, B.T., Webster, D.E., Khavari, P.A., and Chang, H.Y. (2014). Dicer-microRNA-Myc circuit promotes transcription of hundreds of long noncoding RNAs. Nat. Struct. Mol. Biol. *21*, 585–590.

Epigenetic Mechanisms in Rett Syndrome

10

Janine M. LaSalle

Abstract

Mutations in the X-linked gene *MECP2* cause the neurodevelopmental disorder, Rett syndrome, an epigenetic disease at two levels. First, *MECP2* is subject to X-chromosome inactivation and *MECP2* heterozygous female Rett patients are mosaics of both mutant and wild-type expressing cells. Differences in X-chromosome inactivation patterns can therefore affect the phenotypic severity and pathogenesis of Rett syndrome and related disorders of *MECP2* mutation or duplications in females. Second, *MECP2* encodes methyl CpG binding protein 2, a member of the methyl binding domain (MBD) family of epigenetic readers of DNA methylation marks in chromatin. MeCP2 is an inherently disordered protein composed of different isoforms, post-translational modifications, and cofactor associations that dynamically change and mediate a variety of different functions in both neurons and non-neuronal tissues. Investigating both of these levels of epigenetic complexity of *MECP2* is likely to be critical in understanding the molecular pathogenesis and disease progression, as well as developing effective therapies for Rett syndrome.

Introduction to Rett syndrome

Rett syndrome (RTT) is a severe X-linked neurodevelopmental disorder affecting approximately 1 in 10,000 (Armstrong, 1997; Chahrour *et al.*, 2008; Dunn, 2001; Ellaway and Christodoulou, 2001). A hallmark of RTT is the presence of a period of relatively normal developmental milestones in early infancy, followed by a period of rapid regression in motor, social, and language skills between 6 and 18 months of age. The regressive stage is also characterized by an emergence of autistic behaviours, including loss of language and a gain in repetitive behaviours, specifically hand-wringing (Chahrour and Zoghbi, 2007). Girls with RTT frequently develop severe seizures, breathing and feeding difficulties, scoliosis, and loss of motor skills including purposeful hand use and ambulation. The primary neurological feature of RTT is microcephaly, primarily in the impaired growth in head circumference from infancy through the first decade of life (Allanson *et al.*, 2011; Huppke *et al.*, 2003). Most girls with RTT have severe intellectual disability, although some are characterized as a milder 'preserved speech variant' because of their retained ability for limited speech (De Bona *et al.*, 2000; Marschik *et al.*, 2012; Zappella *et al.*, 2001).

RTT is caused by mutations in the gene *MECP2* on Xq28, encoding methyl CpG binding protein 2 (MeCP2) (Amir *et al.*, 1999). RTT is primarily inherited in a X-linked dominant manner arising from *de novo* mutations, where females are heterozygous and affected (Takagi, 2001). However, some rare cases of *MECP2* mutations in males inherited from an unaffected or mildly affected carrier mother have been observed (Meloni *et al.*, 2000; Orrico *et al.*, 2000). These cases will be discussed in greater detail in the section on the influence of X-chromosome inactivation (XCI) in RTT. Because of the influence of XCI in RTT, females with RTT are mosaics for mutant expressing cells, with approximately half of their cells expressing the mutant allele in brain (Ananiev *et al.*, 2011; LaSalle *et al.*, 2001; Shahbazian *et al.*, 2002c).

In addition to *MECP2* mutations, RTT-like features are also caused by mutations in other genes. Mutations in the X-linked *CDKL5* gene cause RTT-like characteristics similar to *MECP2* mutations, with the major distinguishing feature being the presence of early onset developmental delay and seizure onset (Fehr *et al.*, 2013; Hagebeuk *et al.*, 2013). *CDKL5* encodes cyclin-dependent kinase-like 5, that shares several intersecting pathways with MeCP2 functions (Carouge *et al.*, 2010; Kameshita *et al.*, 2008; Lin *et al.*, 2005; Livide *et al.*, 2014; Mari *et al.*, 2005). Mutations in *FOXG1* also cause RTT-like features with the main differentiating feature being infantile spasms and abnormal development in the first 4–5 months (Ariani *et al.*, 2008; Kortum *et al.*, 2011; Mencarelli *et al.*, 2010). *Foxg1* encodes a transcription factor critical for cortical development in mouse models (Kumamoto *et al.*, 2013; MuhChyi *et al.*, 2013; Naidu and Johnston, 2011). Some additional genetic neurodevelopmental disorders with strong overlap with RTT include Angelman syndrome caused by maternal *UBE3A* mutation or deficiency (Tan *et al.*, 2014) or cerebral folate deficiency caused by mutations in a folate receptor gene (Ramaekers *et al.*, 2013). When RTT-like female patients are limited to those showing a period of normal infancy followed by loss of purposeful hand use and language, together with a gain of stereotypical hand wringing behaviour, *MECP2* mutations can be identified in ~95% (Neul *et al.*, 2010). However, the other gene mutations with overlapping features of RTT suggest overlapping molecular pathways of MeCP2 with other critical factors.

Furthermore, *MECP2* mutations, copy number variations (CNVs), or expression deficits have been observed in other disorders beyond RTT. In rare cases of autism or intellectual disability with the other characteristics of RTT, *MECP2* mutations have been described (Hammer *et al.*, 2002; Hoffbuhr *et al.*, 2001; Schanen, 2001; Watson *et al.*, 2001). Duplication of *MECP2* within different sized structural variants has been frequently described in males with autism spectrum disorders or intellectual disability (del Gaudio *et al.*, 2006; Froyen *et al.*, 2007; Lugtenberg *et al.*, 2006; Van Esch *et al.*, 2005). *MECP2* duplication in males is usually inherited from maternal carriers, who are either unaffected or mildly affected with neuropsychiatric conditions (Ramocki *et al.*, 2009). Milder mutations of *MECP2* not observed in RTT females have also been observed in individuals with neuropsychiatric disorders, including aggression, hyperphagia, parkinsonism, and bipolar symptoms (Adegbola *et al.*, 2009; Hanchard *et al.*, 2012; Venkateswaran *et al.*, 2014). Lastly, 79% of postmortem cortical samples from ASDs showed reduced levels of MeCP2 compared with age-matched controls (Nagarajan *et al.*, 2006), suggesting that obtaining the optimal level of MeCP2 expression in the developing brain is a critical common pathway for multiple neurodevelopmental and psychiatric conditions.

Introduction to MeCP2

MeCP2 is an extensively investigated epigenetic regulator of methylated CpG sites in the genome, as a member of the family of methyl binding domain (MBD) proteins (Fan and Hutnick, 2005). Mutations throughout the *MECP2* coding region have been observed in RTT cases, however, the most frequent mutations occur as either missense mutations within the MBD or truncation mutations after the MBD but prior to the transcriptional repression domain (TRD) or C-terminal domain (CTD) (Fig. 10.1). While definitive genotype-phenotype correlations have not been apparent in RTT, C-terminal truncations and the missense mutation R133C are generally milder in severity than other mutation types observed in RTT (Bebbington *et al.*, 2008; Cuddapah *et al.*, 2014). *MECP2* has two alternatively spliced isoforms resulting from the inclusion or exclusion of exon 2, which encodes both a stop codon and an alternative downstream translation initiation codon (Fig. 10.2) (Kriaucionis and Bird, 2004; Mnatzakanian *et al.*, 2004). MeCP2_e1, the ancestrally conserved and most prominent isoform in brain, is translated when exon 2 is included. MeCP2_e2 isoform is created when exon 2 is included in splicing, resulting in a protein that differs from MeCP2_e1 by only a few amino acids in the N-terminus. In addition to alternative splicing, *MECP2* also exhibits

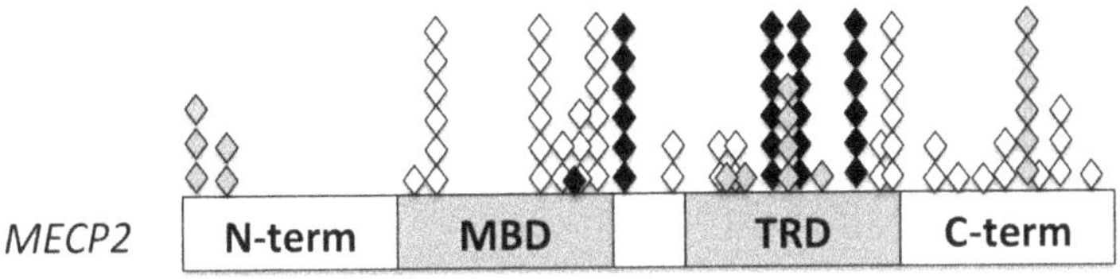

Figure 10.1 Map of Rett syndrome mutations in *MECP2*. The protein structure of known functional domains of MeCP2 is shown with a map of the locations and frequency of the most common mutations. The methyl binding domain (MBD), transcriptional repression domain (TRD), N-terminal, and C-terminal domains are shown. Missense mutations (white diamonds) in RTT are found most frequently in the MBD, and throughout the gene with the exception of the N-terminus. In contrast, frameshift and in frame mutations (grey diamonds) are rarely observed in the MBD, and most frequent in the N-terminal and C-terminal domain. Truncating nonsense mutations (black diamonds) are primarily observed in the TRD and interdomain between the MBD and TRD. The relative frequency of mutation positions are simplified as the number of diamonds, based on the RTT mutation database RettBASE (http://mecp2.chw.edu.au).

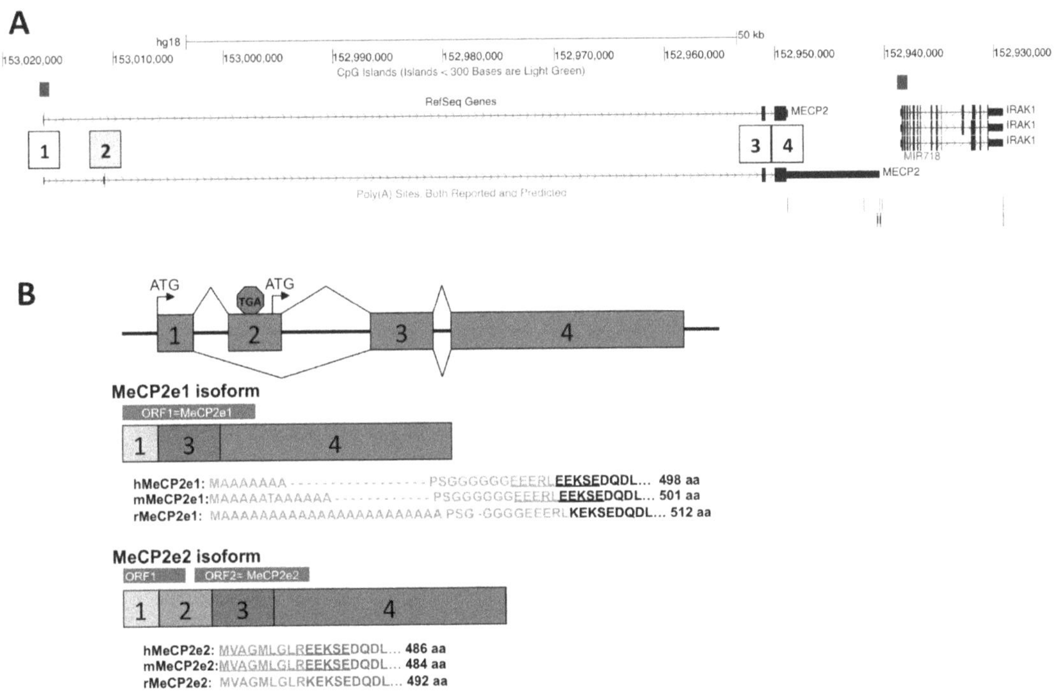

Figure 10.2 Alternative splicing and polyadenylation of *MECP2*. (A) A map from the UCSC Genome Browser of the chromosome Xq28 MECP2 locus showing promoter CpG island, direction of transcription, position of exons and the two major spliced isoforms that either exclude or include exon 2 (grey box), designated *MECP2_e1* or *MECP2_e2*, respectively. In addition to alternative splicing, alternative polyadenylation sites (bottom grey bars) are observed, resulting in both a short and long 3′ untranslated region. The long 3′ UTR is most commonly observed with the *MECP2_e2* transcript in the RefSeq database of spliced mammalian genes. (B) The alternative spliced isoforms of *MECP2* encode proteins that differ in their N-terminal protein sequences due to differential use of translation start sites. The MeCP2e1 isoform utilizes the strong translation start site (ATG) encoded in exon 1, resulting in a highly hydrophobic N-terminal amino acid sequence. In contrast, exon 2 contains a stop codon (TGA), so when exon 2 is included, an alternative downstream ATG is utilized, resulting in a shorter N-terminal peptide sequence. Three mammalian species (human, mouse, and rat) differ in the length and sequence of the N-terminal peptide of MeCP2e1 (light-grey peptide sequence).

alternative polyadenylation of the 3′ untranslated region (UTR), resulting in an extremely long 10kb and relatively shorter 1.9kb transcript (Balmer *et al.*, 2003; Coy *et al.*, 1999; Reichwald *et al.*, 2000).

Approximately 60% of MeCP2 has been experimentally determined to lack a specific structure, and was thereby classified as an intrinsically disordered protein (Adams *et al.*, 2007; Ghosh *et al.*, 2010). Upon interaction with protein partners or substrates, intrinsically disordered proteins can adopt defined structures which are unique to each interaction. This feature fits with the emerging pleiotropic roles of MeCP2 in transcriptional modulation, long-range chromatin structure, and activity-dependent epigenetic responses. MeCP2 contains at least four autonomous DNA binding domains that are capable of cooperatively binding to DNA, but only the methyl binding domain (MBD) is capable of specifically binding to methylated DNA. Because of this, MeCP2 only has an approximately 3-fold greater affinity for DNA containing a single methylated CpG site than for unmethylated DNA. X-ray crystallographic analyses have been performed on the MBD complexed to methylated DNA, revealing a dependence of hydration of the major groove of methylated DNA (Ho *et al.*, 2008). While the MBD of MeCP2 exists as a monomer in solution, it forms a dimer when bound to DNA (Khrapunov *et al.*, 2014). In addition to double-stranded DNA, MeCP2 can bind both single stranded DNA and RNA (Jeffery and Nakielny, 2004; Khrapunov *et al.*, 2014).

MeCP2 has multiple ascribed functions, likely stemming from its inherently disordered structure and the ability to bind RNA as well as DNA (Khrapunov *et al.*, 2014). While MeCP2 has a higher affinity for DNA containing methylated compared to unmethylated CpG sites, it can also bind unmethylated DNA as well as hydroxymethylated DNA. While MeCP2 was first described as a transcriptional repressor because of its ability to repress methylated constructs by recruitment of histone deacetylase and transcriptional repressor Sin3A (Nan *et al.*, 1998), the functions more recently described *in vivo* are much more diverse and will be discussed more thoroughly below.

The influence of X-chromosome inactivation on *MECP2* and RTT pathogenesis

The pathogenesis of RTT involves epigenetic mechanisms in at least two major ways. First, *MECP2* is encoded on the X-chromosome and subject to the epigenetic process of XCI specifically in females. Since RTT overwhelmingly affects females who are heterozygous for *MECP2* mutations, understanding the functional consequences of XCI in the pathogenesis and development of RTT is critical for designing genetically based treatments.

Because females have two X-chromosomes compared to one in males, XCI is established to be the major mechanism to ensure proper dosage compensation of X-linked genes between the sexes. In mammals, a random pattern of X-chromosome inactivation is established around the time of implantation through the differential expression of antisense non-coding RNAs, *Xist* and *Tsix*, expressed from the inactive or active X-chromosome, respectively (Chaligne and Heard, 2014). Once established in early development, each daughter cell maintains the choice of which parental X-chromosome is inactivated. Non-random XCI can occur during the establishment of XCI patterns, resulting in extreme skewing throughout all tissues (Robinson *et al.*, 2001). However, skewed XCI patterns observed in somatic tissues are more likely due to minor allelic differences in cell proliferation or cell survival that can be tissue-specific or of a milder skew (<75/25%).

Most girls with clinical features of RTT exhibit random XCI patterns in blood and brain (Amir *et al.*, 2000; LaSalle *et al.*, 2001; Shahbazian *et al.*, 2002c; Smeets *et al.*, 2003) (Table 10.1). In contrast, for the rare cases of maternal inheritance of *MECP2* mutations, the asymptomatic carrier mothers usually show favourably skewed XCI (Augenstein *et al.*, 2009). Furthermore, in a study with a large enough RTT patient size with defined common mutations (p.R168X, p.T158M), the percentage of mutant expressing cells correlated with clinical severity (Archer *et al.*, 2007). However, despite these findings, there are clear

Table 10.1 Influence of X-chromosome inactivation of *MECP2* related phenotypes

	Skewed XCI?	Tissue	Allele preferentially expressed	Effect on phenotype	Reference
Human					
RTT de novo *MECP2* mutation	No	Blood, brain	ND	ND	Amir *et al*. (2000), LaSalle *et al*. (2001), Shahbazian *et al*. (2002), Smeets *et al*. (2003)
RTT p.R168X and p.T158M	No	Blood	16–65% range of mutant expression	Correlation of mutant expression with clinical severity	Archer *et al*. (2007)
RTT *de novo* *MECP2* mutation	Few	Blood	Maternal	No correlation with severity	Xinhua *et al*. (2008)
Asymptomatic carrier mother	Yes	Blood	Mutant	No correlation with severity	Takahashi *et al*. (2008)
Carrier *MECP2* mutant females	Yes	Blood	ND	Skewed XCI correlated with lower severity	Augenstein *et al*. (2009)
Discordant RTT MZ twins c.806delG	No	Blood, fibroblasts, hair	NA	No correlation with severity	Miyake *et al*. (2013)
MECP2 duplication	No	Blood	ND	Distinct but not less severe than males	Bijlsma *et al*. (2012), Grasshoff *et al*. (2011), Makrythanasis *et al*. (2010), Mayo *et al*. (2011), Reardon *et al*. (2010)
Females	Yes (3 cases)		Duplicated	Similar to male duplications	Fieremans *et al*. (2014), Shimada *et al*. (2013)
MECP2 duplication carriers	Yes	Blood	Wild-type (110%)	Neuropsychiatric phenotypes	Ramocki *et al*. (2009)
	No (1 case)			No apparent phenotype	Shimada *et al*. (2013)
Mouse					
Mecp2$^{308/x}$	Yes	Purkinje neurons	Wild-type (paternal)	Per cent of mutant neurons correlated with severity	Young and Zoghbi (2004)
Mecp2$^{tm1.1Bird/+}$ *Mecp2*$^{tm1.1Jae/+}$	Yes	Cerebral cortex	Wild-type (paternal)	Per cent of mutant cells correlated with reduced MeCP2 in wild-type cells	Braunschweig *et al*. (2004)
Mecp2$^{308/x}$	No	Embryonic brain	Wild-type (maternal)	ND	Watson *et al*. (2005)
Mecp2$^{308/x}$	No	Adult cerebellum			
Mecp2$^{x/308}$	Yes	Adult cerebellum			
Mecp2$^{tm1.1Jae/+}$	No	3 month hippocampus	Wild-type (GFP paternal)	ND	Smrt *et al*. (2011)
	Yes	9 month hippocampus			

GFP, green fluorescent protein; MZ, monozygotic; NA, not applicable; ND, not determined.

specific exceptions to the rule of XCI patterns as the predominant factor in clinical severity. First, a study showing a few RTT patients with highly (>80/<20%) skewed XCI in blood did not correlate with classic versus preserved speech RTT phenotypes (Xinhua *et al.*, 2008). Second, in a familial case of RTT, the *MECP2* mutation was observed in the asymptomatic mother together with an XCI pattern favouring expression of the mutant allele (Takahashi *et al.*, 2008). Lastly, and most puzzling, is the study of monozygotic twins with *MECP2* c806delG mutations discordant for RTT severity in which both twins showed random XCI in three different tissues (blood, fibroblast, buccal cells) (Miyake *et al.*, 2013). No difference in copy number variants could explain the phenotypic differences between the twins, but a genome-wide search for methylation differences revealed three genes with differential methylation (Miyake *et al.*, 2013). These rare exceptions have begun to question the previous assumption that clinical severity could be predicted by the combination of *MECP2* mutation and XCI ratio determination.

MECP2 duplications also do not always follow a strict correlation with XCI and phenotype. While *MECP2* duplications are primarily found in males, females with intellectual disability and *MECP2* duplication have also been identified with a distinct, but not necessarily less severe phenotype than duplication males (Bijlsma *et al.*, 2012; Grasshoff *et al.*, 2011; Makrythanasis *et al.*, 2010; Mayo *et al.*, 2011; Reardon *et al.*, 2010). In rare cases of XCI skewing in blood from *MECP2* duplication females, the phenotype was described as similar to males with *MECP2* in the same study (Shimada *et al.*, 2013) and skewed in favour of expression of the duplicated allele in another (Fieremans *et al.*, 2014). While female carriers of *MECP2* duplication males are generally asymptomatic, a range of milder neuropsychiatric phenotypes were observed in a systematic clinical assessment in one study (Ramocki *et al.*, 2009). Perplexingly, highly skewed XCI in favour of expression of the non-duplicated (wild-type) allele was observed in all of the carrier mothers in this study, despite the phenotypes ascribed to *MECP2* duplication status. These findings suggest that the XCI-phenotype correlation is complex, perhaps due to early developmental effects of the *MECP2* duplications that precede a skewed selection event in peripheral blood. Environmental factors can also interact in complex ways with *Mecp2* mutations in females, potentially explain the clinical exceptions (Woods *et al.*, 2012). Therefore, while determining XCI ratios from blood is still useful in clinical evaluations of *MECP2* mutations and duplications, results should be interpreted with caution.

Determining XCI in *Mecp2* mutant mouse models has been both interesting and informative for investigation of multiple tissues and developmental time points, but the results are equally as complex as the human studies (Table 10.1). Unlike RTT females who largely exhibit random XCI, most mouse models of *Mecp2* mutation or deficiency reveal moderately skewed XCI patterns in favour of expression of the wild-type allele in brain or neurons (Braunschweig *et al.*, 2004; Young and Zoghbi, 2004). The percentage of mutant-expressing neurons correlated with phenotypic severity in adult females (Young and Zoghbi, 2004), as well as with decreased protein levels of MeCP2 in wild-type expressing cells in adult cortex, due to non-cell autonomous effects (Braunschweig *et al.*, 2004). But the observation of skewed XCI depends on both age and parental origin. Unlike adult brain tissues, XCI skewing is not observed in *Mecp2* mutant mouse models in embryonic brain or the 3 month hippocampus, but emerges in adult (Smrt *et al.*, 2011; Watson *et al.*, 2005). Parental origin of the preferentially expressed allele influenced XCI skewing, but was actually opposite the earlier study of the same mutant line (Smrt *et al.*, 2011; Young and Zoghbi, 2004). In mouse but not human, preferential inactivation of the paternal X-chromosome is observed in early development, potentially explaining the differences with human XCI observations (Chaligne and Heard, 2014). These mouse studies also point to skewed XCI patterns associated with *Mecp2* mutations being due to a later selection process, rather than an early random XCI establishment process.

Human stem cell models of RTT have also showed major differences from mouse embryonic stem cells (mESC) in the stability and timing of XCI patterns in culture (Cheung *et al.*, 2012).

mESCs, derived from preimplantation blastocyst at e3.5, are epigenetically pre-XCI and express *Xist* biallelically. In contrast most human (hESC) lines are post-XCI and resemble the post-implantation stage observed in mouse epiblast stem cells more than mESCs. Pre-XCI hESCs have been derived through special culture conditions such as physiological oxygen conditions (5% vs. atmospheric 20% O_2) (Lengner *et al.*, 2010). Since induced pluripotent stem cells (iPSCs) derived from RTT fibroblasts also have the potential to be either pre-XCI or post-XCI, the analysis of allelic expression and XCI ratios is critical for the interpretation of these studies. Post-XCI RTT iPSCs have the distinct advantage of maintaining their XCI states and pairs of isogenic lines expressing either the mutant or wild-type *MECP2* alleles can be obtained from the same individual (Cheung *et al.*, 2011). However, pre-XCI RTT iPSCs may better model the early mosaicism of *MECP2* mutant to wild type ratios that exist in the disease state (Marchetto *et al.*, 2010). Interestingly, a defect in neuronal maturation was observed using multiple RTT iPSCs in both the pre- and post-XCI state (Kim *et al.*, 2011). The determination of XCI ratios in RTT iPSCs will continue to be an important component of understanding the complexity of mosaic cell populations in the pathogenesis of RTT.

MeCP2 as reader of the methylome

The second major influence of epigenetics in the pathogenesis of RTT is in the major functional characteristic of the MBD within MeCP2 is to recognize methylated CpG sites in the genome. MeCP2 binds with higher affinity to methylated compared to unmethylated CpG, but an understanding of what that function means *in vivo* has changed considerably over the last decade. In the past, DNA methylation analyses were mostly limited to the analysis of individual gene promoters and CpG islands through time-consuming and low-throughput bisulfite sequencing-based methods. Recent next-generation sequencing efforts have now given us access to the whole human DNA methylome at base resolution, revealing striking differences in the epigenomic landscapes of pluripotent and lineage-committed human cells (Hawkins *et al.*, 2010; Lister *et al.*, 2009). In brief, these studies have confirmed that CpG island promoters are strongly depleted for DNA methylation and gene promoter methylation is inversely correlated with gene expression. Furthermore, it is now clear that the 'shores' of CpG islands, located up to 2 kb distant from the promoter itself, carry informative marks relevant to tissue-type discrimination (Irizarry *et al.*, 2009). While MeCP2 is associated with a large number of active gene promoters genome-wide, it is actually the shores, not the CpG islands that are the methylated binding sites for MeCP2 on active genes (Chahrour *et al.*, 2008; Yasui *et al.*, 2007).

Interestingly, gene bodies (spanning exons and introns) and intergenic regions show high (>75%) methylation in many human tissues, including embryonic stem cells and cortex (Irizarry *et al.*, 2009; Lister *et al.*, 2009; Rauch *et al.*, 2009; Rollins *et al.*, 2006). However, gene body methylation levels and patterns vary considerably between cell types and developmental stages (Irizarry *et al.*, 2009; Ladd-Acosta *et al.*, 2007; Meissner *et al.*, 2008). Gene body methylation of active genes is also highly enriched for 5hmC as well as mC and genomic sites bound by MeCP2 (Mellen *et al.*, 2012; Tsagaratou *et al.*, 2014). Alternative splicing of exons is predicted to be one function of gene body methylation, and included exons are more highly methylated and bound by MeCP2 than excluded exons (Maunakea *et al.*, 2013).

In the context of neurons, increased transcription of neurogenic genes during neuronal differentiation requires the DNA methyltransferase *Dnmt3a* and the deposition of DNA methylation patterns at regions *flanking* promoter regions (Wu *et al.*, 2010). This suggests that neurogenic transcription requires an unmethylated promoter flanked by highly methylated shores. Multiple studies indicate that the proper deposition of methylation patterns is important for brain function, as indicated by findings that DNMT1 and/or DNMT3A regulate synaptic function (Carvalho *et al.*), memory formation (Chan *et al.*, 2007) and behavioural plasticity (LaPlant and Nestler, 2011). Furthermore, neuronal activity modified specific sites of DNA methylation with significant enrichment in low-CpG density regions, and with brain-specific genes related to neuronal plasticity (Guo *et al.*, 2011).

The low-CpG density regions of the human genome also correspond to a distinct methylation landscape pattern of partially methylated domains (PMDs) that are observed in fetal lung fibroblasts, multiple tumour cell lines, and placenta, but not hESCs, or adult tissues (Hon *et al.*, 2012; Schroeder *et al.*, 2011, 2013). PMDs are defined as large-scale domains (>100 kb) that have lower levels of DNA methylation than the rest of the genome, and can be used to divide the methylome into highly methylated domains (HMDs) versus PMDs. Paradoxically, PMDs span regions that are transcriptionally repressed, low in CpG and CpG island content, and relatively gene-poor compared to HMDs (Lister *et al.*, 2009). PMDs also correspond to the repressive chromatin marks of H3K27me3, H3K9me3, as well as nuclear lamina association, late replication domains, and reduced DNAse hypersensitivity (Berman *et al.*, 2012; Hon *et al.*, 2012). PMDs cover ~40% of the human genome in placenta and IMR-90 fetal fibroblasts, but only ~19% in neuronal SH-SY5Y cells (Schroeder *et al.*, 2011, 2013). What is potentially most interesting is that PMD genes that are repressed in one tissue type are within HMDs in cell types where they are expressed, so the PMD/HMD methylome maps can correctly identify the tissue source. Genes encoding proteins acting at the neuronal synapse are enriched within PMDs in placenta and/or fetal fibroblast, but are within HMD and expressed within neuronal cells (Schroeder *et al.*, 2013). While the reasons and consequences of the PMD/HMD methylome landscape is not yet fully understood, these new views of DNA methylation in neuronal development are likely to be critical to understanding the critical targets of MeCP2 in the pathogenesis of RTT. Since MeCP2 binds globally at near histone occupancy to DNA in neuronal nuclei (Skene *et al.*, 2010), understanding how dynamic changes in the methylome impact MeCP2 functions will be important.

A large number of gene targets of MeCP2 binding have been reported and have been reviewed previously (Singh *et al.*, 2008; Zachariah and Rastegar, 2012). MeCP2 appears to modulate the transcript levels of a class of immediate early genes in response to neuronal activity, including *Bdnf*, *Egf2*, *Arc* and *Junb* (Gonzales *et al.*, 2012; Martinowich *et al.*, 2003; Su *et al.*, 2012; Swanberg *et al.*, 2009). The challenge to a target gene-based approach for understanding the role of MeCP2 in RTT, however, is that there are a large number of genes dysregulated by MeCP2 deficiency or mutation and they vary considerably between tissue type. Also confounding the potential relevance of specific MeCP2 target genes is the visualization of high abundance of MeCP2 binding by genomic sequencing (ChIP-seq) approaches. However, a recent meta-analysis of four different mouse expression studies using similar ages and brain regions identified several overlapping canonical pathways associated with MeCP2 deficiency, including cellular metabolism, intracellular signalling, and cytoskeleton (Bedogni *et al.*, 2014).

Global cellular and nuclear phenotypes associated with MeCP2 deficiency have also been described. While MeCP2 deficient neurons have been long established to have smaller soma and nuclei than wild-type controls, the reasons for this morphological differences are just beginning to be understood. MeCP2 deficient neurons exhibit elevated levels of histone H3 and H4 acetylation (Shahbazian *et al.*, 2002a; Thatcher and LaSalle, 2006; Wan *et al.*, 2001), but paradoxically are smaller, more heterochromatic and have smaller nucleoli than wild-type. Furthermore, two recent analyses of human stem cell lines with *MECP2* mutation demonstrated that both transcription and translation were globally reduced in mutant stem cell-derived neurons (Li *et al.*, 2013; Yazdani *et al.*, 2012). These results are also consistent with earlier mouse studies showing global reduction in the mTOR pathway, reduced phosphorylated ribosomal S6, and reduced protein translation in *Mecp2*-deficient brain (Ricciardi *et al.*, 2011). What remains unanswered is how the global epigenetic findings of increased histone acetylation correspond to global decreases in transcription and translation in MeCP2 deficient neurons. Perhaps since acetylated histones can also accumulate within gene bodies, hampering transcriptional progression in the absence of H3K36 methylation (Smolle *et al.*, 2013), there could be a transcriptionally activating role for MeCP2/HDAC complexes.

Multiple ascribed functions and partners for MeCP2

Since the discovery of *MECP2* mutations as the genetic cause of RTT in 1999, the ascribed functions and roles for MeCP2 have grown considerably. While historically viewed as a repressor of methylated gene promoters, the functions currently predicted for MeCP2 include transcriptional activator (Chahrour *et al.*, 2008; Li *et al.*, 2013), chromatin compactor (Georgel *et al.*, 2003; Ghosh *et al.*, 2010; Hansen *et al.*, 2010), histone H1 variant (Skene *et al.*, 2010), splicing factor (Maunakea *et al.*, 2013; Young *et al.*, 2005), nuclear matrix component (Yasui *et al.*, 2014), cellular metabolism regulator (Li *et al.*, 2013; Ricciardi *et al.*, 2011; Yazdani *et al.*, 2012), and chromatin loop mediator (Horike *et al.*, 2005; Yasui *et al.*, 2011). The complexity of MeCP2 functions is expected to be due to the complexity of the multiple isoforms, post-translational modifications, and cofactors in the nucleus. But instead of taking the defeated view that MeCP2 'binds everywhere and does everything', this section will try to parse out the specific functions ascribed to specific functional domains of MeCP2 and how they are impacted by RTT causing *MECP2* mutations and post-translational modifications.

The N-terminal domain of MeCP2 is the least understood functionally, but potentially the most interesting for the specificity of RTT causing *MECP2* mutations. As shown in Fig. 10.2, alternative splicing of *MECP2* results in two isoforms with distinct N-termini. The MeCP2-e1 isoform has a highly hydrophobic N-terminal peptide that varies in length between different mammalian species and is the ancestrally inherited form, while the MeCP2-e2 N-terminus is only 9 amino acids (Kriaucionis and Bird, 2004). Interestingly, there have been no RTT causing mutations observed in exon 2, but multiple mutations observed in exon 1 (Amir *et al.*, 2005; Quenard *et al.*, 2006; Saunders *et al.*, 2009). Furthermore, a mouse model with exon 2 deleted showed no neurological deficits, confirming clinical findings (Itoh *et al.*, 2012). Finally, a mouse model that mutated only the exon 1 translation start site was sufficient to recapitulate the phenotypes of RTT seen in deletion models, even though the MeCP2-e2 isoform was expressed at twice its normal levels (Yasui *et al.*, 2014). MeCP2-e1 has both a higher translation rate and a greater stability compared to MeCP2-e2 (Kriaucionis and Bird, 2004; Yasui *et al.*, 2014). In addition, when MeCP2 is unphosphorylated at the S30 site, it associates with nuclear matrix factors such as SFPQ and YB-1, but when phosphorylated at S80, it associates with chromocenters, similar to MeCP2-e2 (Gonzales *et al.*, 2012; Yasui *et al.*, 2014). Therefore, despite the 96% similarity between MeCP2-e1 and MeCP2-e2 isoforms, these studies have ascribed unique functions for MeCP2-e1 that cannot be compensated for by MeCP2-e2 in preventing RTT.

The MBD of MeCP2 has clearly defined functions as the DNA portion of the molecule, as well as clear functions relevant to RTT, since missense mutations are most frequently found in the MBD (Fig. 10.1). Classic RTT-causing MBD mutations in R106W, P152R, or T158M dramatically alter the nuclear localization of MeCP2 to heterochromatic chromocenters, although surprisingly R133C mutation does not (Adegbola *et al.*, 2009; Baubec *et al.*, 2013). Interestingly, a milder mutation to P152A that is not RTT causing but cause some neurological phenotypes, also shows an intermediate phenotype in the nuclear localization assay. More recent genomic strategies of examining the MBD of MeCP2 in mouse ESCs has shown that while the predominant predictor of MeCP2 binding was methylated CpG density, MeCP2 was unique among the MBD family members in its retained binding to chromatin in triple methylation knockout (TKO) ESCs (Baubec *et al.*, 2013). In wild-type ESCs, MeCP2 MBD binding sites were enriched for chromatin indicators of active genes (H3K4 methylation, H3K27 acetylation, DNAse hypersensitive sites, and 5hmc), but MeCP2 MBD binding in TKO ESCs was shifted towards even more active promoter regions. These results are consistent with prior MeCP2 ChIP-chip analyses in hypomethylated and PMD-containing SH-SY5Y neuroblastoma cells showing MeCP2 preferentially binding to active gene promoters (Yasui *et al.*, 2007). The MBD of MeCP2 also has at least 2 phosphorylation sites (pS80 and pS149), and the pS80 modification subtly alters chromatin localization and activity dependent gene regulation of MeCP2 when complexed with Sin3a and YB-1 cofactors (Gonzales *et al.*, 2012; Tao *et al.*, 2009).

The TRD of MeCP2 is primarily affected by truncation mutations in RTT (Fig. 10.1) and is functionally important for a large number of cofactor associations with MeCP2. The TRD was originally defined in the canonical MeCP2 complex of MeCP2 with the corepressor Sin3a and HDAC (Nan *et al.*, 1998). Phoshorylation of S229, within the TRD, shifts cofactor association away from Sin3a and in favour of association with HP1 heterochromatin proteins and matrix factor SMC (Gonzales *et al.*, 2012). Phosphorylation of another TRD site, T308, inhibits the association of MeCP2 with the nuclear receptor corepressor (NCoR), affecting its transcriptional repressive functions (Ebert *et al.*, 2013). Common RTT causing mutations in the C-terminal side of the TRD (Fig. 10.1) also impair the association of MeCP2 with NCoR, suggesting a relevant function for RTT (Lyst *et al.*, 2013). The TRD also contains 3 AT-hook motifs and truncation mutations in males at either R270X versus G273X modify severity through affecting recruitment of cofactor ATRX to chromatin (Baker *et al.*, 2013). The TRD also interacts with lamin B receptors at the nuclear periphery (Guarda *et al.*, 2009). Interestingly, both the MBD and TRD need to be wild-type to get toxicity associated with MeCP2 duplication in mouse models, and TRD mutations can have a dominant negative effect on MeCP2 function in this system, perhaps due to the dimerization of MeCP2 and the importance of the TRD in cofactor associations (Heckman *et al.*, 2014).

The C-terminal domain of MeCP2 contains a variety of mutation types in RTT (Fig. 10.1). While mutations of this domain are generally found to be milder, the C-terminal tail of MeCP2 is required for normal neurological function. A mouse model of a human T308X truncation mutation that eliminates the entire C-terminus exhibits many features of RTT including decreased sociability and decreased activity in both male and heterozygous females. The C-terminal domain also contains the best characterized activity dependent phosphorylation site at S421 (Cohen *et al.*, 2011; Zhou *et al.*, 2006). While mutation in the S421 site does not apparently affect genomic MeCP2 binding sites or transcript levels, it is important in activity dependent responses to cocaine or anti-depressants (Hutchinson *et al.*, 2012; Mao *et al.*, 2011). While a WW domain with the C-terminus of MeCP2 predicted cofactor associations (Buschdorf and Stratling, 2004), to date only the nuclear export protein Sdccag1 has been shown to bind only the C-terminus of MeCP2 in brain (Long *et al.*, 2011).

All of the individual functions of the various parts of the MeCP2 molecule come together in critical functions within neurons for axon guidance, dendritic branching, activity-dependent gene modulation, and excitatory to inhibitory balance of neurotransmitters and their receptors (Cohen *et al.*, 2011; Dani *et al.*, 2005; Degano *et al.*, 2009; Gonzales and LaSalle, 2010; Lee *et al.*, 2014; Matijevic *et al.*, 2009; Palmer *et al.*, 2012; Yazdani *et al.*, 2012). While mouse models have shown that MeCP2 is required in the postnatal brain to prevent death and maintain neuronal function, these results do not necessarily imply that MeCP2 does not also have functions in the developing neonatal brain (Guy *et al.*, 2007; McGraw *et al.*, 2011). Recent evident from both mouse models and human ESC derived neurons have demonstrated critical roles for MeCP2 in development prior to the onset of RTT associated symptoms (Baj *et al.*, 2014; Singleton *et al.*, 2011; Tomassy *et al.*, 2014).

The functional consequences of *MECP2* mutations and implications for RTT therapies

In order to develop effective RTT therapies based on the advances in understanding MeCP2 structure and function, consideration for the cell types and developmental stages of action are important. While MeCP2 is at its highest levels in tissues of the central nervous system and specifically in neurons (Balmer *et al.*, 2003; Shahbazian *et al.*, 2002b), there is a growing appreciation for MeCP2 functions in non-neuronal cell types and tissues and their relevance for RTT therapies. Non-neuronal cells in the brain include astrocytes and microglia and both of these cell types are implicated in disease pathogenesis, particularly in heterozygous female mouse models (Ballas *et al.*, 2009; Derecki *et al.*, 2012; Maezawa and Jin,

2010; Maezawa *et al.*, 2009; Nguyen *et al.*, 2013). Interestingly, both astrocytes and microglia deficient in MeCP2 exhibit non-cell-autonomous effects on wild-type neurons (Braunschweig *et al.*, 2004; Maezawa *et al.*, 2009; Nguyen *et al.*, 2013). Astrocytes differentiated from mutant RTT iPSC lines adversely affect wild-type neurons (Williams *et al.*, 2014). Bone marrow transplantation improved phenotypes in a mouse model of RTT through the action of bone marrow derived microglia (Derecki *et al.*, 2012). Consideration of XCI ratios and non-cell autonomous effects in the mosaic female RTT patients is likely going to be important for optimizing the promise of bone marrow transplantation strategies for clinical use. To date, RTT translational studies have focused on three basic strategies.

The initial studies performed the genetic rescue of *Mecp2* in RTT mouse models using a variety of genetic engineering and gene therapy strategies. The restoration of MeCP2 in previously *Mecp2* null mice through tamoxifen-cre deletion of a stop codon has been effective in rescuing neurological deficits in both male and female mice, providing hope for postnatal restoration of MeCP2 in RTT therapies (Guy *et al.*, 2007). Improvements in viral packaging vectors have improved brain delivery of *Mecp2* expression constructs, providing promise for gene therapy (Gray *et al.*, 2011). Recently, the systemic administration of a AAV9 vector carrying the *Mecp2-e1* isoform was able to rescue several neurological deficits in female RTT mice, although XCI ratios and the effects of MeCP2 overexpression on wild-type expressing cells was not investigated (Gadalla *et al.*, 2013; Garg *et al.*, 2013). The major complications still to be overcome in gene therapy strategies for RTT are in mitigating the potential for adverse effects from duplication in wild-type expressing cells and dominant negative effect in mutant expressing cells in mosaic female RTT patients.

The other main area of RTT therapy development involves targeting the molecules upstream of the mTOR pathway in neurons. Currently, the growth factor IGF1 is showing promise in phase II studies at overcoming the cellular metabolism deficits in the AKT/mTOR pathway in RTT. IGF1 was originally shown to rescue RTT phenotypes in a mouse model (Tropea *et al.*, 2009). IGF1 has also been effective in restoring the deficits in neuronal soma size and dendritic branching in human iPSC derived neurons with *MECP2* mutations (Li *et al.*, 2013; Marchetto *et al.*, 2010). In a preliminary clinical assessment of the phase I clinical trial on 12 girls with *MECP2* mutations, IGF1 treatment was safely administered and treatment improved breathing, anxiety, and mood (Khwaja *et al.*, 2014). Combined treatment of a β2-adrenergic receptor agonist clenbuterol together with IGF1 further improved survival in male *Mecp2* null mice, suggesting that future combined therapies may be in the pipeline for future clinical trials. An alternative IGF1-based therapy currently in phase II clinical trials, involves NNZ-2566, an IGF1 tri-peptide that is also undergoing human trials for treatment of traumatic brain injury (TBI) and fragile X syndrome.

BDNF, another extracellular signalling molecule in the mTOR pathway has been the focus of pharmacologic intervention RTT mouse models. *Bdnf* expression has been shown to be epigenetically regulated by MeCP2 (Chen *et al.*, 2003; Martinowich *et al.*, 2003) in mice. Specifically, MeCP2 binding to CpG methylated *Bdnf* promoter IV can recruit factors that modulate transcription in response to extracellular signalling in brain (Chen *et al.*, 2003; Martinowich *et al.*, 2003; Zhou *et al.*, 2006). While it remains unclear the effect that MeCP2 binding has on *Bdnf* transcription in neurons, it is clear that BDNF levels are significantly reduced in the brains of RTT mice. As BDNF itself has poor pharmacologic properties, translational strategies have focused on improving the effects of existing protein levels of attenuation of the BDNF receptor TrkB. LM22A-4, a partial agonist of TrkB has been shown to ameliorate respiratory symptoms in RTT model mice (Schmid *et al.*, 2012).

Despite the current promise of multiple RTT therapies, a better understanding of the epigenetic complexities regulating and involving MeCP2 are likely to be important in treating and improving the lives of girls with RTT. Because of the complications of XCI ratios that vary between mouse models and human patients, it will be important to test potential therapies in female mice and to characterize the effects on the expression levels and phenotypic effects on both mutant and wild-type expressing cells. In addition, further understanding of the critical functions of MeCP2 as an epigenetic regulator of dynamic methylation patterns in the brain and other tissues will likely improve understanding of RTT pathogenesis and treatments.

Acknowledgements

I thank Dr Dag Yasui for critical reading of the manuscript and NIH/NINDS R01NS081913 and the International Rett Syndrome foundation for ongoing support of Rett syndrome research.

References

Adams, V.H., McBryant, S.J., Wade, P.A., Woodcock, C.L., and Hansen, J.C. (2007). Intrinsic disorder and autonomous domain function in the multifunctional nuclear protein, MeCP2. J. Biol. Chem. *282*, 15057–15064.

Adegbola, A.A., Gonzales, M.L., Chess, A., LaSalle, J.M., and Cox, G.F. (2009). A novel hypomorphic MECP2 point mutation is associated with a neuropsychiatric phenotype. Hum. Genet. *124*, 615–623.

Allanson, J.E., Hennekam, R.C., Moog, U., and Smeets, E.E. (2011). Rett syndrome: a study of the face. Am. J. Med. Genet. A 155A, 1563–1567.

Amir, R.E., Van den Veyver, I.B., Wan, M., Tran, C.Q., Francke, U., and Zoghbi, H.Y. (1999). Rett syndrome is caused by mutations in X-linked MECP2, encoding methyl-CpG-binding protein 2. Nat. Genet. *23*, 185–188.

Amir, R.E., Van den Veyver, I.B., Schultz, R., Malicki, D.M., Tran, C.Q., Dahle, E.J., Philippi, A., Timar, L., Percy, A.K., Motil, K.J., et al. (2000). Influence of mutation type and X-chromosome inactivation on Rett syndrome phenotypes. Ann. Neurol. *47*, 670–679.

Amir, R.E., Fang, P., Yu, Z., Glaze, D.G., Percy, A.K., Zoghbi, H.Y., Roa, B.B., and Van den Veyver, I.B. (2005). Mutations in exon 1 of MECP2 are a rare cause of Rett syndrome. J. Med. Genet. 42, e15.

Ananiev, G., Williams, E.C., Li, H., and Chang, Q. (2011). Isogenic pairs of wild type and mutant induced pluripotent stem cell (iPSC) lines from Rett syndrome patients as *in vitro* disease model. PLoS ONE 6, e25255.

Archer, H., Evans, J., Leonard, H., Colvin, L., Ravine, D., Christodoulou, J., Williamson, S., Charman, T., Bailey, M.E., Sampson, J., et al. (2007). Correlation between clinical severity in patients with Rett syndrome with a p.R168X or p.T158M MECP2 mutation, and the direction and degree of skewing of X-chromosome inactivation. J. Med. Genet. *44*, 148–152.

Ariani, F., Hayek, G., Rondinella, D., Artuso, R., Mencarelli, M.A., Spanhol-Rosseto, A., Pollazzon, M., Buoni, S., Spiga, O., Ricciardi, S., et al. (2008). FOXG1 is responsible for the congenital variant of Rett syndrome. Am. J. Hum. Genet. *83*, 89–93.

Armstrong, D.D. (1997). Review of Rett syndrome. J. Neuropathol. Exp. Neurol. *56*, 843–849.

Augenstein, K., Lane, J.B., Horton, A., Schanen, C., and Percy, A.K. (2009). Variable phenotypic expression of a MECP2 mutation in a family. J. Neurodev. Disord. *1*, 313.

Baj, G., Patrizio, A., Montalbano, A., Sciancalepore, M., and Tongiorgi, E. (2014). Developmental and maintenance defects in Rett syndrome neurons identified by a new mouse staging system *in vitro*. Front. Cell. Neurosci. *8*, 18.

Baker, S.A., Chen, L., Wilkins, A.D., Yu, P., Lichtarge, O., and Zoghbi, H.Y. (2013). An AT-hook domain in MeCP2 determines the clinical course of Rett syndrome and related disorders. Cell *152*, 984–996.

Ballas, N., Lioy, D.T., Grunseich, C., and Mandel, G. (2009). Non-cell autonomous influence of MeCP2-deficient glia on neuronal dendritic morphology. Nat. Neurosci. *12*, 311–317.

Balmer, D., Goldstine, J., Rao, Y.M., and LaSalle, J.M. (2003). Elevated methyl-CpG-binding protein 2 expression is acquired during postnatal human brain development and is correlated with alternative polyadenylation. J. Mol. Med. *81*, 61–68.

Baubec, T., Ivanek, R., Lienert, F., and Schubeler, D. (2013). Methylation-dependent and -independent genomic targeting principles of the MBD protein family. Cell *153*, 480–492.

Bebbington, A., Anderson, A., Ravine, D., Fyfe, S., Pineda, M., de Klerk, N., Ben-Zeev, B., Yatawara, N., Percy, A., Kaufmann, W.E., et al. (2008). Investigating genotype–phenotype relationships in Rett syndrome using an international data set. Neurology *70*, 868–875.

Bedogni, F., Rossi, R.L., Galli, F., Cobolli Gigli, C., Gandaglia, A., Kilstrup-Nielsen, C., and Landsberger, N. (2014). Rett syndrome and the urge of novel approaches to study MeCP2 functions and mechanisms of action. Neurosci. Biobehav. Rev. *46*, 187–201.

Berman, B.P., Weisenberger, D.J., Aman, J.F., Hinoue, T., Ramjan, Z., Liu, Y., Noushmehr, H., Lange, C.P., van Dijk, C.M., Tollenaar, R.A., et al. (2012). Regions of focal DNA hypermethylation and long-range hypomethylation in colorectal cancer coincide with nuclear lamina-associated domains. Nat. Genet. *44*, 40–46.

Bijlsma, E.K., Collins, A., Papa, F.T., Tejada, M.I., Wheeler, P., Peeters, E.A., Gijsbers, A.C., van de Kamp, J.M., Kriek, M., Losekoot, M., et al. (2012). Xq28 duplications including MECP2 in five females: Expanding the phenotype to severe mental retardation. Eur. J. Med. Genet. 55, 404–413.

Braunschweig, D., Simcox, T., Samaco, R.C., and LaSalle, J.M. (2004). X-Chromosome inactivation ratios affect wild-type MeCP2 expression within mosaic Rett syndrome and Mecp2–/+ mouse brain. Hum. Mol. Genet. *13*, 1275–1286.

Buschdorf, J.P., and Stratling, W.H. (2004). A WW domain binding region in methyl-CpG-binding protein MeCP2: impact on Rett syndrome. J. Mol. Med. (Berl) 82, 135–143.

Carouge, D., Host, L., Aunis, D., Zwiller, J., and Anglard, P. (2010). CDKL5 is a brain MeCP2 target gene regulated by DNA methylation. Neurobiol. Dis. *38*, 414–424.

Carvalho, C.M., Zhang, F., Liu, P., Patel, A., Sahoo, T., Bacino, C.A., Shaw, C., Peacock, S., Pursley, A., Tavyev, Y.J., et al. (2009). Complex rearrangements in patients with duplications of MECP2 can occur by fork stalling and template switching. Hum. Mol. Genet. *18*, 2188–2203.

Chahrour, M., and Zoghbi, H.Y. (2007). The story of Rett syndrome: from clinic to neurobiology. Neuron *56*, 422–437.

Chahrour, M., Jung, S.Y., Shaw, C., Zhou, X., Wong, S.T., Qin, J., and Zoghbi, H.Y. (2008). MeCP2, a key contributor to neurological disease, activates and represses transcription. Science *320*, 1224–1229.

Chaligne, R., and Heard, E. (2014). X-chromosome inactivation in development and cancer. FEBS Lett. *588*, 2514–2522.

Chan, P.A., Duraisamy, S., Miller, P.J., Newell, J.A., McBride, C., Bond, J.P., Raevaara, T., Ollila, S., Nystrom, M., Grimm, A.J., et al. (2007). Interpreting missense variants: comparing computational methods in human disease genes CDKN2A, MLH1, MSH2, MECP2, and tyrosinase (TYR). Hum. Mutat. *28*, 683–693.

Chen, W.G., Chang, Q., Lin, Y., Meissner, A., West, A.E., Griffith, E.C., Jaenisch, R., and Greenberg, M.E. (2003). Derepression of BDNF transcription involves calcium-dependent phosphorylation of MeCP2. Science *302*, 885–889.

Cheung, A.Y., Horvath, L.M., Grafodatskaya, D., Pasceri, P., Weksberg, R., Hotta, A., Carrel, L., and Ellis, J. (2011). Isolation of MECP2-null Rett Syndrome patient hiPS cells and isogenic controls through X-chromosome inactivation. Hum. Mol. Genet. *20*, 2103–2115.

Cheung, A.Y., Horvath, L.M., Carrel, L., and Ellis, J. (2012). X-chromosome inactivation in rett syndrome human induced pluripotent stem cells. Front. Psychiatry 3, 24.

Cohen, S., Gabel, H.W., Hemberg, M., Hutchinson, A.N., Sadacca, L.A., Ebert, D.H., Harmin, D.A., Greenberg, R.S., Verdine, V.K., Zhou, Z., et al. (2011). Genome-wide activity-dependent MeCP2 phosphorylation regulates nervous system development and function. Neuron 72, 72–85.

Coy, J.F., Sedlacek, Z., Bachner, D., Delius, H., and Poustka, A. (1999). A complex pattern of evolutionary conservation and alternative polyadenylation within the long 3′-untranslated region of the methyl-CpG-binding protein 2 gene (MeCP2) suggests a regulatory role in gene expression. Hum. Mol. Genet. *8*, 1253–1262.

Cuddapah, V.A., Pillai, R.B., Shekar, K.V., Lane, J.B., Motil, K.J., Skinner, S.A., Tarquinio, D.C., Glaze, D.G., McGwin, G., Kaufmann, W.E., et al. (2014). Methyl-CpG-binding protein 2 (MECP2) mutation type is associated with disease severity in Rett syndrome. J. Med. Genet. *51*, 152–158.

Dani, V.S., Chang, Q., Maffei, A., Turrigiano, G.G., Jaenisch, R., and Nelson, S.B. (2005). Reduced cortical activity due to a shift in the balance between excitation and inhibition in a mouse model of Rett syndrome. Proc. Natl. Acad. Sci. U.S.A. *102*, 12560–12565.

De Bona, C., Zappella, M., Hayek, G., Meloni, I., Vitelli, F., Bruttini, M., Cusano, R., Loffredo, P., Longo, I., and Renieri, A. (2000). Preserved speech variant is allelic of classic Rett syndrome. Eur. J. Hum. Genet. *8*, 325–330.

Degano, A.L., Pasterkamp, R.J., and Ronnett, G.V. (2009). MeCP2 deficiency disrupts axonal guidance, fasciculation, and targeting by altering Semaphorin 3F function. Mol. Cell. Neurosci. *42*, 243–254.

Derecki, N.C., Cronk, J.C., Lu, Z., Xu, E., Abbott, S.B., Guyenet, P.G., and Kipnis, J. (2012). Wild-type microglia arrest pathology in a mouse model of Rett syndrome. Nature *484*, 105–109.

Dunn, H.G. (2001). Importance of Rett syndrome in child neurology. Brain Dev. 23(Suppl. 1), S38–43.

Ebert, D.H., Gabel, H.W., Robinson, N.D., Kastan, N.R., Hu, L.S., Cohen, S., Navarro, A.J., Lyst, M.J., Ekiert, R., Bird, A.P., et al. (2013). Activity-dependent phosphorylation of MeCP2 threonine 308 regulates interaction with NCoR. Nature *499*, 341–345.

Ellaway, C., and Christodoulou, J. (2001). Rett syndrome: clinical characteristics and recent genetic advances. Disabil. Rehabil. *23*, 98–106.

Fan, G., and Hutnick, L. (2005). Methyl-CpG binding proteins in the nervous system. Cell Res. *15*, 255–261.

Fehr, S., Wilson, M., Downs, J., Williams, S., Murgia, A., Sartori, S., Vecchi, M., Ho, G., Polli, R., Psoni, S., et al. (2013). The CDKL5 disorder is an independent clinical entity associated with early onset encephalopathy. Eur. J. Hum. Genet. *21*, 266–273.

Fieremans, N., Bauters, M., Belet, S., Verbeeck, J., Jansen, A.C., Seneca, S., Roelens, F., De Baere, E., Marynen, P., and Froyen, G. (2014). *De novo* MECP2 duplications in two females with intellectual disability and unfavorable complete skewed X-inactivation. Hum. Genet. *133*, 1359–1367.

Froyen, G., Van Esch, H., Bauters, M., Hollanders, K., Frints, S.G., Vermeesch, J.R., Devriendt, K., Fryns, J.P., and Marynen, P. (2007). Detection of genomic copy number changes in patients with idiopathic mental retardation by high-resolution X-array-CGH: important role for increased gene dosage of XLMR genes. Hum. Mutat. *28*, 1034–1042.

Gadalla, K.K., Bailey, M.E., Spike, R.C., Ross, P.D., Woodard, K.T., Kalburgi, S.N., Bachaboina, L., Deng, J.V., West, A.E., Samulski, R.J., et al. (2013). Improved survival and reduced phenotypic severity following AAV9/MECP2 gene transfer to neonatal and juvenile male Mecp2 knockout mice. Mol. Ther. *21*, 18–30.

Garg, S.K., Lioy, D.T., Cheval, H., McGann, J.C., Bissonnette, J.M., Murtha, M.J., Foust, K.D., Kaspar, B.K., Bird, A., and Mandel, G. (2013). Systemic delivery of MeCP2 rescues behavioral and cellular deficits in female mouse models of Rett syndrome. J. Neurosci. *33*, 13612–13620.

del Gaudio, D., Fang, P., Scaglia, F., Ward, P.A., Craigen, W.J., Glaze, D.G., Neul, J.L., Patel, A., Lee, J.A., Irons, M., et al. (2006). Increased MECP2 gene copy number as the result of genomic duplication in neurodevelopmentally delayed males. Genet. Med. *8*, 784–792.

Georgel, P.T., Horowitz-Scherer, R.A., Adkins, N., Woodcock, C.L., Wade, P.A., and Hansen, J.C. (2003). Chromatin compaction by human MeCP2. Assembly of novel secondary chromatin structures in the absence of DNA methylation. J. Biol. Chem. *278*, 32181–32188.

Ghosh, R.P., Nikitina, T., Horowitz-Scherer, R.A., Gierasch, L.M., Uversky, V.N., Hite, K., Hansen, J.C., and Woodcock, C.L. (2010). Unique physical properties and interactions of the domains of methylated DNA binding protein 2. Biochemistry *49*, 4395–4410.

Gonzales, M.L., and LaSalle, J.M. (2010). The role of MeCP2 in brain development and neurodevelopmental disorders. Curr. Psychiatry Rep. *12*, 127–134.

Gonzales, M.L., Adams, S., Dunaway, K.W., and LaSalle, J.M. (2012). Phosphorylation of distinct sites in MeCP2 modifies cofactor associations and the dynamics of transcriptional regulation. Mol. Cell. Biol. *32*, 2894–2903.

Grasshoff, U., Bonin, M., Goehring, I., Ekici, A., Dufke, A., Cremer, K., Wagner, N., Rossier, E., Jauch, A., Walter, M., et al. (2011). *De novo* MECP2 duplication in two females with random X-inactivation and moderate mental retardation. Eur. J. Hum. Genet. *19*, 507–512.

Gray, S.J., Foti, S.B., Schwartz, J.W., Bachaboina, L., Taylor-Blake, B., Coleman, J., Ehlers, M.D., Zylka, M.J., McCown, T.J., and Samulski, R.J. (2011). Optimizing promoters for recombinant adeno-associated virus-mediated gene expression in the peripheral and central nervous system using self-complementary vectors. Hum. Gene Ther. *22*, 1143–1153.

Guarda, A., Bolognese, F., Bonapace, I.M., and Badaracco, G. (2009). Interaction between the inner nuclear membrane lamin B receptor and the heterochromatic methyl binding protein, MeCP2. Exp. Cell Res. *315*, 1895–1903.

Guo, J.U., Ma, D.K., Mo, H., Ball, M.P., Jang, M.H., Bonaguidi, M.A., Balazer, J.A., Eaves, H.L., Xie, B., Ford, E., et al. (2011). Neuronal activity modifies the DNA methylation landscape in the adult brain. Nat. Neurosci. *14*, 1345–1351.

Guy, J., Gan, J., Selfridge, J., Cobb, S., and Bird, A. (2007). Reversal of neurological defects in a mouse model of Rett syndrome. Science *315*, 1143–1147.

Hagebeuk, E.E., van den Bossche, R.A., and de Weerd, A.W. (2013). Respiratory and sleep disorders in female children with atypical Rett syndrome caused by mutations in the CDKL5 gene. Dev. Med. Child Neurol. *55*, 480–484.

Hammer, S., Dorrani, N., Dragich, J., Kudo, S., and Schanen, C. (2002). The phenotypic consequences of MECP2 mutations extend beyond Rett syndrome. Ment. Retard. Dev. Disabil. Res. Rev. *8*, 94–98.

Hanchard, N.A., Carvalho, C.M., Bader, P., Thome, A., Omo-Griffith, L., del Gaudio, D., Pehlivan, D., Fang, P., Schaaf, C.P., Ramocki, M.B., et al. (2012). A partial MECP2 duplication in a mildly affected adult male: a putative role for the 3′ untranslated region in the MECP2 duplication phenotype. BMC Med. Genet. *13*, 71.

Hansen, J.C., Ghosh, R.P., and Woodcock, C.L. (2010). Binding of the Rett syndrome protein, MeCP2, to methylated and unmethylated DNA and chromatin. IUBMB Life *62*, 732–738.

Hawkins, R.D., Hon, G.C., Lee, L.K., Ngo, Q., Lister, R., Pelizzola, M., Edsall, L.E., Kuan, S., Luu, Y., Klugman, S., et al. (2010). Distinct epigenomic landscapes of pluripotent and lineage-committed human cells. Cell Stem Cell *6*, 479–491.

Heckman, L.D., Chahrour, M.H., and Zoghbi, H.Y. (2014). Rett-causing mutations reveal two domains critical for MeCP2 function and for toxicity in MECP2 duplication syndrome mice. Elife (Cambridge) *3*, e02676.

Ho, K.L., McNae, I.W., Schmiedeberg, L., Klose, R.J., Bird, A.P., and Walkinshaw, M.D. (2008). MeCP2 binding to DNA depends upon hydration at methyl-CpG. Mol. Cell *29*, 525–531.

Hoffbuhr, K., Devaney, J.M., LaFleur, B., Sirianni, N., Scacheri, C., Giron, J., Schuette, J., Innis, J., Marino, M., Philippart, M., et al. (2001). MeCP2 mutations in children with and without the phenotype of Rett syndrome. Neurology *56*, 1486–1495.

Hon, G.C., Hawkins, R.D., Caballero, O.L., Lo, C., Lister, R., Pelizzola, M., Valsesia, A., Ye, Z., Kuan, S., Edsall, L.E., et al. (2012). Global DNA hypomethylation coupled to repressive chromatin domain formation and gene silencing in breast cancer. Genome Res. *22*, 246–258.

Horike, S., Cai, S., Miyano, M., Cheng, J.F., and Kohwi-Shigematsu, T. (2005). Loss of silent-chromatin looping and impaired imprinting of DLX5 in Rett syndrome. Nat. Genet. 37, 31–40.

Huppke, P., Held, M., Laccone, F., and Hanefeld, F. (2003). The spectrum of phenotypes in females with Rett Syndrome. Brain Dev. *25*, 346–351.

Hutchinson, A.N., Deng, J.V., Cohen, S., and West, A.E. (2012). Phosphorylation of MeCP2 at Ser421 contributes to chronic antidepressant action. J. Neurosci. 32, 14355–14363.

Irizarry, R.A., Ladd-Acosta, C., Wen, B., Wu, Z., Montano, C., Onyango, P., Cui, H., Gabo, K., Rongione, M., Webster, M., et al. (2009). The human colon cancer methylome shows similar hypo- and hypermethylation at conserved tissue-specific CpG island shores. Nat. Genet. *41*, 178–186.

Itoh, M., Tahimic, C.G., Ide, S., Otsuki, A., Sasaoka, T., Noguchi, S., Oshimura, M., Goto, Y.I., and Kurimasa, A. (2012). Methyl CpG-binding protein isoform MeCP2_e2 is dispensable for Rett syndrome phenotypes but essential for embryo viability and placenta development. J. Biol. Chem. *287*, 13859–13867.

Jeffery, L., and Nakielny, S. (2004). Components of the DNA methylation system of chromatin control are RNA-binding proteins. J. Biol. Chem. *279*, 49479–49487.

Kameshita, I., Sekiguchi, M., Hamasaki, D., Sugiyama, Y., Hatano, N., Suetake, I., Tajima, S., and Sueyoshi, N. (2008). Cyclin-dependent kinase-like 5 binds and phosphorylates DNA methyltransferase 1. Biochem. Biophys. Res. Commun. *377*, 1162–1167.

Khrapunov, S., Warren, C., Cheng, H., Berko, E.R., Greally, J.M., and Brenowitz, M. (2014). Unusual characteristics of the DNA binding domain of epigenetic regulatory protein MeCP2 determine its binding specificity. Biochemistry *53*, 3379–3391.

Khwaja, O.S., Ho, E., Barnes, K.V., O'Leary, H.M., Pereira, L.M., Finkelstein, Y., Nelson, C.A. 3rd, Vogel-Farley, V., DeGregorio, G., Holm, I.A., et al. (2014). Safety, pharmacokinetics, and preliminary assessment of efficacy of mecasermin (recombinant human IGF-1) for the treatment of Rett syndrome. Proc. Natl. Acad. Sci. U.S.A. *111*, 4596–4601.

Kim, K.Y., Hysolli, E., and Park, I.H. (2011). Neuronal maturation defect in induced pluripotent stem cells from patients with Rett syndrome. Proc. Natl. Acad. Sci. U.S.A. *108*, 14169–14174.

Kortum, F., Das, S., Flindt, M., Morris-Rosendahl, D.J., Stefanova, I., Goldstein, A., Horn, D., Klopocki, E., Kluger, G., Martin, P., et al. (2011). The core FOXG1 syndrome phenotype consists of postnatal microcephaly, severe mental retardation, absent language, dyskinesia, and corpus callosum hypogenesis. J. Med. Genet. *48*, 396–406.

Kriaucionis, S., and Bird, A. (2004). The major form of MeCP2 has a novel N-terminus generated by alternative splicing. Nucleic Acids Res. *32*, 1818–1823.

Kumamoto, T., Toma, K., Gunadi, McKenna, W.L., Kasukawa, T., Katzman, S., Chen, B., and Hanashima, C. (2013). Foxg1 coordinates the switch from nonradially to radially migrating glutamatergic subtypes in the neocortex through spatiotemporal repression. Cell Rep. 3, 931–945.

Ladd-Acosta, C., Pevsner, J., Sabunciyan, S., Yolken, R.H., Webster, M.J., Dinkins, T., Callinan, P.A., Fan, J.B., Potash, J.B., and Feinberg, A.P. (2007). DNA methylation signatures within the human brain. Am. J. Hum. Genet. *81*, 1304–1315.

LaPlant, Q., and Nestler, E.J. (2011). CRACKing the histone code: cocaine's effects on chromatin structure and function. Horm. Behav. *59*, 321–330.

LaSalle, J.M., Goldstine, J., Balmer, D., and Greco, C.M. (2001). Quantitative localization of heterogeneous methyl-CpG-binding protein 2 (MeCP2) expression phenotypes in normal and Rett syndrome brain by laser scanning cytometry. Hum. Mol. Genet. *10*, 1729–1740.

Lee, W., Yun, J.M., Woods, R., Dunaway, K., Yasui, D.H., LaSalle, J.M., and Gong, Q. (2014). MeCP2 regulates activity-dependent transcriptional responses in olfactory sensory neurons. Hum. Mol. Genet. *23*, 6366–6374.

Lengner, C.J., Gimelbrant, A.A., Erwin, J.A., Cheng, A.W., Guenther, M.G., Welstead, G.G., Alagappan, R., Frampton, G.M., Xu, P., Muffat, J., et al. (2010). Derivation of pre-X inactivation human embryonic stem cells under physiological oxygen concentrations. Cell *141*, 872–883.

Li, Y., Wang, H., Muffat, J., Cheng, A.W., Orlando, D.A., Loven, J., Kwok, S.M., Feldman, D.A., Bateup, H.S., Gao, Q., et al. (2013). Global transcriptional and translational repression in human-embryonic-stem-cell-derived Rett syndrome neurons. Cell Stem Cell *13*, 446–458.

Lin, C., Franco, B., and Rosner, M.R. (2005). CDKL5/Stk9 kinase inactivation is associated with neuronal developmental disorders. Hum. Mol. Genet. *14*, 3775–3786.

Lister, R., Pelizzola, M., Dowen, R.H., Hawkins, R.D., Hon, G., Tonti-Filippini, J., Nery, J.R., Lee, L., Ye, Z., Ngo, Q.M., et al. (2009). Human DNA methylomes at base resolution show widespread epigenomic differences. Nature *462*, 315–322.

Livide, G., Patriarchi, T., Amenduni, M., Amabile, S., Yasui, D., Calcagno, E., Lo Rizzo, C., De Falco, G., Ulivieri, C., Ariani, F., et al. (2014). GluD1 is a common altered player in neuronal differentiation from both MECP2-mutated and CDKL5-mutated iPS cells. Eur. J. Hum. Genet. *23*, 195–201.

Long, S.W., Ooi, J.Y., Yau, P.M., and Jones, P.L. (2011). A brain-derived MeCP2 complex supports a role for MeCP2 in RNA processing. Biosci. Rep. *31*, 333–343.

Lugtenberg, D., de Brouwer, A.P., Kleefstra, T., Oudakker, A.R., Frints, S.G., Schrander-Stumpel, C.T., Fryns, J.P., Jensen, L.R., Chelly, J., Moraine, C., et al. (2006). Chromosomal copy number changes in patients with non-syndromic X linked mental retardation detected by array CGH. J. Med. Genet. *43*, 362–370.

Lyst, M.J., Ekiert, R., Ebert, D.H., Merusi, C., Nowak, J., Selfridge, J., Guy, J., Kastan, N.R., Robinson, N.D., de Lima Alves, F., et al. (2013). Rett syndrome mutations abolish the interaction of MeCP2 with the NCoR/SMRT co-repressor. Nat. Neurosci. *16*, 898–902.

McGraw, C.M., Samaco, R.C., and Zoghbi, H.Y. (2011). Adult neural function requires MeCP2. Science *333*, 186.

Maezawa, I., and Jin, L.W. (2010). Rett syndrome microglia damage dendrites and synapses by the elevated release of glutamate. J. Neurosci. *30*, 5346–5356.

Maezawa, I., Swanberg, S., Harvey, D., LaSalle, J.M., and Jin, L.W. (2009). Rett syndrome astrocytes are abnormal and spread MeCP2 deficiency through gap junctions. J. Neurosci. *29*, 5051–5061.

Makrythanasis, P., Moix, I., Gimelli, S., Fluss, J., Aliferis, K., Antonarakis, S.E., Morris, M.A., Bena, F., and Bottani, A. (2010). *De novo* duplication of MECP2 in a girl with mental retardation and no obvious dysmorphic features. Clin. Genet. *78*, 175–180.

Mao, L.M., Horton, E., Guo, M.L., Xue, B., Jin, D.Z., Fibuch, E.E., and Wang, J.Q. (2011). Cocaine increases phosphorylation of MeCP2 in the rat striatum *in vivo*: a differential role of NMDA receptors. Neurochem. Int. *59*, 610–617.

Marchetto, M.C., Carromeu, C., Acab, A., Yu, D., Yeo, G.W., Mu, Y., Chen, G., Gage, F.H., and Muotri, A.R. (2010). A model for neural development and treatment of Rett syndrome using human induced pluripotent stem cells. Cell *143*, 527–539.

Mari, F., Azimonti, S., Bertani, I., Bolognese, F., Colombo, E., Caselli, R., Scala, E., Longo, I., Grosso, S., Pescucci, C., et al. (2005). CDKL5 belongs to the same molecular pathway of MeCP2 and it is responsible for the early onset seizure variant of Rett syndrome. Hum. Mol. Genet. *14*, 1935–1946.

Marschik, P.B., Pini, G., Bartl-Pokorny, K.D., Duckworth, M., Gugatschka, M., Vollmann, R., Zappella, M., and Einspieler, C. (2012). Early speech-language development in females with Rett syndrome: focusing on the preserved speech variant. Dev. Med. Child Neurol. *54*, 451–456.

Martinowich, K., Hattori, D., Wu, H., Fouse, S., He, F., Hu, Y., Fan, G., and Sun, Y.E. (2003). DNA methylation-related chromatin remodeling in activity-dependent BDNF gene regulation. Science *302*, 890–893.

Matijevic, T., Knezevic, J., Slavica, M., and Pavelic, J. (2009). Rett syndrome: from the gene to the disease. Eur. Neurol. *61*, 3–10.

Maunakea, A.K., Chepelev, I., Cui, K., and Zhao, K. (2013). Intragenic DNA methylation modulates alternative splicing by recruiting MeCP2 to promote exon recognition. Cell Res. *23*, 1256–1269.

Mayo, S., Monfort, S., Rosello, M., Orellana, C., Oltra, S., Armstrong, J., Catala, V., and Martinez, F. (2011). *De novo* interstitial triplication of MECP2 in a girl with neurodevelopmental disorder and random X-chromosome inactivation. Cytogenet. Genome Res. *135*, 93–101.

Meissner, A., Mikkelsen, T.S., Gu, H., Wernig, M., Hanna, J., Sivachenko, A., Zhang, X., Bernstein, B.E., Nusbaum, C., Jaffe, D.B., et al. (2008). Genome-scale DNA methylation maps of pluripotent and differentiated cells. Nature *454*, 766–770.

Mellen, M., Ayata, P., Dewell, S., Kriaucionis, S., and Heintz, N. (2012). MeCP2 binds to 5hmC enriched within active genes and accessible chromatin in the nervous system. Cell *151*, 1417–1430.

Meloni, I., Bruttini, M., Longo, I., Mari, F., Rizzolio, F., D'Adamo, P., Denvriendt, K., Fryns, J.P., Toniolo, D., and Renieri, A. (2000). A mutation in the rett syndrome gene, MECP2, causes X-linked mental retardation and progressive spasticity in males [In Process Citation]. Am. J. Hum. Genet. *67*, 982–985.

Mencarelli, M.A., Spanhol-Rosseto, A., Artuso, R., Rondinella, D., De Filippis, R., Bahi-Buisson, N., Nectoux, J., Rubinsztajn, R., Bienvenu, T., Moncla, A., et al. (2010). Novel FOXG1 mutations associated with the congenital variant of Rett syndrome. J. Med. Genet. *47*, 49–53.

Miyake, K., Yang, C., Minakuchi, Y., Ohori, K., Soutome, M., Hirasawa, T., Kazuki, Y., Adachi, N., Suzuki, S., Itoh, M., et al. (2013). Comparison of genomic and epigenomic expression in monozygotic twins discordant for Rett syndrome. PLoS ONE 8, e66729.

Mnatzakanian, G.N., Lohi, H., Munteanu, I., Alfred, S.E., Yamada, T., MacLeod, P.J., Jones, J.R., Scherer, S.W., Schanen, N.C., Friez, M.J., et al. (2004). A previously unidentified MECP2 open reading frame defines a new protein isoform relevant to Rett syndrome. Nat. Genet. *36*, 339–341.

MuhChyi, C., Juliandi, B., Matsuda, T., and Nakashima, K. (2013). Epigenetic regulation of neural stem cell fate during corticogenesis. Int. J. Dev. Neurosci. *31*, 424–433.

Nagarajan, R.P., Hogart, A.R., Gwye, Y., Martin, M.R., and Lasalle, J.M. (2006). Reduced MeCP2 expression is frequent in autism frontal cortex and correlates with aberrant MECP2 promoter methylation. Epigenetics *1*, 172–182.

Naidu, S., and Johnston, M.V. (2011). Neurodevelopmental disorders: clinical criteria for Rett syndrome. Nat. Rev. Neurol. 7, 312–314.

Nan, X., Ng, H.H., Johnson, C.A., Laherty, C.D., Turner, B.M., Eisenman, R.N., and Bird, A. (1998). Transcriptional repression by the methyl-CpG-binding protein MeCP2 involves a histone deacetylase complex. Nature *393*, 386–389.

Neul, J.L., Kaufmann, W.E., Glaze, D.G., Christodoulou, J., Clarke, A.J., Bahi-Buisson, N., Leonard, H., Bailey, M.E., Schanen, N.C., Zappella, M., et al. (2010). Rett syndrome: revised diagnostic criteria and nomenclature. Ann. Neurol. *68*, 944–950.

Nguyen, M.V., Felice, C.A., Du, F., Covey, M.V., Robinson, J.K., Mandel, G., and Ballas, N. (2013). Oligodendrocyte lineage cells contribute unique features to Rett syndrome neuropathology. J. Neurosci. *33*, 18764–18774.

Orrico, A., Lam, C., Galli, L., Dotti, M.T., Hayek, G., Tong, S.F., Poon, P.M., Zappella, M., Federico, A., and Sorrentino, V. (2000). MECP2 mutation in male patients with non-specific X-linked mental retardation. FEBS Lett. *481*, 285–288.

Palmer, A.M., Degano, A.L., Park, M.J., Ramamurthy, S., and Ronnett, G.V. (2012). Normal mitral cell dendritic development in the setting of Mecp2 mutation. Neuroscience *202*, 108–116.

Quenard, A., Yilmaz, S., Fontaine, H., Bienvenu, T., Moncla, A., des Portes, V., Rivier, F., Mathieu, M., Raux, G., Jonveaux, P., et al. (2006). Deleterious mutations in exon 1 of MECP2 in Rett syndrome. Eur. J. Med. Genet. *49*, 313–322.

Ramaekers, V., Sequeira, J.M., and Quadros, E.V. (2013). Clinical recognition and aspects of the cerebral folate deficiency syndromes. Clin. Chem. Lab. Med. *51*, 497–511.

Ramocki, M.B., Peters, S.U., Tavyev, Y.J., Zhang, F., Carvalho, C.M., Schaaf, C.P., Richman, R., Fang, P., Glaze, D.G., Lupski, J.R., et al. (2009). Autism and other neuropsychiatric symptoms are prevalent in individuals with MeCP2 duplication syndrome. Ann. Neurol. *66*, 771–782.

Rauch, T.A., Wu, X., Zhong, X., Riggs, A.D., and Pfeifer, G.P. (2009). A human B cell methylome at 100 base pair resolution. Proc. Natl. Acad. Sci. U.S.A. *106*, 671–678.

Reardon, W., Donoghue, V., Murphy, A.M., King, M.D., Mayne, P.D., Horn, N., and Birk Moller, L. (2010). Progressive cerebellar degenerative changes in the severe mental retardation syndrome caused by duplication of MECP2 and adjacent loci on Xq28. Eur. J. Pediatr. *169*, 941–949.

Reichwald, K., Thiesen, J., Wiehe, T., Weitzel, J., Poustka, W.A., Rosenthal, A., Platzer, M., Stratling, W.H., and Kioschis, P. (2000). Comparative sequence analysis of the MECP2-locus in human and mouse reveals new transcribed regions. Mamm. Genome *11*, 182–190.

Ricciardi, S., Boggio, E.M., Grosso, S., Lonetti, G., Forlani, G., Stefanelli, G., Calcagno, E., Morello, N., Landsberger, N., Biffo, S., et al. (2011). Reduced AKT/mTOR signaling and protein synthesis dysregulation in a Rett syndrome animal model. Hum. Mol. Genet. *20*, 1182–1196.

Robinson, W.P., Beever, C., Brown, C.J., and Stephenson, M.D. (2001). Skewed X inactivation and recurrent spontaneous abortion. Semin. Reprod. Med. *19*, 175–181.

Rollins, R.A., Haghighi, F., Edwards, J.R., Das, R., Zhang, M.Q., Ju, J., and Bestor, T.H. (2006). Large-scale structure of genomic methylation patterns. Genome Res. *16*, 157–163.

Saunders, C.J., Minassian, B.E., Chow, E.W., Zhao, W., and Vincent, J.B. (2009). Novel exon 1 mutations in MECP2 implicate isoform MeCP2_e1 in classical Rett syndrome. Am. J. Med. Genet. A 149A, 1019–1023.

Schanen, C. (2001). Rethinking the fate of males with mutations in the gene that causes Rett syndrome. Brain Dev. 23(Suppl. 1), S144–146.

Schmid, D.A., Yang, T., Ogier, M., Adams, I., Mirakhur, Y., Wang, Q., Massa, S.M., Longo, F.M., and Katz, D.M. (2012). A TrkB small molecule partial agonist rescues TrkB phosphorylation deficits and improves respiratory function in a mouse model of Rett syndrome. J. Neurosci. *32*, 1803–1810.

Schroeder, D.I., Lott, P., Korf, I., and LaSalle, J.M. (2011). Large-scale methylation domains mark a functional subset of neuronally expressed genes. Genome Res. *21*, 1583–1591.

Schroeder, D.I., Blair, J.D., Lott, P., Yu, H.O., Hong, D., Crary, F., Ashwood, P., Walker, C., Korf, I., Robinson, W.P., et al. (2013). The human placenta methylome. Proc. Natl. Acad. Sci. U.S.A. *110*, 6037–6042.

Shahbazian, M., Young, J., Yuva-Paylor, L., Spencer, C., Antalffy, B., Noebels, J., Armstrong, D., Paylor, R., and Zoghbi, H. (2002a). Mice with truncated MeCP2 recapitulate many Rett syndrome features and display hyperacetylation of histone H3. Neuron 35, 243–254.

Shahbazian, M.D., Antalffy, B., Armstrong, D.L., and Zoghbi, H.Y. (2002b). Insight into Rett syndrome: MeCP2 levels display tissue- and cell-specific differences and correlate with neuronal maturation. Hum. Mol. Genet. *11*, 115–124.

Shahbazian, M.D., Sun, Y., and Zoghbi, H.Y. (2002c). Balanced X-chromosome inactivation patterns in the Rett syndrome brain. Am. J. Med. Genet. *111*, 164–168.

Shimada, S., Okamoto, N., Ito, M., Arai, Y., Momosaki, K., Togawa, M., Maegaki, Y., Sugawara, M., Shimojima, K., Osawa, M., et al. (2013). MECP2 duplication syndrome in both genders. Brain Dev. 35, 411–419.

Singh, J., Saxena, A., Christodoulou, J., and Ravine, D. (2008). MECP2 genomic structure and function: insights from ENCODE. Nucleic Acids Res. *36*, 6035–6047.

Singleton, M.K., Gonzales, M.L., Leung, K.N., Yasui, D.H., Schroeder, D.I., Dunaway, K., and LaSalle, J.M. (2011). MeCP2 is required for global heterochromatic and nucleolar changes during activity-dependent neuronal maturation. Neurobiol. Dis. *43*, 190–200.

Skene, P.J., Illingworth, R.S., Webb, S., Kerr, A.R., James, K.D., Turner, D.J., Andrews, R., and Bird, A.P. (2010). Neuronal MeCP2 is expressed at near histone-octamer levels and globally alters the chromatin state. Mol. Cell *37*, 457–468.

Smeets, E., Schollen, E., Moog, U., Matthijs, G., Herbergs, J., Smeets, H., Curfs, L., Schrander-Stumpel, C., and Fryns, J.P. (2003). Rett syndrome in adolescent and adult females: clinical and molecular genetic findings. Am. J. Med. Genet. A 122A, 227–233.

Smolle, M., Workman, J.L., and Venkatesh, S. (2013). reSETting chromatin during transcription elongation. Epigenetics *8*, 10–15.

Smrt, R.D., Pfeiffer, R.L., and Zhao, X. (2011). Age-dependent expression of MeCP2 in a heterozygous mosaic mouse model. Hum. Mol. Genet. *20*, 1834–1843.

Su, D., Cha, Y.M., and West, A.E. (2012). Mutation of MeCP2 alters transcriptional regulation of select immediate-early genes. Epigenetics 7, 146–154.

Swanberg, S.E., Nagarajan, R.P., Peddada, S., Yasui, D.H., and LaSalle, J.M. (2009). Reciprocal co-regulation of EGR2 and MECP2 is disrupted in Rett syndrome and autism. Hum. Mol. Genet. *18*, 525–534.

Takagi, N. (2001). The role of X-chromosome inactivation in the manifestation of Rett syndrome. Brain Dev. 23(Suppl. 1), S182–185.

Takahashi, S., Ohinata, J., Makita, Y., Suzuki, N., Araki, A., Sasaki, A., Murono, K., Tanaka, H., and Fujieda, K. (2008). Skewed X-chromosome inactivation failed to explain the normal phenotype of a carrier female with MECP2 mutation resulting in Rett syndrome. Clin. Genet. 73, 257–261.

Tan, W.H., Bird, L.M., Thibert, R.L., and Williams, C.A. (2014). If not Angelman, what is it? A review of Angelman-like syndromes. Am. J. Med. Genet. A 164A, 975–992.

Tao, J., Hu, K., Chang, Q., Wu, H., Sherman, N.E., Martinowich, K., Klose, R.J., Schanen, C., Jaenisch, R., Wang, W., et al. (2009). Phosphorylation of MeCP2 at Serine 80 regulates its chromatin association and neurological function. Proc. Natl. Acad. Sci. U.S.A. *106*, 4882–4887.

Thatcher, K.N., and LaSalle, J.M. (2006). Dynamic changes in Histone H3 lysine 9 acetylation localization patterns during neuronal maturation require MeCP2. Epigenetics *1*, 24–31.

Tomassy, G.S., Morello, N., Calcagno, E., and Giustetto, M. (2014). Developmental abnormalities of cortical interneurons precede symptoms onset in a mouse model of Rett syndrome. J. Neurochem. *131*, 115–127.

Tropea, D., Giacometti, E., Wilson, N.R., Beard, C., McCurry, C., Fu, D.D., Flannery, R., Jaenisch, R., and

Sur, M. (2009). Partial reversal of Rett Syndrome-like symptoms in MeCP2 mutant mice. Proc. Natl. Acad. Sci. U.S.A. *106*, 2029–2034.

Tsagaratou, A., Aijo, T., Lio, C.W., Yue, X., Huang, Y., Jacobsen, S.E., Lahdesmaki, H., and Rao, A. (2014). Dissecting the dynamic changes of 5-hydroxymethylcytosine in T-cell development and differentiation. Proc. Natl. Acad. Sci. U.S.A. *111*, E3306–E3315.

Van Esch, H., Bauters, M., Ignatius, J., Jansen, M., Raynaud, M., Hollanders, K., Lugtenberg, D., Bienvenu, T., Jensen, L.R., Gecz, J., et al. (2005). Duplication of the MECP2 region is a frequent cause of severe mental retardation and progressive neurological symptoms in males. Am. J. Hum. Genet. *77*, 442–453.

Venkateswaran, S., McMillan, H.J., Doja, A., and Humphreys, P. (2014). Adolescent onset cognitive regression and neuropsychiatric symptoms associated with the A140V MECP2 mutation. Dev. Med. Child Neurol. *56*, 91–94.

Wan, M., Zhao, K., Lee, S.S., and Francke, U. (2001). MECP2 truncating mutations cause histone H4 hyperacetylation in Rett syndrome. Hum. Mol. Genet. *10*, 1085–1092.

Watson, C.M., Pelka, G.J., Radziewic, T., Shahbazian, M.D., Christodoulou, J., Williamson, S.L., and Tam, P.P. (2005). Reduced proportion of Purkinje cells expressing paternally derived mutant Mecp2308 allele in female mouse cerebellum is not due to a skewed primary pattern of X-chromosome inactivation. Hum. Mol. Genet. *14*, 1851–1861.

Watson, P., Black, G., Ramsden, S., Barrow, M., Super, M., Kerr, B., and Clayton-Smith, J. (2001). Angelman syndrome phenotype associated with mutations in MECP2, a gene encoding a methyl CpG binding protein. J. Med. Genet. *38*, 224–228.

Williams, E.C., Zhong, X., Mohamed, A., Li, R., Liu, Y., Dong, Q., Ananiev, G.E., Mok, J.C., Lin, B.R., Lu, J., et al. (2014). Mutant astrocytes differentiated from Rett syndrome patients-specific iPSCs have adverse effects on wild-type neurons. Hum. Mol. Genet. *23*, 2968–2980.

Woods, R., Vallero, R.O., Golub, M.S., Suarez, J.K., Ta, T.A., Yasui, D.H., Chi, L.H., Kostyniak, P.J., Pessah, I.N., Berman, R.F., et al. (2012). Long-lived epigenetic interactions between perinatal PBDE exposure and Mecp2308 mutation. Hum. Mol. Genet. *21*, 2399–2411.

Wu, H., Coskun, V., Tao, J., Xie, W., Ge, W., Yoshikawa, K., Li, E., Zhang, Y., and Sun, Y.E. (2010). Dnmt3a-dependent nonpromoter DNA methylation facilitates transcription of neurogenic genes. Science *329*, 444–448.

Xinhua, B., Shengling, J., Fuying, S., Hong, P., Meirong, L., and Wu, X.R. (2008). X-chromosome inactivation in Rett Syndrome and its correlations with MECP2 mutations and phenotype. J. Child Neuro. *23*, 22–25.

Yasui, D.H., Peddada, S., Bieda, M.C., Vallero, R.O., Hogart, A., Nagarajan, R.P., Thatcher, K.N., Farnham, P.J., and Lasalle, J.M. (2007). Integrated epigenomic analyses of neuronal MeCP2 reveal a role for long-range interaction with active genes. Proc. Natl. Acad. Sci. U.S.A. *104*, 19416–19421.

Yasui, D.H., Scoles, H.A., Horike, S., Meguro-Horike, M., Dunaway, K.W., Schroeder, D.I., and Lasalle, J.M. (2011). 15q11.2-13.3 chromatin analysis reveals epigenetic regulation of CHRNA7 with deficiencies in Rett and autism brain. Hum. Mol. Genet. *20*, 4311–4323.

Yasui, D.H., Gonzales, M.L., Aflatooni, J.O., Crary, F.K., Hu, D.J., Gavino, B.J., Golub, M.S., Vincent, J.B., Carolyn Schanen, N., Olson, C.O., et al. (2014). Mice with an isoform-ablating Mecp2 exon 1 mutation recapitulate the neurologic deficits of Rett syndrome. Hum. Mol. Genet. *23*, 2447–2458.

Yazdani, M., Deogracias, R., Guy, J., Poot, R.A., Bird, A., and Barde, Y.A. (2012). Disease modeling using embryonic stem cells: MeCP2 regulates nuclear size and RNA synthesis in neurons. Stem Cells *30*, 2128–2139.

Young, J.I., and Zoghbi, H.Y. (2004). X-chromosome inactivation patterns are unbalanced and affect the phenotypic outcome in a mouse model of rett syndrome. Am. J. Hum. Genet. *74*, 511–520.

Young, J.I., Hong, E.P., Castle, J.C., Crespo-Barreto, J., Bowman, A.B., Rose, M.F., Kang, D., Richman, R., Johnson, J.M., Berget, S., et al. (2005). Regulation of RNA splicing by the methylation-dependent transcriptional repressor methyl-CpG binding protein 2. Proc. Natl. Acad. Sci. U.S.A. *102*, 17551–17558.

Zachariah, R.M., and Rastegar, M. (2012). Linking epigenetics to human disease and Rett syndrome: the emerging novel and challenging concepts in MeCP2 research. Neural. Plast. 2012, 415825.

Zappella, M., Meloni, I., Longo, I., Hayek, G., and Renieri, A. (2001). Preserved speech variants of the Rett syndrome: molecular and clinical analysis. Am. J. Med. Genet. *104*, 14–22.

Zhou, Z., Hong, E.J., Cohen, S., Zhao, W.N., Ho, H.Y., Schmidt, L., Chen, W.G., Lin, Y., Savner, E., Griffith, E.C., et al. (2006). Brain-specific phosphorylation of MeCP2 regulates activity-dependent Bdnf transcription, dendritic growth, and spine maturation. Neuron *52*, 255–269.

The Long and Short of Facioscapulohumeral Muscular Dystrophy

11

Sunny Das and Brian P. Chadwick

Abstract

This chapter integrates research over the last few decades from various groups across the globe in order to provide a comprehensive account of different aspects of the debilitating muscle disorder Facioscapulohumeral muscular dystrophy (FSHD). FSHD is the third most common inherited form of muscular dystrophy, it is the first example of a macrosatellite-linked disease, displaying a complex pathogenesis due to interplay between genetic and epigenetic components. The chapter starts with the history of the disease and its clinical features. Various genetic and epigenetic determinants of the disease are subsequently detailed, aimed at understanding how they might be contributing to the phenotype. Finally, we discuss treatment of the disease and translational research for developing therapeutic approaches, along with associated concerns.

History and prevalence

FSHD is a debilitating muscular disorder that was first described by two French physicians, Louis Landouzy and Joseph Dejerine, in 1884. Referring to the condition as facioscapulohumeral type of progressive myopathy, they described muscular atrophy in a male patient who displayed initial weakness of facial muscles, followed by that of shoulder girdle, upper arm and subsequent slow progression of atrophy into trunk and pelvic girdle muscles. Study of the patient's family revealed that the disease had an autosomal dominant inheritance pattern and showed clinical heterogeneity, which was subsequently corroborated in studies with other patients. Hence, this disorder is also known as the 'Landouzy–Dejerine disease'. Alternative names include facioscapulohumeral myopathy. In 1950, a study constructed a massive family pedigree involving a kindred of 1249 individuals spanning six generations (Tyler and Stephens, 1950). All these individuals were descendants of a single patient who had emigrated to Utah (USA) from England in 1840. This study examined 240 members of the kindred and detailed the inheritance pattern and clinical features in 58 affected members of this group.

In spite of this prior knowledge, this disease did not receive much attention until about three decades ago when a landmark study of 107 patients by Dr George W.A.M. Padberg, published in 1982, provided a detailed document of presenting signs and symptoms, precipitating factors, clinical features and progression, differential diagnosis, biochemical studies, mode of inheritance, prevalence and incidence, along with a summary of personal findings (Padberg, 1982). He also coined the term 'facioscapulohumeral disease' for this disorder. Our understanding of the disease has increased manifold in the last three decades. Cumulative research from different groups has contributed to the development of what are currently the most accepted models of FSHD aetiology. These models, along with the research that led to their formulation, will be discussed in subsequent sections.

The prevalence of a disease, defined as the frequency of affected individuals in a population at a given time, is quite variable for FSHD and depends on the geographical region being surveyed. The estimated range for prevalence is

0.4–6.7 per 100,000 individuals of the population, depending on the geographical region being surveyed (Emery, 1991). Similarly, the incidence of the disease, which is defined as the frequency of new occurrences of a disease among individuals of a specific population within a certain period of time, or as the frequency of individuals born in a certain population who will become affected, is also variable. The accepted estimate at about 1 in every 15,000 live births, as calculated from a study of an additional, previously unidentified 971 individuals of the Utah kindred along with 208 patients partly from the kindred and partly from other families (Flanigan *et al.*, 2001). Nevertheless, with the application of improved diagnostic techniques and a more thorough study of populations, the true incidence of the disease could well be found to be lower.

Classification

Elements in D4Z4, a highly polymorphic macrosatellite array in the subtelomeric region of chromosome 4q35 (Fig. 11.1A), are causal to FSHD (Wijmenga *et al.*, 1990). Normal individuals contain 11–150 repeating units in this array, each of which is 3.3 kilobases (kb) in size (Fig. 11.1B). A combination of genetic and epigenetic changes to the transcriptional state of this array results in expression of a gene called double homeobox 4 (*DUX4*) (Hewitt *et al.*, 1994; Winokur *et al.*, 1994; Ding *et al.* 1998; Gabriels *et al.*, 1999), coded by each unit of D4Z4. The stochastic mis-expression of a full-length isoform of this gene from the most distal repeat in patient myoblasts and translation of the protein (Dixit *et al.*, 2007; Lemmers *et al.*, 2010b) results in disease phenotype by signalling a downstream cascade of gene expression that inhibits muscle

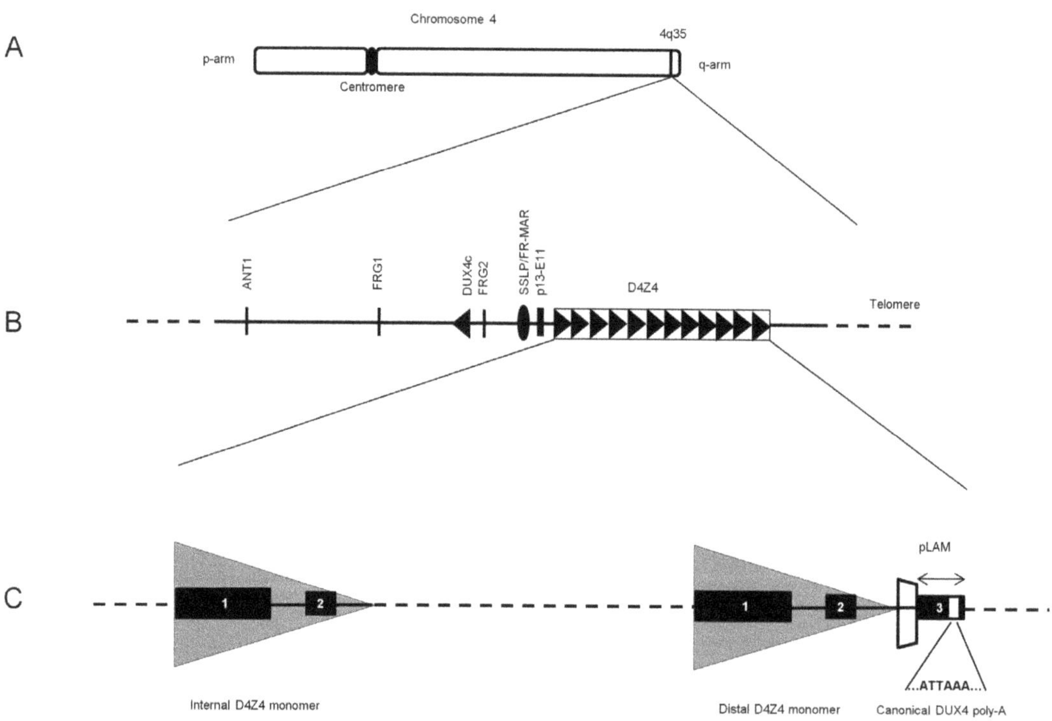

Figure 11.1 The genetic locus of FSHD. (A) shows a schematic diagram of human chromosome 4. Subtelomeric region 4q35 containing D4Z4 and other upstream genes are magnified and shown in (B). The FSHD Related Matrix Attachment Region (FR-MAR) that contains the SSLP found in FSHD-permissive haplotypes, along with the region where the p13-E11 probe hybridizes, is also shown. (C) shows a representation of any one internal D4Z4 monomer and the most distal (telomeric) D4Z4 monomer in a 4qA permissive haplotype. The telomeric monomer differs from the internal ones in that it contains an immediately distal pLAM region that contains a canonical DUX4 polyA sequence (ATTAAA).

differentiation and replenishment (Tupler *et al.*, 2000; Dixit *et al.*, 2007; Belayew, 2010; Geng *et al.*, 2012; Zhao *et al.*, 2014). Occurring in 1 out of every 20,000 individuals, FSHD is the third most common inherited form of muscular dystrophy, after Duchenne muscular dystrophy (DMD) and myotonic dystrophy (MD).

Currently, two forms of the disease FSHD1 (OMIM # 158900) and FSHD2 (OMIM # 158901) are known. Both forms of the disease have some common and some unique determinants but the end result is the same i.e. expression of DUX4, leading to disease phenotype. FSHD1, occurring in the majority (90–95%) of patients, is caused by a contraction in the number of repeats of D4Z4 to fewer than 10 units (Wijmenga, 1992; van Deutekom *et al.*, 1993) whereas the less prevalent FSHD2 (5–10%) is contraction independent (Wijmenga *et al.*, 1991; Gilbert *et al.*, 1992; Bakker *et al.*, 1995).

Both forms require at least one unit of D4Z4 as individuals with a deletion of the entire array on one allele are phenotypically normal (Tupler *et al.*, 1996; Goto *et al.*, 1995; Rossi *et al.*, 2007). Additionally, the presence of a canonical polyadenylation signal in the pLAM region immediately distal to the most telomeric D4Z4 repeat on a 'permissive' 4qA allele (Fig. 11.2A) is required to

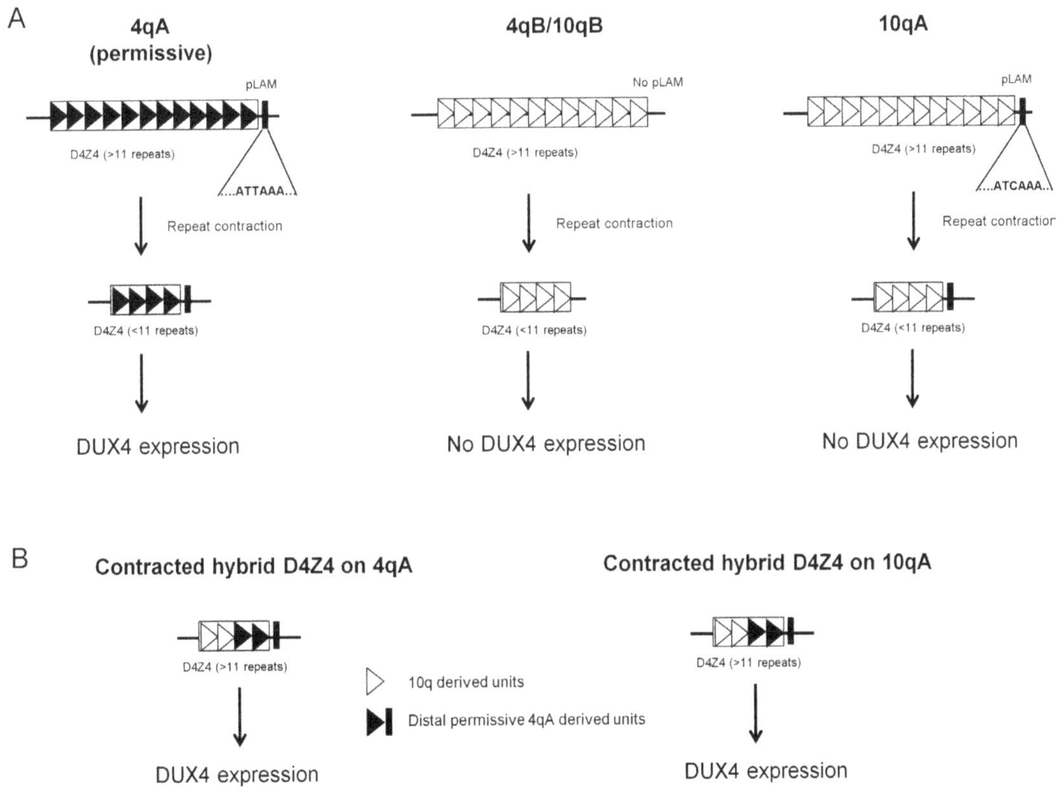

Figure 11.2 DUX4 expression from allelic variants of subtelomeric 4q35 and hybrid repeats due to D4Z4 contraction. (A) shows exemplary D4Z4 arrays on 4qA, 4qB and 10qA alleles. While 4qA and 10qA contain the pLAM region, 4qB does not. For convenience, only a permissive haplotype of 4qA (containing canonical DUX4 polyA sequence ATTAAA) has been shown. 10qA contains a non-canonical (ATCAAA) sequence, which, like 4qB, does not result in DUX4 expression. Any array containing the most distal unit from a permissive 4qA haplotype is capable of expressing DUX4 on repeat contraction, if other epigenetic criteria are favourable. (B) (left) shows one such case where a hybrid array residing on 4q35 may be derived from proximal 10q D4Z4 units and distal permissive 4qA units. Alternatively, a hybrid array residing on 10q26 may be derived from proximal 10q D4Z4 units and distal permissive 4qA units (right). Both these arrays are capable of DUX4 expression and are usually a result of translocations between highly homologous D4Z4 arrays of chromosomes 4 and 10. The 4qC allelic variant is rare and non-permissive (not shown).

stabilize full-length *DUX4* (*DUX4-fl*) mRNA and result in production of DUX4 protein in patients (Fig. 11.1C) (Dixit *et al.*, 2007; Lemmers *et al.*, 2010b). However, unlike FSHD1, most FSHD2 cases also involves the independent inheritance of hemizygous mutations in a second locus, structural maintenance of chromosomes flexible hinge domain 1 (*SMCHD1*; OMIM 614982) on Chromosome 18p11, along with the inheritance of a permissive 4qA allele (Lemmers *et al.*, 2012). *SMCHD1* mutations are not a requirement for FSHD1 but incidence of such mutations affects the severity of phenotype in such patients (Sacconi *et al.*, 2013). Finally, both forms are characterized by CpG hypomethylation (van Overveld *et al.*, 2003; de Greef *et al.*, 2009), a loss of heterochromatic histone markers with a concomitant shift to a more euchromatic state at D4Z4 (Zeng *et al.*, 2009). However, while in FSHD1, methylation loss is restricted to the contracted 4q allele (de Greef *et al.*, 2007), in FSHD2, hypomethylation at D4Z4 occur on both alleles of chromosomes 4 and both alleles of a homologous macrosatellite repeat on chromosome 10 (de Greef *et al.*, 2009, 2010).

FSHD is usually a late-onset disease (Padberg, 1982; Tawil *et al.*, 1998) but there are cases of early onset or infantile FSHD, where the disease manifests from infancy to 4.5 years (Brouwer *et al.*, 1994). The genetic changes causing infantile FSHD are similar to FSHD1 except that the symptoms are more severe (Klinge *et al.*, 2006). The early appearance of features has been linked to larger deletions of D4Z4 that result in fewer residual repeats. This is not surprising, as the inverse relationship between residual D4Z4 repeats and severity of phenotype has been documented earlier (Lunt *et al.*, 1995; Goto *et al.*, 1995; Zatz *et al.*, 1995; Tawil *et al.*, 1996; Ricci *et al.*, 1999).

Clinical features

Both types of adult-onset FSHD have a similar, relatively well-defined set of characteristic features that are used for diagnosis. In his review, Padberg observed that a majority of symptoms were facial muscle weakness, followed by shoulder girdle weakness, ankle dorsiflexor muscle weakness, and pelvic girdle weakness. A uniform set of guidelines for diagnostic criteria of FSHD was first established at an international consortium and published in 1991.

Symptomatic presentation

- Facial muscle weakness: *Orbicularis oculi and orbicularis oris* are the most common facial muscles affected along with an asymmetrical involvement around the lips. Weakness of orofacial muscles may also lead to difficulty in swallowing (Wohlgemuth *et al.*, 2006).
- Shoulder girdle weakness: two characteristic features are the internal rotation of arms and an elevated position of the scapulae due to weakness of middle and lower trapezius muscles. Flattening of the anterior chest wall has been observed and atrophy of pectoralis major and minor along with biceps and triceps, have been reported to varying degrees (Padberg, 1982; Tyler and Stephens, 1950). Significantly, there is marked winging of the scapulae. Bulbar, extraocular, and respiratory muscles are spared.
- Abdominal muscle weakness: a positive Beevor's sign, an FSHD specific feature, has been reported in a majority of patients with FSHD1 (Shahrizaila *et al.*, 2005; Eger *et al.*, 2010) and about 67% of patients with FSHD2 (de Greef *et al.*, 2010).
- Lower extremity muscle weakness: ankle dorsiflexor muscles and pelvic girdle weakness have been reported. In some cases, involvement of muscles of hamstring and quadriceps femoris has also been observed. In patients with FSHD2, scapular, facial and dorsiflexor weakness is present in a majority of patients while pelvic girdle weakness seems to be less common (de Greef *et al.*, 2010).
- Extramuscular features: retinal vasculopathies (Fitzsimons *et al.*, 1987) and hearing loss (Padberg *et al.*, 1995) are also associated with FSHD. The lack of retinal vasculopathies in FSHD2 patients might have been due to the absence of a systematic evaluation of retinal vessels in a majority of the patients under such study (de Greef *et al.*, 2010). Especially, high tone hearing loss has been described in a number of patients in both FSHD1 and

FSHD2. Visual loss has also been reported in a small number of patients with FSHD1. Along with the above-mentioned symptoms, patients also complain of pain and fatigue, which may be caused by factors like exertion and faulty posture due to muscle weakness but reasons behind these are not well understood.

- Cardiac and pulmonary features: some studies have suggested roles of pulmonary and cardiac involvement in FSHD. Respiratory insufficiency has been documented in a small minority of patients who were on nocturnal ventilator support and their risk factors included severe involvement with wheelchair confinement, moderate to severe kyphoscoliosis, and the presence of pectus excavatum (Wohlgemuth *et al.*, 2004). Subclinical cardiac involvement, suggested by a preclinical reduction of left ventricular function and abnormal myocardial activity has also been reported (Galetta *et al.*, 2005). A small number of patients exhibit mild arrhythmias (Stevenson *et al.*, 1990; de Visser *et al.*, 1992) but no cardiovascular risk has been documented (Trevisan *et al.*, 2006).
- Infantile FSHD features: in cases of infantile FSHD (Brouwer *et al.*, 1994), features are similar to the adult onset form but they also develop severe lumbar lordosis and exhibit foot drop deformities along with sensorineural deafness. These children are usually wheelchair dependent by the age of 8–9 years, and by the third decade, thoracic and respiratory compromise becomes clinically significant.

Asymmetry of weakness

One aspect that distinguishes FSHD from other muscular dystrophies is the asymmetry of weakness (Kilmer *et al.*, 1995). Although the reason behind this is not well understood, initial explanations proposed over-work weakness and handedness as the causal reasons. There were also suggestions of asymmetry being linked to an intrinsic disease process or linked to a genetic mechanism. Recent reports have shown that this might well be the case (see 'Pathophysiology'). Despite this asymmetry, the incidence of contractures and scoliosis is low.

Gender bias

Interestingly, the FSHD phenotype shows a gender bias even though there is no molecular evidence for autosomal or X-linked recessive inheritance. An initial study reported a statistically significant difference in the mean age at onset in males and females (Chung and Morton, 1959). A significantly greater proportion of females than males among asymptomatic cases and a significantly greater proportion of affected sons than daughters observed in the offspring of asymptomatic mothers indicate that males are more affected (Zatz *et al.*, 1998). It has been speculated that hormonal differences between males and females may account for this difference in onset or severity. It remains to be seen if this difference is due to the 'protective' effect of a modifier gene in females or the 'pathogenic' effect of a modifier gene in males.

Age at onset and disease progression

Age at onset can vary from infancy to late life, but most affected individuals become symptomatic in the second decade (Padberg, 1982; Tawil *et al.*, 1998). A majority of patients report difficulties with overhead activities in the second or third decade of their life (Tawil *et al.*, 1998). The rate of disease progression is variable but is usually very slow. Most affected individuals are still able to work, often adapting remarkably well to muscle weakness. Up to one-third of patients have no symptoms, but weakness can be found in over 95% of affected individuals by age 20 (Tawil *et al.*, 1998). However, approximately 10% of all patients and 20% of those more than 50 years of age eventually become wheelchair dependent for outdoor activities (Pandya *et al.*, 2008). Lifespan in patients is usually normal (Padberg, 1982; Gardner-Medwin, 1980), although about some of them are severely disabled because of weakness (Padberg, 1982; Tawil *et al.*, 1998; Pandya *et al.*, 2008).

Clinical diagnosis

FSHD is diagnosed by physical examination for features discussed above. Creatine phosphokinase (CPK) levels are normal in FSHD patients and help to differentiate it from DMD patients who have elevated levels for the enzyme. Patient muscles also show pseudohypertrophy and fatty infiltration. Advances in our understanding of the pathophysiology of the disease in the last two decades now also allows for molecular diagnosis (see 'Genetics and epigenetics') of the disease not only in familial and sporadic cases but also in prenatal testing.

Genetics

Mode of inheritance and occurrence

Studies of familial cases in various kindred across the world have reinforced that, with rare exceptions, almost all cases of reported FSHD1 have been found to have an autosomal dominant mode of disease transmission. Most initial studies describing an autosomal recessive mode of inheritance were not well documented (Brown, 1952; Stevenson; 1953). These patients possibly had a myopathy that is similar, but not identical, to FSHD. In FSHD1, the majority of cases are familial, with new mutations occurring in up to one-third of cases (Padberg, 1982; Lunt *et al.*, 1995; Padberg *et al.*, 1995; Zatz *et al.*, 1995, 1998; Tawil *et al.*, 1996). However, in FSHD2, the observed ratio of familial to sporadic cases seems to be the inverse to that of FSHD1 (de Greef *et al.*, 2010).

The disease involves a very complex interplay of genetic and epigenetic factors, which may account for individuals carrying the genetic defect yet no FSHD phenotype i.e. asymptomatic carriers (also see 'Penetrance'). It has been seen that about one-fifth of relatives of FSHD patients carry a D4Z4 deletion but are asymptomatic or have very mild features (Zatz *et al.*, 1998; Tonini *et al.*, 2004). This occurs in a significantly higher proportion of females than males. Biological fitness also seems to be reduced, with no gender bias (Zatz *et al.*, 1995; Tawil *et al.*, 1993). Approximately 40% of the *de novo* cases result from a post-zygotic mutation that leads to somatic mosaicism (Griggs *et al.*, 1993; Weiffenbach *et al.*, 1993; Upadhyaya *et al.*, 1995; van der Maarel *et al.*, 2000). Affected mosaic individuals are typically male while mosaic females are often the unaffected parent of a non-mosaic *de novo* patient (van der Maarel *et al.*, 2000). Mosaic females also have a significantly higher proportion of somatic mosaicism than mosaic males (Zatz *et al.*, 1998; van der Maarel *et al.*, 2000). It has also been shown that severely affected cases originate more often through new mutations or are transmitted through maternal than through paternal lines (Zatz *et al.*, 1998).

Highly homologous chromosome 4-type D4Z4 repeat arrays are present on chromosome 10, in about 10% of normal chromosomes (van Deutekom *et al.*, 1996). In mosaic individuals, 4-type repeats on chromosome 10 are almost five times more frequent. The reverse configuration, however, is less prevalent. This suggests that mutations may arise from interchromosomal interactions, and increase in chromosome 4 repeat clusters biases the more prevalent configuration. The phenomenon of somatic mosaicism primarily suggests a mitotic origin; interchromosomal gene conversion or translocation between fully homologous chromosome 4-type repeat arrays has been proposed as a probable mechanism for new FSHD mutations (van der Maarel *et al.*, 2000). Another interesting phenomenon in FSHD is the relationship between severity of the disease and age at onset. Patients presenting severe phenotype and lesser number of residual D4Z4 repeats display symptoms at an earlier age than those with less severe phenotype and higher number of repeats (Zatz *et al.*, 1995; Goto *et al.*, 1995; Lunt *et al.*, 1995; Tawil *et al.*, 1996). The phenomenon in which signs and symptoms of a genetic disorder tends to become more severe and appear at an earlier age as the disorder is passed from one generation to the next, is called anticipation. Like some other diseases, such as Huntington's disease (MacDonald *et al.*, 1993) and myotonic dystrophy (Aslandis *et al.*, 1992; Buxton *et al.*, 1992; Harley *et al.*, 1992), it has been shown that anticipation occurs for FSHD in the majority of multigenerational families (Zatz *et al.*, 1995).

Penetrance

Initial observations in small pedigrees suggested complete penetrance for FSHD (Tyler and Stephens, 1950). Even with some subsequent early studies, the rate of progression of disease and variability in phenotype was not well documented, leading to an assumption that penetrance is 100%.

However, over the years, as the understanding of the disease evolved and more detailed studies with larger sample size were conducted, it became evident that there is considerable inter- and intra-familial heterogeneity in phenotype. Clinical non-penetrance is observed in asymptomatic carriers. Since monozygotic twins serve as a platform to understand the relative contributions of genetic and epigenetic components in a disease, several groups have reported cases of monozygotic twins showing identical restriction fragment length polymorphism (RFLP) patterns for the *Eco*RI fragment that defines the extent of the D4Z4 repeats (see 'Identification and evolution of the FSHD locus'), but display extreme variability in clinical expression (Tawil *et al.*, 1993; Griggs *et al.*, 1995; Hsu *et al.*, 1997; Tupler *et al.*, 1998). This lack of concordance in phenotype between monozygotic twins points to a strong epigenetic component in the disease. In a study comprising 172 patients, the penetrance of the FSHD1 gene until age 30 was estimated as 83% for both sexes but was significantly greater for males (95%) than for females (69%) (Zatz *et al.*, 1998).

Linkage with other traits

Some early studies examined the possibility of linkage of FSHD with the loci of the ABO, MN, RH, P, FY, and JK blood groups in patients (Tyler and Stephens, 1950; Chung and Morton, 1959) but no association with a known trait or disease has been observed in FSHD to date.

Identification and evolution of the FSHD locus

Central to FSHD is the D4Z4 macrosatellite, which is composed of a uninterrupted tandem repeat of 3.3 kb individual repeat units (Hewitt *et al.*, 1994; Winokur *et al.*, 1994) at subtelomeric 4q35 (Wijmenga *et al.*, 1992; van Deutekom *et al.*, 1993). The array is highly polymorphic in the general population and its size varies from 35 to 350 kb in healthy individuals who have 11–150 repeating units. However, an important feature of FSHD1 is that patients have a contraction of the repeat array, bringing down repeat number to between 1–10 units in patients, resulting in arrays shorter than 35 kb (Wijmenga *et al.*, 1992; van Deutekom *et al.*, 1993). Interestingly, the complete loss of the 4q subtelomeric region is not associated with FSHD, as individuals lacking the entire D4Z4 array from one allele of chromosome 4 are reported to be phenotypically unaffected (Tupler *et al.*, 1996). Such individuals, monosomic for D4Z4 have been found to harbour a deletion of the chromosome 4-subtelomeric region due to an unbalanced translocation (Tupler *et al.*, 1996).

The first report of mapping a gene for FSHD focused on narrowing down the disease locus based on its proximity to known markers. Studies on 10 Dutch families revealed that the FSHD locus maps to the long arm of chromosome 4 (4qter), based on linkage analysis with the marker Mfd 22 (D4S171) with a LOD score greater than 6 (LOD score >3 indicates linkage; Wijmenga *et al.*, 1990). Subsequently, different groups reported more tightly linked markers, namely pH30 (D4S139), D4S163 and the locus for factor XI (F11), with higher LOD scores (Upadyaya *et al.*, 1990, 1991, 1992; Wijmenga *et al.*, 1991). These studies, involving multipoint linkage analysis and *in situ* hybridization, firmly established the location of the FSHD locus to subtelomeric region of chromosome 4q. Additionally, information from these studies helped ascertain the order of these markers with respect to the FSHD gene as Cen-D4S171-F11-D4S163-D4S139-FSHD-Tel, where Cen and Tel represent the chromosome 4 centromere and q-arm telomere, respectively (Wijmenga *et al.*, 1992). In an effort to identify additional distal markers, a 0.8 kb probe called p13E-11 was generated (Wijmenga *et al.*, 1992), derived from subcloning a cosmid clone 13E. The cosmid 13E had previously been isolated in a separate search for homeobox genes (Callen *et al.*, 1990). Using this probe that maps to a locus called D4F104S1, two independent cosmid clones were isolated from a chromosome 4 specific library, in which p13E-11 detected polymorphic *Eco*RI restriction fragments of different sizes. Investigation of *Bam*HI and *Kpn*I sites in these cosmid clones revealed a structure consisting of repetitive DNA sequences (Wijmenga *et al.*, 1992). In healthy individuals, the majority of *Eco*RI fragments were greater than 28 kb whereas in patients they were significantly shorter, ranging between 14–28 kb (Wijmenga *et al.*, 1992). Shorter *Eco*RI fragments emerged in new cases due to *de novo* DNA rearrangements, while in familial cases, FSHD was found to co-segregate exclusively with the *Eco*RI fragments of shorter size (Wijmenga *et al.*, 1992). This implies that virtually all FSHD mutations emerge through rearrangements at the locus near the sequence detected by p13E-11. This was corroborated by another study showing that proximal and distal parts of the *Eco*RI fragments, flanking a region of tandemly repeated 3.3 kb units, are identical in both patients and normal individuals (van Deutekom *et al.*, 1993). This strongly supports the hypothesis that DNA rearrangements observed in FHSD are due to deletions of integral copies of these repeating units, also implying that the number of units might have an effect on the expression of the gene. Sequence analysis of each 3.3 kb repetitive unit of D4Z4 showed that it contains two homeoboxes with a putative open reading frame (ORF) (Hewitt *et al.*, 1994; Ding *et al.*, 1998; Gabriels *et al.*, 1999). Two additional elements, LSau, a repetitive element associated with heterochromatic regions and hhspm3, a low copy number GC-repeat were detected downstream of the homeoboxes (Hewitt *et al.*, 1994). Fluorescence *in situ* hybridization (FISH) with D4Z4 specific probes resulted in cross-hybridization to several heterochromatic regions of the genome (Hewitt *et al.*, 1994; Winokur *et al.*, 1994). Moreover, D4Z4 was found to be subtelomeric, in close proximity to 4q telomere, as previously discussed.

Subsequently, a number of D4Z4-like sequences, organized into subfamilies, have been found on acrocentric chromosomes (Lyle *et al.*, 1995; Beckers *et al.*, 2001), suggesting that D4Z4 is a member of a dispersed family of homeobox containing repeats, distributed across the genome with significant copy number variation (CNV). Although there are similar sequences with almost 90% homology on the Y-chromosome, the one with the maximum sequence similarity (~99%) to D4Z4 is located on chromosome 10q26 (Bakker *et al.*, 1995; Deidda *et al.*, 1996; Winokur *et al.*, 1996; Cacurri *et al.*, 1998). The homology between 4q and 10q subtelomeric regions extends both proximally and distally beyond the D4Z4 repeats (van Geel *et al.*, 2002). Sequence variations and single nucleotide polymorphisms (SNPs) between these otherwise highly homologous 4q35 and 10q26 associated D4Z4 arrays serve as useful markers to differentiate between the two. Canonical 4q35 D4Z4 repeats are exclusively sensitive to *Xap*I digestion (Lemmers *et al.*, 2001) while 10q26 arrays are sensitive to *Bln*I digestion (Deidda *et al.*, 1996; Cacurri *et al.*, 1998). These two restriction enzymes are routinely used in conjunction with *Eco*RI, followed by separation using pulsed field gel electrophoresis (PFGE) and hybridization with the p13-E11 probe, to distinguish D4Z4 repeats from one chromosomal origin from the other. This can be used in conjunction with sequencing studies to dig out allelic information for 4q and 10q variants (Lemmers *et al.*, 2001). The repetitive nature of 4q-associated D4Z4, along with significant sequence similarity to the arrays on 10q26 had for long confounded FSHD research. Additionally, the extensive homology results in frequent exchanges between the 4q35 and 10q26 arrays, as discussed before. This results in formation of 'hybrid' arrays consisting of repeat units derived from both 4q35 and 10q26, which cannot be distinguished from the standard 4q- or 10q-only arrays using routine molecular analysis. 4q is currently known to have three allelic variants, namely 4qA, 4qB and 4qC (Lemmers *et al.*, 2010a). While 4qA and 4qB are equally prevalent in the population, FSHD is exclusively associated with the 4qA variant (Fig. 11.2A) (Lemmers *et al.*, 2002; Thomas *et al.*, 2007). These two alleles share almost 92% sequence identity with considerable variation in organization of sequence elements within these two alleles (van Geel *et al.*, 2002). The most telomeric D4Z4 repeat in 4qA is followed by an immediately distal 260 base pair (bp) sequence called pLAM (van Deutekom *et al.*, 1993; Gabriels *et al.*, 1999) that is absent from the internal or distal repeats, whereas the most telomeric D4Z4 repeat in 4qB contains only the first 570 bp of the 3.3 kb unit (Fig. 11.2A) (van Geel *et al.*, 2002). Additionally, the presence of a 6.2 kb region composed of 68 bp β-satellite sequences immediately distal to D4Z4 along with an adjacent 1 kb divergent (TTAGGG)n array exclusively in the 4qA allele, distinguishes it from the 4qB allele which lacks both these sequences (van Geel *et al.*, 2002). Finally, the A/B sequences also differ in a G/C SNP in the most proximal D4Z4 repeat (Lemmers *et al.*, 2004a). Although D4Z4 in both allelic variants display no difference in propensity to undergo contractions, deletions of the array on 4qB are non-pathogenic (Lemmers *et al.*, 2004b). A simple sequence length polymorphism (SSLP), 3.5 kb proximal to D4Z4, and upstream of the p13–E11 probe hybridization region (Fig. 11.1B), helps determine the FSHD associated haplotypes. The third 4q variant, 4qC is rare and is also not associated with FSHD. 10q is known to have two allelic variants, namely 10qA and 10qB. Contractions of the D4Z4 array on chromosome 10q26, occurring in ~10% of the population, are also non-pathogenic. However, certain hybrid arrays consisting of partial 10q and 4q units have been found to be FSHD associated (Fig. 11.2B).

Initial studies led to the belief that disease pathology is governed by its chromosomal origin i.e. genes responsible for FSHD were only on chromosome 4 and not on chromosome 10 (some of these genes are discussed later) and their transcriptional mis-regulation is responsible for FSHD. However, the existence of certain 4qA haplotypes that are not associated with FSHD argue against this logic (Lemmers *et al.*, 2007). Additionally, it was assumed that the sequence variation between 4qA and 4qB in the region distal to D4Z4 (van Geel *et al.*, 2002) was responsible for the disease to be associated with the 4qA variant exclusively (Lemmers *et al.*, 2002; Thomas *et al.*, 2007). This has indeed been found to be one of the factors governing pathogenicity. Sequencing studies on four types of polymorphisms in 4qter i.e. size of D4Z4, the SSLP, the G/C SNP between A/B and the pLAM region have been used to mine out allelic information at D4Z4 (Lemmers *et al.*, 2007; Lemmers *et al.*, 2010a). 4qter currently has seventeen identified haplotypes, out of which, 4qA161 is the most common FSHD 'permissive' haplotype. In brief, contractions on 4qA161, 4qA161L, 4qA159 and 4qA168 chromosomes are pathogenic, whereas contractions on other 4qA, and all 4qB, 4qC, 10qA and 10qB alleles are non-pathogenic (Lemmers *et al.*, 2010a,b; Spurlock *et al.*, 2010).

The exclusive association of only certain allelic variants with the disease and not others poses a significant question: What determines pathogenicity of these alleles of 4q35 D4Z4 repeats or of certain 'hybrid' repeats?

Each unit of D4Z4 contains the ORF for a double homeobox gene *DUX4* (Hewitt *et al.*, 1994; Ding *et al.*, 1998; Gabriels *et al.*, 1999). It is to be noted that the transcriptional profile of the most distal D4Z4 unit is different from that of the internal units (Dixit *et al.*, 2007; Sinder *et al.*, 2009) but none of the *DUX4* transcripts from the latter seem to be stable, presumably due to the lack of a polyadenylation signal that otherwise stabilizes *DUX4* mRNA (Kojaljow *et al.*, 2007; Dixit *et al.*, 2007; Lemmers *et al.*, 2010b). This gene contains alternatively spliced introns in its 3′ untranslated region (3′ UTR), which maps downstream of the ORF, outside of the most telomeric D4Z4 unit (Dixit *et al.*, 2007; Snider *et al.*, 2010). Detailed sequence analysis of permissive alleles has revealed that the pLAM region, immediately distal to the most telomeric D4Z4 unit, harbours a SNP that confers pathogenicity to permissive alleles. Transcripts, both spliced and unspliced, use the unique 3′ UTR in pLAM and are stabilized only when a canonical polyadenylation (poly A) signal ATTAAA is present (Fig. 11.1C) (Lemmers *et al.*, 2010b). The stability of *DUX4* transcripts is strictly dependent on the sequence of this polyA signal, meaning that any other polymorphism within the sequence makes transcripts unstable and hence undetectable due to decreased mRNA half-life. *DUX4* mRNA and the corresponding protein have been detected in patient primary myoblasts (Coppee *et al.*, 2004) and only the most distal D4Z4 unit and pLAM region containing a canonical polyA signal are sufficient to generate stable DUX4 transcripts *in vitro*, as demonstrated by transfection studies in C2C12 mouse myoblast cells (Lemmers *et al.*, 2010b). However, it has not been established that this is the case *in vivo* as well. Any of the internal D4Z4 units into the pLAM region could in theory, generate stable transcripts, if it splices into the polyA containing pLAM region downstream of the ORF. Stable transcripts are also generated from non-permissive alleles when a (T/C) SNP in the non-canonical ATCAAA found in such alleles is changed to ATTAAA (Lemmers *et al.*, 2010b). Conversely, changing the 'T' to a 'C' on a permissive chromosome results in decreased stability of transcripts, confirming that the polyA signal is instrumental in DUX4 transcript stabilization (Lemmers *et al.*, 2010b).

Three other observations further support the idea of polyA mediated DUX4 stabilization being crucial to the disease. Patients containing a contracted hybrid D4Z4 arrays that have 10q-derived units at the proximal end and 4q-derived units at the distal end on a permissive 4q allele, have been identified (Lemmers *et al.*, 2010b). Another patient has a contracted hybrid array on chromosome 10 that contained distal 4q units derived from a permissive allele (Lemmers *et al.*, 2010b). Disease manifestation in this patient with a chromosome 10 contraction reduces the likelihood that genes previously proposed to be FSHD candidates that are located upstream of D4Z4 on Chromosome 4 contribute to the disease phenotype, since chromosome 10 does not contain any of these genes. These include FSHD region gene 1 (*FRG1*), adenine nucleotide translocator 1 (*ANT1*) and double homeobox protein 4, centromeric (*DUX4c*) (Fig. 11.1B). Analysis of another patient harbouring a massive 4q35 D4Z4 deletion showed that the deleted allele was missing upstream genes *DUX4c* and FSHD region gene 2 (*FRG2*), implying that these were not candidate genes either (Deak *et al.*, 2007). All these hybrid alleles retained the polyA containing pLAM region, reinforcing the notion that *DUX4* expression is central to FSHD pathology.

The pLAM sequence, containing a polyA sequence, is present in in all 4qA and 10qA alleles but absent from 4qB and 10qB alleles (Fig. 11.2A) (Lemmers *et al.*, 2010b), which is why these B-variants are not associated with FSHD. However, out of all haplotypes of 4qA, only the ones that do contain the ATTAAA sequence, such as 4qA161, 4qA161L, 4qA159 and 4qA168 are permissive whereas the others, such as 4qA166 are non-permissive. Additionally, even though 10qA haplotypes contain the pLAM region, the corresponding sequence at the polyA stretch is ATCAAA, which is not known to be a canonical polyA signal for *DUX4*, making these haplotypes non-permissive as well (Lemmers *et al.*, 2010b). Analysis of sequences from various organisms has led to a hypothesis that 4qA arose from an ancestral duplication event involving 4p and 4q (van Geel *et al.*, 2002). The different haplotypes of this variant, referred to as the 4qA cluster, seem to be the most primitive haplotypes with the highest sequence similarity to the Chimpanzee sequence (Lemmers *et al.*, 2010a). The occurrence of D4Z4-like repeats on other chromosomes were due to additional amplification events. Studies in Old and New World monkeys have supported an evolutionary route where an ancestral 4q allele gave rise to the permissive 4qA haplotypes, 4qA159 and 4qA168, the latter evolving further to give rise to 4qA161 (Lemmers *et al.*, 2010a). The permissive alleles then underwent sequence divergence between the subtelomeric regions of 4q, leading to formation of other non-permissive haplotypes of the 4qA cluster. Two independent transfer events of 4qA distal elements on to 10q26, followed by divergence in the canonical polyA sequence gave rise to the non-permissive 10qA cluster. The second allelic variant, 4qB seems to have arisen from a relatively recent duplication and transfer of 4p subtelomeric sequence on 4q, resulting in loss of pLAM in these alleles. 10qB arose from a transfer of D4Z4 distal B sequences on to 10q. Distal B sequence duplication also led to the formation of other haplotypes of the 4qB cluster. The 4qC variant seems to have arisen from a sequence divergence in one of the 4qA haplotypes. In summary, 4qA, 4qC and 10qA clusters possibly evolved from a single ancestral chromosome through three transfer events from 4 to 10 whereas the 4qB and 10qB clusters evolved from a single transfer of distal B variation within chromosome 4 (Lemmers *et al.*, 2010a). Existence of newer, non-permissive haplotypes such as some of the 4qB and 10qA haplotypes suggests that there has been selection towards these haplotypes over evolutionary time in humans.

Pathophysiology

DUX4

As mentioned earlier, the requirement of at least one D4Z4 repeat for the disease indicates that there is something integral to the repeat itself or sequences immediately flanking it, which plays a pivotal role in FSHD aetiology. Additionally, sequence analysis in patients with hybrid arrays containing interchromosomal transfers between 4q and 10q and deletions encompassing some upstream genes, negates the possibility of other candidate genes upstream of the D4Z4 array being central to the disease. In this section we will look at how the identification, characterization and gene expression analysis of *DUX4*, along with understanding of factors contributing to the stability of its various isoforms, have greatly advanced our understanding of the disease in the previous decade.

In spite of evidence hinting that D4Z4 might harbour a functional gene, the identification of full-length *DUX4* mRNA and protein remained a technical challenge for the longest time. Multiple reasons such as rare, low abundant expression, short half-life, high cytotoxicity, existence of similar arrays on other chromosomes scattered throughout the genome and the heterochromatic nature of subtelomeric 4q35 containing the array contributed to this problem. The *DUX4* ORF but not flanking sequences have been conserved for over a hundred million years (Clapp *et al.*, 2007). Moreover, selection of the protein-coding region in primates without the introduction of a stop codon in the ORF, along with a conserved head-to-tail arrangement of repeats, and transcription from a mouse *Dux* array, strongly suggested a coding function for D4Z4 (Clapp *et al.*, 2007). Evolutionary studies propose that the intronless *DUX4* arrays originated from the amplification of an ancestral retrotransposed pseudogene. Phylogenetic analysis and conservation of the C terminal domain (CTD) indicates that *DUX4* probably arose from the retrotransposition of the intron containing *DUXC* gene, which was subsequently lost in primates (Clapp *et al.*, 2007).

Initial attempts at isolating a gene from 4q35 FSHD associated region revealed that each 3.3 kb D4Z4 repeat unit contains a 405 bp homeodomain (HD) sequence within a continuous ORF (Hewitt *et al.*, 1994; Winokur *et al.*, 1994). Additionally the repeat was also found to contain heterochromatic repetitive elements LSau and hhspm3, outside of the ORF. Although this 405 bp region is only a part of the whole *DUX4* gene (as now known) and no active promoter was identified, the fact that developmental defects are a signature of FSHD hinted at the involvement of a HD-containing transcription factor. A putative promoter containing a TACAA box instead of the conventional TATAA box, along with a 30 bp upstream GC box was subsequently assigned 149 bp upstream of this ORF (Ding *et al.*, 1998; Gabriels *et al.*, 1999). This ORF was theoretically capable of coding for a 391 amino acid protein containing the two HDs. A 38 kDa protein corresponding to the partial *DUX4* gene was also generated from *in vitro* translation experiments (Gabriels *et al.*, 1999). The sequence of a known transcription factor *DUX1*, was found to be similar to the D4Z4 ORF, strengthening the notion that the putative gene from D4Z4 is a transcription factor (Ding *et al.*, 1998). Interestingly, the fusion of C-terminal domain (CTD) of *DUX4* to the Capicua transcriptional repressor (*CIC*) gene (encoding a high mobility group transcription factor) resulted in enhanced transcriptional activity in *CIC-DUX4* chimeras (Kawamura-Saito *et al.*, 2006). *DUX4-fl* mRNA and protein have been subsequently identified and characterized using improved RT-PCR techniques and more specific antibodies. Each D4Z4 repeat contains two exons of *DUX4* interrupted by an intron (Fig. 11.3). The LSau and hsspm3 repeats are upstream of the first exon. The first exon (1275 bp) contains the entire ORF while the second exon (92 bp) constitutes a part of the 3′-UTR. Immediately distal to the most telomeric unit of D4Z4 lies an additional intron and the pLAM sequence (Exon 3; 199 bp). *DUX4-fl* has two isoforms, one that retains the first intron and another which splices it out (Fig. 11.3). Both isoforms of *DUX4-fl* are pathogenic (Snider *et al.*, 2010). Apart from these three exons in *DUX4-fl*, downstream exons and polyA signals that are found in alternative transcripts have also been identified (Snider *et al.*, 2009). An alternative isoform, short *DUX4* (Fig. 11.3) (*DUX4-s*), which utilizes a cryptic non-canonical splice donor site within the ORF such that it excludes the CTD of the corresponding protein, is non-pathogenic but dominant negative to DUX4-fl (Snider *et al.*, 2009, 2010). Usage of this cryptic splice donor site seems to be governed by the distribution pattern of local repressive histone modifications (Snider *et al.*, 2010; see Epigenetics)

Control fibroblasts express *DUX4-s* but not *DUX4-fl*. However, induced pluripotent stem (iPS) cells derived from these fibroblasts show expression of *DUX4-fl*, which is suppressed again

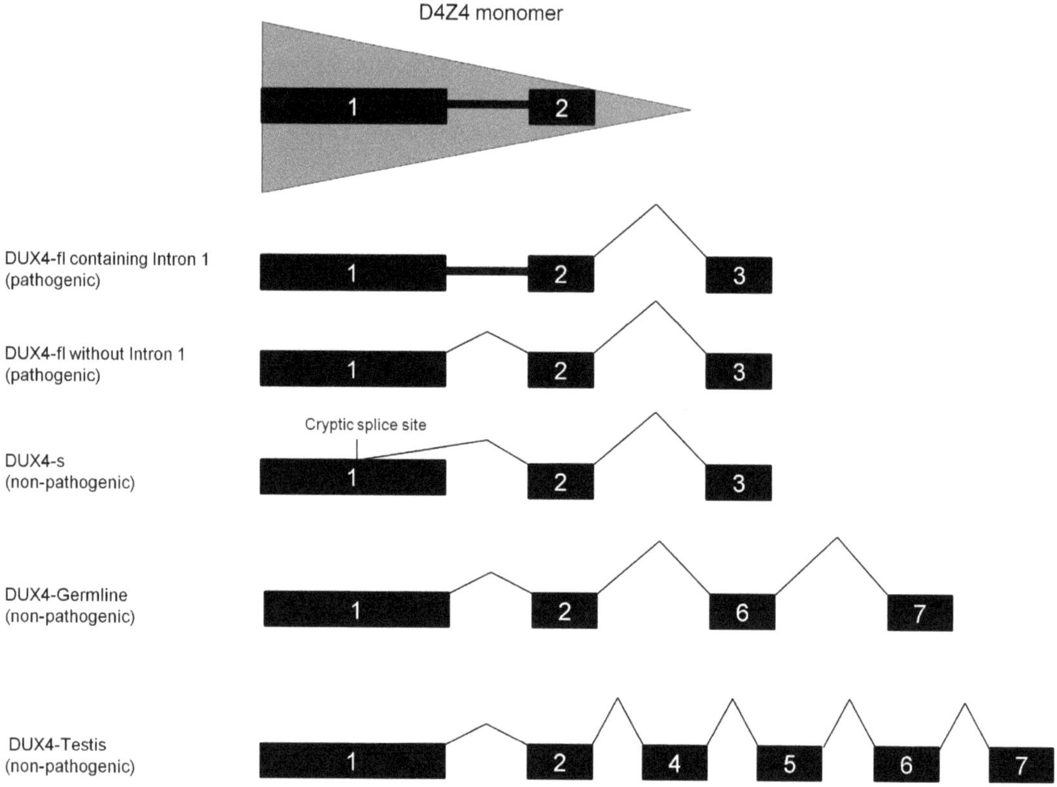

Figure 11.3 Alternative splicing at D4Z4 resulting in various isoforms of DUX4. All alternatively spliced transcripts of DUX4 contain the first two exons (except DUX4-s that contains partial exon 1) that are found in each monomer of D4Z4, as shown on top. Pathogenic DUX4-fl results from splicing into Exon 3 found immediately distal to the most distal D4Z4 unit and can exhibit two isoforms, one that contains Intron 1 and the other that does not. DUX4-s results from alternative splicing from a cryptic splice donor site within Exon 1 (ORF) of DUX4 that results in a truncated protein lacking the transactivator C-terminal domain, making the protein non-pathogenic. Alternative downstream exons are also used in DUX4 transcripts that may play a role in normal germline development, as shown by the last two isoforms. While the germline transcript contains exons 6 and 7 in addition to 1 and 2, the testis-specific transcript contains exons 4, 5, 6 and 7.

(reverting to *DUX4-s* expression) on differentiation of these iPS cells into embryoid bodies (EB). FSHD cells, however, show continued DUX4-fl expression at the fibroblast, iPS cell and EB stages (Snider *et al.*, 2010). These observations suggest a role for *DUX4-fl* in normal germline development with subsequent silencing upon differentiation into normal somatic tissues. Insufficient developmental silencing results in continued expression of *DUX4-fl* in somatic muscle of FSHD patients. It is worth noting that *DUX4* transcripts containing the full ORF originating from chromosome 4 preferably use the polyA signal in the pLAM region for expression in somatic tissue whereas an alternative polyA signal (6.5 kb downstream of the pLAM polyA) is used for expression in the germline (testis) (Fig. 11.3) (Snider *et al.*, 2010). Transcripts from chromosome 10 however, are expressed only in the testis (utilizing the alternative polyA signal) and not in somatic tissues. *DUX4-s* expression is seen in certain normal somatic tissues. It has been observed that only a very small fraction (1 in 1000 myonuclei) produce large amounts of *DUX4-fl* (Snider *et al.*, 2010), which is sufficient to cause the phenotype. Both sense and antisense transcripts have been detected from two promoters in D4Z4 (Block *et al.*, 2012). Transcription is asymmetrical, i.e. antisense transcription is predominant. The directional switching in transcription is temporally dynamic and differentially regulated in human myoblasts and human embryonic stem cells (ESCs). While the promoter generating sense transcripts lies just upstream of the *DUX4* ORF, a second TATA-less promoter lies in a region 33 bp from the 3′ end of the ORF. A regulatory region within the first promoter helps maintain the levels of sense and antisense transcripts. Presence of this regulatory region favours generation of antisense transcripts, while its deletion biases the expression of the sense transcript (Block *et al.*, 2012). The most distal D4Z4 unit has predominantly antisense transcripts in both permissive and non-permissive alleles. Regulation of directionality might explain why not all myonuclei express *DUX4* in FSHD muscle in spite of the presence of a permissive chromosome in all cells (Block *et al.*, 2012).

Although FSHD is attributed exclusively to *DUX4-fl* expression, there are studies which have shown that expression of *DUX4-fl* from a permissive allele at both mRNA and protein levels in myogenic cells and muscle biopsies derived from biceps and deltoid muscles not only in FSHD patients but also in their unaffected first-degree relatives (containing normal sized D4Z4 repeats), although at significantly lower levels (Jones *et al.*, 2012). Additionally, some members in this study showed no clinical symptoms (asymptomatic carriers), in spite of displaying *DUX4-fl* expression. *DUX4-fl* expression is also seen in fetal (non-infantile cases) FSHD quadriceps and trapezius muscles (Ferreboeuf *et al.*, 2014). The gene expression cascade downstream of DUX4 in these cases was similar to that of adult onset FSHD, suggesting that there are additional factors that govern the delayed onset i.e. manifestation of symptoms later in life even when the molecular signatures of the disease are present from fetal life. These observations suggest that additional components, possibly epigenetic modifiers are critical in modulating *DUX4-fl* expression and penetrance. Classical epigenetic factors such as DNA CpG methylation, histone modifications and telomere position effects are already known to be differentially associated with the array between patients and controls and will be discussed in the Epigenetics section in detail. Additionally, modifier genes such as *SMCHD1* may play a vital role in modulating *DUX4-fl* expression (Fig. 11.4A) (Sacconi *et al.*, 2013). Alternatively, proteins such as CCCTC binding Factor (CTCF), known to mediate long-range chromosomal interactions, and bind to D4Z4 in patients, might play an important role. Recent identification of two myogenic enhancers, DME1 and DME2 upstream of the 4q35 D4Z4 array containing CTCF binding motifs lend support to this idea (Himeda *et al.*, 2014). These enhancer sequences, whose activity seems to be dependent on the epigenetic (hypomethylation) status of the *DUX4* promoter, interact specifically with the *DUX4* promoter in muscle cells but not in fibroblasts, indicating that modifiers may act in a tissue-specific manner. Finally, bidirectional transcription and presence of si/miRNA sized transcripts derived from D4Z4 could also be important in mediating D4Z4 silencing (Snider *et al.*, 2009; Block *et al.*, 2012).

The 52 kDa DUX4 protein, resulting from *DUX4-fl* translation, contains two HD towards the N-terminus. Basic residues at the end of these HDs play important roles in nuclear targeting (Ostlund *et al.*, 2005), similar to what has been observed with some other HD containing proteins (Ploski *et al.*, 2004). The CTD seems to have transient interacting partners in the nucleus, with strong transcriptional activity. Being a protein that is actively localized to the nucleus, three *in-silico* nuclear localization signals (NLS) have been annotated within the two HDs of DUX4 (Corona *et al.*, 2013). Triple deletion mutants for these three NLSs are still partially localized to the nucleus, suggesting that additional factors are important for localization. Indeed, deletion of a region from the CTD on a triple NLS deletion mutant background localizes the protein completely to the cytoplasm, implying that the CTD could be recruiting additional factors for nuclear localization via its transactivator region (Corona *et al.*, 2013). Nuclear import of the protein does not seem to utilize the classical α/β importin pathway. The DUX4 binding site has two canonical HD binding motifs TAAT separated by a nucleotide and arranged in tandem. Additionally, although *DUX4-fl* and *DUX4-s* protein products bind to the same motif, the latter cannot activate downstream genes involved in FSHD, suggesting that the CTD that DUX4-s lacks, must act as a transactivator domain.

RNA-seq analysis of normal human myoblasts transduced with *DUX4* shows gene expression

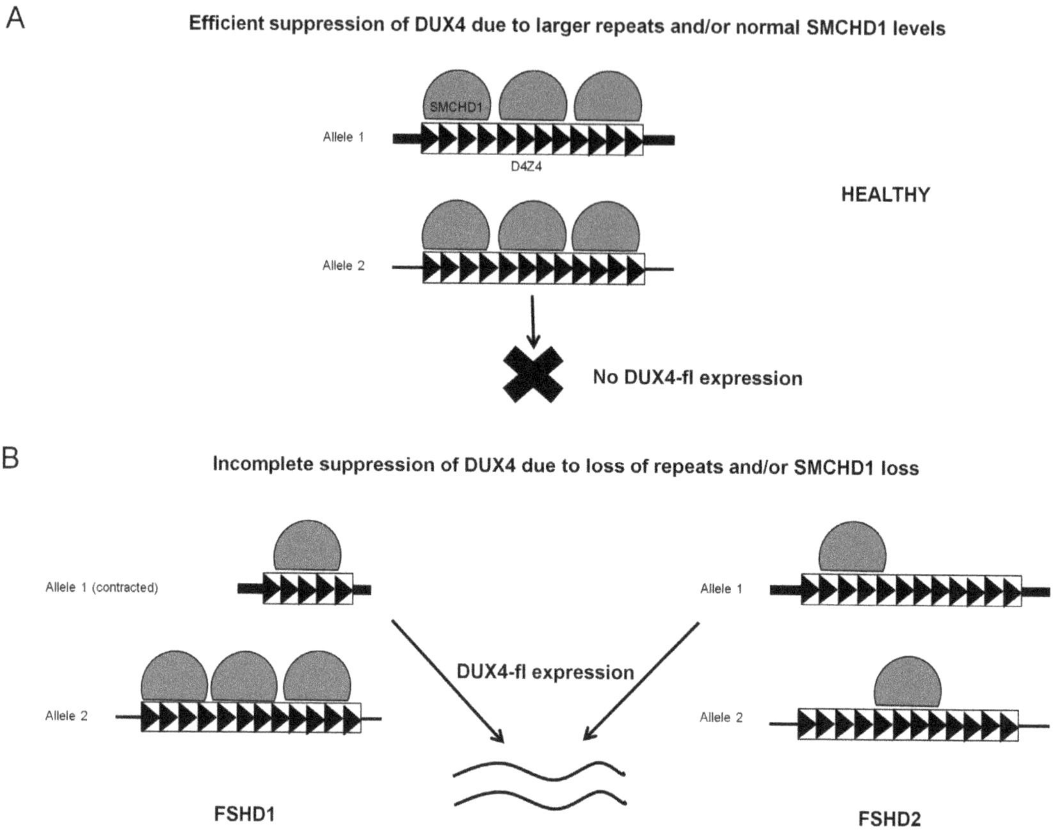

Figure 11.4 Contraction-dependent and independent modes of FSHD. (A) shows subtelomeric D4Z4 on two alleles of chromosome 4, one of which (bold black line) is a permissive 4qA haplotype. SMCHD1, which is believed to play a role in D4Z4 methylation, is intact in these healthy individuals and keeps the array hypermethylated state along with other repressive factors such as histone modifications (not shown), resulting in efficient DUX4 repression. This efficient repression could be lost in two ways as shown in (B). Contraction of the permissive allele in FSHD1 (left) may result in loss of CpG methylation from the contracted allele or due to hemizygous mutations in SMCHD1 in FSHD2 (right) may result in loss of CpG methylation from both alleles of 4q (and also 10q; not shown). Both these instances would lead to expression of DUX4-fl from a permissive allele, as observed in patients.

changes that are consistent with *DUX4* overexpression (Yao *et al.*, 2014). Similar gene expression patterns are also observed in FSHD biopsies, reinforcing that *DUX4* induced misexpression of target genes is important for FSHD. Similar expression profiles between FSHD1 and FSHD2 patients confirm that *DUX4* expression is common to all forms of the disease (Yao *et al.*, 2014).

Although the entire array of downstream genes activated by DUX4 is not yet known, there is considerable information on the group of genes it activates, revealing how its mis-expression relates to clinical features. Some of these target genes, such as leucine twenty homeobox (*LEUTX*), preferentially expressed antigen in melanoma family member 2 (*PRAMEF2*), tripartite motif containing 43 (*TRIM43*) and KH homology domain containing 1-Like (*KHDC1L*) have been proposed as candidate biomarkers for FSHD (Ferrebouef *et al.*, 2014; Geng *et al.*, 2012; Yao *et al.*, 2014). *DUX4-fl* expression in human primary myoblasts induces expression of genes that are normally not detected in skeletal muscle. These genes, also expressed in patient muscle, are primate specific and are normally expressed in the germline and during early development but silenced in normal somatic muscle. DUX4 binds to many long terminal repeats (LTRs) of the Mammalian apparent LTR-Retrotransposon (MaLR) and Endogenous Retroviral sequence (ERV) families, initiating transcription from a number of these sites (Geng *et al.*, 2012; Young *et al.*, 2013). Interestingly, some of these DUX4-bound LTRs form alternative promoters for protein coding genes, long non-coding RNAs and antisense transcripts. This activity of DUX4 at some of these elements was observed in both patient myoblasts and in testis of a control, suggesting a role in normal germline biology (Young *et al.*, 2013). This also demonstrates how repetitive elements derived from retrotransposons can play a role in regulating host gene expression over evolutionary time. The reactivation of germline specific genes in somatic cells could result in a conflict in cellular transcriptional programs with serious consequences on cell physiology, the extreme end of which would be apoptosis. Not surprisingly, *DUX4* overexpression has been found to be cytotoxic, involving a *p53* mediated apoptotic pathway and causing the induction of *caspase3/7* along with altered emerin distribution at the nuclear envelope (Kojaljow *et al.*, 2007). DUX4 overexpression causes apoptosis in most tissues, except for certain tumour cells lines that are *p53* inactive and in *p53*-null mice (Wallace *et al.*, 2011). Through down-regulation of genes providing resistance to oxidative stress, *DUX4* also renders cells sensitive to oxidative damage-induced stress (seen in FSHD myoblasts), contributing to cytotoxicity (Winokur *et al.*, 2003; Macaione *et al.*, 2007; Bosnakovski *et al.*, 2008a, 2014). *DUX4* directly activates paired homeodomains transcription factor 1 (*PITX1*), a gene often up-regulated in patients and involved in specification of hind-limb identity and left–right symmetry (Dixit *et al.*, 2007). Dysregulation of *PITX1* by *DUX4* might explain asymmetry of muscle weakness that distinguishes FSHD from other muscular dystrophies. Interesting, *PITX1* is known to activate components of the *p53* pathway (Liu and Lobie, 2007), suggesting its contribution to *DUX4*-mediated apoptosis. *DUX4-fl* also regulates expression of members of the *PRAME* family, which are known cancer-testis antigens. It has been suggested that mis-regulation of these cancer-testis loci by *DUX4-fl* might trigger an adaptive immune response often seen in FSHD skeletal muscle but not in the germline, since the latter is an immune-privileged site (Geng *et al.*, 2012). Defensin Beta 3 (*DEFB103*), another gene induced by *DUX4*, inhibits innate immune response to lentiviral infection in skeletal muscle and also plays a pro-inflammatory role in adaptive immune response. *DEFB103* induction represses genes involved in myogenic differentiation and results in mild myostatin induction that prevents differentiation of myoblasts, suggesting a mechanism of how *DUX4* expression might cause muscle atrophy (Geng *et al.*, 2012). Genes involved in RNA splicing, regulatory components of the RNA PolII transcription complex and ubiquitin-mediated proteasome degradation pathways are also mis-regulated upon *DUX4-fl* expression (Geng *et al.*, 2012). In mouse C2C12 myoblasts, *DUX4* induction results in down-regulation of Myogenic Differentiation gene (*MyoD*) expression and up-regulation of *p21* in patient muscle biopsies and patient myoblast cultures, respectively (Bosnakovski *et al*, 2008a). MyoD is a transcription factor involved in myogenic commitment and skeletal muscle differentiation. The HDs of DUX4 are similar to members of the paired box (*Pax*) family, *Pax3* and *Pax7*, which play important roles in maintenance of satellite muscle cells (Seale *et al.*, 2000). DUX4 competes with both Pax proteins for binding and inappropriate regulation of their target myogenic genes (Bosnakovski *et al.*, 2008a).

Thus, *DUX4* seems to affect skeletal muscle physiology in multiple ways. Its cytotoxic effect reduces some of the myoblast population through apoptosis and compromises the capacity of the rest to differentiate into myotubes. Additionally, its expression seems to reduce the muscle stem cell population, resulting in lowered muscle regeneration and repair capacity. One intriguing aspect of FSHD muscle pathology is how = the mis-expression of *DUX4-fl* in such a low fraction of myonuclei (~1 in 1000) causes atrophy in relatively large regions of muscle fibre. For this, a nuclear protein-spreading model of DUX4 protein has been proposed (Tassin *et al.*, 2013; Ferreboeuf *et al.*, 2014). In this model, a stochastic reactivation of *DUX4-fl* in one myonucleus within the syncytial myotube is followed by export of *DUX4* mRNA into the cytoplasm where it is translated into several molecules of DUX4 protein. The protein molecules diffuse in the cytoplasm and are imported into neighbouring myonuclei, guided by their nuclear localization signals. Additionally, these DUX4 protein molecules activate their cascade of downstream genes, many of which are transcription factors themselves. This sets up an amplification of the initial trigger that results in atrophy, immune responses, oxidative stress and inhibited differentiation of regions throughout the muscle fibre.

Other candidate/contributing genes

Apart from DUX4, a number of other genes located at varying distances upstream of the array on chromosome 4 have been proposed as candidates for FSHD since they have been found to be up-regulated in patients and seem to recapitulate certain disease features. Though *DUX4* is believed to be central to FSHD pathology, we will briefly discuss two of these genes and understand if they help modulate the phenotype in addition to *DUX4*.

DUX4c

Located 42 kb upstream of the D4Z4 array on 4q35, this gene containing a truncated and inverted D4Z4 element, encodes a 47 kDa protein (Wright *et al.*, 1993). Its double HDs are identical to that of DUX4 but has a divergent CTD (Ansseau *et al.*, 2009). Though both DUX4 and DUX4c bind to the mouse *Pitx1* and Xenopus *pitx1* promoter (Dixit *et al.*, 2007; Wuebbles *et al*, 2010), DUX4c shows weaker transcriptional activity due to absence of the transcriptionally active DUX4-type CTD (Dixit *et al.*, 2007). Forced expression of *DUX4c* does not cause cytotoxicity in proliferating mouse myoblasts but does inhibit differentiation even at very low levels of expression. This inhibitory effect is mediated by the down-regulation of *MyoD* and Myogenic factor 5 (*Myf5*) with similar down-regulation of *MYOD1* in human myoblasts as well (Ansseau *et al.*, 2009). In contrast, *DUX4c* (but not DUX4) induces and probably stabilizes the myogenic regulator MYF5 protein, which increases myoblast proliferation but inhibits differentiation in human myoblasts (Ansseau *et al.*, 2009; Bosnakovski *et al.*, 2008b).

FRG1

Located 125 kb upstream of the 4q D4Z4 array, this gene was shown to be up-regulated during myogenic differentiation of human myoblasts (Bodega *et al*, 2009), although other studies did not agree with this observation. *Frg1*, but not *Frg2* or *Ant1* transgenic mice that overexpress *Frg1* in skeletal muscle develop skeletal muscle abnormalities similar to what is seen in patients (Gabellini *et al.*, 2006). Furthermore, *frg1* overexpression in *Xenopus* results in disruption of vasculature (Wuebbles *et al*, 2009) and musculature (Hanel *et al.*, 2009), two features of FSHD. This is consistent with the suggestions that in muscle, *FRG1* is a developmentally regulated sarcomeric protein with a possible role in muscle contraction (Hanel *et al.*, 2011). Inconsistent reports on *FRG1* overexpression (Bodega *et al.*, 2002; Gabellini *et al.*, 2002, 2006; Klooster *et al.*, 2009; Osborne *et al.*, 2007) in patients, along with the characterization of *DUX4* and all the experimental evidence in its favour, leave little doubt that *DUX4* is the primary FSHD gene. However, in addition to *DUX4*, it is possible that up-regulation of *FRG1* and other non-DUX4 genes contribute to the phenotype, through an epigenetic de-repression mechanism (discussed later).

As discussed previously, the identification of a patient with a contracted hybrid allele and proximal D4Z4 deletion reduce the probability of these genes being directly involved in pathogenesis. However, it is possible that up-regulation of these genes in patients are epigenetically facilitated from the non-permissive chromosome 4 allele and they act in *trans* on the permissive contracted allele on 4q or the hybrid contracted allele on 10q to activate their downstream genes and contribute to the development and heterogeneity of phenotype in FSHD. *FRG2*, for example, has a homologue on chromosome 10, which might be the source of up-regulated *FRG2* mRNA in patient myoblasts.

Epigenetics

Complex disorders such as FSHD often display a multifactorial aetiology where various epigenetic mechanisms play a critical role in modulating the manifestation and severity of disease phenotype. Although not a comprehensive list, we will discuss some of the important epigenetic contributors in FSHD.

Copy number variation in macrosatellites

Packaged mostly into heterochromatin, repetitive DNA accounts for more than half of the human genome (Schmid and Deininger, 1975). Tandem repeats consist of a significant fraction of such repetitive DNA and are organized as multiple copies of a DNA sequence repeated and arranged in a head-to-tail manner. These regions are classified on the basis of the size of the individual repeating units and exhibit considerable variation in repeat size and number. Historically regarded as 'junk' DNA (Ohno, 1972; Orgel and Crick, 1980), these repeats have only recently started receiving attention in light of findings that imply that tandem repeats might be playing important regulatory roles.

Macrosatellites are one such category of tandem repeats where the repeating units can vary from a few hundred base pairs to a several kilobases and often span hundreds of kilobases of DNA in the genome. These represent an extreme case of copy number variation since the number of individual repeat units within the macrosatellites are highly variable from one individual to the next, meaning that they are polymorphic in the general population. Various studies have identified and characterized different macrosatellites scattered throughout the human genome. Some of these, such as the GAGE family (Gjerstorff, 2008) and CT47 (Chen *et al.*, 2006), are cancer-testis loci, with readily detectable transcription from their arrays in testis and certain forms of cancer, as their name suggests. The X-linked macrosatellite DXZ4 (Giacalone *et al.*, 1992), located at Xq23 shows differential chromatin conformation between the two X-chromosomes in females, primarily adopting a heterochromatic organization on the active X-chromosome, whereas the allele on the inactive X-chromosome (Xi) is primarily arranged into euchromatin that is bound by CTCF (Chadwick, 2008). The functional consequence of this is that the Xi-allele makes CTCF-dependent long-range intrachromosomal interactions between DXZ4 and other similarly packaged tandem repeats exclusively on the Xi (Horakova *et al.*, 2012), and is perhaps involved in facilitating chromosome folding.

Though some macrosatellites have been characterized and many others remain to be explored, D4Z4 is the only macrosatellite repeat array currently known that has a direct disease association, although small alleles of a macrosatellite unique to chromosome 5p15.1 have potentially been linked to schizophrenia (Bruce *et al.*, 2009). Nevertheless, D4Z4 is a great example of how CNVs can contribute to disease pathology. We already know how SNPs in *DUX4* allelic variants stabilize transcripts and lead to DUX4 protein-mediated pathology. Additionally, CNVs at D4Z4 in FSHD1 represent another class of determinant polymorphism in this disease. That CNVs have a significant contribution to the evolution of human genome is well known (Conrad *et al.*, 2010). They are also known to change coding function of genes and have regulatory roles. We have already discussed how *de-novo* contractions might arise in FSHD. An increasing number of Mendelian disorders and complex diseases have a CNV component to them (Stankiewiez and Lupski, 2010).

CpG methylation

Repetitive sequences and macrosatellites are typically packaged into heterochromatin. Loss of heterochromatin at such regions may be regulated and intentional, such as DXZ4 on Xi, or may result *de novo*, such as seen on contracted D4Z4 alleles. Such loss of heterochromatin at these repeats is likely to have functional consequences. DNA methylation and histone modifications are believed to mediate chromatin structure, in addition to other factors such as associated proteins and different classes of regulatory RNAs. Although DNA methylation shows a genome-wide distribution pattern, the presence or absence of methylation marks on cytosine residues at CpG sites are crucial determinants of gene regulation. Clusters of CpG dinucleotides, known as CpG islands (CGIs) are found at both promoters and intragenic regions. Transcriptionally active genes have unmethylated promoters whereas methylation results in stable gene silencing and chromatin condensation, assisted by the distribution of specific histone modifications. Methylation of gene bodies is believed to suppress initiation of non-specific transcription (Suzuki and Bird, 2008). These 5-methylcytosine marks, laid down by different DNA methyltransferases (DNMTs), also block access of regulatory proteins to genomic regions containing or flanking these marks (Chen and Li, 2004). Proteins such as CTCF (Hark *et al.*, 2000) and Ying Yang 1 (YY1) (Kim *et al.*, 2003), both involved in FSHD, are methylation-sensitive and show differential localization around D4Z4 between patients and normal individuals (Bodega *et al.*, 2009; Gabellini *et al.*, 2002; Ottaviani *et al.*, 2009). As mentioned earlier, each unit of D4Z4 contains repetitive LSau and hhspm3 sequences. Moreover, the D4Z4 repeat units are extremely GC rich (70% GC) (Gaillard *et al.*, 2014) and are normally packaged into heterochromatin. In unaffected individuals, D4Z4 arrays are hypermethylated at CpG dinucleotides (Tsien *et al.*, 2001), whereas in FSHD1 patients, the array is hypomethylated on the contracted allele only (van Overveld *et al.*, 2003). It is to be noted that although D4Z4 contractions are always associated with hypomethylation, this reduction in methylation is not chromosome or haplotype specific, as demonstrated in normal individuals with deletions on non-pathogenic chromosome 4 and 10 alleles (de Greef *et al.*, 2009). It will be interesting to see if contraction associated hypomethylation is a general phenomenon for GC rich macrosatellites such as DXZ4 and RNU2. In contrast, FSHD2 patients, carrying non-contracted D4Z4 arrays exhibit array hypomethylation on both alleles on chromosome 4 and 10. This suggests that hypomethylation of D4Z4 is a common feature of both forms of the disease.

Although the exact methyltransferase(s) responsible for methylation of D4Z4 are not known, knocking out both *DNMT1* and *DNMT3B* (but not *DNMT1* or *DNMT3B* alone) in the cell line HCT116 that contains a permissive 4qA allele, results in reactivation of *DUX4-fl*, suggesting that these methyltransferases might act cooperatively in maintaining array hypermethylation (Das and Chadwick, unpublished data). SMCHD1 is believed to facilitate hypermethylation of the array in normal individuals (Fig. 11.4A) (Lemmers *et al.*, 2012). Hemizygous mutations in *SMCHD1* in FSHD2 patients are associated with loss of hypermethylation at the array (Lemmers *et al.*, 2012). Since SMCHD1 binding to D4Z4 is not allele or chromosome specific, its haploinsufficiency results in hypomethylation of both alleles of chromosomes 4 and 10 but *DUX4* reactivation occurs only from the permissive 4qA allele (Fig. 11.4B).

Initial studies assessed FSHD specific methylation levels of both proximal and internal D4Z4 repeats using Southern blot analysis post digestion of genomic DNA with methylation sensitive restriction enzymes, such as *Fse*I, *Bsa*AI, and *Cpo*I (de Greef *et al.*, 2009, 2010; van Overveld *et al.*, 2003, 2005). However, the number of CpG sites assessed by these enzymes is limited to those associated with the restriction enzyme recognition sequences. Nevertheless, it was found that the proximal 4q D4Z4 repeats in patients were hypomethylated compared to controls and methylation levels in the proximal unit was also less than that of internal 4q units (de Greef *et al.*, 2009). On chromosome 10, however, there was no methylation differences between proximal and internal units, suggesting that hypomethylation within individual arrays might be influenced by its chromosomal context (de Greef *et al.*, 2009). Data obtained through this approach have given us an idea of general methylation patterns at the array. However, as mentioned above, since restriction endonuclease analysis assesses a very limited and specific set of CpG dinucleotides at each of the enzyme recognition sites (for which D4Z4 contains only a single *Cpo*I, *Bsa*AI and *Fse*I per D4Z4 repeat unit), this assay does not provide information for methylation at the more than 300 other CpG dinucleotides in a single repeat monomer and therefore provides a restricted methylation assessment. More informative has been the simultaneous analysis of the methylation state at multiple CpG sites in a single monomer, which has indicated that hypomethylation of some parts of D4Z4 might be more important than others. Using bisulfite sequencing (BiS) to scan 74 CpG sites in three different regions across D4Z4, a study reported the methylation status of D4Z4 in a large cohort of FSHD2 individuals (Hartweck *et al.*, 2013). This revealed that instead of global changes in CpG methylation, only specific regions of each D4Z4 unit undergo methylation changes. Three such regions (DR1,

DR2 and DR3) were hypomethylated in patients compared to controls, with hypomethylation of DR1 being the most severe. While DR3 is located within the *DUX4* ORF, DR1 and DR2 are at varying distances upstream of the *DUX4* promoter. This pattern was consistent throughout all the units of the array. Also, DR1 hypomethylation was more severe in FSHD2 myoblasts compared with FSHD2 peripheral blood lymphocytes, indicating that DNA methylation levels are influenced in a tissue-specific manner. In light of these findings, it has been proposed that in patients, the array does not undergo uniform demethylation throughout its entire length; instead it shows focused hypomethylation at these three domains. The extreme hypomethylation of DR1 upstream of the DUX4 ORF as opposed to DR3 within the ORF suggests, that DR1 might encompass a regulatory element critical to the expression of *DUX4*. Also, the presence of a CTCF binding site upstream of DR1 suggests that the region could play a role in CTCF-mediated insulation of *DUX4* from heterochromatin, further contributing to DUX4 expression. This function would only be possible when DR1 is hypomethylated, as the binding of CTCF to DNA is compromised if its target sequence contains methylated CpG.

Another study (Gaillard *et al.*, 2014) examined D4Z4 methylation at three specific regions (D4Z4-5′, D4Z4-Mid, and D4Z4-3′) at both global and single-base resolution using methylated DNA immunoprecipitation (Me-DIP) and BiS respectively. The analysis included unaffected individuals, FSHD1 and FSHD2 patients as well as individuals that have pathogenic-sized D4Z4 alleles on a permissive background but are phenotypically normal and therefore are considered asymptomatic carriers. In addition to confirming previous findings, this study revealed several other important pieces of information. First, in this study, no significant methylation differences were found between controls and asymptomatic carriers containing short D4Z4 arrays (< 11 units), suggesting that the manifestation of clinical features is strongly linked to hypomethylation. Therefore, asymptomatic carriers likely represent individuals with non-penetrance of disease. The fact that D4Z4 contractions on permissive alleles is insufficient to elicit disease onset, strongly supports involvement of an epigenetic component. This study also revealed that FSHD1 and FSHD2 patients have similar levels of reduced methylation at D4Z4-5′, which is significantly lower than that observed in controls or asymptomatic carriers. Finally, FSHD2 patients also showed hypomethylation at D4Z4-3′. The existence of asymptomatic carriers with D4Z4 contractions and FSHD2 patients without D4Z4 contractions, coupled with these methylation data suggest that clinical features are not correlated with D4Z4 repeat size but rather with the epigenomic organization, such as hypomethylation and SMCHD1 haploinsufficiency. Differences in hypomethylation levels may also explain why patients with 1–3 repeats display more severe phenotype than those with 4–10 repeats. The presence of specific hypomethylated regions in D4Z4 that are absent in other macrosatellites surveyed suggests the involvement of a D4Z4-specific methylation maintenance mechanism that selectively acts at some but not all CpG sites.

In summary these findings suggest that D4Z4 hypomethylation is necessary for FSHD. The next question is whether it is sufficient. Patients with immunodeficiency, centromere instability and facial anomalies syndrome 1 (ICF1; OMIM # 242860) (Maraschio *et al.*, 1998) who have mutations in the DNMT3B gene (Hansen *et al.*, 1999; Okano *et al.*, 1999; Xu *et al.*, 1999) display hypomethylation of specific repeat sequences in their genome such as Sat2, Sat3, NBL2 (chromosomes 1 and 16) and D4Z4 (Kondo *et al.*, 2000). Besides hypomethylated D4Z4, there are no clinical similarities between ICF1 and FSHD patients. A recent study showed that a female ICF1 patient, who possesses a hypomethylated 4qA allele, does not express DUX4-fl (Das and Chadwick, unpublished data). Thus, it seems that while CpG methylation is necessary and may be a modifier of penetrance and variable clinical heterogeneity observed in the disease, it is not sufficient to cause FSHD.

Histone modifications and non-coding RNAs

Eukaryotic DNA is organized into nucleosomes in order to efficiently 'package' DNA into the nucleus. A nucleosome consists of approximately

146 bp of DNA wrapped around a core histone octamer made up of two copies each of histones H2A, H2B, H3 and H4 (Campos and Reinberg, 2009). Histones are covalently modified by a variety of enzymes that attach or remove certain chemical groups at specific amino acid side chains, primarily at the amino-terminal tails that project out from the nucleosome core. These post-translational modifications such as methylation, acetylation, phosphorylation, ubiquitination and SUMOylation can be commonly found on lysine, arginine, serine and threonine residues (Bernstein *et al.*, 2007). Like DNA methylation, these modifications influence local chromatin organization through the recruitment or eviction of chromatin associated proteins, and is referred to as the 'histone code hypothesis' (Jenuwein and Allis, 2001). Coordinated action of CpG methylation and histone modifications can cooperate to regulate gene expression. For example, methyl binding domain (MBD) proteins recognize and bind to methylated DNA. In turn, these proteins can recruit histone deacetylases (HDACs) and histone methyltransferases (HMTases), as well as chromatin proteins such as heterochromatin protein 1 (HP1), which recognize and bind to methylated histones, locking in a repressive chromatin environment (Bannister *et al.*, 2001; Lachner *et al.*, 2001). A comprehensive list of effects of specific histone modifications is widely available in the literature, but we will focus only on some common modifications in the context of FSHD. Trimethylation of lysine residues 9 or 27 on histone H3 (H3K9me3 or H3K27me3) is linked to heterochromatin formation and transcriptional repression, while dimethylation or trimethylation of lysine residue 4 (H3K4me2 or H3K4me3) or acetylation on histone H3 (H3Ac) is associated with euchromatin and transcriptional activity (Bernstein *et al.*, 2005; Peters *et al.*, 2003; Rice *et al.*, 2003). In this section we will examine how various histone modifications govern repeat mediated heterochromatin formation in normal individuals and how loss of heterochromatin in patients may result in *DUX4* reactivation.

Owing to DNA hypermethylation and recruitment of a repressor complex to the array, it is easy to assume that D4Z4 is uniformly heterochromatic (Fig. 11.5). Transcriptional inactivity can result from either constitutive heterochromatin or from unexpressed euchromatin. Initial chromatin immunoprecipitation (ChIP) studies in regions around D4Z4 arrays on chromosomes 4 and 10 showed histone 4 acetylation (H4Ac) levels that were characteristic of unexpressed euchromatin rather than constitutive heterochromatin (Jiang *et al.*, 2003; Yang *et al.*, 2004). Subsequently, it was found that barring the most proximal repeat, which is uniformly euchromatic, all D4Z4 repeats consist of distinct heterochromatic and euchromatic regions within each unit. H3K9me3 and H3K27me3 characterize heterochromatic domains while the euchromatic domains are enriched for H3K4me2 and H3Ac. Loss of H3K9me3 but not H3K27me3 or H3K4me2, is seen in both FSHD1 and FSHD2 patients, compared to controls (Zeng *et al.*, 2009), and therefore D4Z4 arrays in patients tend to adopt a more euchromatic conformation (Fig. 11.5). Interestingly, unlike CpG hypomethylation, loss of H3K9me3 occurs on both alleles of chromosomes 4 and 10 and is not restricted to the contracted allele in FSHD1 patients (Zeng *et al.*, 2009). Similar loss was also seen in FSHD2 on both chromosome pairs (Zeng *et al.*, 2009). This might allow for transvection effects, as seen in *Drosophila* where components of one chromosome might act in *trans* to mis-regulate genes on another chromosome. This also indicates that retention of H3K9me3 is independent of CpG methylation, further supported by the fact that H3K9me3 levels at D4Z4 appear unchanged in ICF1 patient cell lines despite CpG hypomethylation (Zeng *et al.*, 2009). Given that *DUX4-fl* remains silent in an ICF1 patient that carries a hypomethylated 4qA allele, this suggests that H3K9me3 is central to maintaining D4Z4 in a transcriptionally repressive state. Furthermore, ICF1 patients have abnormally short telomeres, so maintenance of silencing is unlikely to be attributed to telomere proximity, supporting H3K9me3 loss at D4Z4 as a triggering factor in reactivation of *DUX4-fl* in FSHD (Yehezkel *et al.*, 2008).

Additionally, CpG hypomethylation might be an accessory epigenetic modifier of DUX4

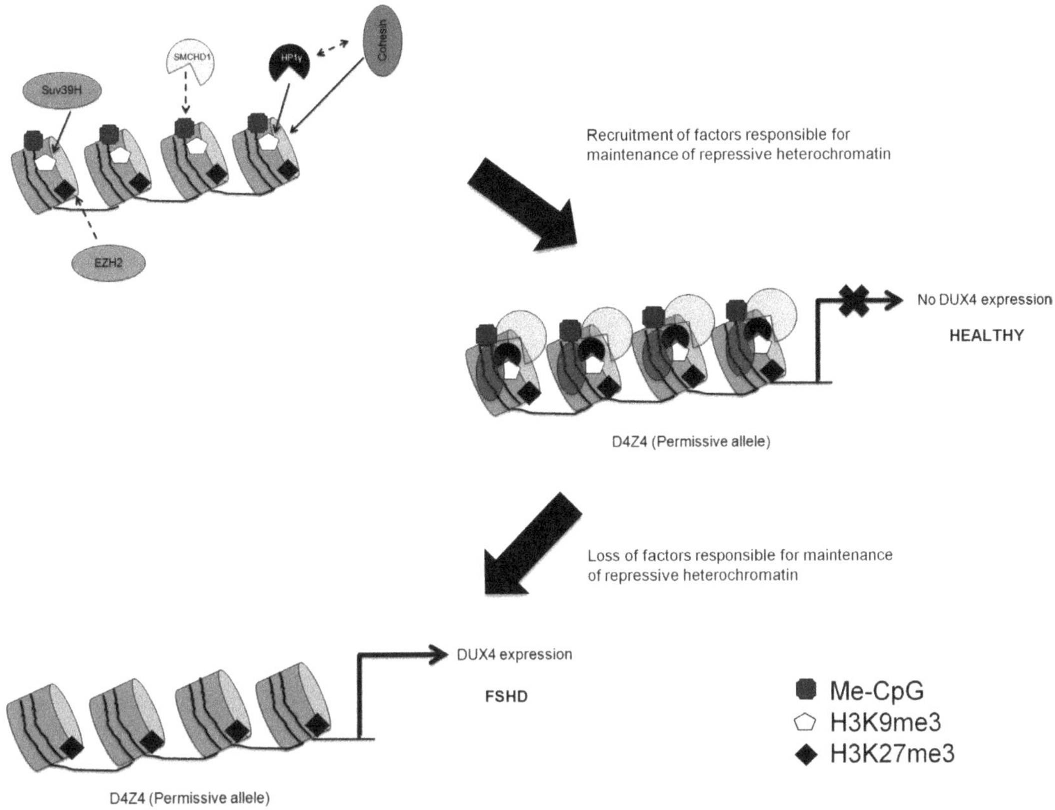

Figure 11.5 Loss-of-heterochromatin mediated de-repression of DUX4. Top panel shows recruitment of various factors mediating heterochromatin formation at D4Z4. DNA (black) wrapped around nucleosomes (grey) is marked by methylated CpGs. SMCHD1, recruited in an H3K9me3-dependent manner, may contribute to CpG methylation and keep the array hypermethylated. HP1γ and Cohesin are recruited in a mutually dependent manner and this requires H3K9me3, which is recruited by SUV39H1. An interaction between SMCH1 and HP1γ has also been speculated. H3K27me3 is possibly recruited by EZH2. Together, these repressive protein complexes, histone modifications and CpG hypermethylation at D4Z4 keeps the array repressed in healthy individuals, as shown in the middle panel. These repressive proteins are lost in FSHD that results in de-repression of the array and expression of DUX4-fl in patients. Loss of H3K9me3 but not H3K27me3 in specifically observed in patients with both FSHD1 and FSHD2.

expression and phenotype. More evidence suggesting that this indeed might be the case comes from the identification of Suppressor of Variegation 3-9 Homologue 1 (*SUV39H1*), as the histone methyltransferase that is responsible for H3K9me3 at D4Z4 (Zeng *et al.*, 2014). Experimental evidence pointing to the importance of H3K9me3 loss in FSHD came from a recent study that demonstrated up-regulation of *DUX4-fl* in immortalized normal myoblasts that were subjected to either chemical inhibition of or short hairpin RNA (shRNA)-mediated knockdown of *SUV39H1*. Importantly, this also showed that suppression of H3K9me3 reduces SMCHD1 association with D4Z4, suggesting that SMCHD1 is a H3K9me3-dependent component of D4Z4 heterochromatin. The role of SMCHD1 as an epigenetic modifier will be discussed shortly but the above observation suggests that H3K9me3 contributes to the recruitment of SMCHD1, which itself directs DNA hypermethylation at D4Z4. HP1γ and cohesin are also recruited to D4Z4 in an H3K9me-dependent manner in normal myoblasts and are specifically lost in FSHD patient myoblasts, resulting in a transcriptionally permissive chromatin structure (Zeng *et al.*, 2009). Normal lymphoblasts, however, do not show HP1γ and cohesion binding, indicating cell-type specific factor(s) might be responsible for retention of these two proteins in myoblasts. Experiments also suggest that HP1γ and cohesin are downstream effectors of H3K9me3. Cohesin, a sister chromatid cohesion complex, is not involved in heterochromatin formation at D4Z4 *per se* but its presence in myoblasts as well as in non-proliferating myotubes indicates that it plays a role beyond mitosis at D4Z4. HP1, on the other hand, is known to play a role in formation of repressive heterochromatin.

A multiprotein complex known as D4Z4 repressor complex (DRC) binds a sequence called the D4Z4 Binding Element (DBE) within each D4Z4 unit (Gabellini *et al.*, 2002). DRC is composed of YY1, High-Mobility Group Protein 2 (HMGB2) and nucleolin; proteins that collectively mediate heterochromatin formation and facilitate gene silencing. Contraction of repeats leads to reduced association of this complex in FSHD1 and may contribute to loss of silencing at D4Z4 or upstream genes. The localization dynamics of this complex in FSHD2 patients has not been defined and it remains to be seen if DRC is lost from FSHD2 D4Z4 as well. If so, it might indicate that recruitment of DRC (or its components) is contingent on H3K9me3 presence (and not on the number of repeats). In addition to SUV39H1-mediated H3K9me3, normal D4Z4 arrays are also marked by the Enhancer of Zeste 2 (*EZH2*) mediated H3K27me3 modification. EZH2 is a component of the polycomb repressor complex 2 (PRC2). In contrast to the previous study (Zeng *et al.*, 2009) where no loss in H3K27me3 was seen in FSHD patients, another study (Cabianca *et al.*, 2012) showed significant loss of this mark in patients, compared to controls. Additionally, transcription of a long non-coding RNA from the DBE region, termed DBE-T was observed in patients but not in controls. It was hypothesized that loss of PRC2-mediated silencing at D4Z4 results in DBE-T transcription, which consequently activates a Trithorax protein called ASH1L, that antagonizes the effects of PRC2 and helps de-repress 4q35 genes in patients. This apparent contradiction in H3K27me3 distribution in patients between the two studies can potentially be resolved by understanding where the primers used for ChIP analysis were located. In the first study (Zeng *et al.*, 2009), primers were located within each D4Z4 repeat, whereas in the second study (Cabianca *et al.*, 2012) primers were located both within and upstream of the most proximal D4Z4 repeat. The significant difference in H3K27me3 levels observed between patients and controls were only for ChIP with the primers upstream of the array. No significant loss in H3K27me3 was seen within the D4Z4 unit, which corroborates the earlier finding. However, difference in H3K27me3 levels between patients and controls in a region upstream of the array but not within the repeats suggests that this region may play a regulatory role. Since H3K9me3 levels were not assessed in this second study, based on the earlier finding, it could be possible that transcription of DBE-T is actually because of H3K9me3 loss in patients. This would link H3K9me3 and not H3K27me3 to ASH1L mediated up-regulation of 4q35 genes, though this speculation needs thorough experimental verification.

Apart from governing transcriptional regulation of the array as a whole, local H3K9me3 enrichment might play a role in dictating *DUX4-fl/DUX4-s* expression by influencing usage of specific splice sites. As discussed before, *DUX4-s* is generated through the usage of a non-canonical splice donor site internal to the *DUX4* ORF that is normally not favoured for splicing. Repressive chromatin modifications are known to govern splice donor usage. ChIP in the vicinity of this cryptic splice site show that the control fibroblasts and EBs expressing *DUX4-s* are enriched for H3K9me3, compared to iPS cells, which express *DUX4-fl*. FSHD cells, however, maintain low levels of H3K9me3 in both iPS and differentiated cells, suggesting a correlation between the relatively higher levels of repressive chromatin modifications and the use of the cryptic splice donor to produce *DUX4-s* (Snider *et al.*, 2010). Although the mechanism for this is not known, it is suggested that this might be facilitated through either preferential targeting of one of the two transcripts by endogenous siRNAs generated from D4Z4 or by slowing down RNA polymerase around H3K9me3 accumulated sites.

Role of CTCF, subnuclear localization and chromosomal cross-talk in FSHD

The subnuclear location of genomic sequences has been linked to their transcriptional state and time of replication during S-phase (Schneider and Grossched, 2007). Although replication timing of the D4Z4 locus does not vary between patient and control myoblasts (Pope *et al.*, 2011), subnuclear localization of 4qter might have an important effect on transcription. It is intriguing that unlike most telomeres on human chromosomes, the 4q telomere is normally associated with a heterochromatin-rich compartment at the nuclear periphery and is anchored to the nuclear matrix (Masny *et al.*, 2004). Association with the nuclear periphery is an intrinsic property of the 4q subtelomere, which does not differ between controls and FSHD patients (Tam *et al.*, 2004). This localization is also restricted to D4Z4 arrays on chromosome 4 and is consistent between cell-types and cell-cycle stages in both patients and controls. Nuclear anchorage seems to be mediated by more than one mechanism. An 80 bp sequence in the 5′ end of each D4Z4 repeat guides positioning to the nuclear periphery in a CTCF and Lamin-A dependent manner in artificial telomeric constructs with just a single copy of D4Z4 (Ottaviani *et al.*, 2009b). This positioning capacity is lost upon multimerization of D4Z4 due to the inability of CTCF to bind to such arrays. Binding of CTCF to D4Z4 is believed to confer insulator activity to the latter that may protect genes upstream of 4q D4Z4 from position effect of the heterochromatic subtelomeric 4q35 region (Ottaviani *et al.*, 2009a). CTCF might facilitate this by mediating long-range chromosomal interactions to bring regions with similar transcriptional states together, creating local 'protected' subdomains within which genes are not influenced by other subdomains. Protection against silencing, like subnuclear positioning, is mediated in part by CTCF and A-type lamins. Both CTCF binding and insulator activity are lost upon multimerization of D4Z4, as found in normal individuals, suggesting that FSHD is associated with a gain of function of CTCF (Ottaviani *et al.*, 2009a). However, since the subnuclear location of 4q35 does not change between patients and controls, it seems that other means of nuclear periphery association might exist. Not surprisingly, it has been found that a 215 bp region identified upstream of the array (Masny *et al.*, 2004) also governs nuclear periphery positioning of the 4q telomere and the positioning capacity of this region is strictly dependent on Lamin A, but not on CTCF. A lamin-associated domain, identified approximately 50 kb proximal to D4Z4, might be critical in mediating this localization (Guelen *et al.*, 2008). It seems that the CTCF-dependent mechanism maintains subnuclear localization of 4q35 in FSHD patients with fewer D4Z4 repeats that are capable of CTCF binding while the CTCF independent mechanism achieves the same goal in normal individuals with non-contracted D4Z4 arrays (Ottaviani *et al.*, 2009b). Deregulation of genes at and upstream of D4Z4 in patients in the absence of change in subnuclear position leaves the possibility that this locus may be relocated exclusively in patients to a different nuclear periphery subdomain, which might facilitate transcription. Alternatively, changes in higher order chromatin structure at the subtelomeric region might be affected in patients, significantly contributing to mis-expression of genes. An epigenetic model for up-regulation of genes upstream of D4Z4 in patients has been proposed based on the results of chromosome capture conformation (3C) studies. In this model, an FSHD-related nuclear matrix attachment region (FR-MAR) near the 5′ end of the array blocks a potential enhancer (identified in the same 5′ region) in normal myoblasts. This allows the chromatin to form a D4Z4 containing chromatin loop by anchoring the FR-MAR to the nuclear matrix, which is transcriptionally insulated from a second loop containing the upstream genes. However, in FSHD myoblasts, weakening of FR-MAR-nuclear matrix interaction results in a single loop that brings the upstream genes in close proximity to the D4Z4 4qA enhancer, which up-regulates these genes (Petrov *et al.*, 2006, 2008). A physical interaction of the *FRG1* promoter with the D4Z4 array has also been reported (Pirozhkova *et al.*, 2008). Additionally, change to *cis* interactions between *DUX4* and *DUX4c*, as well as *DUX4c* interacting with *FRG1* and *FRG2* suggest that long-range chromatin interactions could impact local gene regulation (Pirozhkova *et al.*, 2008). Hence, it is possible that up-regulation of these non-DUX4 genes is independent of DUX4 expression, but results from the same set of epigenetic phenomena that make the D4Z4 array and flanking regions transcriptionally permissive.

SMCHD1 and telomere position effect as additional modifiers of phenotype

Haploinsufficiency of *SMCHD1*, along with inheritance of the permissive allele, accounts for most cases of FSHD2 (Lemmers *et al.*, 2012). *SMCHD1* belongs to the structural maintenance of chromosomes (SMC) gene superfamily, whose members are integral to chromosome condensation and segregation; two events that are fundamental to the cell cycle (Ball and Yokomori, 2001). SMC proteins are largely conserved in eukaryotes and at least six different proteins are found in individual species. Each of them binds to a specific partner, creating three different heterodimers (Hirano, 2005). SMC protein monomers have a modular structure and contain a central hinge domain flanked by two coiled-coiled domains and two ATP binding motifs which suggest that their ATPase activity could play critical roles in SMC–DNA interactions seen in cohesin and condensin complexes. The 226 kDa SMCHD1 protein contains an N-terminal ATPase and a C-terminal hinge domain. SMCHD1 was first identified in mouse as an epigenetic modifier of transgene expression in an ENU mutagenesis screen (Blewitt *et al.*, 2005). *SMCHD1* is involved in maintaining the methylation of some, but not all CGIs during X-chromosome inactivation (XCI) in mouse (Gendre *et al.*, 2012) and other suggested roles are as a tumour suppressor (Leong *et al.*, 2013) and DNA damage response gene (Coker and Brockdorff *et al.*, 2014). To date, 17 mutations in *SMCHD1* have been associated with cases of FSHD2 (Lemmers *et al.*, 2012; Mituhashi *et al.*, 2013; Winston *et al.*, 2014).

Apart from naturally occurring mutations in FSHD2 patients, shRNA-mediated knockdown of *SMCHD1* in normal myoblasts also results in the up-regulation of *DUX4-fl* transcription. It has been postulated that SMCHD1 maintains somatic repression of *DUX4* through facilitating the retention of CpG hypermethylation of the array and that haploinsufficiency of *SMCHD1* results in loss-of-methylation induced reactivation of *DUX4-fl* in FSHD2 patient myoblasts. *SMCHD1* has also been reported to be an epigenetic modifier of FSHD1, where mutation of this gene in FSHD1 patients increases expression of *DUX4-fl* and its target genes beyond levels induced by repeat contraction alone, impacting the severity of the phenotype (Sacconi *et al.*, 2013).

Although SMCHD1 has been shown to localize at D4Z4 and its modifier effect cannot be disputed, currently, there is no resolution on *how* the protein assists in maintaining array hypermethylation. Also, it is not clear, whether it has a direct causal role or is a downstream effector of some other causal factor.

A SMCHD1-like protein is required for RNA directed DNA methylation (RdDM) and gene silencing in *Arabidopsis* (Kanno *et al.*, 2008). The gene in question, Defective in Meristem Silencing 3 (*DMS3*), encodes a putative chromosome architectural protein whose domains are similar to that of members of the SMC family, but lacks the ATPase domain. The DNA methylation mechanism is dependent on the production of primary and secondary short interfering (siRNA), whose silencing effect spreads beyond the targeted region. *DMS3* might be a general component of the RdDM machinery and it is postulated that DMS3 homodimers function by binding to DNA-RNA or primary siRNA–RNA hybrids, thereby stabilizing them. These hybrids are subsequently targeted by a host of other proteins (such as Argonaute, DRD1, PolIVb and DCL3) to generate primary or secondary siRNAs that direct DRM2, a *de novo* methyltransferase, towards methylation and spreading of silencing beyond the targeted locus. Small RNAs emanating from D4Z4 could be involved in a similar mechanism (Snider *et al.*, 2009).

Alternatively, SMCHD1 could recruit additional proteins that directly or indirectly methylate D4Z4. Mouse *Smchd1* plays a critical role in XCI (Blewitt *et al.*, 2008; Mould *et al.*, 2013). There exists a specific requirement of Smchd1 for some, but not all CGI methylation on the Xi. Haploinsufficiency of *Smchd1* retards CGI methylation initially, but they reach wild-type methylated levels with time. Validating this notion, two parallel pathways of CGI methylation on the Xi have been reported in mouse ESCs (Gendrel *et al.*, 2012). The secondary *Smchd1*-independent pathway rapidly methylates a subset of CGI on the Xi, shortly after the initiation of XCI, while the primary *Smchd1* dependent pathway methylates the remaining gradually in association with Smchd1. Both pathways require Dnmt3b. Speculation is rife on whether Smchd1 directly recruits Dmnt3b or makes the DNA at CGI more accessible for Dnmt3b by manipulating chromatin structure along with other histone modifying enzymes. The fact that Smchd1 has an ATPase domain suggests that it could be a potential chromatin remodelling enzyme, or a part of such a complex. Nevertheless, existence of a *Smchd1* independent CGI methylation pathway in mouse raises a possibility of an analogous mechanism in humans. Co-localization studies on SMCHD1 could help identify its potential interacting partners.

As mentioned previously, reduced H3K9me3 displaces SMCHD1 from D4Z4, implying that SMCHD1 is a component of D4Z4 heterochromatin. H3K9me3 may directly recruit SMCHD1, which subsequently facilitates DNA methylation at D4Z4 (Zeng *et al.*, 2014). The fact that H3K9me3 loss occurs at all D4Z4 alleles in FSHD1 and FSHD2, raises the possibility that SMCHD1 binding to D4Z4 is already compromised in FSHD cases in which the SMCHD1 gene is not mutated, but that mutation of the gene further compounds the effect. Comparison of D4Z4-associated SMCHD1 levels in FSHD1/FSHD2 patient myoblasts without SMCHD1 mutations relative to controls and patients with mutations, would test this hypothesis.

Results from a recent study show that while SMCHD1 loss might increase transcription from D4Z4 and result in generation of unspliced *DUX4* transcripts, spliced *DUX4-fl* is generated only when accompanied by CpG hypomethylation and associated loss of H3K9me3 (Das and Chadwick, unpublished data). Thus, *SMCHD1* loss makes cells more vulnerable to *DUX4-fl* expression, and splicing is influenced by an altered epigenetic state. Although this study was performed in a non-myoblast cell line, the finding strengthens an idea we have discussed previously, that local chromatin environment might influence splicing of *DUX4*. Additionally, it was observed that shortening of telomeres has an effect partially redundant to SMCHD1 loss. This observation adds to the speculation about the mechanism of SMCHD1-mediated repression of *DUX4*. SMCHD1 is a known telomeric component that associates with repeats packaged into H3K9me3 mediated heterochromatin (Arnoult *et al.*, 2012). SMCHD1, along with its interacting partner, the HP1-binding protein HBiX1 (Nozawa *et al.*, 2013), is preferentially enriched at longer telomeres (Grolimund *et al.*, 2013). Given that SUV39H1 establishes H3K9me3 at telomeric repeats in addition to D4Z4, suggests that chromatin at these two repeat regions may involve a common mechanistic pathway, and that longer telomeres might facilitate SUV39H–H3K9me3 mediated heterochromatin formation at D4Z4 through SMCHD1 binding. As SMCHD1 is tethered to both the telomere and D4Z4, it is feasible that it could bridge the two regions, bringing them into close proximity. This would mediate a telomere position effect and keep *DUX4-fl* in a more repressive environment. This is a strong possibility, since telomere position effect on *DUX4-fl* expression has been demonstrated in immortalized human myoblasts, with telomere length having an inverse correlation with *DUX4-fl* reactivation (Stadler *et al.*, 2013). Chromosome capture studies involving D4Z4 and regions of the 4q telomere might reveal novel interactions between these regions.

The existence of rare FSHD patients without D4Z4 contractions or *SMCHD1* mutations (Lemmers *et al.*, 2012) suggests that additional modifier genes might exist, which regulate *DUX4* expression in these patients, and may include SMCHD1 associated protein such as HBiX1.

A muscle-specific deregulation of the protocadherin gene *FAT1*, located 3.6 megabases (Mb) upstream of D4Z4, segregates with FSHD and this mis-regulation is a possible consequence of *DUX4* overexpression (Caruso *et al.*, 2013). *Fat1* deficient mice mimic some phenotypic FSHD features and an alteration of muscle shape is observed. Interestingly, some patients with no D4Z4 contractions carry a deletion of a putative *FAT1* enhancer and it has been proposed that loss of this regulatory region could be a potential disease modifier in these contraction-independent FSHD patients, although a detailed analysis of more such cases would be required to validate this hypothesis.

Translational therapeutics

The ultimate goal of understanding a disease is the development of an effective therapy or cure. While a lot of diseases have been effectively managed or treated, developing resources for treating multifactorial diseases such as FSHD poses a major challenge. Translational research in FSHD has been hindered by its inherent complexity. Until recently, most treatments are aimed at improving existing conditions in patients or are in order to provide relief, rather than address the root cause, since the lack of a clear consensus on a causative gene in FSHD made it difficult to develop biomedical approaches that somehow alleviate the effect of the gene or its downstream targets.

In general, an active lifestyle and light aerobic exercises are recommended (Olsen *et al.*, 2005), among other things. However, the benefits offered by different kinds of exercise remain inconclusive due to small sample size and other inconsistencies, as highlighted by the a Cochrane review of muscle strength (Voet *et al.*, 2013). Different corticosteroid drugs, such as albuterol (Kissel *et al.*, 1998; van der Kooi *et al.*, 2004) and prednisolone (Tawil *et al.*, 1997) have been shown to offer no therapeutic benefit in FSHD. Surgical treatments, such as scapulothoracic fusion help increase the functionality of the shoulder by allowing a greater range of motion, and also alleviate discomfort of the shoulder (Cooney *et al.*, 2014; Diab *et al.*, 2005).

The emergence of a unifying model of FSHD, initially triggered by *DUX4-fl* reactivation, along with some resolution on its downstream targets now offer a new hope for patients. In the last decade, an improved understanding of the genetic and epigenetic components of the disease and better molecular diagnostic techniques, along with parallel progress in the fields of targeted gene therapy has given a much-needed push to the possibility of translational medicine in FSHD. These tools, such as chemical inhibitors, small molecules, morpholinos, RNA interference (RNAi) using small RNAs and custom-built nucleases have the potential to be used to target *DUX4* or its downstream effectors to develop therapeutic approaches. In addition, gene targeting strategies could be coupled to pluripotent stem cell technology to develop customized patient-specific treatments that overcome issues of immune rejection of non-self samples.

The availability of an animal model is another factor that plays a vital role in bridging the gap between laboratory research and clinical trials. Although, the mere presence of an animal model does not guarantee quick development of an effective therapy, animal models do offer an important *in vivo* platform on which to test the feasibility, efficacy and safety of newly developed approaches, before progressing to clinical trials in humans.

DUX4 transgenic FSHD mouse model

Transgenic mouse models of genes upstream of 4q35 such as *FRG1*, *ANT1* and *FRG2* were generated in the past decade (Gabellini *et al.*, 2006), testing if overexpression of these genes recapitulates features of FSHD. However, based on the mounting evidence, the consensus points to *DUX4* as the FSHD gene, necessitating a DUX4 transgenic mouse model for the disease.

An important development in understanding *DUX4* mediated pathology of FSHD was the recent generation of two *DUX4* transgenic mouse models containing both 12.5 (normal sized) and 2.5 (FSHD contracted) 4q35 D4Z4 repeats (Krom *et al.*, 2013). Although transgenic mice have been made for other FSHD candidate genes, one with human D4Z4 is ideal for studying *in vivo* effects of *DUX4* de-repression along with an understanding of epigenetic mechanisms that regulate transcription from D4Z4. It must be mentioned however, that some targets of DUX4 are primate-specific (Leidenroth and Hewitt, 2010) and only a fraction of these genes are present in mouse (Geng *et al.*, 2012; Krom *et al.*, 2013). Hence, even if we discount intrinsic differences between human and mouse genomic background, the recapitulation of all patient features will likely not be possible in a mouse model. Nevertheless, these transgenic mice mirror quite a few features of the disease. *DUX4* transgene is expressed in both normal and FSHD mice germline cells whereas its expression is exclusive to FSHD mouse somatic cells, reflecting the observation in humans. Similar to humans, *DUX4* is expressed in a small fraction of FSHD mouse myonuclei but interestingly, its expression is not restricted to somatic skeletal muscle in FSHD mice, suggesting that *DUX4* expression may not be muscle-specific. The downstream transcriptional cascade of *DUX4* is also similar but not comprehensively identical to that of humans such as regulation of genes closely related to known human *DUX4* targets, MaLR LTRs. Around a quarter of genes mis-regulated in mouse models, including ones involved in germline development and innate immunity are also mis-regulated in humans. Along with the gene expression profiles, human epigenetic patterns at D4Z4 such as CpG methylation and histone distribution are also conserved in normal and FSHD mice. Probably due to a lack of the complete complement of human *DUX4* targets, the exact skeletal muscle phenotype is not recapitulated in mouse but a trend towards muscle weakness is observed, including smaller fibres, and delayed response to artificially induced muscle damage. However, a lot of the FSHD mice develop an abnormal eye phenotype that leads to blindness. Though not identical, appearance of such a phenotype in mouse that is on the lines of weakness of eyelid muscles and retinal telangiectasis observed in some patients is encouraging.

In summary, these transgenic mouse models tend to mirror some phenotypes of patients, which make them a good platform for investigative studies on pathophysiology and therapeutics in FSHD. It will be interesting to see how some of the existing approaches can be used in these DUX4 transgenic mouse models to demonstrate their therapeutic potential.

RNAi

Two 'proof-of-principle' studies devised a gene therapy strategy using adeno-associated virus (AAV)-delivered micro RNAs (miRNAs) or shRNAs to selectively target the *FRG1* gene in a transgenic mouse model overexpressing the gene (Bortolanza *et al.*, 2011; Wallace *et al.*, 2011). This resulted in marked reduction in FRG1 protein levels, demonstrating the effectiveness of this targeting approach. Moreover, alleviation of symptoms induced by *FRG1* overexpression in these RNAi targeted mice, such as lowered fibrosis and fat deposition along with decreased myofibre degeneration, was observed in control mice (overexpressing *FRG1*). Muscle histology in targeted mice was characteristic of wild type, along with improved muscle function and strength, showing therapeutic promise for this approach. Subsequently, after *DUX4* was established as the causal gene in FSHD, two research groups also tested this approach *in vivo* in mouse muscles overexpressing transduced *DUX4* (Vanderplanck *et al.*, 2011; Wallace *et al.*, 2012), since a *DUX4* transgenic mouse model was not available at that time. *DUX4* microRNA (miRNA) was found to protect muscles from *DUX4* induced degeneration which showed normal muscle histology and a reversal of myopathic phenotype, characterized by reduced myofibre degeneration. An increase in muscle strength was also observed, compared to controls. *caspase-3*, a gene induced by *DUX4* and present in the myofibres of *DUX4*-overexpressing control mice, was found to be absent in these knockdown mice, hinting that RNAi was successful. Importantly, reversal of myopathy strongly correlated with reduction in *DUX4* mRNA and protein levels and down-regulation of FSHD-specific markers that are activated by *DUX4*, demonstrating the effectiveness of targeting and causality of this gene. The miRNA did not seem to demonstrate any significant toxicity or myopathy in mouse muscle, suggesting that this might be a safe approach.

Morpholinos

Downstream effectors of *DUX4* that are known to contribute to disease phenotype serve as alternate targets for gene therapy. *PITX1* is one such gene that is directly bound by DUX4 and is up-regulated in patients (Dixit *et al.*, 2007). It is believed to contribute to *DUX4* induced myopathy involving a *p53* pathway. Additionally, overexpression of *Pitx1* in a muscle-specific tetracycline repressible *Pitx1* transgenic mouse model leads to recapitulation of some phenotypes of FSHD, such as muscle wasting in selective muscle groups and atrophy of myofibres (Pandey *et al.*, 2012). A subsequent study from the same research group showed that blocking the translation of the Pitx1 protein *in vivo* in the abovementioned *Pitx1* transgenic mice using morpholinos, results in reversal of myopathic features (Pandey *et al.*, 2014). These mice were administered with intravenous doses of two different morpholinos and it was seen that one of them (vivo-morpholino), effectively reduced Pitx1 expression at the protein level but not mRNA level. A reduction in the number of atrophic myofibres along with increased muscle strength was also observed. At experimental concentrations, neither of the two morpholinos showed any significant cytotoxicity. Morpholinos are known to be stable in cells and do not induce immune responses, as confirmed by some other studies. This study provides a proof-of-principle for blocking FSHD related protein translation *in vivo*, using morpholinos.

Alternatives and concerns

It has been demonstrated that activation of the Wnt/β-catenin signalling pathway suppresses *DUX4* expression in both FSHD1 and FSHD2 myotubes and rescues them from DUX4-mediated apoptosis, suggesting that this pathway plays a role in the suppression of *DUX4* (Block *et al.*, 2013). Additionally, reduction of components of this pathway, such as β-catenin, *Wnt3A* and *Wnt9B* at the mRNA level results in *DUX4* up-regulation. Thus, it is possible that reactivation of DUX4 in patients is possibly facilitated by escape from one or more components of this repressive pathway. Mutations in Wnt signalling receptors are known to result in retinopathies, a feature also observed in some FSHD patients. Treatment of a mouse model of DMD using *Wnt7a*, to ameliorate myofibre hypertrophy and increase muscle strength (von Maltzahn *et al.*, 2012) suggests that a similar approach involving the Wnt/β-catenin pathway could be taken in the available DUX4 transgenic FSHD mouse model.

Another study has recently identified numerous small molecule inhibitors that reduce *DUX4* induced toxicity in mouse C2C12 myoblasts (Bosnakovski *et al.*, 2014). Although the exact mechanism of their action is not known, a subset of these compounds protects muscles against oxidative stress induced by *DUX4*. Thus, further investigation of these and similar inhibitors might help develop an effective oxidative stress response to *DUX4* activation, providing a therapeutic benefit in FSHD.

Apart from the experiments that have already been discussed, alternative gene targeting approaches using custom-built nucleases also offer a potential therapeutic route for FSHD. These nucleases, such as Zinc Finger Nucleases (ZFNs), transcription activator-like effector nucleases (TALENs) and clustered regularly interspaced short palindromic repeats (CRISPRs), are proteins or ribonucleoproteins that can bind to specific desired regions of the genome and disrupt these sequences, allowing genes to be manipulated and engineered (Gaj *et al.*, 2013). These nucleases have already been used to 'correct' various gene defects involved in other diseases, and could be developed to incapacitate the *DUX4* gene, possibly by mutating the canonical polyA signal to a non-permissive sequence. Alternatively, the entire D4Z4 array could be removed as proof-of-principle verification of causality of disease, by observing alleviation of downstream effects in its absence. If successful, the integration of these approaches in patient iPS cells could offer a much-needed patient-specific muscle transplantation therapy for FSHD.

However, any technique developed in the lab aimed at offering therapeutic benefit must be carefully evaluated for efficacy and more importantly, safety concerns. The dominant nature of the disease poses a problem because its effect will be manifested even with the presence of one defective allele. Additionally, FSHD results from a toxic gain-of-function of *DUX4*, rather than a loss-of-function. If the latter were the case, then ways to increase expression of the normal allele could have been looked into, which is not possible here. However, since *DUX4-s* seems to be dominant negative to *DUX4-fl* an avenue of suppression of *DUX4-fl* by increasing *DUX4-s* expression, could be explored. Local histone modifications that might be playing a role in splice donor choice may be regulated to bias this outcome. Morpholinos are also known to modify splicing outcomes by regulating splice junctions and splice regulatory sites. Alternatively, since *SMCHD1* haploinsufficiency is an epigenetic modifier of disease, increasing levels of this protein, either endogenously or through overexpression of the transduced gene could be explored to see if there is a reversal of phenotype.

The safety of all the tools discussed above is not substantiated at clinical trial levels. Inhibitory RNAs may exhibit non-specific off-targeting effects, interfere or compete with normal cellular pathways involving short RNAs, resulting in potential cytotoxicity (Grimm *et al.*, 2006). Incomplete suppression by RNAi has also remains a concern (Pecot *et al.*, 2011).

Off-targeting is also a concern with custom-built nucleases, but techniques to improve specificity are being developed (Miller *et al.*, 2007; Mussolino *et al.*, 2014; Shen *et al.*, 2014). In addition, targeting of the disease-specific allele is also an important requirement. *DUX4-s* is expressed in some cell types of healthy individuals, suggesting some normal function. Additionally, with detection of transcripts (not *DUX4-fl*) containing *DUX4* ORF that are theoretically capable of producing DUX4 protein in some human somatic tissues, indicate that any effective therapy should be muscle-specific or selectively target *DUX4-fl* and not any other isoform in order to avoid disruption of normal cellular function.

Specificity of small molecule inhibitors is also an area that needs to be carefully assessed while implementing drug development. Delivery of antisense oligonucleotides is another concern, although AAV-mediated delivery have been shown to be efficient and safe in clinical trials (Mendell *et al.*, 2009, 2010). Also, lipid-based transfection or electroporation of plasmids encoding custom-built nucleases can be used to effectively deliver these nucleases into cultured mammalian cells.

If patient-specific corrected myogenic cells can be developed, there still lies a challenge in subsequent muscle transplantation. Before implementation in the clinic, there are a few issues that need to be addressed. Post cell transplantation, there can be issues such as cell loss due to rejection or immune response to the genetic manipulation. Therefore, immune suppression might be required for transplantation. Increased risk of tumour formation in transplanted cells is also something that needs to be minimized. More importantly, scalability of corrected myogenic cells to produce sufficient mass, along with optimization of tissue sites that might be best suited for such procedures also needs to be looked into in detail. Use of patient-specific differentiated cells hold great promise in cell-based therapies (Fox *et al.*, 2014). With increased research on aspects catering to muscle-specific lineage, it might be possible to offer a promising solution in FSHD. Although a mouse model offers the convenience of testing the feasibility of different therapeutic techniques, it seems that a platform that is more closely related to humans, such as a primate model, would best help understand and gauge the outcome of these therapies, before human clinical trials begin.

While RNAi, morpholinos and small molecule inhibitors aim to counter the adverse effects of defective genes, cell-based therapies look to remove causality. Which of these approaches will be the most effective and the safest, is yet to be determined. It is possible that a combination of more than one approach may be required to successfully treat patients.

Concluding remarks

FSHD has emerged as a classic example of a multifactorial complex disease and its D4Z4 linkage is crucial in showing that repetitive elements such as macrosatellites are not 'junk' DNA. Mapping of the disease to the D4Z4 macrosatellite, identification and characterization of a causative gene *DUX4*, and SNPs that make an allele permissive are landmark findings in FSHD research. Studies from various groups have shown that while *DUX4* expression might be necessary for the disease, it is not sufficient. Incomplete penetrance and variable heterogeneity brings into the picture, various other factors that govern manifestation of symptoms and severity. While both forms of the disease are characterized by D4Z4 CpG hypomethylation, it appears that loss of repressive histone modifications play a more deterministic role in *DUX4* expression. Telomere position effect also seems to play a role in conjunction with modifier genes such as *SMCHD1* in regulating DUX4 expression and this relationship needs to be probed further. A comprehensive, though not exhaustive, list of downstream targets of DUX4 are now known and studies seem to indicate a DUX4 protein-spreading model for the disease. Existence of individuals with FSHD but no D4Z4 contractions or *SMCHD1* mutations strongly suggest the possibility of additional modifier loci. If discovered, this may help classify patients into an additional category of FSHD that is independent of contractions and *SMCHD1* mutations. The role of regulatory proteins such as CTCF, which facilitate chromosomal crosstalk, along with the 3-D organization of chromatin around D4Z4 needs to be investigated further in order to highlight the role of the array as a possible regulatory element that shields upstream genes from subtelomeric silencing in patients. Additionally, the role of small and long non-coding RNAs emanating from D4Z4 needs to be further explored. Although significant copy number variation is observed in this disease, the link between D4Z4 repeat number and phenotype is not well established. It seems unlikely that a decrease in number of repeats from 11 to 10 would decrease methylation levels and change chromatin conformation substantially enough, 'making the cut' for *DUX4* reactivation and disease progression. Eleven, the minimum number of repeats in a healthy individual, is an empirical number assigned on the basis of all patients who have been studied so far, and it is known that the number of such patients is limited. It is therefore possible that no 'absolute' value for the threshold number of repeats exists. As an alternative, it might be helpful to categorize individuals on the basis of reduction of some epigenetic factor(s) such as H3K9me3 levels (and CpG methylation), below some statistically determined threshold, rather than simply on the number of repeats. Something along these lines is already done with the use of chromatin compaction scores (ChCS) that calculate the ratio of H3K9me to H3K4me2 and correlating it to clinical severity scores (CSS) (Balog *et al.*, 2012). A similar approach may serve as a tighter indicator of chromatin permissiveness. This is indeed an exciting time for FSHD research. The plethora of information on the disease that has been revealed in the last few decades, availability of a DUX4 transgenic mouse model, along with emergence of latest therapeutic tools, offers immense hope for patients afflicted with this disorder. However, as with any other disease, comprehensive exchange of information between laboratory and clinical scientists along with increased collaboration between academia and industry will be crucial in clinical treatment of FSHD.

References

Ansseau, E., Laoudj-Chenivesse, D., Marcowycz, A., Tassin, A., Vanderplanck, C., Sauvage, S., Barro, M., Mahieu, I., Leroy, A., Leclercq, I., *et al.* (2009). DUX4c is up-regulated in FSHD. It induces the MYF5 protein and human myoblast proliferation. PLoS ONE *4*, e7482.

Arnoult, N., Van Beneden, A., and Decottignies, A. (2012). Telomere length regulates TERRA levels through increased trimethylation of telomeric H3K9 and HP1alpha. Nat. Struct. Mol. Biol. *19*, 948–956.

Aslanidis, C., Jansen, G., Amemiya, C., Shutler, G., Mahadevan, M., Tsilfidis, C., Chen, C., Alleman, J., Wormskamp, N.G., and Vooijs, M. (1992). Cloning of the essential myotonic dystrophy region and mapping of the putative defect. Nature *355*, 548–551.

Bakker, E., Wijmenga, C., Vossen, R.H., Padberg, G.W., Hewitt, J., van der Wielen, M., Rasmussen, K., and Frants, R.R. (1995). The FSHD-linked locus D4F104S1 (p13E-11) on 4q35 has a homologue on 10qter. Muscle Nerve Suppl. *2*, S39–44.

Ball, A.R. Jr, and Yokomori, K. (2001). The structural maintenance of chromosomes (SMC) family of proteins in mammals. Chromosome Res. *9*, 85–96.

Balog, J., Thijssen, P.E., de Greef, J.C., Shah, B., van Engelen, B.G.M., Yokomori, K., Tapscott, S.J., Tawil, R., and van der Maarel, S.M. (2012). Correlation analysis of clinical parameters with epigenetic modifications in the DUX4 promoter in FSHD. Epigenetics *7*, 579–584.

Bannister, A.J., Zegerman, P., Partridge, J.F., Miska, E.A., Thomas, J.O., Allshire, R.C., and Kouzarides, T. (2001). Selective recognition of methylated lysine 9 on histone H3 by the HP1 chromo domain. Nature *410*, 120–124.

Beckers, M., Gabriels, J., van der Maarel, S., De Vriese, A., Frants, R.R., Collen, D., and Belayew, A. (2001). Active genes in junk DNA? Characterization of DUX genes embedded within 3.3 kb repeated elements. Gene *264*, 51–57.

Belayew, A. (2010). [Cascade of gene activation in Landouzy Dejerine muscular dystrophy]. Bull. Mem. Acad. R. Med. Belg. *165*, 87–95; discussion 95–97.

Bernstein, B.E., Kamal, M., Lindblad-Toh, K., Bekiranov, S., Bailey, D.K., Huebert, D.J., McMahon, S., Karlsson, E.K., Kulbokas, E.J. 3rd, Gingeras, T.R., *et al.* (2005). Genomic maps and comparative analysis of histone modifications in human and mouse. Cell *120*, 169–181.

Bernstein, B.E., Meissner, A., and Lander, E.S. (2007). The mammalian epigenome. Cell *128*, 669–681.

Blewitt, M.E., Vickaryous, N.K., Hemley, S.J., Ashe, A., Bruxner, T.J., Preis, J.I., Arkell, R., and Whitelaw, E. (2005). An N-ethyl-N-nitrosourea screen for genes involved in variegation in the mouse. Proc. Natl. Acad. Sci. U.S.A. *102*, 7629–7634.

Blewitt, M.E., Gendrel, A.-V., Pang, Z., Sparrow, D.B., Whitelaw, N., Craig, J.M., Apedaile, A., Hilton, D.J., Dunwoodie, S.L., Brockdorff, N., *et al.* (2008).

SmcHD1, containing a structural-maintenance-of-chromosomes hinge domain, has a critical role in X inactivation. Nat. Genet. *40*, 663–669.

Block, G.J., Petek, L.M., Narayanan, D., Amell, A.M., Moore, J.M., Rabaia, N.A., Tyler, A., van der Maarel, S.M., Tawil, R., Filippova, G.N., *et al.* (2012). Asymmetric bidirectional transcription from the FSHD-causing D4Z4 array modulates DUX4 production. PLoS ONE *7*, 1–12.

Block, G.J., Narayanan, D., Amell, A.M., Petek, L.M., Davidson, K.C., Bird, T.D., Tawil, R., Moon, R.T., and Miller, D.G. (2013). Wnt/beta-catenin signaling suppresses DUX4 expression and prevents apoptosis of FSHD muscle cells. Hum. Mol. Genet. *22*, 4661–4672.

Bodega, B., Ramirez, G.D.C., Grasser, F., Cheli, S., Brunelli, S., Mora, M., Meneveri, R., Marozzi, A., Mueller, S., Battaglioli, E., *et al.* (2009). Remodeling of the chromatin structure of the facioscapulohumeral muscular dystrophy (FSHD) locus and upregulation of FSHD-related gene 1 (FRG1) expression during human myogenic differentiation. BMC Biol. *7*, 1–15.

Bortolanza, S., Nonis, A., Sanvito, F., Maciotta, S., Sitia, G., Wei, J., Torrente, Y., Di Serio, C., Chamberlain, J.R., and Gabellini, D. (2011). AAV6-mediated systemic shRNA delivery reverses disease in a mouse model of facioscapulohumeral muscular dystrophy. Mol. Ther. *19*, 2055–2064.

Bosnakovski, D., Xu, Z., Gang, E.J., Galindo, C.L., Liu, M., Simsek, T., Garner, H.R., Agha-Mohammadi, S., Tassin, A., Coppee, F., *et al.* (2008a). An isogenetic myoblast expression screen identifies DUX4-mediated FSHD-associated molecular pathologies. EMBO J. *27*, 2766–2779.

Bosnakovski, D., Lamb, S., Simsek, T., Xu, Z., Belayew, A., Perlingeiro, R., and Kyba, M. (2008b). DUX4c, an FSHD candidate gene, interferes with myogenic regulators and abolishes myoblast differentiation. Exp. Neurol. *214*, 87–96.

Bosnakovski, D., Choi, S.H., Strasser, J.M., Toso, E.A., Walters, M.A., and Kyba, M. (2014). High-throughput screening identifies inhibitors of DUX4-induced myoblast toxicity. Skelet. Muscle. *4*, 1–11.

Brouwer, O.F., Padberg, G.W., Wijmenga, C., and Frants, R.R. (1994). Facioscapulohumeral muscular dystrophy in early childhood. Arch. Neurol. *51*, 387–394.

Brown, M.R. (1951). The incidence and heredity of muscular dystrophy; a study of 71 patients admitted to the Massachusetts General Hospital. N. Engl. J. Med. *244*, 88–90.

Bruce, H.A., Sachs, N., Rudnicki, D.D., Lin, S.G., Willour, V.L., Cowell, J.K., Conroy, J., McQuaid, D.E., Rossi, M., Gaile, D.P., *et al.* (2009). Long tandem repeats as a form of genomic copy number variation: structure and length polymorphism of a chromosome 5p repeat in control and schizophrenia populations. Psychiatr. Genet. *19*, 64–71.

Buxton, J., Shelbourne, P., Davies, J., Jones, C., Van Tongeren, T., Aslanidis, C., de Jong, P., Jansen, G., Anvret, M., and Riley, B. (1992). Detection of an unstable fragment of DNA specific to individuals with myotonic dystrophy. Nature 355, 547–548.

Cabianca, D.S., Casa, V., Bodega, B., Xynos, A., Ginelli, E., Tanaka, Y., and Gabellini, D. (2012). A long ncRNA links copy number variation to a polycomb/trithorax epigenetic switch in FSHD muscular dystrophy. Cell *149*, 819–831.

Cacurri, S., Piazzo, N., Deidda, G., Vigneti, E., Galluzzi, G., Colantoni, L., Merico, B., Ricci, E., and Felicetti, L. (1998). Sequence homology between 4qter and 10qter loci facilitates the instability of subtelomeric KpnI repeat units implicated in facioscapulohumeral muscular dystrophy. Am. J. Hum. Genet. *63*, 181–190.

Callen, D.F., Baker, E.G., and Lane, S.A. (1990). Re-evaluation of GM2346 from a del (16)(q22) to t(4;16)(q35;q22.1). Clin. Genet. *38*, 466–468.

Campos, E.I., and Reinberg, D. (2009). Histones: annotating chromatin. Annu. Rev. Genet. *43*, 559–599.

Caruso, N., Herberth, B., Bartoli, M., Puppo, F., Dumonceaux, J., Zimmermann, A., Denadai, S., Lebosse, M., Roche, S., Geng, L., *et al.* (2013). Deregulation of the protocadherin gene FAT1 alters muscle shapes: implications for the pathogenesis of facioscapulohumeral dystrophy. PLoS Genet. *9*, 1–29.

Chadwick, B.P. (2008). DXZ4 chromatin adopts an opposing conformation to that of the surrounding chromosome and acquires a novel inactive X-specific role involving CTCF and antisense transcripts. Genome Res. *18*, 1259–1269.

Chen, T., and Li, E. (2004). Structure and function of eukaryotic DNA methyltransferases. Curr. Top. Dev. Biol. *60*, 55–89.

Chen, Y.T., Iseli, C., Venditti, C.A., Old, L.J., Simpson, A.J.G., and Jongeneel, C.V. (2006). Identification of a new cancer/testis gene family, CT47, among expressed multicopy genes on the human X-chromosome. Genes Chromosomes Cancer 45, 392–400.

Chung, C.S., and Morton, N.E. (1959). Discrimination of genetic entities in muscular dystrophy. Am. J. Hum. Genet. *11*, 339–359.

Clapp, J., Mitchell, L.M., Bolland, D.J., Fantes, J., Corcoran, A.E., Scotting, P.J., Armour, J.A.L., and Hewitt, J.E. (2007). Evolutionary conservation of a coding function for D4Z4, the tandem DNA repeat mutated in facioscapulohumeral muscular dystrophy. Am. J. Hum. Genet. *81*, 264–279.

Coker, H., and Brockdorff, N. (2014). SMCHD1 accumulates at DNA damage sites and facilitates the repair of DNA double-strand breaks. J. Cell. Sci. *127*, 1869–1874.

Conrad, D.F., Pinto, D., Redon, R., Feuk, L., Gokcumen, O., Zhang, Y., Aerts, J., Andrews, T.D., Barnes, C., Campbell, P., *et al.* (2010). Origins and functional impact of copy number variation in the human genome. Nature *464*, 704–712.

Cooney, A.D., Gill, I., and Stuart, P.R. (2014). The outcome of scapulothoracic arthrodesis using cerclage wires, plates, and allograft for facioscapulohumeral dystrophy. J. Shoulder Elbow Surg. *23*, e8–13.

Corona, E.D., Jacquelin, D., Gatica, L., and Rosa, A.L. (2013). Multiple protein domains contribute to nuclear import and cell toxicity of DUX4, a candidate pathogenic protein for facioscapulohumeral muscular dystrophy. PLoS ONE *8*, 1–11.

Deak, K.L., Lemmers, R.J.L.F., Stajich, J.M., Klooster, R., Tawil, R., Frants, R.R., Speer, M.C., van der Maarel, S.M., and Gilbert, J.R. (2007). Genotype-phenotype study in an FSHD family with a proximal deletion encompassing p13E-11 and D4Z4. Neurology *68*, 578–582.

Deidda, G., Cacurri, S., Piazzo, N., and Felicetti, L. (1996). Direct detection of 4q35 rearrangements implicated in facioscapulohumeral muscular dystrophy (FSHD). J. Med. Genet. *33*, 361–365.

van Deutekom, J.C., Wijmenga, C., van Tienhoven, E.A., Gruter, A.M., Hewitt, J.E., Padberg, G.W., van Ommen, G.J., Hofker, M.H., and Frants, R.R. (1993). FSHD associated DNA rearrangements are due to deletions of integral copies of a 3.2 kb tandemly repeated unit. Hum. Mol. Genet. *2*, 2037–2042.

van Deutekom, J.C., Bakker, E., Lemmers, R.J., van der Wielen, M.J., Bik, E., Hofker, M.H., Padberg, G.W., and Frants, R.R. (1996). Evidence for subtelomeric exchange of 3.3 kb tandemly repeated units between chromosomes 4q35 and 10q26: implications for genetic counselling and etiology of FSHD1. Hum. Mol. Genet. *5*, 1997–2003.

Diab, M., Darras, B.T., and Shapiro, F. (2005). Scapulothoracic fusion for facioscapulohumeral muscular dystrophy. J. Bone. Joint. Surg Am. *87*, 2267–2275.

Ding, H., Beckers, M.C., Plaisance, S., Marynen, P., Collen, D., and Belayew, A. (1998). Characterization of a double homeodomain protein (DUX1) encoded by a cDNA homologous to 3.3 kb dispersed repeated elements. Hum. Mol. Genet. *7*, 1681–1694.

Dixit, M., Ansseau, E., Tassin, A., Winokur, S., Shi, R., Qian, H., Sauvage, S., Matteotti, C., van Acker, A.M., Leo, O., *et al.* (2007). DUX4, a candidate gene of facioscapulohumeral muscular dystrophy, encodes a transcriptional activator of PITX1. Proc. Natl. Acad. Sci. U.S.A. *104*, 18157–18162.

Eger, K., Jordan, B., Habermann, S., and Zierz, S. (2010). Beevor's sign in facioscapulohumeral muscular dystrophy: an old sign with new implications. J Neurol. *257*, 436–438.

Emery, A.E. (1991). Population frequencies of inherited neuromuscular diseases-a world survey. Neuromuscul. Disord. *1*, 19–29.

Ferreboeuf, M., Mariot, V., Bessieres, B., Vasiljevic, A., Attie-Bitach, T., Collardeau, S., Morere, J., Roche, S., Magdinier, F., Robin-Ducellier, J., *et al.* (2014a). DUX4 and DUX4 downstream target genes are expressed in fetal FSHD muscles. Hum. Mol. Genet. *23*, 171–181.

Ferreboeuf, M., Mariot, V., Furling, D., Butler-Browne, G., Mouly, V., and Dumonceaux, J. (2014b). Nuclear protein spreading: implication for pathophysiology of neuromuscular diseases. Hum. Mol. Genet. *23*, 4125–4133.

Fitzsimons, R.B., Gurwin, E.B., and Bird, A.C. (1987). Retinal vascular abnormalities in facioscapulohumeral muscular dystrophy. A general association with genetic and therapeutic implications. Brain *110* (Part 3), 631–648.

Flanigan, K.M., Coffeen, C.M., Sexton, L., Stauffer, D., Brunner, S., and Leppert, M.F. (2001). Genetic characterization of a large, historically significant Utah kindred with facioscapulohumeral dystrophy. Neuromuscul. Disord. *11*, 525–529.

Fox, I.J., Daley, G.Q., Goldman, S.A., Huard, J., Kamp, T.J., and Trucco, M. (2014). Stem cell therapy. Use of differentiated pluripotent stem cells as replacement therapy for treating disease. Science *345*, 889.

Gabellini, D., Green, M.R., and Tupler, R. (2002). Inappropriate gene activation in FSHD: a repressor complex binds a chromosomal repeat deleted in dystrophic muscle. Cell *110*, 339–348.

Gabellini, D., D'Antona, G., Moggio, M., Prelle, A., Zecca, C., Adami, R., Angeletti, B., Ciscato, P., Pellegrino, M.A., Bottinelli, R., *et al.* (2006). Facioscapulohumeral muscular dystrophy in mice overexpressing FRG1. Nature *439*, 973–977.

Gabriels, J., Beckers, M.C., Ding, H., De Vriese, A., Plaisance, S., van der Maarel, S.M., Padberg, G.W., Frants, R.R., Hewitt, J.E., Collen, D., *et al.* (1999). Nucleotide sequence of the partially deleted D4Z4 locus in a patient with FSHD identifies a putative gene within each 3.3 kb element. Gene *236*, 25–32.

Gaillard, M.-C., Roche, S., Dion, C., Tasmadjian, A., Bouget, G., Salort-Campana, E., Vovan, C., Chaix, C., Broucqsault, N., Morere, J., *et al.* (2014). Differential DNA methylation of the D4Z4 repeat in patients with FSHD and asymptomatic carriers. Neurology *83*, 733–742.

Gaj, T., Gersbach, C.A., and Barbas, C.F. 3rd (2013). ZFN, TALEN, and CRISPR/Cas-based methods for genome engineering. Trends Biotechnol. *31*, 397–405.

Galetta, F., Franzoni, F., Sposito, R., Plantinga, Y., Femia, F.R., Galluzzi, F., Rocchi, A., Santoro, G., and Siciliano, G. (2005). Subclinical cardiac involvement in patients with facioscapulohumeral muscular dystrophy. Neuromuscul. Disord. *15*, 403–408.

Gardner-Medwin, D. (1980). Clinical features and classification of the muscular dystrophies. Br. Med. Bull *36*, 109–115.

van Geel, M., Dickson, M.C., Beck, A.F., Bolland, D.J., Frants, R.R., van der Maarel, S.M., de Jong, P.J., and Hewitt, J.E. (2002). Genomic analysis of human chromosome 10q and 4q telomeres suggests a common origin. Genomics *79*, 210–217.

Gendrel, A.-V., Apedaile, A., Coker, H., Termanis, A., Zvetkova, I., Godwin, J., Tang, Y.A., Huntley, D., Montana, G., Taylor, S., *et al.* (2012). Smchd1-dependent and -independent pathways determine developmental dynamics of CpG island methylation on the inactive X-chromosome. Dev. Cell *23*, 265–279.

Geng, L.N., Yao, Z., Snider, L., Fong, A.P., Cech, J.N., Young, J.M., van der Maarel, S.M., Ruzzo, W.L., Gentleman, R.C., Tawil, R., *et al.* (2012). DUX4 activates germline genes, retroelements, and immune mediators: implications for facioscapulohumeral dystrophy. Dev. Cell *22*, 1–14.

Giacalone, J., Friedes, J., and Francke, U. (1992). A novel GC-rich human macrosatellite VNTR in Xq24 is differentially methylated on active and inactive X-chromosomes. Nat. Genet. *1*, 137–143.

Gilbert, J.R., Stajich, J.M., Wall, S., Carter, S.C., Qiu, H., Vance, J.M., Stewart, C.S., Speer, M.C., Pufky, J., and

Yamaoka, L.H. (1993). Evidence for heterogeneity in facioscapulohumeral muscular dystrophy (FSHD). Am. J. Hum. Genet. 53, 401–408.

Gjerstorff, M.F., and Ditzel, H.J. (2008). An overview of the GAGE cancer/testis antigen family with the inclusion of newly identified members. Tissue Antigens 71, 187–192.

Goto, K., Lee, J.H., Matsuda, C., Hirabayashi, K., Kojo, T., Nakamura, A., Mitsunaga, Y., Furukawa, T., Sahashi, K., and Arahata, K. (1995). DNA rearrangements in Japanese facioscapulohumeral muscular dystrophy patients: clinical correlations. Neuromuscul. Disord. 5, 201–208.

de Greef, J.C., Wohlgemuth, M., Chan, O.A., Hansson, K.B., Smeets, D., Frants, R.R., Weemaes, C.M., Padberg, G.W., and van der Maarel, S.M. (2007). Hypomethylation is restricted to the D4Z4 repeat array in phenotypic FSHD. Neurology 69, 1018–1026.

de Greef, J.C., Lemmers, R.J.L.F., van Engelen, B.G.M., Sacconi, S., Venance, S.L., Frants, R.R., Tawil, R., and van der Maarel, S.M. (2009). Common epigenetic changes of D4Z4 in contraction-dependent and contraction-independent FSHD. Hum. Mutat. 30, 1449–1459.

de Greef, J.C., Lemmers, R.J.L.F., Camano, P., Day, J.W., Sacconi, S., Dunand, M., van Engelen, B.G.M., Kiuru-Enari, S., Padberg, G.W., Rosa, A.L., *et al.* (2010). Clinical features of facioscapulohumeral muscular dystrophy 2. Neurology 75, 1548–1554.

Griggs, R.C., Tawil, R., Storvick, D., Mendell, J.R., and Altherr, M.R. (1993). Genetics of facioscapulohumeral muscular dystrophy: new mutations in sporadic cases. Neurology 43, 2369–2372.

Grimm, D., Streetz, K.L., Jopling, C.L., Storm, T.A., Pandey, K., Davis, C.R., Marion, P., Salazar, F., and Kay, M.A. (2006). Fatality in mice due to oversaturation of cellular microRNA/short hairpin RNA pathways. Nature 441, 537–541.

Grolimund, L., Aeby, E., Hamelin, R., Armand, F., Chiappe, D., Moniatte, M., and Lingner, J. (2013). A quantitative telomeric chromatin isolation protocol identifies different telomeric states. Nat. Commun. 4, 1–12.

Guelen, L., Pagie, L., Brasset, E., Meuleman, W., Faza, M.B., Talhout, W., Eussen, B.H., de Klein, A., Wessels, L., de Laat, W., *et al.* (2008). Domain organization of human chromosomes revealed by mapping of nuclear lamina interactions. Nature 453, 948–951.

Hanel, M.L., Wuebbles, R.D., and Jones, P.L. (2009). Muscular dystrophy candidate gene FRG1 is critical for muscle development. Dev. Dyn. 238, 1502–1512.

Hanel, M.L., Sun, C.-Y.J., Jones, T.I., Long, S.W., Zanotti, S., Milner, D., and Jones, P.L. (2011). Facioscapulohumeral muscular dystrophy (FSHD) region gene 1 (FRG1) is a dynamic nuclear and sarcomeric protein. Differentiation 81, 107–118.

Hansen, R.S., Wijmenga, C., Luo, P., Stanek, A.M., Canfield, T.K., Weemaes, C.M., and Gartler, S.M. (1999). The DNMT3B DNA methyltransferase gene is mutated in the ICF immunodeficiency syndrome. Proc. Natl. Acad. Sci. U.S.A. 96, 14412–14417.

Hark, A.T., Schoenherr, C.J., Katz, D.J., Ingram, R.S., Levorse, J.M., and Tilghman, S.M. (2000). CTCF mediates methylation-sensitive enhancer-blocking activity at the H19/Igf2 locus. Nature 405, 486–489.

Harley, H.G., Rundle, S.A., Reardon, W., Myring, J., Crow, S., Brook, J.D., Harper, P.S., and Shaw, D.J. (1992). Unstable DNA sequence in myotonic dystrophy. Lancet 339, 1125–1128.

Hartweck, L.M., Anderson, L.J., Lemmers, R.J., Dandapat, A., Toso, E.A., Dalton, J.C., Tawil, R., Day, J.W., van der Maarel, S.M., and Kyba, M. (2013). A focal domain of extreme demethylation within D4Z4 in FSHD2. Neurology 80, 392–399.

Hewitt, J.E., Lyle, R., Clark, L.N., Valleley, E.M., Wright, T.J., Wijmenga, C., van Deutekom, J.C., Francis, F., Sharpe, P.T., and Hofker, M. (1994). Analysis of the tandem repeat locus D4Z4 associated with facioscapulohumeral muscular dystrophy. Hum. Mol. Genet. 3, 1287–1295.

Himeda, C.L., Debarnot, C., Homma, S., Beermann, M.L., Miller, J.B., Jones, P.L., and Jones, T.I. (2014). Myogenic enhancers regulate expression of the facioscapulohumeral muscular dystrophy-associated DUX4 gene. Mol. Cell. Biol. 34, 1942–1955.

Hirano, T. (2005). SMC proteins and chromosome mechanics: from bacteria to humans. Philos. Trans. R. Soc. Lond. B. Biol. Sci. 360, 507–514.

Horakova, A.H., Moseley, S.C., McLaughlin, C.R., Tremblay, D.C., and Chadwick, B.P. (2012). The macrosatellite DXZ4 mediates CTCF-dependent long-range intrachromosomal interactions on the human inactive X-chromosome. Hum. Mol. Genet. 21, 4367–4377.

Hsu, Y.D., Kao, M.C., Shyu, W.C., Lin, J.C., Huang, N.E., Sun, H.F., Yang, K.D., and Tsao, W.L. (1997). Application of chromosome 4q35-qter marker (pFR-1) for DNA rearrangement of facioscapulohumeral muscular dystrophy patients in Taiwan. J. Neurol. Sci. 149, 73–79.

Jenuwein, T., and Allis, C.D. (2001). Translating the histone code. Science 293, 1074–1080.

Jiang, G., Yang, F., van Overveld, P.G.M., Vedanarayanan, V., van der Maarel, S., and Ehrlich, M. (2003). Testing the position-effect variegation hypothesis for facioscapulohumeral muscular dystrophy by analysis of histone modification and gene expression in subtelomeric 4q. Hum. Mol. Genet. 12, 2909–2921.

Jones, T.I., Chen, J.C.J., Rahimov, F., Homma, S., Arashiro, P., Beermann, M.L., King, O.D., Miller, J.B., Kunkel, L.M., Emerson, C.P.J., *et al.* (2012). Facioscapulohumeral muscular dystrophy family studies of DUX4 expression: evidence for disease modifiers and a quantitative model of pathogenesis. Hum Mol. Genet. 21, 4419–4430.

Kanno, T., Bucher, E., Daxinger, L., Huettel, B., Bohmdorfer, G., Gregor, W., Kreil, D.P., Matzke, M., and Matzke, A.J.M. (2008). A structural-maintenance-of-chromosomes hinge domain-containing protein is required for RNA-directed DNA methylation. Nat. Genet. 40, 670–675.

Kawamura-Saito, M., Yamazaki, Y., Kaneko, K., Kawaguchi, N., Kanda, H., Mukai, H., Gotoh, T., Motoi, T.,

Fukayama, M., Aburatani, H., *et al.* (2006). Fusion between CIC and DUX4 up-regulates PEA3 family genes in Ewing-like sarcomas with t(4;19)(q35;q13) translocation. Hum. Mol. Genet. *15*, 2125–2137.

Kilmer, D.D., Abresch, R.T., McCrory, M.A., Carter, G.T., Fowler, W.M.J., Johnson, E.R., and McDonald, C.M. (1995). Profiles of neuromuscular diseases. Facioscapulohumeral muscular dystrophy. Am. J. Phys. Med. Rehabil. *74*, S131–S139.

Kim, J., Kollhoff, A., Bergmann, A., and Stubbs, L. (2003). Methylation-sensitive binding of transcription factor YY1 to an insulator sequence within the paternally expressed imprinted gene, Peg3. Hum. Mol. Genet.*12*, 233–245.

Kissel, J.T., McDermott, M.P., Natarajan, R., Mendell, J.R., Pandya, S., King, W.M., Griggs, R.C., and Tawil, R. (1998). Pilot trial of albuterol in facioscapulohumeral muscular dystrophy. FSH-DY Group. Neurology *50*, 1402–1406.

Klinge, L., Eagle, M., Haggerty, I.D., Roberts, C.E., Straub, V., and Bushby, K.M. (2006). Severe phenotype in infantile facioscapulohumeral muscular dystrophy. Neuromuscul. Disord.*16*, 553–558.

Klooster, R., Straasheijm, K., Shah, B., Sowden, J., Frants, R., Thornton, C., Tawil, R., and van der Maarel, S. (2009). Comprehensive expression analysis of FSHD candidate genes at the mRNA and protein level. Eur. J. Hum. Genet. *17*, 1615–1624.

Kondo, T., Bobek, M.P., Kuick, R., Lamb, B., Zhu, X., Narayan, A., Bourc'his, D., Viegas-Pequignot, E., Ehrlich, M., and Hanash, S.M. (2000). Whole-genome methylation scan in ICF syndrome: hypomethylation of non-satellite DNA repeats D4Z4 and NBL2. Hum. Mol. Genet. *9*, 597–604.

van der Kooi, E.L., Vogels, O.J.M., van Asseldonk, R.J.G.P., Lindeman, E., Hendriks, J.C.M., Wohlgemuth, M., van der Maarel, S.M., and Padberg, G.W. (2004). Strength training and albuterol in facioscapulohumeral muscular dystrophy. Neurology *63*, 702–708.

Kowaljow, V., Marcowycz, A., Ansseau, E., Conde, C.B., Sauvage, S., Matteotti, C., Arias, C., Corona, E.D., Nunez, N.G., Leo, O., *et al.* (2007). The DUX4 gene at the FSHD1A locus encodes a pro-apoptotic protein. Neuromuscul. Disord. *17*, 611–623.

Krom, Y.D., Thijssen, P.E., Young, J.M., den Hamer, B., Balog, J., Yao, Z., Maves, L., Snider, L., Knopp, P., Zammit, P.S., *et al.* (2013). Intrinsic epigenetic regulation of the D4Z4 macrosatellite repeat in a transgenic mouse model for FSHD. PLoS Genet. *9*, 1–17.

Lachner, M., O'Carroll, D., Rea, S., Mechtler, K., and Jenuwein, T. (2001). Methylation of histone H3 lysine 9 creates a binding site for HP1 proteins. Nature *410*, 116–120.

Leidenroth, A., and Hewitt, J.E. (2010). A family history of DUX4: phylogenetic analysis of DUXA, B, C and Duxbl reveals the ancestral DUX gene. BMC Evol. Biol. *10*, 1–13.

Lemmers, R.J.L., de Kievit, P., van Geel, M., van der Wielen, M.J., Bakker, E., Padberg, G.W., Frants, R.R., and van der Maarel, S.M. (2001). Complete allele information in the diagnosis of facioscapulohumeral muscular dystrophy by triple DNA analysis. Ann. Neurol. *50*, 816–819.

Lemmers, R.J.L.F., de Kievit, P., Sandkuijl, L., Padberg, G.W., van Ommen, G.-J.B., Frants, R.R., and van der Maarel, S.M. (2002). Facioscapulohumeral muscular dystrophy is uniquely associated with one of the two variants of the 4q subtelomere. Nat. Genet. *32*, 235–236.

Lemmers, R.J.F.L., Wohlgemuth, M., Frants, R.R., Padberg, G.W., Morava, E., and van der Maarel, S.M. (2004a). Contractions of D4Z4 on 4qB subtelomeres do not cause facioscapulohumeral muscular dystrophy. Am. J. Hum. Genet. 75, 1124–1130.

Lemmers, R.J.L.F., Van Overveld, P.G.M., Sandkuijl, L.A., Vrieling, H., Padberg, G.W., Frants, R.R., and van der Maarel, S.M. (2004b). Mechanism and timing of mitotic rearrangements in the subtelomeric D4Z4 repeat involved in facioscapulohumeral muscular dystrophy. Am. J. Hum. Genet. 75, 44–53.

Lemmers, R.J.L.F., Wohlgemuth, M., van der Gaag, K.J., van der Vliet, P.J., van Teijlingen, C.M.M., de Knijff, P., Padberg, G.W., Frants, R.R., and van der Maarel, S.M. (2007). Specific sequence variations within the 4q35 region are associated with facioscapulohumeral muscular dystrophy. Am. J. Hum. Genet. *81*, 884–894.

Lemmers, R.J.L.F., van der Vliet, P.J., Klooster, R., Sacconi, S., Camano, P., Dauwerse, J.G., Snider, L., Straasheijm, K.R., van Ommen, G.J., Padberg, G.W., *et al.* (2010a). A unifying genetic model for facioscapulohumeral muscular dystrophy. Science *329*, 1650–1653.

Lemmers, R.J.L.F., van der Vliet, P.J., van der Gaag, K.J., Zuniga, S., Frants, R.R., de Knijff, P., and van der Maarel, S.M. (2010b). Worldwide population analysis of the 4q and 10q subtelomeres identifies only four discrete interchromosomal sequence transfers in human evolution. Am. J. Hum. Genet. *86*, 364–377.

Lemmers, R.J.L.F., Tawil, R., Petek, L.M., Balog, J., Block, G.J., Santen, G.W.E., Amell, A.M., van der Vliet, P.J., Almomani, R., Straasheijm, K.R., *et al.* (2012). Digenic inheritance of an SMCHD1 mutation and an FSHD-permissive D4Z4 allele causes facioscapulohumeral muscular dystrophy type 2. Nat. Genet. *44*, 1370–1374.

Leong, H.S., Chen, K., Hu, Y., Lee, S., Corbin, J., Pakusch, M., Murphy, J.M., Majewski, I.J., Smyth, G.K., Alexander, W.S., *et al.* (2013). Epigenetic regulator Smchd1 functions as a tumor suppressor. Cancer Res. *73*, 1591–1599.

Liu, D.X., and Lobie, P.E. (2007). Transcriptional activation of p53 by Pitx1. Cell Death Differ. *14*, 1893–1907.

Lunt, P.W., Jardine, P.E., Koch, M.C., Maynard, J., Osborn, M., Williams, M., Harper, P.S., and Upadhyaya, M. (1995). Correlation between fragment size at D4F104S1 and age at onset or at wheelchair use, with a possible generational effect, accounts for much phenotypic variation in 4q35-facioscapulohumeral muscular dystrophy (FSHD). Hum. Mol. Genet. *4*, 951–958.

Lyle, R., Wright, T.J., Clark, L.N., and Hewitt, J.E. (1995). The FSHD-associated repeat, D4Z4, is a member of a dispersed family of homeobox-containing repeats, subsets of which are clustered on the short arms of the acrocentric chromosomes. Genomics *28*, 389–397.

van der Maarel, S.M., Deidda, G., Lemmers, R.J., van Overveld, P.G., van der Wielen, M., Hewitt, J.E., Sandkuijl, L., Bakker, B., van Ommen, G.J., Padberg, G.W., *et al.* (2000). De novo facioscapulohumeral muscular dystrophy: frequent somatic mosaicism, sex-dependent phenotype, and the role of mitotic transchromosomal repeat interaction between chromosomes 4 and 10. Am. J. Hum. Genet. *66*, 26–35.

Macaione, V., Aguennouz, M., Rodolico, C., Mazzeo, A., Patti, A., Cannistraci, E., Colantone, L., Di Giorgio, R.M., De Luca, G., and Vita, G. (2007). RAGE-NF-kappaB pathway activation in response to oxidative stress in facioscapulohumeral muscular dystrophy. Acta. Neurol. Scand. *115*, 115–121.

MacDonald, M.E., Barnes, G., Srinidhi, J., Duyao, M.P., Ambrose, C.M., Myers, R.H., Gray, J., Conneally, P.M., Young, A., and Penney, J. (1993). Gametic but not somatic instability of CAG repeat length in Huntington's disease. J. Med. Genet. *30*, 982–986.

von Maltzahn, J., Renaud, J.-M., Parise, G., and Rudnicki, M.A. (2012). Wnt7a treatment ameliorates muscular dystrophy. Proc. Natl. Acad. Sci. U.S.A. *109*, 20614–20619.

Maraschio, P., Zuffardi, O., Dalla Fior, T., and Tiepolo, L. (1988). Immunodeficiency, centromeric heterochromatin instability of chromosomes 1, 9, and 16, and facial anomalies: the ICF syndrome. J. Med. Genet. *25*, 173–180.

Masny, P.S., Bengtsson, U., Chung, S.-A., Martin, J.H., van Engelen, B., van der Maarel, S.M., and Winokur, S.T. (2004). Localization of 4q35.2 to the nuclear periphery: is FSHD a nuclear envelope disease? Hum. Mol. Genet. *13*, 1857–1871.

Mendell, J.R., Rodino-Klapac, L.R., Rosales-Quintero, X., Kota, J., Coley, B.D., Galloway, G., Craenen, J.M., Lewis, S., Malik, V., Shilling, C., *et al.* (2009). Limb-girdle muscular dystrophy type 2D gene therapy restores alpha-sarcoglycan and associated proteins. Ann. Neurol. *66*, 290–297.

Mendell, J.R., Rodino-Klapac, L.R., Rosales, X.Q., Coley, B.D., Galloway, G., Lewis, S., Malik, V., Shilling, C., Byrne, B.J., Conlon, T., *et al.* (2010). Sustained alpha-sarcoglycan gene expression after gene transfer in limb-girdle muscular dystrophy, type 2D. Ann. Neurol. *68*, 629–638.

Miller, J.C., Holmes, M.C., Wang, J., Guschin, D.Y., Lee, Y.-L., Rupniewski, I., Beausejour, C.M., Waite, A.J., Wang, N.S., Kim, K.A., *et al.* (2007). An improved zinc-finger nuclease architecture for highly specific genome editing. Nat. Biotechnol. *25*, 778–785.

Mitsuhashi, S., Boyden, S.E., Estrella, E.A., Jones, T.I., Rahimov, F., Yu, T.W., Darras, B.T., Amato, A.A., Folkerth, R.D., Jones, P.L., *et al.* (2013). Exome sequencing identifies a novel SMCHD1 mutation in facioscapulohumeral muscular dystrophy 2. Neuromuscul. Disord. *23*, 975–980.

Mould, A.W., Pang, Z., Pakusch, M., Tonks, I.D., Stark, M., Carrie, D., Mukhopadhyay, P., Seidel, A., Ellis, J.J., Deakin, J., *et al.* (2013). Smchd1 regulates a subset of autosomal genes subject to monoallelic expression in addition to being critical for X inactivation. Epigenetics Chromatin *6*, 1–16.

Mussolino, C., Alzubi, J., Fine, E.J., Morbitzer, R., Cradick, T.J., Lahaye, T., Bao, G., and Cathomen, T. (2014). TALENs facilitate targeted genome editing in human cells with high specificity and low cytotoxicity. Nucleic Acids Res. *42*, 6762–6773.

Nozawa, R.-S., Nagao, K., Igami, K.-T., Shibata, S., Shirai, N., Nozaki, N., Sado, T., Kimura, H., and Obuse, C. (2013). Human inactive X-chromosome is compacted through a PRC2-independent SMCHD1-HBiX1 pathway. Nat. Struct. Mol. Biol. *20*, 566–573.

Ohno, S. (1972). So much "junk" DNA in our genome. Brookhaven Symp. Biol. *23*, 366–370.

Okano, M., Bell, D.W., Haber, D.A., and Li, E. (1999). DNA methyltransferases Dnmt3a and Dnmt3b are essential for de novo methylation and mammalian development. Cell *99*, 247–257.

Olsen, D.B., Orngreen, M.C., and Vissing, J. (2005). Aerobic training improves exercise performance in facioscapulohumeral muscular dystrophy. Neurology *64*, 1064–1066.

Orgel, L.E., and Crick, F.H. (1980). Selfish DNA: the ultimate parasite. Nature *284*, 604–607.

Osborne, R.J., Welle, S., Venance, S.L., Thornton, C.A., and Tawil, R. (2007). Expression profile of FSHD supports a link between retinal vasculopathy and muscular dystrophy. Neurology *68*, 569–577.

Ostlund, C., Garcia-Carrasquillo, R.M., Belayew, A., and Worman, H.J. (2005). Intracellular trafficking and dynamics of double homeodomain proteins. Biochemistry *44*, 2378–2384.

Ottaviani, A., Rival-Gervier, S., Boussouar, A., Foerster, A.M., Rondier, D., Sacconi, S., Desnuelle, C., Gilson, E., and Magdinier, F. (2009). The D4Z4 macrosatellite repeat acts as a CTCF and A-type lamins-dependent insulator in facio-scapulo-humeral dystrophy. PLoS Genet. *5*, 1–11.

van Overveld, P.G.M., Lemmers, R.J.F.L., Sandkuijl, L.A., Enthoven, L., Winokur, S.T., Bakels, F., Padberg, G.W., van Ommen, G.-J.B., Frants, R.R., and van der Maarel, S.M. (2003). Hypomethylation of D4Z4 in 4q-linked and non-4q-linked facioscapulohumeral muscular dystrophy. Nat. Genet. *35*, 315–317.

van Overveld, P.G.M., Enthoven, L., Ricci, E., Rossi, M., Felicetti, L., Jeanpierre, M., Winokur, S.T., Frants, R.R., Padberg, G.W., and van der Maarel, S.M. (2005). Variable hypomethylation of D4Z4 in facioscapulohumeral muscular dystrophy. Ann. Neurol. *58*, 569–576.

Padberg, G. (1982). Facioscapulohumeral disease. Thesis. University of Leiden, Leiden, Netherlands.

Padberg, G.W., Frants, R.R., Brouwer, O.F., Wijmenga, C., Bakker, E., and Sandkuijl, L.A. (1995a). Facioscapulohumeral muscular dystrophy in the Dutch population. Muscle Nerve Suppl. *2*, S81–S84.

Padberg, G.W., Brouwer, O.F., de Keizer, R.J., Dijkman, G., Wijmenga, C., Grote, J.J., and Frants, R.R. (1995b). On the significance of retinal vascular disease and hearing loss in facioscapulohumeral muscular dystrophy. Muscle Nerve Suppl. 2, S73–80

Pandey, S.N., Cabotage, J., Shi, R., Dixit, M., Sutherland, M., Liu, J., Muger, S., Harper, S.Q., Nagaraju, K., and Chen, Y.-W. (2012). Conditional over-expression of PITX1 causes skeletal muscle dystrophy in mice. Biol. Open. *1*, 629–639.

Pandey, S.N., Lee, Y.-C., Yokota, T., and Chen, Y.-W. (2014). Morpholino treatment improves muscle function and pathology of Pitx1 transgenic mice. Mol. Ther. *22*, 390–396.

Pandya, S., King, W.M., and Tawil, R. (2008). Facioscapulohumeral dystrophy. Phys. Ther. *88*, 105–113.

Pecot, C.V., Calin, G.A., Coleman, R.L., Lopez-Berestein, G., and Sood, A.K. (2011). RNA interference in the clinic: challenges and future directions. Nat. Rev. Cancer *11*, 59–67.

Peters, A.H.F.M., Kubicek, S., Mechtler, K., O'Sullivan, R.J., Derijck, A.A.H.A., Perez-Burgos, L., Kohlmaier, A., Opravil, S., Tachibana, M., Shinkai, Y., *et al.* (2003). Partitioning and plasticity of repressive histone methylation states in mammalian chromatin. Mol. Cell.*12*, 1577–1589.

Petrov, A., Pirozhkova, I., Carnac, G., Laoudj, D., Lipinski, M., and Vassetzky, Y.S. (2006). Chromatin loop domain organization within the 4q35 locus in facioscapulohumeral dystrophy patients versus normal human myoblasts. Proc. Natl. Acad. Sci. U.S.A. *103*, 6982–6987.

Petrov, A., Allinne, J., Pirozhkova, I., Laoudj, D., Lipinski, M., and Vassetzky, Y.S. (2008). A nuclear matrix attachment site in the 4q35 locus has an enhancer-blocking activity *in vivo*: implications for the facio-scapulo-humeral dystrophy. Genome Res. *18*, 39–45.

Pirozhkova, I., Petrov, A., Dmitriev, P., Laoudj, D., Lipinski, M., and Vassetzky, Y. (2008). A functional role for 4qA/B in the structural rearrangement of the 4q35 region and in the regulation of FRG1 and ANT1 in facioscapulohumeral dystrophy. PLoS ONE 3, 1–9.

Ploski, J.E., Shamsher, M.K., and Radu, A. (2004). Paired-type homeodomain transcription factors are imported into the nucleus by karyopherin 13. Mol. Cell. Biol. *24*, 4824–4834.

Pope, B.D., Tsumagari, K., Battaglia, D., Ryba, T., Hiratani, I., Ehrlich, M., and Gilbert, D.M. (2011). DNA replication timing is maintained genome-wide in primary human myoblasts independent of D4Z4 contraction in FSH muscular dystrophy. PLoS ONE *6*, 1–9.

Ricci, E., Galluzzi, G., Deidda, G., Cacurri, S., Colantoni, L., Merico, B., Piazzo, N., Servidei, S., Vigneti, E., Pasceri, V., *et al.* (1999). Progress in the molecular diagnosis of facioscapulohumeral muscular dystrophy and correlation between the number of KpnI repeats at the 4q35 locus and clinical phenotype. Ann. Neurol.*45*, 751–757.

Rice, J.C., Briggs, S.D., Ueberheide, B., Barber, C.M., Shabanowitz, J., Hunt, D.F., Shinkai, Y., and Allis, C.D. (2003). Histone methyltransferases direct different degrees of methylation to define distinct chromatin domains. Mol. Cell. *12*, 1591–1598.

Rossi, M., Ricci, E., Colantoni, L., Galluzzi, G., Frusciante, R., Tonali, P.A., and Felicetti, L. (2007). The Facioscapulohumeral muscular dystrophy region on 4qter and the homologous locus on 10qter evolved independently under different evolutionary pressure. BMC Med. Genet. *8*, 1–8.

Sacconi, S., Lemmers, R.J.L.F., Balog, J., van der Vliet, P.J., Lahaut, P., van Nieuwenhuizen, M.P., Straasheijm, K.R., Debipersad, R.D., Vos-Versteeg, M., Salviati, L., *et al.* (2013). The FSHD2 gene SMCHD1 is a modifier of disease severity in families affected by FSHD1. Am. J. Hum. Genet. 93, 744–751.

Schmid, C.W., and Deininger, P.L. (1975). Sequence organization of the human genome. Cell 6, 345–358.

Schneider, R., and Grosschedl, R. (2007). Dynamics and interplay of nuclear architecture, genome organization, and gene expression. Genes Dev. *21*, 3027–3043.

Seale, P., Sabourin, L.A., Girgis-Gabardo, A., Mansouri, A., Gruss, P., and Rudnicki, M.A. (2000). Pax7 is required for the specification of myogenic satellite cells. Cell *102*, 777–786.

Shahrizaila, N., and Wills, A.J. (2005). Significance of Beevor's sign in facioscapulohumeral dystrophy and other neuromuscular diseases. J. Neurol. Neurosurg. Psychiatry 76, 869–870.

Shen, B., Zhang, W., Zhang, J., Zhou, J., Wang, J., Chen, L., Wang, L., Hodgkins, A., Iyer, V., Huang, X., *et al.* (2014). Efficient genome modification by CRISPR-Cas9 nickase with minimal off-target effects. Nat. Methods. *11*, 399–402.

Snider, L., Asawachaicharn, A., Tyler, A.E., Geng, L.N., Petek, L.M., Maves, L., Miller, D.G., Lemmers, R.J.L.F., Winokur, S.T., Tawil, R., *et al.* (2009). RNA transcripts, miRNA-sized fragments and proteins produced from D4Z4 units: new candidates for the pathophysiology of facioscapulohumeral dystrophy. Hum. Mol. Genet. *18*, 2414–2430.

Snider, L., Geng, L.N., Lemmers, R.J.L.F., Kyba, M., Ware, C.B., Nelson, A.M., Tawil, R., Filippova, G.N., van der Maarel, S.M., Tapscott, S.J., *et al.* (2010). Facioscapulohumeral dystrophy: incomplete suppression of a retrotransposed gene. PLoS Genet. *6*, 1–14.

Spurlock, G., Jim, H.-P., and Upadhyaya, M. (2010). Confirmation that the specific SSLP microsatellite allele 4qA161 segregates with fascioscapulohumeral muscular dystrophy (FSHD) in a cohort of multiplex and simplex FSHD families. Muscle Nerve 42, 820–821.

Stadler, G., Rahimov, F., King, O.D., Chen, J.C.J., Robin, J.D., Wagner, K.R., Shay, J.W., Emerson, C.P.J., and Wright, W.E. (2013). Telomere position effect regulates DUX4 in human facioscapulohumeral muscular dystrophy. Nat. Struct. Mol. Biol. *20*, 671–678.

Stankiewicz, P., and Lupski, J.R. (2010). Structural variation in the human genome and its role in disease. Annu. Rev. Med. *61*, 437–455.

Stevenson, A.C. (1953). Muscular dystrophy in Northern Ireland, I. An account of the condition in fifty-one families. Ann. Eugen. *18*, 50–93.

Stevenson, W.G., Perloff, J.K., Weiss, J.N., and Anderson, T.L. (1990). Facioscapulohumeral muscular dystrophy: evidence for selective, genetic electrophysiologic cardiac involvement. J. Am. Coll. Cardiol. *15*, 292–299.

Suzuki, M.M., and Bird, A. (2008). DNA methylation landscapes: provocative insights from epigenomics. Nat. Rev. Genet. *9*, 465–476.

Tam, R., Smith, K.P., and Lawrence, J.B. (2004). The 4q subtelomere harboring the FSHD locus is specifically anchored with peripheral heterochromatin unlike most human telomeres. J. Cell. Biol. *167*, 269–279.

Tassin, A., Laoudj-Chenivesse, D., Vanderplanck, C., Barro, M., Charron, S., Ansseau, E., Chen, Y.-W., Mercier, J., Coppee, F., and Belayew, A. (2013). DUX4 expression in FSHD muscle cells: how could such a rare protein cause a myopathy? J. Cell. Mol. Med. *17*, 76–89.

Tawil, R., Storvick, D., Feasby, T.E., Weiffenbach, B., and Griggs, R.C. (1993). Extreme variability of expression in monozygotic twins with FSH muscular dystrophy. Neurology *43*, 345–348.

Tawil, R., Forrester, J., Griggs, R.C., Mendell, J., Kissel, J., McDermott, M., King, W., Weiffenbach, B., and Figlewicz, D. (1996). Evidence for anticipation and association of deletion size with severity in facioscapulohumeral muscular dystrophy. The FSH-DY Group. Ann. Neurol. *39*, 744–748.

Tawil, R., McDermott, M.P., Pandya, S., King, W., Kissel, J., Mendell, J.R., and Griggs, R.C. (1997). A pilot trial of prednisone in facioscapulohumeral muscular dystrophy. FSH-DY Group. Neurology *48*, 46–49.

Tawil, R., Figlewicz, D.A., Griggs, R.C., and Weiffenbach, B. (1998). Facioscapulohumeral dystrophy: a distinct regional myopathy with a novel molecular pathogenesis. FSH Consortium. Ann. Neurol. *43*, 279–282.

Tonini, M.M.O., Passos-Bueno, M.R., Cerqueira, A., Matioli, S.R., Pavanello, R., and Zatz, M. (2004). Asymptomatic carriers and gender differences in facioscapulohumeral muscular dystrophy (FSHD). Neuromuscul. Disord. *14*, 33–38.

Trevisan, C.P., Pastorello, E., Armani, M., Angelini, C., Nante, G., Tomelleri, G., Tonin, P., Mongini, T., Palmucci, L., Galluzzi, G., *et al.* (2006). Facioscapulohumeral muscular dystrophy and occurrence of heart arrhythmia. Eur. Neurol. *56*, 1–5.

Tupler, R., Berardinelli, A., Barbierato, L., Frants, R., Hewitt, J.E., Lanzi, G., Maraschio, P., and Tiepolo, L. (1996). Monosomy of distal 4q does not cause facioscapulohumeral muscular dystrophy. J. Med. Genet. *33*, 366–370.

Tupler, R., Barbierato, L., Memmi, M., Sewry, C.A., De Grandis, D., Maraschio, P., Tiepolo, L., and Ferlini, A. (1998). Identical de novo mutation at the D4F104S1 locus in monozygotic male twins affected by facioscapulohumeral muscular dystrophy (FSHD) with different clinical expression. J. Med. Genet. 35, 778–783.

Tyler, F.H., and Stephens, F.E. (1950). Studies in disorders of muscle. II Clinical manifestations and inheritance of facioscapulohumeral dystrophy in a large family. Ann. Intern. Med. *32*, 640–660.

Upadhyaya, M., Lunt, P.W., Sarfarazi, M., Broadhead, W., Daniels, J., Owen, M., and Harper, P.S. (1990). DNA marker applicable to presymptomatic and prenatal diagnosis of facioscapulohumeral disease. Lancet *336*, 1320–1321.

Upadhyaya, M., Lunt, P.W., Sarfarazi, M., Broadhead, W., Daniels, J., Owen, M., and Harper, P.S. (1991). A closely linked DNA marker for facioscapulohumeral disease on chromosome 4q. J. Med. Genet. *28*, 665–671.

Upadhyaya, M., Lunt, P., Sarfarazi, M., Broadhead, W., Farnham, J., and Harper, P.S. (1992). The mapping of chromosome 4q markers in relation to facioscapulohumeral muscular dystrophy (FSHD). Am. J. Hum. Genet. *51*, 404–410.

Vanderplanck, C., Ansseau, E., Charron, S., Stricwant, N., Tassin, A., Laoudj-Chenivesse, D., Wilton, S.D., Coppee, F., and Belayew, A. (2011). The FSHD atrophic myotube phenotype is caused by DUX4 expression. PLoS ONE *6*, 1–14.

de Visser, M., de Voogt, W.G., and la Riviere, G.V. (1992). The heart in Becker muscular dystrophy, facioscapulohumeral dystrophy, and Bethlem myopathy. Muscle Nerve *15*, 591–596.

Voet, N.B.M., van der Kooi, E.L., Riphagen, I.I., Lindeman, E., van Engelen, B.G.M., and Geurts, A.C.H. (2013). Strength training and aerobic exercise training for muscle disease. Cochrane Database Syst. Rev. *7*, 1–50.

Wallace, L.M., Garwick, S.E., Mei, W., Belayew, A., Coppee, F., Ladner, K.J., Guttridge, D., Yang, J., and Harper, S.Q. (2011a). DUX4, a candidate gene for facioscapulohumeral muscular dystrophy, causes p53-dependent myopathy *in vivo*. Ann. Neurol. *69*, 540–552.

Wallace, L.M., Garwick-Coppens, S.E., Tupler, R., and Harper, S.Q. (2011b). RNA interference improves myopathic phenotypes in mice over-expressing FSHD region gene 1 (FRG1). Mol. Ther. *19*, 2048–2054.

Wallace, L.M., Liu, J., Domire, J.S., Garwick-Coppens, S.E., Guckes, S.M., Mendell, J.R., Flanigan, K.M., and Harper, S.Q. (2012). RNA interference inhibits DUX4-induced muscle toxicity *in vivo*: implications for a targeted FSHD therapy. Mol. Ther. *20*, 1417–1423.

Wijmenga, C., Frants, R.R., Brouwer, O.F., Moerer, P., Weber, J.L., and Padberg, G.W. (1990). Location of facioscapulohumeral muscular dystrophy gene on chromosome 4. Lancet *336*, 651–653.

Wijmenga, C., Padberg, G.W., Moerer, P., Wiegant, J., Liem, L., Brouwer, O.F., Milner, E.C., Weber, J.L., van Ommen, G.B., and Sandkuyl, L.A. (1991). Mapping of facioscapulohumeral muscular dystrophy gene to chromosome 4q35-qter by multipoint linkage analysis and in situ hybridization. Genomics *9*, 570–575.

Wijmenga, C., Hewitt, J.E., Sandkuijl, L.A., Clark, L.N., Wright, T.J., Dauwerse, H.G., Gruter, A.M., Hofker, M.H., Moerer, P., and Williamson, R. (1992). Chromosome 4q DNA rearrangements associated with facioscapulohumeral muscular dystrophy. Nat. Genet. *2*, 26–30.

Winokur, S.T., Bengtsson, U., Feddersen, J., Mathews, K.D., Weiffenbach, B., Bailey, H., Markovich, R.P., Murray, J.C., Wasmuth, J.J., and Altherr, M.R. (1994). The DNA rearrangement associated with facioscapulohumeral muscular dystrophy involves a heterochromatin-associated repetitive element: implications for a role of chromatin structure in the pathogenesis of the disease. Chromosome Res. *2*, 225–234.

Winokur, S.T., Bengtsson, U., Vargas, J.C., Wasmuth, J.J., Altherr, M.R., Weiffenbach, B., and Jacobsen, S.J. (1996). The evolutionary distribution and structural organization of the homeobox-containing repeat D4Z4 indicates a functional role for the ancestral copy in the FSHD region. Hum. Mol. Genet. *5*, 1567–1575.

Winokur, S.T., Barrett, K., Martin, J.H., Forrester, J.R., Simon, M., Tawil, R., Chung, S.-A., Masny, P.S., and Figlewicz, D.A. (2003). Facioscapulohumeral muscular dystrophy (FSHD) myoblasts demonstrate increased susceptibility to oxidative stress. Neuromuscul. Disord. *13*, 322–333.

Winston, J., Duerden, L., Mort, M., Frayling, I.M., Rogers, M.T., and Upadhyaya, M. (2014). Identification of two novel SMCHD1 sequence variants in families with FSHD-like muscular dystrophy. Eur. J. Hum. Genet. *23*, 67–71

Wohlgemuth, M., van der Kooi, E.L., van Kesteren, R.G., van der Maarel, S.M., and Padberg, G.W. (2004). Ventilatory support in facioscapulohumeral muscular dystrophy. Neurology *63*, 176–178.

Wohlgemuth, M., de Swart, B.J.M., Kalf, J.G., Joosten, F.B.M., Van der Vliet, A.M., and Padberg, G.W. (2006). Dysphagia in facioscapulohumeral muscular dystrophy. Neurology *66*, 1926–1928.

Wright, T.J., Wijmenga, C., Clark, L.N., Frants, R.R., Williamson, R., and Hewitt, J.E. (1993). Fine mapping of the FSHD gene region orientates the rearranged fragment detected by the probe p13E-11. Hum. Mol. Genet. *2*, 1673–1678.

Wuebbles, R.D., Hanel, M.L., and Jones, P.L. (2009). FSHD region gene 1 (FRG1) is crucial for angiogenesis linking FRG1 to facioscapulohumeral muscular dystrophy-associated vasculopathy. Dis. Model Mech. *2*, 267–274.

Wuebbles, R.D., Long, S.W., Hanel, M.L., and Jones, P.L. (2010). Testing the effects of FSHD candidate gene expression in vertebrate muscle development. Int. J. Clin. Exp. Pathol. *3*, 386–400.

Xu, G.L., Bestor, T.H., Bourc'his, D., Hsieh, C.L., Tommerup, N., Bugge, M., Hulten, M., Qu, X., Russo, J.J., and Viegas-Pequignot, E. (1999). Chromosome instability and immunodeficiency syndrome caused by mutations in a DNA methyltransferase gene. Nature *402*, 187–191.

Yang, F., Shao, C., Vedanarayanan, V., and Ehrlich, M. (2004). Cytogenetic and immuno-FISH analysis of the 4q subtelomeric region, which is associated with facioscapulohumeral muscular dystrophy. Chromosoma *112*, 350–359.

Yao, Z., Snider, L., Balog, J., Lemmers, R.J.L.F., Van Der Maarel, S.M., Tawil, R., and Tapscott, S.J. (2014). DUX4-induced gene expression is the major molecular signature in FSHD skeletal muscle. Hum. Mol. Genet. *23*, 5342–5352.

Yehezkel, S., Segev, Y., Viegas-Pequignot, E., Skorecki, K., and Selig, S. (2008). Hypomethylation of subtelomeric regions in ICF syndrome is associated with abnormally short telomeres and enhanced transcription from telomeric regions. Hum. Mol. Genet. *17*, 2776–2789.

Young, J.M., Whiddon, J.L., Yao, Z., Kasinathan, B., Snider, L., Geng, L.N., Balog, J., Tawil, R., van der Maarel, S.M., and Tapscott, S.J. (2013). DUX4 binding to retroelements creates promoters that are active in FSHD muscle and testis. PLoS Genet. *9*, 1–19.

Zatz, M., Marie, S.K., Passos-Bueno, M.R., Vainzof, M., Campiotto, S., Cerqueira, A., Wijmenga, C., Padberg, G., and Frants, R. (1995). High proportion of new mutations and possible anticipation in Brazilian facioscapulohumeral muscular dystrophy families. Am. J. Hum. Genet. *56*, 99–105.

Zatz, M., Marie, S.K., Cerqueira, A., Vainzof, M., Pavanello, R.C., and Passos-Bueno, M.R. (1998). The facioscapulohumeral muscular dystrophy (FSHD1) gene affects males more severely and more frequently than females. Am. J. Med. Genet. *77*, 155–161.

Zeng, W., de Greef, J.C., Chen, Y.Y., Chien, R., Kong, X., Gregson, H.C., Winokur, S.T., Pyle, A., Robertson, K.D., Schmiesing, J.A., *et al.* (2009). Specific loss of histone H3 lysine 9 trimethylation and HP1gamma/cohesin binding at D4Z4 repeats is associated with facioscapulohumeral dystrophy (FSHD). PLoS Genet. *5*, 1–14.

Zeng, W., Chen, Y.Y., Newkirk, D.A., Wu, B., Balog, J., Kong, X., Ball, A.R.J., Zanotti, S., Tawil, R., Hashimoto, N., *et al.* (2014). Genetic and epigenetic characteristics of FSHD-associated 4q and 10q D4Z4 that are distinct from non-4q/10q D4Z4 homologs. Hum. Mutat. *35*, 998–1010.

The Epigenetics of Nuclear Reprogramming to Pluripotency

12

Theodore P. Rasmussen

Abstract

Recent exciting advances in cellular biology prove that it is possible to change virtually any cell type into another cell type. This is possible due to reprogramming, whereby the epigenome and transcriptome of cells is coaxed to assume the identity of another type of cell, often a pluripotent cell. In all cases, profound epigenetic changes are associated with successful reprogramming. This chapter discusses natural and artificial means of reprogramming, the chromatin-based mechanisms involved, and the problem of incomplete reprogramming and epigenetic memory. In particular, natural processes in which reprogramming is central are considered, including fertilization, preimplantation development, and reprogramming in the germ line. Artificially induced reprogramming is also considered, to include somatic cell nuclear transfer (cloning), pluripotent stem cell derivation, fusion-mediated reprogramming, and the production of induced pluripotent stem cells. Many of these methods suffer from the problem of somatic cell epigenetic memory, and this is also considered.

Introduction

All differentiated multicellular organisms are faced with a fundamental problem, namely how to establish two or more distinct cell types while using only a single fixed genome. Once development is complete, all differentiated somatic cell types contain identical DNA content, yet a large number of distinct and specialized cell types are present in each adult body plan. Different cells types are possible in multicellular organisms only because a *unique and specific subset* of the total available genes within the larger genome are expressed in each cell type. Developmental biology is replete with hundreds of examples of transcription factors being induced at key points during development that are involved in the specification of unique cell types, often in response to morphogen gradients or other external signalling mechanisms. Together, extracellular signalling and downstream cell-intrinsic responses, including the induction of tissue-specific transcription factors, collude to establish a cell type-specific transcriptome that is unique to each type of differentiated cell. In this way, a liver hepatocyte expresses a set of genes that is quite different than the genes expressed in a glial cell within the brain, yet both cell types contain identical DNA content.

Though the role of transcription factors in specifying distinct cell types is a clearly established principle, the long-term maintenance of patterns of gene expression is intimately related to chromatin structure, and chromatin constitutes the molecular machinery responsible for epigenetics. Chromatin is remarkably diverse in its chemical structure, and the molecular state of chromatin is intimately related to the transcriptional activity or silence of genes and regulatory loci with which it is physically associated. Chromatin modifications of importance include patterns of 5-methylcytosine (5-meC) bases within CpG dinucleotides (in mammalian cells especially) and post-translational modifications of any of the histones present within nucleosomes, which constitute the core subunit structure of chromatin. In general, 5-meC-modified CpG dinucleotides are associated with transcriptional silencing, and regions dense in CpG dinucleotides often occur upstream from gene promoters (CpG islands) or in DNA that is firmly inactive transcriptionally (constitutive heterochromatin). CpG islands often display tissue-specific patterns of methylation that are correlated with the transcriptional activity or silence of the genes they regulate. CpG island methylation or demethylation is therefore central to tissue-specific gene expression and the maintenance of cell type. Therefore, CpG methylation can serve as a 'lock' for cell type-specific patterns of gene expression, and just as each cell type contains a unique transcriptome, they also contain a unique methylome, which helps to stabilize the transcriptional state of cells.

A second major class of epigenetic molecular machinery is that of histone modifications. An octamer of histones (2H2A, 2H2B, 2H3 and 2H4) is assembled by non-covalent protein–protein interactions between histones, and around this is wrapped 146 bases of DNA to form the core nucleosome. A short region of non-nucleosomal spacer DNA separates each nucleosome from the next, and this can be bound by linker histone H1, which affects the overall level of chromatin compaction. Within the core nucleosome, each histone contains an N-terminal tail that is rich in lysines (K) and arginines (R), and these tails protrude from the core nucleosome, where they are subject to post-translational modification by acetylation (Ac), methylation (me), phosphorylation (P) and others. A nomenclature for histone modification was established at a conference in Brno, Czechoslovakia (Turner, 2005), in which the histone is denoted first, followed by the histone residue (numbered relative to the N-terminus), followed by the modification abbreviation and the number of such modifications at this residue, which can vary between 1 and 3 in the case of methylation. For instance, histone H3 can be modified by trimethlyation of lysine 9, and this is denoted H3K9me3. Lysines can also be acetylated (with only one acetyl group possible), and while lysines can be either acetylated or methylated, these are mutually exclusive modification events. In summary, mono-, di-, and trimethylated states of lysine and arginine residues are common, in addition to acetylation, and more rarely, phosphorylation, monoubiquitination and other more rare modifications. In total, more than 100 residue-specific modification states are possible, and when one considers the total possible combinations of nucleosome modification that are theoretically possible, the sum is a vast number, far exceeding the number of bases in the genome.

Chromatin modification is dynamically regulated in cells, even in post-mitotic cells. Histone modifications can be added and removed by histone acetyltransferases, (HATs) and histone deacetylases (HDACs) in the case of acetyl groups, and histone methyltransferases (HMTs) and demethylases (HDMs) in the case of methyl groups. Likewise, cytosine methyl groups within CpG dinucleotides can be added by *de novo* methyltransferases (notably DNMT3a/b) and re-established immediately after DNA replication by the maintenance DNA CpG methyltransferase DNMT1. Similar mechanisms for the maintenance of histone methyl and acetylation patterns after DNA replication also exist for nucleosomes, but the mechanisms are far more complex and only recently partially elucidated.

The connection between post-translational modifications of histones and transcriptional activity is more complex than the case for DNA cytosine methylation. In general, histone acetylation is associated with the transcriptional activity of associated genes. However, histone methylation is more complex. For example, H3K9me3 and H3K27me3 are transcriptionally repressive marks, while H3K4me3 promotes transcription. H3K36me3 is found in gene bodies of expressed genes and is associated with regions of active transcription. All in all, genes are assembled upon a series of nucleosomes from end to end, with specific patterns of DNA methylation and histone modification present within enhancer, promoter, and gene body regions. These patterns of chromatin vary markedly between differing cell types and help to specify tissue-specific patterns of gene expression.

This introduction briefly describes the molecular basis of epigenetics (modified chromatin). This chapter focuses in part on the epigenetics of artificial reprogramming to a state of pluripotency, which can be experimentally induced by several methods (Fig. 12.1). However, a brief consideration of natural forms of reprogramming is warranted prior to the consideration of artificially induced reprogramming methods, since many of the epigenetic reprogramming activities that operate during artificially induced reprogramming methods probably co-opt endogenous activities that normally function during development (Table 12.1).

Epigenetic reprogramming in development and germ cell maturation

During development, completely new cell types are established, and these each acquire characteristic patterns of chromatin modification (the cellular epigenome). In the adult organism (after the completion of development), each somatic cell has its own pattern of chromatin modifications. In

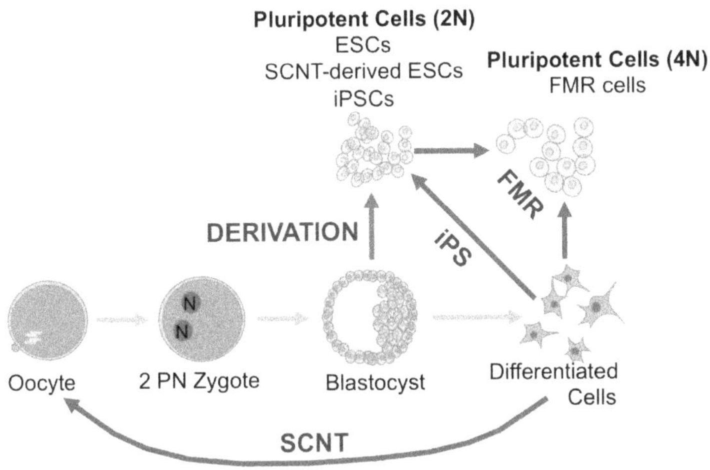

Figure 12.1 Natural forms of reprogramming, and artificial approaches in developmental context. The oocyte contains reprogramming activities that remodel the male pronucleus soon after fertilization. In addition, reprogramming of the zygotic genome occurs during subsequent divisions of preimplantation development. SCNT makes use of endogenous reprogramming activities of the oocyte, and FMR makes use of endogenous reprogramming activities of ESCs (which are very similar to the ICM of blastocysts). iPS is unique in that it does not directly utilize living cells as vehicles for the delivery of reprogramming factors. Rather, transcription factors that maintain the pluripotent state of ESCs are ectopically expressed in somatic cells, leading to their change in developmental state to one of pluripotency.

Table 12.1 Reprogramming modes

Name	Context	Description
Pre-implantation development	Natural	Resetting of zygote epigenome to achieve pluripotency
Post implantation development	Natural	Specification of tissue-specific epigenomes
Germline establishment and maturation	Natural	Resetting of epigenome for totipotency Erasure and establishment of imprints
Somatic cell nuclear transfer	Artificial	Transfer of differentiated nucleus into an oocyte Establishment of totipotency
Pluripotent stem cell derivation and culture	Artificial	Adaptation of pluripotent cells to cell culture conditions
Fusion-mediated reprogramming	Artificial	Reprogramming of epigenome between nuclei of heterokaryons
Factor-induced pluripotency	Artificial	Induction of pluripotency by defined factors

essence, multicellular life begins with a single epigenome, that is present in a single cell (the zygote), which is successively modified during the course of development as cellular lineages are specified and allocated. Therefore, development occurs in conjunction with a series of naturally occurring and carefully regulated chromatin reprogramming events, and these transitions in chromatin structure that occur in the course of developmentally regulated asymmetrical cell divisions can each be considered as cellular reprogramming events at the level of chromatin and epigenetic states. The notion that oocytes contain potent reprogramming activities has been greatly supported by experiments with nuclear transfer, that show that ooplasm can reprogram the epigenome of somatic cells (Chan *et al.*, 2012). CpG demethylation is a major reprogramming event during early embryonic development. Recently, it has been shown that the 5-meC demethylating Tet3 enzyme (leading to hydroxymethylation and later demethylation) acts robustly on the male pronucleus shortly after fertilization (Wossidlo *et al.*, 2011). Reprogramming of the histone content of the zygote genome has been directly observed for the histone variant macroH2A, which is present in oocyte chromatin, but rapidly removed upon fertilization (Chang *et al.*, 2005). MacroH2A of embryonic origin appears only later during preimplantation development.

The epigenome can be viewed as metastable, in that it can be altered at key times (during a developmental fate decision for instance), but once a terminal epigenome is established, it can be maintained through subsequent cell division in proliferative tissues, or for an exceedingly long time (essentially a lifetime) in post-mitotic cells such as most neurons. Thus, epigenetic alterations during development can be viewed as naturally induced reprogramming events, and it seems reasonable that many of the mechanisms involved in such developmental reprogramming may be co-opted during the artificial reprogramming methods that are described in subsequent sections of this chapter.

In addition to development, the composition of the epigenome changes during the course of germ line specification and during the maturation of germ cells. Therefore, it is clear that reprogramming events are a key part of reproductive biology, and the life cycles of all eukaryotes. Significant differences in the state of the epigenome are evident when the germ line and gametes of males and females are compared. Though germ cells may have earlier developmental origins, they become prominent during gonadal development after migration to the genital ridge. The H3K27-specific histone demethylases UTX has been shown to function in the germ line. This demethylases acts in PGCs at embryonic day 10.5, and loss of UTX function leads PGC failure at day 12.5 (Mansour *et al.*, 2012). After gonadal sex determination, primordial germ cells (PGCs) subsequently develop into spermatogonia, spermatids, and mature sperm (males), or oogonia, germinal vesicles, and mature oocytes (females). Imprinting is a specialized form of monoallelic gene expression whereby only one of two copies of a gene (of paternal or maternal origin) is expressed in somatic cells. Imprints are established as CpG methylation marks, and these differ male and female germ lines. Therefore, imprinted methylation marks must be erased and re-established (in a sex-specific pattern) during the course of germ cell maturation and gamete production. The establishment of imprints is essentially a reprogramming event, since an epigenetic methylation state is first removed, and then a sex-specific pattern is established prior to the production of gametes. It is likely that there is a vast amount of reprogramming of non-imprinted genes that is involved in the establishment of PGCs and their subsequent maturation in both male and female germ lines. However, since it has been difficult to identify and analyse the epigenomes of early stages of PGC development (such as embryonic germ cells of the epiblast and hindgut prior to their migration to the genital ridge), this has been difficult to assess. However, it is clear that PGCs have their own unique and plastic epigenome, which is distinct from that of mature gametes and subsequent epigenomes of differentiated cells of the conceptus.

Together, analyses of chromatin changes during the establishment of cellular lineages in development and dynamic changes of chromatin during germ line maturation demonstrate that reprogramming processes (broadly defined as identifiable changes in the epigenome) are an important and prevalent part of normal biology. The most significant contexts for massive epigenomic reprogramming events are the period following fertilization and subsequent preimplantation development during normal development, and a likely sequence of epigenetic erasure and the establishment of new epigenetic signatures during germ cell development and gametogenesis. The latter process is sexually dimorphic as evidenced by imprinting, and the vastly differing processing of chromosomes during male and female gametogenesis.

Somatic cell nuclear transfer

The oocyte can be viewed as a very large cell with a store of enzymes that have reprogramming activities, including a variety of DNA methyltransferases, HMTs, HDMs, HATs and HDACs. The presence of a highly active cocktail of reprogramming enzymes within oocytes likely exists for two important reasons: (1) the egg must rapidly confer a developmentally competent epigenetic state on the male genome, which arrives in the sperm as a highly compacted structure that is packaged mostly with protamines rather than histones and has high levels of CpG methylation; (2) the epigenomes of the male and female pronuclei differ in epigenetic content, and must be remodelled in the course of a few cell divisions to yield cleavage-stage blastomeres, which possess a 'ground state' epigenetic configuration that is functionally compatible with the condition of totipotent developmental potential that can lead to the later production of the entirety of the embryo proper, and the embryonic portion of the placenta. The notion that oocytes contain a pool of reprogramming enzymes is supported by the direct demonstration that oocyte extracts can reprogramme fibroblasts into pluripotent cells that can differentiate into all three germ layers (Bui *et al.*, 2012).

Since oocytes are the repositories of a cocktail of reprogramming factors, they have been used successfully to achieve artificial reprogramming by transferring the nuclei of somatic cells that have been made devoid of their own genetic material. This feat was first achieved with *Xenopus* oocytes, in which nuclei from somatic cells were transferred into recipient oocytes (Gurdon, 1962). The reconstructed embryos, resembling zygotes, were then allowed to develop and led to the production of live cloned offspring. The overall procedure of animal cloning is best described by the term somatic cell nuclear transfer (SCNT), and this procedure results in the massive reprogramming of differentiated somatic cell epigenomes to a state resembling that of the early embryo. Subsequently, SCNT has been successfully carried out in sheep (Wilmut *et al.*, 1997), mice (Wakayama *et al.*, 1998; Wakayama and Yanagimachi, 1999), a variety of agricultural mammals, and even in humans, where ESCs were derived from the cloned blastocysts (Chung *et al.*, 2014). The reprogramming activities of oocytes seem to be highly conserved through evolution. One study showed that *Xenopus* oocyte extracts can reactivate Oct4 in mammalian nuclei, and also cause extensive CpG demethylation events, that occur prior to transcriptional activation (Simonsson and Gurdon, 2004). In practice, SCNT procedures carried out by micromanipulation of oocytes and the nuclei of somatic cells using micropipets. In mammalian SCNT, the mitotic spindle of mature oocytes that are arrested at metaphase II (MII) is removed (a procedure that is often inaccurately deemed 'enucleation', since no nuclear envelope exists at this stage). Then, either individual nuclei, or entire somatic cells from a donor animal are injected into the oocytes, and reprogramming ensues quickly during the culture of the artificially constructed embryos. In this way, an animal with genetic content identical to the somatic cell donor can be produced, and live cloned offspring have been obtained after embryo transfer and implantation of cloned blastocysts.

From the standpoint of epigenetic reprogramming, the epigenome of a differentiated cell (often a fibroblast or cumulus cell for mammalian cloning) is converted to that of an early blastomere, probably after one or two divisions of the artificially assembled 'zygote'. Indeed, a number of studies have compared the epigenome of SCNT embryos (usually at preimplantation stages) to normal embryos. Though it is clear that the epigenome of somatic cells is greatly reprogrammed during the SCNT process, this reprogramming is often only partial, and associated with epigenetic memory of the somatic cell epigenome (Armstrong *et al.*, 2006). The problem of epigenetic memory exists for SCNT and other artificial reprogramming methods, but always involves the retention of somatic chromatin marks after the reprogramming method is completed (Fig. 12.2). Thus, SCNT reprogramming is an inefficient and often incomplete process, and only the rare transferred SCNT blastocyst develops to term and the production of live offspring. However, the fact that cloned live offspring can be recovered at all is remarkable, given the massive epigenetic reprogramming events that must occur within the transplanted somatic cell nucleus. Poor quality embryos that exhibit only partial reprogramming after SCNT often exhibit high levels of both DNA and histone methylation, reminiscent of the state in somatic donor cells (Santos *et al.*, 2003). More recent studies using epigenomic sequencing approaches have supported this earlier finding (Chan *et al.*, 2012). In addition, a study that examined the expression of the key pluripotency factor Oct4 in SCNT blastocysts found variability from clone to clone even though all were derived from genetically identical cumulus cells (Boiani *et al.*, 2005), a finding that indicates that SCNT reprogramming is an imperfect and stochastic process that is subject to variations in epigenomic integrity. Also, imprints are not always faithfully maintained in SCNT embryos, and the imprinted gene H19 was found to be biallelically expressed in many cloned cells (Suzuki *et al.*, 2011), which is clearly the result of inappropriate reprogramming events. Another study using bovine SCNT assessed the ability of oocytes to reprogramme α-satellite repeats of centromeres, which are highly methylated at CpG dinucleotides in somatic cells. This repeat retains CpG methylation (at least to some degree) and thus preserves a somatic cell epigenetic state even after the SCNT process (Couldrey and Wells, 2013). Even in mammals that survive to term, epigenetic heterogeneity exists, possibly due to incomplete reprogramming. For instance, in fibroblasts derived from the ears of cloned sows, methylation patterns of differentially methylated regions (DMRs) of imprinted genes were detected (Shen *et al.*, 2012). Failures in reprogramming also exist for histone methylation, where it has been observed that H3K27 methylation levels are higher in the ICM of cloned mouse blastocysts as compared to normal blastocysts (Zhang *et al.*, 2009). Together, these studies provide ample evidence that although SCNT can achieve the remarkable feat of the cloning of animals from somatic cells, incomplete reprogramming and reprogramming errors are commonplace.

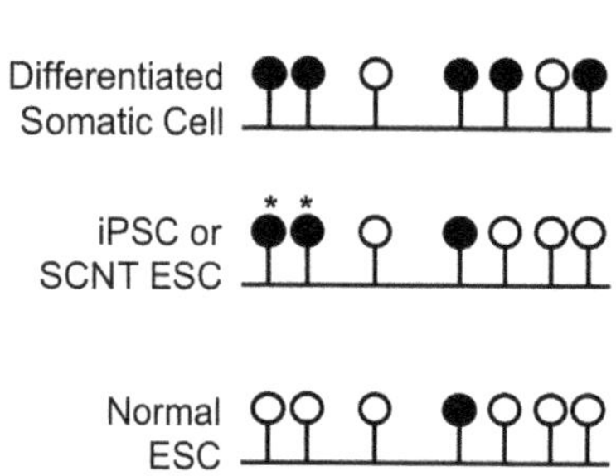

Figure 12.2 Epigenetic memory from the standpoint of chromatin. During iPS or the production of ESCs derived from SCNT embryos, the goal is to produce pluripotent cells that are identical to ESCs. However, some somatic epigenetic marks (indicated with filled circles) are retained in the final reprogrammed cells (indicated by asterisks). These can affect the expression of genes in the 'reprogrammed cells'. This diagram uses a commonly used convention to indicate CpG methylation, but, in principle, histone modifications may also be retained after reprogramming.

For the reasons just outlined, attempts have been made to artificially reset the epigenetic states of somatic cells prior to nuclear transfer. One study investigated the effects of pre-treating somatic cells with an epigenetic modifying agent prior to SCNT. In this case nuclear donor fibroblasts were pre-treated with valproic acid, a histone deacetylase inhibitor. Though hyperacetylation was observed in the treated cells prior to SCNT, this had no effect on the rates of development to the blastocyst stage *in vitro*, or to term *in vivo* (Sangalli *et al.*, 2014). However, another study showed that the differentiation state of somatic cell donor nuclei is quite important for successful cloning frequencies, where the rates of deriving ESCs from cloned embryos made by using neural stem cell nuclei are greater than that of nuclei from terminally differentiated neurons (Blelloch *et al.*, 2006). In addition, the use of nuclei from cells bearing a hypomorphic allele of *Dnmt1* improved cloning efficiency. Additional studies will be required to discern more accurately the influence of chromatin variations within somatic cell nuclei on SCNT outcomes, and the degree to which chemically and genetically induced alterations in the somatic cell epigenome can make somatic cells more amenable to the cloning process.

As outlined in the introduction of this chapter, it is clear that naturally occurring reprogramming events exist during the course of implantation development. Yet in SCNT, a somatic nucleus with its differentiated epigenome is artificially introduced into the ooplasm, which asks the oocyte to perform reprogramming on a much different substrate than the normal substrate of pronuclei. Therefore, many attempts to 'assist' the early SCNT embryo with epigenetic reprogramming have been attempted, most by exposing early SCNT preimplantation embryos to chemical agents with epigenetic modulatory activities. In one such study, an HDAC inhibitor, oxamflatin, was used to treat porcine embryos soon after SCNT. This treatment resulted in increased rates of development to blastocyst and to term upon implantation (Hou *et al.*, 2014). This study, which treated SCNT embryos soon after activation, found increased levels of H3K9Ac, and H4K4Ac, possibly acting through suppression of HCAC1 and HDAC6. In addition, DNMT1 activity and CpG methylation was decreased, showing possible cross-talk between histone acetylation and DNA methylation. OCT4 expression in the SCNT embryos was also increased. In another study, SCNT embryos were treated with Trichostatin A (TSA), another potent HDAC inhibitor (Bui *et al.*, 2010). The treated embryos exhibited improved chromosome decondensation after egg activation, increased histone acetylation, alterations in histone methylation (again indicating cross-talk between chromatin modification activities within the cell), and increased induction of transcription in SCNT preimplantation embryos. In a porcine study, an HDAC inhibitor, CBHA, caused increased rates of development to the blastocyst stage upon treatment immediately following activation of SCNT embryos, with attendant increases in histone acetylation and the expression of imprinted genes (Song *et al.*, 2014), though no improvement of development to term was noted. Reductions in repressive histone methylation levels are also correlated with improved SCNT outcomes. In one study, an inducible allele of JMJD2B (an H3K9 demethylase) was used to demethylate chromatin in ESCs harbouring this inducible allele (Antony *et al.*, 2013). Nuclei from the resulting cells exhibited increases rates of successful SCNT development to term, though H3K9me levels were quickly normalized after nuclear transfer. Reprogramming involving histone variants has been demonstrated in studies involving the histone variant macroH2A. Oocytes have little or no macroH2A, but somatic cells have high levels of this histone (Chang *et al.*, 2005). After SCNT, macroH2A of somatic origin is quickly removed from the transplanted nucleus in a process that is quite rapid, suggesting an active chromatin remodelling process during reprogramming (Chang *et al.*, 2010).

Overall, many experiments that have attempted to induce the removal of heterochromatin during the development of SCNT embryos to the blastocyst stage have had moderate to excellent success. These findings suggest that a great deal of reprogramming occurs during the cleavage stages of preimplantation embryonic development, and that the removal of silencing chromatin marks is crucial for successful reprogramming during the cloning process. Furthermore, it seems likely that extensive epigenetic reprogramming events occur during the course of normal preimplantation embryo development, and it is likely these same mechanisms that are utilized during SCNT. However, it is clear that cloning by SCNT is a highly inefficient and often incomplete process, as evidenced by low rates of development to term of cloned offspring, and the prevalence of epigenetic abnormalities in those that survive. Therefore, the reprogramming environment of the oocyte and early embryo are likely taxed to their limits after SCNT is performed.

Reprogramming in pluripotent stem cell culture

Embryonic stem cells (ESCs) are self-renewing pluripotent cells that are derived from the inner cell mass (ICM) of blastocysts. However, the epigenome of ESCs likely differs subtly from the ICM, as ESCs must undergo a cell culture adaptation process during derivation. Mouse ESCs are exquisitely pluripotent, as evidenced by their ability to contribute to the entire embryo proper after injection into a host blastocyst and implantation into a pseudopregnant recipient mother. The colony morphology of mouse ESCs is such that ESCs grow in a distinct colony that is several cells in depth. In contrast, human ESCs and induced pluripotent stem cells (iPSCs, which are described in detail in the last section of this chapter) grow in monolayer, epithelial-like colonies with distinct edges, and seem to be somewhat more restricted in their degree of pluripotency as compared to mouse ESCs. However, it has recently become possible to induce changes in human ESCs that render them more similar to naïve mouse ESCs (Gafni *et al.*, 2013). In addition, human epiblast stem cells (EpiSCs) have a unique developmental potential that differs from ESCs. Since it is possible to induce alterations in human pluripotent stem cells to convert them between these states *in vitro*, it seems possible that these conversions are indeed reprogramming events, and this notion has been borne out by assessments of the epigenomes of these related, but developmentally distinct states of pluripotent cells.

Direct physical evidence for the endogenous reprogramming capacity of ESCs exists since extracts of ESCs can induce changes in somatic cells that have much in common with epigenetic reprogramming to a pluripotent state (Collas and Taranger, 2006). In addition, the genetic material and chromatin of somatic cells can be reprogrammed upon fusion with ESCs, which is described in detail in a section below. Together, these results show that ESCs have endogenous enzymatic activities that can facilitate reprogramming events *in vitro*.

ESCs are normally derived from the ICM of blastocysts. However, it is also possible to derive ESC-like cells from the epiblast of post-implantation embryos (Bao *et al.*, 2009). In order to do this, epiblast stem cells (EpiSCs) are first derived from day 5.5 to 7.5 mouse embryos in response to LIF/Stat3 signalling. These can then be reverted to an ESC state *in vitro*, thus demonstrating that reprogramming-like events can occur during stem cell derivation and cell culture. In the case of human pluripotent cells, it has become possible to revert developmentally primed human ESCs (the usual state of most human ESC lines, which in some regards resemble EpiSCs) to developmentally naïve cells in culture with a cocktail defined signalling components (Gafni *et al.*, 2013).

The evidence shows that when pluripotent cells are derived either from the ICM or the epiblast, reprogramming-like events occur during the process of derivation and adaptation to the cell culture environment. Indeed, some of the signalling pathways and culture conditions that vary between ESCs, EpiSCs, and ground state naïve pluripotent cells have recently been elucidated, and used to produce pluripotent cells that differ in their epigenetic qualities and developmental potential. However, the intrinsic reprogramming activities resident within pluripotent cells may also contribute to their relative epigenetic instability. Especially in human ESCs, the epigenetic state is somewhat unstable in culture as evidenced by the propensity of these cells to undergo changes in the status of imprinted gene expression and X-chromosome inactivation (in female ESC lines) (Tanasijevic *et al.*, 2009; Wutz, 2012).

Fusion-mediated reprogramming

ESCs are derived from the ICM, a set of cells that are developmentally removed from oocytes by only a few cell divisions. Therefore, it seems reasonable that some biochemical reprogramming activities might be retained in ESCs. In addition, the ICM *in vivo* is poised to undergo massive epigenomic reorganizations during gastrulation in which ectoderm, endoderm, and mesoderm are established, the three principle somatic germ layers from which all tissues and organs are derived during subsequent development. For these reasons, attempts have been made to reprogram the differentiated somatic epigenome by fusion of somatic cells with ESCs. The results have been convincing, and demonstrate that potent reprogramming activities do indeed reside within ESCs.

When somatic cells such as fibroblasts are fused with ESCs or even PGCs or embryonic carcinoma cells, roughly tetraploid hybrid cells are formed that have extensive self renewal potential and a moderate to excellent degree of pluripotency (Ambrosi and Rasmussen, 2005; Serov *et al.*, 2011; Soza-Ried and Fisher, 2012). Cellular fusion to yield heterokaryons can be achieved either with polyethylene glycol (Ambrosi *et al.*, 2007), or by electrofusion (Tada and Tada, 2006a,b). Another approach involves the exposure of fusion partner cells to a fusogenic engineered measles haemagglutinin protein (Long and Rossi, 2011). Though laboratory experiments demonstrate the pluripotency of such cells, their clinical use is restricted by their tetraploid DNA content (Ambrosi and Rasmussen, 2005; Do *et al.*, 2006). However, possible approaches to produce diploid reprogrammed cells after fusion have been proposed (Pralong *et al.*, 2006).

Analysis of the transcriptomes of ESCs, fibroblasts, and tetraploid fusion cell lines indicated that many, but not all genes of the fibroblast fusion partner become reprogrammed to an ESC-like state (Ambrosi *et al.*, 2007). Though not all fibroblast genes are reprogrammed by this strategy, key pluripotency genes, including those that encode ESC-specific transcription factors are among those that are reprogrammed (Ambrosi *et al.*, 2007). However, it is clear that the somatic cell epigenome exhibits some degree of epigenetic memory after the FMR process is completed (Choi and Do, 2010). Another interesting study fused embryonic germ cells with somatic cells (thymic cells) to investigate the intrinsic reprogramming activities of germ cells at the level of epigenetics. Interestingly, many genes, including imprinted genes became demethylated in the hybrids, and this resulted in derepression and transcription of genes from thymic cell chromosomes that are silent in normal thymic cells (Tada *et al.*, 1997). This system, which uses embryonic germ cells may constitute a cell culture model for endogenous reprogramming in the germ line. The factors involved in FMR have not been well characterized, but in one study it was shown that Nanog (a key pluripotency transcription factor) participates in the reprogramming process and conversion of the neural stem cell epigenome upon fusion with ESCs (Silva *et al.*, 2006). In addition, extracts prepared from ESCs have some ability to alter the epigenome of somatic cells exposed to them (Collas and Taranger, 2006).

An interesting feature of all reprogrammed fusion cells is that the genetic characteristics of one of the two fusion partners is dominant over the other. In fusions of fibroblasts and ESCs, the ESC transcriptional program is mostly (but not completely) dominant (Ambrosi *et al.*, 2007). Another study showed that when C2C12 myotube cells are fused with lymphocytes, the muscle-specific pattern of gene expression is dominant. These experiments also showed that activation and repression of genes in the hybrids are separable events. Interestingly, the inhibition of HDAC function prevents the silencing of lymphocyte-specific genes (Terranova *et al.*, 2006). The dominance effect is also observed even when two cells with only restricted developmental potential are fused. For instance, in fusions between human fibroblasts and fetal haematopoietic progenitor cells ($CD34^+$ cells present in the fetal liver), the resulting hybrids resemble the fetal haematopoietic progenitor cells most closely (Sandler *et al.*, 2011). In short, several studies have shown that the transcriptional characteristics of one cell type is dominant over the other, though the dominant cell type is usually the most developmental primitive of the two. The cause of this dominance is unknown, but may be a reflection of the relative 'power' of the transcription factors present in one cell type as compared to the other.

FMR is a reprogramming method that has largely been supplanted by other methods, likely because of the roughly tetraploid content of the reprogrammed cells, a ploidy that is incompatible with future stem cell therapies. However, it is a scientifically interesting method, because it shows that ESC, like oocytes, contain potent reprogramming activities leading to pluripotency. The oocyte's reprogramming activities likely exist because massive epigenetic changes must be performed during the course of normal preimplantation development. In an analogous fashion, the ESC (like the ICM) is poised to differentiate into cells of all three germ layers, in processes that likely recapitulate aspects of gastrulation. Thus, the ESC likely harbours reprogramming factors that ordinarily are necessary to prime the execution of tissue-specific patterns of gene expression, and it may be these factors that participate in the FMR process.

Factor-induced reprogramming

In 2006 Shinya Yamanaka and coworkers showed that a somatic cell such as a skin fibroblast could be directly converted to a cell type nearly identical to ESCs (Takahashi and Yamanaka, 2006). These were deemed induced pluripotent stem cells (iPSCs), and this achievement was recognized by the award of the Nobel Prize in 2012, along with recognition of the work of John Gurdon, the first person to clone a vertebrate. To achieve this feat, powerful master transcription factors that normally are expressed in ESCs and the ICM were expressed ectopically in skin fibroblasts after retroviral delivery. These factors were Oct4, Sox2, Klf4, and c-Myc (often abbreviated OSKM). In particular, Oct4 and Sox2 (which collectively transactivate thousands of genes in ESCs) are involved in the maintenance of the pluripotency-related transcriptional state in ESCs. Thus, the expression of these key transcription factors in fibroblasts was sufficient to convert (reprogram) them to an early developmental state, with a transcriptome and an epigenome nearly identical to that of ESCs (Wernig *et al.*, 2007). Whether transcriptional reprogramming precedes epigenetic reprogramming, or vice versa, once established, patterns of chromatin modification can be retained for numerous cell divisions with excellent fidelity. In addition, it has been possible to produce iPS cells by using small molecules rather than the expression of iPS factors from genetic constructs. These molecules likely inhibit enzymes that otherwise would block reprogramming. In one such example, mouse iPSCs were produced by adding a combination of seven small molecules that were identified using a chemical screen utilizing an Oct4-GFP reporter gene (Hou *et al.*, 2013). In a variation on induced pluripotency, EpiSCs can be generated using a standard set of iPS factors for reprogramming of fibroblasts, followed by culture under EpiSC conditions instead of ESC conditions (Han *et al.*, 2011). There have been other successes using factor-based reprogramming strategies in which one differentiated somatic cell type is directly transdifferentiated into another. In short, it is now clear that in theory any cell type can be morphed into any other, so long as an intact genome is present. Therefore, the epigenome must possess the property of metastability, in that it is sufficiently stable to preserve cellular identity in the absence of major external signals, but it is sufficiently malleable that reprogramming to other cell types (or even cell types that do not occur *in vivo*) is possible.

Ultimately, a key step in iPS reprogramming is the activation of key genes encoding pluripotency transcription factors including Oct4, Sox2, Klf4, Nanog, and others. Interestingly, an early step in Nanog derepression during iPS is mediated by poly (ADP-ribose) polymerase I (Parp1) and ten-eleven tranlocation-2 (Tet2), which are both involved in modifying histones and 5-meC, respectively (Doege *et al.*, 2012). These enzymes are recruited to the *Nanog* locus only 4 days after iPS is initiated by OSKM, yet the entire iPS procedure when using mouse cells takes about 2 weeks. Thus, it appears that erasure of somatic chromatin marks precedes transcriptional activation of key genes such as *Nanog*. Other chromatin remodelling enzymes have been implicated in the iPS reprogramming process. An shRNA screen that knocked down the expression of candidate factors shows that reduction of polycomb repressive complex 1 and 2 components (PRC1/2) reduced reprogramming efficiency, indicating that H3K27 methylation and histone ubiquitination activities (normally performed by PRC1/2 complexes) is important for the iPS process to proceed efficiently (Onder *et al.*, 2012).

In many regards, microRNA function resembles classical epigenetic pathways in that gene expression can be altered (by microRNA activity) independently of DNA sequence, based on the availability of expressed microRNAs in a specific cell type. In keeping with this concept, it has been found that some microRNA sets enhance iPS reprogramming. For instance miR 302–367 cluster miRNAs enhance iPS, likely by accelerating a transition from a mesenchymal to epithelial state (the reverse of normal development), possibly by inducing the expression of E-cadherins, which are normally expressed in pluripotent cells and involved in cell–cell contacts that are required for the maintenance of pluripotency (Liao *et al.*, 2011). Numerous other examples of the involvement of microRNAs in the establishment and exit from pluripotency exist.

In contrast, some chromatin-modifying

factors inhibit the iPS reprogramming process. For instance, down-regulation of the nucleosome remodelling and deacetylation complex (NuRD) is necessary for high efficiency iPS reprogramming. This has been shown by manipulating 5-meC-binding protein 3 (Mbd3), which when overexpressed blocks iPS, likely by stabilizing CpG methylation at key genomic loci including the regulatory regions of pluripotency genes (Luo *et al.*, 2013). Another example is that of the *Ink4/Arf* locus, which contains cyclin-dependent kinase genes. These genes are robustly expressed in most somatic differentiated cells, but are silenced in both ESCs and iPSCs, where they are marked by bivalent domains (both H3K4me3 and H3K27me3), which silences genes, but renders them poised for later developmentally regulated activation. Knock-down of the *Ink4/Arf* genes greatly facilitates iPS reprogramming efficiency (Li *et al.*, 2009). Other chromatin remodelling factors have also been shown to inhibit iPS reprogramming, including SUV39H1, DOT1L, and YY1. In fact, inhibition of DOT1L, an H3K79 HMT protein can functionally substitute for KLF4 and c-Myc in the iPS procedure (Onder *et al.*, 2012). Yet another study found that experimentally induced deficiency of the histone variant macroH2A1 or both macroH2A1 and macroH2A2 can increase iPS reprogramming efficiency, which suggests that these histone variants (which are abundantly expressed in differentiated cells and substitute for histone H2A in specialized nucleosomes) act as a barrier to reprogramming (Gaspar-Maia *et al.*, 2013; Pasque *et al.*, 2012). These examples show that some epigenetic modifying enzymes are barriers to reprogramming, and their inhibition (either intentionally or spontaneously during the course of iPS reprogramming) leads to successful outcomes in the iPS reprogramming procedure. The chromatin of ESCs and iPSCs is in general reduced in the content of heterochromatin. In one study that used mass spectrometry, it was found that H3K9 methylation (a marker of constitutive heterochromatin) is reduced in iPSCs. Heterochromatin protein 1 (HP1) binds methylated forms of H3K9, and depletion of HP1 increased iPS efficiency (Sridharan *et al.*, 2013). The results outlined above suggest that a general principle with regards to epigenetic changes that are involved in iPS reprogramming is the removal of heterochromatin patterns that are present in differentiated somatic cells. Overall, it is clear that during the iPS reprogramming process, certain factors and pathways assist in the reprogramming process, while others are inhibitory (Fig. 12.3). Successful iPS likely is possible because pro-reprogramming activities are able to overcome inhibitory mechanisms during the reprogramming process.

After iPS-mediated reprogramming is complete, the resulting iPSCs resemble ESCs in many ways. Analyses of the methylome of iPSCs reveal that overall the CpG methylation patterns of iPS cells are vastly different from those of differentiated somatic cells from which they are derived (Doi *et al.*, 2009). CpG methylation patterns of iPS cells closely resemble ESCs as judged by a genome-wide study that compared the methylome of human iPSCs derived from human fibroblasts or endothelial cells to human ESCs (Lagarkova *et al.*, 2010). Female XX somatic cells contain a single, stably inactivated X-chromosome. A recent iPS study showed that female cells subjected to iPS acquire two active X-chromosomes (which resembles that pre-X inactivation status of the ICM and female mouse ESC lines), and then maintain this state in some instances, or undergo spontaneous X-chromosome inactivation after continued cell culture (Kim *et al.*, 2014). Since the inactive X-chromosome is a well-established model for heterochromatin, its reactivation is an excellent demonstration of reprogramming of heterochromatin to euchromatin, but in female iPSC and ESCs (especially human lines) this state is unstable, for unknown reasons. The kinetics of epigenetic reprogramming in iPSCs have also been investigated. In another study, histone modification profiles were assessed in iPS cells. It was found that H3K4me2 was rapidly reprogrammed in nascent iPS cells, while H3K27me3 kinetics were later, and the chromatin changes preceded changes in transcriptional profiles (Koche *et al.*, 2011).

Though the epigenome of iPSCs broadly resembles that of ESCs, subtle but potentially important differences occur that may arise from incomplete or inappropriate reprogramming. Many of these differences seem to stem from

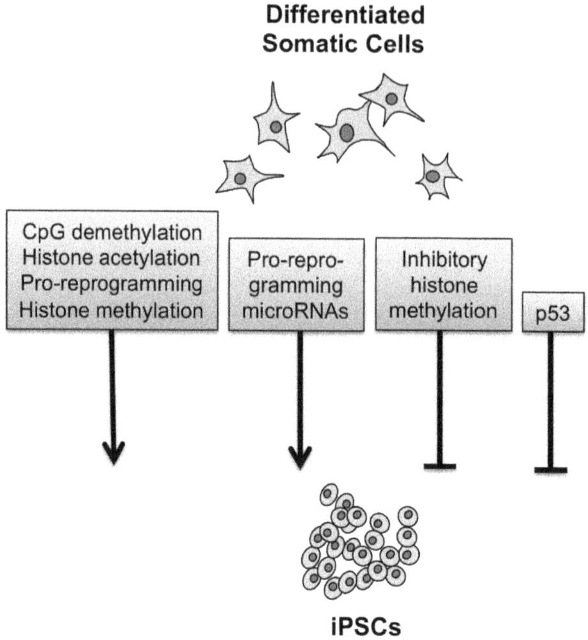

Figure 12.3 Pro- and anti-reprogramming activities within somatic cells. Somatic cells in general have a higher content of heterochromatin than pluripotent cells. Therefore, it has been found that activities that promote CpG demethylation, histone acetylation, and the removal of transcriptionally repressive histone modifications all participate in the reprogramming process. In addition, certain microRNAs are involved in maintaining the pluripotent state. In contrast, certain histone modifications (such as H3K9me3) strongly inhibit reprogramming. In addition, p53 expression inhibits reprogramming by complex means likely involved in monitoring the integrity of the genome.

epigenetic memory, which is likely caused by the retention of chromatin modification states that were originally present in the differentiated somatic cells prior to their reprogramming. A comprehensive study of the methylation pattern of 22 human iPSC lines derived from multiple somatic cell types showed that some methylation differences do exist in comparisons between early passage human iPSC and ESC cell lines (Nishino *et al.*, 2011). However, this study also showed that methylation patterns in iPSC lines became more similar to those of ESCs with continued cell culture, suggesting that reprogramming associated with adaptation to cell culture conditions exists. The iPS procedure, like other reprogramming methods, can often be incomplete. 'Epigenetic memory' has been detected in iPS cell lines that harbour residual traces of 5-meC at loci that are methylated in various somatic cell types used for iPS (Kim *et al.*, 2010).

Residual CpG methylation traces have been found in several iPSC lines derived from different somatic cell sources. One study found that nine genes were consistently aberrantly methylated in iPSCs regardless of the type of somatic cell from which they were derived (Ruiz *et al.*, 2012). Another example of epigenetic memory in iPSCs is exemplified in a study that showed that higher levels of CpG methylation in iPSCs exist at key loci, and in DNA as a whole (Zhou *et al.*, 2013). In another study that addressed the transcriptional consequences of incomplete reprogramming, human iPSCs were derived from cells of endoderm, ectoderm, and mesoderm. All the resulting low passage iPSCs exhibited some degree of transcriptional memory reminiscent of the somatic cell from which they were derived (Ohi *et al.*, 2011). It has also been shown that microRNAs can contribute to the transcriptional memory of gene expression patterns in iPSCs that are remnants of transcriptional patterns within somatic cells (Vitaloni *et al.*, 2014). Imprinted genes are expressed monoallelically in somatic cells and the imprinted loci exhibit heterozygous patterns of CpG methylation. For the most part, imprints are maintained during iPS-mediated reprogramming, but some errors do occur, usually resulting in biallelic expression of imprinted genes with corresponding loss of controlling methylation patterns (Hiura *et al.*, 2013). Lastly, transposable retroelements present in the genome become activated during iPS, likely by the formation of transcriptionally permissive chromatin at their regulatory elements (Friedli *et al.*, 2014). This finding suggests that there is some risk of the activation of endogenous mutagenesis mechanisms during the production of iPSCs.

In considering the rich set of research that focuses on the epigenetics of iPSCs, it is clear that iPS reprogramming is intricately entwined with the modification of chromatin, and enzymes with epigenetic modifying activities have been implicated in iPS reprogramming. Indeed, some epigenetic enzymes have been found which participate in iPS reprogramming, and others have been found to be inhibitory. In terms of the fidelity of the iPS reprogramming procedure, overall, the extent and fidelity is quite good. However, there are some convincing examples of the memory of somatic chromatin and transcriptional states of some genes. Furthermore, imprinted gene expression is only quasi-stable in iPSCs as are X-chromosome inactivation states. iPSCs are promising because they can be made from human subjects with specific genetic disorders, thus leading to the development of 'disease in a dish' cell culture models of human diseases. However, should they be used as sources of differentiated cells for stem cell therapy, caution is indicated because of subtly altered epigenetic profiles, which could lead to undesirable cellular behaviour including the activation of immunogenic transplantation antigens and the production of neoplastic cells.

Conclusions

This chapter shows that nuclear reprogramming is intimately associated with changes in epigenetic state that are rooted in chromatin structure. In addition, it is clear that reprogramming-like events occur in normal biology, to include epigenetic changes during cellular lineage allocation during developmental biology, and management of the germ line through successive generations of organisms. Reprogramming also occurs during the initial stages of preimplantation development, prior to the formation of the epiblast. Similar reprogramming events can be induced artificially by SCNT, FMR, and iPS, all methods that convert somatic cells to a pluripotent state. Recently, there have also been successes in the direct conversion of one type of somatic cell into another, without transiting through a pluripotent state and subsequent differentiation. These successes are transdifferentiation events. A collective analysis of all these cell type conversion events now makes it clear that any cell type with a complete and intact genome can be converted into any other by epigenetic means, provided proper induction strategies are employed.

Acknowledgement

This review was supported in part by an award from the Connecticut State Stem Cell Research Program, grant number 11SCDIS02.

References

Ambrosi, D.J., and Rasmussen, T.P. (2005). Reprogramming mediated by stem cell fusion. J Cell. Mol. Med. 9, 320–330.

Ambrosi, D.J., Tanasijevic, B., Kaur, A., Obergfell, C., O'Neill, R.J., Krueger, W., and Rasmussen, T.P. (2007). Genome-wide reprogramming in hybrids of somatic cells and embryonic stem cells. Stem Cells 25, 1104–1113.

Antony, J., Oback, F., Chamley, L.W., Oback, B., and Laible, G. (2013). Transient JMJD2B-mediated reduction of H3K9me3 levels improves reprogramming of embryonic stem cells into cloned embryos. Mol. Cell. Biol. 33, 974–983.

Armstrong, L., Lako, M., Dean, W., and Stojkovic, M. (2006). Epigenetic modification is central to genome reprogramming in somatic cell nuclear transfer. Stem Cells 24, 805–814.

Bao, S., Tang, F., Li, X., Hayashi, K., Gillich, A., Lao, K., and Surani, M.A. (2009). Epigenetic reversion of post-implantation epiblast to pluripotent embryonic stem cells. Nature *461*, 1292–1295.

Blelloch, R., Wang, Z., Meissner, A., Pollard, S., Smith, A., and Jaenisch, R. (2006). Reprogramming efficiency following somatic cell nuclear transfer is influenced by the differentiation and methylation state of the donor nucleus. Stem Cells 24, 2007–2013.

Boiani, M., Gentile, L., Gambles, V.V., Cavaleri, F., Redi, C.A., and Scholer, H.R. (2005). Variable reprogramming of the pluripotent stem cell marker Oct4 in mouse clones: distinct developmental potentials in different culture environments. Stem Cells *23*, 1089–1104.

Bui, H.T., Wakayama, S., Kishigami, S., Park, K.K., Kim, J.H., Thuan, N.V., and Wakayama, T. (2010). Effect of trichostatin A on chromatin remodeling, histone modifications, DNA replication, and transcriptional activity in cloned mouse embryos. Biol. Reprod. *83*, 454–463.

Bui, H.T., Kwon, D.N., Kang, M.H., Oh, M.H., Park, M.R., Park, W.J., Paik, S.S., Van Thuan, N., and Kim, J.H. (2012). Epigenetic reprogramming in somatic cells induced by extract from germinal vesicle stage pig oocytes. Development *139*, 4330–4340.

Chan, M.M., Smith, Z.D., Egli, D., Regev, A., and Meissner, A. (2012). Mouse ooplasm confers context-specific reprogramming capacity. Nat. Genet. *44*, 978–980.

Chang, C.C., Ma, Y., Jacobs, S., Tian, X.C., Yang, X., and Rasmussen, T.P. (2005). A maternal store of macroH2A is removed from pronuclei prior to onset of somatic macroH2A expression in preimplantation embryos. Dev. Biol. *278*, 367–380.

Chang, C.C., Gao, S., Sung, L.Y., Corry, G.N., Ma, Y., Nagy, Z.P., Tian, X.C., and Rasmussen, T.P. (2010). Rapid elimination of the histone variant MacroH2A from somatic cell heterochromatin after nuclear transfer. Cell. Reprogram. *12*, 43–53.

Choi, H.W., and Do, J.T. (2010). Kinetics of reprogramming in cell fusion hybrids. Int. J. Dev. Biol. *54*, 1697–1702.

Chung, Y.G., Eum, J.H., Lee, J.E., Shim, S.H., Sepilian, V., Hong, S.W., Lee, Y., Treff, N.R., Choi, Y.H., Kimbrel, E.A., *et al.* (2014). Human somatic cell nuclear transfer using adult cells. Cell Stem Cell *14*, 777–780.

Collas, P., and Taranger, C.K. (2006). Epigenetic reprogramming of nuclei using cell extracts. Stem Cell Rev. *2*, 309–317.

Couldrey, C., and Wells, D.N. (2013). DNA methylation at a bovine alpha satellite I repeat CpG site during development following fertilization and somatic cell nuclear transfer. PLoS ONE *8*, e55153.

Do, J.T., Han, D.W., and Scholer, H.R. (2006). Reprogramming somatic gene activity by fusion with pluripotent cells. Stem Cell Rev. *2*, 257–264.

Doege, C.A., Inoue, K., Yamashita, T., Rhee, D.B., Travis, S., Fujita, R., Guarnieri, P., Bhagat, G., Vanti, W.B., Shih, A., *et al.* (2012). Early-stage epigenetic modification during somatic cell reprogramming by Parp1 and Tet2. Nature *488*, 652–655.

Doi, A., Park, I.H., Wen, B., Murakami, P., Aryee, M.J., Irizarry, R., Herb, B., Ladd-Acosta, C., Rho, J., Loewer, S., *et al.* (2009). Differential methylation of tissue- and cancer-specific CpG island shores distinguishes human induced pluripotent stem cells, embryonic stem cells and fibroblasts. Nat. Genet. *41*, 1350–1353.

Friedli, M., Turelli, P., Kapopoulou, A., Rauwel, B., Castro-Diaz, N., Rowe, H.M., Ecco, G., Unzu, C., Planet, E., Lombardo, A., *et al.* (2014). Loss of transcriptional control over endogenous retroelements during reprogramming to pluripotency. Genome Res. *24*, 1251–1259.

Gafni, O., Weinberger, L., Mansour, A.A., Manor, Y.S., Chomsky, E., Ben-Yosef, D., Kalma, Y., Viukov, S., Maza, I., Zviran, A., *et al.* (2013). Derivation of novel human ground state naive pluripotent stem cells. Nature *504*, 282–286.

Gaspar-Maia, A., Qadeer, Z.A., Hasson, D., Ratnakumar, K., Leu, N.A., Leroy, G., Liu, S., Costanzi, C., Valle-Garcia, D., Schaniel, C., *et al.* (2013). MacroH2A histone variants act as a barrier upon reprogramming towards pluripotency. Nat. Commun. *4*, 1565.

Gurdon, J.B. (1962). The developmental capacity of nuclei taken from intestinal epithelium cells of feeding tadpoles. J. Embryol. Exp. Morphol. *10*, 622–640.

Han, D.W., Greber, B., Wu, G., Tapia, N., Arauzo-Bravo, M.J., Ko, K., Bernemann, C., Stehling, M., and Scholer, H.R. (2011). Direct reprogramming of fibroblasts into epiblast stem cells. Nat. Cell Biol. *13*, 66–71.

Hiura, H., Toyoda, M., Okae, H., Sakurai, M., Miyauchi, N., Sato, A., Kiyokawa, N., Okita, H., Miyagawa, Y., Akutsu, H., *et al.* (2013). Stability of genomic imprinting in human induced pluripotent stem cells. BMC Genetics *14*, 32.

Hou, L., Ma, F., Yang, J., Riaz, H., Wang, Y., Wu, W., Xia, X., Ma, Z., Zhou, Y., Zhang, L., *et al.* (2014). Effects of histone deacetylase inhibitor oxamflatin on *in vitro* porcine somatic cell nuclear transfer embryos. Cell. Reprogram. *16*, 253–265.

Hou, P., Li, Y., Zhang, X., Liu, C., Guan, J., Li, H., Zhao, T., Ye, J., Yang, W., Liu, K., *et al.* (2013). Pluripotent stem cells induced from mouse somatic cells by small-molecule compounds. Science *341*, 651–654.

Kim, K., Doi, A., Wen, B., Ng, K., Zhao, R., Cahan, P., Kim, J., Aryee, M.J., Ji, H., Ehrlich, L.I., *et al.* (2010). Epigenetic memory in induced pluripotent stem cells. Nature *467*, 285–290.

Kim, K.Y., Hysolli, E., Tanaka, Y., Wang, B., Jung, Y.W., Pan, X., Weissman, S.M., and Park, I.H. (2014). X-chromosome of female cells shows dynamic changes in status during human somatic cell reprogramming. Stem Cell Reports *2*, 896–909.

Koche, R.P., Smith, Z.D., Adli, M., Gu, H., Ku, M., Gnirke, A., Bernstein, B.E., and Meissner, A. (2011). Reprogramming factor expression initiates widespread targeted chromatin remodeling. Cell Stem Cell *8*, 96–105.

Lagarkova, M.A., Shutova, M.V., Bogomazova, A.N., Vassina, E.M., Glazov, E.A., Zhang, P., Rizvanov, A.A., Chestkov, I.V., and Kiselev, S.L. (2010). Induction of pluripotency in human endothelial cells resets epigenetic profile on genome scale. Cell Cycle *9*, 937–946.

Li, H., Collado, M., Villasante, A., Strati, K., Ortega, S., Canamero, M., Blasco, M.A., and Serrano, M. (2009). The Ink4/Arf locus is a barrier for iPS cell reprogramming. Nature *460*, 1136–1139.

Liao, B., Bao, X., Liu, L., Feng, S., Zovoilis, A., Liu, W., Xue, Y., Cai, J., Guo, X., Qin, B., *et al.* (2011). MicroRNA cluster 302–367 enhances somatic cell reprogramming by accelerating a mesenchymal-to-epithelial transition. J. Biol. Chem. *286*, 17359–17364.

Long, M.A., and Rossi, F.M. (2011). Targeted cell fusion facilitates stable heterokaryon generation *in vitro* and *in vivo*. PLoS ONE *6*, e26381.

Luo, M., Ling, T., Xie, W., Sun, H., Zhou, Y., Zhu, Q., Shen, M., Zong, L., Lyu, G., Zhao, Y., *et al.* (2013). NuRD blocks reprogramming of mouse somatic cells into pluripotent stem cells. Stem Cells *31*, 1278–1286.

Mansour, A.A., Gafni, O., Weinberger, L., Zviran, A., Ayyash, M., Rais, Y., Krupalnik, V., Zerbib, M., Amann-Zalcenstein, D., Maza, I., *et al.* (2012). The H3K27 demethylase Utx regulates somatic and germ cell epigenetic reprogramming. Nature *488*, 409–413.

Nishino, K., Toyoda, M., Yamazaki-Inoue, M., Fukawatase, Y., Chikazawa, E., Sakaguchi, H., Akutsu, H., and Umezawa, A. (2011). DNA methylation dynamics in human induced pluripotent stem cells over time. PLoS Genet. *7*, e1002085.

Ohi, Y., Qin, H., Hong, C., Blouin, L., Polo, J.M., Guo, T., Qi, Z., Downey, S.L., Manos, P.D., Rossi, D.J., *et al.* (2011). Incomplete DNA methylation underlies a transcriptional memory of somatic cells in human iPS cells. Nat. Cell Biol. *13*, 541–549.

Onder, T.T., Kara, N., Cherry, A., Sinha, A.U., Zhu, N., Bernt, K.M., Cahan, P., Marcarci, B.O., Unternaehrer, J., Gupta, P.B., *et al.* (2012). Chromatin-modifying enzymes as modulators of reprogramming. Nature *483*, 598–602.

Pasque, V., Radzisheuskaya, A., Gillich, A., Halley-Stott, R.P., Panamarova, M., Zernicka-Goetz, M., Surani, M.A., and Silva, J.C. (2012). Histone variant macroH2A marks embryonic differentiation *in vivo* and acts as an epigenetic barrier to induced pluripotency. J. Cell Sci. *125*, 6094–6104.

Pralong, D., Trounson, A.O., and Verma, P.J. (2006). Cell fusion for reprogramming pluripotency: toward elimination of the pluripotent genome. Stem Cell Rev. *2*, 331–340.

Ruiz, S., Diep, D., Gore, A., Panopoulos, A.D., Montserrat, N., Plongthongkum, N., Kumar, S., Fung, H.L., Giorgetti, A., Bilic, J., *et al.* (2012). Identification of a specific reprogramming-associated epigenetic signature in human induced pluripotent stem cells. Proc. Natl. Acad. Sci. U.S.A. *109*, 16196–16201.

Sandler, V.M., Lailler, N., and Bouhassira, E.E. (2011). Reprogramming of embryonic human fibroblasts into fetal hematopoietic progenitors by fusion with human fetal liver CD34+ cells. PLoS ONE *6*, e18265.

Sangalli, J.R., Chiaratti, M.R., De Bem, T.H., de Araujo, R.R., Bressan, F.F., Sampaio, R.V., Perecin, F., Smith, L.C., King, W.A., and Meirelles, F.V. (2014). Development to term of cloned cattle derived from donor cells treated with valproic Acid. PLoS ONE *9*, e101022.

Santos, F., Zakhartchenko, V., Stojkovic, M., Peters, A., Jenuwein, T., Wolf, E., Reik, W., and Dean, W. (2003). Epigenetic marking correlates with developmental potential in cloned bovine preimplantation embryos. Curr. Biol. *13*, 1116–1121.

Serov, O.L., Matveeva, N.M., and Khabarova, A.A. (2011). Reprogramming mediated by cell fusion technology. Int. Rev. Cell Mol. Biol. *291*, 155–190.

Shen, C.J., Cheng, W.T., Wu, S.C., Chen, H.L., Tsai, T.C., Yang, S.H., and Chen, C.M. (2012). Differential differences in methylation status of putative imprinted genes among cloned swine genomes. PLoS ONE *7*, e32812.

Silva, J., Chambers, I., Pollard, S., and Smith, A. (2006). Nanog promotes transfer of pluripotency after cell fusion. Nature *441*, 997–1001.

Simonsson, S., and Gurdon, J. (2004). DNA demethylation is necessary for the epigenetic reprogramming of somatic cell nuclei. Nat. Cell Biol. *6*, 984–990.

Song, Y., Hai, T., Wang, Y., Guo, R., Li, W., Wang, L., and Zhou, Q. (2014). Epigenetic reprogramming, gene expression and *in vitro* development of porcine SCNT embryos are significantly improved by a histone deacetylase inhibitor – m-carboxycinnamic acid bishydroxamide (CBHA). Protein Cell *5*, 382–393.

Soza-Ried, J., and Fisher, A.G. (2012). Reprogramming somatic cells towards pluripotency by cellular fusion. Curr. Opin. Genet. Dev. *22*, 459–465.

Sridharan, R., Gonzales-Cope, M., Chronis, C., Bonora, G., McKee, R., Huang, C., Patel, S., Lopez, D., Mishra, N., Pellegrini, M., *et al.* (2013). Proteomic and genomic approaches reveal critical functions of H3K9 methylation and heterochromatin protein-1gamma in reprogramming to pluripotency. Nat. Cell Biol. *15*, 872–882.

Suzuki, J., Jr., Therrien, J., Filion, F., Lefebvre, R., Goff, A.K., Perecin, F., Meirelles, F.V., and Smith, L.C. (2011). Loss of methylation at H19 DMD is associated with biallelic expression and reduced development in cattle derived by somatic cell nuclear transfer. Biol. Reprod. *84*, 947–956.

Tada, M., and Tada, T. (2006a). Epigenetic reprogramming of somatic genomes by electrofusion with embryonic stem cells. Methods Mol. Biol. *325*, 67–79.

Tada, M., and Tada, T. (2006b). Nuclear reprogramming of somatic nucleus hybridized with embryonic stem cells by electrofusion. Methods Mol. Biol. *329*, 411–420.

Tada, M., Tada, T., Lefebvre, L., Barton, S.C., and Surani, M.A. (1997). Embryonic germ cells induce epigenetic reprogramming of somatic nucleus in hybrid cells. EMBO J. *16*, 6510–6520.

Takahashi, K., and Yamanaka, S. (2006). Induction of pluripotent stem cells from mouse embryonic and adult fibroblast cultures by defined factors. Cell *126*, 663–676.

Tanasijevic, B., Dai, B., Ezashi, T., Livingston, K., Roberts, R.M., and Rasmussen, T.P. (2009). Progressive accumulation of epigenetic heterogeneity during human ES cell culture. Epigenetics *4*, 330–338.

Terranova, R., Pereira, C.F., Du Roure, C., Merkenschlager, M., and Fisher, A.G. (2006). Acquisition and extinction of gene expression programs are separable events in heterokaryon reprogramming. J. Cell Sci. *119*, 2065–2072.

Turner, B.M. (2005). Reading signals on the nucleosome with a new nomenclature for modified histones. Nat. Struct. Mol. Biol. *12*, 110–112.

Vitaloni, M., Pulecio, J., Bilic, J., Kuebler, B., Laricchia-Robbio, L., and Izpisua Belmonte, J.C. (2014). MicroRNAs contribute to induced pluripotent stem cell somatic donor memory. J. Biol. Chem. *289*, 2084–2098.

Wakayama, T., and Yanagimachi, R. (1999). Cloning of male mice from adult tail-tip cells. Nat. Genet. *22*, 127–128.

Wakayama, T., Perry, A.C., Zuccotti, M., Johnson, K.R., and Yanagimachi, R. (1998). Full-term development of mice from enucleated oocytes injected with cumulus cell nuclei. Nature *394*, 369–374.

Wernig, M., Meissner, A., Foreman, R., Brambrink, T., Ku, M., Hochedlinger, K., Bernstein, B.E., and Jaenisch, R. (2007). *In vitro* reprogramming of fibroblasts into a pluripotent ES-cell-like state. Nature *448*, 318–324.

Wilmut, I., Schnieke, A.E., McWhir, J., Kind, A.J., and Campbell, K.H. (1997). Viable offspring derived from fetal and adult mammalian cells. Nature *385*, 810–813.

Wossidlo, M., Nakamura, T., Lepikhov, K., Marques, C.J., Zakhartchenko, V., Boiani, M., Arand, J., Nakano, T., Reik, W., and Walter, J. (2011). 5-Hydroxymethylcytosine in the mammalian zygote is linked with epigenetic reprogramming. Nat. Commun. *2*, 241.

Wutz, A. (2012). Epigenetic alterations in human pluripotent stem cells: a tale of two cultures. Cell Stem Cell *11*, 9–15.

Zhang, M., Wang, F., Kou, Z., Zhang, Y., and Gao, S. (2009). Defective chromatin structure in somatic cell cloned mouse embryos. J. Biol. Chem. *284*, 24981–24987.

Zhou, W., Wang, K., Ruan, W., Bo, Z., Liu, L., Cao, Z., Chai, L., and Cao, G. (2013). Higher methylation in genomic DNA indicates incomplete reprogramming in induced pluripotent stem cells. Cell. Reprogram. *15*, 92–99.

Emerging Role of the Guanine-quadruplex DNA Secondary Structure in Epigenetics

13

Aradhita Baral, Dhurjhoti Saha and Shantanu Chowdhury

Abstract

Chromosomal DNA packaged as chromatin controls an intriguing variety of biological functions. Notably, a largely similar template of DNA sequence, comprising sequential arrangement of the four nucleic acid moieties, manages these functions. This is striking. In addition, this phrases an interesting question about whether the DNA sequence template plays a 'bystander' role in genome function or do locally attained structural formations also exert control. Recent studies directly implicate the chromatin in many processes including chromosome organization, transcription, replication and genomic stability (Gray *et al.*, 2014; Katapadi *et al.*, 2012; Paeschke *et al.*, 2013). To some extent this is attributed to the supramolecular flexible structure of genomic DNA, which allows it to bend, twist and compress. Together these studies have pioneered a school of thought that DNA secondary structure derived from a particular encoded arrangement of DNA bases could participate in function (Bacolla and Wells, 2004; Wadkins, 2000). In this chapter we will focus on a particular type of DNA secondary structure – the G4 motif or G Quadruplexes formed from a specific arrangement of guanine bases and discuss studies that together implicate a larger role of DNA structures in regulating genome-wide function.

G Quadruplex motif: an introduction

Inherent flexibility of the DNA sequence can lead to formation of various inter- and intra-chromosomal secondary and tertiary structures that are distinct from the Watson and Crick DNA double helix. Increasing evidence points towards functional relevance of non-B DNA structural motifs, including guanine quadruplex or G4 motifs, triplexes, cruciform and Z-DNA – notable among these is the conservation of sequence with potential to form such structures across many species (Zhao *et al.*, 2010), and particularly, enriched presence in regulatory regions within genomes of several species (Bacolla and Wells, 2004).

G4 motifs (see Fig. 13.1 for details of the structure) were found to be prevalent throughout genomes of prokaryotes and eukaryotes (Rawal *et al.*, 2006; Huppert and Balasubramanian, 2005). The first indication of a more direct genome wide functional role of G4 motifs in gene regulation came from studies *in E. coli* and other bacteria where authors found sharply enriched presence of such structures in promoters (Rawal *et al.*, 2006; Yadav *et al.*, 2007). Subsequent studies pointed out a possible role for the G4 motif in many other functions including recombination (Mani *et al.*, 2009), replication (Lopes *et al.*, 2011), initiation of double strand breaks (De and Michor, 2011), telomere maintenance (Paeschke *et al.*, 2010), and in expression quantitative trait loci (Baral *et al.*, 2012). Interaction of G4 motif with proteins, their role in regulation or modulation of the genome and therapeutic potential of such interactions have been reviewed recently (Sissi *et al.*, 2011; Thakur *et al.*, 2009). Apart from the DNA G4 motif an interesting arena of quadruplex biology describes RNA G4 motifs and DNA–RNA hybrid G4 motifs (Agarwal *et al.*, 2012; Bugaut and Balasubramanian, 2012; Millevoi *et al.*, 2012). Interestingly, a possible role of the G4 motif has

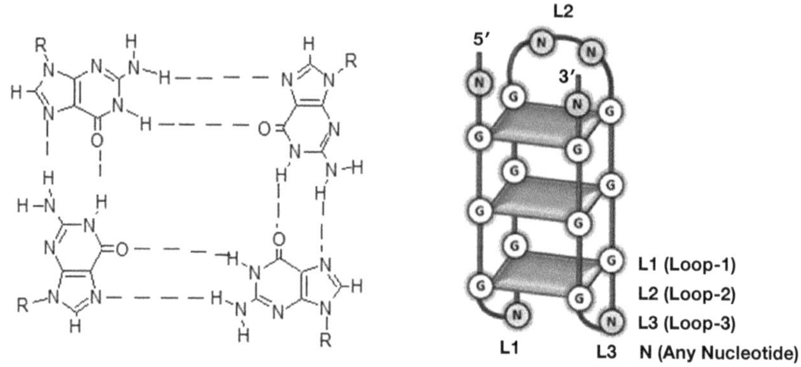

Figure 13.1 The Guanine-quadruplex or G4 motif DNA secondary structure. From intractable guanine gels to the discovery of gene regulatory role, the G-quadruplex or G4 motif story has traversed a long road. In 1962, Gellert *et al.* reported 3′ GMP and 5′ GMP can self-assemble using hydrogen bonds into guanine tetrads that stack upon one another in the form of layers (Gellert *et al.*, 1962). X-ray and fibre-diffraction studies in the 1970s established the four-stranded conformation adopted by polyguanilic acid and polyinosinic acid (Arnott *et al.*, 1974; Zimmerman, 1975). Around the same time Sasisekharan *et al.* (1975) studied 5′ GMP gels, and based on intricate analysis proposed a continuous hydrogen-bonded helix involving 15 nucleotides in four turns, setting the stage for discovery of the biologically relevant intramolecular guanine-quadruplex form that followed almost two decades later. Taking cue from previous reports of the four-stranded inosinic acid residues being stabilized by cations (Howard and Miles, 1982), the Tetrahymena and Oxytricha telomere sequences were synthesized and their *in vitro* structural alterations were noted in the presence of various cations (Williamson *et al.*, 1989). Based on this, the 'G-Quartet model' of the telomeres was put forward in 1989 (Williamson *et al.*, 1989). The tetrads use both Watson and Crick and Hoogsteen bond formations to adopt a stacked G4 motif that is interlinked by a sugar phosphate backbone. Following this, in 1994, Balagurumoorthy and Brahmachari studied the human telomere (TTAGGG) repeat sequence and demonstrated that it forms a stable G4 motif in solution (Balagurumoorthy and Brahmachari, 1994). In the current literature, not only are *in vitro* evidence using synthetic oligonucleotides well established, *in vivo* evidence have also been reported (Guedin *et al.*, 2010; Hazel *et al.*, 2004; Henderson *et al.*, 2014). The continuous expanse of guanine residues providing the stacked tetrad form is known as the stem, and the stretch of DNA comprising of variable bases interspersing the guanine repeats form the loop as shown in the figure (Balagurumoorthy and Brahmachari, 1994). Many structural and in turn topological variations of the G4 motif are known. The primary mediators of this conformational diversity are strand orientation, glycosidic conformation, nature of the binding cation (K^+ or Na^+ generally) and loop length and/or sequence (Bourdoncle *et al.*, 2006; Burge *et al.*, 2006; Risitano and Fox, 2004); e.g. grouping according to the strand orientation gives parallel or anti-parallel quadruplexes (Phan *et al.*, 2004, 2007).

also been noted in DNA methylation and nucleosome positioning (reviewed in Baral *et al.*, 2013). Herein we focus primarily on describing recent literature on methylation and nucleosomal rearrangements that are linked to the presence of G4 motifs, and discuss how protein-G4 motif interactions may impact our understanding. Together these portend a novel role of G4 motifs in epigenetic processes.

Biological relevance of G quadruplex

DNA sequence with potential to form G4 (PG4) motifs occur in clusters within biologically relevant regions like telomeres, promoters, 5′UTRs, 3′UTRs, replication origins, exons and introns. Based on enriched presence within these regions PG4 motifs have been implicated to have possible roles in multiple biological functions (reviewed in Bochman *et al.*, 2012) (Fig. 13.2). Telomeric DNA consists of repeats of a short sequence motif, GGGTTA, in humans. This propels the sequence, particularly the single-stranded end of telomeric DNA, to fold into the quadruplex motif due to hydrogen bonding interaction between the consecutive 'GGG' repeats. This arrangement has been implicated to inhibit telomere elongation and hence has been a focus for anti-cancer drug development (Han and Hurley,

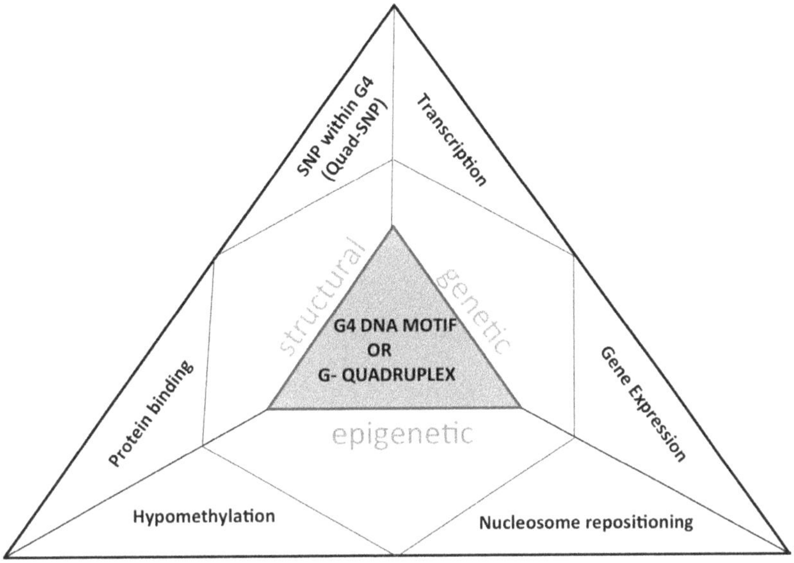

Figure 13.2 Various G4 motif related modules that can control cellular processes.

2000; Raymond *et al.*, 2000; Smith *et al.*, 2011). Recently, *KEM1*, a G4 motif-specific nuclease, was found to block meiosis in yeast thereby indicating G4 motifs in synapsis (Liu and Gilbert, 1994; Sen and Gilbert, 1988). Moreover, it was noted that recombination was compromised by mutation in *MutSa*, which reportedly interacts with the G4 motif (Larson *et al.*, 2005). Reports also showed that recombinogenic regions were enriched in G4 motifs and significantly associated with recombination factors such as c-Rel and NF-kB (Mani *et al.*, 2009), supporting earlier studies suggesting that the immunoglobulin class switch recombination region were activated by intronic repetitive elements containing G4 motifs where formation of G-loops facilitates the process resulting in deletion of several kilobases (Duquette *et al.*, 2004; Manis *et al.*, 2002).

Interestingly, increased incidence of PG4 motifs was not only found to be enriched within promoters, particularly transcription factor binding sites (TFBS), but also conserved across human, chimpanzee, mouse and rat promoters (Verma *et al.*, 2008; Rawal *et al.*, 2006; Kumar *et al.*, 2011). Though initial studies indicated that a G4 motif within the nuclease hypersensitive element in the *c-myc* promoter was required for regulation of *c-myc* (Siddiqui-Jain *et al.*, 2002; Simonsson *et al.*, 1998), evidence of a genome-wide role in gene expression came from comparison of more than 60,000 prokaryote open reading frames across 18 organisms that found enrichment of the G4 motif within promoters (Rawal *et al.*, 2006). Based on detailed studies on *E. coli* promoters, which indicated association of specific TFBS with G4 rich sequences, this study predicted a wide spread role of G4 motifs as regulatory elements. Subsequent studies found G4 motifs within several human promoters including, *BCL2, VEGF, HIF1a, MYB, RB1* and *TERT* (reviewed in Brooks and Hurley, 2009). Like *c-myc*, in several other cases, the presence of promoter G4 motifs were found to influence gene regulation (Cogoi and Xodo, 2006; Wei *et al.*, 2012), including the *Thymidine kinase 1* promoter where a novel quadruplex with repeats consisting of two guanines was found to regulate gene expression (Basundra *et al.*, 2010). These studies were further substantiated by the finding that synthetic G4 motifs when inserted upstream of constitutive promoters give significant alteration in gene expression (Baral *et al.*, 2012). Furthermore, on comparing genome-wide gene expression change as a result of treatment with intracellular G4-binding ligand, it was found to significantly influence regulation of genes that had PG4 motifs within promoters (Verma *et al.*, 2008). By far the most convincing findings of G4 motif mediated transcription were reported when a transcription factor was found to specifically interact with the G4 motif for transcription regulation (Thakur *et al.*, 2009). Subsequent studies found other transcription factors that were implicated to interact with G4 motifs in regulatory modes (Cogoi and Xodo, 2006; Cogoi *et al.*, 2010). Association between G4 motifs and transcription factors were also found from genome wide computational studies – interestingly many of these were zinc finger proteins (Kumar *et al.*, 2011).

In addition to these, promoter PG4 motifs have been observed to be conserved across diverse species. A study on seven species of yeast (five of which are *sensu stricto*), which considered PG4 motifs in 2-kb regions flanking transcriptions start sites, showed conservation of 64% of the motif in at least two of the strains and 7% of the motif in all the *sensu stricto* strains (Capra *et al.*, 2010). At the nucleotide level, G4 containing sequences have a greater conservation score as compared to flanking sequences, indicating importance for retention of the motifs (Capra *et al.*, 2010). In yet another study comparing PG4 motifs in a 1-kb flanking region of transcription start sites in human, chimp, mouse and rat, it was seen that 1414 PG4 motifs in humans were conserved (Verma *et al.*, 2008). G4 motifs were also found to be sensitive to their nucleotide composition. Studies on single nucleotide polymorphisms (SNP) within G4 motifs showed depleted presence of SNPs within the motifs; a SNP within a quadruplex, or Quad-SNP, was found to strongly correlate to differential gene expression within populations (Baral *et al.*, 2012).

Epigenetic modulation and DNA structure

While change in arrangement of DNA bases can influence function, in many cases cellular processes are determined or modulated by alterations not in base-sequence but other modifications of the chromosomal DNA. These modifications are classified as epigenetic changes. Though many factors are known to influence epigenetic processes, two major players are nucleosome positioning, or the basic packaging machinery of the chromosomes, and methylation of DNA residues (Richards and Elgin, 2002). In addition to the effect of epigenetic alterations on the basic DNA backbone this can also impact binding of transcription factors or other DNA binding proteins (Taqi *et al.*, 2012). Given this, it is apparent that DNA secondary structure forms that either impacts the DNA backbone, and/or the ability of DNA residues to be methylated, would also impact epigenetic processes. Some evidence in literature exists about Z-DNA forms and other repetitive DNA that could adopt unusual structures (Neguembor and Gabellini, 2010; Zacharias *et al.*, 1990); herein we will focus on the literature that asserts a possible role of the G4-motif in epigenetic modifications.

CpG methylation and quadruplexes

CpG islands, 'CG' nucleotide doublets present in tandem, where the cytosine residue can get methylated typically occur within GC rich regions of the genome. Cytosine methylation is aided by DNA methyltransferases, it can occur *de novo* and be inherited and generally results in silencing of genes when CpG islands are methylated within promoters (Bird, 2002). Furthermore, studies have shown CpG methylation reduces dynamics of the DNA phosphate-sugar backbone, and this in turn decreases DNA flexibility due to steric hindrance and hydrophobicity (Severin *et al.*, 2011). Interestingly, specific cytosine methylations were also found to influence formation/deformation of alternative non-B DNA motifs (Temiz *et al.*, 2012). Together these findings built a framework for exploring CpG methylations vis-à-vis G4 motif-forming potential and its possible role in biological functions.

Depletion of PG4 motifs in methylated genomic regions was established from multiple genome wide experiments. Comparison of PG4 motifs and methylation status of CpG revealed highly methylated CpGs had relatively reduced occurrence of PG4 motifs. This was also supported by the finding that CpG islands within gene promoters had depleted methylation status and enriched presence of PG4 motifs (Halder *et al.*, 2010) (Fig. 13.3). Moreover the depletion of PG4 motifs in highly methylated regions was not limited to promoters, this was found to be

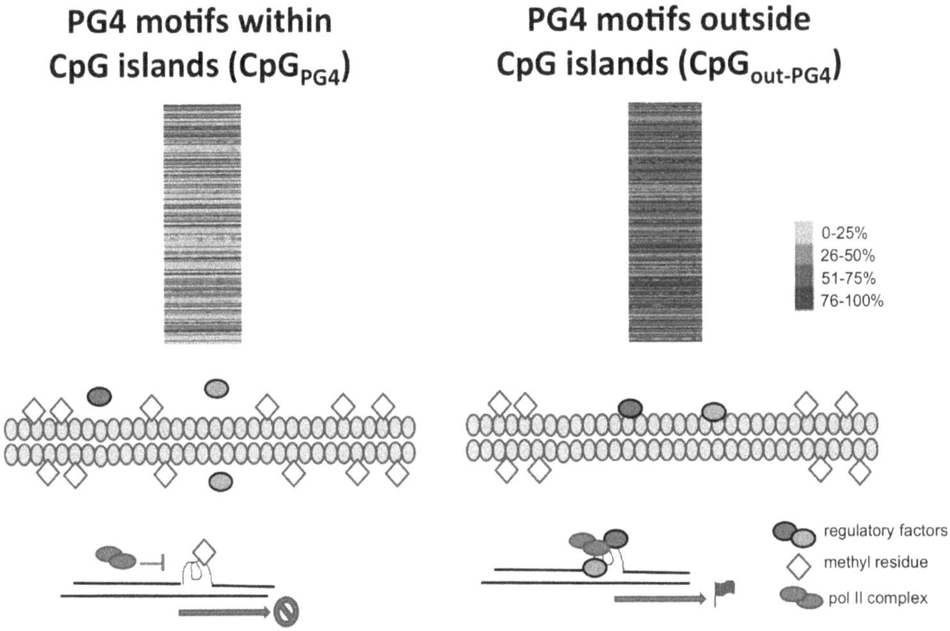

Figure 13.3 CpG methylation and G4 motifs. Top panel shows data from 18 individuals, which indicated that G4 motifs are hypomethylated within CpG islands (Halder *et al.*, 2010). Lower left panel shows a putative model where in case of methylated CpGPG4 the motif is stabilized. This could inhibit Pol II from binding, resulting in stalled or aborted transcription. On the other hand, G4 motif stabilizing proteins may bind unmethylated G4 motifs and help Pol II occupancy for transcription (lower right panel). A colour version of this figure is available in the plate section at the back of the book.

the case also in untranslated regions (UTR), exons and introns (Halder *et al.*, 2010). A relatively large-scale analysis in humans studied 2.1 million CpGs across 12 different tissues from the Human Epigenome Project. On average, CpGs with low methylation had significantly more PG4 motifs when compared to highly methylated CpG islands (Halder *et al.*, 2010). Additionally, to further test their analyses, the authors experimentally determined methylation status of >600,000 CpGs across 18 individuals using bisulfate sequencing – results found significant depletion of PG4 motifs in methylated CpGs relative to CpGs with low methylation levels. Notably, similar observations were made from methylation data obtained from human embryonic stem cells (H1) and differentiated fibroblasts (IMR90) (Lister *et al.*, 2009). Together these studies implicated presence of G4 motifs with hypomethylated regions.

Hypomethylation has been reported as a hallmark of many cancers (Hanahan and Weinberg, 2011). De and Michor (2011) analysed 663,446 DNA breakpoints associated with somatic copy-number alterations (SCNAs) in 26 cancer types. They noticed two striking features: SCNAs were enriched in PG4 motifs and hypomethylated (Li *et al.*, 2012). An interesting study on the G4 motif present within the *bcl-2* promoter found methylation of a cytosine residue in the CpG island can induce a conformational change in the G4 motif (Lin *et al.*, 2013). It was postulated that methylation increased stability of the motif by reducing competitive inhibition from the complimentary strand, and the stable G4 motif in turn led to inactivation of transcription.

Nucleosome repositioning and quadruplexes

The topology of chromosomal DNA and how this can influence nucleosome organization, including positioning has been discussed in the literature (Ioshikhes *et al.*, 2011; Patterton and Simpson, 1995). Though a role for nucleosome packaging in gene regulation is well documented, recent studies suggesting nucleosome positioning can be influenced by G4 motifs appear to add a new dimension to our understanding of how local DNA structure impacts higher order chromatin arrangement and consequent functional implications.

Chromatin at the transcription starts sites of active genes is generally found to have nucleosome-depleted regions (NDR) (Bai and Morozov, 2010). Interestingly, initial genome wide analyses in the yeast *S. cerevisiae* revealed low nucleosome occupancy regions or NDR were also regions with relatively high density of PG4 motifs (Halder *et al.*, 2009). This was further tested in humans using experimentally determined nucleosome positioning and DNA hypersensitive sites – results further substantiated the view that PG4 motifs were relatively enriched in nucleosome-free regions genome wide (Halder *et al.*, 2009) (Fig. 13.4A). Similar observations were also made from analyses in the ciliate *C. elegans* (Wong and Huppert, 2009).

It has been noted that DNA sequence with potential to form secondary structures are more fragile during DNA replication and histone re-assembly (Burrow *et al.*, 2010). It was further observed that members of the Y family polymerases like *REV1*, aid in replication at G4 motifs – experiments with *REV1* deficient cells, that lack the translesion protein required for G4 replication, resulted in stalled replication forks (Edmunds *et al.*, 2008). Similarly, genome instability was also noted in mutants of *FANCJ*, *WRN* and *BLM* helicase, which have been implicated in resolution of G4 motifs inside cells (Sarkies *et al.*, 2012). Together these studies underline the notion that the presence of epigenetic factors/proteins could be essential for G4 metabolism in order to provide genomic stability, particularly for functions that require folding/unfolding or repackaging of chromatin.

Interaction between G4 motifs and proteins

Many recent studies have indicated direct association of G4 motifs with proteins, and implicated

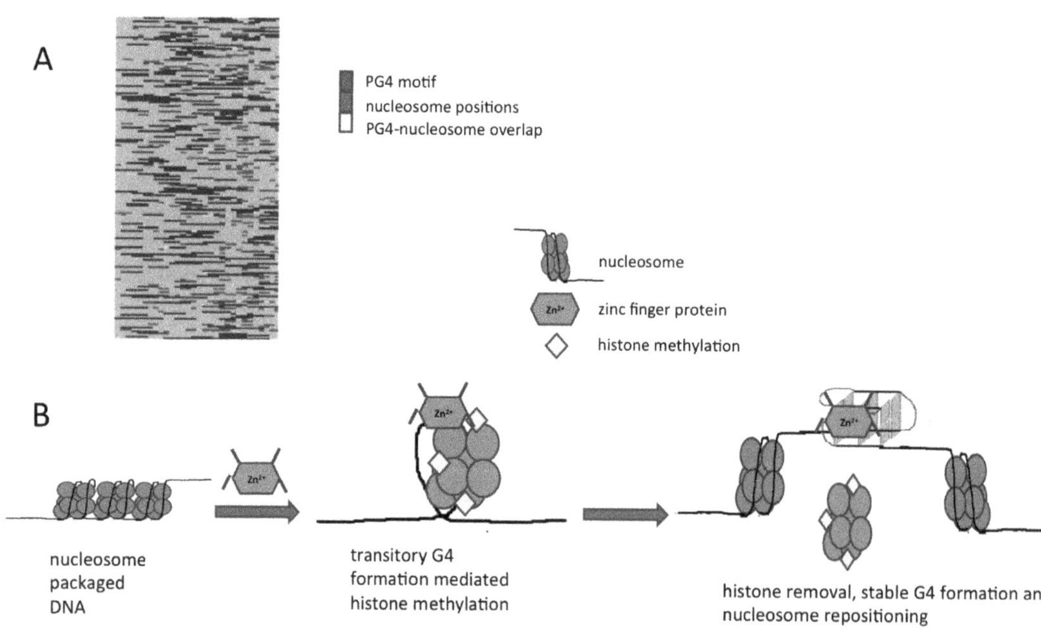

Figure 13.4 (A) Data from 180 promoters highlighting regions of nucleosome occupancy, G4 motifs and nucleosome-G4 overlap, suggests depletion within regions of G4-nucleosome overlap (Halder *et al.*, 2009). (B) Scheme showing speculative model for G4-mediated nucleosome repositioning – zinc finger proteins, which specially bind to G4 motifs could be responsible for histone methylation, thereby G4 motifs could act as a scaffold for histone methylation and nucleosome remodelling. A colour version of this figure is available in the plate section at the back of the book.

a role of G4–protein interactions in cellular function. Following initial studies that predicted G4 motif interaction with transcription factors in *E. coli* gene regulation (Rawal *et al.*, 2006), Thakur *et al.* (2009) provided experimental evidence arguing for G4-mediated transcription of *c-myc*. In this paper, the authors demonstrated an intact G4 motif was necessary within the *c-myc* promoter for transcriptional activation by the regulatory protein NDPK-B (also known as NME2). These findings supported earlier studies implicating G4 motif–protein interactions: for example, between HIV-1 zinc finger nucleocapsid and the G4 motifs present within the compact assembly of the viral nucleocapsid (Lyonnais *et al.*, 2003); interaction of nucleolin and hnRNP D subunits, as well as association of the Activation Induced Cytosine Deaminase (AID) protein with G4 structures of the immunoglobulin class switch region (Dempsey *et al.*, 1999; Duquette *et al.*, 2005; Hanakahi *et al.*, 1999). Interaction with G4 motifs was also noted with several DNA binding proteins in human, chimpanzee, mouse and rat including Sp1, MAZ (Myc-associated zinc finger), hnRNP A1, MyoD, BLM and nucleolin (Cogoi *et al.*, 2010; Cohen *et al.*, 2004; Raiber *et*

Table 13.1 Quadruplex interacting proteins

Protein	Source	Reference
Thrombin	Human	Bock *et al*. (1992), Macaya *et al*. (1993)
MyoD	Human	Etzioni *et al*. (2005)
QUAD	Murine	Weismanshomer and Fry (1994)
Macrophage scavenger receptor	Bovine	Pearson *et al*. (1993)
Tetrahymena binding protein (TBP)	*Tetrahymena thermophila*	Schierer and Henderson (1994)
G4p1	*Saccharomyces cerevisiae*	Frantz and Gilbert (1995)
Gene 5 protein (g5p)	Bacteriophage fd/Ff	Gilbert (1995), Hayashi and Murakami (2002)
Ku protein	Human	Uliel *et al*. (2000)
Nucleocapsid protein (NCp)	HIV-1	Kankia (2005)
MutSα	Human	Larson *et al*. (2005)
Cytoplasmic intermediate filament proteins (cIF)	Murine	Tolstonog *et al*. (2005)
Subunit β of telomere binding protein	*Oxytricha nova*	Fang and Cech (1993)
RAP1	*Saccharomyces cerevisiae*	Giraldo and Rhodes (1994)
DNA topoisomerase I	Human	Arimondo (2000)
Hop1	*Saccharomyces cerevisiae*	Muniyappa *et al*. (2000)
Telomere end-binding proteins α and β	*Stylonychia lemnae*	Paeschke *et al*. (2005)
WRN	Human	Mohaghegh (2001)
BLM	Human	Mohaghegh (2001)
Sgs1	*Saccharomyces cerevisiae*	Sun *et al*. (1999)
hnRNP D	Murine	Enokizono *et al*. (2005)
Cdc13p	*Saccharomyces cerevisiae*	Lin *et al*. (2001)
Unwinding Protein 1 (UP1)	Murine	Fukuda *et al*. (2005)
Protection Of Telomere 1 (POT1)	Human	Zaug *et al*. (2005)
Replication protein A (RPA)	Human	Salas *et al*. (2006)
KEM1 gene product	*Saccharomyces cerevisiae*	Liu and Gilbert (1994)
G Quartet Nuclease 1 (GQN1)	Human	Sun *et al*. (2001)
Mre11	*Saccharomyces cerevisiae*	Ghosal and Muniyappa (2005)

al., 2012; Shklover *et al.*, 2010). A summary of various reported G4 motif protein interactions is given in Table 13.1.

Interestingly, the human telomeric repeat factor 2 (TRF2) and TBP25 were found to bind intramolecular forms of the G4 motif in the ciliate tetrahymena, and this was proposed as a mode of protection of the 3′ end of telomeres (Paeschke *et al.*, 2005). Likewise G4 motif binding proteins, TEB β, RAP1, Topo I were observed to promote *in vivo* formation of G4 motifs (Arimondo *et al.*, 2000; Giraldo and Rhodes, 1994; Paeschke *et al.*, 2005). In addition, it was demonstrated that RecQ family helicases, MutSα and AID participated in DNA unwinding and recombination using G4 motif binding (Fry, 2007). Other than the RecQ family, Pif1 helicases, WRN and BLM helicases were also shown to bind G4 motifs and unwind the folded architecture, and thereby contribute to replication and genomic instability (Kamath-Loeb *et al.*, 2001; Paeschke *et al.*, 2013; Sun *et al.*, 1998).

It is possible that activity of some cellular proteins that can destabilize the G4 motif exploit the dynamic nature of G4 motif formation/deformation. Proteins of this class can disrupt different types of tetrahelical DNA in a non-catalytic fashion – hnRNPD and UP1 were found to destabilize intra- and intermolecular quadruplex structures of the telomeric sequence d(TTAGGG)$_{1-4}$ *in vitro* (Enokizono *et al.*, 2005; Eversole and Maizels, 2000; Fukuda *et al.*, 2002). It was also found that the proteins protection of telomere 1 (POT1) and replication protein A (RPA), which bind to the guanine-rich single-stranded telomeric overhang were essential for the maintenance of telomeres (Smith *et al.*, 2000), possibly by destabilizing the tetraplex structure and thereby stimulating telomerase activity as noted *in vitro* (Cohen *et al.*, 2004; Schramke *et al.*, 2004). Another interesting G4 motif–protein interaction emerges from recent findings related to transcription from telomere ends that produce RNA known as telomeric repeat containing RNA (TERRA) (Azzalin *et al.*, 2007; Luke and Lingner, 2009). The C-terminal region of a telomere interacting protein, TLS, was found to associate with telomeric DNA and the transcribed RNA in a ternary complex. Here, the authors found that the DNA G4 motif provides the required scaffold for TLS binding, which in turn affects chromatin dynamics by changing histone marks and thereby regulates telomere length *in vivo* (Takahama *et al.*, 2013)

Future perspective – a model involving G4 motifs in epigenetic changes

Based on what we know about nucleosome positioning, methylation and protein interactions in connection with G4 motifs, a framework appears to be emerging. This envisages a model where G4 motifs (or possibly other local small non-B DNA structures) can render epigenetic changes that involve cis elements (that is DNA fragments on the same chromosome) where the structural form or architecture attained by the cis element plays a more central role. Though current studies only implicate such models, there are interesting possibilities that can be tested. For example: a role for G4 motifs in hypo- or hypermethylation may be significant in transcription regulation. While reports described here suggest association of PG4 motifs with DNA methylation, how methylated or un-methylated DNA influences occupancy of regulatory factors within promoters, which is required for polymerase II binding is well studied. Taking these together, it is possible that G4 motif-mediated DNA methylation is an important factor in transcription control (Fig. 13.3). Second, as described in this article, G4 motifs were found to be enriched in linker regions (the stretch of DNA between two nucleosomes) – which are known to be protein associated, particularly by transcription factors within gene regulatory regions, and therefore it is possible that G4-transcription factor associations are also crucial in determining nucleosome positioning or remodelling. This gains support from multiple studies demonstrating biologically relevant G4 motif-transcription factor associations (described above). The finding that G4 motifs preferably associate with zinc finger protein is interesting to consider in a putative model – where G4 motifs recruit zinc finger factors on to DNA (Kumar *et al.*, 2011) and the recruited factors in turn modify histone methylation. Involvement of zinc finger proteins in histone methylation has been recently reviewed (Blackledge *et al.*, 2013; Long *et al.*, 2013). A speculative model along these lines is shown in Fig. 13.4B. Taken together, it is possible that G4 motifs in association with protein counterparts may emerge as an important link between cis elements and DNA methylation and other epigenetic changes. While recent findings are helping to bridge the gap in our understanding between DNA secondary structure and transcription (Bochman *et al.*, 2012), recombination (Maizels and Gray, 2013), and replication (Paeschke *et al.*, 2011) using the G4 motif as a 'model' architecture, it remains to be seen how G4 motif–protein interactions emerge in connection with epigenetic changes in the genome.

Acknowledgements

SC is a senior fellow of the Wellcome Trust/DBT India Alliance. AB and DJ acknowledge senior research fellowships from CSIR, Government of India.

References

Agarwal, T., Jayaraj, G., Pandey, S.P., Agarwala, P., and Maiti, S. (2012). RNA G-quadruplexes: G-quadruplexes with "U" turns. Curr. Pharm. Des *18*, 2102–2111.

Arimondo, P.B. (2000). Interaction of human DNA topoisomerase I with G-quartet structures. Nucleic Acids Res. *28*, 4832–4838.

Arimondo, P.B., Riou, J.F., Mergny, J.L., Tazi, J., Sun, J.S., Garestier, T., and Hélène, C. (2000). Interaction of human DNA topoisomerase I with G-quartet structures. Nucleic Acids Res. *28*, 4832–4838.

Azzalin, C.M., Reichenbach, P., Khoriauli, L., Giulotto, E., and Lingner, J. (2007). Telomeric repeat containing RNA and RNA surveillance factors at mammalian chromosome ends. Science *318*, 798–801.

Bacolla, A., and Wells, R.D. (2004). Non-B DNA conformations, genomic rearrangements, and human disease. J. Biol. Chem. *279*, 47411–47414.

Bai, L., and Morozov, A.V. (2010). Gene regulation by nucleosome positioning. Trends Genet. *26*, 476–483.

Balagurumoorthy, P., and Brahmachari, S.K. (1994). Structure and stability of human telomeric sequence. J. Biol. Chem. *269*, 21858–21869.

Baral, A., Kumar, P., Halder, R., Mani, P., Yadav, V.K., Singh, A., Das, S.K., and Chowdhury, S. (2012). Quadruplex-single nucleotide polymorphisms (Quad-SNP) influence gene expression difference among individuals. Nucleic Acids Res. *40*, 3800–3811.

Baral, A., Kumar, P., Pathak, R., and Chowdhury, S. (2013). Emerging trends in G-quadruplex biology – role in epigenetic and evolutionary events. Mol. Biosyst. *9*, 1568–1575.

Basundra, R., Kumar, A., Amrane, S., Verma, A., Phan, A.T., and Chowdhury, S. (2010). A novel G-quadruplex motif modulates promoter activity of human thymidine kinase 1. FEBS J. *277*, 4254–4264.

Bird, A. (2002). DNA methylation patterns and epigenetic memory. Genes Dev. *16*, 6–21.

Blackledge, N.P., Thomson, J.P., and Skene, P.J. (2013). CpG island chromatin is shaped by recruitment of ZF-CxxC proteins. Cold Spring Harb. Perspect. Biol. *5*, a018648.

Bochman, M.L., Paeschke, K., and Zakian, V.A. (2012). DNA secondary structures: stability and function of G-quadruplex structures. Nat. Rev. Genet. *13*, 770–780.

Bock, L.C., Griffin, L.C., Latham, J.A., Vermaas, E.H., and Toole, J.J. (1992). Selection of single-stranded DNA molecules that bind and inhibit human thrombin. Nature *355*, 564–566.

Bourdoncle, A., Estevez, T.A., Gosse, C., Lacroix, L., Vekhoff, P., Le, S.T., Jullien, L., and Mergny, J.L. (2006). Quadruplex-based molecular beacons as tunable DNA probes. J. Am. Chem. Soc. *128*, 11094–11105.

Brooks, T.A., and Hurley, L.H. (2009). The role of supercoiling in transcriptional control of MYC and its importance in molecular therapeutics. Nat. Rev. Cancer *9*, 849–861.

Bugaut, A., and Balasubramanian, S. (2012). 5′-UTR RNA G-quadruplexes: translation regulation and targeting. Nucleic Acids Res. *40*, 4727–4741.

Burge, S., Parkinson, G.N., Hazel, P., Todd, A.K., and Neidle, S. (2006). Quadruplex DNA: sequence, topology and structure. Nucleic Acids Res. *34*, 5402–5415.

Burrow, A.A., Marullo, A., Holder, L.R., and Wang, Y.H. (2010). Secondary structure formation and DNA instability at fragile site FRA16B. Nucleic Acids Res. *38*, 2865–2877.

Capra, J.A., Paeschke, K., Singh, M., and Zakian, V.A. (2010). G-quadruplex DNA sequences are evolutionarily conserved and associated with distinct genomic features in *Saccharomyces cerevisiae*. PLoS. Comput. Biol. *6*. DOI: 10.1371/pcbi.1000861

Cogoi, S., and Xodo, L.E. (2006). G-quadruplex formation within the promoter of the KRAS proto-oncogene and its effect on transcription. Nucleic Acids Res. *34*, 2536–2549.

Cogoi, S., Paramasivam, M., Membrino, A., Yokoyama, K.K., and Xodo, L.E. (2010). The KRAS promoter responds to Myc-associated zinc finger and poly(ADP-ribose) polymerase 1 proteins, which recognize a critical quadruplex-forming GA-element. J. Biol. Chem. *285*, 22003–22016.

Cohen, S., Jacob, E., and Manor, H. (2004). Effects of single-stranded DNA binding proteins on primer extension by telomerase. Biochim. Biophys. Acta *1679*, 129–140.

De, S., and Michor, F. (2011). DNA secondary structures and epigenetic determinants of cancer genome evolution. Nat. Struct. Mol. Biol. *18*, 950–955.

Dempsey, L.A., Sun, H., Hanakahi, L.A., and Maizels, N. (1999). G4 DNA binding by LR1 and its subunits, nucleolin and hnRNP D, A role for G–G pairing in immunoglobulin switch recombination. J. Biol. Chem. *274*, 1066–1071.

Duquette, M.L., Handa, P., Vincent, J.A., Taylor, A.F., and Maizels, N. (2004). Intracellular transcription of G-rich DNAs induces formation of G-loops, novel structures containing G4 DNA. Genes Dev. *18*, 1618–1629.

Duquette, M.L., Pham, P., Goodman, M.F., and Maizels, N. (2005). AID binds to transcription-induced structures in c-MYC that map to regions associated with translocation and hypermutation. Oncogene *24*, 5791–5798.

Edmunds, C.E., Simpson, L.J., and Sale, J.E. (2008). PCNA ubiquitination and REV1 define temporally distinct mechanisms for controlling translesion synthesis in the avian cell line DT40. Mol. Cell *30*, 519–529.

Enokizono, Y., Konishi, Y., Nagata, K., Ouhashi, K., Uesugi, S., Ishikawa, F., and Katahira, M. (2005). Structure of hnRNP D complexed with single-stranded telomere DNA and unfolding of the quadruplex by heterogeneous nuclear ribonucleoprotein D. J. Biol. Chem. *280*, 18862–18870.

Etzioni, S., Yafe, A., Khateb, S., Weisman-Shomer, P., Bengal, E., and Fry, M. (2005). Homodimeric MyoD preferentially binds tetraplex structures of regulatory sequences of muscle-specific genes. J. Biol. Chem. *280*, 26805–26812.

Eversole, A., and Maizels, N. (2000). *In vitro* properties of the conserved mammalian protein hnRNP D suggest a role in telomere maintenance. Mol. Cell. Biol. *20*, 5425–5432.

Fang, G., and Cech, T.R. (1993). The β subunit of Oxytricha telomere-binding protein promotes G-quartet formation by telomeric DNA. Cell *74*, 875–885.

Frantz, J.D., and Gilbert, W. (1995). A novel yeast gene product, G4p1, with a specific affinity for quadruplex nucleic acids. J. Biol. Chem. *270*, 20692–20697.

Fry, M. (2007). Tetraplex DNA and its interacting proteins. Front. Biosci. J. Virtual Libr. *12*, 4336–4351.

Fukuda, H., Katahira, M., Tsuchiya, N., Enokizono, Y., Sugimura, T., Nagao, M., and Nakagama, H. (2002). Unfolding of quadruplex structure in the G-rich strand of the minisatellite repeat by the binding protein UP1. Proc. Natl. Acad. Sci. U.S.A. *99*, 12685–12690.

Fukuda, H., Katahira, M., Tanaka, E., Enokizono, Y., Tsuchiya, N., Higuchi, K., Nagao, M., and Nakagama, H. (2005). Unfolding of higher DNA structures formed by the d(CGG) triplet repeat by UP1 protein: UP1 unfolds secondary structure of CGG repeat. Genes Cells *10*, 953–962.

Gellert, M., Lipsett, M.N., and Davies, D.R. (1962). Helix formation by guanylic acid. Proc. Natl. Acad. Sci. U.S.A. *48*, 2013–2018.

Ghosal, G., and Muniyappa, K. (2005). *Saccharomyces cerevisiae* Mre11 is a high-affinity G4 DNA-binding protein and a G-rich DNA-specific endonuclease: implications for replication of telomeric DNA. Nucleic Acids Res. *33*, 4692–4703.

Gilbert, W. (1995). A yeast gene product, G4p2, with a specific affinity for quadruplex nucleic acids. J. Biol. Chem. *270*, 9413–9419.

Giraldo, R., and Rhodes, D. (1994). The yeast telomere-binding protein RAP1 binds to and promotes the formation of DNA quadruplexes in telomeric DNA. EMBO J. *13*, 2411–2420.

Gray, L.T., Vallur, A.C., Eddy, J., and Maizels, N. (2014). G quadruplexes are genomewide targets of transcriptional helicases XPB and XPD. Nat. Chem. Biol. *10*, 313–318.

Guedin, A., Gros, J., Alberti, P., and Mergny, J.L. (2010). How long is too long? Effects of loop size on G-quadruplex stability. Nucleic Acids Res. *38*, 7858–7868.

Halder, K., Halder, R., and Chowdhury, S. (2009). Genome-wide analysis predicts DNA structural motifs as nucleosome exclusion signals. Mol. Biosyst. *5*, 1703–1712.

Halder, R., Halder, K., Sharma, P., Garg, G., Sengupta, S., and Chowdhury, S. (2010). Guanine quadruplex DNA structure restricts methylation of CpG dinucleotides genome-wide. Mol. Biosyst. *6*, 2439–2447.

Han, H., and Hurley, L.H. (2000). G-quadruplex DNA: a potential target for anti-cancer drug design. Trends Pharmacol. Sci. *21*, 136–142.

Hanahan, D., and Weinberg, R.A. (2011). Hallmarks of cancer: the next generation. Cell *144*, 646–674.

Hanakahi, L.A., Sun, H., and Maizels, N. (1999). High affinity interactions of nucleolin with G-G-paired rDNA. J. Biol. Chem. *274*, 15908–15912.

Hayashi, N., and Murakami, S. (2002). STM1, a gene which encodes a guanine quadruplex binding protein, interacts with CDC13 in *Saccharomyces cerevisiae*. Mol. Genet. Genomics *267*, 806–813.

Hazel, P., Huppert, J., Balasubramanian, S., and Neidle, S. (2004). Loop-length-dependent folding of G-quadruplexes. J. Am. Chem. Soc. *126*, 16405–16415.

Henderson, A., Wu, Y., Huang, Y.C., Chavez, E.A., Platt, J., Johnson, F.B., Brosh, R.M., Jr., Sen, D., and Lansdorp, P.M. (2014). Detection of G-quadruplex DNA in mammalian cells. Nucleic Acids Res. *42*, 860–869.

Howard, F.B., and Miles, H.T. (1982). Poly(inosinic acid) helices: essential chelation of alkali metal ions in the axial channel. Biochemistry (Mosc.) *21*, 6736–6745.

Huppert, J.L., and Balasubramanian, S. (2005). Prevalence of quadruplexes in the human genome. Nucleic Acids Res. *33*, 2908–2916.

Ioshikhes, I., Hosid, S., and Pugh, B.F. (2011). Variety of genomic DNA patterns for nucleosome positioning. Genome Res. *21*, 1863–1871.

Kamath-Loeb, A.S., Loeb, L.A., Johansson, E., Burgers, P.M., and Fry, M. (2001). Interactions between the Werner syndrome helicase and DNA polymerase delta specifically facilitate copying of tetraplex and hairpin structures of the d(CGG)n trinucleotide repeat sequence. J. Biol. Chem. *276*, 16439–16446.

Kankia, B.I. (2005). Unfolding of DNA quadruplexes induced by HIV-1 nucleocapsid protein. Nucleic Acids Res. *33*, 4395–4403.

Katapadi, V.K., Nambiar, M., and Raghavan, S.C. (2012). Potential G-quadruplex formation at breakpoint regions of chromosomal translocations in cancer may explain their fragility. Genomics *100*, 72–80.

Kumar, P., Yadav, V.K., Baral, A., Kumar, P., Saha, D., and Chowdhury, S. (2011). Zinc-finger transcription factors are associated with guanine quadruplex motifs in human, chimpanzee, mouse and rat promoters genome-wide. Nucleic Acids Res. *39*, 8005–8016.

Larson, E.D., Duquette, M.L., Cummings, W.J., Streiff, R.J., and Maizels, N. (2005). MutSalpha binds to and promotes synapsis of transcriptionally activated immunoglobulin switch regions. Curr. Biol. *15*, 470–474.

Li, Y., Zhang, L., Ball, R.L., Liang, X., Li, J., Lin, Z., and Liang, H. (2012). Comparative analysis of somatic copy-number alterations across different human cancer types reveals two distinct classes of breakpoint hotspots. Hum. Mol. Genet. *21*, 4957–4965.

Lin, J., Hou, J.Q., Xiang, H.D., Yan, Y.Y., Gu, Y.C., Tan, J.H., Li, D., Gu, L.Q., Ou, T.M., and Huang, Z.S. (2013). Stabilization of G-quadruplex DNA by C-5-methyl-cytosine in bcl-2 promoter: implications for epigenetic regulation. Biochem. Biophys. Res. Commun. *433*, 368–373.

Lin, Y.-C., Shih, J.-W., Hsu, C.-L., and Lin, J.-J. (2001). Binding and Partial Denaturing of G-quartet DNA by Cdc13p of *Saccharomyces cerevisiae*. J. Biol. Chem. *276*, 47671–47674.

Lister, R., Pelizzola, M., Dowen, R.H., Hawkins, R.D., Hon, G., Tonti-Filippini, J., Nery, J.R., Lee, L., Ye, Z., Ngo, Q.M., *et al.* (2009). Human DNA methylomes at base resolution show widespread epigenomic differences. Nature *462*, 315–322.

Liu, Z., and Gilbert, W. (1994). The yeast KEM1 gene encodes a nuclease specific for G4 tetraplex DNA: implication of *in vivo* functions for this novel DNA structure. Cell 77, 1083–1092.

Long, H.K., Blackledge, N.P., and Klose, R.J. (2013). ZF-CxxC domain-containing proteins, CpG islands and the chromatin connection. Biochem. Soc. Trans. *41*, 727–740.

Lopes, J., Piazza, A., Bermejo, R., Kriegsman, B., Colosio, A., Teulade-Fichou, M.P., Foiani, M., and Nicolas, A. (2011). G-quadruplex-induced instability during leading-strand replication. EMBO J. *30*, 4033–4046.

Luke, B., and Lingner, J. (2009). TERRA: telomeric repeat-containing RNA. EMBO J. *28*, 2503–2510.

Lyonnais, S., Gorelick, R.J., Mergny, J.L., Le, C.E., and Mirambeau, G. (2003). G-quartets direct assembly of HIV-1 nucleocapsid protein along single-stranded DNA. Nucleic Acids Res. *31*, 5754–5763.

Macaya, R.F., Schultze, P., Smith, F.W., Roe, J.A., and Feigon, J. (1993). Thrombin-binding DNA aptamer forms a unimolecular quadruplex structure in solution. Proc. Natl. Acad. Sci. U.S.A. *90*, 3745–3749.

Maizels, N., and Gray, L.T. (2013). The G4 Genome. PLoS Genet. *9*, e1003468.

Mani, P., Yadav, V.K., Das, S.K., and Chowdhury, S. (2009). Genome-wide analyses of recombination prone regions predict role of DNA structural motif in recombination. PLoS ONE *4*, e4399.

Manis, J.P., Tian, M., and Alt, F.W. (2002). Mechanism and control of class-switch recombination. Trends Immunol. *23*, 31–39.

Millevoi, S., Moine, H., and Vagner, S. (2012). G-quadruplexes in RNA biology. Wiley Interdiscip. Rev. RNA *3*, 495–507.

Mohaghegh, P. (2001). The Bloom's and Werner's syndrome proteins are DNA structure-specific helicases. Nucleic Acids Res. *29*, 2843–2849.

Muniyappa, K., Anuradha, S., and Byers, B. (2000). Yeast meiosis-specific protein Hop1 binds to G4 DNA and promotes its formation. Mol. Cell. Biol. *20*, 1361–1369.

Neguembor, M.V., and Gabellini, D. (2010). In junk we trust: repetitive DNA, epigenetics and facioscapulohumeral muscular dystrophy. Epigenomics. *2*, 271–287.

Paeschke, K., Simonsson, T., Postberg, J., Rhodes, D., and Lipps, H.J. (2005). Telomere end-binding proteins control the formation of G-quadruplex DNA structures *in vivo*. Nat. Struct. Mol. Biol. *12*, 847–854.

Paeschke, K., McDonald, K.R., and Zakian, V.A. (2010). Telomeres: structures in need of unwinding. FEBS Lett. *584*, 3760–3772.

Paeschke, K., Capra, J.A., and Zakian, V.A. (2011). DNA replication through G-quadruplex motifs is promoted by the *Saccharomyces cerevisiae* Pif1 DNA helicase. Cell *145*, 678–691.

Paeschke, K., Bochman, M.L., Garcia, P.D., Cejka, P., Friedman, K.L., Kowalczykowski, S.C., and Zakian, V.A. (2013). Pif1 family helicases suppress genome instability at G-quadruplex motifs. Nature *497*, 458–462.

Patterton, H.G., and Simpson, R.T. (1995). Modified curved DNA that could allow local DNA underwinding at the nucleosomal pseudodyad fails to position a nucleosome *in vivo*. Nucleic Acids Res. *23*, 4170–4179.

Pearson, A.M., Rich, A., and Krieger, M. (1993). Polynucleotide binding to macrophage scavenger receptors depends on the formation of base-quartet-stabilized four-stranded helices. J. Biol. Chem. *268*, 3546–3554.

Phan, A.T., Modi, Y.S., and Patel, D.J. (2004). Propeller-type parallel-stranded G-quadruplexes in the human c-myc promoter. J. Am. Chem. Soc. *126*, 8710–8716.

Phan, A.T., Kuryavyi, V., Luu, K.N., and Patel, D.J. (2007). Structure of two intramolecular G-quadruplexes formed by natural human telomere sequences in K+ solution. Nucleic Acids Res. *35*, 6517–6525.

Raiber, E.A., Kranaster, R., Lam, E., Nikan, M., and Balasubramanian, S. (2012). A non-canonical DNA structure is a binding motif for the transcription factor SP1 *in vitro*. Nucleic Acids Res. *40*, 1499–1508.

Rawal, P., Kummarasetti, V.B., Ravindran, J., Kumar, N., Halder, K., Sharma, R., Mukerji, M., Das, S.K., and Chowdhury, S. (2006). Genome-wide prediction of G4 DNA as regulatory motifs: role in Escherichia coli global regulation. Genome Res. *16*, 644–655.

Raymond, E., Soria, J.C., Izbicka, E., Boussin, F., Hurley, L., and Von Hoff, D.D. (2000). DNA G-quadruplexes, telomere-specific proteins and telomere-associated enzymes as potential targets for new anticancer drugs. Invest. New Drugs *18*, 123–137.

Richards, E.J., and Elgin, S.C. (2002). Epigenetic codes for heterochromatin formation and silencing: rounding up the usual suspects. Cell *108*, 489–500.

Risitano, A., and Fox, K.R. (2004). Influence of loop size on the stability of intramolecular DNA quadruplexes. Nucleic Acids Res. *32*, 2598–2606.

Salas, T.R., Petruseva, I., Lavrik, O., Bourdoncle, A., Mergny, J.-L., Favre, A., and Saintome, C. (2006). Human replication protein A unfolds telomeric G-quadruplexes. Nucleic Acids Res. *34*, 4857–4865.

Sarkies, P., Murat, P., Phillips, L.G., Patel, K.J., Balasubramanian, S., and Sale, J.E. (2012). FANCJ coordinates two pathways that maintain epigenetic stability at G-quadruplex DNA. Nucleic Acids Res. *40*, 1485–1498.

Sasisekharan, V., Zimmerman, S., and Davies, D.R. (1975). The structure of helical 5′-guanosine monophosphate. J. Mol. Biol. *92*, 171–179.

Schierer, T., and Henderson, E. (1994). A protein from *Tetrahymena thermophila* that specifically binds parallel-stranded G4-DNA. Biochemistry (Mosc.) *33*, 2240–2246.

Schramke, V., Luciano, P., Brevet, V., Guillot, S., Corda, Y., Longhese, M.P., Gilson, E., and Géli, V. (2004). RPA regulates telomerase action by providing Est1p access to chromosome ends. Nat. Genet. *36*, 46–54.

Sen, D., and Gilbert, W. (1988). Formation of parallel four-stranded complexes by guanine-rich motifs in DNA and its implications for meiosis. Nature *334*, 364–366.

Severin, P.M.D., Zou, X., Gaub, H.E., and Schulten, K. (2011). Cytosine methylation alters DNA mechanical properties. Nucleic Acids Res. *39*, 8740–8751.

Shklover, J., Weisman-Shomer, P., Yafe, A., and Fry, M. (2010). Quadruplex structures of muscle gene promoter sequences enhance *in vivo* MyoD-dependent gene expression. Nucleic Acids Res. *38*, 2369–2377.

Siddiqui-Jain, A., Grand, C.L., Bearss, D.J., and Hurley, L.H. (2002). Direct evidence for a G-quadruplex in a promoter region and its targeting with a small molecule to repress c-MYC transcription. Proc. Natl. Acad. Sci. U.S.A. *99*, 11593–11598.

Simonsson, T., Pecinka, P., and Kubista, M. (1998). DNA tetraplex formation in the control region of c-myc. Nucleic Acids Res. *26*, 1167–1172.

Sissi, C., Gatto, B., and Palumbo, M. (2011). The evolving world of protein-G-quadruplex recognition: a medicinal chemist's perspective. Biochimie *93*, 1219–1230.

Smith, J., Zou, H., and Rothstein, R. (2000). Characterization of genetic interactions with RFA1: the role of RPA in DNA replication and telomere maintenance. Biochimie *82*, 71–78.

Smith, J.S., Chen, Q., Yatsunyk, L.A., Nicoludis, J.M., Garcia, M.S., Kranaster, R., Balasubramanian, S., Monchaud, D., Teulade-Fichou, M.P., Abramowitz, L., *et al.* (2011). Rudimentary G-quadruplex-based telomere capping in *Saccharomyces cerevisiae*. Nat. Struct. Mol. Biol. *18*, 478–485.

Sun, H., Karow, J.K., Hickson, I.D., and Maizels, N. (1998). The Bloom's syndrome helicase unwinds G4 DNA. J. Biol. Chem. *273*, 27587–27592.

Sun, H., Bennett, R.J., and Maizels, N. (1999). The *Saccharomyces cerevisiae* Sgs1 helicase efficiently unwinds G-G paired DNAs. Nucleic Acids Res. *27*, 1978–1984.

Sun, H., Yabuki, A., and Maizels, N. (2001). A human nuclease specific for G4 DNA. Proc. Natl. Acad. Sci. U.S.A. *98*, 12444–12449.

Takahama, K., Takada, A., Tada, S., Shimizu, M., Sayama, K., Kurokawa, R., and Oyoshi, T. (2013). Regulation of telomere length by G-quadruplex telomere DNA- and TERRA-binding protein TLS/FUS. Chem. Biol. *20*, 341–350.

Taqi, M.M., Warmlander, S.K., Yamskova, O., Madani, F., Bazov, I., Luo, J., Zubarev, R., Verbeek, D., Graslund, A., and Bakalkin, G. (2012). Conformation effects of CpG methylation on single-stranded DNA oligonucleotides: analysis of the opioid peptide dynorphin-coding sequences. PLoS ONE 7. DOI: 10.1371/journal.pone.0039605

Temiz, N.A., Donohue, D.E., Bacolla, A., Luke, B.T., and Collins, J.R. (2012). The role of methylation in the intrinsic dynamics of B- and Z-DNA. PLoS ONE 7. DOI: 10.1371/journal.pone.0035558

Thakur, R.K., Kumar, P., Halder, K., Verma, A., Kar, A., Parent, J.L., Basundra, R., Kumar, A., and Chowdhury, S. (2009). Metastases suppressor NM23-H2 interaction with G-quadruplex DNA within c-MYC promoter nuclease hypersensitive element induces c-MYC expression. Nucleic Acids Res. *37*, 172–183.

Tolstonog, G.V., Li, G., Shoeman, R.L., and Traub, P. (2005). Interaction *in vitro* of type III intermediate filament proteins with higher order structures of single-stranded DNA, particularly with G-quadruplex DNA. DNA Cell Biol. *24*, 85–110.

Uliel, L., Weisman-Shomer, P., Oren-Jazan, H., Newcomb, T., Loeb, L.A., and Fry, M. (2000). Human Ku antigen tightly binds and stabilizes a tetrahelical form of the Fragile X syndrome d(CGG) in expanded sequence. J. Biol. Chem. *275*, 33134–33141.

Verma, A., Halder, K., Halder, R., Yadav, V.K., Rawal, P., Thakur, R.K., Mohd, F., Sharma, A., and Chowdhury, S. (2008). Genome-wide computational and expression analyses reveal G-quadruplex DNA motifs as conserved cis-regulatory elements in human and related species. J. Med. Chem. *51*, 5641–5649.

Wadkins, R.M. (2000). Targeting DNA secondary structures. Curr. Med. Chem. *7*, 1–15.

Wei, D., Parkinson, G.N., Reszka, A.P., and Neidle, S. (2012). Crystal structure of a c-kit promoter quadruplex reveals the structural role of metal ions and water molecules in maintaining loop conformation. Nucleic Acids Res. *40*, 4691–4700.

Weismanshomer, P., and Fry, M. (1994). Stabilization of tetrahelical DNA by the quadruplex DNA binding protein QUAD. Biochem. Biophys. Res. Commun. *205*, 305–311.

Williamson, J.R., Raghuraman, M.K., and Cech, T.R. (1989). Monovalent cation-induced structure of telomeric DNA: the G-quartet model. Cell *59*, 871–880.

Wong, H.M., and Huppert, J.L. (2009). Stable G-quadruplexes are found outside nucleosome-bound regions. Mol. Biosyst. *5*, 1713–1719.

Yadav, V.K., Abraham, J.K., Mani, P., Kulshrestha, R., and Chowdhury, S. (2007). QuadBase: genome-wide database of G4 DNA occurrence and conservation in human, chimpanzee, mouse and rat promoters and 146 microbes. Nucleic Acids Res. *36*, D381–D385.

Zacharias, W., Jaworski, A., and Wells, R.D. (1990). Cytosine methylation enhances Z-DNA formation *in vivo*. J. Bacteriol. *172*, 3278–3283.

Zaug, A.J., Podell, E.R., and Cech, T.R. (2005). Human POT1 disrupts telomeric G-quadruplexes allowing telomerase extension *in vitro*. Proc. Natl. Acad. Sci. U.S.A. *102*, 10864–10869.

Zhao, J., Bacolla, A., Wang, G., and Vasquez, K.M. (2010). Non-B DNA structure-induced genetic instability and evolution. Cell. Mol. Life Sci. *67*, 43–62.

Clinical Epigenetics in Cancer

14

María G. García*, Estela G. Toraño*, Agustín F. Fernández and Mario F. Fraga

Abstract

Tumours harbour many epigenetic alterations affecting, among other things, genomic DNA methylation and histone post-translational modifications. Different epigenetic alterations are associated with the specific characteristics of each type of tumour, which has led to exploration of their potential use as clinical tumoral markers. Indeed, recent evidence increasingly points to epigenetic markers being a promising clinical tool for diagnosis, prognosis and therapy in cancer. Future challenges in the field include the routine use of genome-wide technologies and the development of new epigenetic drugs.

Introduction

It is well known that epigenetics plays an important role in the development of a multitude of diseases, including cancer (Dawson and Kouzarides, 2012; Ogino *et al.*, 2013; Tsai and Baylin, 2011). Next generation sequence approaches have proved to be a valuable tool with which to study alterations in the epigenome during the human disease process. The identification of these changes can be used for the identification of specific biomarkers for disease detection, diagnosis and prognosis. Also, it could be useful for the development of new personalized therapies which increase efficiency and have fewer side effects.

The best-known epigenetic modification is *DNA methylation*. This epigenetic mark plays a role in genomic imprinting, X-chromosome inactivation and in the silencing of repetitive elements and specific genes. It consists in a covalent addition of a methyl group at the 5′ position of a cytosine that precedes a guanine (CpG). The distribution of these CpGs is not random, they are clustered in specific regions which are located in repetitive sequences, gene bodies, and the 5′ region (promoter, untranslated region and exon 1) of many genes and are known as 'CpG islands' (CGIs). Most CpGs are methylated except those located in the promoter gene regions, which are usually unmethylated (Bird, 1986; Esteller, 2007).

Another important epigenetic mechanism is *histone modifications*. Histone tails are susceptible to post-translational modifications (PTMs) such as methylation, acetylation, ubiquitination and phosphorylation, among others. These covalent modifications have effects on chromatin organization and, as a consequence, play a role in gene activation and inactivation depending on the degree of accessibility of the transcriptional machinery involved (Delcuve *et al.*, 2009). Together, the different PTMs, which can occur simultaneously, and the effects they exert are referred to as the 'Histone Code' (Kouzarides, 2007).

The central role that epigenetic processes play in the formation of a large number of diseases suggests the importance of finding new epigenetic biomarkers to improve diagnostic, prognostic and therapeutic approaches. Currently, there are few biomarkers that have proved appropriate for clinical applications (Heyn *et al.*, 2013). For many years, genetic disorders, such as familial hypercholesterolaemia, human hypertension, and other disorders, have been the main molecular target for improving understanding of the causes of pathogenesis and for creating personalized medicine (Novelli *et al.*, 2008). In some cases

*These authors contributed equally to this chapter

there are genetic mutations that affect epigenome organization which directly participate in abnormal epigenetic events in many types of tumour or other human diseases (Egger *et al.*, 2004). For example, genetic aberrations in DNA methyltransferases can modify the DNA's functional activity and cause several human disorders, such as mutations in *DNMT3A* which have been associated with a strong loss of methylation activity (Yamashita *et al.*, 2010) and poor outcome in patients with acute myeloid leukaemia (AML) (Ley *et al.*, 2010). Another condition, ICF (immunodeficiency centromeric region instability and facial anomalies), is due to mutations in the *DNMT3B* gene (Ehrlich, 2003).

Mutations have also been identified in chromatin-modifying enzymes in various cancers. The compactional status of chromatin is influenced by the activity of enzymes such as HAT and HDAC (Kouzarides, 1999), which form part of different protein complexes and whose recruitment to the site of action is brought about by co-repressor molecules. In various types of leukaemia, certain fusion proteins, originating through different translocations, can become bound to these protein complexes and guide their activity (Hake *et al.*, 2004). Acute myeloid leukaemia gene 1 (*AML1*) is often implicated in chromosomal translocations in this liquid tumour type and encodes the AML1 protein. When the latter translocates with ETO protein, the fusion protein interacts with the repressor protein complex N-CoR–Sin3–HDAC1 and suppresses transcription by maintaining histones in a deacetylated state. The result is the blocking of haematopoietic cell differentiation and leukaemia develops (Fernandez *et al.*, 2012; Volkmar *et al.*, 2012).

The use of recently emerged high-throughput approaches such as DNA methylation arrays, genome-wide bisulfite sequencing, meDIP-seq, and ChIP-seq (Heyn and Esteller, 2012) could open a gateway to the development of new specific candidates for clinical situations. Besides, the reversibility of methylation processes through the use, for example, of demethylating agents gives epigenetic biomarkers an advantage over genetic biomarkers (Esteller, 2005, 2006).

The aim of this chapter is to summarize the different epigenetic modifications that are or would be useful for clinical trials. We shall focus mainly on DNA methylation marks since histone PTMs are more complex to analyse and the changes are less stable than those resulting from DNA methylation (Feinberg, 2007).

Epigenetics for disease diagnosis

The early detection of disease is one of the most important goals in the development of biomarkers. What is more, the use of non-invasive methods, such as blood, urine, saliva, or stool analysis, for disease detection possesses a clear advantage for patients compared to the use of invasive methods such as biopsy.

Of all common diseases, the influence of epigenetic alterations in cancer has been the most studied (Esteller, 2008; Feinberg and Tycko, 2004). Indeed, it was the first human disease linked to epigenetics, in 1983 when researchers found changes in DNA methylation between normal and tumoral tissues in colon cancer (Feinberg and Vogelstein, 1983). Currently, epigenetics is one of most promising fields of biomedical studies in terms of increasing our understanding of the aetiology of cancer as well as other diseases.

Aberrant changes in methylation patterns or chromatin modifications are common in tumour cells, and may even take place before symptoms are present. The early detection of such differences compared with normal cells could be helpful in the diagnosis and classification of tumours, and reduce the high mortality related with late diagnosis. The novel, quick and sensitive techniques used in epigenetics contribute to the early identification of genetic targets in cancer.

Epigenetic alterations deregulate important cellular pathways and steps involved in carcinogenesis. Altered epigenetic patterns and chromosomal instability may affect transcriptional activity and can give rise to tumour initiation and progression (Jones, 2012; Jones and Baylin, 2002; Sincic and Herceg, 2011).

Interest in the study and analysis of genome DNA methylation has increased due to the link between alterations in DNA methylation patterns and early-stage cancers (Egger *et al.*, 2004; Irizarry *et al.*, 2009; Jones and Baylin, 2002; Robertson,

2002; Weisenberger *et al.*, 2006). Likewise, epigenetic alterations have the potential to become valuable biomarkers for early detection screening and diagnosis of tumorigenesis. The stability of DNA allows aberrant methylation patterns in this molecule to be analysed using non-invasive, automated and semiquantitative methods (Esteller *et al.*, 1999; Mayor *et al.*, 2009). Tumour DNA can be present in body fluids like blood, plasma, sputum (Palmisano *et al.*, 2000), urine (Cairns *et al.*, 2001), stools, ductal lavage (Evron *et al.*, 2001), semen (Goessl *et al.*, 2000), bronchoalveolar lavage (Ahrendt *et al.*, 1999) or saliva (Rosas *et al.*, 2001). For example, tumour blood cells can be directly isolated from the primary tissue (Sandoval *et al.*, 2013); colon cancer cell DNA can be detected in stools (Glockner *et al.*, 2009; Muller *et al.*, 2004); and urine analysis is useful in prostate cancer (Payne *et al.*, 2009).

Circulating free DNA (CFDNA) of tumour cells can be found in blood. Its molecular weight is lower than that of genomic DNA, as well as its levels being higher in cancer patients than in healthy individuals (Nogueira da Costa and Herceg, 2012). While in normal individuals most free DNA is haematopoietic in origin, in cancer patients it may also derive from apoptotic and necrotic cancer cells (Jahr *et al.*, 2001). Depending on the type of cancer, its stage, grade and location, the amount of CFDNA can oscillate such that changes in its concentration in blood may potentially be used as a biomarker in tumour diagnosis. In addition, analysing tumoral CFDNA it is possible to detect epigenetic alterations (Begum *et al.*, 2011; Chang *et al.*, 2008; Huang *et al.*, 2011; Iyer *et al.*, 2010), and some of these aberrant profiles have clinical applications. For example, higher percentages of methylated vimentin have been found in tumour DNA serum of patients with gastric cancer (Shirahata *et al.*, 2012) and an increased detection of susceptibility to colon cancer detection was found using septin 9 methylation biomarker (Warren *et al.*, 2011). In fact, using a panel of only a few genes, it has proved possible to obtain a particular diagnosis for cancer patients (Chan *et al.*, 2002; Dulaimi *et al.*, 2004). However, although there is a strong association between promoter DNA methylation of a certain few genes and cancer, improvement in the sensitivity of such tools is needed in order to assess multiple epigenetic markers at the same time (Herceg and Hainaut, 2007).

In cancer processes, global hypomethylation often accompanies local CGI hypermethylation of promoter regions of specific genes involved in various cellular pathways (cell cycle, apoptosis or genome maintenance) (Baylin *et al.*, 2001; Laird, 2005). Both mechanisms may promote malignant cell formation and tumour development. Hypomethylation is, generally, characterized by a decline in genome-wide methylation levels in CpGs of repetitive regions and in CGIs outside of promoter regions (Jones, 2012) which leads to genomic instability and the re-expression of silenced genes (Berdasco and Esteller, 2010; Jones and Baylin, 2007).

As mention above, hypermethylation consist in the addition of methyl groups, normally in CGIs located in promoter regions of tumour suppressor genes (Esteller, 2002). These regions are normally unmethylated in normal cells and become hypermethylated in cancer cells. This excess of methyl groups in promoter regions, which impedes the binding of transcription factor, could produce gene inactivation of tumour suppressor genes.

Although global hypomethylation is a common event in different types of cancer, the detection of DNA hypomethylation is more complicated than promoter-specific DNA methylation detection, despite there being many different techniques to detect it (Toraño *et al.*, 2012). Furthermore, it is currently very difficult to establish a relationship between global DNA hypomethylation and cancer, since hypomethylation is also a common feature of ageing and other diseases (Johnson *et al.*, 2012). Recent studies using methylation microarrays have identified the aberrant DNA hypomethylation of specific genes, such as *CCND1*, *SPRR3*, *BTC*, EGF and *FGF6* in neuroblastoma pathogenesis, as being suitable as biomarkers (Mayol *et al.*, 2012). Loss of imprinting, which is a common event in cancer disease, has been seen on the insulin-like growth factor 2 (*IGF2*) gene. It has been linked to hypomethylation and is considered useful in colon cancer diagnosis (Cui *et al.*, 2002; Ito *et al.*, 2008).

However, it is in fact more frequent to find aberrant hypermethylation profiles associated with cancer. In some cases these findings have allowed the location of target genes which are suitable for early diagnosis.

One example is p16INK4a (*p16*). This tumour suppressor gene, involved in the cell cycle, is aberrantly hypermethylated in a wide range of human cancers (Merlo *et al.*, 1995; Rocco and Sidransky, 2001). In lung cancer, *p16* inactivation is necessary not only for tumour progression, but also for the formation of precursor lesions and consequently the detection of *p16* methylation changes in sputum may be used as a biomarker for early diagnosis, detection and risk assessment (Belinsky, 2004; Belinsky *et al.*, 1998). Abnormal promoter DNA hypermethylation of this tumour suppressor gene can also be detected in the serum of cancer patients for diagnostic purposes. A significant proportion of aberrant methylation of the *p16* gene has also been detected in the peripheral circulation of colorectal cancer patients (Zou *et al.*, 2002), in the plasma and serum of hepatocellular carcinoma patients (Wong *et al.*, 1999), and in early stages of lung cancer (Belinsky *et al.*, 1998). Proteins such as MLH1 and BRCA1, belonging to DNA repair machinery (Fukui, 2010; Kantelinen *et al.*, 2010; Venkitaraman, 2002) are associated with aberrant methylation in colon (Menigatti *et al.*, 2001), endometrial (Bischoff *et al.*, 2012), ovarian (Esteller *et al.*, 2000b; Ozdemir *et al.*, 2012) and breast cancer (Dobrovic and Simpfendorfer, 1997).

In contrast, the determination of changes in histone PTMs in disease in order to develop new biomarkers is virtually unexplored. There is no doubt that they play an important role in epigenetic regulation and their mis-regulation may lead to disease. Acetylation and methylation are the main histone PTMs that have been shown to be clearly involved in cancer development (Fraga *et al.*, 2005). The loss of either mark has been identified as a cancer biomarker (Fraga *et al.*, 2005). Also, alterations of the enzymes responsible for histone methylation and histone acetylation have been implicated in various diseases. Fusion proteins which involve HATs have been identified in many haematological malignancies (Ellis *et al.*, 2009). The development of new and more accurate approaches is necessary for the detection of histone PTM biomarkers. This would be of huge clinical benefit because, as mentioned above, epigenetic alterations take place early and at high frequency in cancer; thus, it could be possible to detect the disease even before symptoms appearance. However, there are currently very few epigenetic trials in clinical practice approved by the US Food and Drug Administration (FDA). Further investigation is necessary to include more clinical routine screenings using epigenetic biomarkers.

Epigenetics for disease prognosis

Apart from aiding diagnosis, epigenetic features could also help to determine the status of the disease, the clinical outcome and, in the case of tumours, the recurrence probability and chemotherapy response.

There are many different promoter hypermethylation events in cancer which are often associated with poor prognosis (Bhalla, 2005; Das and Singal, 2004; Herranz and Esteller, 2007). For example, *CDH1* (cadherin 1) is associated with poor clinical outcome and Dkk-3 with poor survival in gastric cancer (Graziano *et al.*, 2004; Tamura *et al.*, 2000; Yu *et al.*, 2009). In acute leukaemia, *CDKN2A* and *CDKN1A* have been associated with poor clinical outcome (Bhalla, 2005). The hypermethylation of *NSD1* in neuroblastomas, *DAPK1* in lung cancer, *EMP3* in brain tumours and *CDKN2A* in colon cancer provide some examples of the link between aberrant epigenetics and poor outcome in cancer (Rodriguez-Paredes and Esteller, 2011). In prostate disease, methylation promoter detection of the glutathione-S-transferase P1 (*GSTP1*) gene in serum can be helpful in the distinguishing between malignant and non-malignant prostate neoplasias (Hashad *et al.*, 2011).

The identification and removal of metastatic tumour cells are crucial in cancer prognosis. DNA methylation alterations in the cadherin family have been found firstly, in breast cancer (Graff *et al.*, 2000) and, later, in melanoma and head and neck cancers (Carmona *et al.*, 2012). In other cancer types they have been identified as markers of metastasis (Fang *et al.*, 2011; Lujambio *et al.*, 2008).

In addition to single gene prognostic biomarkers, it is possible to identify sets of methylated genes through the use of genome-wide methylation approaches and DNA methylation panels. The study of the hypermethylation status of *p16*, *SOCS1*, *GSTP1* and *CDH1* genes has been associated with poor prognosis in hepatocellular carcinoma (Mah and Lee, 2014). Through hypermethylome screening, Jeschke and colleagues found a set of hypermethylated genes (*CDO1*, *CKM*, *CRIP1*, *KL* and *TAC1*) which predicts poor overall survival in breast cancer (Jeschke *et al.*, 2012). Another methylation panel has been designed which detects metastatic breast cancer (Fackler *et al.*, 2014).

The prognostic potential of DNA methylation analysis was first reported for childhood acute lymphoblastic leukaemia (Milani *et al.*, 2010; Stumpel *et al.*, 2009). In this study, DNA methylation profiles allowed the identification of different subgroups which stratified patients with different clinical outcome. Specific methylation profiles have also been established in different subtypes of acute myeloid leukaemia (AML) (Figueroa *et al.*, 2010) and breast cancer (Bediaga *et al.*, 2010). In contrast, additional studies are needed to determine the potential clinical interest of the identification of thyroid cancer subtypes associated with promoter aberrant methylation (Rodriguez-Rodero *et al.*, 2013).

Apart from DNA hypermethylation, another modification linked to poor survival in many tumours is *LINE-1* hypomethylation (Baba *et al.*, 2014; Benard *et al.*, 2013; Ogino *et al.*, 2008; Zhu *et al.*, 2014), and also promoter specific DNA hypomethylation of some genes has been related to poor prognosis. In addition, genes which possess high DNA methylation levels in embryonic stem cells are shown hypomethylated in many types of cancer and related with poor prognosis (Heyn and Esteller, 2012; Zhuang *et al.*, 2012). In non-small cell lung cancer, hypomethylation of the *CHRNB4* gene is an unfavourable prognostic factor (Yoo *et al.*, 2014), and in epithelial ovarian cancer DNA hypomethylation at 6p21.3 is an important predictor of recurrence (Wang *et al.*, 2014).

Histone PTMs have also been found to be related to clinical outcome. For example, reduced histone H4 acetylation and H3K9 methylation have been found to be correlated with poor prognosis in gastric cancer patients (Ono *et al.*, 2002; Park *et al.*, 2008; Yasui *et al.*, 2003), and tumour recurrence in prostate cancer (Seligson *et al.*, 2005).

The ability to identify tumour subtypes, their differing response to treatments and the possibility of relapse greatly improves the accuracy of prognosis. However, for the future, DNA methylation and histone PTMs screenings using next generation sequence approaches need to be performed in order to find sets of epigenetic marks which give indications of patient survival, cancer subtype, and the possibility of recurrence or metastasis.

Epigenetics for drug response

Apart from prognosis and diagnosis, biomarkers could provide support to decisions regarding the best treatment for the patient. Some epigenetic features could give an explanation of why some patients respond to a specific treatment and why others not (Liu *et al.*, 2014). Knowing in advance if the patient is going to respond to the treatment is the key to effective personalized treatment.

The DNA-repair enzyme O6-methylguanine-DNA methyltransferase (MGMT) protects cancer cells against alkylating agents by reversing cross-link formations in the DNA. Several studies indicate that its levels are lower in a large group of gliomas (Silber *et al.*, 1993, 1998). The lack of expression of this gene, due to promoter methylation, has been associated not only with better responsiveness of these tumours to alkylating agents, but also with a favourable outcome in some subtypes (Esteller *et al.*, 2000a; Weller *et al.*, 2010).

There are a number of other known examples of DNA hypermethylation of DNA-repair enzyme promoters which predict response to chemotherapy. One is the hypermethylation of the mismatch repair of the *MLH1* gene and *BRCA1* which are related to cisplatin response. While the first has been found to be related to in vitro resistance to cisplatin in ovary cancer cells (Strathdee *et al.*, 1999), the second has been associated with increased sensitivity to this chemotherapeutic agent in breast as well as ovarian cancer (Stefansson *et al.*, 2012).

Another example of the epigenetic effect on therapeutic response to chemotherapeutic substances in a clinical setting is the case of *GSTP1*. This metabolic enzyme takes part in cell detoxification, removing toxic substances or anticancer drugs through the conjugation of glutathione (Coles and Kadlubar, 2003). In vitro studies have demonstrated that the reduction of *GSTP1* expression by promoter hypermethylation in tumour cells is correlated with response to chemotherapy (Miyake *et al.*, 2012; Nakamichi *et al.*, 2007; Townsend and Tew, 2003; Zhang *et al.*, 2009). In contrast, in CRCs, DNA hypomethylation, which is measured by *LINE-1* methylation, has been described as a biomarker of poor prognosis, with respect to response to oral fluoropyrimidines (Kawakami *et al.*, 2011).

The identification of further candidates implicated in a better or worse response to specific treatment has high potential clinical applications. In addition, knowing beforehand which treatment is the best in each patient's case could reduce the economic costs of cancer treatment.

Epigenetic therapy

The possibility of reversing epigenetic abnormalities makes the development of drugs which can bring about these changes of great interest for clinical practice (Helin and Dhanak, 2013). In this respect, there are a number of DNA methylation and histone deacetylase inhibitors (HDACis) which are currently under investigation.

There are two DNA methylation inhibitors approved by the FDA in the USA for the treatment of certain myelodysplastic syndromes; 5-azacytidine (azacitidine) and 5-aza-2′-deoxycytidine (decitabine). The basis of the use of these kinds of drugs is that they are able to reactivate the expression of tumour suppressor genes which have been silenced by promoter DNA hypermethylation. However, while only the reactivation of specific genes is necessary in these cases, the action of these drugs is unspecific, and the long-term effects of their use still need to be evaluated.

In the case of HDACis, they are able to inhibit HDAC activity (Kretsovali *et al.*, 2012), and are classified into various families: the hydroxamic acids [trichostatin A (TSA)], the short-chain fatty acids [sodium butyrate, phenylbutyrate, and valproic acid (VPA)], suberoylanilide hydroxamic acid (SAHA), the epoxy ketones (trapoxin), and the benzamides (Abel and Zukin, 2008). The biological responses they are involved in are very varied and affect different molecular processes (Bolden *et al.*, 2006). Apart from causing an increase in histone acetylation, there are many non-histone targets for HDACs, including p53, NF-kB and E2F1 (Johnstone and Licht, 2003; Johnstone *et al.*, 2002). A better understanding of the pathways of HDACi activity is necessary to decide when use of HDACi treatment is indicated.

Demethylating agents used alone are not able to completely reverse aberrant epigenetic alterations in hematologic tumours (McGarvey *et al.*, 2006), and it has been demonstrated that there is no response to their use as treatment for many solid tumours (Abele *et al.*, 1987). Epigenetic drugs are thus often best employed in combination with other drugs, one of the best examples being their combination with antineoplastic treatments in chemotherapy or with HDACis. For instance, IL-2 combined with 5-aza-2′-deoxycytidine improves response in malignant melanoma (Gollob *et al.*, 2006). Also, both azacitidine and decitabine have shown promising results when combined with chemotherapeutic agents approved by the FDA, such as lenalidomide in the case of myelodysplastic syndromes, imatinib mesylate for myelogenous leukaemia, gemtuzumab ozogamicin for acute myeloid leukaemia, and carboplatin for ovarian cancer (Chowdhury *et al.*, 2009; Fang *et al.*, 2014; Sekeres *et al.*, 2010). Also, the combination of DNA demethylating agents with HDACis causes a synergistic (or additive) effect in cancer treatment (Herranz and Esteller, 2007; Soriano *et al.*, 2007). For instance, they increase the apoptotic effect of chemotherapeutic agents in different types of cancer, such as bladder cancer (Wang *et al.*, 2013a), neuroblastoma (Wang *et al.*, 2013b), and glioblastoma (Hamed *et al.*, 2013), among others. Further investigations, however, need to be carried out with a view to improving the use of epigenetic drugs as an adjuvant therapy in cancer.

Conclusions

To date, there are few validated epigenetic biomarkers currently in clinical use. Furthermore, all those being used are obtained through single-gene approaches. For the improvement and development of personalized medicine, automation is necessary, and a huge reduction in the cost of high-throughput approaches.

Acknowledgements

We thank Ronnie Lendrum for editorial assistance. This work has been financially supported by FICYT to (M.G.G. and E.G.T.), the Spanish National Research Council (CSIC; 2008201172 to M.F.F.); Fundación Ramón Areces (to M.F.F); the Fondo de Investigaciones Sanitarias FIS/FEDER (PI11/01728 to A.F.F., PI12/01080 to M.F.F.). A.F.F. is sponsored by ISCIII-Subdirección General de Evaluación y Fomento de la Investigación (CP11/00131). The IUOPA is supported by the Obra Social Cajastur, Spain.

References

Abel, T., and Zukin, R.S. (2008). Epigenetic targets of HDAC inhibition in neurodegenerative and psychiatric disorders. Curr. Opin. Pharmacol. *8*, 57–64.

Abele, R., Clavel, M., Dodion, P., Bruntsch, U., Gundersen, S., Smyth, J., Renard, J., van Glabbeke, M., and Pinedo, H.M. (1987). The EORTC Early Clinical Trials Cooperative Group experience with 5-aza-2′-deoxycytidine (NSC 127716) in patients with colo-rectal, head and neck, renal carcinomas and malignant melanomas. Eur. J. Cancer Clin. Oncol. *23*, 1921–1924.

Ahrendt, S.A., Chow, J.T., Xu, L.H., Yang, S.C., Eisenberger, C.F., Esteller, M., Herman, J.G., Wu, L., Decker, P.A., Jen, J., *et al.* (1999). Molecular detection of tumor cells in bronchoalveolar lavage fluid from patients with early stage lung cancer. J. Natl. Cancer Inst. *91*, 332–339.

Baba, Y., Watanabe, M., Murata, A., Shigaki, H., Miyake, K., Ishimoto, T., Iwatsuki, M., Iwagami, S., Yoshida, N., Oki, E., *et al.* (2014). LINE-1 hypomethylation, DNA copy number alterations, and CDK6 amplification in esophageal squamous cell carcinoma. Clin. Cancer Res. *20*, 1114–1124.

Baylin, S.B., Esteller, M., Rountree, M.R., Bachman, K.E., Schuebel, K., and Herman, J.G. (2001). Aberrant patterns of DNA methylation, chromatin formation and gene expression in cancer. Hum. Mol. Genet. *10*, 687–692.

Bediaga, N.G., Acha-Sagredo, A., Guerra, I., Viguri, A., Albaina, C., Ruiz Diaz, I., Rezola, R., Alberdi, M.J., Dopazo, J., Montaner, D., *et al.* (2010). DNA methylation epigenotypes in breast cancer molecular subtypes. Breast Cancer Res. *12*, R77.

Begum, S., Brait, M., Dasgupta, S., Ostrow, K.L., Zahurak, M., Carvalho, A.L., Califano, J.A., Goodman, S.N., Westra, W.H., Hoque, M.O., *et al.* (2011). An epigenetic marker panel for detection of lung cancer using cell-free serum DNA. Clin. Cancer Res. *17*, 4494–4503.

Belinsky, S.A. (2004). Gene-promoter hypermethylation as a biomarker in lung cancer. Nat. Rev. Cancer *4*, 707–717.

Belinsky, S.A., Nikula, K.J., Palmisano, W.A., Michels, R., Saccomanno, G., Gabrielson, E., Baylin, S.B., and Herman, J.G. (1998). Aberrant methylation of p16(INK4a) is an early event in lung cancer and a potential biomarker for early diagnosis. Proc. Natl. Acad. Sci. U.S.A. *95*, 11891–11896.

Benard, A., van de Velde, C.J., Lessard, L., Putter, H., Takeshima, L., Kuppen, P.J., and Hoon, D.S. (2013). Epigenetic status of LINE-1 predicts clinical outcome in early-stage rectal cancer. Br. J. Cancer *109*, 3073–3083.

Berdasco, M., and Esteller, M. (2010). Aberrant epigenetic landscape in cancer: how cellular identity goes awry. Dev. Cell *19*, 698–711.

Bhalla, K.N. (2005). Epigenetic and chromatin modifiers as targeted therapy of hematologic malignancies. J. Clin. Oncol. *23*, 3971–3993.

Bird, A.P. (1986). CpG-rich islands and the function of DNA methylation. Nature *321*, 209–213.

Bischoff, J., Ignatov, A., Semczuk, A., Schwarzenau, C., Ignatov, T., Krebs, T., Kuster, D., Przadka-Rabaniuk, D., Roessner, A., Costa, S.D., *et al.* (2012). hMLH1 promoter hypermethylation and MSI status in human endometrial carcinomas with and without metastases. Clin. Exp. Metastasis *29*, 889–900.

Bolden, J.E., Peart, M.J., and Johnstone, R.W. (2006). Anticancer activities of histone deacetylase inhibitors. Nat. Rev. Drug Discov. *5*, 769–784.

Cairns, P., Esteller, M., Herman, J.G., Schoenberg, M., Jeronimo, C., Sanchez-Cespedes, M., Chow, N.H., Grasso, M., Wu, L., Westra, W.B., *et al.* (2001). Molecular detection of prostate cancer in urine by GSTP1 hypermethylation. Clin. Cancer Res. *7*, 2727–2730.

Carmona, F.J., Villanueva, A., Vidal, A., Munoz, C., Puertas, S., Penin, R.M., Goma, M., Lujambio, A., Piulats, J.M., Mesia, R., *et al.* (2012). Epigenetic disruption of cadherin-11 in human cancer metastasis. J. Pathol. *228*, 230–240.

Chan, M.W., Chan, L.W., Tang, N.L., Tong, J.H., Lo, K.W., Lee, T.L., Cheung, H.Y., Wong, W.S., Chan, P.S., Lai, F.M., *et al.* (2002). Hypermethylation of multiple genes in tumor tissues and voided urine in urinary bladder cancer patients. Clin. Cancer Res. *8*, 464–470.

Chang, H., Yi, B., Li, L., Zhang, H.Y., Sun, F., Dong, S.Q., and Cao, Y. (2008). Methylation of tumor associated genes in tissue and plasma samples from liver disease patients. Exp. Mol. Pathol. *85*, 96–100.

Chowdhury, S., Seropian, S., and Marks, P.W. (2009). Decitabine combined with fractionated gemtuzumab ozogamicin therapy in patients with relapsed or refractory acute myeloid leukemia. Am. J. Hematol. *84*, 599–600.

Coles, B.F., and Kadlubar, F.F. (2003). Detoxification of electrophilic compounds by glutathione S-transferase catalysis: determinants of individual response to chemical carcinogens and chemotherapeutic drugs? Biofactors *17*, 115–130.

Cui, H., Onyango, P., Brandenburg, S., Wu, Y., Hsieh, C.L., and Feinberg, A.P. (2002). Loss of imprinting in colorectal cancer linked to hypomethylation of H19 and IGF2. Cancer Res. *62*, 6442–6446.

Das, P.M., and Singal, R. (2004). DNA methylation and cancer. J. Clin. Oncol. *22*, 4632–4642.

Dawson, M.A., and Kouzarides, T. (2012). Cancer epigenetics: from mechanism to therapy. Cell *150*, 12–27.

Delcuve, G.P., Rastegar, M., and Davie, J.R. (2009). Epigenetic control. J. Cell Physiol. *219*, 243–250.

Dobrovic, A., and Simpfendorfer, D. (1997). Methylation of the BRCA1 gene in sporadic breast cancer. Cancer Res. *57*, 3347–3350.

Dulaimi, E., Uzzo, R.G., Greenberg, R.E., Al-Saleem, T., and Cairns, P. (2004). Detection of bladder cancer in urine by a tumor suppressor gene hypermethylation panel. Clin. Cancer Res. *10*, 1887–1893.

Egger, G., Liang, G., Aparicio, A., and Jones, P.A. (2004). Epigenetics in human disease and prospects for epigenetic therapy. Nature *429*, 457–463.

Ehrlich, M. (2003). The ICF syndrome, a DNA methyltransferase 3B deficiency and immunodeficiency disease. Clin. Immunol. *109*, 17–28.

Ellis, L., Atadja, P.W., and Johnstone, R.W. (2009). Epigenetics in cancer: targeting chromatin modifications. Mol. Cancer Ther. *8*, 1409–1420.

Esteller, M. (2002). CpG island hypermethylation and tumor suppressor genes: a booming present, a brighter future. Oncogene *21*, 5427–5440.

Esteller, M. (2005). DNA methylation and cancer therapy: new developments and expectations. Curr. Opin. Oncol. *17*, 55–60.

Esteller, M. (2006). Epigenetics provides a new generation of oncogenes and tumour-suppressor genes. Br. J. Cancer *94*, 179–183.

Esteller, M. (2007). Cancer epigenomics: DNA methylomes and histone-modification maps. Nat. Rev. Genet. *8*, 286–298.

Esteller, M. (2008). Epigenetics in cancer. N Engl J Med *358*, 1148–1159.

Esteller, M., Sanchez-Cespedes, M., Rosell, R., Sidransky, D., Baylin, S.B., and Herman, J.G. (1999). Detection of aberrant promoter hypermethylation of tumour suppressor genes in serum DNA from non-small cell lung cancer patients. Cancer Res *59*, 67–70.

Esteller, M., Garcia-Foncillas, J., Andion, E., Goodman, S.N., Hidalgo, O.F., Vanaclocha, V., Baylin, S.B., and Herman, J.G. (2000a). Inactivation of the DNA-repair gene MGMT and the clinical response of gliomas to alkylating agents. N Engl J Med *343*, 1350–1354.

Esteller, M., Silva, J.M., Dominguez, G., Bonilla, F., Matias-Guiu, X., Lerma, E., Bussaglia, E., Prat, J., Harkes, I.C., Repasky, E.A., *et al.* (2000b). Promoter hypermethylation and BRCA1 inactivation in sporadic breast and ovarian tumours. J Natl Cancer Inst *92*, 564–569.

Evron, E., Dooley, W.C., Umbricht, C.B., Rosenthal, D., Sacchi, N., Gabrielson, E., Soito, A.B., Hung, D.T., Ljung, B., Davidson, N.E., *et al.* (2001). Detection of breast cancer cells in ductal lavage fluid by methylation-specific PCR. Lancet *357*, 1335–1336.

Fackler, M.J., Lopez Bujanda, Z., Umbricht, C., Teo, W.W., Cho, S., Zhang, Z., Visvanathan, K., Jeter, S., Argani, P., Wang, C., *et al.* (2014). Novel methylated biomarkers and a robust assay to detect circulating tumour DNA in metastatic breast cancer. Cancer Res *74*, 2160–2170.

Fang, F., Turcan, S., Rimner, A., Kaufman, A., Giri, D., Morris, L.G., Shen, R., Seshan, V., Mo, Q., Heguy, A., *et al.* (2011). Breast cancer methylomes establish an epigenomic foundation for metastasis. Sci Transl Med *3*, 75ra25.

Fang, F., Zuo, Q., Pilrose, J., Wang, Y., Shen, C., Li, M., Wulfridge, P., Matei, D., and Nephew, K.P. (2014). Decitabine reactivated pathways in platinum resistant ovarian cancer. Oncotarget *5*, 3579–3589.

Feinberg, A.P. (2007). Phenotypic plasticity and the epigenetics of human disease. Nature *447*, 433–440.

Feinberg, A.P., and Tycko, B. (2004). The history of cancer epigenetics. Nat. Rev. Cancer *4*, 143–153.

Feinberg, A.P., and Vogelstein, B. (1983). Hypomethylation of ras oncogenes in primary human cancers. Biochem. Biophys. Res. Commun. *111*, 47–54.

Fernandez, A.F., Assenov, Y., Martin-Subero, J.I., Balint, B., Siebert, R., Taniguchi, H., Yamamoto, H., Hidalgo, M., Tan, A.C., Galm, O., *et al.* (2012). A DNA methylation fingerprint of 1628 human samples. Genome Res. *22*, 407–419.

Figueroa, M.E., Lugthart, S., Li, Y., Erpelinck-Verschueren, C., Deng, X., Christos, P.J., Schifano, E., Booth, J., van Putten, W., Skrabanek, L., *et al.* (2010). DNA methylation signatures identify biologically distinct subtypes in acute myeloid leukaemia. Cancer Cell *17*, 13–27.

Fraga, M.F., Ballestar, E., Villar-Garea, A., Boix-Chornet, M., Espada, J., Schotta, G., Bonaldi, T., Haydon, C., Ropero, S., Petrie, K., *et al.* (2005). Loss of acetylation at Lys16 and trimethylation at Lys20 of histone H4 is a common hallmark of human cancer. Nat. Genet. *37*, 391–400.

Fukui, K. (2010). DNA mismatch repair in eukaryotes and bacteria. J Nucleic Acids *2010*.

Glockner, S.C., Dhir, M., Yi, J.M., McGarvey, K.E., Van Neste, L., Louwagie, J., Chan, T.A., Kleeberger, W., de Bruine, A.P., Smits, K.M., *et al.* (2009). Methylation of TFPI2 in stool DNA: a potential novel biomarker for the detection of colorectal cancer. Cancer Res *69*, 4691–4699.

Goessl, C., Krause, H., Muller, M., Heicappell, R., Schrader, M., Sachsinger, J., and Miller, K. (2000). Fluorescent methylation-specific polymerase chain reaction for DNA-based detection of prostate cancer in bodily fluids. Cancer Res *60*, 5941–5945.

Gollob, J.A., Sciambi, C.J., Peterson, B.L., Richmond, T., Thoreson, M., Moran, K., Dressman, H.K., Jelinek, J., and Issa, J.P. (2006). Phase I trial of sequential low-dose 5-aza-2′-deoxycytidine plus high-dose intravenous bolus interleukin-2 in patients with melanoma or renal cell carcinoma. Clin Cancer Res *12*, 4619–4627.

Graff, J.R., Gabrielson, E., Fujii, H., Baylin, S.B., and Herman, J.G. (2000). Methylation patterns of the E-cadherin 5′ CpG island are unstable and reflect the dynamic, heterogeneous loss of E-cadherin expression during metastatic progression. J. Biol. Chem. *275*, 2727–2732.

Graziano, F., Arduini, F., Ruzzo, A., Bearzi, I., Humar, B., More, H., Silva, R., Muretto, P., Guilford, P., Testa, E., *et al.* (2004). Prognostic analysis of E-cadherin gene promoter hypermethylation in patients with surgically resected, node-positive, diffuse gastric cancer. Clin Cancer Res *10*, 2784–2789.

Hake, S.B., Xiao, A., and Allis, C.D. (2004). Linking the epigenetic 'language' of covalent histone modifications to cancer. Br. J. Cancer *90*, 761–769.

Hamed, H.A., Yacoub, A., Park, M.A., Archer, K., Das, S.K., Sarkar, D., Grant, S., Fisher, P.B., and Dent, P. (2013). Histone deacetylase inhibitors interact with melanoma differentiation associated-7/interleukin-24 to kill primary human glioblastoma cells. Mol. Pharmacol. *84*, 171–181.

Hashad, D.I., Hashad, M.M., Talaat, I.M., and Ibrahim, M.A. (2011). Role of glutathione-S-transferase P1 hypermethylation in molecular detection of prostate cancer. Genet. Test Mol. Biomarkers *15*, 667–670.

Helin, K., and Dhanak, D. (2013). Chromatin proteins and modifications as drug targets. Nature *502*, 480–488.

Herceg, Z., and Hainaut, P. (2007). Genetic and epigenetic alterations as biomarkers for cancer detection, diagnosis and prognosis. Mol. Oncol. *1*, 26–41.

Herranz, M., and Esteller, M. (2007). DNA methylation and histone modifications in patients with cancer: potential prognostic and therapeutic targets. Methods Mol. Biol. *361*, 25–62.

Heyn, H., and Esteller, M. (2012). DNA methylation profiling in the clinic: applications and challenges. Nat. Rev. Genet. *13*, 679–692.

Heyn, H., Mendez-Gonzalez, J., and Esteller, M. (2013). Epigenetic profiling joins personalized cancer medicine. Expert Rev. Mol. Diagn. *13*, 473–479.

Huang, Z.H., Hu, Y., Hua, D., Wu, Y.Y., Song, M.X., and Cheng, Z.H. (2011). Quantitative analysis of multiple methylated genes in plasma for the diagnosis and prognosis of hepatocellular carcinoma. Exp. Mol. Pathol. *91*, 702–707.

Irizarry, R.A., Ladd-Acosta, C., Wen, B., Wu, Z., Montano, C., Onyango, P., Cui, H., Gabo, K., Rongione, M., Webster, M., *et al.* (2009). The human colon cancer methylome shows similar hypo- and hypermethylation at conserved tissue-specific CpG island shores. Nat. Genet. *41*, 178–186.

Ito, Y., Koessler, T., Ibrahim, A.E., Rai, S., Vowler, S.L., Abu-Amero, S., Silva, A.L., Maia, A.T., Huddleston, J.E., Uribe-Lewis, S., *et al.* (2008). Somatically acquired hypomethylation of IGF2 in breast and colorectal cancer. Hum. Mol. Genet. *17*, 2633–2643.

Iyer, P., Zekri, A.R., Hung, C.W., Schiefelbein, E., Ismail, K., Hablas, A., Seifeldin, I.A., and Soliman, A.S. (2010). Concordance of DNA methylation pattern in plasma and tumor DNA of Egyptian hepatocellular carcinoma patients. Exp. Mol. Pathol. *88*, 107–111.

Jahr, S., Hentze, H., Englisch, S., Hardt, D., Fackelmayer, F.O., Hesch, R.D., and Knippers, R. (2001). DNA fragments in the blood plasma of cancer patients: quantitations and evidence for their origin from apoptotic and necrotic cells. Cancer Res. *61*, 1659–1665.

Jeschke, J., Van Neste, L., Glockner, S.C., Dhir, M., Calmon, M.F., Deregowski, V., Van Criekinge, W., Vlassenbroeck, I., Koch, A., Chan, T.A., *et al.* (2012). Biomarkers for detection and prognosis of breast cancer identified by a functional hypermethylome screen. Epigenetics 7, 701–709.

Johnson, A.A., Akman, K., Calimport, S.R., Wuttke, D., Stolzing, A., and de Magalhaes, J.P. (2012). The role of DNA methylation in aging, rejuvenation, and age-related disease. Rejuvenation Res. *15*, 483–494.

Johnstone, R.W., and Licht, J.D. (2003). Histone deacetylase inhibitors in cancer therapy: is transcription the primary target? Cancer Cell *4*, 13–18.

Johnstone, R.W., Ruefli, A.A., and Lowe, S.W. (2002). Apoptosis: a link between cancer genetics and chemotherapy. Cell *108*, 153–164.

Jones, P.A. (2012). Functions of DNA methylation: islands, start sites, gene bodies and beyond. Nat. Rev. Genet. *13*, 484–492.

Jones, P.A., and Baylin, S.B. (2002). The fundamental role of epigenetic events in cancer. Nat. Rev. Genet. 3, 415–428.

Jones, P.A., and Baylin, S.B. (2007). The epigenomics of cancer. Cell *128*, 683–692.

Kantelinen, J., Kansikas, M., Korhonen, M.K., Ollila, S., Heinimann, K., Kariola, R., and Nystrom, M. (2010). MutSbeta exceeds MutSalpha in dinucleotide loop repair. Br. J. Cancer *102*, 1068–1073.

Kawakami, K., Matsunoki, A., Kaneko, M., Saito, K., Watanabe, G., and Minamoto, T. (2011). Long interspersed nuclear element-1 hypomethylation is a potential biomarker for the prediction of response to oral fluoropyrimidines in microsatellite stable and CpG island methylator phenotype-negative colorectal cancer. Cancer Sci. *102*, 166–174.

Kouzarides, T. (1999). Histone acetylases and deacetylases in cell proliferation. Curr. Opin. Genet. Dev. *9*, 40–48.

Kouzarides, T. (2007). Chromatin modifications and their function. Cell *128*, 693–705.

Kretsovali, A., Hadjimichael, C., and Charmpilas, N. (2012). Histone deacetylase inhibitors in cell pluripotency, differentiation, and reprogramming. Stem Cells Int. *2012*, 184154.

Laird, P.W. (2005). Cancer epigenetics. Hum. Mol. Genet. *14*(Spec. No 1), R65–76.

Ley, T.J., Ding, L., Walter, M.J., McLellan, M.D., Lamprecht, T., Larson, D.E., Kandoth, C., Payton, J.E., Baty, J., Welch, J., *et al.* (2010). DNMT3A mutations in acute myeloid leukemia. N. Engl. J. Med. *363*, 2424–2433.

Liu, M.Z., McLeod, H.L., He, F.Z., Chen, X.P., Zhou, H.H., Shu, Y., and Zhang, W. (2014). Epigenetic perspectives on cancer chemotherapy response. Pharmacogenomics *15*, 699–715.

Lujambio, A., Calin, G.A., Villanueva, A., Ropero, S., Sanchez-Cespedes, M., Blanco, D., Montuenga, L.M.,

Rossi, S., Nicoloso, M.S., Faller, W.J., *et al.* (2008). A microRNA DNA methylation signature for human cancer metastasis. Proc. Natl. Acad. Sci. U.S.A. *105*, 13556–13561.

McGarvey, K.M., Fahrner, J.A., Greene, E., Martens, J., Jenuwein, T., and Baylin, S.B. (2006). Silenced tumor suppressor genes reactivated by DNA demethylation do not return to a fully euchromatic chromatin state. Cancer Res. *66*, 3541–3549.

Mah, W.C., and Lee, C.G. (2014). DNA methylation: potential biomarker in hepatocellular carcinoma. Biomark Res. *2*, 5.

Mayol, G., Martin-Subero, J.I., Rios, J., Queiros, A., Kulis, M., Sunol, M., Esteller, M., Gomez, S., Garcia, I., de Torres, C., *et al.* (2012). DNA hypomethylation affects cancer-related biological functions and genes relevant in neuroblastoma pathogenesis. PLoS ONE 7, e48401.

Mayor, R., Casadome, L., Azuara, D., Moreno, V., Clark, S.J., Capella, G., and Peinado, M.A. (2009). Long-range epigenetic silencing at 2q14.2 affects most human colorectal cancers and may have application as a non-invasive biomarker of disease. Br. J. Cancer *100*, 1534–1539.

Menigatti, M., Di Gregorio, C., Borghi, F., Sala, E., Scarselli, A., Pedroni, M., Foroni, M., Benatti, P., Roncucci, L., Ponz de Leon, M., *et al.* (2001). Methylation pattern of different regions of the MLH1 promoter and silencing of gene expression in hereditary and sporadic colorectal cancer. Genes Chromosomes Cancer *31*, 357–361.

Merlo, A., Herman, J.G., Mao, L., Lee, D.J., Gabrielson, E., Burger, P.C., Baylin, S.B., and Sidransky, D. (1995). 5′ CpG island methylation is associated with transcriptional silencing of the tumour suppressor p16/CDKN2/MTS1 in human cancers. Nat. Med. *1*, 686–692.

Milani, L., Lundmark, A., Kiialainen, A., Nordlund, J., Flaegstad, T., Forestier, E., Heyman, M., Jonmundsson, G., Kanerva, J., Schmiegelow, K., *et al.* (2010). DNA methylation for subtype classification and prediction of treatment outcome in patients with childhood acute lymphoblastic leukemia. Blood *115*, 1214–1225.

Miyake, T., Nakayama, T., Naoi, Y., Yamamoto, N., Otani, Y., Kim, S.J., Shimazu, K., Shimomura, A., Maruyama, N., Tamaki, Y., *et al.* (2012). GSTP1 expression predicts poor pathological complete response to neoadjuvant chemotherapy in ER-negative breast cancer. Cancer Sci. *103*, 913–920.

Muller, H.M., Oberwalder, M., Fiegl, H., Morandell, M., Goebel, G., Zitt, M., Muhlthaler, M., Ofner, D., Margreiter, R., and Widschwendter, M. (2004). Methylation changes in faecal DNA: a marker for colorectal cancer screening? Lancet *363*, 1283–1285.

Nakamichi, I., Tomita, Y., Zhang, B., Sugiyama, H., Kanakura, Y., Fukuhara, S., Hino, M., Kanamaru, A., Ogawa, H., and Aozasa, K. (2007). Correlation between promoter hypermethylation of GSTP1 and response to chemotherapy in diffuse large B cell lymphoma. Ann. Hematol. *86*, 557–564.

Nogueira da Costa, A., and Herceg, Z. (2012). Detection of cancer-specific epigenomic changes in biofluids: powerful tools in biomarker discovery and application. Mol. Oncol. *6*, 704–715.

Novelli, G., Ciccacci, C., Borgiani, P., Papaluca Amati, M., and Abadie, E. (2008). Genetic tests and genomic biomarkers: regulation, qualification and validation. Clin. Cases Miner Bone Metab. *5*, 149–154.

Ogino, S., Nosho, K., Kirkner, G.J., Kawasaki, T., Chan, A.T., Schernhammer, E.S., Giovannucci, E.L., and Fuchs, C.S. (2008). A cohort study of tumoral LINE-1 hypomethylation and prognosis in colon cancer. J. Natl. Cancer Inst. *100*, 1734–1738.

Ogino, S., Lochhead, P., Chan, A.T., Nishihara, R., Cho, E., Wolpin, B.M., Meyerhardt, J.A., Meissner, A., Schernhammer, E.S., Fuchs, C.S., *et al.* (2013). Molecular pathological epidemiology of epigenetics: emerging integrative science to analyze environment, host, and disease. Mod. Pathol. *26*, 465–484.

Ono, S., Oue, N., Kuniyasu, H., Suzuki, T., Ito, R., Matsusaki, K., Ishikawa, T., Tahara, E., and Yasui, W. (2002). Acetylated histone H4 is reduced in human gastric adenomas and carcinomas. J. Exp. Clin. Cancer Res. *21*, 377–382.

Ozdemir, F., Altinisik, J., Karateke, A., Coksuer, H., and Buyru, N. (2012). Methylation of tumor suppressor genes in ovarian cancer. Exp. Ther. Med. *4*, 1092–1096.

Palmisano, W.A., Divine, K.K., Saccomanno, G., Gilliland, F.D., Baylin, S.B., Herman, J.G., and Belinsky, S.A. (2000). Predicting lung cancer by detecting aberrant promoter methylation in sputum. Cancer Res. *60*, 5954–5958.

Park, Y.S., Jin, M.Y., Kim, Y.J., Yook, J.H., Kim, B.S., and Jang, S.J. (2008). The global histone modification pattern correlates with cancer recurrence and overall survival in gastric adenocarcinoma. Ann. Surg. Oncol. *15*, 1968–1976.

Payne, S.R., Serth, J., Schostak, M., Kamradt, J., Strauss, A., Thelen, P., Model, F., Day, J.K., Liebenberg, V., Morotti, A., *et al.* (2009). DNA methylation biomarkers of prostate cancer: confirmation of candidates and evidence urine is the most sensitive body fluid for non-invasive detection. Prostate *69*, 1257–1269.

Robertson, K.D. (2002). DNA methylation and chromatin – unraveling the tangled web. Oncogene *21*, 5361–5379.

Rocco, J.W., and Sidransky, D. (2001). p16(MTS-1/CDKN2/INK4a) in cancer progression. Exp. Cell Res. *264*, 42–55.

Rodriguez-Paredes, M., and Esteller, M. (2011). Cancer epigenetics reaches mainstream oncology. Nat. Med. *17*, 330–339.

Rodriguez-Rodero, S., Fernandez, A.F., Fernandez-Morera, J.L., Castro-Santos, P., Bayon, G.F., Ferrero, C., Urdinguio, R.G., Gonzalez-Marquez, R., Suarez, C., Fernandez-Vega, I., *et al.* (2013). DNA methylation signatures identify biologically distinct thyroid cancer subtypes. J. Clin. Endocrinol. Metab. *98*, 2811–2821.

Rosas, S.L., Koch, W., da Costa Carvalho, M.G., Wu, L., Califano, J., Westra, W., Jen, J., and Sidransky, D. (2001). Promoter hypermethylation patterns of p16, O6-methylguanine-DNA-methyltransferase, and death-associated protein kinase in tumors and saliva of head and neck cancer patients. Cancer Res. *61*, 939–942.

Sandoval, J., Heyn, H., Mendez-Gonzalez, J., Gomez, A., Moran, S., Baiget, M., Melo, M., Badell, I., Nomdedeu, J.F., and Esteller, M. (2013). Genome-wide DNA methylation profiling predicts relapse in childhood B-cell acute lymphoblastic leukaemia. Br. J. Haematol. *160*, 406–409.

Sekeres, M.A., List, A.F., Cuthbertson, D., Paquette, R., Ganetzky, R., Latham, D., Paulic, K., Afable, M., Saba, H.I., Loughran, T.P. Jr., *et al.* (2010). Phase I combination trial of lenalidomide and azacitidine in patients with higher-risk myelodysplastic syndromes. J. Clin. Oncol. *28*, 2253–2258.

Seligson, D.B., Horvath, S., Shi, T., Yu, H., Tze, S., Grunstein, M., and Kurdistani, S.K. (2005). Global histone modification patterns predict risk of prostate cancer recurrence. Nature *435*, 1262–1266.

Shirahata, A., Sakuraba, K., Kitamura, Y., Yokomizo, K., Gotou, T., Saitou, M., Kigawa, G., Nemoto, H., Sanada, Y., and Hibi, K. (2012). Detection of vimentin methylation in the serum of patients with gastric cancer. Anticancer Res. *32*, 791–794.

Silber, J.R., Mueller, B.A., Ewers, T.G., and Berger, M.S. (1993). Comparison of O6-methylguanine-DNA methyltransferase activity in brain tumors and adjacent normal brain. Cancer Res. *53*, 3416–3420.

Silber, J.R., Bobola, M.S., Ghatan, S., Blank, A., Kolstoe, D.D., and Berger, M.S. (1998). O6-methylguanine-DNA methyltransferase activity in adult gliomas: relation to patient and tumor characteristics. Cancer Res. *58*, 1068–1073.

Sincic, N., and Herceg, Z. (2011). DNA methylation and cancer: ghosts and angels above the genes. Curr. Opin. Oncol. *23*, 69–76.

Soriano, A.O., Yang, H., Faderl, S., Estrov, Z., Giles, F., Ravandi, F., Cortes, J., Wierda, W.G., Ouzounian, S., Quezada, A., *et al.* (2007). Safety and clinical activity of the combination of 5-azacytidine, valproic acid, and all-trans retinoic acid in acute myeloid leukemia and myelodysplastic syndrome. Blood *110*, 2302–2308.

Stefansson, O.A., Villanueva, A., Vidal, A., Marti, L., and Esteller, M. (2012). BRCA1 epigenetic inactivation predicts sensitivity to platinum-based chemotherapy in breast and ovarian cancer. Epigenetics 7, 1225–1229.

Strathdee, G., MacKean, M.J., Illand, M., and Brown, R. (1999). A role for methylation of the hMLH1 promoter in loss of hMLH1 expression and drug resistance in ovarian cancer. Oncogene *18*, 2335–2341.

Stumpel, D.J., Schneider, P., van Roon, E.H., Boer, J.M., de Lorenzo, P., Valsecchi, M.G., de Menezes, R.X., Pieters, R., and Stam, R.W. (2009). Specific promoter methylation identifies different subgroups of MLL-rearranged infant acute lymphoblastic leukemia, influences clinical outcome, and provides therapeutic options. Blood *114*, 5490–5498.

Tamura, G., Yin, J., Wang, S., Fleisher, A.S., Zou, T., Abraham, J.M., Kong, D., Smolinski, K.N., Wilson, K.T., James, S.P., *et al.* (2000). E-Cadherin gene promoter hypermethylation in primary human gastric carcinomas. J. Natl. Cancer Inst. *92*, 569–573.

Toraño, E.G., Petrus, S., Fernandez, A.F., and Fraga, M.F. (2012). Global DNA hypomethylation in cancer: review of validated methods and clinical significance. Clin. Chem. Lab. Med. *50*, 1733–1742.

Townsend, D.M., and Tew, K.D. (2003). The role of glutathione-S-transferase in anti-cancer drug resistance. Oncogene *22*, 7369–7375.

Tsai, H.C., and Baylin, S.B. (2011). Cancer epigenetics: linking basic biology to clinical medicine. Cell Res. *21*, 502–517.

Venkitaraman, A.R. (2002). Cancer susceptibility and the functions of BRCA1 and BRCA2. Cell *108*, 171–182.

Volkmar, M., Dedeurwaerder, S., Cunha, D.A., Ndlovu, M.N., Defrance, M., Deplus, R., Calonne, E., Volkmar, U., Igoillo-Esteve, M., Naamane, N., *et al.* (2012). DNA methylation profiling identifies epigenetic dysregulation in pancreatic islets from type 2 diabetic patients. EMBO J. *31*, 1405–1426.

Wang, C., Cicek, M.S., Charbonneau, B., Kalli, K.R., Armasu, S.M., Larson, M.C., Konecny, G.E., Winterhoff, B., Fan, J.B., Bibikova, M., *et al.* (2014). Tumor hypomethylation at 6p21.3 associates with longer time to recurrence of high-grade serous epithelial ovarian cancer. Cancer Res. *74*, 3084–3091.

Wang, D., Jing, Y., Ouyang, S., Liu, B., Zhu, T., Niu, H., and Tian, Y. (2013a). Inhibitory effect of valproic acid on bladder cancer in combination with chemotherapeutic agents and. Oncol. Lett. *6*, 1492–1498.

Wang, G., Edwards, H., Caldwell, J.T., Buck, S.A., Qing, W.Y., Taub, J.W., Ge, Y., and Wang, Z. (2013b). Panobinostat synergistically enhances the cytotoxic effects of cisplatin, doxorubicin or etoposide on high-risk neuroblastoma cells. PLoS ONE *8*, e76662.

Warren, J.D., Xiong, W., Bunker, A.M., Vaughn, C.P., Furtado, L.V., Roberts, W.L., Fang, J.C., Samowitz, W.S., and Heichman, K.A. (2011). Septin 9 methylated DNA is a sensitive and specific blood test for colorectal cancer. BMC Med. *9*, 133.

Weisenberger, D.J., Siegmund, K.D., Campan, M., Young, J., Long, T.I., Faasse, M.A., Kang, G.H., Widschwendter, M., Weener, D., Buchanan, D., *et al.* (2006). CpG island methylator phenotype underlies sporadic microsatellite instability and is tightly associated with BRAF mutation in colorectal cancer. Nat. Genet. *38*, 787–793.

Weller, M., Stupp, R., Reifenberger, G., Brandes, A.A., van den Bent, M.J., Wick, W., and Hegi, M.E. (2010). MGMT promoter methylation in malignant gliomas: ready for personalized medicine? Nat. Rev. Neurol. *6*, 39–51.

Wong, I.H., Lo, Y.M., Zhang, J., Liew, C.T., Ng, M.H., Wong, N., Lai, P.B., Lau, W.Y., Hjelm, N.M., and Johnson, P.J. (1999). Detection of aberrant p16 methylation in the plasma and serum of liver cancer patients. Cancer Res. *59*, 71–73.

Yamashita, Y., Yuan, J., Suetake, I., Suzuki, H., Ishikawa, Y., Choi, Y.L., Ueno, T., Soda, M., Hamada, T., Haruta, H., *et al.* (2010). Array-based genomic resequencing of human leukemia. Oncogene *29*, 3723–3731.

Yasui, W., Oue, N., Ono, S., Mitani, Y., Ito, R., and Nakayama, H. (2003). Histone acetylation and gastrointestinal carcinogenesis. Ann. N. Y. Acad. Sci. *983*, 220–231.

Yoo, S.S., Lee, S.M., Do, S.K., Lee, W.K., Kim, D.S., and Park, J.Y. (2014). Unmethylation of the CHRNB4 gene is an unfavorable prognostic factor in non-small cell lung cancer. Lung Cancer *86*, 85–90.

Yu, J., Tao, Q., Cheng, Y.Y., Lee, K.Y., Ng, S.S., Cheung, K.F., Tian, L., Rha, S.Y., Neumann, U., Rocken, C., *et al.* (2009). Promoter methylation of the Wnt/beta-catenin signaling antagonist Dkk-3 is associated with poor survival in gastric cancer. Cancer *115*, 49–60.

Zhang, Y., Qu, X., Jing, W., Hu, X., Yang, X., Hou, K., Teng, Y., Zhang, J., and Liu, Y. (2009). GSTP1 determines cis-platinum cytotoxicity in gastric adenocarcinoma MGC803 cells: regulation by promoter methylation and extracellular regulated kinase signaling. Anticancer Drugs *20*, 208–214.

Zhu, C., Utsunomiya, T., Ikemoto, T., Yamada, S., Morine, Y., Imura, S., Arakawa, Y., Takasu, C., Ishikawa, D., Imoto, I., *et al.* (2014). Hypomethylation of long interspersed nuclear element-1 (LINE-1) is associated with poor prognosis via activation of c-MET in hepatocellular carcinoma. Ann. Surg. Oncol. *21* (Suppl. 4), S729–735.

Zhuang, J., Jones, A., Lee, S.H., Ng, E., Fiegl, H., Zikan, M., Cibula, D., Sargent, A., Salvesen, H.B., Jacobs, I.J., *et al.* (2012). The dynamics and prognostic potential of DNA methylation changes at stem cell gene loci in women's cancer. PLoS Genet. *8*, e1002517.

Zou, H.Z., Yu, B.M., Wang, Z.W., Sun, J.Y., Cang, H., Gao, F., Li, D.H., Zhao, R., Feng, G.G., and Yi, J. (2002). Detection of aberrant p16 methylation in the serum of colorectal cancer patients. Clin. Cancer Res. *8*, 188–191.

Environment and the Epigenetic Transgenerational Inheritance of Disease

15

Ingrid Sadler-Riggleman and Michael K. Skinner

Abstract

Environmental factors can lead to transgenerational epigenetic changes in generations originating from ancestors exposed to environmental insults such as toxicants, abnormal nutrition or stress. These epigenetic germline changes can in turn increase disease susceptibility in offspring of exposed ancestors. The focus of this chapter is on environmentally induced epigenetic transgenerational inheritance of disease. The molecular mechanisms that allow these inherited epimutations to increase disease susceptibility will be reviewed. Environmental exposure specificity and exposure-specific disease development are also discussed.

Introduction

The term epigenetics was originally coined by C.H. Waddington in the 1940s in relation to his studies of gene–environment interactions involving non-Mendelian genetic phenomena (Van Speybroeck, 2002; Waddington, 1940). In the 1970s the first of the molecular factors enabling epigenetic processes – DNA methylation – was discovered (Holliday and Pugh, 1975). Other epigenetic modifications, like histone modification (Turner, 1998) and action of non-coding RNAs (Jodar *et al.*, 2013; Mattick, 2009) were later described. The complex integration of epigenetic modifications is referred to as the 'epigenome' with the first whole epigenome analysis accomplished in 2005 (Pokholok *et al.*, 2005). Epigenetics has dramatic impacts on developmental processes and disease aetiologies.

The definition of epigenetic transgenerational inheritance is 'germline-mediated inheritance of epigenetic information between generations in the *absence* of direct environmental influences, that leads to phenotypic variation' (Skinner, 2011). Multigenerational exposures on the other hand means that the observed effects are the result of direct exposure and may disappear when exposure is no longer present. The term transgenerational is often misused as multigenerational and transgenerational being the same, so the following has to be taken into consideration to use the term correctly. Direct environmental exposure of the parents (either male or female = F0 generation) will also affect the germline of either parent, Fig. 15.1. Therefore, the next generation (F1) derived from this germline is still considered exposed, so not truly transgenerational. The F2 generation offspring would be considered the first transgenerational unexposed generation. This is different when a pregnant female is exposed, because then the fetus and the fetus's germline are directly exposed as well. In this case the F3 generation is the first unexposed transgenerational offspring (Fig. 15.1).

The Agouti mouse model would be an example of multigenerational inheritance (Blewitt *et al.*, 2006; Waterland *et al.*, 2007). When pregnant Agouti mice are exposed to a methyl donor in their diet, they will experience increased methylation on an allele of their Agouti gene, which will lead to a coat colour change in their offspring. But this change will not be passed on to future generations. Instead the normal process of epigenetic reprogramming in the germline and early

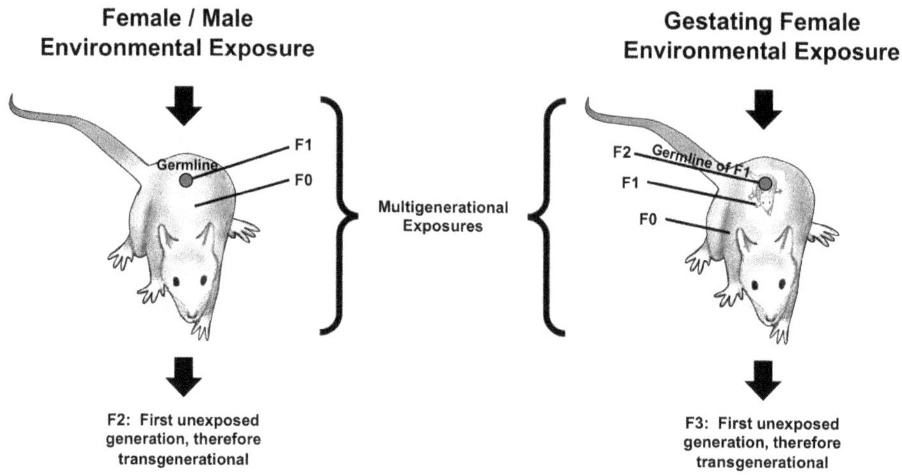

Figure 15.1 Environmentally induced transgenerational epigenetic inheritance: schematic of environmental exposure and affected generations.

embryo will return the DNA methylation state to its original setting. Similarly, exposure of human pregnant women to diethylstilbestrol (DES) did result in abnormalities and increased disease risk in only their direct offspring and grandchildren. In an animal model these multigenerational effects of DES have been shown to be caused by epigenetic alterations (Bromer *et al.*, 2009; Harris and Waring, 2012; Li *et al.*, 2014).

There are a large number of examples of transgenerational inheritance in the literature. The authors' laboratory has examined several environmental toxicants (e.g. DDT, vinclozolin, methoxychlor, plastics, hydrocarbons) and shown transgenerational inheritance of disease after exposure to these toxicants (Anway *et al.*, 2005; Guerrero-Bosagna *et al.*, 2013; Manikkam *et al.*, 2012a; Skinner *et al.*, 2013), Table 15.1. Other examples of transgenerational inheritance are caused by caloric restriction or high fat diet. The Överkalix study by Bygren *et al.* (2014) shows how cardiovascular mortality can be influenced by childhood food supply. Effects will reach into the second generation. Maternal high fat diet in mice (Dunn and Bale, 2011) will increase body size and reduce insulin sensitivity in F3 generation female offspring. A paternal prediabetic condition in mice will be inherited transgenerationally as shown by impaired glucose tolerance and decreased insulin sensitivity (Wei *et al.*, 2014). A number of studies from other laboratories have shown transgenerational inheritance after exposure to environmental factors such as stress and toxicants. Wolstenholme *et al.* (2012) showed that ancestral exposure to bisphenol A caused changes in social behaviour in juvenile mice and changes in expression of neural genes like oxytocin and vasopressin. Franklin *et al.* (2011) found that the stress of maternal separation in mice will transgenerationally impair social interaction and cognition but will also make the mice more stress resilient. Transgenerational inheritance has been shown in a variety of organism: vertebrates (Anway *et al.*, 2005), invertebrates like *C. elegans* (Kelly, 2014) or *Artemia* (Norouzitallab *et al.*, 2014) and plants (Zheng *et al.*, 2013). Table 15.1 shows a select list of studies of transgenerational inheritance after various environmental exposures.

Phenoptypic diversity of transgenerationally inherited diseases

The transgenerational inheritance of germline epigenetic modifications and its effects on disease, reproduction and behaviour has been a focus in the author's laboratory for the past decade and it was one of the first laboratories to study transgenerational effects on the epigenome. The initial research involved studying impacts of the agricultural agents vinclozolin and methoxychlor (Anway *et al.*, 2005) on pregnant rats. These

Table 15.1 Examples of transgenerational inheritance and its specific effects

Exposure	Effects	Reference
Environmental toxicants		
Vinclozolin	Impaired male fertility; prostate, kidney disease, tumours, immune and reproductive pathologies	Anway *et al.* (2005, 2006)
Vinclozolin	Gender-specific changes in anxiety-like behaviour	Skinner *et al.* (2008)
Vinclozolin	Immune and reproductive pathologies	Nilsson *et al.* (2008)
Methoxychlor	Impaired male fertility	Anway *et al.* (2005)
Permethrin/DEET	Prostate, kidney disease	Manikkam *et al.* (2012b)
Dioxin	Prostate, kidney disease; reduced fertility, negative effects on pregnancy outcome	Manikkam *et al.* (2012c), Bruner-Tran (2011)
BPA/phthalates	Prostate, kidney disease; obesity	Manikkam *et al.* (2013)
Hydrocarbon mixture (jet fuel)	Prostate, kidney disease; obesity; immune and reproductive pathologies	Tracey *et al.* (2013)
Vinclozolin, permethrin/DEET, plastics, dioxin, jet fuel	Polycystic ovaries, reduced primordial follicle pool	Nilsson *et al.* (2012)
DDT	Obesity	Skinner *et al.* (2013)
Phthalate	Disruption of testicular germ cell organization and spermatogonial stem cell function	Doyle *et al.* (2013)
Tributyltin	Increase in fat depot size	Chamorro-Garcia *et al.* (2013)
BPA	Changes in social behaviour and neural gene expression; increased post-implantation loss, decreased litter size; decrease in sperm count and motility	Wolstenholme *et al.* (2012), Salian *et al.* (2009)
Other exposures		
Caloric restriction	Cardiovascular mortality	Bygren *et al.* (2014)
High-fat diet	Increased body size; reduced insulin sensitivity	Dunn and Bale (2011)
Folate	Congenital malformations	Padmanabhan *et al.* (2013)
Stress	Reduced social interaction; increased stress resilience	Franklin *et al.* (2011)
Traumatic stress	Change in behavioural and metabolic responses	Gapp *et al.* (2014)
Drought	DNA methylation changes	Zheng *et al.* (2013)
Heat/salt stress	Accelerated flowering, increased salt tolerance	Suter and Widmer (2013)
Prediabetes	Impaired glucose tolerance; reduced insulin sensitivity	Wei *et al.* (2014)
Smoking	Abnormal pulmonary function	Rehan (2013)
Alcohol consumption	Deficit in POMC (proopiomelanocortin) neuronal function	Govorko (2012)
Heat stress	Increased Hsp70 production and tolerance to heat stress	Norouzitallab *et al.* (2014)

early findings were fortuitous observations in animals bred beyond the originally planned F1 generation. Effects in the F2 generation animals (increased spermatogenic apoptosis) were found to persist through the F3 and F4 generation and in outcrossed offspring, thus showing true transgenerational inheritance in a non-Mendelian manner with 90% of the male population affected at each generation.

The authors' initial research on epigenetic transgenerational inheritance showed a rise in spermatogenic cell apoptosis in the F3 generation after exposure to vinclozolin in the F0 generation. In later experiments, after treatment of the F0 generation with an array of additional toxicants (Table 15.1), a number of other problems were detected, for example prostate and kidney disease (Anway *et al.*, 2006; Anway and Skinner, 2008; Manikkam *et al.*, 2012a–c, 2013; Tracey *et al.*, 2013), mammary tumours (Anway *et al.*, 2006), immune and reproductive pathologies (Anway *et al.*, 2006; Nilsson *et al.*, 2008; Tracey *et al.*, 2013), obesity (Skinner *et al.*, 2013; Tracey *et al.*, 2013), as well as behavioural effects (Skinner *et al.*, 2008). Therefore, between the authors' and other laboratories (Chamorro-Garcia *et al.*, 2013; Doyle *et al.*, 2013), evidence has been given that a variety of diseases and behavioural abnormalities will occur transgenerationally after exposure to a large number of environmental toxicants and other environmental exposures, (Table 15.1). The disease phenotypes observed in these experiments often depend on the specific exposure of the F0 generation. For example, increased obesity risk is inherited transgenerationally after ancestral exposure to DDT, plastic compounds, hydrocarbons and methoxychlor (Manikkam *et al.*, 2013, 2014; Skinner *et al.*, 2013) but not others. Jet fuel hydrocarbons will induce an elevated rate of luteal ovarian cyst formation in F3 females (Manikkam *et al.*, 2012a; Tracey *et al.*, 2013), a phenotype not observed with other exposures. On the other hand, some ovarian disorders like polycystic ovaries and reduction of the primordial follicle pool size have been shown to be inherited transgenerationally after exposure of the F0 generation to all the toxicants studied so far in the authors' laboratory (Nilsson *et al.*, 2012). The explanation for this phenomenon may be that some developmental mechanisms, in this case ovarian follicle development, are more sensitive to gene expression changes and will be affected more often than other cells and tissues.

Germline epimutations and specificity of environmental toxicants

The term epimutations is defined as differential presence of epigenetic marks that lead to altered genome activity, see Glossary. It is a prerequisite for environmentally induced epigenetic transgenerational inheritance to have epimutations in the germline, because the germ cells (sperm and egg) are the only cells that can transmit information from the parents to their offspring. DNA methylation is the most thoroughly investigated molecular mechanism involved in the transmission of epigenetic changes via the germ cells to future generations, but other factors like histone modification and actions by non-coding RNAs appear to also be involved. Normally, environmentally induced methylation changes will be reset by a genome wide DNA methylation reprogramming after fertilization in the early embryo (Hackett and Surani, 2013; Smith *et al.*, 2012). This phenomenon allows embryonic stem cells to be developed, but also one would not want parental epigenetic changes to be established in the offspring. These epigenetic reprogramming events take place at two developmental periods, during germ cell specification at gonadal sex determination and in the early embryo. After this DNA methylation erasure parent-specific imprints are established (imprinted genes) (Constancia *et al.*, 1998). It has been found that certain regions in the genome called differential DNA methylation regions (DMRs) (Skinner *et al.*, 2010) appear to behave as imprinted-like genes in the way that their methylation pattern persists. By definition, true imprinted genes display 'parent-of-origin allelic transmission with monoallelic gene expression'. DMRs often demonstrate parent-of-origin allelic transmission, but monoallelic gene expression may not be involved. Differentially methylated sites connected with transgenerational inheritance are called 'imprinted-like' (Jirtle and Skinner, 2007).

When epigenetic information is passed on to future generations via germ cells, the epigenome of the developing embryonic stem cells would be expected to have an altered epigenome that will change the epigenetic and transcriptome programming of all subsequent somatic cells developed (Guerrero-Bosagna *et al.*, 2013). All developing tissues will have cell and tissue specific effects on the transgenerational transcriptome. Therefore, one would expect a different transcriptome stemming from the same germline epigenome in each somatic cell type. Those tissues that have a sensitivity to alterations in their transcriptomes may also show increased sensitivity to disease development, which will be inherited because of the causative epigenetic changes in the germline (Skinner *et al.*, 2010, 2012).

The first examples of transgenerational epimutations come from studies from the authors' laboratory. The studies investigated Vinclozolin, a dicarboximide fungicide, used especially in vineyards and on fruits and vegetables (Anway *et al.*, 2005). A genome-wide promoter analysis was applied to look for DNA modifications in the sperm DNA and approximately 50 differential DNA methylated regions (epimutations) were identified in vinclozolin lineage F3 generation sperm DNA versus control lineage (Guerrero-Bosagna *et al.*, 2010).

After having established a specific set of epimutations in vinclozolin lineage animals, the authors' laboratory pursued the question if treatment with different environmental toxicants would also lead to germline epimutations. Experiments similar to the original vinclozolin experiment were performed with a number of additional toxicants: dioxin (Manikkam *et al.*, 2012c), a mixture of permethrin and DEET (Manikkam *et al.*, 2012b), bisphenol A and phthalates (Manikkam *et al.*, 2013) and jet fuel (hydrocarbons) (Tracey *et al.*, 2013). All these toxicants were found to promote transgenerational inheritance of disease and sperm epimutations. Intriguingly, it was observed that each toxicant produced an exposure specific set of epigenetic changes in the sperm DNA and comparison between the individual exposures demonstrated negligible overlap (Manikkam *et al.*, 2012a). The possibility of using these unique 'epimutation signatures' as a diagnostic tool in the future is speculated.

Molecular basis of transgenerationally inherited disease

Transgenerational inheritance of environmentally induced epigenetic changes requires transmission through the germ line from parents to future generations. Ensuing disease such as cancer, prostate or kidney abnormalities and obesity are brought on by disturbances in gene expression in the pertinent somatic cells. The hypothesis is that the epimutations in the germline alter the epigenome of the embryonic stem cells that then affect all subsequent somatic cell epigenomes and transcriptomes (Guerrero-Bosagna *et al.*, 2013; Nilsson *et al.*, 2012). These cell and tissue specific epimutations promote tissue-specific alterations in transcriptomes (Skinner *et al.*, 2012). These aberrant transcriptomes could then lead to physiological abnormalities and disease.

Experiments have shown that F0 generation exposure to environmental toxicants will affect and change the transcriptomes of multiple tissues in future generations (Skinner *et al.*, 2012). It was observed that the epigenomic changes of one cell type could be very different from another cell type in the body, despite the fact that they all result from a common germ line. The authors' laboratory examined the transcriptomes of 11 different tissue types from adult male and female animals who were of the vinclozolin lineage mentioned earlier. They found that (a) there were (not-unexpectedly) gene expression differences between control and vinclozolin lineage animals and (b) there was only minor overlap in the differentially expressed genes between tissues. These findings warranted a closer look at the genomic locations of epimutations and differentially expressed genes. Regions of the genome were found to have statistically over-represented clusters of gene expression changes. These regions in the genome were called epigenetic control regions (ECR). These epigenetic control regions are 2–5 MB in size and have gene clusters of statistically over-represented regulated genes. These ECR are often located close to epimutations (Skinner *et al.*, 2012) and to regions of long non-coding (nc) RNA expression (Kung *et al.*, 2013). It is well known that long ncRNAs play a role in regulation of distal gene transcription and epigenetic regulation. Further research will reveal if the long ncRNAs have a part in the management of the genes located in the epigenetic control regions.

Several studies in the authors' laboratory have elucidated at least partially how the molecular mechanisms of environmentally induced transgenerational inheritance may lead to tissue specific disease appearance. As mentioned earlier, two ovarian disorders, polycystic ovarian syndrome and primary ovarian insufficiency (premature reduction of the primary follicle pool), were both induced transgenerationally by a number of environmental toxicants (Nilsson *et al.*, 2008). In order to study this phenomenon as related to the vinclozolin lineage animals, granulosa cells were isolated from ovarian follicles from young female rats before disease onset. The epigenomes and transcriptomes of these granulosa cells (control and vinclozolin lineage) were analysed (Nilsson *et al.*, 2012). Granulosa cells from vinclozolin lineage rats had differences in both the epigenome and the transcriptome compared to control lineage. Closer examination revealed that specific signalling pathways were affected, and some of the affected genes had been shown previously to be associated to polycystic ovarian syndrome and primary ovarian insufficiency. Similar results were obtained when the molecular basis of vinclozolin lineage male infertility in rats was examined. As above, changes in the epigenome and transcriptome were found in testicular Sertoli cells of F3 generation rats (Guerrero-Bosagna *et al.*, 2013). Several of the differentially regulated genes were known to be involved in male infertility.

Impact on evolution

The fact that studies have shown that environmental factors can induce transgenerational changes in organisms leads to the assumption that epigenetic transgenerational inheritance may also play a role in evolution. Evolutionary change through DNA mutations is a slow process and not always suitable for an aggressively changing environment. Epigenetic inheritance would provide a quick-response solution and adaptation in a rapidly shifting setting.

Darwin (1871) considered sexual selection one of the important processes driving evolution. He defined sexual selection as either male–male competition or females' choice of a male mate (mate selection). One of the authors' studies has shown that F3 generation rats from the vinclozolin lineage mentioned earlier have a significant change in mate preference behaviour. Female rats of both control and vinclozolin F3 generation lineages were found to preferentially chose males from the control group instead of males from the vinclozolin lineage (Crews *et al.*, 2007). Considering that sexual selection is a major element in evolutionary biology, the fact that environmental factors can change mate preference in a transgenerational way would support a critical role of epigenetic transgenerational inheritance in evolution. Generally, epigenetic changes in the germline that are passed on to offspring can produce inherited changes in gene expression in all cells and tissues, and if those gene expression changes result in an increased phenotypic diversity, one might expect that natural selection will act upon these phenotypic variations and lead to evolutionary change.

The rats used in the mate preference study are from a lineage that had been shown to have epimutations in the sperm genome that alters epigenomes in adult tissues transgenerationally (Skinner *et al.*, 2010, 2012). The mate preference study investigated male and female brain regions and showed distinct differentially expressed gene sets in each region (Skinner *et al.*, 2014b). Gene bionetworks were found in various brain regions that statistically correlated to the altered mate preference behaviours (Skinner *et al.*, 2014b). Observations indicate the environmentally induced changes in mate preference appear to impact a major determinant of evolution, sexual selection.

A more recent experiment investigated the phylogenetic relationships of epimutations and genetic mutations in Darwin's finches in the Galapagos Islands (Skinner *et al.*, 2014a). The epigenetic changes, epimutations, among five different species demonstrated a statistically significant correlation with the phylogenetic association of these species. Interestingly, the number of epigenetic changes were generally greater than genetic mutations and more correlated with the phylogenetics (Skinner *et al.*, 2014a). Although additional research is needed to establish a functional role of environmental epigenetics in evolution, the ability of environment to promote the epigenetic transgenerational inheritance of phenotypic variation will clearly have a significant impact on natural selection.

Conclusions

Research in the area of environmentally induced epigenetic transgenerational inheritance of disease and phenotypic variation has provided evidence of transgenerational inheritance of epimutations in plants, worms, flies, fish, pigs, mice, rats and humans. Ancestral exposure to environmental influences such as toxicants or nutritional stress can transmit changes in the epigenome to their descendants. These epimutations caused by individual exposures are required to be transmitted through the germline. They subsequently will induce somatic cell changes in the epigenome and transcriptome in the offspring. There are several mechanisms that will be involved in the creation of epigenetic change, DNA methylation being the most thoroughly researched. A major mechanism involved in the epigenetic transgenerational inheritance of disease is through changes in DNA methylation which are transferred through the germline to somatic cells causing the observed phenotypical changes. Therefore, these ancestral exposures to environmental toxicants can lead to transgenerational changes in the epigenome and transcriptome of future generation and lead to increased incidence of disease. The potential role of these ancestral exposures and epigenetic transgenerational inheritance in disease aetiology needs to be seriously considered.

Since it has been shown in initial experiments that the majority of the environmental influences tested have a unique pattern of epigenetic changes in the offspring, one might consider the possibility for epigenetic diagnostics in the future, testing for ancestral exposure and associated disease risk in a population.

Glossary

Epigenetics: molecular factors/processes around the DNA that regulate genome activity independent of DNA sequence, and are mitotically stable.

Epigenetic transgenerational inheritance: germline-mediated inheritance of epigenetic information between generations in the absence of direct environmental influences, that leads to phenotypic variation.

Epimutation: differential presence of epigenetic marks that lead to altered genome activity.

F0, F1, F2, F3: various generations – parents, offspring, grand-offspring, great-grand-offspring.

Acknowledgements

We thank Dr Eric Nilsson for helpful comments and Ms Heather Johnson for assistance in preparation of the manuscript. This research was supported by NIH grants to MKS. Drs Ingrid Sadler-Riggleman and Michael K. Skinner are at the School of Biological Sciences at Washington State University, Pullman, WA, USA.

References

Anway, M.D., and Skinner, M.K. (2008). Transgenerational effects of the endocrine disruptor vinclozolin on the prostate transcriptome and adult onset disease. Prostate *68*, 517–529.

Anway, M.D., Cupp, A.S., Uzumcu, M., and Skinner, M.K. (2005). Epigenetic transgenerational actions of endocrine disruptors and male fertility. Science *308*, 1466–1469.

Anway, M.D., Leathers, C., and Skinner, M.K. (2006). Endocrine disruptor vinclozolin induced epigenetic transgenerational adult-onset disease. Endocrinology *147*, 5515–5523.

Blewitt, M.E., Vickaryous, N.K., Paldi, A., Koseki, H., and Whitelaw, E. (2006). Dynamic reprogramming of DNA methylation at an epigenetically sensitive allele in mice. PLoS Genet. *2*, e49.

Bromer, J.G., Wu, J., Zhou, Y., and Taylor, H.S. (2009). Hypermethylation of homeobox A10 by in utero diethylstilbestrol exposure: an epigenetic mechanism for altered developmental programming. Endocrinology *150*, 3376–3382.

Bruner-Tran, K.L. and Osteen, K.G. (2011). Developmental exposure to TCDD reduces fertility and negatively affects pregnancy outcomes across multiple generations. Reprod. Toxicol. *31*, 344–350.

Bygren, L.O., Tinghog, P., Carstensen, J., Edvinsson, S., Kaati, G., Pembrey, M.E., and Sjostrom, M. (2014). Change in paternal grandmothers' early food supply influenced cardiovascular mortality of the female grandchildren. BMC Genetics *15*, 12.

Chamorro-Garcia, R., Sahu, M., Abbey, R.J., Laude, J., Pham, N., and Blumberg, B. (2013). Transgenerational inheritance of increased fat depot size, stem cell reprogramming, and hepatic steatosis elicited by prenatal exposure to the obesogen tributyltin in mice. Environ. Health Perspect. *121*, 359–366.

Constancia, M., Pickard, B., Kelsey, G., and Reik, W. (1998). Imprinting mechanisms. Genome Res. *8*, 881–900.

Crews, D., Gore, A.C., Hsu, T.S., Dangleben, N.L., Spinetta, M., Schallert, T., Anway, M.D., and Skinner, M.K. (2007). Transgenerational epigenetic imprints on mate preference. Proc. Natl. Acad. Sci. U.S.A. *104*, 5942–5946.

Darwin, C. (1871). The Descent of Man, and Selection in Relation to Sex (T. Murray, London).

Doyle, T.J., Bowman, J.L., Windell, V.L., McLean, D.J., and Kim, K.H. (2013). Transgenerational effects of di-(2-ethylhexyl) phthalate on testicular germ cell associations and spermatogonial stem cells in mice. Biol. Reprod. *88*, 112.

Dunn, G.A., and Bale, T.L. (2011). Maternal high-fat diet effects on third-generation female body size via the paternal lineage. Endocrinology *152*, 2228–2236.

Franklin, T.B., Linder, N., Russig, H., Thony, B., and Mansuy, I.M. (2011). Influence of early stress on social abilities and serotonergic functions across generations in mice. PLoS ONE *6*, e21842.

Gapp, K., Jawaid, A., Sarkies, P., Bohacek, J., Pelczar, P., Prados, J., Farinelli, L., Miska, E., and Mansuy, I.M. (2014). Implication of sperm RNAs in transgenerational inheritance of the effects of early trauma in mice. Nat. Neurosci. *17*, 667–669.

Govorko, D., Bekdash, R.A., Zhang, C., and Sarkar, D.K. (2012). Male germline transmits fetal alcohol adverse effect on hypothalamic proopiomelanocortin gene across generations. Biol. Psychiatry *72*, 378–388.

Guerrero-Bosagna, C., Settles, M., Lucker, B., and Skinner, M. (2010). Epigenetic transgenerational actions of vinclozolin on promoter regions of the sperm epigenome. PLoS ONE *5*, e13100.

Guerrero-Bosagna, C., Savenkova, M., Haque, M.M., Sadler-Riggleman, I., and Skinner, M.K. (2013). Environmentally induced epigenetic transgenerational inheritance of altered sertoli cell transcriptome and epigenome: molecular etiology of male infertility. PLoS ONE *8*, e59922.

Hackett, J.A., and Surani, M.A. (2013). Beyond DNA: programming and inheritance of parental methylomes. Cell *153*, 737–739.

Harris, R.M., and Waring, R.H. (2012). Diethylstilboestrol – a long-term legacy. Maturitas *72*, 108–112.

Holliday, R., and Pugh, J.E. (1975). DNA modification mechanisms and gene activity during development. Science *187*, 226–232.

Jirtle, R.L., and Skinner, M.K. (2007). Environmental epigenomics and disease susceptibility. Nat. Rev. Genet. *8*, 253–262.

Jodar, M., Selvaraju, S., Sendler, E., Diamond, M.P., and Krawetz, S.A. (2013). The presence, role and clinical use of spermatozoal RNAs. Hum. Reprod. Update *19*, 604–624.

Kelly, W.G. (2014). Transgenerational epigenetics in the germline cycle of *Caenorhabditis elegans*. Epigenetics Chromatin *7*, 6.

Kung, J.T., Colognori, D., and Lee, J.T. (2013). Long noncoding RNAs: past, present, and future. Genetics *193*, 651–669.

Li, Y., Hamilton, K.J., Lai, A.Y., Burns, K.A., Li, L., Wade, P.A., and Korach, K.S. (2014). Diethylstilbestrol (DES)-stimulated hormonal toxicity is mediated by ERalpha alteration of target gene methylation patterns and epigenetic modifiers (DNMT3A, MBD2, and HDAC2) in the mouse seminal vesicle. Environ. Health Perspect. *122*, 262–268.

Manikkam, M., Guerrero-Bosagna, C., Tracey, R., Haque, M.M., and Skinner, M.K. (2012a). Transgenerational actions of environmental compounds on reproductive disease and epigenetic biomarkers of ancestral exposures. PLoS ONE *7*, e31901.

Manikkam, M., Tracey, R., Guerrero-Bosagna, C., and Skinner, M. (2012b). Pesticide and insect repellent mixture (Permethrin and DEET) induces epigenetic transgenerational inheritance of disease and sperm epimutations. Reproduct. Toxicol. *34*, 708–719.

Manikkam, M., Tracey, R., Guerrero-Bosagna, C., and Skinner, M.K. (2012c). Dioxin (TCDD) induces epigenetic transgenerational inheritance of adult onset disease and sperm epimutations. PLoS ONE *7*, e46249.

Manikkam, M., Tracey, R., Guerrero-Bosagna, C., and Skinner, M. (2013). Plastics derived endocrine disruptors (BPA, DEHP and DBP) induce epigenetic transgenerational inheritance of adult-onset disease and sperm epimutations. PLoS ONE *8*, e55387.

Manikkam, M., Haque, M.M., Guerrero-Bosagna, C., Nilsson, E., and Skinner, M. (2014). Pesticide methoxychlor promotes the epigenetic transgenerational inheritance of adult onset disease through the female germline. PLoS ONE *9*, e102091.

Mattick, J.S. (2009). The genetic signatures of noncoding RNAs. PLoS Genet. *5*, e1000459.

Nilsson, E.E., Anway, M.D., Stanfield, J., and Skinner, M.K. (2008). Transgenerational epigenetic effects of the endocrine disruptor vinclozolin on pregnancies and female adult onset disease. Reproduction *135*, 713–721.

Nilsson, E., Larsen, G., Manikkam, M., Guerrero-Bosagna, C., Savenkova, M., and Skinner, M. (2012). environmentally induced epigenetic transgenerational inheritance of ovarian disease. PLoS ONE *7*, e36129.

Norouzitallab, P., Baruah, K., Vandegehuchte, M., Van Stappen, G., Catania, F., Vanden Bussche, J., Vanhaecke, L., Sorgeloos, P., and Bossier, P. (2014). Environmental heat stress induces epigenetic transgenerational inheritance of robustness in parthenogenetic Artemia model. FASEB J. *28*, 355–3563.

Padmanabhan, N. and Watson, E.D. (2013). Lessons from the one-carbon metabolism: passing it along to the next generation. Reprod. Biomed. Online *27*, 637–643.

Pokholok, D.K., Harbison, C.T., Levine, S., Cole, M., Hannett, N.M., Lee, T.I., Bell, G.W., Walker, K., Rolfe, P.A., Herbolsheimer, E., *et al.* (2005). Genome-wide map of nucleosome acetylation and methylation in yeast. Cell *122*, 517–527.

Rehan, V.K., Liu, J., Sakurai, R., and Torday, J.S. (2013). Perinatal nicotine-induced transgenerational asthma. Am. J. Physiol. Lung Cell. Mol. Physiol. *305*, L501–507.

Salian, S., Doshi, T. and Vanage, G. (2009). Perinatal exposure of rats to Bisphenol A affects the fertility of male offspring. Life Sci. *85*, 742–752.

Skinner, M.K. (2011). Environmental epigenetic transgenerational inheritance and somatic epigenetic mitotic stability. Epigenetics *6*, 838–842.

Skinner, M.K., Anway, M.D., Savenkova, M.I., Gore, A.C., and Crews, D. (2008). Transgenerational epigenetic programming of the brain transcriptome and anxiety behavior. PLoS ONE 3, e3745.

Skinner, M.K., Manikkam, M., and Guerrero-Bosagna, C. (2010). Epigenetic transgenerational actions of environmental factors in disease etiology. Trends Endocrinol. Metab. *21*, 214–222.

Skinner, M.K., Manikkam, M., Haque, M.M., Zhang, B., and Savenkova, M. (2012). Epigenetic transgenerational inheritance of somatic transcriptomes and epigenetic control regions. Genome Biol. *13*, R91.

Skinner, M.K., Manikkam, M., Tracey, R., Nilsson, E., Haque, M.M., and Guerrero-Bosagna, C. (2013). Ancestral dichlorodiphenyltrichloroethane (DDT) exposure promotes epigenetic transgenerational inheritance of obesity. BMC Medicine *11*, 228.

Skinner, M.K., Gurerrero-Bosagna, C., Haque, M.M., Nilsson, E.E., Koop, J.A., Knutie, S.A., and Clayton, D.H. (2014). Epigenetics and the evolution of Darwin's Finches. Genome Biol. Evol. *6*, 1972–1989.

Skinner, M.K., Savenkova, M., Zhang, B., and Crews, D. (2014b). Gene bionetworks involved in epigenetic transgenerational inheritance of altered mate preference: environmental epigenetics and evolutionary biology. BMC Genomics *15*, 377.

Smith, Z.D., Chan, M.M., Mikkelsen, T.S., Gu, H., Gnirke, A., Regev, A., and Meissner, A. (2012). A unique regulatory phase of DNA methylation in the early mammalian embryo. Nature *484*, 339–344.

Suter, L. and Widmer, A. (2013). Environmental heat and salt stress induce transgenerational phenotypic changes in *Arabidopsis thaliana*. PLoS One *8*, e60364.

Tracey, R., Manikkam, M., Guerrero-Bosagna, C., and Skinner, M. (2013). Hydrocarbon (jet fuel JP-8) induces epigenetic transgenerational inheritance of adult-onset disease and sperm epimutations. Reprod. Toxicol. *36*, 104–116.

Turner, B.M. (1998). Histone acetylation as an epigenetic determinant of long-term transcriptional competence. Cell. Mol. Life Sci. *54*, 21–31.

Van Speybroeck, L. (2002). From epigenesis to epigenetics: the case of C. H. Waddington. Ann. N. Y. Acad. Sci. *981*, 61–81.

Waddington, C.H. (1940). Organisers and Genes (The University Press, Cambridge).

Waterland, R.A., Travisano, M., and Tahiliani, K.G. (2007). Diet-induced hypermethylation at agouti viable yellow is not inherited transgenerationally through the female. FASEB J. *21*, 3380–3385.

Wei, Y., Yang, C.R., Wei, Y.P., Zhao, Z.A., Hou, Y., Schatten, H., and Sun, Q.Y. (2014). Paternally induced transgenerational inheritance of susceptibility to diabetes in mammals. Proc. Natl. Acad. Sci. U.S.A. *111*, 1873–1878.

Wolstenholme, J.T., Edwards, M., Shetty, S.R., Gatewood, J.D., Taylor, J.A., Rissman, E.F., and Connelly, J.J. (2012). Gestational exposure to bisphenol a produces transgenerational changes in behaviors and gene expression. Endocrinology *153*, 3828–3838.

Zheng, X., Chen, L., Li, M., Lou, Q., Xia, H., Wang, P., Li, T., Liu, H., and Luo, L. (2013). Transgenerational variations in DNA methylation induced by drought stress in two rice varieties with distinguished difference to drought resistance. PLoS ONE *8*, e80253.

Metabolic Inputs into Epigenetics

16

Scott J. Bultman

Abstract

Gene–environment interactions converge at the level of the epigenome, which, in turn, influences transcriptome profiles and phenotypic outcomes (Fig. 16.1). In this context, we must consider our diet as an amalgam of environmental factors that we are continuously exposed to and as something not restricted to caloric content. Certain bioactive food components and energy metabolites are known to modulate epigenomic marks and gene expression. In fact, many epigenetic enzymes require intermediary energy metabolites as essential co-factors (Table 16.1). Collectively, this represents a suite of mechanisms that link nutrient availability with regulation of gene expression to maintain homeostasis. These mechanisms are important for human health, and perturbations can alter an individual's susceptibility to various disease states including cancer. This chapter discusses some of the better-understood relationships between diet/energy metabolites and DNA methylation, histone post-translational modifications, and ATPase chromatin-remodelling complexes in mammals. Each section includes specific examples of 'metaboloepigenetics' that have been documented in humans and mouse models that are relevant to disease.

1C metabolism, SAM and DNA methylation

DNA methylation occurs at cytosines, primarily in the context of CpG dinucleotides, in mammals (Bird, 2002). It is a widespread and relatively stable epigenomic mark of considerable importance because promoter hypermethylation confers transcriptional silence (Bird, 2002). Both of the *de novo* DNA methyltransferase enzymes (DNMT3A and DNMT3B) and the only maintenance enzyme (DNMT1) require *S*-adenosylmethionine (SAM) as a co-factor because it serves as the methyl-group donor

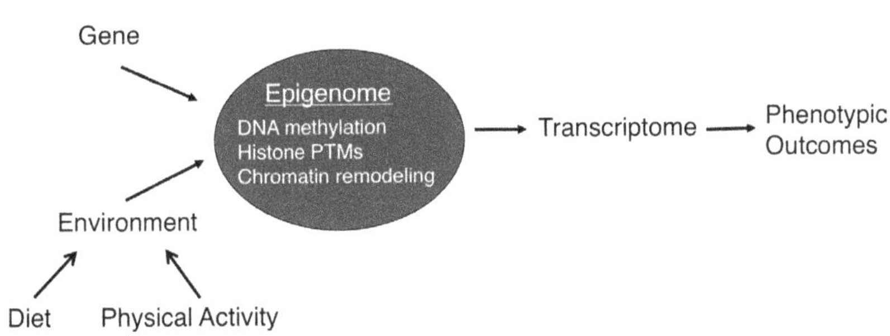

Figure 16.1 Energy balance affects epigenetics. Many gene–environment interactions converge on the epigenome, which regulates the transcriptome to influence phenotypic outcomes. Diet and physical activity are environmental factors that modulate energy metabolites that can affect DNA methylation, histone post-translational modifications (PTMs), and ATPase chromatin–remodelling complexes.

Table 16.1 Energy metabolites that are co-factors for epigenetic enzymes

Energy metabolite	Source (energetic process or pathway)	Affected epigenetic enzyme(s)	Effect on metabolite on enzymatic activity
SAM	1C metabolism	DNA methyltransferases (DNMTs)	Increase (methyl group donor)
SAM	1C metabolism	Histone methyltransferases	Increase (methyl group donor)
α-KG	TCA cycle	DNA 'demethylases' (TETs)	Increase (R-2HG metabolite decreases)
α-KG	TCA cycle	Histone demethylases with JmjC domain	Increase (R-2HG metabolite decreases)
Succinate	TCA cycle	DNA 'demethylases' (TETs)	Decrease
Succinate	TCA cycle	Histone demethylases with JmjC domain	Decrease
Fumarate	TCA cycle	DNA 'demethylases' (TETs)	Decrease
Fumarate	TCA cycle	Histone demethylases with JmjC domain	Decrease
Acetyl-CoA	TCA cycle-citrate shuttle-ACL; ACS; nuclear PDH	Histone acetyltransferases (HATs)	Increase (acetyl group donor)
NAD+	TCA cycle/electron transport chain	Class III histone deacetylases (sirtuins)	Increase
NAD+	TCA cycle/electron transport chain	Poly-ADP-ribose polymerases (PARPs)	Increase (ADP-ribose group donor)
FADH2	TCA cycle/electron transport chain	LSD histone demethylases	Increase
ATP	Electron transport chain/ oxidative phosphorylation	Chromatin-remodelling complexes with ATPase catalytic subunits	Increase
ATP	Electron transport chain/ oxidative phosphorylation	AMP kinase (AMPK)	Decrease (AMP/ATP ratio)

(Table 16.1). For each methyltransferase reaction, the methyl group of SAM is catalytically transferred by a particular DNMT (usually DNMT1) to the 5 position of cytosine to generate 5-methylcytosine. SAM is a product of one-carbon (1C) metabolism that involves a number of micronutrients including zinc, methionine, and vitamin B family members such as choline and folate/folic acid (Fig. 16.2). Folate is a natural component of leafy vegetables and certain other foods, whereas folic acid is a synthetic form used as a dietary supplement. They are considered equivalent because each one is metabolized to 5-methyltetrahydrofolate (5-methylTHF).

The epigenetic effects of 1C micronutrients can be readily observed in the viable-yellow agouti mouse model where genetically identical mice (when they are maintained on an inbred genetic background) can range in coat colour from yellow, through various degrees of mottling where patches of yellow are mixed with agouti (brown), to completely agouti depending on the extent of DNA methylation near one of the transcriptional start sites at the agouti locus (Morgan *et al.*, 1999). At the two extremes, yellow coat colour is associated with hypomethylation and higher levels of transcriptional activity, while agouti is associated with hypermethylation and lower levels of transcription. When viable-yellow females are pregnant and provided a diet with a 3-fold increase in several 1C methyl donors including folic acid and choline, the coat-colour distribution of their progeny shifts from yellow towards agouti (Fig. 16.3) (Cooney *et al.*, 2002; Waterland and Jirtle, 2003). These studies clearly demonstrate that micronutrients such as vitamin B family members can be rate-limiting factors in SAM production and DNA methylation reactions *in vivo*. The viable-yellow phenotype is pleiotropic, and the shift in coat colour is correlated with decreased incidence of obesity, type-2 diabetes, and cancer that is observed in all-yellow mice

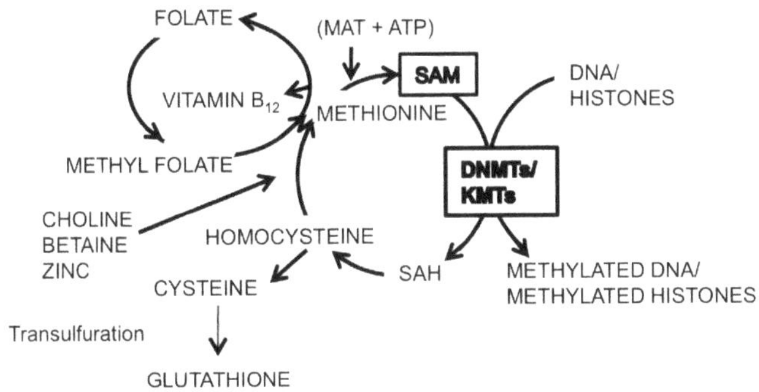

Figure 16.2 1C metabolism and SAM influences the methylation of DNA and histones. A schematic of one-carbon (1C) metabolism. A number of micronutrients including certain B vitamins are used to synthesize methionine, which is converted to S-adenosylmethionine (SAM) by the enzyme MAT in an ATP-dependent manner. SAM is an essential co-factor for DNA methyltransferases (DNMTs) and histone methyltransferases [also known as lysine methyltransferases (KMTs)] because it is the methyl-group donor. In methyltransferase reactions, SAM is converted to *S*-adenosylhomocysteine (SAH). High levels of SAH can inhibit methyltransferase reactions via feedback inhibition, and SAH is converted to homocysteine, which is recycled to make more methionine. Carbons from SAH can also be redirected to glutathione (via transulfuration).

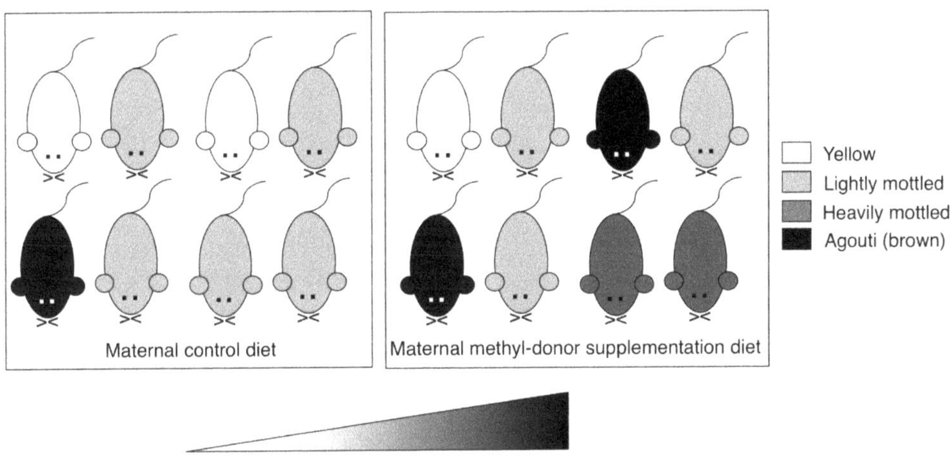

Figure 16.3 Maternal diet can influence the coat-colour phenotype of agouti viable-yellow mice. Shown is a schematic of two litters of viable-yellow mice with their coat colours indicated. On the left are offspring of a mother provided a control diet. The yellow and lightly mottled coat colours are well represented (see key on far right). On the right are offspring of a mother provided a diet with a 3-fold increase in several methyl donors such as folic acid and choline while pregnant. The heavily mottled and agouti coat colours are well represented (see key on far right). At the very bottom, a gradient-coloured triangle shows that the coat-colour distribution of the offspring shifts based on maternal diet.

but not all-agouti mice. Dietary supplementation in viable-yellow mice also has transgenerational epigenetic effects because yellow-to-agouti skewing can be apparent in the next generation without any further supplementation although the effect size is smaller (Cropley *et al.*, 2006). The basis for this transgenerational effect is that increased levels of SAM *in utero* result in increased methylation of the agouti gene in developing germ cells as well as developing skin/hair follicles within the fetus (Cropley *et al.*, 2006).

The viable-yellow mouse model serves as a particularly good reporter of DNA methylation, but it is not unique. Folic acid or choline supplementation during pregnancy has been reported to alter the DNA methylation status and expression of other genes in both mice and humans including a differentially methylated region (DMR) of the imprinted gene *IGF2* (insulin-like growth factor 2) (Steegers-Theunissen *et al.*, 2009; Waterland *et al.*, 2006; Zeisel, 2009). DNA methylation profiling and transcriptome profiling experiments are beginning to identify diverse genes (i.e. not associated with transposable elements as are viable-yellow agouti and axin-fused) that are regulated by these factors via DNA methylation in a tissue-specific manner (Amarasekera *et al.*, 2014; Binder and Michels, 2013; Fryer *et al.*, 2011; Schaible *et al.*, 2011). However, the affected genes can be either hypermethylated or hypomethylated, which indicates the situation is not as simple as methyl donors increasing SAM and DNA methylation at all loci. One potential explanation for differential effects is that SAM might not be randomly distributed throughout the nucleus. Chromatin immunoprecipitation (ChIP) experiments have demonstrated that MAT (methionine adenosyltransferase), which catalyses the final step of SAM production, is physically associated with chromatin and enriched at certain loci (Katoh *et al.*, 2011), suggesting that SAM levels spike at specific sites that are methylated and silenced.

Developmental, genetic, and mechanistic aspects of 1C metabolism

1C metabolism is important for human health. Epidemiological experiments have demonstrated reproducibly that adequate folate/folic acid intake is important during pregnancy and can prevent neural tube defects such as spina bifida from occurring (Czeizel and Dudas, 1992). The importance of these findings from a public health perspective led to flour being fortified with folic acid because the neural tube closes in the fetus during the first trimester before many women realize they are pregnant and have had an opportunity to modify their diets. However, this approach is potentially complicated by other epidemiologic findings suggesting that increased folate/folic acid intake during pregnancy is associated with several disease states in children after they are born including atopic dermatitis and asthma (Dunstan *et al.*, 2012; Haberg *et al.*, 2009; Kiefte-de Jong *et al.*, 2012; Whitrow *et al.*, 2009; Smith *et al.*, 2008). If confirmed, this discrepancy would be compatible with folate/folic acid affecting many genes in a tissue-specific manner.

Genetics will also be an important variable when it comes to supplementation with different individuals having different dietary requirements. For example, women who are carriers of a *MTHFD1* (5,10-methylenetetrahydrofolate dehydrogenase) single-nucleotide polymorphism (SNP) have >15-fold increased risk of developing signs of choline deficiency when on a low-choline diet plus an increased risk of having children with neural tube defects (Zeisel, 2009). This non-synonymous SNP (resulting in G1958A) is predicted to affect the availability of 5-methylTHF for homocysteine remethylation and may therefore increase demand for choline as a methyl-group donor. Additionally, a C677T SNP in *MTHFR* (5,10-methylenetetrahydrofolate reductase), which results in a single amino-acid substitution (A222V), also results in neural tube defects (Friso and Choi, 2005). MTHFR is a key enzyme in 1C metabolism that catalyses the production of 5-methyl-THF and maintains the flux of methyl groups for the re-methylation of homocysteine to methionine, and the A222V variant decreases this activity. To add another layer of complexity on top of genetic heterogeneity, certain pharmaceuticals, including some drugs used to treat seizures, are associated with folate deficiency suggesting that folate depletion is an adverse side effect. It seems likely that other environmental exposures will affect SAM levels and DNA methylation.

Despite the large body of literature regarding 1C metabolism, the precise mechanism of how it impacts human health is unclear. Although methyl-donor depletion/supplementation and SNPs in 1C enzymes modulate global DNA methylation levels (Friso and Choi, 2005; Zeisel, 2009), the target genes that contribute to neural tube defects and other disease states have not been identified. More importantly, some studies have been underpowered in terms of the number of samples and loci analysed, which has led to some findings not being replicated in subsequent studies (Crider *et al.*, 2012). One challenge is that DNA methylation is highly cell type dependent so cellular heterogeneity within samples can be a confounding factor. Another challenge is that there is a considerable amount of normal variation that occurs even in homogeneous samples such as cell lines and flow-sorted cells (which is evident from bisulfite sequencing analyses). Finally, the mechanism might not be entirely due to DNA methylation because SAM is also a co-factor for enzymes that methylate histones and non-histone proteins as well as RNAs and certain lipids. It should also be noted that 1C metabolism influences nucleotide pools for DNA synthesis and also affects amino-acid metabolism. Determining the relative importance of these different processes that are dependent on 1C metabolism will probably not be trivial.

Maternal diet, developmental windows of epigenetic susceptibility to disease, and transgenerational effects

The importance of maternal diet, which was discussed in the two previous sections, is supported by cohort studies that evaluated the health status of >2000 people at ~50 years of age who were conceived during the Dutch famine of 1944–1945 (Painter *et al.*, 2005). In Nazi-occupied Netherlands, calorie consumption dropped to 400–800 per day, and children who were conceived during this period became adults who were more likely to suffer from schizophrenia and heart disease than their siblings. Although it is unclear whether this higher incidence was mediated through epigenetic alterations, it is known that dynamic changes in DNA methylation occur during the earliest stages of mammalian embryogenesis (Messerschmidt *et al.*, 2014; Saitou *et al.*, 2012) and these studies identified a correlation between disease state and altered CpG methylation in certain genes including *IGF2* (Painter *et al.*, 2005; Tobi *et al.*, 2009). It is plausible that periconceptual folate deficiency contributed to this change (Lumey *et al.*, 2011; Steegers-Theunissen *et al.*, 2009; Sinclair *et al.*, 2007). One caveat of the Dutch famine studies is that CpG methylation profiles of blood cells are being correlated with schizophrenia, and it is not clear whether these cells are functionally relevant or if they serve as a reliable proxy for CpG methylation profiles in the CNS. Prenatal malnutrition during the Great Leap Forward famine of 1959–1961 in China has also been associated with increased risk of adult schizophrenia (Xu *et al.*, 2009), but this study did not analyse DNA methylation. A number of similar studies have reported increased incidence of cardiovascular problems, metabolic syndrome, and other health ailments due to the effects of various famines during different stages of pregnancy (Lumey *et al.*, 2011).

Many dietary factors and environmental agents will undoubtedly affect 1C metabolism and/or DNA methylation in general and during fetal development in particular. In fact, it is tempting to speculate that many teratogens perturb DNA methylation and other epigenomic marks. As a possible precedent, high levels of alcohol consumption can cause genome-wide changes in CpG methylation in mouse embryos that are associated with neural tube defects and other teratogenic effects (Liu *et al.*, 2009). Therefore, fetal alcohol syndrome may have an epigenetic component (Haycock, 2009). Although the mechanism is not clear, one possibility is that alcohol-induced oxidative stress leads to glutathione turnover, which is replenished by the transulfuration pathway that pulls carbons away from the 1C cycle at the level of cysteine and homocysteine (Fig. 16.2). This idea is supported by work where acute glutathione depletion affected methionine and led to CpG hypomethylation (Lertratanangkoon *et al.*, 1997).

Recent mouse studies have also revealed links between paternal diet and DNA methylation in their offspring. One study showed that a paternal low-protein diet led to many modest (~10–20%) changes in DNA methylation status in the liver with altered expression of genes involved in lipid metabolism (Carone *et al.*, 2010). A second study showed that paternal high-fat diets diminished the DNA methylation status of the *Il13ra2* gene in the pancreatic islets of female offspring (Ng *et al.*, 2010). Yet another study demonstrated that maternal calorie restriction during pregnancy gave rise to males with an aberrant DNA methylation of *Lxra* (a gene involved in lipogenesis) in their sperm that was transmitted to a somatic tissue (liver) the next generation (Martinez *et al.*,

2014). This is an example of a transgenerational effect persisting in animals that were not exposed as fetuses to the stimulus (caloric restriction in this case). Another example of this is a gene-trap line of mice carrying a hypomorphic mutation of *Mtrr* (methionine synthase reductase), which encodes an enzyme involved in 1C metabolism (Padmanabhan *et al.*, 2013). This mutation results in global hypomethylated DNA profiles and confers intrauterine growth restriction, developmental delay, and congenital malformations including cardiac and neural tube defects. These defects are dependent on the *Mtrr* genotypes of the maternal grandparents, rather than the maternal or embryonic genotypes, and persist for 5 generations, possibly through transgenerational epigenetic inheritance.

The viable-yellow mouse model demonstrates that what mom ate while pregnant or even what grandma ate while pregnant can influence the epigenetics and phenotype of children/grandchildren. These observations have led to the idea that there are windows of susceptibility for dietary factors or environmental exposures during development that contribute to subsequent heath outcomes and transgenerational effects. Support for this idea comes from a study of bisphenol A (BPA), which is a chemical used to manufacture polycarbonate plastic that has been implicated in autism. When viable-yellow mice are exposed to BPA during fetal or neonatal development, their coat-colour distribution is shifted towards yellow, opposite of methyl-donor supplementation, due to decreased CpG methylation (Dolinoy *et al.*, 2007). Interestingly, maternal supplementation of methyl donors including folic acid or the phytooestrogen genistein (an isoflavine derived from soy beans that is abundant in soy-based food) negates the effect of BPA in viable-yellow mice (Dolinoy *et al.*, 2007). This result suggests that maternal diet might counteract certain adverse environmental exposures. Human case–control studies are currently being performed to investigate whether exposure to dietary factors or environmental agents during pregnancy or postnatal development can alter the epigenome to influence the incidence of autism spectrum disorders and other conditions where the aetiology is poorly understood.

Methyl-donor supplementation beyond pregnancy

The influence of methyl donor supplementation on 1C metabolism is likely to be important beyond pregnancy, but this has been more difficult to establish. For example, there is widespread interest in reversing the changes in DNA methylation (and other epigenomic marks) that occur during carcinogenesis. Tumours often exhibit global hypomethylation, particularly in pericentric repeats which can lead to chromosomal instability, while also displaying focal hypermethylation that can silence tumour-suppressor genes. Unfortunately, studies that have evaluated the effect of folic acid supplementation on DNA methylation status in tumours have been inconsistent (Crider *et al.*, 2012). This could be due to limited sample sizes, different tissues/tumour types, and different DNA methylation assays performed at different loci. Genetics could also be a complicating factor as discussed above for *MTHFR* CC versus TT genotypes. Folic acid supplementation has been implicated in colorectal cancer prevention, but this is somewhat controversial and some evidence suggests that it may actually increase the progression of existing colonic adenomas or precancerous lesions in the prostate and possibly other sites (Kim, 2007a,b; Smith *et al.*, 2008; Ulrich and Potter, 2007). Therefore, genetic background, dose (optimal vs. too high), and the timing of supplementation may be crucial in distinguishing beneficial and detrimental effects. Both genetics and diet can influence the basal levels of a particular micronutrient, and this is an important factor based on a study of 672 individuals with histories of benign colorectal tumours where half of the participants received 1 mg of folic acid each day for ≥3 years and the other half received placebo (Wu *et al.*, 2009). Although folic acid supplementation did not have a significant effect when all of the participants were analysed, it did have a significant effect reducing the risk of recurrence in those individuals with the lowest levels of folate/folic acid intake prior to the onset of the trial (Wu *et al.*, 2009). Additional work will be necessary to determine whether methyl-donor supplementation and 1C metabolism can modify CpG methylation profiles in adult tissues with cancer-preventative or other beneficial properties. Generally speaking, stratifying cases and controls will probably reveal novel relationships between diet (including methyl-donor supplementation), epigenomic profiles, and health outcomes in human studies.

A number of environmental agents including air pollution, benzene exposure, and particulate pollution have been reported to change DNA methylation in adults (Bollati *et al.*, 2007; Crider *et al.*, 2012; Madrigano *et al.*, 2011). Although the mechanism(s) is not clear, it could involve reactive oxygen species (ROS), glutathione depletion, and the transulfuration pathway to affect 1C flux and DNA methylation as speculated above for alcohol exposure. Oestrogen-replacement therapy has been associated with decreased plasma homocysteine levels and increased DNA methylation (Friso *et al.*, 2007). An important area of future research will be to learn more about how the 'exposome' impacts the epigenome to influence health outcomes.

Oncometabolites and DNA demethylation

CpG methylation profiles are determined by DNMT-mediated methylation as well as TET-mediated demethylation. There are three TET (ten-eleven translocation) proteins that convert 5-methylcytosine (5mC) to 5-hydroxylmethylcytosine (5hmC) and can further oxidize this base to 5-formylcytosine (5fC) and 5-carboxylcytosine (5caC) (Tan and Shi, 2012). Although these 5-methylcytosine derivatives may or may not have their own epigenetic properties, they are clearly intermediates in the process of demethylation as they can be removed by glycosylases/base-excision repair enzymes and then be replaced by unmethylated cytosines (Bhutani *et al.*, 2011; Nabel and Kohli, 2011).

The TETs are dioxygenase enzymes that require oxygen, iron (Fe^{2+}) and a-ketoglutarate (a-KG) (Table 16.1). a–KG plays a crucial role in oxidative metabolism as an intermediate in the tricarboxylic acid (TCA) cycle (Fig. 16.4). In addition to being produced by isocitrate dehydrogenase (IDH) in the TCA cycle, a-KG can also be generated from glutamine, which is the most abundant amino acid in the body, where it enters the TCA via anapleurosis mediated by two enzymatic reactions (Fig. 16.4). Intracellular a-KG levels are estimated to be in the low millimolar range, well above the TET K_m values, suggesting that they are not normally rate limiting (Chowdhury *et al.*, 2011; Thirstrup *et al.*, 2011). However, neomorphic IDH mutations, which are common in certain tumours, deplete a–KG (also known as 2-oxyglutarate) and convert it to the R enantiomer of 2-hydroxyglutarate (R-2HG) (Kaelin and McKnight, 2013; Yang *et al.*, 2013). R-2HG is referred to as an oncometabolite because it is present at trace levels in normal tissues but accumulates to millimolar levels in IDH mutant tumours. Importantly, R-2HG inhibits TET activity and is both necessary and sufficient to promote certain aspects of tumorigenesis (Losman *et al.*, 2013; Xu *et al.*, 2011). Therefore, normal tissues have high levels of a-KG, which stimulates TET activity, whereas IDH mutant tumours have high levels of R-2HG that inhibit the same enzymes (Kaelin and McKnight, 2013) (Table 16.1 and Fig. 16.4). As one would expect, this imbalance results in IDH mutant tumours having significantly increased global levels of DNA methylation (and histone methylation as mentioned in the next section) (Figueroa *et al.*, 2010; Turcan *et al.*, 2012; Xu *et al.*, 2011).

There is additional evidence to support the above mechanism. In other tumours, loss-of-function mutations in two other genes encoding TCA cycle enzymes (Fig. 16.4), succinate dehydrogenase (SDH) and fumarate hydrase (FH), result in the accumulation of succinate and fumarate, respectively (Kaelin and McKnight, 2013; Yang *et al.*, 2013). These TCA cycle intermediates are also oncometabolites that inhibit TETs and cause global hypermethylation of CpGs (Killian *et al.*, 2013; Letouze *et al.*, 2013; Xiao *et al.*, 2012) (Fig. 16.4).

Energy metabolites involved in histone methylation and demethylation

Genomic DNA is packaged as chromatin, and the nucleosome is the most fundamental unit. Each nucleosome consists of a 147-bp segment of DNA wrapped around a histone octamer (two copies each of histones H2A, H2B, H3, and H4).

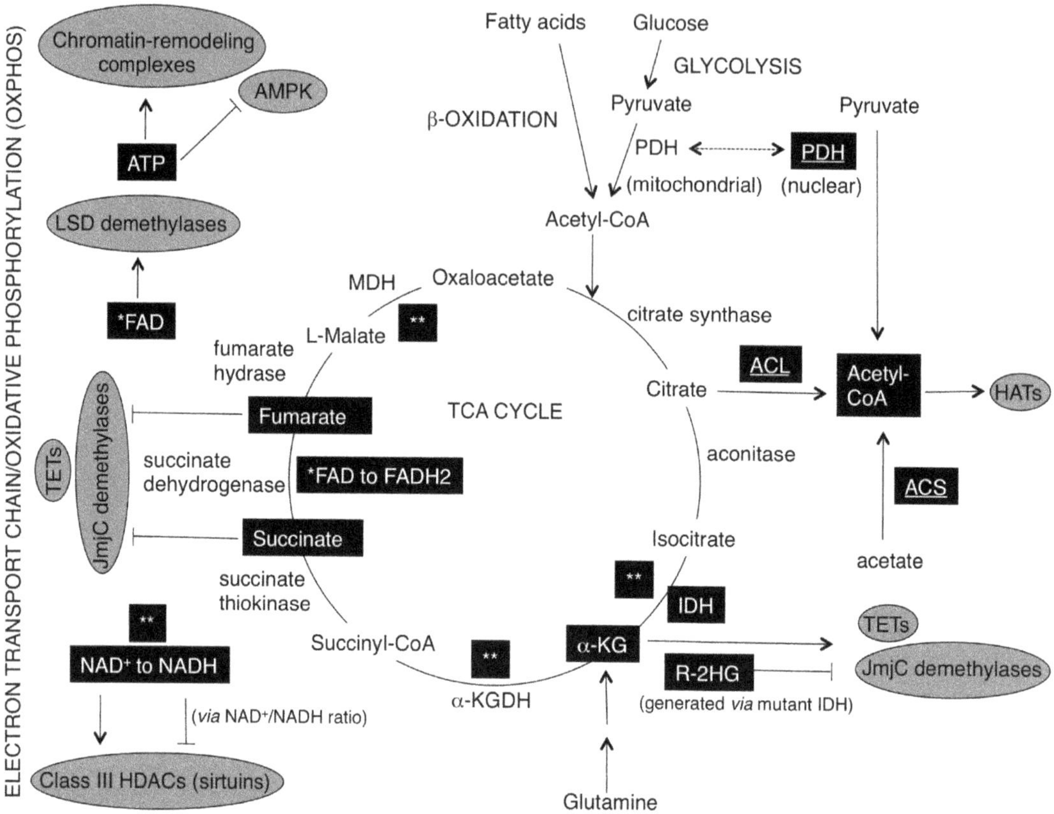

Figure 16.4 The TCA cycle can influence the activity of epigenetic enzymes. The TCA cycle is shown along with inputs and outputs at different positions as indicated. Labelling of the TCA cycle and other metabolic processes/pathways are capitalized. Individual enzymes (IDH, ACL, ACS, PDH), energy metabolites (_-KG, succinate, fumarate, acetyl-CoA, ATP), and electron carriers modulated by redox state (NAD+, NADH, FAD) that activate (arrows) or inhibit (perpendicular lines) epigenetic enzymes are highlighted by white text in black boxes. All of the metabolic enzymes that produce these energy metabolites and electron carries are mitochondrial (TCA cycle and OXPHOS) except for several nucleocytoplasmic enzymes that are underlines (ACL, ACS, PDH). The epigenetic enzymes that are regulated by energy metabolites and electron carries are shown in grey ovals. See text for details.

Histones undergo a variety of post-translational modifications (PTMs) including methylation and acetylation that alter nucleosome function. Methylation occurs at arginine (R) and lysine (K) residues and has differential effects on transcription. For example, histone 3, lysine 4 methylation (H3K4me) is associated with active transcription, whereas H3K9me and H3K27me are associated with transcriptional repression. Protein arginine methyltransferases (PRMTs) and protein lysine methyltransferases (PKMTs) catalyse histone methylation and also methylate other proteins. The nine PRMTs are relatively promiscuous because they methylate multiple R residues on multiple histones, whereas the ~50 PKMTs exhibit target specificity for histone substrates (Greer and Shi, 2012). For example, MLL family members methylate H3K4, G9a and SUV39H1 methylate H3K9, and EZH2 methylates H3K27. Although PKMTs do exhibit target specificity, there are many instances of overlap (e.g. G9a and SUV39H1 both target H3K9) with differences in their ability to mono/di/trimethylate (me1, me2, me3) or to function in pericentric heterochromatic regions and different euchromatic loci. All of the PKMTs except DOT1L (which methylates H3K79) contain a SET domain that catalyses the methyltransferase reaction.

All of the PRMT and PKMT enzymes are dependent on SAM as a methyl donor co-factor similar to DNMTs (Table 16.1 and Fig. 16.2). Methyl-donor supplementation/deficiency and genetic variants of enzymes involved in 1C metabolism such as MTHFR will undoubtedly affect histone methylation levels, but relatively few studies have been published so far (Dobosy *et al.*, 2008; Mehedint *et al.*, 2010). The metabolism of the amino acid threonine and SAM are coupled to regulate H3K4me3 and pluripotency in mouse embryonic stem (ES) cells (Shyh-Chang *et al.*, 2013). Although it is not clear why other histone methylation marks are apparently not regulated by this pathway, H3K4me3 is particularly abundant and turns over rapidly in this cell type. This finding suggests that other dietary factors and metabolic pathways will impinge on 1C metabolism. However, as mentioned above, it will be a challenge to determine the relative importance of methyl donor availability, 1C genetics, and SAM levels on DNA methylation versus histone methylation, especially since there is crosstalk between these epigenomic marks.

Similar to DNA, histone methylation is reversed by demethylation. This represents a newer area of research so our understanding lags behind other epigenetic processes. Two families of lysine demethylase (KDM) enzymes have been characterized, but the identification of arginine demethylases has been more elusive. The larger family of KDMs contain a catalytic jumonji C domain (JMJD) and consist of ~30 members that demethylate specific K residues. Multiple KDMs with JMJDs can act on the same residue with differences in their ability to cleave me1/me2/me3 as well as which loci they act upon in the genome. They also can demethylate non-histone proteins. Similar to TETs, these KDM/JMJD enzymes are dioxygenases that require oxygen, iron (Fe^{2+}) and the TCA cycle intermediate a-ketoglutarate (a-KG) (Table 16.1 and Fig. 16.3). Consequently, neomorphic IDH mutations that generate the oncometabolite R-2HG in tumours result in increased histone methylation in addition to increased CpG methylation (Lu *et al.*, 2012).

The second family of histone demethylase enzymes (although discovered first) consists of KDM1A and KDM1B (also known as lysine-specific demethylases LSD1 and LSD2). The former acts on HeK4me1/2 and H3K9me1/2, while the later has only been shown to act on H3K4me1/2. Both can also demethylate non-histone proteins as well as DNA. These enzymes are structurally distinct from the JMJD family and are catalysed by an amine oxidase domain. KDM1A and KDM1B require flavin adenine dinucleotide (FAD) as a co-factor (Table 16.1), which is an energy carrier derived from riboflavin (vitamin B2) and ADP that is dependent on the redox state of the cell. Their activities are stimulated by FAD when it is oxidized but not when it is reduced to $FADH_2$ such as occurs in the TCA cycle to carry high-energy electrons used in the electron transport chain of mitochondria that culminates in oxidative phosphorylation and ATP production (Fig. 16.4). KDM1A/LSD1 directly regulates PPAR function and mitochondrial respiration (Hino *et al.*, 2012), which could potentially contribute to additional epigenomic alterations.

Acetyl-CoA and histone acetylation

Another intensively studied post-translational modification is histone acetylation. Unlike methylation, it only occurs at K residues and is usually associated with transcriptional activity. Histone acetyltransferases (HATs) (also called lysine acetyl transferases or KATs because they acetylate non-histone protein substrates) require acetyl-CoA as a co-factor because it is the acetyl group donor (Table 16.1). Acetyl-CoA is central to energy metabolism. It is generated in the mitochondria as product of both glucose metabolism (glycolysis followed by pyruvate dehydrogenase) and lipid metabolism (b oxidation) and condenses with oxaloacetate to form citrate in the first step of the TCA cycle (Fig. 16.4). Although acetyl-CoA cannot exit the mitochondria by diffusing across membranes, a citrate antiporter system can shuttle citrate from the mitochondria to the cytosol, where ATP citrate lyase (ACL) reconverts it to acetyl-CoA (Fig. 16.4). ATP and CoA are consumed by this ACL-catalysed reaction, and oxaloacetate is another reaction product. Cytosolic acetyl-CoA is the building block of *de novo* lipid biogenesis that involves reactions catalysed by acetyl-CoA carboxylase and fatty acid synthase. Phospholipids are important for doubling the biomass of the plasma membrane every time a cell divides, and triglycerides are utilized for energy storage, especially in adipocytes and hepatocytes. ACL is also present in the nucleus, and citrate can diffuse through nuclear pores into the nucleus (Wellen *et al.*, 2009). Acetyl-CoA generated by ACL in the cytosol can also enter the nucleus through nuclear pores. Thus, the citrate shuttle and ACL represent a major pathway for the production of nuclear acetyl-CoA. Depending on nutritional/metabolic conditions, intracellular acetyl-CoA levels can fluctuate by ~10-fold and are within the K_m range of many HATs, suggesting that cetyl-CoA availability can be rate-limiting for histone acetylation (Cai *et al.*, 2011; Takamura and Nomura, 1988). The importance of this pathway is supported by RNAi-mediated knockdown studies demonstrating that ACL is crucial for maintaining global histone acetylation levels (Wellen *et al.*, 2009). It is believed that the citrate shuttle-ACL pathway is utilized more prominently when nutrients meet or surpass the energy needs of the cell, and ACL is overexpressed in several tumour types and important for tumour growth (Bauer *et al.*, 2005; Hatzivassiliou *et al.*, 2005). Not surprisingly, this pathway is subject to feed-forward activation as glucose and insulin signalling up-regulate ACL activity via phosphorylation by AKT.

The importance of nucleocytosolic acetyl-CoA is underscored by it being produced not only by ACL but also by other mechanisms. Two acetyl-CoA synthetase (ACS) enzymes condense acetate and coenzyme A (Table 16.1, Fig. 16.4). As one might expect, RNAi-mediated knockdown of one of these enzymes in addition to ACL results in particularly low levels of global histone acetylation (Wellen *et al.*, 2009). A more recent discovery is that pyruvate dehydrogenase (PDH) can translocate to the nucleus instead of being exclusively localized to the mitochondria. This mitochondrial-to-nuclear translocation of PDH occurs when cells are stimulated with serum or epidermal growth factor (EGF), and RNAi knockdown studies demonstrate that PDH also contributes to nuclear pools of acetyl-CoA histone acetylation (Sutendra *et al.*, 2014) (Table 16.1 and Fig. 16.4).

It is tempting to speculate that acetyl-CoA links energy status and gene expression to regulate cell-cycle progression. When cells are serum starved or lack enough energy metabolites to complete the next round of cell division, they do not pass a G_1 checkpoint called the restriction point and become quiescent because it would be deleterious to begin DNA replication without being able to complete it and the ensuing mitosis. This checkpoint protects against genomic instability and cancer. Re-addition of serum or nutrients allows cells to move beyond the restriction point and the G_1/S phase transition and also induces histone acetylation. RB phosphorylation is involved in this process, but acetyl-CoA is also a good candidate because it is influenced by energy balance, helps to build new cells via lipid biogenesis, and regulates gene expression as a HAT co-factor. In support of this idea, acetyl-CoA levels can fluctuate by 6-fold in yeast depending on their metabolic cycle (oxidative vs. reductive growth), and this correlates with H3K9ac levels at certain loci (Kaelin and McKnight, 2013). Furthermore, when carbon sources are added back to nutrient-deprived yeast that are quiescent, acetyl-CoA levels increase and induce histone acetylation (via the GCN5 HAT) and transcription of growth-related genes (Cai *et al.*, 2011). It is also likely that acetyl-CoA-driven HAT activity triggers the G_1/S phase transition by facilitating DNA replication because origins of replication are enriched in histone acetylation (Goren *et al.*, 2008). Finally, it should be reiterated that HATs/KATs acetylate non-histone proteins, and the acetylated proteome is enriched for metabolic enzymes and cell-cycle regulators including ACL (Lin *et al.*, 2013; Zhao *et al.*, 2010).

NAD^+ and sirtuins

Histone deacetylases (HDACs) are divided into several classes, and the class III HDACs (also known as sirtuins) require nicotinamide adenine dinucleotide (NAD^+) as an essential co-factor (Table 16.1). NAD^+ is an energy carrier that accepts electrons from other molecules as it is reduced to NADH. Similar to FAD and $FADH_2$, this change in redox status links the TCA cycle to the electron transport chain and oxidative phosphorylation for ATP production in mitochondria (Fig. 16.4). A relatively low NAD^+/NADH ratio is therefore a measure of favourable energetics that is 'sensed' by sirtuins (SIRTs) to down-regulate their deacetylase activity (Guarente, 2011). However, in response to prolonged fasting and diminished glucose intake, this ratio is elevated, which leads to increased sirtuin activity including SIRT1-mediated deacetylation of H3K9ac and H4K16ac at ribosomal DNA (rDNA) repeats that down-regulates rDNA transcription, ribosome biogenesis, and consequently protein synthesis. SIRT1 and other sirtuins also regulate FoxO transcription factors and various aspects of autophagy (Hariharan *et al.*, 2010; Lee *et al.*, 2008; Ng and Tang, 2013). These findings are compatible with sirtuins playing a role in cellular growth when nutrients are abundant and autophagy when nutrients are scarce.

Not only can diet can elevate NAD^+ levels and the NAD^+/NADH ratio (Canto *et al.*, 2012), but alcohol intake can have the opposite effect when it is metabolized by alcohol dehydrogenase (ADH) to acetaldehyde and then by acetaldehyde dehydrogenase (ALDH) to acetate because both reactions involve the reduction of NAD^+ to NADH (Shepard and Tuma, 2009). This is expected to increase the NAD^+/NADH ratio and inhibit sirtuins, which is consistent with observations that that excessive alcohol exposure can increase histone acetylation (Haycock, 2009). In this scenario, increased histone acetylation could also be driven by a second mechanism where the acetate end product is converted to acetyl-CoA by acetyl-CoA synthetase (ACS) to increase HAT activity. Although alcohol-induced histone acetylation should occur primarily in the liver, it does have effects in the CNS (Pascual *et al.*, 2012) and might contribute to the teratogenic effects of alcohol in addition to decreased DNA methylation (and epigenetic-independent mechanisms).

Similar to other HDACs, the sirtuins also deacetylate non-histone proteins. In fact, most of the seven sirtuins are localized outside of the nucleus with SIRT2 being cytosolic and SIRT3, SIRT4, and SIRT5 being mitochondrial (Houtkooper *et al.*, 2012). This subcellular localization is consistent with the sirtuins regulating energy metabolism and supports proteomic data that metabolic enzymes are regulated by acetylation to a greater extent than other proteins. Only SIRT1, SIRT6, and SIRT7 are localized to the nucleus, and SIRT1 is also localized in the cytosol while SIRT7 is primarily found in the nucleolus (Houtkooper *et al.*, 2012). SIRT1 plays an important role in the maintenance of circadian rhythms, and part of this driven by the oscillation of NAD^+ levels (Nakahata *et al.*, 2009; Ramsey *et al.*, 2009) (Sassone-Corsi, 2012). Energy metabolism and autophagy play a central role in carcinogenesis, and several sirtuins including SIRT6 are tumour suppressors that coordinate glycolytic and TCA cycle activity (Sebastian *et al.*, 2012). Part of this mechanism involves SIRT6 deacetylation of H3K9 at the promoters genes involved in glycolysis (Zhong *et al.*, 2010).

Caloric restriction, lifespan and histone acetylation

Reducing caloric intake to ~60% of *ad libitum* is known to diminish certain age-related diseases and increase lifespan up to 150% in organisms ranging from yeast to humans, and much of this effect has been attributed to SIRT1 activity. Although there has been some controversy regarding SIRT1 and lifespan extension in *C. elegans* and *Drosophila* (Burnett *et al.*, 2011), it has not been challenged in mammals. For example, caloric restriction increases NAD^+ levels in certain tissues (muscle, liver, white adipose tissue) and extends the lifespan of wild-type mice, but it does not have this effect in SIRT1-deficient mice (Herranz and Serrano, 2010). Although this indicates that the effect of caloric restriction is SIRT1-dependent, a caveat is that SIRT1-deficient mice are short-lived regardless of diet, with many dying in the neonatal period. Further support comes from the observation that transgenic mice overexpressing SIRT1 are protected from glucose intolerance and hyperinsulinaemia in response to a high-fat diet although longevity is unaffected (Herranz and Serrano, 2010). Similarly, resveratrol, which is a naturally occurring phenol present in the skin of red grapes (but present at low levels in red wine) and other fruits, activates SIRT1 and can improve insulin sensitivity and extend lifespan of mice provided a high-fat diet. However, resveratrol probably activates SIRT1 indirectly via AMPK, and the relative importance of SIRT1 compared to other targets is not known. Nevertheless, these findings support the idea that SIRT1 protects against metabolic stress and promotes 'healthspan'. It is not clear to what extent this is due to deacetylation of histones versus non-histone targets. However, caloric restriction does lead to changes in histone acetylation and other epigenomic marks at various loci (Li *et al.*, 2011). Other energy metabolites involved in epigenetics will probably be involved in mammalian ageing as suggested by work in more experimentally tractable eukaryotic model systems. For example, a-KG extends the lifespan of *C. elegans* by inhibiting ATP synthase, and this is dependent on target of rapamycin (TOR) and also associated with increased autophagy (Chin *et al.*, 2014). Furthermore, RNAi-mediated depletion of acetyl-CoA synthetase in yeast (*S. cerevisiae*) and fruit flies (*D. melanogaster*) alters nucleocytosolic acetyl-CoA levels to stimulate autophagy and prolong lifespan, and part of this is mediated by H4K16ac and up-regulation of autophagy (atg) genes (Eisenberg *et al.*, 2014).

Prolonged fasting during caloric restriction results in the liver producing ketone bodies such as b-hydroxybutyrate (b-OHB) as an energy source. This also occurs when glucose levels drop due to ketogenic diets (i.e. high fat/low carbohydrate diets such as the Atkins diet) or prolonged exercise. Serum levels of b-OHB are induced to the millimolar range compared to basal levels in the micromolar range. In addition to being important for energy metabolism, millimolar concentrations of b-OHB function as an inhibitor of class I HDACs (distinct from the sirtuins) and increases H3K9ac and H3K14ac (Shimazu *et al.*, 2013). This is the case globally and at specific genes such as FOXO3A that are up-regulated and may account for some of the homeostatic effects (Shimazu *et al.*, 2013). It should be noted that b-OHB production is activated in response to fasting by mitochondrial SIRT3 deacetylating an enzyme (HMGCS2) that regulates ketone body production (Shimazu *et al.*, 2010).

Additional post-translational modifications of histones

Histones undergo additional PTMs that are derived from NAD^+ and other energy metabolites. One or more ADP-ribose moieties can be added to K, R, and several other amino-acid residues by 18 different poly-ADP-ribose polymerase (PARP) enzymes (Messner and Hottiger, 2011). PARP1 has been investigated in the most detail and catalyses the majority of ADP-ribose chains. Similar to sirtuins, the PARPs require NAD^+. NAD^+ is consumed because it donates the ADP-ribose group (Table 16.1). The poly-ADP-ribose chain length can be determined by NAD^+ concentrations (Huletsky *et al.*, 1985), which is influenced by nutritional status and redox state (Fig. 16.4). Histone poly-ADP ribosylation plays a role in transcriptional regulation and DNA replication but is particularly important for DNA repair (Messner and Hottiger, 2011). Because PARPs and sirtuins compete for the same NAD^+ pools, extensive DNA damage may result in PARP-mediated depletion of NAD^+ to an extent that attenuates sirtuin-mediated histone deacetylation. Similarly, it is tempting speculate that PARP inhibitors, which are used as chemotherapeutic agents, might affect sirtuin activity. Either of these scenarios could involve non-histone proteins as supported by the observation that DNA damage and mitochondrial SIRT4 tumour-suppressive function involve glutamine metabolism/anaplerosis and carbon flux into the TCA cycle (Jeong *et al.*, 2013). Like other epigenomic marks, poly-ADP-ribosylation is reversible and PARPs are counteracted by ADP-ribosyl hydrolases and PAR glycohydrolases.

Glucose induces histone GlcNacylation, and GlcNAc is derived from glycolysis via the hexosamine pathway (Fujiki *et al.*, 2011) Glucose also induces histone monoubiquitination (H2B-K120ub), which is an active mark that is not targeted for proteosome-mediated degradation. New marks have recently been identified by mass spectrometry, and it will not be surprising if some are influenced by energy metabolism.

ATP levels, chromatin-remodelling complexes and AMPK

Four families of chromatin-remodelling complexes (SWI/SNF, ISWI, NuRD/Mi2/CHD, and INO80) have been characterized that utilize catalytic subunits with ATPase activity to slide or evict nucleosomes at many sites in the genome. Much attention has focused on their occupancy at promoters and their role in transcriptional initiation. Because each one has a catalytic ATPase subunit, these complexes have activities that are dependent on adenosine triphosphate (ATP) (Table 16.1), which is produced by oxidative phosphorylation in the mitochondria and serves as the energy currency of the cell (Fig. 16.4). Theoretically, suboptimal energetic conditions could result in such low levels of ATP that it becomes rate limiting for chromatin remodelling. However, there is not any evidence to support this notion at the current time. For example, SWI/SNF complexes have robust activity in cell-free assays when 1–2 mmol/l ATP is used, and this ATP concentration is towards the lower end of the normal intracellular concentration range (0.5–10 mmol/l). However, ATP levels could conceivably drop below this range in the nucleus under extreme conditions (without causing cell death), and it is not clear whether dose–response experiments have been performed that titrate ATP to determine the threshold concentration necessary for maximal ATPase and chromatin-remodelling activities. It is also possible that there are differences in the local concentration of ATP within different regions of the nucleus (e.g. nucleolus, nuclear periphery, pericentric heterochromatin), which could diminish chromatin remodelling at a subset of genomic loci.

In contrast to chromatin-remodelling complexes, the enzyme AMPK [5′ adenosine monophosphate (AMP)-activated protein kinase] is sensitive to ATP levels in the cell. It is activated by high AMP/ATP ratios and activates many substrates in multiple pathways to adapt to energetic stress and enable cell survival (Table 16.1 and Fig. 16.4). Among its direct targets is H2B-S36. AMPK-mediated phosphorylation of this residue is important for transcriptional regulation and homeostasis (Bungard *et al.*, 2010). Activation of AMPK alters the NAD^+/NADH ratio to modulate SIRT1 activity and the deacetylation of PPARs and FOXO transcription factors as another way to restore energy homeostasis (Canto *et al.*, 2009). This suggests that AMPK may deacetylate histones indirectly via SIRT1 or SIRT6.

Lack of physical activity, the obesity epidemic, cancer, and epigenetics

Physical activity as well as diet influence energy metabolites and epigenetics (Fig. 16.1). Although our knowledge of the former is not as strong as the latter, strenuous exercise can affect NAD^+ levels, AMPK activation, and certain epigenomic marks including DNA methylation (Kirchner *et al.*, 2013). A more sedentary lifestyle combined with the Western diet (energy-rich, processed foods) is probably responsible for most or all of the increased adiposity of human populations, and the obesity epidemic will continue to have profound deleterious consequences on cardiovascular disease, diabetes, cancer, and other disease states. Many of the underlying disease aetiologies will involve direct effects on energy balance that culminate in increased adiposity and metabolic syndrome, which are associated with atherosclerosis, insulin resistance, inflammation, and a tumour-permissive environment. However, chronic ‘overnutrition’ is likely to have other effects including altered gene expression due to perturbed epigenomic marks that arise because of altered energy metabolites that function as co-factors for epigenetic enzymes (via mechanisms discussed in above sections). For example, the enzyme NNMT (nicotinamide *N*-methyltransferase) methylates nicotinamide (vitamin B3) using SAM as a methyl donor and is up-regulated in white adipose tissue and liver from obese humans and rodents. NNMT can impact sirtuins because nicotinamide is a precursor of NAD^+, while it can simultaneously affect DNA methylation and histone methylation because SAM is converted to SAH. NNMT-depleted mice are resistant to diet-induced obesity and insulin resistance, and this phenotype is associated with altered SAM, NAD^+ and H3K4me levels in white adipose tissue (Kraus *et al.*, 2014).

In human epidemiological studies, a strong association between obesity and increased incidence of various cancers has been very reproducible (Willett, 2008). This link is compatible with metformin, which targets AMPK and has long used to treat type 2 diabetes, being repurposed for cancer treatment with considerable success (Quinn *et al.*, 2013). This link may involve obesity leading to epigenomic perturbations that precede oncogenic transformation and contribute to carcinogenesis. For example, diet-induced obesity in mice lacking any oncogenic stimuli alters H3K27ac profiles at putative enhancers and transcriptome profiles in colonocytes, and these obesity-associated profiles resemble a colorectal cancer signature (Li *et al.*, 2014). However, the relative contribution of epigenomics must still be determined for the obesity-cancer connection. Another challenge is that, unlike obesity, epidemiologic studies that have evaluated individual dietary factors such as fruits, vegetables, fibre, and red meat in cancer prevention have yielded inconsistent findings (Willett, 2008).

Owing to the Warburg effect (Vander Heiden *et al.*, 2009), tumours primarily utilize aerobic glycolysis instead of oxidative metabolism. Although this is inefficient from an energetic perspective (2–4 ATPs vs. 36 ATPs per glucose molecule, respectively), carbons and nitrogens from glycolytic intermediates and the pentose phosphate shunt are diverted into nucleotide and amino acid salvage pathways (instead of the carbon 'escaping' as CO_2), which enables the rapidly dividing cell to double its biomass in as little as 6–8 hours. Similarly, the citrate shuttle-ACL pathway is crucial for increased for lipid biogenesis (to double the plasma membrane content), but this also increases histone acetylation. This represents one way in which the transcriptome profile of the cancer cell diverges from the original cell-of-origin without requiring mutations in the promoter regions of the affected genes. To compensate for their inefficient ATP production, tumours up-regulate the expression of their glucose transporters (GLUTs) and take up >10′ glucose than normal cells, and this is the basis for tumour imaging in the clinic (via 2-FDG and PET scanning). Mutations in proto-oncogenes and tumour-suppressor genes play an important role in the overexpression of GLUTs and other aspects of metabolic reprogramming (Soga, 2013; Lee *et al.*, 2014).

Bioactive food components that inhibit epigenetic enzymes

Certain naturally occurring bioactive food components can inhibit epigenetic enzymes by mechanisms other than 1C metabolism (Huang *et al.*, 2011; Ong *et al.*, 2011; Supic *et al.*, 2013). Biochemical analysis of plant-derived extracts, particularly those used as traditional medicines in China and India, has led to a number of bioactive molecules. For example, several polyphenols can inhibit DNMTs: epigallocatechin-3-gallate (EGCG), a catechin abundant in green tea; genistein, an isoflavone from soybeans; curcumin from the roots of tumeric herb; resveratrol from grape skins; and caffeic acid and chlorogenic acid from coffee. Naturally occurring HDAC inhibitors include sulforaphane, an isothiocyanate abundant in cruciferous vegetables, diallyl disulfide, as well as allyl mercaptan, organosulfur compounds abundant in garlic; butyrate, a short-chain fatty acid produced by bacterial fermentation of fibre in the gut. The gut microbiome has prodigious metabolic capacity and will likely produce many metabolites other than butyrate and choline that affect host epigenetics (Donohoe *et al.*, 2011, 2012; Spencer *et al.*, 2011). Other dietary compounds inhibit HATs such as anacardic acid, salicylic acids with alkyl chains present in certain nuts; curcumin (listed above for DNMTs); plumbagin, a hydroxynaphthoquinone from the roots of a leadwort plant. Other epigenetic enzymes can also be targeted. For example, omega-3 fatty acids, which are common in fish oils, can inhibit the EZH2 histone methyltransferase.

Depending on which specific HDACs or HATs (or other epigenetic enzyme) are affected by each dietary agent, different target genes will be affected in different tissues. This is relevant because many studies have been performed in cell lines and need to be extended to mouse models and human subjects. A benefit of *in vivo* studies is that they will more rigorously test the hypothesis that some of these factors are beneficial for human health, especially for cancer prevention. However, even in cases where these links are confirmed, it will be important to determine how much of the chemoprotective effect is mediated by altering epigenomic marks versus other possible mechanisms such as affecting non-histone proteins or antioxidant activity. Many other bioactive food components and micronutrients may affect histone acetylation or other epigenomic marks that still await identification. Considering the size of the nutritional supplement industry and how little we know about many agents, it would be useful to learn more about their potential epigenetic effects by performing large-scale ChIP-seq, BS-seq, and FAIRE-seq screens to evaluate histone PTMs, DNA methylation, and nucleosome occupancy/density, respectively.

References

Amarasekera, M., Martino, D., Ashley, S., Harb, H., Kesper, D., Strickland, D., Saffery, R., and Prescott, S.L. (2014). Genome-wide DNA methylation profiling identifies a folate-sensitive region of differential methylation upstream of ZFP57-imprinting regulator in humans. FASEB J. [Epub ahead of print]

Bauer, D.E., Hatzivassiliou, G., Zhao, F., Andreadis, C., and Thompson, C.B. (2005). ATP citrate lyase is an important component of cell growth and transformation. Oncogene *24*, 6314–6322.

Bhutani, N., Burns, D.M., and Blau, H.M. (2011). DNA demethylation dynamics. Cell *146*, 866–872.

Binder, A.M., and Michels, K.B. (2013). The causal effect of red blood cell folate on genome-wide methylation in cord blood: a Mendelian randomization approach. BMC Bioinformatics *14*, 353.

Bird, A. (2002). DNA methylation patterns and epigenetic memory. Genes Dev. *16*, 6–21.

Bollati, V., Baccarelli, A., Hou, L., Bonzini, M., Fustinoni, S., Cavallo, D., Byun, H.M., Jiang, J., Marinelli, B., Pesatori, A.C., et al. (2007). Changes in DNA methylation patterns in subjects exposed to low-dose benzene. Cancer Res. *67*, 876–880.

Bungard, D., Fuerth, B.J., Zeng, P.Y., Faubert, B., Maas, N.L., Viollet, B., Carling, D., Thompson, C.B., Jones, R.G., and Berger, S.L. (2010). Signaling kinase AMPK activates stress-promoted transcription via histone H2B phosphorylation. Science *329*, 1201–1205.

Burnett, C., Valentini, S., Cabreiro, F., Goss, M., Somogyvari, M., Piper, M.D., Hoddinott, M., Sutphin, G.L., Leko, V., McElwee, J.J., et al. (2011). Absence of effects of Sir2 overexpression on lifespan in *C. elegans* and *Drosophila*. Nature *477*, 482–485.

Cai, L., Sutter, B.M., Li, B., and Tu, B.P. (2011). Acetyl-CoA induces cell growth and proliferation by promoting the acetylation of histones at growth genes. Mol. Cell *42*, 426–437.

Canto, C., Gerhart-Hines, Z., Feige, J.N., Lagouge, M., Noriega, L., Milne, J.C., Elliott, P.J., Puigserver, P., and Auwerx, J. (2009). AMPK regulates energy expenditure by modulating NAD+ metabolism and SIRT1 activity. Nature *458*, 1056–1060.

Canto, C., Houtkooper, R.H., Pirinen, E., Youn, D.Y., Oosterveer, M.H., Cen, Y., Fernandez-Marcos, P.J., Yamamoto, H., Andreux, P.A., Cettour-Rose, P., et al. (2012). The NAD(+) precursor nicotinamide riboside enhances oxidative metabolism and protects against high-fat diet-induced obesity. Cell Metab. *15*, 838–847.

Carone, B.R., Fauquier, L., Habib, N., Shea, J.M., Hart, C.E., Li, R., Bock, C., Li, C., Gu, H., Zamore, P.D., et al. (2010). Paternally induced transgenerational environmental reprogramming of metabolic gene expression in mammals. Cell *143*, 1084–1096.

Chin, R.M., Fu, X., Pai, M.Y., Vergnes, L., Hwang, H., Deng, G., Diep, S., Lomenick, B., Meli, V.S., Monsalve, G.C., et al. (2014). The metabolite alpha-ketoglutarate extends lifespan by inhibiting ATP synthase and TOR. Nature *509*, 397–401.

Chowdhury, R., Yeoh, K.K., Tian, Y.M., Hillringhaus, L., Bagg, E.A., Rose, N.R., Leung, I.K., Li, X.S., Woon, E.C., Yang, M., et al. (2011). The oncometabolite 2-hydroxyglutarate inhibits histone lysine demethylases. EMBO Rep. *12*, 463–469.

Cooney, C.A., Dave, A.A., and Wolff, G.L. (2002). Maternal methyl supplements in mice affect epigenetic variation and DNA methylation of offspring. J. Nutr. *132*, 2393S-2400S.

Crider, K.S., Yang, T.P., Berry, R.J., and Bailey, L.B. (2012). Folate and DNA methylation: a review of molecular mechanisms and the evidence for folate's role. Adv. Nutr. *3*, 21–38.

Cropley, J.E., Suter, C.M., Beckman, K.B., and Martin, D.I. (2006). Germ-line epigenetic modification of the murine A vy allele by nutritional supplementation. Proc. Natl. Acad. Sci. U.S.A. *103*, 17308–17312.

Czeizel, A.E., and Dudas, I. (1992). Prevention of the first occurrence of neural-tube defects by periconceptional vitamin supplementation. N. Eng. J. Med. *327*, 1832–1835.

Dobosy, J.R., Fu, V.X., Desotelle, J.A., Srinivasan, R., Kenowski, M.L., Almassi, N., Weindruch, R., Svaren, J., and Jarrard, D.F. (2008). A methyl-deficient diet modifies histone methylation and alters Igf2 and H19 repression in the prostate. Prostate *68*, 1187–1195.

Dolinoy, D.C., Huang, D., and Jirtle, R.L. (2007). Maternal nutrient supplementation counteracts bisphenol A-induced DNA hypomethylation in early development. Proc. Natl. Acad. Sci. U.S.A. *104*, 13056–13061.

Donohoe, D.R., Garge, N., Zhang, X., Sun, W., O'Connell, T.M., Bunger, M.K., and Bultman, S.J. (2011). The microbiome and butyrate regulate energy metabolism and autophagy in the mammalian colon. Cell Metab. *13*, 517–526.

Donohoe, D.R., Collins, L.B., Wali, A., Bigler, R., Sun, W., and Bultman, S.J. (2012). The Warburg effect dictates the mechanism of butyrate-mediated histone acetylation and cell proliferation. Mol. Cell *48*, 612–626.

Dunstan, J.A., West, C., McCarthy, S., Metcalfe, J., Meldrum, S., Oddy, W.H., Tulic, M.K., D'Vaz, N., and Prescott, S.L. (2012). The relationship between maternal folate status in pregnancy, cord blood folate levels, and allergic outcomes in early childhood. Allergy *67*, 50–57.

Eisenberg, T., Schroeder, S., Andryushkova, A., Pendl, T., Kuttner, V., Bhukel, A., Marino, G., Pietrocola, F., Harger, A., Zimmermann, A., et al. (2014). Nucleocytosolic depletion of the energy metabolite acetyl-coenzyme a stimulates autophagy and prolongs lifespan. Cell Metab. *19*, 431–444.

Figueroa, M.E., Abdel-Wahab, O., Lu, C., Ward, P.S., Patel, J., Shih, A., Li, Y., Bhagwat, N., Vasanthakumar, A., Fernandez, H.F., et al. (2010). Leukemic IDH1 and IDH2 mutations result in a hypermethylation phenotype, disrupt TET2 function, and impair hematopoietic differentiation. Cancer Cell *18*, 553–567.

Friso, S., and Choi, S.W. (2005). Gene–nutrient interactions in one-carbon metabolism. Curr. Drug Metab. *6*, 37–46.

Friso, S., Lamon-Fava, S., Jang, H., Schaefer, E.J., Corrocher, R., and Choi, S.W. (2007). Oestrogen replacement therapy reduces total plasma homocysteine and enhances genomic DNA methylation in postmenopausal women. Br. J. Nutr. *97*, 617–621.

Fryer, A.A., Emes, R.D., Ismail, K.M., Haworth, K.E., Mein, C., Carroll, W.D., and Farrell, W.E. (2011). Quantitative, high-resolution epigenetic profiling of CpG loci identifies associations with cord blood plasma homocysteine and birth weight in humans. Epigenetics *6*, 86–94.

Fujiki, R., Hashiba, W., Sekine, H., Yokoyama, A., Chikanishi, T., Ito, S., Imai, Y., Kim, J., He, H.H., Igarashi, K., et al. (2011). GlcNAcylation of histone H2B facilitates its monoubiquitination. Nature *480*, 557–560.

Goren, A., Tabib, A., Hecht, M., and Cedar, H. (2008). DNA replication timing of the human beta-globin domain is controlled by histone modification at the origin. Genes Dev. *22*, 1319–1324.

Greer, E.L., and Shi, Y. (2012). Histone methylation: a dynamic mark in health, disease and inheritance. Nat. Rev. Genet. *13*, 343–357.

Guarente, L. (2011). The logic linking protein acetylation and metabolism. Cell Metab. *14*, 151–153.

Haberg, S.E., London, S.J., Stigum, H., Nafstad, P., and Nystad, W. (2009). Folic acid supplements in pregnancy and early childhood respiratory health. Arch. Dis. Child. *94*, 180–184.

Hariharan, N., Maejima, Y., Nakae, J., Paik, J., Depinho, R.A., and Sadoshima, J. (2010). Deacetylation of FoxO by Sirt1 plays an essential role in mediating starvation-induced autophagy in cardiac myocytes. Circ. Res. *107*, 1470–1482.

Hatzivassiliou, G., Zhao, F., Bauer, D.E., Andreadis, C., Shaw, A.N., Dhanak, D., Hingorani, S.R., Tuveson, D.A., and Thompson, C.B. (2005). ATP citrate lyase inhibition can suppress tumor cell growth. Cancer Cell *8*, 311–321.

Haycock, P.C. (2009). Fetal alcohol spectrum disorders: the epigenetic perspective. Biol. Reprod. *81*, 607–617.

Herranz, D., and Serrano, M. (2010). SIRT1: recent lessons from mouse models. Nat. Rev. Cancer *10*, 819–823.

Hino, S., Sakamoto, A., Nagaoka, K., Anan, K., Wang, Y., Mimasu, S., Umehara, T., Yokoyama, S., Kosai, K., and Nakao, M. (2012). FAD-dependent lysine-specific demethylase-1 regulates cellular energy expenditure. Nat. Commun. *3*, 758.

Houtkooper, R.H., Pirinen, E., and Auwerx, J. (2012). Sirtuins as regulators of metabolism and healthspan. Nat. Rev. Mol. Cell Biol. *13*, 225–238.

Huang, J., Plass, C., and Gerhauser, C. (2011). Cancer chemoprevention by targeting the epigenome. Curr. Drug Targets *12*, 1925–1956.

Huletsky, A., Niedergang, C., Frechette, A., Aubin, R., Gaudreau, A., and Poirier, G.G. (1985). Sequential ADP-ribosylation pattern of nucleosomal histones. ADP-ribosylation of nucleosomal histones. Eur. J. Biochem. *146*, 277–285.

Jeong, S.M., Xiao, C., Finley, L.W., Lahusen, T., Souza, A.L., Pierce, K., Li, Y.H., Wang, X., Laurent, G., German, N.J., et al. (2013). SIRT4 has tumor-suppressive activity and regulates the cellular metabolic response to DNA damage by inhibiting mitochondrial glutamine metabolism. Cancer Cell *23*, 450–463.

Kaelin, W.G. Jr., and McKnight, S.L. (2013). Influence of metabolism on epigenetics and disease. Cell *153*, 56–69.

Katoh, Y., Ikura, T., Hoshikawa, Y., Tashiro, S., Ito, T., Ohta, M., Kera, Y., Noda, T., and Igarashi, K. (2011). Methionine adenosyltransferase II serves as a transcriptional corepressor of Maf oncoprotein. Mol. Cell *41*, 554–566.

Kiefte-de Jong, J.C., Timmermans, S., Jaddoe, V.W., Hofman, A., Tiemeier, H., Steegers, E.A., de Jongste, J.C., and Moll, H.A. (2012). High circulating folate and vitamin B-12 concentrations in women during pregnancy are associated with increased prevalence of atopic dermatitis in their offspring. J. Nutr. *142*, 731–738.

Killian, J.K., Kim, S.Y., Miettinen, M., Smith, C., Merino, M., Tsokos, M., Quezado, M., Smith, W.I., Jr., Jahromi, M.S., Xekouki, P., et al. (2013). Succinate dehydrogenase mutation underlies global epigenomic divergence in gastrointestinal stromal tumor. Cancer Discov. *3*, 648–657.

Kim, Y.I. (2007a). Folate and colorectal cancer: an evidence-based critical review. Mol. Nutr. Food Res. *51*, 267–292.

Kim, Y.I. (2007b). Folic acid fortification and supplementation – good for some but not so good for others. Nutr. Rev. *65*, 504–511.

Kirchner, H., Osler, M.E., Krook, A., and Zierath, J.R. (2013). Epigenetic flexibility in metabolic regulation: disease cause and prevention? Trends Cell Biol. *23*, 203–209.

Kraus, D., Yang, Q., Kong, D., Banks, A.S., Zhang, L., Rodgers, J.T., Pirinen, E., Pulinilkunnil, T.C., Gong, F., Wang, Y.C., et al. (2014). Nicotinamide N-methyltransferase knockdown protects against diet-induced obesity. Nature *508*, 258–262.

Lee, I.H., Cao, L., Mostoslavsky, R., Lombard, D.B., Liu, J., Bruns, N.E., Tsokos, M., Alt, F.W., and Finkel, T. (2008). A role for the NAD-dependent deacetylase Sirt1 in the regulation of autophagy. Proc. Natl. Acad. Sci. U.S.A. *105*, 3374–3379.

Lee, Y., Dominy, J.E., Choi, Y.J., Jurczak, M., Tolliday, N., Camporez, J.P., Chim, H., Lim, J.H., Ruan, H.B., Yang, X., et al. (2014). Cyclin D1-Cdk4 controls glucose metabolism independently of cell cycle progression. Nature *510*, 547–551.

Lertratanangkoon, K., Wu, C.J., Savaraj, N., and Thomas, M.L. (1997). Alterations of DNA methylation by glutathione depletion. Cancer Lett. *120*, 149–156.

Letouze, E., Martinelli, C., Loriot, C., Burnichon, N., Abermil, N., Ottolenghi, C., Janin, M., Menara, M., Nguyen, A.T., Benit, P., et al. (2013). SDH mutations establish a hypermethylator phenotype in paraganglioma. Cancer Cell *23*, 739–752.

Li, R., Grimm, S.A., Chrysovergis, K., Kosak, J., Wang, X., Du, Y., Burkholder, A., Janardhan, K., Mav, D., Shah, R., et al. (2014). Obesity, rather than diet, drives epigenomic alterations in colonic epithelium resembling cancer progression. Cell Metab. *19*, 702–711.

Li, Y., Daniel, M., and Tollefsbol, T.O. (2011). Epigenetic regulation of caloric restriction in aging. BMC Medicine *9*, 98.

Lin, R., Tao, R., Gao, X., Li, T., Zhou, X., Guan, K.L., Xiong, Y., and Lei, Q.Y. (2013). Acetylation stabilizes ATP-citrate lyase to promote lipid biosynthesis and tumor growth. Mol. Cell *51*, 506–518.

Liu, Y., Balaraman, Y., Wang, G., Nephew, K.P., and Zhou, F.C. (2009). Alcohol exposure alters DNA methylation profiles in mouse embryos at early neurulation. Epigenetics *4*, 500–511.

Losman, J.A., Looper, R.E., Koivunen, P., Lee, S., Schneider, R.K., McMahon, C., Cowley, G.S., Root, D.E., Ebert, B.L., and Kaelin, W.G., Jr. (2013). (R)-2-hydroxyglutarate is sufficient to promote leukemogenesis and its effects are reversible. Science *339*, 1621–1625.

Lu, C., Ward, P.S., Kapoor, G.S., Rohle, D., Turcan, S., Abdel-Wahab, O., Edwards, C.R., Khanin, R., Figueroa, M.E., Melnick, A., et al. (2012). IDH mutation impairs histone demethylation and results in a block to cell differentiation. Nature *483*, 474–478.

Lumey, L.H., Stein, A.D., and Susser, E. (2011). Prenatal famine and adult health. Annu. Rev. Public Health *32*, 237–262.

Madrigano, J., Baccarelli, A., Mittleman, M.A., Wright, R.O., Sparrow, D., Vokonas, P.S., Tarantini, L., and Schwartz, J. (2011). Prolonged exposure to particulate pollution, genes associated with glutathione pathways, and DNA methylation in a cohort of older men. Environ. Health Perspect. *119*, 977–982.

Martinez, D., Pentinat, T., Ribo, S., Daviaud, C., Bloks, V.W., Cebria, J., Villalmanzo, N., Kalko, S.G., Ramon-Krauel, M., Diaz, R., et al. (2014). In utero undernutrition in male mice programs liver lipid metabolism in the second-generation offspring involving altered lxra DNA methylation. Cell Metab. *19*, 941–951.

Mehedint, M.G., Niculescu, M.D., Craciunescu, C.N., and Zeisel, S.H. (2010). Choline deficiency alters global histone methylation and epigenetic marking at the Re1 site of the calbindin 1 gene. FASEB J. *24*, 184–195.

Messerschmidt, D.M., Knowles, B.B., and Solter, D. (2014). DNA methylation dynamics during epigenetic reprogramming in the germline and preimplantation embryos. Genes Dev. *28*, 812–828.

Messner, S., and Hottiger, M.O. (2011). Histone ADP-ribosylation in DNA repair, replication and transcription. Trends Cell Biol. *21*, 534–542.

Morgan, H.D., Sutherland, H.G., Martin, D.I., and Whitelaw, E. (1999). Epigenetic inheritance at the agouti locus in the mouse. Nat. Genet. *23*, 314–318.

Nabel, C.S., and Kohli, R.M. (2011). Molecular biology. Demystifying DNA demethylation. Science *333*, 1229–1230.

Nakahata, Y., Sahar, S., Astarita, G., Kaluzova, M., and Sassone-Corsi, P. (2009). Circadian control of the NAD+ salvage pathway by CLOCK-SIRT1. Science *324*, 654–657.

Ng, F., and Tang, B.L. (2013). Sirtuins' modulation of autophagy. J. Cell. Physiol. *228*, 2262–2270.

Ng, S.F., Lin, R.C., Laybutt, D.R., Barres, R., Owens, J.A., and Morris, M.J. (2010). Chronic high-fat diet in fathers programs beta-cell dysfunction in female rat offspring. Nature *467*, 963–966.

Ong, T.P., Moreno, F.S., and Ross, S.A. (2011). Targeting the epigenome with bioactive food components for cancer prevention. J. Nutrigenet. Nutrigenomics *4*, 275–292.

Padmanabhan, N., Jia, D., Geary-Joo, C., Wu, X., Ferguson-Smith, A.C., Fung, E., Bieda, M.C., Snyder, F.F., Gravel, R.A., Cross, J.C., et al. (2013). Mutation in folate metabolism causes epigenetic instability and transgenerational effects on development. Cell *155*, 81–93.

Painter, R.C., Roseboom, T.J., and Bleker, O.P. (2005). Prenatal exposure to the Dutch famine and disease in later life: an overview. Reprod. Toxicol. *20*, 345–352.

Pascual, M., Do Couto, B.R., Alfonso-Loeches, S., Aguilar, M.A., Rodriguez-Arias, M., and Guerri, C. (2012). Changes in histone acetylation in the prefrontal cortex of ethanol-exposed adolescent rats are associated with ethanol-induced place conditioning. Neuropharmacology *62*, 2309–2319.

Quinn, B.J., Kitagawa, H., Memmott, R.M., Gills, J.J., and Dennis, P.A. (2013). Repositioning metformin for cancer prevention and treatment. Trends Endocrinol. Metab. *24*, 469–480.

Ramsey, K.M., Yoshino, J., Brace, C.S., Abrassart, D., Kobayashi, Y., Marcheva, B., Hong, H.K., Chong, J.L., Buhr, E.D., Lee, C., et al. (2009). Circadian clock feedback cycle through NAMPT-mediated NAD^+ biosynthesis. Science *324*, 651–654.

Saitou, M., Kagiwada, S., and Kurimoto, K. (2012). Epigenetic reprogramming in mouse pre-implantation development and primordial germ cells. Development *139*, 15–31.

Sassone-Corsi, P. (2012). Minireview: NAD+, a circadian metabolite with an epigenetic twist. Endocrinology *153*, 1–5.

Schaible, T.D., Harris, R.A., Dowd, S.E., Smith, C.W., and Kellermayer, R. (2011). Maternal methyl-donor supplementation induces prolonged murine offspring colitis susceptibility in association with mucosal epigenetic and microbiomic changes. Hum. Mol. Genet. *20*, 1687–1696.

Sebastian, C., Zwaans, B.M., Silberman, D.M., Gymrek, M., Goren, A., Zhong, L., Ram, O., Truelove, J.,

Guimaraes, A.R., Toiber, D., et al. (2012). The histone deacetylase SIRT6 is a tumor suppressor that controls cancer metabolism. Cell *151*, 1185–1199.

Shepard, B.D., and Tuma, P.L. (2009). Alcohol-induced protein hyperacetylation: mechanisms and consequences. World J. Gastroenterol. *15*, 1219–1230.

Shimazu, T., Hirschey, M.D., Hua, L., Dittenhafer-Reed, K.E., Schwer, B., Lombard, D.B., Li, Y., Bunkenborg, J., Alt, F.W., Denu, J.M., et al. (2010). SIRT3 deacetylates mitochondrial 3-hydroxy-3-methylglutaryl CoA synthase 2 and regulates ketone body production. Cell Metab. *12*, 654–661.

Shimazu, T., Hirschey, M.D., Newman, J., He, W., Shirakawa, K., Le Moan, N., Grueter, C.A., Lim, H., Saunders, L.R., Stevens, R.D., et al. (2013). Suppression of oxidative stress by beta-hydroxybutyrate, an endogenous histone deacetylase inhibitor. Science *339*, 211–214.

Shyh-Chang, N., Locasale, J.W., Lyssiotis, C.A., Zheng, Y., Teo, R.Y., Ratanasirintrawoot, S., Zhang, J., Onder, T., Unternaehrer, J.J., Zhu, H., et al. (2013). Influence of threonine metabolism on S-adenosylmethionine and histone methylation. Science *339*, 222–226.

Sinclair, K.D., Allegrucci, C., Singh, R., Gardner, D.S., Sebastian, S., Bispham, J., Thurston, A., Huntley, J.F., Rees, W.D., Maloney, C.A., et al. (2007). DNA methylation, insulin resistance, and blood pressure in offspring determined by maternal periconceptional B vitamin and methionine status. Proc. Natl. Acad. Sci. U.S.A. *104*, 19351–19356.

Smith, A.D., Kim, Y.I., and Refsum, H. (2008). Is folic acid good for everyone? Am. J. Clin. Nutr. *87*, 517–533.

Soga, T. (2013). Cancer metabolism: key players in metabolic reprogramming. Cancer Sci. *104*, 275–281.

Spencer, M.D., Hamp, T.J., Reid, R.W., Fischer, L.M., Zeisel, S.H., and Fodor, A.A. (2011). Association between composition of the human gastrointestinal microbiome and development of fatty liver with choline deficiency. Gastroenterology *140*, 976–986.

Steegers-Theunissen, R.P., Obermann-Borst, S.A., Kremer, D., Lindemans, J., Siebel, C., Steegers, E.A., Slagboom, P.E., and Heijmans, B.T. (2009). Periconceptional maternal folic acid use of 400 microg per day is related to increased methylation of the IGF2 gene in the very young child. PLoS ONE 4, e7845.

Supic, G., Jagodic, M., and Magic, Z. (2013). Epigenetics: a new link between nutrition and cancer. Nutr. Cancer *65*, 781–792.

Sutendra, G., Kinnaird, A., Dromparis, P., Paulin, R., Stenson, T.H., Haromy, A., Hashimoto, K., Zhang, N., Flaim, E., and Michelakis, E.D. (2014). A nuclear pyruvate dehydrogenase complex is important for the generation of acetyl-CoA and histone acetylation. Cell *158*, 84–97.

Takamura, Y., and Nomura, G. (1988). Changes in the intracellular concentration of acetyl-CoA and malonyl-CoA in relation to the carbon and energy metabolism of Escherichia coli K12. J. Gen. Microbiol. *134*, 2249–2253.

Tan, L., and Shi, Y.G. (2012). Tet family proteins and 5-hydroxymethylcytosine in development and disease. Development *139*, 1895–1902.

Thirstrup, K., Christensen, S., Moller, H.A., Ritzen, A., Bergstrom, A.L., Sager, T.N., and Jensen, H.S. (2011). Endogenous 2-oxoglutarate levels impact potencies of competitive HIF prolyl hydroxylase inhibitors. Pharmacol. Res. *64*, 268–273.

Tobi, E.W., Lumey, L.H., Talens, R.P., Kremer, D., Putter, H., Stein, A.D., Slagboom, P.E., and Heijmans, B.T. (2009). DNA methylation differences after exposure to prenatal famine are common and timing- and sex-specific. Hum. Mol. Genet. *18*, 4046–4053.

Turcan, S., Rohle, D., Goenka, A., Walsh, L.A., Fang, F., Yilmaz, E., Campos, C., Fabius, A.W., Lu, C., Ward, P.S., et al. (2012). IDH1 mutation is sufficient to establish the glioma hypermethylator phenotype. Nature *483*, 479–483.

Ulrich, C.M., and Potter, J.D. (2007). Folate and cancer – timing is everything. JAMA 297, 2408–2409.

Vander Heiden, M.G., Cantley, L.C., and Thompson, C.B. (2009). Understanding the Warburg effect: the metabolic requirements of cell proliferation. Science *324*, 1029–1033.

Waterland, R.A., and Jirtle, R.L. (2003). Transposable elements: targets for early nutritional effects on epigenetic gene regulation. Mol. Cell. Biol. *23*, 5293–5300.

Waterland, R.A., Dolinoy, D.C., Lin, J.R., Smith, C.A., Shi, X., and Tahiliani, K.G. (2006). Maternal methyl supplements increase offspring DNA methylation at Axin Fused. Genesis *44*, 401–406.

Wellen, K.E., Hatzivassiliou, G., Sachdeva, U.M., Bui, T.V., Cross, J.R., and Thompson, C.B. (2009). ATP-citrate lyase links cellular metabolism to histone acetylation. Science *324*, 1076–1080.

Whitrow, M.J., Moore, V.M., Rumbold, A.R., and Davies, M.J. (2009). Effect of supplemental folic acid in pregnancy on childhood asthma: a prospective birth cohort study. Am. J. Epidemiol. *170*, 1486–1493.

Willett, W. (2008). Nutrition and cancer: the search continues. Nutr. Cancer *60*, 557–559.

Wu, K., Platz, E.A., Willett, W.C., Fuchs, C.S., Selhub, J., Rosner, B.A., Hunter, D.J., and Giovannucci, E. (2009). A randomized trial on folic acid supplementation and risk of recurrent colorectal adenoma. Am. J. Clin. Nutr. *90*, 1623–1631.

Xiao, M., Yang, H., Xu, W., Ma, S., Lin, H., Zhu, H., Liu, L., Liu, Y., Yang, C., Xu, Y., et al. (2012). Inhibition of alpha-KG-dependent histone and DNA demethylases by fumarate and succinate that are accumulated in mutations of FH and SDH tumor suppressors. Genes Dev. *26*, 1326–1338.

Xu, M.Q., Sun, W.S., Liu, B.X., Feng, G.Y., Yu, L., Yang, L., He, G., Sham, P., Susser, E., St Clair, D., et al. (2009). Prenatal malnutrition and adult schizophrenia: further evidence from the 1959–1961 Chinese famine. Schizophr. Bull. *35*, 568–576.

Xu, W., Yang, H., Liu, Y., Yang, Y., Wang, P., Kim, S.H., Ito, S., Yang, C., Wang, P., Xiao, M.T., et al. (2011). Oncometabolite 2-hydroxyglutarate is a competitive inhibitor of alpha-ketoglutarate-dependent dioxygenases. Cancer Cell *19*, 17–30.

Yang, M., Soga, T., and Pollard, P.J. (2013). Oncometabolites: linking altered metabolism with cancer. J. Clin. Invest. *123*, 3652–3658.

Zeisel, S.H. (2009). Epigenetic mechanisms for nutrition determinants of later health outcomes. Am. J. Clin. Nutr. *89*, 1488S–1493S.

Zhao, S., Xu, W., Jiang, W., Yu, W., Lin, Y., Zhang, T., Yao, J., Zhou, L., Zeng, Y., Li, H., et al. (2010). Regulation of cellular metabolism by protein lysine acetylation. Science *327*, 1000–1004.

Zhong, L., D'Urso, A., Toiber, D., Sebastian, C., Henry, R.E., Vadysirisack, D.D., Guimaraes, A., Marinelli, B., Wikstrom, J.D., Nir, T., et al. (2010). The histone deacetylase Sirt6 regulates glucose homeostasis via Hif1alpha. Cell *140*, 280–293.

Environmental Exposures: Impact on the Epigenome

17

Jaclyn M. Goodrich and Dana C. Dolinoy

Abstract

Exposures to environmental toxicants contribute significantly to the global burden of disease and are a prominent public health issue. Environmental toxicants exert their impact on health and disease in part by modifying the epigenome, DNA modifications that do not affect the underlying sequence but can result in altered gene expression and downstream phenotypic changes. While potentially reversible, toxicant-induced epigenetic change at vulnerable life stages such as embryogenesis and adolescence may have a lasting effect on susceptibility to disease later in life and could even impact subsequent generations. In this chapter, we describe evidence for mitotically inheritable epigenetic perturbation by toxicants from four chemical classes – heavy metals, persistent organic pollutants (POPs), endocrine-disrupting chemicals (EDCs), and air pollution, a representative exposure mixture – from animal models and epidemiological cohorts. Research addressing mechanisms by which toxicants perturb the epigenome, the biological and functional significance of small epigenetic changes (e.g. DNA methylation, histone modifications, and non-coding RNA expression), and the impact of epigenetic change longitudinally on health outcomes are discussed. The translation of environmental epigenetics research to environmental policy and public health solutions will enhance chemical risk assessment and enable clinicians to identify at-risk populations prior to disease onset.

Introduction

The health consequences of chronic, low level environmental toxicants are of increasing concern to societies world-wide, as well as to government agencies responsible for regulating the manufacture, use, and environmental impact of such chemicals. Thus, research in basic molecular toxicology continues to underpin key innovations and advances in environmental sciences by striving to elucidate the cellular, biochemical, and genomic mechanisms of action of toxic substances. Over the past several decades, our understanding of the mechanisms by which chemicals and environmental toxicants impact health outcomes via the genome has greatly expanded. Traditional approaches in identifying the molecular underpinnings associated with toxicant exposure relied on observed responses identified via pathology, histology, or phenotypical observation. Advances in molecular technologies have allowed for the evaluation of more subtle biological effects across large systems, generally termed toxicogenomics, which encompasses genomics, transcriptomics, proteomics, and highlighted in this chapter, epigenomics.

Through compelling animal model and human epidemiological evidence, it is now widely accepted that environmental exposures can alter gene expression and affect phenotype by not only mutating promoter and coding regions of genes, but also by modifying the epigenome (Waterland and Jirtle, 2004). Literally meaning 'above the genome,' the epigenome comprises the heritable changes in gene expression that occur in the absence of changes to the DNA sequence itself, including chromatin folding and attachment to the nuclear matrix, packaging of DNA around nucleosomes, covalent modifications of histone tails, and DNA methylation. Unlike genetic mutations, which are static and generally non-modifiable, epigenetic profiles are dynamic and potentially reversible, providing a unique avenue to potentiate the effects of environmental toxicants.

Environmental exposures, epigenetics, and the developmental origins of health and disease

DNA methylation and other epigenetic modifications are prone to change throughout the life course, but are especially vulnerable during reprogramming events associated with embryogenesis and early development (Reik *et al.*, 2001). Consequently, epigenetic modifications have emerged as mechanisms of the 'developmental origins of health and disease' (DOHaD) paradigm, which posits that environment interactions during early life result in long-lasting effects on health and disease risk. Specifically, the DOHaD hypothesis postulates that nutrition and other environmental factors during prenatal and early postnatal development influence developmental plasticity, thereby altering susceptibility to adult chronic diseases, including metabolic syndrome, obesity, cancer, and neurodegenerative disorders (see Fig. 17.1) (Bateson *et al.*, 2004).

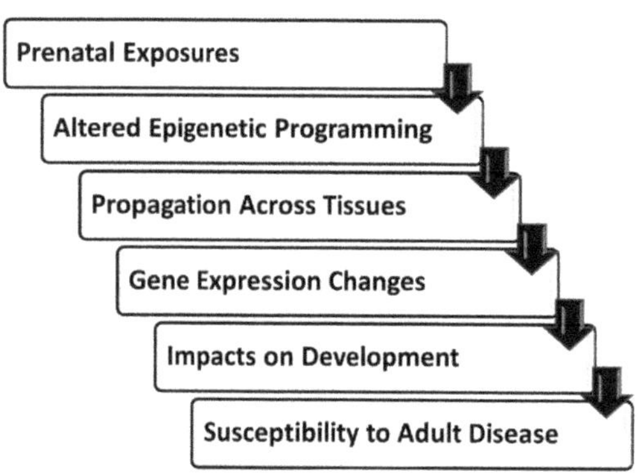

Figure 17.1 Epigenetics and the developmental origins of health and disease (DOHaD). Prenatal exposures to environmental agents, including chemicals discussed in this chapter, can influence developmental plasticity and ultimately alter susceptibility to disease in adulthood. Modification of the fetal epigenome by exposures is one mechanism that can set off this series of events leading to lifelong consequences for health and disease.

Although the epigenome is vulnerable to environmental perturbations during multiple periods, it is particularly susceptible to environmental deregulation during embryogenesis, adolescence, and into late adulthood. First, during early development, the DNA synthetic rate is high, and the elaborate DNA methylation pattern required for normal tissue development is established during early development (Jirtle and Skinner, 2007). Second, during the peripubertal period, mammary and other tissue growth is highly proliferative (Daniel and Silberstein, 1987; Elston and Ellis, 1998; Russo and Russo, 1987), making pre-adolescents especially vulnerable to insults acting via an epigenetic mechanism. For example, chemical exposures during the pre-adolescent and adolescent period may prolong puberty, resulting in increased mammary epithelial cell proliferation and altered epigenetic remodelling of the developing mammary gland (Knower *et al.*, 2014). Finally, as individuals age, gradual DNA hypomethylation occurs at the genome-wide level, concurrent with locus-specific promoter increases in DNA methylation at normally unmethylated genomic regions (Feinberg and Tycko, 2004), leading, for example, to genome instability or gene-specific suppression, respectively.

Animal model and human population approaches in environmental epigenetics

Researchers in the field of environmental toxicology and epigenetic epidemiology aspire to identify epigenetic biomarkers of exposure or disease status that will accelerate the identification of at-risk individuals prior to disease onset. As such, many epigenetic epidemiology analyses investigating the effects of the environment are performed using biologically available DNA (e.g. blood, saliva, stool), which serve as proxies for epigenetic status in tissues targeted by exposures. While methylation of peripheral DNA may not mirror that of all target tissues, it has been used as a biomarker and associated with several cancers, disease states, and environmental exposures (Terry *et al.*, 2011; Wright *et al.*, 2010). As there is no *a priori* reason to expect that peripheral samples will contain biologically relevant DNA methylation data at all loci, it is imperative that surrogate DNA be interpreted carefully.

Pairing human studies with parallel efforts in animal models will increase our understanding of the complex relationship between environmental exposures and epigenetics. Animal models provide several experimental advantages, including access to both target and biologically available tissues. Laboratory settings allow for careful control of exposures and modifying factors such as diet and stress. Thus, animal models are well-poised to carefully elucidate complex relationships among environmental exposures, epigenetic tissue specificity, and time-dependent epigenetic drift. On the other hand, differences in genetics and metabolism of chemicals in animal models compared to humans, limit their applicability to human health. In addition, many early epigenotoxicology studies conducted with animal models failed to carefully consider dose and routes of exposure, limiting their interpretability. Thus many complementary approaches must be undertaken to understand the role of environmental exposures in modifying disease risk via epigenetic alterations.

Goals for this chapter

This chapter focuses on a wide range of relevant environmental toxicants (e.g. heavy metals, persistent organic pollutants, and endocrine disrupting chemicals), paying particular attention to epigenetic alterations at vulnerable stages across the life course. Each section begins with an overview of the chemical class, including sources, modes of action, and well-established toxicological and epidemiological outcome data. Next we review the effects on the epigenome of several individual toxicants from each class, with a focus on *in vivo* animal model and human epidemiological data. We generally focus on effects of toxicants on somatic cell epigenetic programming, and in some cases direct effects on the epigenome in germ line cells. Transgenerational effects of environmental exposures, are of growing interest, and are summarized in Chapter 15. Owing to space limitations, we do not generally focus on non-toxicant environmental factors with potential epigenetic consequences, including cigarette smoke, pharmaceuticals, radiation, diet, and stress.

Metals

Exposure to heavy metals is detrimental to the health of the general public and also certain occupational groups worldwide. In the USA, environmental regulatory bodies (Environmental Protection Agency, EPA, and Agency for Toxic Substances and Disease Registry, ATSDR) rank pollutants of concern based on known toxicity and likelihood of exposure. Four of the top seven pollutants are heavy metals – arsenic, lead, mercury, and cadmium (As, Pb, Hg, Cd) and chromium (Cr) is seventeenth (ATSDR, 2013). Three of these metals are also known to be carcinogenic (As, Cd, Cr). While exposures to some of these top metals are declining in the USA (e.g. Pb), exposures to others remain high worldwide (Hg, Cd, As) stemming from sources such as contaminated fish, rice, and water. Various occupational groups are exposed to high levels of the top metals including small-scale gold miners (Hg) and smelters (As, Pb, Cd, Cr, and more). Additional heavy metals pose a threat to occupationally exposed populations or populations living near contaminated industrial sites (manganese, nickel, zinc, aluminium, uranium – Mn, Ni, Zn, Al, U). Some metals that fall into the micronutrient category exhibit hormesis. The metal is beneficial at optimal doses but can be harmful at both high doses and in cases of deficiency (e.g. Zn, selenium – Se).

While metals share some common pathways of toxicity and adverse effects, each has its own unique profile of toxicokinetics and downstream adverse health outcomes, some of which are still being elucidated. Recent research has shown that the epigenome is one common entity metals impact though specific alterations are unique to each metal (for reviews, see Basu *et al.*, 2014; Cheng *et al.*, 2012; Chervona *et al.*, 2012; Martinez-Zamudio and Ha, 2011; Smeester *et al.*, 2014). For the purposes of this chapter, we will describe examples of epigenetic effects from exposures to three heavy metals – Hg, Cd, and Pb – throughout the life course to demonstrate the types of effects these toxicants have on populations environmentally and/or occupationally exposed.

Mercury (Hg)

Mercury, a potent toxicant that the general population is exposed to through fish consumption (in the form of methylmercury) and dental amalgams (as inorganic Hg), is known to impact various systems including the neurological system, cardiovascular system, and kidneys. As with many toxicants, health effects are dependent upon chemical species, dose and duration of exposure (chronic vs. acute) (Clarkson and Magos, 2006). The discovery of epigenetic alterations from Hg exposure is relatively new. While most studies focus on methylmercury, some epigenetic impacts from inorganic Hg have also been observed (see review by Basu *et al.*, 2014). There is evidence of methylmercury's impact on the epigenome from exposures at different life stages (*in utero* to adulthood) in humans and animal models. Onishchenko *et al.* (2008) demonstrated the potential for DOHaD-like epigenetic effects from early life exposure to methylmercury in mice. Dams were perinatally exposed to methylmercury in drinking water, and mice were tested for epigenetic changes and depression-like symptoms at 9 weeks, 12 weeks and 14 months. The exposed group exhibited evidence of depression and lasting epigenetic changes at the brain-derived neurotrophic factor (Bdnf) promoter IV at later time points while the unexposed controls did not. Specifically, the promoter was in a repressive state noted by increased DNA methylation, increased trimethylation of H3K27, and decreased acetylation of H3. Decreased mRNA levels of Bdnf were apparent in the hippocampal formation at 12 weeks and 14 months, corresponding to the repressive epigenetic changes in the same region and the lasting depression phenotype.

While longitudinal studies examining epigenetic change from prenatal Hg exposure have not yet been performed in human populations, cross-sectional studies reveal the potential for methylmercury to impact the human epigenome. We observed an association between hair Hg levels (indicative of methylmercury exposure) specifically in males and hypomethylation of the selenoprotein p1 (*SEPP1*) promoter, an antioxidant gene whose protein product binds Hg, in a cohort of dental professionals (Goodrich *et al.*, 2013). In a cohort of women of reproductive age, methylation increased at the glutathione *S*-transferase M1/5 (*GSTM1/5*) promoter, another antioxidant gene, with increasing blood Hg levels (Hanna *et al.*, 2012). As with other toxicants, the field is ready to move away from candidate gene methylation studies to epigenome-wide studies, and changes to the histone code, non-coding RNAs, and gene expression merit exploration.

Cadmium (Cd)

Cadmium is a carcinogenic metal that humans are primarily exposed to via diet, lifestyle and occupation. Many plant species readily uptake Cd from the soil. As such, Cd contamination in foods such as leafy vegetables and rice is common in certain parts of the world. Tobacco plants are excellent concentrators of Cd leading to high Cd exposures among cigarette smokers. Certain industrial practices such as smelting and electroplating release Cd in the air which can result in occupational exposure and environmental contamination (ATSDR, 2012). Chronic Cd exposure is associated with neurodevelopmental delays and smaller birth size among children exposed *in utero*, and many effects appear to be sex-specific (ATSDR, 2012; Kippler *et al.*, 2012). In a birth cohort study in Bangladesh, Kippler *et al.* (2013) used epigenome-wide array technology and observed DNA methylation changes that were associated with *in utero* Cd exposure. Interestingly, top differentially methylated sites and their direction of change differed by sex. In a separate group of 4-year-old children from the same region, some of the same sites were also differentially methylated by current Cd exposure. Methylation changes may have implications for Cd-related growth restriction as methylation at some CpG sites increased with Cd exposure and were related to lower birth weight (Kippler *et al.*, 2013). In another mother–newborn pair study, maternal blood Cd levels were associated with distinct sets of differentially methylated genes in the mothers versus the newborns (Sanders *et al.*, 2014).

Among chronically exposed adults, Cd is associated with renal dysfunction, osteoporosis, and several cancers (including lung, breast, and bladder) (ATSDR, 2012). Epigenetic changes, including histone modifications and DNA methylation, are important hallmarks in tumour cell formation and cancer progression and are targets of some current therapeutic options. Specific sets of changes, often called signatures, vary by type of cancer (Ellis *et al.*, 2009; Huang *et al.*, 2011). Given the association of metals such as Cd with both cancers and epigenetic modifications, carcinogenic metals may work in part through epigenetic mechanisms. Animal and *in vitro* models have shown that Cd can alter DNA methylation, histone modifications, and microRNA expression in pathways that lead to downstream malignant transformation of cells (see reviews see Brocato and Costa, 2013; Wang *et al.*, 2012). Long-term Cd exposure in these models is often associated with hypermethylation of specific genes including the tumour suppressor, *P16*. In a cohort of Argentinian women with low-dose Cd exposure, urinary Cd levels were associated with hypomethylation of long interspersed element-1 (LINE-1), repetitive elements used as a marker of global methylation, and genotype of DNA methyltransferase 1 (*DNMT1*) modified this relationship (Hossain *et al.*, 2012). Future studies in Cd epigenetics should assess exposure and epigenetic status in individuals with Cd-associated cancers to gain a better understanding of the connections between results observed in animal, *in vitro*, and human studies to date.

Lead (Pb)

Lead exposure and toxicity is a prominent public health concern worldwide. Tetraethyl Pb, a gasoline additive, and Pb-based paint were major sources of Pb globally for decades. Both have since been banned in most countries, and exposures are decreasing (Laidlaw and Filippelli, 2008). However, Pb exposure remains an issue for socioeconomically disadvantaged groups (e.g. inhabitants of decrepit housing with old Pb-based paint, families residing near contaminated industrial sites). *In utero* development and early childhood are vulnerable life stages to effects from Pb toxicity including growth and neurodevelopment delays. The persistent impact of *in utero* Pb exposure on growth and neurodevelopment in later childhood (Afeiche *et al.*, 2011; Hu *et al.*, 2006) and on Alzheimer's Disease-like pathology in late adulthood (in rhesus monkeys (Wu *et al.*, 2008)) suggests early epigenetic change as a mechanistic link. Rodent studies corroborate this hypothesis. Viable yellow agouti mice, which serve a biosensor for maternal environmental exposures (Dolinoy, 2008), whose mothers consumed leaded drinking water from preconception until weaning exhibited altered dose-dependent DNA methylation at two metastable loci on postnatal day 22 (Faulk *et al.*, 2013), and methylation drift over time at three loci from postnatal 22 to 10 months differed by perinatal Pb exposure group (Faulk *et al.*, 2014).

While longitudinal birth cohort data exploring epigenetic mechanisms as the link between prenatal Pb exposure and later outcomes are forthcoming, several cross-sectional studies have indicated that Pb may perturb the human methylome of adults. For example, among elderly men of the Normative Aging Study, patella Pb – a biomarker of long-term exposure – was associated with global hypomethylation quantified at LINE-1 repetitive elements (Wright *et al.*, 2010). Future epidemiological studies designed to address health implications of observed modifications will be useful in solidifying epigenetic change as a mechanism causing persistent health effects from Pb exposure.

Persistent organic pollutants (POPs)

Persistent organic pollutants (POPs) constitute a broad group of organic chemicals that often serve as pesticides or are used in manufacturing, are resistant to common forms of degradation, and thus persist in the environment. In 2001, the Stockholm Convention mandated a global ban on 12 POPs, selected based on toxicity, called the 'Dirty Dozen': aldrin, chlordane, dieldrin, dieldrin, heptachlor, hexachlorobenzene, mirex, toxaphene, polychlorinated biphenyls (PCBs), dichlorodiphenyltrichloroethane (DDT), dioxins, and polychlorinated dibenzofurans (Secretariat of the Stockholm Convention, 2001). Others have since been added to the list. Health effects vary by POP and include disruption to the endocrine and immune systems, impacts on *in utero* development and reproductive success, and cancers. Exposure to POPs remains a public health problem worldwide but extent of exposure and specific POPs vary by location, occupation, and lifestyle practices. For example, while banned for agricultural use, DDT is used in some countries with high malaria prevalence, and some regions have only recently phased out insecticides such as mirex which can end up in meat.

POPs differ widely in chemical formula and structure, and thus toxicokinetic properties of each are dissimilar (e.g. half-life, target organs and pathways, storage in the body, mechanisms of toxicity). Epigenetic disruption may be one pathway of toxicity that POPs share, though not all POPs have been studied with respect to epigenetics to date. Exact targets likely differ by chemical. In this chapter, we will describe epigenetic perturbations associated with exposures to two widely studied legacy POPs: PCBs and dioxins.

Polychorinated biphenyls (PCBs)

Due to their insulating properties, PCBs were commonly used until growing evidence of toxicity led to a manufacturing ban in the USA in 1979. Today the legacy of PCBs lives on in old equipment (e.g. transformers), hazardous waste sites, and soils and water bodies where PCBs deposited over the years and remain due to resistance to degradation. PCBs bioaccumulate in the food chain, and as such many fish consumption advisories are due to PCB levels (ATSDR, 2000). Exposure to PCBs through maternal diet during *in utero* development and lactation is associated with a multitude of effects including decreased birth weight, learning deficits, and decreased dopamine in the frontal cortex in rodents and monkeys. Many of these changes persist long after cessation of exposure. In epidemiological studies of prenatal and postnatal exposure, evidence for effects are inconsistent though several studies have shown negative impacts on memory, growth, and immune function later in the exposed children's lives (ATSDR, 2000). Several rodent studies have begun to explore epigenetic change following pre- or post-natal exposure to PCBs. Desaulniers *et al.* (2009) observed decreased expression of three DNMT genes, decreased S-adenosylmethionine (SAM), and hypomethylation of $p16^{INK4a}$ in postnatal day 29 livers following perinatal exposure to PCBs. Chromatin remodelling was also observed in postnatal day 12 rat liver samples following perinatal exposure. Reduced levels of two histone modifications – H4K16Ac and H3K4me3 – and increased expression of a deacetylating enzyme, SirT1, and a demethylating enzyme, Jarid1b, were noted (Casati *et al.*, 2012, 2013).

Birth cohort studies investigating PCBs and the epigenome are currently scarce, though evidence for DNA methylation changes in cross-sectional studies is emerging in adult cohorts. Among 399 Japanese women, global blood leucocyte DNA methylation, approximated by the luminometric methylation assay (LUMA), significantly decreased with quartiles of several PCB congeners (Itoh *et al.*, 2014). Hypomethylation of *Alu*, a repetitive element and indicator of global methylation, was observed in blood leucocyte DNA with increasing total PCB concentrations among Greenlandic Inuit (Rusiecki *et al.*, 2008). Levels of other POPs also associated with hypomethylation in this population which is one of the most highly POPs-exposed groups worldwide.

Dioxins

Dioxins are a family of POPs with related chemical structures that include 2,3,7,8-tetrachlorodibenzoparadioxin (TCDD), polychlorinated dibenzoparadioxins (PCDDs), polychlorinated dibenzofurans (PCDFs), and some dioxin-like PCBs. TCDD is the most toxic and the most widely studied dioxin. Human exposure occurs primarily though consumption of contaminated meat, dairy products, and fish (Canady *et al.*, 2001). At certain doses, TCDD is considered to be a human carcinogen (Baan *et al.*, 2009). As with many other chemicals, developing babies are especially sensitive to toxicity from dioxins which readily cross the placenta. In a murine model, preimplantation exposure to TCDD led to greater DNMT activity at the preimplantation stage. At embryonic day 14, the exposed offspring exhibited lower weight, hypermethylation of the imprint control region for *H19*/*Igf2* (a reciprocally imprinted region important for prenatal growth) and reduced *H19* expression (Wu *et al.*, 2004). In a rat model, prenatal exposure to TCDD altered mammary gland architecture later in life (postnatal days 46 and 71). This was accompanied by lasting hypermethylation of the breast cancer 1 early onset (Brca1) promoter with concomitant decreased mRNA and protein BRCA-1 levels, and these changes could increase the offspring's risk for developing breast cancer (Papoutsis *et al.*, 2013).

Along with animal models, *in vitro* models provide evidence for aberrant DNA methylation, histone modifications, or microRNA expression from TCDD exposure (Beedanagari *et al.*, 2010; Elyakim *et al.*, 2010; Nakajima *et al.*, 2003). However, epidemiological data on dioxins and the epigenome remain limited. In peripheral blood mononuclear cells, higher levels of dioxin-like PCBs in serum were associated with increased expression of the microRNA, miR-191 (Guida *et al.*, 2013). Octachlorodibenzo-*p*-dioxin (OCDD) was associated with hypomethylation assessed by LUMA in a cohort of Swedish septuagenarians (Lind *et al.*, 2013). Future epidemiological studies should examine the impact of dioxins and other POPs on the epigenome throughout the life course and determine whether these changes contribute to toxicity.

Endocrine-disrupting chemicals (EDCs)

Endocrine-disrupting chemicals include naturally occurring and man-made chemicals that mimic or interfere with the normal function of hormones in the endocrine system. The endocrine system encompasses female ovaries, male testes, and pituitary, thyroid, and adrenal glands and is responsible for the regulation of multiple and complex biological processes across the life course, including brain and nervous system development, reproductive system maturation and function, and metabolism. EDCs have been shown to affect the normal functions of numerous tissues and organs and have been associated with developmental, reproductive, neurological, and immune effects across *in vitro*, *in vivo*, human epidemiological, and ecotoxicology studies (Vandenberg *et al.*, 2012). Of note, EDCs comprise a wide range of substances, many of which are frequently encountered by humans and wildlife, including pharmaceutical drugs, environmental toxicants, and natural chemicals found in plants. EDCs are present in plastic food and beverage containers, metal food cans, personal care products and detergents, flame retardants, toys, and pesticides.

The ability of EDCs to elicit epigenetic effects was first studied in the eostrogenic pharmaceutical agent diethylstilboestrol (DES). Initial work demonstrated hypomethylation in the oestrogen-responsive lactoferrin promoter and in the promoter and intronic regions of the *c-fos* oncogene in mice exposed *in utero* or in the perinatal period, respectively (Li *et al.*, 1997, 2003). In humans, increased incidence of uncommon disorders was observed in the granddaughters and grandsons of DES exposed women, suggesting epigenetic reprogramming during *in utero* exposure (reviewed in Newbold, 2004). Rodent studies likewise demonstrated that the effects of maternal DES exposure are transmitted through the maternal germ line to offspring via epigenetic mechanisms (Newbold *et al.*, 2006). More recently, the chaperone protein, Hsp90, was implicated mechanistically with DES-induced epigenetic gene alterations via interactions with histone proteins important for active chromatin conformations (Ruden *et al.*, 2005). This compelling work in DES epigenetics promoted research over the past decade to shift focus to endocrine active toxicants, such as bisphenol A (BPA) and phthalates, reviewed in greater detail below.

Bisphenol A

Bisphenol A (BPA), a monomer of polycarbonate plastics and epoxy resins, is produced in high volumes and is found in consumer products including food and water containers, baby bottles, thermal paper, dental composites, and metal can linings. BPA leaching from consumer products presents many opportunities for human exposure via inhalation, ingestion, and dermal routes. Hence, human exposure to BPA is widespread, and several studies have reported detectable levels of total urinary BPA in a majority of individuals in populations around the world, including the USA, China, and Korea (Calafat *et al.*, 2008; Zhang *et al.*, 2011). BPA can mimic or antagonize endogenous hormones, and subsequently perturb endocrine function, by binding weakly to several steroid receptors including the oestrogen receptors (ER α and β) and thyroid hormone receptor (Gould *et al.*, 1998; Kuiper *et al.*, 1998; Moriyama *et al.*, 2002). BPA has a strong binding affinity for the trans-membrane ER, G protein-coupled receptor 30 (GPR30), and the orphan nuclear receptor, oestrogen related receptor gamma (ERRγ) (Takayanagi *et al.*, 2006; Thomas and Dong, 2006). BPA can also activate transcription factors, such as peroxisome x receptor (PXR) and the aryl hydrocarbon receptor (AhR), which are often involved in cross talk with steroid receptors (Kruger *et al.*, 2008; Sui *et al.*, 2012). Many of these receptors play an important role in gene regulation, suggesting that BPA may influence normal differentiation and maturation processes especially during embryonic and fetal development.

Several animal studies have focused on the effects of early life BPA exposure on DNA methylation patterning. In a rat model, following gestational exposure to 10 μg/kg/day, Ho and colleagues observed multiple changes in gene-specific DNA methylation patterns in the adult male prostate, including hypomethylation of the phosphor-diesterase type 4 variant 4 (*Pde4d4*) (Ho *et al.*, 2006). Using the viable yellow agouti mouse model, we have demonstrated that perinatal exposure to 50 ng, 50 μg, and 50 mg BPA/kg diet shifts the distribution of offspring coat colour by altering methylation levels at the A^{vy} allele (Anderson *et al.*, 2012a; Dolinoy *et al.*, 2007). In addition, maternal nutritional supplementation of methyl donors or genistein was able to counteract the shift of the distribution of coat colour in offspring exposed to 50 mg BPA/kg diet (Dolinoy *et al.*, 2007). Using these same dietary BPA exposures and employing epigenome-wide DNA methylation profiling of offspring liver (Kim *et al.*, 2014), we also identified several non-monotonic responses to perinatal BPA exposure, corroborating previous observations of non-monotonic outcomes (Vandenberg, 2013). A series of recent investigations have identified an effect of early BPA exposure on the brain epigenome. Following maternal exposure to 2, 20, and 200 μg BPA/kg body weight (BW)/day in mice, the expression of both DNA methyltransferase 1 (*Dnmt1*), the maintenance methyltransferase enzyme, and DNA methyltransferase 3 (*Dnmt3a*), the enzyme involved in *de novo* DNA methylation in early development, were significantly altered in the prefrontal cortex and the hypothalamus of offspring (Kundakovic *et al.*, 2013). Additionally, exposed male offspring displayed a decrease in DNA methylation at oestrogen receptor 1 (*Esr1*) in the prefrontal cortex while females exposed to 20 μg/kg/day displayed a decrease in methylation at *Esr1* in the hypothalamus (Wolstenholme *et al.*, 2011).

A growing number of human birth cohort studies and cross sectional epidemiological and clinical studies are beginning to investigate BPA levels and DNA methylation profiles. Using healthy 1st to 2nd trimester human fetal liver specimens quantified for internal BPA levels, we examined xenobiotic metabolizing enzyme gene expression and identified 2 phase I and 12 phase II genes with significantly reduced expression with higher BPA levels, including isoforms from the carboxylesterase, catechol O-methyltransferase (COMT), glutathione S-transferase, sulfotransferase (SULT), and UDP-glucuronosyltransferase families (Nahar *et al.*, 2013). Analysis of DNA methylation indicated higher BPA levels were associated with increased site-specific methylation at COMT and increased average methylation at SULT2A1 promoters. Building upon our endocrine-disrupting chemicals exposure studies in Gharbiyah, Egypt (Colacino *et al.*, 2011; Nahar *et al.*, 2012), we correlated urinary concentrations of total (free plus conjugated) species of BPA in spot samples with genome-wide DNA methylation levels in concurrently collected saliva DNA samples (Kim *et al.*, 2013), showing that BPA may affect human health through specific epigenomic modification of genes in relevant pathways.

A number of BPA studies in animal models have examined the effects of exposure on epigenetic marks other than DNA methylation. For example, exposure to 10 nM of BPA during gestation and lactation resulted in decreased binding of acetylated H3K9 to the *Kcc2* promoter in mouse offspring resulting in diminished *Kcc2* expression, a gene that plays an important role in neurodevelopment (Yeo *et al.*, 2013). Additionally, mice offspring displayed an increase in EZH2, a histone methyltransferase, expression in adult mammary tissue after exposure to 10 μg BPA/kg body weight with a concomitant increase in H3 trimethylation (Doherty *et al.*, 2010). MicroRNA (miRNA) expression was measured in fetal ovaries of sheep after *in utero* exposure to 0.5 mg/kg/day of BPA (Veiga-Lopez *et al.*, 2013). BPA exposure resulted in 45 and 11 miRNAs that were differentially expressed at GD 65 and 90 in fetal tissue, respectively. Parallel efforts are under way in human birth and longitudinal studies, using carefully collected and stored samples.

Phthalates

Phthalates are widely used in consumer goods, including building materials, food packaging, cosmetics, cleaning materials, pharmaceuticals, and medical devices (Heudorf *et al.*, 2007). High molecular weight phthalates, such as di-(2-ethylhexyl) phthalate (DEHP), are often used as plasticizers in polyvinyl chloride (PVC), while low molecular weight phthalates, such as dibutyl phthalate (DBP), are used in cosmetics, insecticides, and pharmaceuticals (Schettler, 2006). Phthalates may leach from consumer products into air or food (Hauser *et al.*, 2004) and exposures occur via diet and to some degree via inhalation and dermal absorption from personal care products (Colacino *et al.*, 2010; Green *et al.*, 2005; Hauser *et al.*, 2004). Studies have detected metabolites of DEHP and diethyl phthalate (DEP) in the majority of the US population (Silva *et al.*, 2004). Typically, children show higher urinary concentrations of metabolites than adults (Centers for Disease Control and Prevention, 2009; Koch and Calafat, 2009; Silva *et al.*, 2004). Exposure of rats to high doses of phthalates orally (0.5–2 g/kg/day) was associated with prolonged oestrous cycles and mid-pregnancy abortion (Davis *et al.*, 1994; Gray *et al.*, 2006). Epidemiological studies suggest developmental and reproductive associations, including preterm birth (Meeker *et al.*, 2009) and early breast and pubic hair development with high molecular weight phthalates, and delayed breast and pubic hair development with low molecular weight phthalates (Wolff *et al.*, 2010).

The high molecular weight phthalate DEHP is the most extensively studied phthalate for epigenetic alterations associated with DOHaD. A series of studies in mice assessed the effects of maternal exposure to DEHP (500 mg/kg/day), on offspring testes DNA methylation and gene expression, noting increased global DNA methylation, increased testis-specific gene insulin-like hormone 3 (*Insl3*) DNA methylation, and increased DNA methyltransferase expression (Wu *et al.*, 2010a,b). Recently Li and colleagues examined the effects of maternal DEHP exposure (40 mg/kg/day) on mouse offspring oocyte DNA methylation, and identified altered methylation patterns at imprinted genes (Li *et al.*, 2014). Human birth cohort studies are beginning to evaluate epigenetics as a mediator of phthalates exposures and birth outcomes. Using the Predictors of Preeclampsia Study and the Harvard Epigenetic Birth Cohorts, LaRocca *et al.* (2014) recently showed that maternal first trimester phthalate levels are associated with altered DNA methylation at imprinted genes. Specifically, hypomethylation of *H19* in the placenta was associated with the sum of phthalate metabolites and low molecular weight phthalate metabolites (LaRocca *et al.*, 2014). Additionally, there is growing interest in the effects of EDCs, including phthalates, on allergy and asthma predisposition. When exposed to DEHP or butyl benzyl phthalate (BBP), circulating plasmacytoid DC cells isolated from healthy volunteers exhibited suppressed DNA methylation-induced interferon expression, providing evidence that phthalates may affect immune response via epigenetic mechanisms (Kuo *et al.*, 2013).

Other EDCs

Many other EDCs out there such as phytoestrogens (found in plants, such as genistein and daidzein, both present in soy-derived products) are not discussed here but have been shown to elicit epigenetic effects in animal models (Dolinoy *et al.*, 2006, 2007), with growing interest as cancer chemoprevention agents (McKay and Mathers, 2011). Other EDCs, including PCBs, dioxins, POPs (e.g. some pesticides) and some heavy metals (e.g. Pb), also display endocrine disruption, and are reviewed above. There is growing concern that many of the greater than 700 chemicals used in the hydraulic fracking process to extract natural gas exhibit endocrine disruption (Kassotis *et al.*, 2014). Similarly, endocrine disrupting properties have been observed in composite samples from municipal waste and surface water (Jalova *et al.*, 2013). Thus, considering the effects of chemical mixtures on epigenetic programming is emerging as a major initiative in environmental epigenetics research (see next Section).

Mixtures

Widespread and growing evidence in toxicology points to the totality of exposures and lifestyle factors, or exposure mixtures, as the entities that collectively lead to disease along with genetics. The epigenome is known to be affected (sometimes in opposite directions) by many chemical exposures (described in this chapter), nutritional factors (Anderson *et al.*, 2012b; Park *et al.*, 2012), and sociobehavioural factors (Weaver *et al.*, 2004). Realistically, multiple factors influence the epigenome and subsequent health and disease states simultaneously throughout the life course. The study of mixtures is complicated by need for a larger sample size, statistical modelling procedures that incorporate multiple exposures at once, and analytical methods for determining multiple exposures in bioavailable samples. Few exposure mixtures have been studied to date with regards to the epigenome. Some exceptions include combinations of PCBs, POPs and/or methylmercury (Desaulniers *et al.*, 2009; Huen *et al.*, 2014; Rusiecki *et al.*, 2008) and air pollution.

Air pollution

Particulate matter (PM) air pollution is one exposure mixture which has received attention in toxicology and environmental epigenetics. Air pollution is essentially a mixture of mixtures. The PM that constitutes air pollution is categorized by size – PM_{10}, $PM_{2.5}$, PM_1 – referring to particles less than 10, 2.5, or 1 mm in diameter, respectively. Polycyclic aromatic hydrocarbons (PAHs), which stem from combustion reactions in primarily diesel engines, form another component of air pollution often bound to $PM_{2.5}$ particulates, and other components include sulfates and black carbon. In urban areas, air pollution consists of combustion by-products (from coal-fired power plants or diesel exhaust, for example). Indoor air pollution can be an issue in industrial settings (e.g. metals-rich PM at steel mills) and in rural areas where wood burning cook stoves are staple household items. Early life exposure to air pollution increases risk for developing asthma and allergic disorders in childhood (Patel *et al.*, 2011; Zhou *et al.*, 2013). Several studies have thus examined associations between maternal exposure to PM air pollution and/or PAHs and the offspring's epigenome. A Belgian birth cohort study found reduced levels of global DNA methylation (via measurement of total methylated cytosine) in placenta DNA among mothers with higher $PM_{2.5}$ exposure based on regional data (Janssen *et al.*, 2013). Exposure across the entire pregnancy and the first trimester alone were associated with placental global hypomethylation. Early changes to the placental epigenome from environmental exposures could lead to altered gene expression and placenta function throughout pregnancy and adverse pregnancy or postnatal outcomes. Perera *et al.* (2009) identified differentially methylated genes in umbilical cord blood leucocytes based on maternal PAHs exposure. Decreased expression of one of the hypermethylated genes, acyl-CoA synthetase long-chain family member 3 (*ACSL3*), was noted in placenta tissue, and hypermethylation of this gene was associated with asthma or probable asthma when the children were 5 years old (Perera *et al.*, 2009).

Epidemiological studies have shown evidence for DNA methylation changes following chronic or acute exposure to air pollution. Many of these studies focus on markers of global DNA methylation such as repetitive elements or tandem repeats in blood leucocyte DNA. Among truck drivers in Beijing, $PM_{2.5}$ from personal monitors and ambient PM_{10} levels were associated with hypomethylation of the tandem repeat *SATα*. PM_{10} was also associated with decreased methylation at the tandem repeat *NBL2* (Guo *et al.*, 2014). PM_{10} was associated with hypomethylation of the repetitive elements LINE-1 and Alu among electric furnace steel plant workers with chronic exposure (Tarantini *et al.*, 2009, 2013). Chronic exposure to PAHs among firefighters was associated with hypomethylation of the dual specificity phosphatase 22 (*DUSP22*) gene promoter compared with non-firefighters (Ouyang *et al.*, 2012). Among elderly men participating in the Normative Aging Study, $PM_{2.5}$ and black carbon were both associated with

LINE-1 hypomethylation (Baccarelli *et al.*, 2010). Decreased miRNA expression associated with $PM_{2.5,}$ black carbon, organic carbon, and sulfates was also noted in a subset of the Normative Aging Study cohort (Fossati *et al.*, 2014). The changes to LINE-1 methylation and miRNA differed according to timing of the exposure (e.g. moving averages from the previous 4 hours, 48 hours, 7 days, or 28 days before sample collection). Methylation of repetitive elements and tandem repeats serves as a biomarker of epigenetic status following exposure, and hypomethylation in these regions can lead to genomic instability and undesirable transposition providing a link between exposure and adverse health effects.

The aforementioned air pollution studies reveal the complexity of analysing mixtures. Even within the air pollution field, a given study often focuses on one or two components (e.g. PAHs, $PM_{2.5}$), and the complete picture of the impact of short- and long-term air pollution exposure on the epigenome is unknown. Animal studies addressing the impact of exposure mixtures require careful and thoughtful design. In one example, mice were exposed to diesel exhaust particles for three weeks while being sensitized to fungal allergen. Exposure and sensitization together led to altered methylation at two genes with immune function that correlated with IgE levels, suggesting a link between traffic-related air pollution and asthma development (Liu *et al.*, 2008). While animal studies have great potential to examine links between exposure mixtures and outcomes, components of the mixture, exposure timeframe, and timing of outcome assessment must be limited to provide evidence for causality or gain insight into mechanisms. Tools and methods within epidemiology and toxicology that incorporate exposure mixtures will continue to evolve. In the meantime, the significance of epigenetic change by toxicants in isolation for health and disease needs to be interpreted in the light of the fact that multiple toxicants are acting on the epigenome simultaneously.

Conclusions

Environmental factors have the ability to alter the epigenome at various life stages from conception to old age and across generations. In this chapter, we detailed classes of environmental chemicals that human populations are commonly exposed to and how they perturb the epigenome in epidemiological cohorts, animal models, and *in vitro*: heavy metals, POPs, EDCs, and PM air pollution. Along with the chemicals described here, other chemicals (e.g. cigarette smoke), nutrients (e.g. genistein), sociobehavioural factors (e.g. maternal care), and physical agents (e.g. ultraviolet light) modify the epigenome. Environmental factors have the ability to change mitotic inheritance of epigenetic marks and to exert lasting transgenerational impacts, not described here but noted in rodent models (Manikkam *et al.*, 2012). In the foreseeable future, the field of environmental epigenetics may improve public health by identifying exposures that adversely impact health through epigenetic mechanisms and also beneficial factors (e.g. nutrients; meditation) that counter harmful effects. The field continues to address three areas that are critical to our understanding of chemical impact on the epigenome and its ultimate implication for human health and disease: (1) mechanisms linking toxicants to epigenetic modification, (2) biological significance of results, and (3) best practices for study design.

Mechanisms

The disparate epigenetic effects observed to date for various chemicals suggest that, while some common mechanisms exist, multiple pathways lead to epigenetic modification. A slew of enzymes, binding proteins, co-factors, and donors are involved in creating *de novo* epigenetic patterns shortly after conception and in maintaining epigenetic marks (DNA methylation, histone modifications, and non-coding RNAs) in the 200-plus human cell types following mitotic events throughout the life course. Fig. 17.2 depicts broad pathways chemicals can disrupt that would lead to an altered epigenome (with a focus on DNA methylation and histone modifications). Major mechanistic pathways include but are not limited to: (1) deregulating expression of key epigenetic machinery genes, (2) inhibiting epigenetic machinery enzymes, and (3) disrupting the one-carbon metabolism pathway.

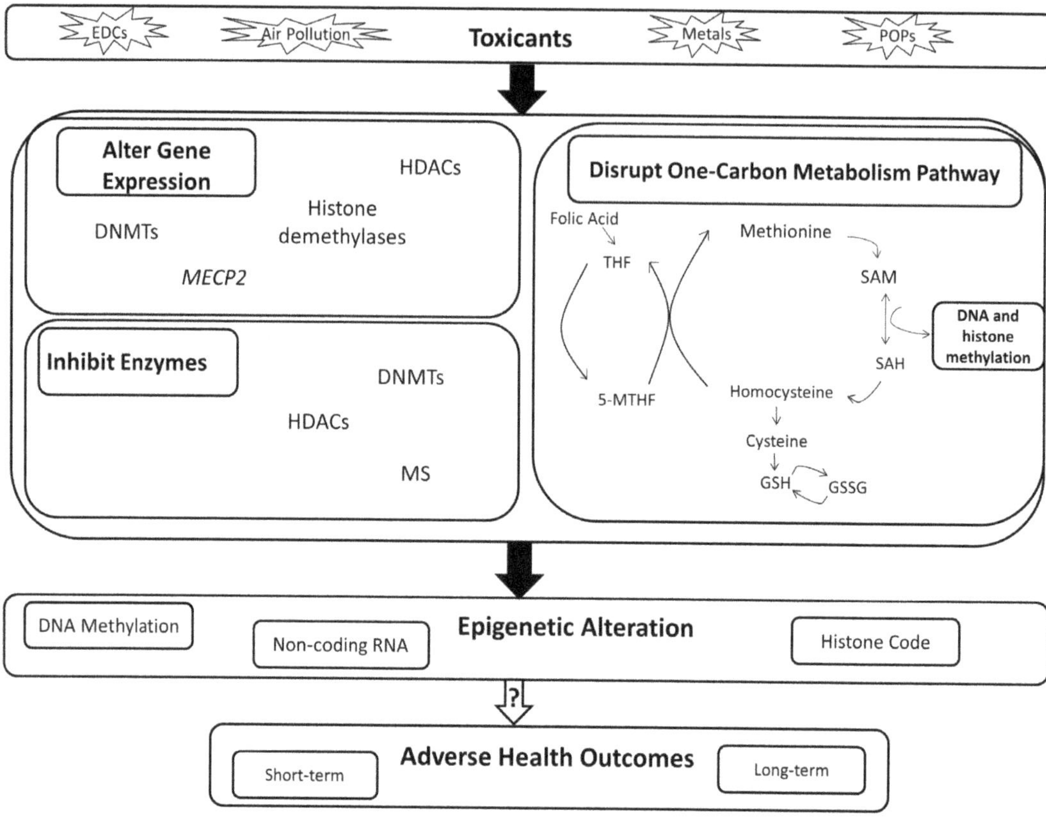

Figure 17.2 Pathways of toxicant-induced epigenetic change. Toxicants including air pollution, EDCs (endocrine disrupting chemicals), heavy metals, and persistent organic pollutants (POPs) have the ability to change epigenetic patterning including DNA methylation, histone tail modifications, and expression of non-coding RNAs. Three broad mechanistic pathways toxicants can work through to cause epigenetic change are depicted: altered expression of epigenetic machinery genes, inhibition of epigenetic machinery enzymes, and disruption of the one-carbon metabolism pathway. Example genes and enzymes are given along with a simple schematic of the one-carbon metabolism pathway. Exposure-induced epigenetic changes have the potential to persist throughout the life course and contribute to adverse health outcomes close to the time of exposure or much later in life. (Additional abbreviations appearing in the figure: DNMTs, DNA methyltransferases; GSH, glutathione; GSSG, glutathione disulfide; HDACs, histone deacetylases; MECP2, methyl-CpG binding protein 2; MS, methionine synthase; 5-MTHF, 5-tetrahydrofolate; THF, tetrahydrofolate; SAH S-adenosyl-homocysteine; SAM, S-adenosyl methionine.)

Dysregulation of gene expression

In order for cells to carry out their biological roles, a distinct epigenome must be maintained in each of the different cell types. These patterns must propagate during cell division, and in some cases, purposefully change to alter function at certain stages in development or later in life. As such, the epigenome is highly regulated and maintained by specialized enzymes, some of which are still being discovered. Toxicologists are well aware that vast gene expression changes accompany acute or prolonged exposure to most environmental agents. Evidence for epigenetic modification by environmental agents has led researchers (primarily those using animal or *in vitro* models) to screen for expression of key genes involved in DNA methylation or propagation of the histone code to determine whether toxicants alter the epigenome by deregulating those particular genes. Indeed, numerous chemicals have been shown to increase or decrease expression of DNMTs or histone regulators such as histone deacetylases (HDACs) and histone demethylases in rodent and cell culture models including Cd, Hg, Pb, organochlorine pesticides, PCBs, and PM_{10} (Casati *et al.*, 2012; Desaulniers *et al.*, 2009; Miousse *et al.*, 2014; Schneider *et al.*, 2013; Yuan *et al.*, 2013). In some cases, the impact of early life exposures on expression of epigenetic pathway genes was maintained until late life. For example, prenatal and/or postnatal Pb exposure led to decreased protein levels of Dnmt1, Dnmt3a, and methyl-CpG binding protein-2 (MeCP2) in rat hippocampi collected on postnatal day 55 (Schneider *et al.*, 2013). Likewise, among elderly rhesus macaques (23 years), those exposed to Pb for the first year of life had significantly decreased levels of DNMT1, DNMT3a, and MeCP2 protein levels in cerebral cortex compared with unexposed age-matched controls (Bihaqi *et al.*, 2011). The decreased protein levels correlated with an enrichment of down-regulated gene expression in exposed animal brains. In cultured mouse macrophages, short-term PM_{10} exposure led to decreased expression of DNMTs and histone machinery genes including *Hdac2*, reduced methylation of repetitive elements, and increased expression of these elements (Miousse *et al.*, 2014).

Inhibition of enzymes

While environmental agents can impact the epigenome via expression changes of key epigenetic pathway genes, they can also directly impact epigenetic machinery by interacting with the enzymes. Chemicals that are able to inhibit epigenetic pathway enzymes via oxidation, direct binding, or molecular mimicry have the potential to alter *de novo* or maintenance epigenetic patterns on a broad scale. DNMTs catalyse the attachment of a methyl group from the donor, SAM, to cytosine residues. Methylmercury levels of mink with dietary exposure correlated with decreased DNMT activity in brain tissue (Basu *et al.*, 2013). In a cell culture model, treatment with metals including Hg reduced methionine synthase (MS) activity, an important enzyme in the one-carbon metabolism pathway which supplies SAM (Waly *et al.*, 2004). The pharmaceutical sciences take advantage of this mechanism to reverse aberrant epigenetic patterning. For example, HDAC inhibitors are part of the treatment regimen for certain cancers (Bolden *et al.*, 2006; Dokmanovic *et al.*, 2007). Toxicants that modify the epigenome via inhibition of epigenetic machinery enzymes would exert their greatest impact during the exposure timeframe until the chemical is depleted from the body.

Disruption of the one-carbon metabolism pathway

Along with the enzymes needed to catalyse reactions that methylate DNA or modify histone tails (e.g. by acetylation, methylation, or phosphorylation), the supply of donors for these modifications is essential to maintain the epigenome. The one-carbon metabolism pathway synthesizes SAM, the donor that provides methyl groups for DNA and histone tail methylation (see a simple schematic in Fig. 17.2, upper right). This pathway, reviewed by Fox and Stover, involves a multitude of enzymes, cofactors, and nutrients, and the balance of these components influences the availability of SAM (Fox and Stover, 2008). Toxicants that inhibit/activate enzymes or otherwise influence the availability of intermediates of the one-carbon metabolism pathway can potentially deplete SAM (Christensen and Marsit, 2011). In human neuroblastoma cells, Hg, thiomerosal, Pb, Al, and ethanol exposure inhibit methionine synthase (MS), the enzyme responsible for converting homocysteine back to methionine in the one-carbon pathway (Waly *et al.*, 2004). Inorganic arsenic competes for SAM as this toxicant is methylated in the body using SAM as the methyl-donor (Howe *et al.*, 2014). Toxicant-induced changes to the one-carbon metabolism pathway thus have the ability to alter methylation capabilities of DNA or histone tails, with the greatest impact expected to be around the time of exposure until the toxicant is excreted.

Understanding the molecular mechanisms linking toxicant exposure to epigenetic change will benefit risk assessment by improving our knowledge of the doses at which epigenetic perturbation occurs. Furthermore, mechanistic insight may provide opportunities to prevent or ameliorate adverse changes. For example, targeted nutrient supplementation to increase SAM levels could counter SAM depletion by toxicants that disrupt the one-carbon metabolism pathway. In some instances, toxicants work through multiple mechanisms to disrupt the epigenome [for an example, see Hg review (Basu *et al.*, 2014)]. Exposure reduction may be the best solution for these toxicants as intervention would be complex.

Biological significance

Since the advent of environmental epigenetics, there has been no shortage of epidemiological, animal model, and *in vitro* studies reporting statistically significant differences in DNA methylation (primarily), microRNA expression, and histone modifications by exposure group or exposure dose. Many of these differences are described in this chapter. However, the effect sizes are often small, and functional and biological significance is not well understood. Animal and epidemiological studies have observed gene expression differences that correlate with epigenetic changes (Desaulniers *et al.*, 2009; Onishchenko *et al.*, 2008; Perera *et al.*, 2009). For example, methylation at four CpG sites of the Bdnf promoter was 2–3% higher among mice perinatally exposed to methylmercury. Exposed mice also had repressive changes to the histone code at Bdnf, and the mRNA levels were reduced by 25% compared to controls (Onishchenko *et al.*, 2008). However, not all studies assessing DNA methylation or histone modification find the expected correlation with gene expression. Thus, the ability of small epigenetic changes (e.g. <5% methylation difference) to produce biologically meaningful gene expression and phenotypic changes is still debated. The impact likely depends on the genic region, type of modification, cell type, and life stage. Studies aimed at identifying epigenetically labile genes with large expression changes and phenotypic consequences following exposure will improve the translation of environmental epigenetics research to benefit public health.

Best practices for study design

Initial environmental epigenetic studies, especially those in epidemiological cohorts, were limited in scope by tissue collection and storage, technologies used, and information on exposures and other lifestyle factors influencing the epigenome. Adding epigenetic inquiries to longstanding cohort studies is complicated by the quality of banked samples which is not necessarily adequate for isolation of DNA, RNA, or protein. Moving forward, new animal and epidemiological epigenetic studies can maximize their potential for meaningful discoveries by carefully planning procedures for sample collection, exposure and outcomes assessment, and epigenome interrogation. Establishing correlations between epigenomes of surrogate tissues (e.g. blood leucocytes and saliva) and target tissues will further maximize the interpretability of associations observed in epidemiological studies.

While longstanding cohorts are limited to DNA isolation from aged frozen blood samples, new molecular epidemiology studies have the opportunity to collect and store biospecimen using methods compatible with the isolation of DNA, RNA, and protein. In most cases, biospecimen available in human cohorts will remain restricted (e.g. blood leucocytes, saliva, placenta, tumours, and in some cases prenatal blood spot cards). Collecting two or more surrogate tissues from the same individuals and comparing the epigenome of all available tissues may improve the identification of environmentally induced epigenetic changes. Animal studies have the distinct advantage of collecting target organs (e.g. brain, heart, liver) at the time of sacrifice. As such, care should be taken to store tissue suitable for DNA, RNA, and protein isolation from as many organs as possible even if immediate research aims will not make use of all tissues. In both human and animal research, collecting samples at multiple time points will help to address questions of epigenetic change over time as it correlates with exposures and health outcomes (Faulk *et al.*, 2014).

Exposure and health outcome assessment should be thorough with adequate timing to address hypotheses. Cross-sectional studies report associations but cannot show causality since exposures, outcomes, and epigenetic marks are assessed simultaneously. While there is evidence that exposures can produce rapid change (Baccarelli *et al.*, 2009), most epigenetic change and resulting adverse health outcomes are likely the product of multiple exposures during susceptible periods (e.g. *in utero* development) or compounded throughout life. Certain toxicants (e.g. Cd, Pb) have long half-lives enabling exposure assessment of preceding decades. Others have short half-lives (e.g. BPA), and researchers rely on assumptions that exposure sources remain consistent over time. Maximizing exposure history will aid environmental epigenetic studies and reduce confounding. Animal and birth cohort studies that track early life exposures and assess epigenetic status and health outcomes later in life are costly and difficult to implement but have the most potential to reveal lifelong consequences of early exposure-induced epigenetic change.

DNA methylation is the most common epigenetic modification studied, but the histone code and non-coding RNAs are equally important modifications to understand. In the ideal research setting, epigenome-wide technologies would be used to interrogate the DNA methylome and the histone code. Technologies such as RNA-seq could then be used to assess non-coding RNA expression and gene expression. Bioinformatics tools would integrate the data, and genes with epigenetic marks and expression modified by variable of interest (e.g. exposure) would be identified. Available technology, cost, and bioinformatics expertise currently prevent most research groups from carrying out the ideal research project though capabilities are rapidly increasing at major research institutes. Identifying interesting genes using an epigenome-wide technology (e.g. DNA methylation array technology or reduced representation bisulfite sequencing), and targeting those genes for further analyses (e.g. assessment of histone modifications, quantitative DNA methylation, and gene expression) is one feasible approach researchers can adapt. As the field advances, the importance of new types of epigenetic marks such as DNA hydroxymethylation may become apparent (Dao *et al.*, 2014).

Blood leucocytes are the most common source of DNA and RNA in epigenetic epidemiology. It is still unclear whether associations between exposures and epigenetic changes in peripheral tissues (e.g. blood leucocytes, saliva) serve as biomarkers of epigenetic effects in inaccessible target tissues. Studies examining epigenetic correlation between peripheral tissues and target tissues in human cohorts or animal models will improve our ability to interpret data from peripheral tissues. For example, Murphy and colleagues compared DNA methylation at seven imprinted genes in up to 18 tissue types and observed good correlation among most, but not all, tissues (Murphy *et al.*, 2012). The use of blood leucocyte DNA is further complicated by the heterogeneity of blood which results in significant inter-individual epigenetic differences based on cell type proportions (Lam *et al.*, 2012). Fortunately, the use of blood leucocytes as a surrogate tissue when studying the impact of environmental factors on the epigenome can be greatly improved by statistically controlling for cell type proportions (Koestler *et al.*, 2013). If cell type differentials were not performed on the blood samples, cell type proportions can be inferred from a panel of epigenetic markers that are unique to each cell type (Accomando *et al.*, 2014). These statistical methods have proved useful in reducing cell-type variability in methylation data from other heterogeneous tissues such as placenta (Houseman *et al.*, 2014).

Environmental epigenetics research has made great strides to date but more work is necessary to better understand, prevent, and ameliorate environmentally induced disease and toxicity. Longitudinal studies that characterize *in utero* and early life exposures and track epigenetic modifications and health outcomes later in life are recommended. Identification of other susceptible life stages to epigenetic change should be determined by longitudinal studies of adults. Interrogating multiple layers of the epigenome (e.g. DNA methylation, non-coding RNA, histone modifications) and coupling these data with gene expression or protein levels will provide a better understanding of the functional consequences of statistical associations. While this chapter focused on mitotic inheritance, multi- and transgenerational effects of environmental exposures are important aspects this field needs to address (see the Transgenerational Inheritance chapter by Sadler-Riggleman and Skinner). Chemicals discussed in this chapter (e.g. BPA, phthalates, TCDD) have been linked in rodent models to transgenerational impacts on the epigenome (Manikkam *et al.*, 2012). Moving forward, it will also be important to evaluate direct effects on the germline following *in utero* exposures.

Acknowledgements

We acknowledge Maxwell Scher for his contribution to figure design and the literature review. We also acknowledge John Francis for his work on the literature review. The authors are supported by the US NIEHS/EPA (National Institute of Environmental Health Sciences/Environmental Protection Agency) Children's Environmental Health and Disease Prevention Center (Grant No P01 ES02284401/RD 83543601), the University of Michigan NIEHS P30 Environmental Health Science Core Center (Grant No P30 ES017885), and the National Center for Advancing Translational Sciences (Award No 2UL1TR00433). The contents of this article are solely the responsibility of the authors and do not necessarily reflect the official views of the NIEHS, the National Institutes of Health, or the US EPA. No conflict of interest is declared.

References

Accomando, W.P., Wiencke, J.K., Houseman, E.A., Nelson, H.H., and Kelsey, K.T. (2014). Quantitative reconstruction of leukocyte subsets using DNA methylation. Genome Biol. *15*, R50.

Afeiche, M., Peterson, K.E., Sanchez, B.N., Cantonwine, D., Lamadrid-Figueroa, H., Schnaas, L., Ettinger, A.S., Hernandez-Avila, M., Hu, H., and Tellez-Rojo, M.M. (2011). Prenatal lead exposure and weight of 0- to 5-year-old children in Mexico city. Environ. Health Perspect. *119*, 1436–1441.

Anderson, O.S., Nahar, M.S., Faulk, C., Jones, T.R., Liao, C., Kannan, K., Weinhouse, C., Rozek, L.S., and Dolinoy, D.C. (2012a). Epigenetic responses following maternal dietary exposure to physiologically relevant levels of bisphenol A. Environ. Mol. Mutagen. *53*, 334–342.

Anderson, O.S., Sant, K.E., and Dolinoy, D.C. (2012b). Nutrition and epigenetics: an interplay of dietary methyl donors, one-carbon metabolism and DNA methylation. J. Nutr. Biochem. *23*, 853–859.

Agency for Toxic Substances and Disease Registry (ATSDR) (2000). Toxicological Profile for Polychlorinated Biphenyls (PCBs). US Department of Health and Human Services, Atlanta, GA.

Agency for Toxic Substances and Disease Registry (ATSDR) (2012). Toxicological profile for cadmium. US Department of Health and Human Services, Atlanta, GA.

Agency for Toxic Substances and Disease Registry (ATSDR) (2013). Priority List of Hazardous Substances. Available at: http://www.atsdr.cdc.gov/spl/ [accessed 10 June 2014].

Baan, R., Grosse, Y., Straif, K., Secretan, B., El Ghissassi, F., Bouvard, V., Benbrahim-Tallaa, L., Guha, N., Freeman, C., Galichet, L., and Cogliano, V. (2009). A review of human carcinogens-Part F: Chemical agents and related occupations. Lancet Oncol. *10*, 1143–1144.

Baccarelli, A., Wright, R.O., Bollati, V., Tarantini, L., Litonjua, A.A., Suh, H.H., Zanobetti, A., Sparrow, D., Vokonas, P.S., and Schwartz, J. (2009). Rapid DNA methylation changes after exposure to traffic particles. Am. J. Respir. Crit. Care Med. *179*, 572–578.

Baccarelli, A., Tarantini, L., Wright, R.O., Bollati, V., Litonjua, A.A., Zanobetti, A., Sparrow, D., Vokonas, P., and Schwartz, J. (2010). Repetitive element DNA methylation and circulating endothelial and inflammation markers in the VA normative aging study. Epigenetics *5*, 222–228.

Basu, N., Head, J.A., Nam, D.H., Pilsner, J.R., Carvan, M.J., Chan, H.M., Goetz, F.W., Murphy, C.A., Rouvinen-Watt, K., and Scheuhammer, A.M. (2013). Effects of methylmercury of epigenetic markers in a mammal, bird, and fish. Comp. Biochem. Physiol. Pt C *157*, 322–322–327.

Basu, N., Goodrich, J.M., and Head, J. (2014). Ecogenetics of mercury: From genetic polymorphisms and epigenetics to risk assessment and decision-making. Environ. Toxicol. Chem. *33*, 1248–1258.

Bateson, P., Barker, D., Clutton-Brock, T., Deb, D., D'Udine, B., Foley, R.A., Gluckman, P., Godfrey, K., Kirkwood, T., Lahr, M.M., *et al.* (2004). Developmental plasticity and human health. Nature *430*, 419–421.

Beedanagari, S.R., Taylor, R.T., Bui, P., Wang, F., Nickerson, D.W., and Hankinson, O. (2010). Role of epigenetic mechanisms in differential regulation of the dioxin-inducible human CYP1A1 and CYP1B1 genes. Mol. Pharmacol. *78*, 608–616.

Bihaqi, S.W., Huang, H., Wu, J., and Zawia, N.H. (2011). Infant exposure to lead (Pb) and epigenetic modifications in the aging primate brain: implications for Alzheimer's disease. J. Alzheimers Dis. *27*, 819–833.

Bolden, J.E., Peart, M.J., and Johnstone, R.W. (2006). Anticancer activities of histone deacetylase inhibitors. Nat. Rev. Drug Discov. *5*, 769–784.

Brocato, J., and Costa, M. (2013). Basic mechanics of DNA methylation and the unique landscape of the DNA methylome in metal-induced carcinogenesis. Crit. Rev. Toxicol. *43*, 493–514.

Calafat, A.M., Ye, X., Wong, L.Y., Reidy, J.A., and Needham, L.L. (2008). Exposure of the U.S. population to bisphenol A and 4-tertiary-octylphenol: 2003–2004. Environ. Health Perspect. *116*, 39–44.

Canady, R., Crump, K., Feeley, M., Freijer, J., Kogevinas, M., Malisch, R., Verger, P., Wilson, J., and Zeilmaker, M. (2001). Safety evaluation of certain food additives and contaminants: polychlorinated dibenzodioxins, polychlorinated dibenzofurans, and coplanar polychlorinated biphenyls. WHO Food Additive Series no. *48*, 1021–1044.

Casati, L., Sendra, R., Colciago, A., Negri-Cesi, P., Berdasco, M., Esteller, M., and Celotti, F. (2012). Polychlorinated biphenyls affect histone modification pattern in early development of rats: a role for androgen receptor-dependent modulation? Epigenomics *4*, 101–112.

Casati, L., Sendra, R., Poletti, A., Negri-Cesi, P., and Celotti, F. (2013). Androgen receptor activation by polychlorinated biphenyls: epigenetic effects mediated by the histone demethylase Jarid1b. Epigenetics *8*, 1061–1068.

Centers for Disease Control and Prevention (2009). Fourth National Report on Human Exposure to Environmental Chemicals. Available at: http://www.cdc.gov/exposurereport/pdf/fourthreport.pdf [accessed 1 June 2014].

Cheng, T.F., Choudhuri, S., and Muldoon-Jacobs, K. (2012). Epigenetic targets of some toxicologically relevant metals: a review of the literature. J. Appl. Toxicol. *32*, 643–653.

Chervona, Y., Arita, A., and Costa, M. (2012). Carcinogenic metals and the epigenome: understanding the effect of nickel, arsenic, and chromium. Metallomics *4*, 619–627.

Christensen, B.C., and Marsit, C.J. (2011). Epigenomics in environmental health. Front. Genet. *2*, 84.

Clarkson, T.W., and Magos, L. (2006). The toxicology of mercury and its chemical compounds. Crit. Rev. Toxicol. *36*, 609–662.

Colacino, J.A., Harris, T.R., and Schecter, A. (2010). Dietary intake is associated with phthalate body burden in a nationally representative sample. Environ. Health Perspect. *118*, 998–1003.

Colacino, J.A., Soliman, A.S., Calafat, A.M., Nahar, M.S., Van Zomeren-Dohm, A., Hablas, A., Seifeldin, I.A., Rozek, L.S., and Dolinoy, D.C. (2011). Exposure to phthalates among premenstrual girls from rural and urban Gharbiah, Egypt: a pilot exposure assessment study. Environ. Health *10*, 40–069X-10-40.

Daniel, C.W., and Silberstein, G.B. (1987). Postnatal development of the rodent mammary gland. In The Mammary Gland: Development, Regulation, And Function, Neville, M.C., and Daniel, C.W. eds. (Springer-Verlag, New York, NY).

Dao, T., Cheng, R.Y., Revelo, M.P., Mitzner, W., and Tang, W. (2014). Hydroxymethylation as a novel environmental biosensor. Curr. Environ. Health. Rep. *1*, 1–10.

Davis, B.J., Maronpot, R.R., and Heindel, J.J. (1994). Di-(2-ethylhexyl) phthalate suppresses estradiol and ovulation in cycling rats. Toxicol. Appl. Pharmacol. *128*, 216–223.

Desaulniers, D., Xiao, G.H., Lian, H., Feng, Y.L., Zhu, J., Nakai, J., and Bowers, W.J. (2009). Effects of mixtures of polychlorinated biphenyls, methylmercury, and organochlorine pesticides on hepatic DNA methylation in prepubertal female Sprague-Dawley rats. Int. J. Toxicol. *28*, 294–307.

Doherty, L.F., Bromer, J.G., Zhou, Y., Aldad, T.S., and Taylor, H.S. (2010). *In utero* exposure to diethylstilbestrol (DES) or bisphenol-A (BPA) increases EZH2 expression in the mammary gland: an epigenetic mechanism linking endocrine disruptors to breast cancer. Horm. Cancer. *1*, 146–155.

Dokmanovic, M., Clarke, C., and Marks, P.A. (2007). Histone deacetylase inhibitors: overview and perspectives. Mol. Cancer. Res. *5*, 981–989.

Dolinoy, D.C. (2008). The agouti mouse model: an epigenetic biosensor for nutritional and environmental alterations on the fetal epigenome. Nutr. Rev. *66* (Suppl. 1), S7–11.

Dolinoy, D.C., Weidman, J.R., Waterland, R.A., and Jirtle, R.L. (2006). Maternal genistein alters coat color and protects Avy mouse offspring from obesity by modifying the fetal epigenome. Environ. Health Perspect. *114*, 567–572.

Dolinoy, D.C., Huang, D., and Jirtle, R.L. (2007). Maternal nutrient supplementation counteracts bisphenol A-induced DNA hypomethylation in early development. Proc. Natl. Acad. Sci. U.S.A. *104*, 13056–13061.

Ellis, L., Atadja, P.W., and Johnstone, R.W. (2009). Epigenetics in cancer: targeting chromatin modifications. Mol. Cancer. Ther. *8*, 1409–1420.

Elston, C.W., and Ellis, I.O. (1998). Normal structure and developmental abnormalities. In The Breast: Systemic Pathology, Elston, C.W., and Ellis, I.O. eds. (Churchill Livingstone, Edinburgh) pp. 1–19.

Elyakim, E., Sitbon, E., Faerman, A., Tabak, S., Montia, E., Belanis, L., Dov, A., Marcusson, E.G., Bennett, C.F., Chajut, A., Cohen, D., and Yerushalmi, N. (2010). hsa-miR-191 is a candidate oncogene target for hepatocellular carcinoma therapy. Cancer Res. *70*, 8077–8087.

Faulk, C., Barks, A., Liu, K., Goodrich, J.M., and Dolinoy, D.C. (2013). Early-life lead exposure results in dose- and sex-specific effects on weight and epigenetic gene regulation in weanling mice. Epigenomics *5*, 487–500.

Faulk, C., Liu, K., Barks, A., Goodrich, J.M., and Dolinoy, D.C. (2014). Longitudinal epigenetic drift in mice perinatally exposed to lead. Epigenetics *9*, 934–941.

Feinberg, A.P., and Tycko, B. (2004). The history of cancer epigenetics. Nat. Rev. Cancer. *4*, 143–153.

Fossati, S., Baccarelli, A., Zanobetti, A., Hoxha, M., Vokonas, P.S., Wright, R.O., and Schwartz, J. (2014). Ambient particulate air pollution and microRNAs in elderly men. Epidemiology *25*, 68–78.

Fox, J.T., and Stover, P.J. (2008). Folate-mediated one-carbon metabolism. Vitam. Horm. *79*, 1–44.

Goodrich, J.M., Basu, N., Franzblau, A., and Dolinoy, D.C. (2013). Mercury biomarkers and DNA methylation among Michigan dental professionals. Environ. Mol. Mutagen. *54*, 195–203.

Gould, J.C., Leonard, L.S., Maness, S.C., Wagner, B.L., Conner, K., Zacharewski, T., Safe, S., McDonnell, D.P., and Gaido, K.W. (1998). Bisphenol A interacts with the estrogen receptor alpha in a distinct manner from estradiol. Mol. Cell. Endocrinol. *142*, 203–214.

Gray, L.E.,Jr, Laskey, J., and Ostby, J. (2006). Chronic di-n-butyl phthalate exposure in rats reduces fertility and alters ovarian function during pregnancy in female Long Evans hooded rats. Toxicol. Sci. *93*, 189–195.

Green, R., Hauser, R., Calafat, A.M., Weuve, J., Schettler, T., Ringer, S., Huttner, K., and Hu, H. (2005). Use of di(2-ethylhexyl) phthalate-containing medical products and urinary levels of mono(2-ethylhexyl) phthalate in neonatal intensive care unit infants. Environ. Health Perspect. *113*, 1222–1225.

Guida, M., Marra, M.L., Zullo, F., Guida, M., Trifuoggi, M., Biffali, E., Borra, M., De Mieri, G., D'Alessandro, R., and De Felice, B. (2013). Association between exposure to dioxin-like polychlorinated biphenyls and miR-191 expression in human peripheral blood mononuclear cells. Mutat. Res. *753*, 36–41.

Guo, L., Byun, H.M., Zhong, J., Motta, V., Barupal, J., Zheng, Y., Dou, C., Zhang, F., McCracken, J.P., Diaz, A., *et al.* (2014). Effects of short-term exposure to inhalable particulate matter on DNA methylation of tandem repeats. Environ. Mol. Mutagen. *55*, 322–335.

Hanna, C.W., Bloom, M.S., Robinson, W.P., Kim, D., Parsons, P.J., Vom Saal, F.S., Taylor, J.A., Steuerwald, A.J., and Fujimoto, V.Y. (2012). DNA methylation changes in whole blood is associated with exposure to the environmental contaminants, mercury, lead, cadmium and bisphenol A, in women undergoing ovarian stimulation for IVF. Hum. Reprod. *27*, 1401–1410.

Hauser, R., Duty, S., Godfrey-Bailey, L., and Calafat, A.M. (2004). Medications as a source of human exposure to phthalates. Environ. Health Perspect. *112*, 751–753.

Heudorf, U., Mersch-Sundermann, V., and Angerer, J. (2007). Phthalates: toxicology and exposure. Int. J. Hyg. Environ. Health *210*, 623–634.

Ho, S.M., Tang, W.Y., Belmonte de Frausto, J., and Prins, G.S. (2006). Developmental exposure to estradiol and bisphenol A increases susceptibility to prostate carcinogenesis and epigenetically regulates phosphodiesterase type 4 variant 4. Cancer Res. *66*, 5624–5632.

Hossain, M.B., Vahter, M., Concha, G., and Broberg, K. (2012). Low-level environmental cadmium exposure is associated with DNA hypomethylation in Argentinean women. Environ. Health Perspect. *120*, 879–884.

Houseman, E.A., Molitor, J., and Marsit, C.J. (2014). Reference-free cell mixture adjustments in analysis of DNA methylation data. Bioinformatics *30*, 1431–1439.

Howe, C.G., Niedzwiecki, M.M., Hall, M.N., Liu, X., Ilievski, V., Slavkovich, V., Alam, S., Siddique, A.B., Graziano, J.H., and Gamble, M.V. (2014). Folate and cobalamin modify associations between S-adenosylmethionine and methylated arsenic metabolites in arsenic-exposed Bangladeshi adults. J. Nutr. *144*, 690–697.

Hu, H., Tellez-Rojo, M.M., Bellinger, D., Smith, D., Ettinger, A.S., Lamadrid-Figueroa, H., Schwartz, J., Schnaas, L., Mercado-Garcia, A., and Hernandez-Avila, M. (2006). Fetal lead exposure at each stage of pregnancy as a predictor of infant mental development. Environ. Health Perspect. *114*, 1730–1735.

Huang, Y., Nayak, S., Jankowitz, R., Davidson, N.E., and Oesterreich, S. (2011). Epigenetics in breast cancer: what's new? Breast Cancer Res. *13*, 225.

Huen, K., Yousefi, P., Bradman, A., Yan, L., Harley, K.G., Kogut, K., Eskenazi, B., and Holland, N. (2014). Effects of age, sex, and persistent organic pollutants on DNA methylation in children. Environ. Mol. Mutagen. *55*, 209–222.

Itoh, H., Iwasaki, M., Kasuga, Y., Yokoyama, S., Onuma, H., Nishimura, H., Kusama, R., Yoshida, T., Yokoyama, K., and Tsugane, S. (2014). Association between serum organochlorines and global methylation level of leukocyte DNA among Japanese women: a cross-sectional study. Sci. Total Environ. *490*, 603–609.

Jalova, V., Jarosova, B., Blaha, L., Giesy, J.P., Ocelka, T., Grabic, R., Jurcikova, J., Vrana, B., and Hilscherova, K. (2013). Estrogen-, androgen- and aryl hydrocarbon receptor mediated activities in passive and composite samples from municipal waste and surface waters. Environ. Int. *59*, 372–383.

Janssen, B.G., Godderis, L., Pieters, N., Poels, K., Kici Ski, M., Cuypers, A., Fierens, F., Penders, J., Plusquin, M., Gyselaers, W., and Nawrot, T.S. (2013). Placental DNA hypomethylation in association with particulate air pollution in early life. Part. Fibre Toxicol. *10*, 22.

Jirtle, R.L., and Skinner, M.K. (2007). Environmental epigenomics and disease susceptibility. Nat. Rev. Genet. *8*, 253–262.

Karlaganis, G., Marioni, R., Sieber, I., and Weber, A. (2001). The elaboration of the Stockholm Convention on Persistent Organic Pollutants (POPs): a negotiation process fraught with obstacles and opportunities. Env. Sci. Poll. Res. *8*, 216–221.

Kassotis, C.D., Tillitt, D.E., Davis, J.W., Hormann, A.M., and Nagel, S.C. (2014). Estrogen and androgen receptor activities of hydraulic fracturing chemicals and surface and ground water in a drilling-dense region. Endocrinology *155*, 897–907.

Kim, J.H., Rozek, L.S., Soliman, A.S., Sartor, M.A., Hablas, A., Seifeldin, I.A., Colacino, J.A., Weinhouse, C., Nahar, M.S., and Dolinoy, D.C. (2013). Bisphenol A-associated epigenomic changes in prepubescent girls: a cross-sectional study in Gharbiah, Egypt. Environ. Health *12*, 33–069X-12–33.

Kim, J.H., Sartor, M.A., Rozek, L.S., Faulk, C., Anderson, O.S., Jones, T.R., Nahar, M.S., and Dolinoy, D.C. (2014). Perinatal bisphenol A exposure promotes dose-dependent alterations of the mouse methylome. BMC Genomics *15*, 30–2164–15–30.

Kippler, M., Tofail, F., Gardner, R., Rahman, A., Hamadani, J.D., Bottai, M., and Vahter, M. (2012). Maternal cadmium exposure during pregnancy and size at birth: a prospective cohort study. Environ. Health Perspect. *120*, 284–289.

Kippler, M., Engstrom, K., Mlakar, S.J., Bottai, M., Ahmed, S., Hossain, M.B., Raqib, R., Vahter, M., and Broberg, K. (2013). Sex-specific effects of early life cadmium exposure on DNA methylation and implications for birth weight. Epigenetics *8*, 494–503.

Knower, K.C., To, S.Q., Leung, Y.K., Ho, S.M., and Clyne, C.D. (2014). Endocrine disruption of the epigenome: a breast cancer link. Endocr. Relat. Cancer *21*, T33–55.

Koch, H.M., and Calafat, A.M. (2009). Human body burdens of chemicals used in plastic manufacture. Philos. Trans. R. Soc. Lond. B. Biol. Sci. *364*, 2063–2078.

Koestler, D.C., Christensen, B., Karagas, M.R., Marsit, C.J., Langevin, S.M., Kelsey, K.T., Wiencke, J.K., and Houseman, E.A. (2013). Blood-based profiles of DNA methylation predict the underlying distribution of cell types: a validation analysis. Epigenetics *8*, 816–826.

Kruger, T., Long, M., and Bonefeld-Jorgensen, E.C. (2008). Plastic components affect the activation of the aryl hydrocarbon and the androgen receptor. Toxicology *246*, 112–123.

Kuiper, G.G., Lemmen, J.G., Carlsson, B., Corton, J.C., Safe, S.H., van der Saag, P.T., van der Burg, B., and Gustafsson, J.A. (1998). Interaction of estrogenic chemicals and phytoestrogens with estrogen receptor beta. Endocrinology *139*, 4252–4263.

Kundakovic, M., Gudsnuk, K., Franks, B., Madrid, J., Miller, R.L., Perera, F.P., and Champagne, F.A. (2013). Sex-specific epigenetic disruption and behavioral changes following low-dose in utero bisphenol A exposure. Proc. Natl. Acad. Sci. U.S.A. *110*, 9956–9961.

Kuo, C.H., Hsieh, C.C., Kuo, H.F., Huang, M.Y., Yang, S.N., Chen, L.C., Huang, S.K., and Hung, C.H. (2013). Phthalates suppress type I interferon in human plasmacytoid dendritic cells via epigenetic regulation. Allergy *68*, 870–879.

Laidlaw, M.A.S., and Filippelli, G.M. (2008). Resuspension of urban soils as a persistent source of lead poisoning in children: A review and new directions. Appl. Geochem. *23*, 2021–2039.

Lam, L.L., Emberly, E., Fraser, H.B., Neumann, S.M., Chen, E., Miller, G.E., and Kobor, M.S. (2012). Factors underlying variable DNA methylation in a human community cohort. Proc. Natl. Acad. Sci. U.S.A. 109 2(Suppl. 2), 17253–17260.

LaRocca, J., Binder, A.M., McElrath, T.F., and Michels, K.B. (2014). The impact of first trimester phthalate and phenol exposure on IGF2/H19 genomic imprinting and birth outcomes. Environ. Res. *133*, 396–406.

Li, L., Zhang, T., Qin, X.S., Ge, W., Ma, H.G., Sun, L.L., Hou, Z.M., Chen, H., Chen, P., Qin, G.Q., Shen, W., and Zhang, X.F. (2014). Exposure to diethylhexyl phthalate (DEHP) results in a heritable modification of imprint genes DNA methylation in mouse oocytes. Mol. Biol. Rep. *41*, 1227–1235.

Li, S., Washburn, K.A., Moore, R., Uno, T., Teng, C., Newbold, R.R., McLachlan, J.A., and Negishi, M. (1997). Developmental exposure to diethylstilbestrol elicits demethylation of estrogen-responsive lactoferrin gene in mouse uterus. Cancer Res. *57*, 4356–4359.

Li, S., Hursting, S.D., Davis, B.J., McLachlan, J.A., and Barrett, J.C. (2003). Environmental exposure, DNA methylation, and gene regulation: lessons from diethylstilbesterol-induced cancers. Ann. N. Y. Acad. Sci. *983*, 161–169.

Lind, L., Penell, J., Luttropp, K., Nordfors, L., Syvanen, A.C., Axelsson, T., Salihovic, S., van Bavel, B., Fall, T., Ingelsson, E., and Lind, P.M. (2013). Global DNA hypermethylation is associated with high serum levels of persistent organic pollutants in an elderly population. Environ. Int. *59*, 456–461.

Liu, J., Ballaney, M., Al-alem, U., Quan, C., Jin, X., Perera, F., Chen, L.C., and Miller, R.L. (2008). Combined inhaled diesel exhaust particles and allergen exposure alter methylation of T helper genes and IgE production *in vivo*. Toxicol. Sci. *102*, 76–81.

McKay, J.A., and Mathers, J.C. (2011). Diet induced epigenetic changes and their implications for health. Acta Physiol. (Oxf) *202*, 103–118.

Manikkam, M., Guerrero-Bosagna, C., Tracey, R., Haque, M.M., and Skinner, M.K. (2012). Transgenerational actions of environmental compounds on reproductive disease and identification of epigenetic biomarkers of ancestral exposures. PLoS ONE *7*, e31901.

Martinez-Zamudio, R., and Ha, H.C. (2011). Environmental epigenetics in metal exposure. Epigenetics *6*, 820–827.

Meeker, J.D., Hu, H., Cantonwine, D.E., Lamadrid-Figueroa, H., Calafat, A.M., Ettinger, A.S., Hernandez-Avila, M., Loch-Caruso, R., and Tellez-Rojo, M.M. (2009). Urinary phthalate metabolites in relation to preterm birth in Mexico city. Environ. Health Perspect. *117*, 1587–1592.

Miousse, I.R., Chalbot, M.C., Aykin-Burns, N., Wang, X., Basnakian, A., Kavouras, I.G., and Koturbash, I. (2014). Epigenetic alterations induced by ambient particulate matter in mouse macrophages. Environ. Mol. Mutagen. *55*, 428–435.

Moriyama, K., Tagami, T., Akamizu, T., Usui, T., Saijo, M., Kanamoto, N., Hataya, Y., Shimatsu, A., Kuzuya, H., and Nakao, K. (2002). Thyroid hormone action is disrupted by bisphenol A as an antagonist. J. Clin. Endocrinol. Metab. *87*, 5185–5190.

Murphy, S.K., Huang, Z., and Hoyo, C. (2012). Differentially methylated regions of imprinted genes in prenatal, perinatal and postnatal human tissues. PLoS ONE *7*, e40924.

Nahar, M.S., Soliman, A.S., Colacino, J.A., Calafat, A.M., Battige, K., Hablas, A., Seifeldin, I.A., Dolinoy, D.C., and Rozek, L.S. (2012). Urinary bisphenol A concentrations in girls from rural and urban Egypt: a pilot study. Environ. Health *11*, 20.

Nahar, M.S., Kim, J.H., Sartor, M.A., and Dolinoy, D.C. (2013). Bisphenol A-associated alterations in the expression and epigenetic regulation of genes encoding xenobiotic metabolizing enzymes in human fetal liver. Environ. Mol. Mutagen. *55*, 184–195.

Nakajima, M., Iwanari, M., and Yokoi, T. (2003). Effects of histone deacetylation and DNA methylation on the constitutive and TCDD-inducible expressions of the human CYP1 family in MCF-7 and HeLa cells. Toxicol. Lett. *144*, 247–256.

Newbold, R.R. (2004). Lessons learned from perinatal exposure to diethylstilbestrol. Toxicol. Appl. Pharmacol. *199*, 142–150.

Newbold, R.R., Padilla-Banks, E., and Jefferson, W.N. (2006). Adverse effects of the model environmental estrogen diethylstilbestrol are transmitted to subsequent generations. Endocrinology *147*, S11–7.

Onishchenko, N., Karpova, N., Sabri, F., Castren, E., and Ceccatelli, S. (2008). Long-lasting depression-like behavior and epigenetic changes of BDNF gene expression induced by perinatal exposure to methylmercury. J. Neurochem. *106*, 1378–1387.

Ouyang, B., Baxter, C.S., Lam, H.M., Yeramaneni, S., Levin, L., Haynes, E., and Ho, S.M. (2012). Hypomethylation of dual specificity phosphatase 22 promoter correlates with duration of service in firefighters and is inducible by low-dose benzo[a]pyrene. J. Occup. Environ. Med. *54*, 774–780.

Papoutsis, A.J., Selmin, O.I., Borg, J.L., and Romagnolo, D.F. (2013). Gestational exposure to the AhR agonist 2,3,7,8-tetrachlorodibenzo-p-dioxin induces BRCA-1 promoter hypermethylation and reduces BRCA-1 expression in mammary tissue of rat offspring: preventive effects of resveratrol. Mol. Carcinog. doi: 10.1002/mc.22095.

Park, L.K., Friso, S., and Choi, S.W. (2012). Nutritional influences on epigenetics and age-related disease. Proc. Nutr. Soc. *71*, 75–83.

Patel, M.M., Quinn, J.W., Jung, K.H., Hoepner, L., Diaz, D., Perzanowski, M., Rundle, A., Kinney, P.L., Perera, F.P., and Miller, R.L. (2011). Traffic density and stationary sources of air pollution associated with wheeze, asthma, and immunoglobulin E from birth to age 5 years among New York City children. Environ. Res. *111*, 1222–1229.

Perera, F., Tang, W.Y., Herbstman, J., Tang, D., Levin, L., Miller, R., and Ho, S.M. (2009). Relation of DNA methylation of 5′-CpG island of ACSL3 to transplacental exposure to airborne polycyclic aromatic hydrocarbons and childhood asthma. PLoS ONE *4*, e4488.

Reik, W., Dean, W., and Walter, J. (2001). Epigenetic reprogramming in mammalian development. Science *293*, 1089–1093.

Ruden, D.M., Xiao, L., Garfinkel, M.D., and Lu, X. (2005). Hsp90 and environmental impacts on epigenetic states: a model for the trans-generational effects of diethylstibesterol on uterine development and cancer. Hum. Mol. Genet. *14*(Spec. No 1), R149–55.

Rusiecki, J.A., Baccarelli, A., Bollati, V., Tarantini, L., Moore, L.E., and Bonefeld-Jorgensen, E.C. (2008). Global DNA hypomethylation is associated with high serum-persistent organic pollutants in Greenlandic Inuit. Environ. Health Perspect. *116*, 1547–1552.

Russo, J., and Russo, I.H. (1987). Development of the human mammary gland. In The Mammary Gland: Development, Regulation, And Function, Neville, M.C., and Daniel, C.W. eds. (Springer-Verlag, New York, NY) pp. 67–93.

Sanders, A.P., Smeester, L., Rojas, D., DeBussycher, T., Wu, M.C., Wright, F.A., Zhou, Y.H., Laine, J.E., Rager, J.E., Swamy, G.K., *et al.* (2014). Cadmium exposure and the epigenome: Exposure-associated patterns of DNA methylation in leukocytes from mother–baby pairs. Epigenetics 9, 212–221.

Schettler, T. (2006). Human exposure to phthalates via consumer products. Int. J. Androl. *29*, 134–139; discussion 181–185.

Schneider, J.S., Kidd, S.K., and Anderson, D.W. (2013). Influence of developmental lead exposure on expression of DNA methyltransferases and methyl cytosine-binding proteins in hippocampus. Toxicol. Lett. *217*, 75–81.

Silva, M.J., Barr, D.B., Reidy, J.A., Malek, N.A., Hodge, C.C., Caudill, S.P., Brock, J.W., Needham, L.L., and Calafat, A.M. (2004). Urinary levels of seven phthalate metabolites in the U.S. population from the National Health and Nutrition Examination Survey (NHANES) 1999-2000. Environ. Health Perspect. *112*, 331–338.

Smeester, L., Yosim, A.E., Nye, M.D., Hoyo, C., Murphy, S.K., and Fry, R.C. (2014). Imprinted genes and the environment: links to the toxic metals arsenic, cadmium, lead and mercury. Genes 5, 477–496.

Sui, Y., Ai, N., Park, S.H., Rios-Pilier, J., Perkins, J.T., Welsh, W.J., and Zhou, C. (2012). Bisphenol A and its analogues activate human pregnane X receptor. Environ. Health Perspect. *120*, 399–405.

Takayanagi, S., Tokunaga, T., Liu, X., Okada, H., Matsushima, A., and Shimohigashi, Y. (2006). Endocrine disruptor bisphenol A strongly binds to human estrogen-related receptor gamma (ERRgamma) with high constitutive activity. Toxicol. Lett. *167*, 95–105.

Tarantini, L., Bonzini, M., Apostoli, P., Pegoraro, V., Bollati, V., Marinelli, B., Cantone, L., Rizzo, G., Hou, L., Schwartz, J., Bertazzi, P.A., and Baccarelli, A. (2009). Effects of particulate matter on genomic DNA methylation content and iNOS promoter methylation. Environ. Health Perspect. *117*, 217–222.

Tarantini, L., Bonzini, M., Tripodi, A., Angelici, L., Nordio, F., Cantone, L., Apostoli, P., Bertazzi, P.A., and Baccarelli, A.A. (2013). Blood hypomethylation of inflammatory genes mediates the effects of metal-rich airborne pollutants on blood coagulation. Occup. Environ. Med. *70*, 418–425.

Terry, M.B., Delgado-Cruzata, L., Vin-Raviv, N., Wu, H.C., and Santella, R.M. (2011). DNA methylation in white blood cells: association with risk factors in epidemiologic studies. Epigenetics. *6*, 828–837.

Thomas, P., and Dong, J. (2006). Binding and activation of the seven-transmembrane estrogen receptor GPR30 by environmental estrogens: a potential novel mechanism of endocrine disruption. J. Steroid Biochem. Mol. Biol. *102*, 175–179.

Vandenberg, L.N. (2013). Non-monotonic dose responses in studies of endocrine disrupting chemicals: bisphenol a as a case study. Dose Response *12*, 259–276.

Vandenberg, L.N., Colborn, T., Hayes, T.B., Heindel, J.J., Jacobs, D.R.,Jr, Lee, D.H., Shioda, T., Soto, A.M., vom Saal, F.S., Welshons, W.V., Zoeller, R.T., and Myers, J.P. (2012). Hormones and endocrine-disrupting chemicals: low-dose effects and nonmonotonic dose responses. Endocr. Rev. 33, 378–455.

Veiga-Lopez, A., Luense, L.J., Christenson, L.K., and Padmanabhan, V. (2013). Developmental programming: gestational bisphenol-A treatment alters trajectory of fetal ovarian gene expression. Endocrinology *154*, 1873–1884.

Waly, M., Olteanu, H., Banerjee, R., Choi, S.W., Mason, J.B., Parker, B.S., Sukumar, S., Shim, S., Sharma, A., Benzecry, J.M., Power-Charnitsky, V.A., and Deth, R.C. (2004). Activation of methionine synthase by insulin-like growth factor-1 and dopamine: a target for neurodevelopmental toxins and thimerosal. Mol. Psychiatry 9, 358–370.

Wang, B., Li, Y., Shao, C., Tan, Y., and Cai, L. (2012). Cadmium and its epigenetic effects. Curr. Med. Chem. *19*, 2611–2620.

Waterland, R.A., and Jirtle, R.L. (2004). Early nutrition, epigenetic changes at transposons and imprinted genes, and enhanced susceptibility to adult chronic diseases. Nutrition *20*, 63–68.

Weaver, I.C., Cervoni, N., Champagne, F.A., D'Alessio, A.C., Sharma, S., Seckl, J.R., Dymov, S., Szyf, M., and

Meaney, M.J. (2004). Epigenetic programming by maternal behavior. Nat. Neurosci. 7, 847–854.

Wolff, M.S., Teitelbaum, S.L., Pinney, S.M., Windham, G., Liao, L., Biro, F., Kushi, L.H., Erdmann, C., Hiatt, R.A., Rybak, M.E., Calafat, A.M., and Breast Cancer and Environment Research Centers. (2010). Investigation of relationships between urinary biomarkers of phytoestrogens, phthalates, and phenols and pubertal stages in girls. Environ. Health Perspect. *118*, 1039–1046.

Wolstenholme, J.T., Taylor, J.A., Shetty, S.R., Edwards, M., Connelly, J.J., and Rissman, E.F. (2011). Gestational exposure to low dose bisphenol A alters social behavior in juvenile mice. PLoS ONE *6*, e25448.

Wright, R.O., Schwartz, J., Wright, R.J., Bollati, V., Tarantini, L., Park, S.K., Hu, H., Sparrow, D., Vokonas, P., and Baccarelli, A. (2010). Biomarkers of lead exposure and DNA methylation within retrotransposons. Environ. Health Perspect. *118*, 790–795.

Wu, J., Basha, M.R., Brock, B., Cox, D.P., Cardozo-Pelaez, F., McPherson, C.A., Harry, J., Rice, D.C., Maloney, B., Chen, D., Lahiri, D.K., and Zawia, N.H. (2008). Alzheimer's disease (AD)-like pathology in aged monkeys after infantile exposure to environmental metal lead (Pb): evidence for a developmental origin and environmental link for AD. J. Neurosci. *28*, 3–9.

Wu, Q., Ohsako, S., Ishimura, R., Suzuki, J.S., and Tohyama, C. (2004). Exposure of mouse preimplantation embryos to 2,3,7,8-tetrachlorodibenzo-p-dioxin (TCDD) alters the methylation status of imprinted genes H19 and Igf2. Biol. Reprod. *70*, 1790–1797.

Wu, S., Zhu, J., Li, Y., Lin, T., Gan, L., Yuan, X., Xiong, J., Liu, X., Xu, M., Zhao, D., *et al.* (2010a). Dynamic epigenetic changes involved in testicular toxicity induced by di-2-(ethylhexyl) phthalate in mice. Basic Clin. Pharmacol. Toxicol. *106*, 118–123.

Wu, S., Zhu, J., Li, Y., Lin, T., Gan, L., Yuan, X., Xu, M., and Wei, G. (2010b). Dynamic effect of di-2-(ethylhexyl) phthalate on testicular toxicity: epigenetic changes and their impact on gene expression. Int. J. Toxicol. *29*, 193–200.

Yeo, M., Berglund, K., Hanna, M., Guo, J.U., Kittur, J., Torres, M.D., Abramowitz, J., Busciglio, J., Gao, Y., Birnbaumer, L., and Liedtke, W.B. (2013). Bisphenol A delays the perinatal chloride shift in cortical neurons by epigenetic effects on the Kcc2 promoter. Proc. Natl. Acad. Sci. U.S.A. *110*, 4315–4320.

Yuan, D., Ye, S., Pan, Y., Bao, Y., Chen, H., and Shao, C. (2013). Long-term cadmium exposure leads to the enhancement of lymphocyte proliferation via down-regulating p16 by DNA hypermethylation. Mutat. Res. *757*, 125–131.

Zhang, Z., Alomirah, H., Cho, H.S., Li, Y.F., Liao, C., Minh, T.B., Mohd, M.A., Nakata, H., Ren, N., and Kannan, K. (2011). Urinary bisphenol A concentrations and their implications for human exposure in several Asian countries. Environ. Sci. Technol. *45*, 7044–7050.

Zhou, C., Baiz, N., Zhang, T., Banerjee, S., Annesi-Maesano, I., and EDEN Mother-Child Cohort Study Group. (2013). Modifiable exposures to air pollutants related to asthma phenotypes in the first year of life in children of the EDEN mother-child cohort study. BMC Public Health *13*, 506.

Index

I

K

L

M

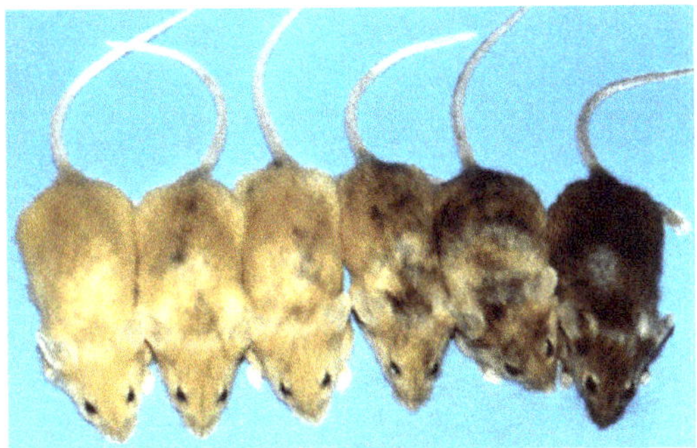

Figure 5.2 The coat colour spectrum in isogenic A^{vy} mice demonstrates the phenotypic outcome of stochastic gene silencing at a locus under the control of an upstream promoter. Low DNA methylation at the upstream promoter is associated with constitutive expression of the *agouti* gene (left). Intermediate methylation states are seen in mottled mice and no expression occurs when the upstream promoter is completely methylated in pseudoagouti mice (right).

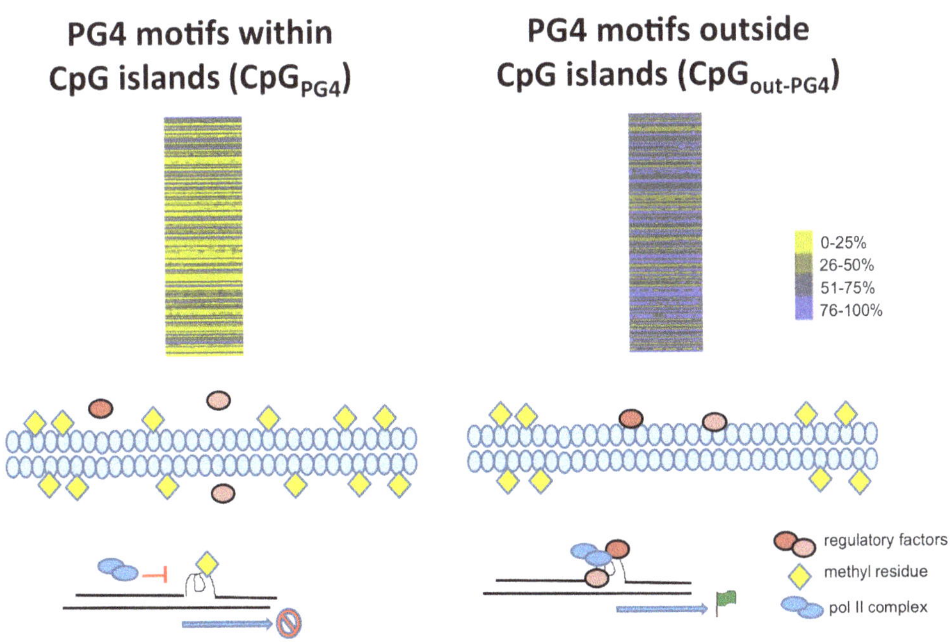

Figure 13.3 CpG methylation and G4 motifs. Top panel shows data from 18 individuals, which indicated that G4 motifs are hypomethylated within CpG islands (Halder *et al.*, 2010). Lower left panel shows a putative model where in case of methylated CpGPG4 the motif is stabilized. This could inhibit Pol II from binding, resulting in stalled or aborted transcription. On the other hand, G4 motif stabilizing proteins may bind unmethylated G4 motifs and help Pol II occupancy for transcription (lower right panel).

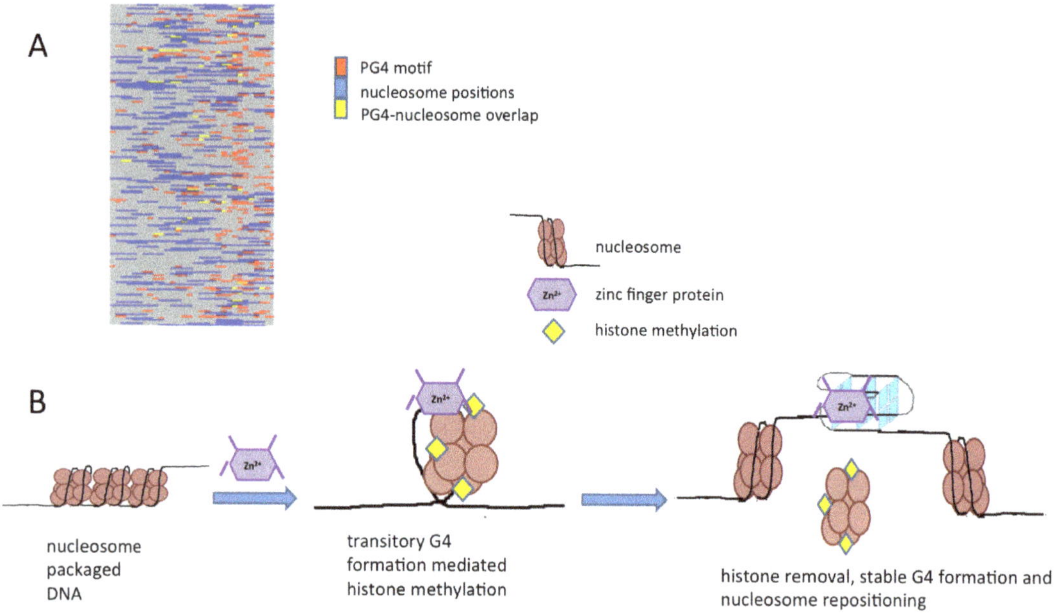

Figure 13.4 (A) Data from 180 promoters highlighting regions of nucleosome occupancy, G4 motifs and nucleosome-G4 overlap, suggests depletion within regions of G4-nucleosome overlap (Halder *et al.*, 2009). (B) Scheme showing speculative model for G4-mediated nucleosome repositioning – zinc finger proteins, which specially bind to G4 motifs could be responsible for histone methylation, thereby G4 motifs could act as a scaffold for histone methylation and nucleosome remodelling.

www.ingramcontent.com/pod-product-compliance
Lightning Source LLC
LaVergne TN
LVHW071411210926
848486LV00004B/244